TELECOM INDUSTRY ALMANAC 2023

The Only Comprehensive Guide to the Telecommunications Industry

Jack W. Plunkett

Published by:
Plunkett Research®, Ltd., Houston, Texas
www.plunkettresearch.com

PLUNKETT'S
TELECOMMUNICATIONS INDUSTRY
ALMANAC 2023

Editor and Publisher:
Jack W. Plunkett

Executive Editor and Database Manager:
Martha Burgher Plunkett

Senior Editor and Researchers:
Isaac Snider
Michael Cappelli

Editors, Researchers and Assistants:
Coby Cho
Bryant Hyunh
Annie Paynter
Gina Sprenkel
Jason Suerte

Information Technology Manager:
Rebeca Tijiboy

Special Thanks to:
Cellular Telecommunications & Internet
Association (CTIA)
Cisco Visual Networking Index (VNI)
eMarketer
Federal Communications Commission (FCC)
Gartner, Inc.
International Telecommunication Union (ITU)
Telecommunications Industry Association
U.S. Bureau of the Census
U.S. Bureau of Economic Analysis
U.S. Bureau of Labor Statistics
U.S. Department of Commerce

Plunkett Research®, Ltd.
P. O. Drawer 541737, Houston, Texas 77254 USA
Phone: 713.932.0000   Fax: 713.932.7080
www.plunkettresearch.com

Copyright © 2022, Plunkett Research®, Ltd. All rights reserved. Except as provided for below, you may not copy, resell, reproduce, distribute, republish, download, display, post, or transmit any portion of this book in any form or by any means, including, but not limited to, electronic, mechanical, photocopying, recording, or otherwise, without the express prior written permission of Plunkett Research, Ltd. Additional copyrights are held by other content providers, including, in certain cases, Morningstar, Inc. The information contained herein is proprietary to its owners and it is not warranted to be accurate, complete or timely. Neither Plunkett Research, Ltd. nor its content providers are responsible for any damages or losses arising from any use of this information. Market and industry statistics, company revenues, profits and other details may be estimates. Financial information, company plans or status, and other data can change quickly and may vary from those stated here. **Past performance is no guarantee of future results**.

Plunkett Research®, Ltd.
P. O. Drawer 541737
Houston, Texas 77254-1737
Phone: 713.932.0000, Fax: 713.932.7080 www.plunkettresearch.com

<u>**ISBN13 #**</u> **978-1-64788-000-2 (eBook Edition # 978-1-62831-992-7)**

Limited Warranty and Terms of Use:

Users' publications in static electronic format containing any portion of the content of this book (and/or the content of any related Plunkett Research, Ltd. online service to which you are granted access, hereinafter collectively referred to as the "Data") or Derived Data (that is, a set of data that is a derivation made by a User from the Data, resulting from the applications of formulas, analytics or any other method) may be resold by the User only for the purpose of providing third-party analysis within an established research platform under the following conditions: (However, Users may not extract or integrate any portion of the Data or Derived Data for any other purpose.)

a) Users may utilize the Data only as described herein. b) User may not export more than an insubstantial portion of the Data or Derived Data, c) Any Data exported by the User may only be distributed if the following conditions are met:

    i) Data must be incorporated in added-value reports or presentations, either of which are part of the regular services offered by the User and not as stand-alone products.
    ii) Data may not be used as part of a general mailing or included in external websites or other mass communication vehicles or formats, including, but not limited to, advertisements.
    iii) Except as provided herein, Data may not be resold by User.

"Insubstantial Portions" shall mean an amount of the Data that (1) has no independent commercial value, (2) could not be used by User, its clients, Authorized Users and/or its agents as a substitute for the Data or any part of it, (3) is not separately marketed by the User, an affiliate of the User or any third-party source (either alone or with other data), and (4) is not retrieved by User, its clients, Authorized Users and/or its Agents via regularly scheduled, systematic batch jobs.

<u>LIMITED WARRANTY; DISCLAIMER OF LIABILITY</u>: While Plunkett Research, Ltd. ("PRL") has made an effort to obtain the Data from sources deemed reliable, PRL makes no warranties, expressed or implied, regarding the Data contained herein. This book and its Data are provided to the End-User "AS IS" without warranty of any kind. No oral or written information or advice given by PRL, its employees, distributors or representatives will create a warranty or in any way increase the scope of this Limited Warranty, and the Customer or End-User may not rely on any such information or advice. <u>Customer Remedies</u>: PRL's entire liability and your exclusive remedy shall be, at PRL's sole discretion, either (a) return of the price paid, if any, or (b) repair or replacement of a book that does not meet PRL's Limited Warranty and that is returned to PRL with sufficient evidence of or receipt for your original purchase.

<u>NO OTHER WARRANTIES</u>: TO THE MAXIMUM EXTENT PERMITTED BY APPLICABLE LAW, PRL AND ITS DISTRIBUTORS DISCLAIM ALL OTHER WARRANTIES AND CONDITIONS, EITHER EXPRESSED OR IMPLIED, INCLUDING, BUT NOT LIMITED TO, IMPLIED WARRANTIES OR CONDITIONS OF MERCHANTABILITY, FITNESS FOR A PARTICULAR PURPOSE, TITLE AND NON-INFRINGEMENT WITH REGARD TO THE BOOK AND ITS DATA, AND THE PROVISION OF OR FAILURE TO PROVIDE SUPPORT SERVICES. <u>LIMITATION OF LIABILITY</u>: TO THE MAXIMUM EXTENT PERMITTED BY APPLICABLE LAW, IN NO EVENT SHALL PRL BE LIABLE FOR ANY SPECIAL, INCIDENTAL OR CONSEQUENTIAL DAMAGES WHATSOEVER (INCLUDING, WITHOUT LIMITATION, DAMAGES FOR LOSS OF BUSINESS PROFITS, BUSINESS INTERRUPTION, ABILITY TO OBTAIN OR RETAIN EMPLOYMENT OR REMUNERATION, ABILITY TO PROFITABLY MAKE AN INVESTMENT, OR ANY OTHER PECUNIARY LOSS) ARISING OUT OF THE USE OF, OR RELIANCE UPON, THE BOOK OR DATA, OR THE INABILITY TO USE THIS DATA OR THE FAILURE OF PRL TO PROVIDE SUPPORT SERVICES, EVEN IF PRL HAS BEEN ADVISED OF THE POSSIBILITY OF SUCH DAMAGES. IN ANY CASE, PRL'S ENTIRE LIABILITY SHALL BE LIMITED TO THE AMOUNT ACTUALLY PAID BY YOU FOR THE BOOK.

PLUNKETT'S TELECOMMUNICATIONS INDUSTRY ALMANAC 2023

CONTENTS

Introduction		1
How To Use This Book		3
<u>Chapter 1:</u> **Major Trends Affecting the Telecommunications Industry**		7
1)	Introduction to the Telecommunications Industry	7
2)	Landline Subscribers Cancel Service/ Bundled Services Pick Up Market Share	9
3)	5G Wireless Networks Rollout Worldwide, Enabling the Internet of Things (IoT)/Massive Investments Required	9
4)	Wi-Fi Enables Wireless Traffic Growth, Including the Internet of Things (IoT)	12
5)	Wireless Service Subscriptions Worldwide Reach 9.1 Billion	12
6)	Handset Makers Adopt Android, Push Advanced 4G and 5G Smartphones	13
7)	VOIP (Telephony over the Internet) Continues To Revolutionize the Telecommunications Industry	13
8)	Telecom Equipment Makers Face Intense Competition from Manufacturers in China/Huawei Faces Controversy	14
9)	Telecom Companies, Including AT&T and Verizon, Compete Fiercely Against Cable in the TV, Internet and Telephone Market	15
10)	Fiber-to-the-Home (FTTH) Gains Traction	16
11)	Global Internet Market Tops 5.2 Billion Users/Ultrafast Broadband Expands, both Fixed and Wireless	16
12)	Telecommunications Systems Move Online Including Unified Communications, Telepresence	17
13)	Carriers Reinstate Unlimited Access Plans for Smartphones/Face Intense Subscription Price Competition	18
14)	Smaller Satellites (SmallSats and CubeSats) and Low Earth Orbit Revolutionize Telecommunications	19
15)	The Internet of Things (IoT) and M2M to Boom, Enhanced by Artificial Intelligence (AI)/ Open New Avenues for Hacking	20
<u>Chapter 2:</u> **Telecommunications Industry Statistics**		23
Telecommunications Industry Statistics and Market Size Overview		24
Wired Telecommunications Carriers, Estimated Sources of Revenue, U.S.: 2016-2020		25
Wired Telecommunications Carriers, Estimated Breakdown of Revenue by Type of Customer, U.S.: 2017-2020		26
ATT & the Bell Companies, Then & Now		27
Total Retail Local Telephone Service Connections, U.S.: 2016-2019		28
Wireless Telecommunications Carriers (except Satellite): Estimated Sources of Revenue & Expenses, U.S.: 2016-2020		29
Top Mobile Operators by Number of Subscribers, Worldwide		30
Satellite Telecommunications: Estimated Sources of Revenue & Expenses, U.S.: 2017-2020		31

Continued on next page

Number of Business & Residential High Speed Internet Lines, U.S.: 2016-2021   **32**
Exports & Imports of Telecommunications Equipment, U.S.: 2016-2021   **33**
Employment in the Telecommunications Industry, U.S.: 1990-May 2022   **34**
<u>Chapter 3:</u> **Important Telecommunications Industry Contacts**   **35**
Addresses, Telephone Numbers and Internet Sites
<u>Chapter 4:</u> **THE TELECOMMUNICATIONS 350:**
Who They Are and How They Were Chosen   **65**
Index of Companies Within Industry Groups   **66**
Alphabetical Index   **76**
Index of U.S. Headquarters Location by State   **79**
Index of Non-U.S. Headquarters Location by Country   **82**
Individual Profiles on each of THE TELECOMMUNICATIONS 350   **85**

<u>Additional Indexes</u>
Index of Hot Spots for Advancement for Women/Minorities   **448**
Index of Subsidiaries, Brand Names and Selected Affiliations   **450**

A Short Telecommunications Industry Glossary   **461**

INTRODUCTION

PLUNKETT'S TELECOMMUNICATIONS INDUSTRY ALMANAC is designed as a general source for researchers of all types.

The data and areas of interest covered are intentionally broad, ranging from the trends relating to landline and wireless telephone service, to emerging technology, to an in-depth look at the major for-profit firms (which we call "THE TELECOMMUNICATIONS 350") within the many industry sectors that make up the telecom sector.

This reference book is especially intended to assist with market research, strategic planning, employment searches, contact or prospect list creation and financial research, and as a data resource for executives and students of all types.

PLUNKETT'S TELECOMMUNICATIONS INDUSTRY ALMANAC takes a rounded approach for the general reader and presents a complete overview of the telecommunications field (see "How To Use This Book"). For example, you will find trends in unified communications, 5G wireless markets, satellites and VOIP, along with easy-to-use charts and tables on all facets of telecommunications in general: from the sales and profits of the leading telecom companies to statistics relating to revenues, cellular communications and internet access.

THE TELECOMMUNICATIONS 350 is our unique grouping of the biggest, most successful corporations in all segments of the telecommunications industry. Tens of thousands of pieces of information, gathered from a wide variety of sources, have been researched and are presented in a unique form that can be easily understood. This section includes thorough indexes to THE TELECOMMUNICATIONS 350, by geography, industry, sales, brand names, subsidiary names and many other topics. (See Chapter 4.)

Especially helpful is the way in which PLUNKETT'S TELECOMMUNICATIONS INDUSTRY ALMANAC enables readers who have no business background to readily compare the financial records and growth plans of telecommunications companies and major industry groups. You'll see the mid-term financial record of each firm, along with the impact of earnings, sales and strategic plans on each company's potential to fuel growth, to serve new markets and to provide investment and employment opportunities.

No other source provides this book's easy-to-understand comparisons of growth, expenditures, technologies, corporations and many other items of great importance to people of all types who may be studying this, one of the largest industries in the world today.

By scanning the data groups and the unique indexes, you can find the best information to fit your personal research needs. The major companies in telecommunications are profiled and then ranked using several different groups of specific criteria. Which firms are the biggest employers? Which companies earn the most profits? These things and much more are easy to find.

In addition to individual company profiles, a thorough analysis of trends in telecommunications sectors is provided. This book's job is to help you sort through easy-to-understand summaries of today's trends and technologies in a quick and effective manner.

Whatever your purpose for researching the telecommunications field, you'll find this book to be a valuable guide. Nonetheless, as is true with all resources, this volume has limitations that the reader should be aware of:

- Financial data and other corporate information can change quickly. A book of this type can be no more current than the data that was available as of the time of editing. Consequently, the financial picture, management and ownership of the firm(s) you are studying may have changed since the date of this book. For example, this almanac includes the most up-to-date sales figures and profits available to the editors as of mid-2022. That means that we have typically used corporate financial data as of late-2021.

- Corporate mergers, acquisitions and downsizing are occurring at a very rapid rate. Such events may have created significant change, subsequent to the publishing of this book, within a company you are studying.

- Some of the companies in THE TELECOMMUNICATIONS 350 are so large in scope and in variety of business endeavors conducted within a parent organization, that we have been unable to completely list all subsidiaries, affiliations, divisions and activities within a firm's corporate structure.

- This volume is intended to be a general guide to a vast industry. That means that researchers should look to this book for an overview and, when conducting in-depth research, should contact the specific corporations or industry associations in

question for the very latest changes and data. Where possible, we have listed contact names, toll-free telephone numbers and internet site addresses for the companies, government agencies and industry associations involved so that the reader may get further details without unnecessary delay.

- Tables of industry data and statistics used in this book include the latest numbers available at the time of printing, generally through late-2021. In a few cases, the only complete data available was for earlier years.

- We have used exhaustive efforts to locate and fairly present accurate and complete data. However, when using this book or any other source for business and industry information, the reader should use caution and diligence by conducting further research where it seems appropriate. We wish you success in your endeavors, and we trust that your experience with this book will be both satisfactory and productive.

Jack W. Plunkett
Houston, Texas
August 2022

HOW TO USE THIS BOOK

The two primary sections of this book are devoted first to the telecommunications industry as a whole and then to the "Individual Data Listings" for THE TELECOMMUNICATIONS 350. If time permits, you should begin your research in the front chapters of this book. Also, you will find lengthy indexes in Chapter 4 and in the back of the book.

THE TELECOMMUNICATIONS INDUSTRY

Chapter 1: Major Trends Affecting the Telecommunications Industry.
This chapter presents an encapsulated view of the major trends that are creating rapid changes in the telecommunications industry today.

Chapter 2: Telecommunications Industry Statistics.
This chapter presents in-depth statistics including an industry overview.

Chapter 3: Important Telecommunications Industry Contacts – Addresses, Telephone Numbers and Internet Sites.
This chapter covers contacts for important government agencies, industry organizations and trade groups. Included are numerous important Internet sites.

THE TELECOMMUNICATIONS 350

Chapter 4: THE TELECOMMUNICATIONS 350: Who They Are and How They Were Chosen.
The companies compared in this book were carefully selected from the telecommunications industry, on a global basis. Many of the firms are based outside the U.S. For a complete description, see THE TELECOMMUNICATIONS 350 indexes in this chapter.

Individual Data Listings:
Look at one of the companies in THE TELECOMMUNICATIONS 350's Individual Data Listings. You'll find the following information fields:

Company Name:
The company profiles are in alphabetical order by company name. If you don't find the company you are seeking, it may be a subsidiary or division of one of the firms covered in this book. Try looking it up in the Index by Subsidiaries, Brand Names and Selected Affiliations in the back of the book.

Industry Code:
Industry Group Code: An NAIC code used to group companies within like segments.

Types of Business:
A listing of the primary types of business specialties conducted by the firm.

Brands/Divisions/Affiliations:

Major brand names, operating divisions or subsidiaries of the firm, as well as major corporate affiliations—such as another firm that owns a significant portion of the company's stock. A complete Index by Subsidiaries, Brand Names and Selected Affiliations is in the back of the book.

Contacts:

The names and titles up to 27 top officers of the company are listed, including human resources contacts.

Growth Plans/ Special Features:

Listed here are observations regarding the firm's strategy, hiring plans, plans for growth and product development, along with general information regarding a company's business and prospects.

Financial Data:

Revenue (2021 or the latest fiscal year available to the editors, plus up to five previous years): This figure represents consolidated worldwide sales from all operations. These numbers may be estimates.

R&D Expense (2021 or the latest fiscal year available to the editors, plus up to five previous years): This figure represents expenses associated with the research and development of a company's goods or services. These numbers may be estimates.

Operating Income (2021 or the latest fiscal year available to the editors, plus up to five previous years): This figure represents the amount of profit realized from annual operations after deducting operating expenses including costs of goods sold, wages and depreciation. These numbers may be estimates.

Operating Margin % (2021 or the latest fiscal year available to the editors, plus up to five previous years): This figure is a ratio derived by dividing operating income by net revenues. It is a measurement of a firm's pricing strategy and operating efficiency. These numbers may be estimates.

SGA Expense (2021 or the latest fiscal year available to the editors, plus up to five previous years): This figure represents the sum of selling, general and administrative expenses of a company, including costs such as warranty, advertising, interest, personnel, utilities, office space rent, etc. These numbers may be estimates.

Net Income (2021 or the latest fiscal year available to the editors, plus up to five previous years): This figure represents consolidated, after-tax net profit from all operations. These numbers may be estimates.

Operating Cash Flow (2021 or the latest fiscal year available to the editors, plus up to five previous years): This figure is a measure of the amount of cash generated by a firm's normal business operations. It is calculated as net income before depreciation and after income taxes, adjusted for working capital. It is a prime indicator of a company's ability to generate enough cash to pay its bills. These numbers may be estimates.

Capital Expenditure (2021 or the latest fiscal year available to the editors, plus up to five previous years): This figure represents funds used for investment in or improvement of physical assets such as offices, equipment or factories and the purchase or creation of new facilities and/or equipment. These numbers may be estimates.

EBITDA (2021 or the latest fiscal year available to the editors, plus up to five previous years): This figure is an acronym for earnings before interest, taxes, depreciation and amortization. It represents a company's financial performance calculated as revenue minus expenses (excluding taxes, depreciation and interest), and is a prime indicator of profitability. These numbers may be estimates.

Return on Assets % (2021 or the latest fiscal year available to the editors, plus up to five previous years): This figure is an indicator of the profitability of a company relative to its total assets. It is calculated by dividing annual net earnings by total assets. These numbers may be estimates.

Return on Equity % (2021 or the latest fiscal year available to the editors, plus up to five previous years): This figure is a measurement of net income as a percentage of shareholders' equity. It is also called the rate of return on the ownership interest. It is a vital indicator of the quality of a company's operations. These numbers may be estimates.

Debt to Equity (2021 or the latest fiscal year available to the editors, plus up to five previous years): A ratio of the company's long-term debt to its shareholders' equity. This is an indicator of the overall financial leverage of the firm. These numbers may be estimates.

Address:

The firm's full headquarters address, the headquarters telephone, plus toll-free and fax numbers where available. Also provided is the internet address.

Stock Ticker, Exchange: When available, the unique stock market symbol used to identify this firm's common stock for trading and tracking purposes is indicated. Where appropriate, this field may contain "private" or "subsidiary" rather than a ticker symbol. If the firm is a publicly-held company headquartered outside of the U.S., its international ticker and exchange are given.

Total Number of Employees: The approximate total number of employees, worldwide, as of the end of 2021 (or the latest data available to the editors).

Parent Company: If the firm is a subsidiary, its parent company is listed.

Salaries/Bonuses:

(The following descriptions generally apply to U.S. employers only.)

Highest Executive Salary: The highest executive salary paid, typically a 2021 amount (or the latest year available to the editors) and typically paid to the Chief Executive Officer.

Highest Executive Bonus: The apparent bonus, if any, paid to the above person.

Second Highest Executive Salary: The next-highest executive salary paid, typically a 2021 amount (or the latest year available to the editors) and typically paid to the President or Chief Operating Officer.

Second Highest Executive Bonus: The apparent bonus, if any, paid to the above person.

Other Thoughts:

Estimated Female Officers or Directors: It is difficult to obtain this information on an exact basis, and employers generally do not disclose the data in a public way. However, we have indicated what our best efforts reveal to be the apparent number of women who either are in the posts of corporate officers or sit on the board of directors. There is a wide variance from company to company.

Hot Spot for Advancement for Women/Minorities: A "Y" in appropriate fields indicates "Yes." These are firms that appear either to have posted a substantial number of women and/or minorities to high posts or that appear to have a good record of going out of their way to recruit, train, promote and retain women or minorities. (See the Index of Hot Spots For Women and Minorities in the back of the book.) This information may change frequently and can be difficult to obtain and verify. Consequently, the reader should use caution and conduct further investigation where appropriate.

Glossary: A short list of telecommunications industry terms.

Chapter 1

MAJOR TRENDS AFFECTING THE TELECOMMUNICATIONS INDUSTRY

Major Trends Affecting the Telecommunications Industry
1) Introduction to the Telecommunications Industry
2) Landline Subscribers Cancel Service/ Bundled Services Pick Up Market Share
3) 5G Wireless Networks Rollout Worldwide, Enabling the Internet of Things (IoT)/Massive Investments Required
4) Wi-Fi Enables Wireless Traffic Growth, Including the Internet of Things (IoT)
5) Wireless Service Subscriptions Worldwide Reach 9.1 Billion
6) Handset Makers Adopt Android, Push Advanced 4G and 5G Smartphones
7) VOIP (Telephony over the Internet) Continues To Revolutionize the Telecommunications Industry
8) Telecom Equipment Makers Face Intense Competition from Manufacturers in China/Huawei Faces Controversy
9) Telecom Companies, Including AT&T and Verizon, Compete Fiercely Against Cable in the TV, Internet and Telephone Market
10) Fiber-to-the-Home (FTTH) Gains Traction
11) Global Internet Market Tops 5.2 Billion Users/Ultrafast Broadband Expands, both Fixed and Wireless
12) Telecommunications Systems Move Online Including Unified Communications, Telepresence
13) Carriers Reinstate Unlimited Access Plans for Smartphones/Face Intense Subscription Price Competition
14) Smaller Satellites (SmallSats and CubeSats) and Low Earth Orbit Revolutionize Telecommunications
15) The Internet of Things (IoT) and M2M to Boom, Enhanced by Artificial Intelligence (AI)/Open New Avenues for Hacking

1) Introduction to the Telecommunications Industry

No other industry touches as many technology-related business sectors as telecommunications, which, by definition, encompasses not only the traditional areas of local and long-distance telephone service, but also advanced technology-based services including wireless communications, the internet, fiber-optics and satellites, as well as the software and hardware that enable these fields. Telecom is also deeply intertwined with entertainment of all types. Cable TV systems, such as Comcast, are aggressively offering local telephone service and high-speed internet access. The relationship between the telecom and cable sectors has become even more complex as traditional telecommunications firms such as AT&T are selling television via the internet and competing directly against cable for consumers' entertainment dollars.

Consequently, the various organizations that monitor the global telecommunications industry have their own ways of estimating total revenues, and their own thoughts on including, or not including, specific business sectors. Does "telecom" include equipment manufacturing and certain types of consulting? Or, should it be considered as services only, such as subscriber lines and data networks?

Information and Communication Technologies (ICT) is a term that is used to help describe the relationship between the myriad types of hardware, software, services and networks that make up the global information and telecommunications system. Sectors considered to be ICT include landlines, private networks, the internet, wireless communications, (including subscriber fees) and satellites. ICT is also generally considered to include cloud computing, computers and software. Globally, in the broadest possible sense, ICT was expected by Plunkett Research to grow to $7.3 trillion by 2022.

Telecommunications remains one of the major providers of employment in the world, with 661,000 employees in the U.S. alone as of mid-2022, and that number reflects only jobs in pure telecommunications service sectors.

There were approximately 9.1 billion wireless service subscriptions worldwide as of 2021. This is immense growth from about 4 billion at the end of 2008 and only 1.41 billion back in 2003. However, the actual number of individuals holding those subscriptions is somewhat less, as many people have more than one wireless device.

Experts at Cisco, with their Cisco Visual Networking Index, estimated that by 2023, the global number of internet-connected mobile devices would reach 3.6 per capita (nearly 28.5 billion devices).

Wireless service providers have set service prices low enough to be affordable for vast numbers of people, even in very low-income nations. Inexpensive cellphones are now indispensable to consumers from Haiti to Africa to New Guinea. Simple handsets can be bought for as little as $20 in such markets, and they can be topped-off with a segment of prepaid minutes for as little as 50 cents. Smartphones are more expensive, but good smartphones that are internet-capable are on the market at modest prices.

The International Telecommunication Union (ITU) estimates global landlines at 915 million as of 2019, down from 1.21 billion in 2009. This is only 11.9 landlines per 100 global population. Clearly, the worldwide boom in telecommunications is in wireless subscriptions and internet access—not in old-fashioned landlines.

Several major factors are creating deep changes in the telecommunications sector today, including: a) a shift in business and commercial telephones to VOIP (Voice Over Internet Protocol) services, that is,

telephone via the internet; b) a shift in residential and personal telephone use from landline services to wireless; c) intense competition between cable, satellite and wired services providers; d) soaring growth in the amount of data and video accessed via the internet and over wireless devices for information and entertainment, especially video; the e) introduction of advanced, much more cost-effective satellites, f) rapid growth in machine-to-machine communications (the Internet of Things) and g) the continuing evolution of advanced wireless technologies, including more powerful smartphones and ultra-high-speed 5G services. Simply put, a vast number of telecommunications service users prefer to make their phone calls, download data, view entertainment and otherwise access the internet via smartphones, not fixed telephones or PCs plugged into the wall.

Ingenuity, innovation, cost control, mergers between large companies and a reasonable approach to spending and investment will help to move the telecom industry ahead while it goes through these evolutionary changes. New technologies promise continuous advancement. The cost of a cellphone call has become a bargain worldwide. Meanwhile, competition among handset makers is more intense than ever. On the higher end, manufacturers are adding advanced new features to smartphones on a regular basis. These phones now contain significant computing power and memory, to the extent that today's smartphones easily have more computer processing power than a PC of 15 years ago, at a fraction of the size and weight of a PC. Improved cellphone service has prompted tens of millions of consumers to cancel their landlines altogether, eating into traditional revenue streams at AT&T and Verizon, among others.

As more consumers recognize the promise, and good value, of phone service using VOIP, millions of households and businesses worldwide have signed up for this less-expensive service as an alternative to landlines, often through their cable providers as part of a bundle of services. A handful of firms, such as Comcast, lead the VOIP market, along with relatively young companies like Skype (a Microsoft subsidiary) and Vonage. Savvy consumers realize that some VOIP services are essentially free, such as certain international calls on Skype or Viber.

At the same time, local phone companies, led by Verizon and AT&T, as well as some cable companies, are laying fiber-optic cable directly to the neighborhood, and even directly into the home and office, in order to retain customers with promises of ultra-high-speed internet connections and enhanced entertainment offerings online. If mobile phone owners are dropping their landlines, while VOIP over cable takes even more landline customers away, then the best weapon that traditional telcos can use in their battle for market share is very high-speed internet. AT&T and its peers are focusing on bundled service packages (combining wireless accounts, very high-speed internet access and TV, in addition to VOIP or landlines).

AT&T took a significant strategic step in mid-2015 by completing its acquisition of satellite TV firm DIRECTV for $48.5 billion. However, in 2021, AT&T made an agreement to spin-off its TV subscription businesses (DIRECTV, AT&T TV and U-Verse) into a joint-venture with private equity firm TPG Capital.

In June 2018, AT&T completed a massive acquisition of Time Warner, Inc., adding such entertainment content as HBO, Warner Bros. and Turner to AT&T's businesses. However, by May 2021, AT&T announced plans to spin off its media assets and merge them with Discovery, Inc. in a $43 billion deal. The spin off was completed in April 2022, with the newly named Warner Bros. Discovery, Inc. owning the Discovery Channel, Warner Bros. Entertainment, CNN, HBO, Cartoon Network, Discovery+ and HBO Max and major franchises including Batman and Harry Potter.

Telephone giant Verizon completed a strategic acquisition of AOL in 2015, largely to acquire AOL's superb online advertising technologies. In mid-2017, Verizon acquired Yahoo!'s online businesses, also to boost Verizon's advertising capabilities.

Meanwhile, government regulations are evolving quickly, which will bring even bigger changes to business strategies. In Europe and the U.S., regulators want much greater control over the practices of internet carriers and the ways in which they provide reliable access to consumers. Overall, the telecommunications industry is in a state of continuous technological and economic flux driven by intense competition and new technologies.

2) Landline Subscribers Cancel Service/ Bundled Services Pick Up Market Share

Massive numbers of American cellphone users have given up their landline telephones. Cutting off the landline saves consumers a significant amount of money each month. According to the National Health Interview Survey (NHIS), 73.9% of U.S. households had wireless-only phone service as of December 2021, up from only 10.5% back in

December 2006. (The NHIS makes this survey each year in order to keep track of what types of telephones that households might use to contact EMS and medical offices.)

There is more at work here than simply saving money. Consumers, particularly younger people, prefer their smartphones to any other communications medium. They use them for phone calls, text messages, email and internet access. It would not be convenient for them to switch back and forth from a cellphone to a landline, and a landline doesn't offer the advanced features and rich experience of a cellphone (particularly a smartphone, which is now carried by more than 85% of U.S. cellphone subscribers.) This same pattern is true throughout the world.

At the same time, large numbers of homes have adopted internet based VOIP phones, often from their cable providers. With new developments in technologies such as VOIP and enhanced, more reliable 4G cellular service (and even faster 5G in many cities), the demand for wired phone service will continue to fall.

AT&T is pushing its AT&T Fiber with DIRECTV STREAM packages (formerly U-verse) paring high speed internet and streaming TV, while Verizon is boosting its own Fios high speed internet service. Both services offer TV via the internet that is similar in nature to the program choices offered by leading cable companies. They have implemented massive investments that are bringing fiber-optic cable directly to the home and office, or at least directly into the neighborhood. Fiber to the Premises (FTTP) enables these firms to add fast access to enhanced entertainment, such as video on demand, as an important part of their bundled services. At the same time, they are able to offer internet access at extremely high speeds.

Verizon offers its Fios bundles to residential customers in a number of major U.S. cities. Consumers can receive a local phone number, unlimited long-distance in the U.S. and Canada, along with several enhanced phone features, plus Fios internet access and Fios Digital TV.

3) 5G Wireless Networks Rollout Worldwide, Enabling the Internet of Things (IoT)/Massive Investments Required

Fast wireless systems make it possible for subscribers to receive high-quality music, video and other features, and to connect interactively with the internet at high speeds. Fast wireless, combined with high-resolution color screens and cameras on smartphones, means that subscribers are using their handsets to play games, access Facebook, get detailed news or financial data, watch mobile TV and shop online. Smartphone owners are filming videos with their phones and then posting them to social media. They are also viewing internet sites featuring store and restaurant listings, as well as advertising, based on the location of the user. Film and video sites Netflix, Hulu and YouTube are immense drivers of traffic.

The next big thing in wireless internet service is 5G technology. The concept enables blinding download speeds from one gigabyte per second (Gbps), to perhaps as high as 10 Gbps. Part of the appeal is 5G's low latency (response time) of as little as one millisecond, compared to 4G's 50 milliseconds.

Ultrafast 5G wireless service will be the critical backbone that will enable IoT on a vast scale. Far beyond simply improving smartphone services, 5G will connect sensors and devices to networks, gathering data and enabling machine-to-machine communications at near instantaneous speeds. For example, it will be of vital, real-time use in robotics and self-driving cars and trucks.

In theory, 5G wireless can be more than 50 times faster than 4G. A test in New York City found that 5G enabled the download of a 2.1 GB movie file in two minutes, 57 seconds, compared to about one hour for a 4G download of the same film. However, actual 5G downloads achieved vary from city-to-city and carrier-to-carrier, and the movie download cited above may take a much longer tie to complete in some places. Actual users of 5G in 2020 often found lower than expected speed increases, as 5G networks and equipment are still in early stages of roll-out.

In the U.S., Sprint and T-Mobile completed a merger in April 2020. The combined businesses, operating under the T-Mobile name, planned to invest approximately $40 billion in infrastructure by 2022, developing what they expect to be one of the first nationwide 5G networks. Sprint and T-Mobile hope that the combined companies, with 90 million total customers, can operate more efficiently while aggressively establishing advanced 5G wireless services.

South Korea's largest mobile carriers (SK Telecom Co. and KT Corp.), manufacturers including Samsung Electronics Co. and LG Electronics, Inc. and the South Korean government have been true 5G pioneers. In April 2019, both SK Telecom and KT Corp. launched their first 5G commercial services.

The EU and South Korea agreed to cooperate to develop systems, establish standards and acquire radio frequencies necessary to implement 5G networks. The partnership will be led by Europe's 5G PPP (backed by Telefonica SA and Nokia Oyj) and South Korea's 5G Forum.

The United Nations' International Telecommunication Union (ITU) agreed on a set of requirements for 5G technology in 2018. Download speeds should be at least 20 gigabits per second, response times should be less than 1 millisecond, and at least 1 million devices should be able to connect within a high-traffic area equal to one square kilometer. In practical terms, a full-length, high-resolution movie should transmit in two seconds. Carriers, equipment manufacturers and tech companies must agree to work together to establish seamless connections around the world, along with methods of managing these connections.

U.S. cellphone service providers have invested billions of dollars in enhancing their networks over recent years. Subscribers wanting 5G were initially limited to a small number of devices capable of using the service. However, by 2021 5G had been rolled out, at least to some extent, in most major metro markets and was in use on a variety of devices.

However, all major wireless device makers are investing heavily in the design of 5G-capable handsets. As services roll out to more and more cities worldwide, consumers' demand for new handsets will accelerate and will generate tremendous new revenues for handset makers. Apple launched its first 5G iPhone (the iPhone 12) in late 2020, and later introduced the iPhone 13 and the iPhone SE, both of which support 5G. The earliest models of true 5G handsets were expensive, and, due to high energy consumption, lacked the long battery endurance enjoyed by typical smartphones. (5G smartphones are designed to work with lower speed 4G networks when necessary. For example, if the handsets reach high temperatures, they will switch to 4G to conserve power.) However, 5G will improve rapidly, and consumers will eventually adopt them in massive quantities, as long as monthly subscription fees are reasonable and handset prices fall to reasonable levels. All-in-all, this is terrific news for handset manufacturers, as it has become more and more daunting to create improvements to today's smartphones that are compelling enough to get consumers to upgrade.

Despite these positives, 5G infrastructure faces a long and expensive rollout. In a recent report, Deloitte Consulting estimated that U.S. conversion to

5G would require carriers to invest a combined $130 billion to $150 billion. Other analysts believe that deployment of 5G (outside of urban areas with the highest population densities) will take more than 10 years due to the expense and the necessity of cooperation among carriers and local municipalities. According to CTIA, in 2020, the U.S. had 154,000 cellular service towers. However, an additional 769,000 transmission must be created by 2026 to achieve full 5G coverage, due to the fact that 5G has shorter range and requires more antennas. Analysts at Evercore estimated in 2021 that it could be two years before two-thirds of Americans can have 5G service that is notably better than 4G.

While 4G mobile phone service can travel fairly long distances and transmissions come and go from tall towers, the fastest-service (known as high-band) 5G antennae must be spaced fairly close together. Carriers will be competing to offer the fastest possible speeds within dense cities. Consequently, carriers will be placing small 5G antennae on streetlights, the sides of buildings and other unique spots. Seeing this boom of new antenna locations as a potential source of tax revenue, some U.S. cities have been attempting to charge high permit fees to the carriers for each new antenna deployed. Unfortunately, the super-fast high-band systems may not be able to penetrate walls, creating the need for even more antennae. Slower low- and mid-band systems are likely to be used in areas that are not as densely populated.

5G may also increase cybersecurity threats since it will eventually connect to vast numbers of devices, pieces of equipment and remote sensors as the Internet of Things grows. Its potential use in autonomous vehicles, factories, transportation systems, utilities and smart-city infrastructure raises security concerns.

Chinese mobile equipment manufacturer Huawei Technologies Co., which was previously a dominant player in the global 5G market, faced cybersecurity doubts among certain foreign governments in recent years, resulting in U.S. and Australian government bans of Huawei 5G equipment. The bans, in addition to other Huawei restrictions in the U.K. and a number of EU countries, have opened doors for other cellular equipment makers to gain market share and play larger roles in establishing global standards. In the U.S., a number of smaller companies such as Airspan Networks (www.airspan.com), Mavenir (www.mavenir.com), JMA Wireless (jmawireless.com), Parallel Wireless (www.parallelwireless.com) and Rakuten Symphony (symphony.rakuten.com), which acquired Altiostar in 2019, are pushing ahead using open-standard 5G software.

Satellite provider Dish is spending an estimated $10 billion to launch a 5G wireless service. The service will be focused on creating wide-area networks for major enterprises, capable of powering the Internet of Things (IoT). Dish will be working with Dell Technologies' RAN (Radio Access Network) architecture and a strategy known as "edge" computing which enables faster IoT data processing. Dish is hoping that its service will cost much less to operate than those of traditional competitors.

Meanwhile, some carriers have been advertising a "lite" version of 5G, optimistically referred to as 5GE—the E stands for "evolution." Such service is technically based on an enhanced 4G system. Another technology is 5G NR ("new radio"), which upgrades existing 4G networks with regard to home internet access. For example, Verizon offers the 5G NR Enhanced Gateway modem made by D-Link.

High speed mobile access is a global phenomenon. China ranks number one in the world in the total number of high-speed internet subscribers. India's high-speed wireless count is soaring, thanks to massive investments in infrastructure, along with intense competition that keeps prices low. China's Huawei Technologies is investing heavily in 5G technology development and plans to compete head-on with the world's other technology giants in hopes to own very significant market share in 5G equipment.

Faster speed may also be achieved through new routers which support Wi-Fi 6 (also known as 802.11ax). They enable more devices in crowded areas to exchange data, such as texts and tweets, more quickly. Peak download speeds max out at 9.6 gigabits per second. In addition, Bluetooth 5 is a new standard with improved wireless range and reliability. Growing numbers of devices are compatible with Bluetooth 5.

SPOTLIGHT: pCell

Artemis Networks, www.artemis.com, may solve wireless network traffic jams with a technology called the pCell mobile wireless data transmission system. Compared to conventional cell towers, which share spectrum capacity and suffer slow performance when many users are on the network simultaneously, pCell uses interference caused by closely placed antennas to create a unique and clear wireless signal for each user on the network. The firm claims that all pCell service is full speed, no matter how many users are involved.

4) Wi-Fi Enables Wireless Traffic Growth, Including the Internet of Things (IoT)

While cellular phone companies are investing billions of dollars in technologies to give their subscribers enhanced services such and 4G mobile internet, Wi-Fi is more vital than ever for wireless access. As the number of cellular device subscriptions for smartphones, tablets, laptops and aircards has soared, so has the demand placed upon cellular networks. Wi-Fi acts as a vital relief valve. Wireless device owners increasingly want to access immense files, such as Netflix movies on demand. If the world's rapidly increasing wireless data traffic relied solely on cellular networks, the system would be under severe stress. However, since wireless device owners frequently have access to Wi-Fi as an alternative for a large portion of the day, they can switch to Wi-Fi from cellular as needed, reducing their total cellular subscription costs and dramatically reducing the load placed on cellular networks. The role played by Wi-Fi will remain vitally important, even as ultrafast 5G cellular networks are rolled out. At the same time, Wi-Fi will become even more important to the world's technology users as connected devices proliferate in the rapidly growing Internet of Things (IoT).

Experts at Cisco, with their Cisco Visual Networking Index, estimated that by 2023, the global number of internet-connected mobile devices will reach 3.6 per capita (nearly 28.5 billion devices). Smartphones will represent about 44% of these devices and connections. More than 50% of mobile traffic is offloaded to Wi-Fi rather than remaining on cellular networks.

Wi-Fi routers have very high theoretical data transfer speeds, but actual speeds rely on local internet connections. On the fixed end, each Wi-Fi network is tied into an internet router. This means that the actual download speed enjoyed by the Wi-Fi user is limited to the speed of the internet service connected to the router. If a user has a local internet connection with a 50 meg download speed via a cable modem, then the local Wi-Fi system will also be limited to 50 meg.

Wi-Fi is now advancing through enhanced technologies. Recent enhancements include MU-MIMO (multi-user, multiple-input, multiple-output), which allows a Wi-Fi device to handle data requests from multiple sources at once. Another recent technology known as OFDMA (orthogonal frequency-division multiple access) can split a Wi-Fi channel into many data pipes simultaneously. Kumu Networks, Inc., a startup based in Santa Clara, California (kumunetworks.com), has developed "full duplex" technology that enables Wi-Fi to transmit and receive simultaneously, effectively doubling the speed and capability of the network.

In early 2019, Amazon acquired Eero, a Wi-Fi startup that makes small wireless routers for home use, promising no Wi-Fi dead zones within the home. The acquisition enables Amazon to compete with Google's OnHubs.

Amazon and Apple are using devices such as iPhones, smartwatches, "smart" speakers and personal digital assistants such as Alexa to provide connectivity and power wireless networks. The companies use their wireless networks, such as Amazon's Sidewalk, as well as Apple's AirTag and Find My Network, to allow the devices to transmit tiny bits of data from any available wireless connection, thereby supplementing Wi-Fi networks and reducing wireless communication problems. Data is encrypted for security.

5) Wireless Service Subscriptions Worldwide Reach 9.1 Billion

Around the world, the wireless market continues to grow. As of the end of 2021, there were 371.6 million mobile internet connections of all types in use in the U.S. (according to Plunkett Research estimates). On a global basis, Plunkett Research estimates there were 9.1 billion subscriptions as of the end of 2021.

However, the actual number of individuals who subscribe to wireless connections is considerably lower than the total number of subscriptions. There are several reasons for this. In emerging nations, many people have multiple service subscriptions. Subscribers own and swap multiple SIM, or network subscription cards, on a frequent basis within their handsets, depending on the level of service quality available in a given location, or the balance left in a prepaid subscription account. In developed nations, subscribers may own multiple devices that require subscriptions, such as aircards, smartphones and tablets.

Cellphone manufacturers are continuously adding new features, cellphone service providers are offering more minutes per dollar spent and a growing number of consumers are discontinuing their landline phones. In 2021, 68.7% of U.S. adults lived in households that were wireless-only, according to the National Health Interview Survey, up from 31.6% in 2011.

Cellular companies are facing increased competition that is resulting in lower prices. The companies are also competing to purchase increased wireless spectrum in order to serve ever-growing downloads of data and video via wireless devices.

6) Handset Makers Adopt Android, Push Advanced 4G and 5G Smartphones

For 2021, analysts at Gartner reported that 1.434 billion smartphone units were sold worldwide (as well as much smaller number of basic cellphones). Advanced smartphone features on modestly-priced handsets are helping to fuel sales.

Upstart competitors have been growing in the mobile handset business. However, this is a very challenging, highly competitive business. Profit margins can be very thin (outside of Apple's enormously profitable iPhone business).

A firm in China called Xiaomi Corp. rapidly grew to be a major competitor in the handset manufacturer race. Xiaomi's phones are full featured but sold at near cost. The company is shipping its phones worldwide, with English versions bearing the brand "Mi." The firm's strategy is to build long-term revenues by selling services, such as music, video and games. Mi also quickly became one of India's leading brands in terms of market share. Meanwhile, India's Micromax, the largest smartphone vendor in India, is also making its phones available to Western consumers, at modest prices.

Phones are also featuring larger screens, especially in China. Large screens had been slow to catch on in the U.S. until the launch of the very popular Apple iPhone "Plus" models, which feature large screens.

Google forced transparency and open software on the cellular market in one bold act during 2007, with the launch of its Android wireless device operating system. Android is readily usable, at no fee, by any or all of the world's manufacturers. At the same time, it is open to the world's software and mobile app developers, free of charge. Google's highly successful strategy was to offer an open platform for mobile devices that is flexible and free, thus setting itself up to be the leader in mobile search and related services. The result is that Google is the envy of all other mobile search providers and all other mobile operating system creators. Google's Android is the most popular smartphone operating system worldwide.

Google's Android and Apple's iOS put additional competitive pressure on Microsoft in the mobile arena. Microsoft offers its Windows mobile operating system, which combines potential access to its Office suite of applications, Xbox LIVE gaming and a wide array of apps. However, Microsoft is running far behind Apple and Android in the mobile market.

The biggest news in the handset industry is the massive investments that manufacturers are making in order to offer state-of-the-art handsets capable of accessing ultra-fast 5G cellular networks. Such networks can offer ultrafast service, including internet downloads, at speeds substantially higher than those of 4G networks. 5G service is steadily rolling out in major cities worldwide, and consumers are excited about the prospect of near-instantaneous download of full-length movies, higher resolution for images and fast access to all online services.

Pioneers in 5G handsets include Samsung, OnePlus, Motorola, Huawei, Xiaomi and LG. Apple's 5G iPhone was first released in October 2020.

7) VOIP (Telephony over the Internet) Continues to Revolutionize the Telecommunications Industry

Local exchange telephone companies continue to face a dismal market for landlines. The popularity of smartphones is one reason, but another big threat to traditional phone lines is phone calls made over the internet. The rapid proliferation of high-speed access to the internet in homes and offices makes internet-based telephony a logical next step for a wide variety of businesses and consumers. Most newly installed telephone systems in mid-size to large business offices are based on Voice over Internet Protocol (VOIP), with virtually all telephone equipment manufacturers now offering VOIP equipment. Likewise, massive numbers of consumers around the globe are adopting VOIP for their long-distance services, particularly via the extremely popular Skype service, which is owned by Microsoft.

The fact that phone service can be provided online enables cable companies to compete head-on with telephone companies, since cable offers internet service as well as entertainment programming. Consequently, many households are subscribing to cable bundles that include VOIP, internet and entertainment/TV (and dropping their traditional landlines).

This makes cable companies major, direct competitors to traditional telephone companies. The competitive landscape has become even more complicated now that software companies are also getting into VOIP-type services. For example, WhatsApp, a global text messaging service owned by Facebook, offers VOIP features, putting it in direct competition with Skype. Viber, part of Japan's Rakuten, Inc., is another popular service used worldwide.

VOIP transforms phone calls into digital format, which is transmitted across the internet almost effortlessly and, more tellingly, cheaply. Rapid growth is occurring outside the U.S., particularly in nations where very high-speed broadband is commonplace in homes and offices. In emerging nations, VOIP growth has been fueled by its low cost.

Some VOIP services are essentially free of cost, or extremely inexpensive, but they offer limited services that often make or receive calls only with other members of the same service. This is how Skype got started. The strategy of Skype and similar firms has been to first offer free services, and then sell value-added subscriptions.

What a full-featured VOIP subscription offers is really a local telephone number bundled with long-distance service, at rates far cheaper than plans sold by other carriers, especially for international calls. In addition, many VOIP providers do not charge extra fees for services such as voice mail, call-waiting, caller ID and do-not-disturb, which temporarily blocks calls from certain parties and directs them to voice mail. Additional features are three-way calling as well as message retrieval and system settings via phone, web site or e-mail from any location worldwide.

A particularly interesting feature of VOIP service gives subscribers the ability to choose almost any area code as the prefix to their phone numbers. This means that a business or home can be located in Portland, Oregon, for example, but have an area code for Miami, Florida. This can be convenient for businesses that want to create local phone numbers for virtual locations in several cities or countries. Both businesses and consumers have the ability to set up toll-free numbers via VOIP as well.

Google offers its Voice app, with all calls within the U.S. free of charge and international calls at low cost without roaming fees. The service works using Wi-Fi networks.

Not surprisingly, there are a large number of players in the VOIP service market. A leader in this technology is the Vonage, a New Jersey-based firm, which was acquired by Ericsson in July 2022. In addition to relatively new firms like Vonage and Net2Phone, however, are the traditional telephone companies, such as Verizon and AT&T, as well as cable companies like Comcast. These companies followed Vonage's lead into VOIP, and all offer some type of VOIP service.

One of the more interesting players in VOIP is Skype. Skype works on peer-to-peer technology; that is, users must download the free Skype software to their computers or smartphones. They can then make VOIP calls to other computers that contain Skype software. Several users may participate in a conference call at once. Microsoft has integrated Skype's features into many of its products, such as its extremely popular Xbox game machines. Skype's challenge was to move beyond the distribution of free software, in order to generate significant subscription-based fees.

VOIP service is already a highly competitive market. Consequently, consumers can look forward to more new features and better service while prices drop. It can be vastly cheaper to operate an internet-based phone company than a traditional landline provider, because VOIP service providers don't have to invest in billions of dollars' worth of telephone equipment—instead, they rely on the internet as the backbone of their services. (Of course, this means that your VOIP telephone service is only as reliable as your internet connection.)

8) Telecom Equipment Makers Face Intense Competition from Manufacturers in China/Huawei Faces Controversy

Telephone service providers around the world have been facing up to vast changes in telecom technology. They have invested hundreds of billions of dollars in advanced switching equipment capable of providing ultra-fast internet service, wireless service and VOIP. Such investments involve massive purchases of state-of-the-art equipment. In the not-too-distant past, American companies were the world's leading makers of telecom equipment. Today however, there is intense global competition.

China-based telecom equipment manufacturers are booming. For example, Huawei Technologies Co., Ltd. has become a global leader. Chinese firms such as Datang Telecom Technology and ZTE Corp. are winning major contracts with telecom service providers around the world, including large sales of VOIP equipment, wireless infrastructure and network equipment. China has gained respect as a base for the manufacture of technology gear of very high quality. ZTE historically had sold to customers in China, elsewhere in Asia, and Africa. However, the firm is now aggressively marketing to telecom companies on a worldwide basis. ZTE has opened offices in U.S. cities and in Canada, as well as dozens of offices throughout Europe, Latin America, Africa and Asia/Pacific.

Nonetheless, some corporations and governments are uncomfortable installing and relying upon vital telecom equipment that has been designed and manufactured by Chinese firms. Particularly in today's environment of concerns about Chinese hackers gaining access to proprietary and highly sensitive corporate and government via the internet, some observers are worried that security breaches may be built into equipment made by companies like Huawei and its competitor ZTE, with or without the knowledge of corporate management.

In addition to a growing dependence upon wireless devices and ever-faster internet access, major corporations are committing to VOIP for their future telecommunications needs. VOIP not only saves on their voice communication costs, it also saves on costs for data and video. This is due to the fact that voice, video and data are able to run on the same network under VOIP.

Chinese competition was particularly hard on equipment maker Alcatel-Lucent. The France-based firm was the result of the merger of a leading European firm, Alcatel, and a leading U.S. telecom gear manufacturer, Lucent. The company was acquired by Nokia in 2016.

While Chinese telecom equipment makers have gained market share, many U.S. firms continue to do well in particular sectors. An industry giant is Cisco Systems, Inc., one of the world's leading makers of IP-based networking and communications technology. Additional leading firms include F5 Networks (internet traffic management tools), Ciena Corp. (network infrastructure equipment) and Viavi Solutions (formerly part of JDS Uniphase, which has split into two firms including Viavi). The next phase of global telecom equipment purchases will focus on upgrading services to ultra-fast 5G networks.

9) Telecom Companies, Including AT&T and Verizon, Compete Fiercely Against Cable in the TV, Internet and Telephone Market

Telecommunications companies like AT&T and Verizon are competing fiercely to take television viewers away from their cable and satellite competitors. Telecom companies are competing directly against cable companies by offering television programming in addition to telephone service and high-speed internet access.

Broadcast, cable and satellite TV providers are competing by positioning themselves with value-added content and services, such as enhanced interactive TV, video on demand and expensive programming such as made-for-TV movies and exclusive major league sports coverage.

AT&T took a significant strategic step in mid-2015 by completing its acquisition of satellite TV firm DIRECTV for $48.5 billion. In February 2021, the company agreed to sell 30% of its pay TV services (including Dish, U-Verse and AT&T TV) to TPG Capital, a private equity firm, for $1.8 billion. Operations of the three services are now jointly run under a new company called DIRECTV.

Verizon is also investing heavily in the delivery of TV over its internet connections. The firm offers a full menu of hundreds of TV channels (including HD channels) plus an on-demand library of thousands of movies. The service, called Fios, also provides premium channel packages for additional fees, as well as single or multi-room DVR. It provides subscribers with a very high-speed internet access service for speedy downloads.

The prospect of selling television subscriptions is a boon to telephone companies, since it gives them a highly coveted entertainment service to add to their phone service offerings. They have the ability to bundle services that might include any or all of the following: internet access, landline telephone, long-distance, VOIP telephony, cellular telephone and television—all on one discounted bill. The major telephone service providers have been in a tough spot in the past few years, with revenues from landline customers declining as more consumers switch to cellular phones and VOIP as their standard methods of communication. Delivering entertainment via the internet lines gives them a way to develop significant new revenues.

10) Fiber-to-the-Home (FTTH) Gains Traction

The major telephone firms are looking for ways to increase revenues through enhanced services while retaining their customer bases. One such way is through the delivery of ultra-high-speed internet access, combined with enhanced entertainment and telephone options, often by installing true fiber-to-the-home (FTTH) networks.

Under traditional telephone and internet service, homes are served by copper wires, which are limited to relatively slow service. However, old-fashioned copper networks are not up to the demands of today's always-on internet consumers. Fiber-optic cable might be used in trunk lines to connect regions and cities to each other at the switch and network level, but speed drops significantly once the service hits local copper lines.

Things are much more advanced in major markets today. In an AT&T project, fiber is being brought into special hubs in existing neighborhoods. From those hubs, fast services are delivered into the home with advanced technologies on short, final runs of copper wire.

FTTH, in contrast, delivers fiber-optic cable all the way from the network to the local switch directly into the living room—with this system, internet delivery speeds and the types of entertainment and data delivered can be astounding. Google's "Google Fiber" service offers extremely fast download speeds of 1,000 Mbps (one gigabit), roughly 50 to 100 times faster than typical DSL or cable service. Recently, Google offered the service in cities including Atlanta, Georgia; Austin, Texas; Charlotte, North Carolina; Kansas City, Missouri; Kansas City, Kansas; Nashville, Tennessee; and Provo Utah, among others. This puts tremendous competitive pressure on other internet access companies. It also makes consumers wonder why they have been paying high monthly rates for slow service for many years. In response, AT&T has begun providing 1,000 Mbps service in several major cities. The long-term result is going to be much better, faster internet access in major U.S. cities, at a reasonable cost.

Verizon has completed a multi-year FTTH program that makes it a U.S. leader in FTTH. Verizon's FTTH (called Fios) offers exceptional speeds up to 1 gigabit.

The Fiber-to-the-Home Council (www.ftthcouncil.org) tracks FTTH trends. In many cases, FTTH has been provided by local government or by subdivision developers who are determined to provide leading-edge connectivity as a value-added feature to new homes.

FTTH technologies, though expensive, may save the Bells from being trampled by the cable companies. Fiber-optic networks can give consumers extremely fast internet connections. Such ultra-high-speeds will also allow consumers to download movies in seconds and make videoconferencing a meaningful reality for businesses. (Additional fiber terms used in the industry include FTTP for Fiber-to-the-premises and FTTO for fiber-to-the-office.)

FTTH has been widely adopted in South Korea, Hong Kong, Japan, the United Arab Emirates and Taiwan, according to the FTTH Councils of Asia-Pacific, Europe and North America.

11) Global Internet Market Tops 5.2 Billion Users/Ultrafast Broadband Expands, both Fixed and Wireless

The majority of American cellphones are now smartphones. Big improvements in the devices, such as the latest iPhones and Android-based units, along with enhanced high-speed access via 4G networks, are fueling this growth. In addition, most major e-commerce, news and entertainment sites have carefully designed their web pages to perform reasonably well on the "third screen," that is, cellphones (with TV being the first screen and desktop or laptop computers being the second screen). Globally, the number of internet users was 5.2 billion as of March 2021, (including wireless) according to Internetworldstats.com.

Internet access speeds continue to increase dramatically. Google launched its "Google Fiber" ultra-high-speed internet service in Kansas City, Kansas in 2012, and soon expanded into Austin, Texas; Provo, Utah; and Atlanta, Georgia. This system allows homes and businesses to have 1 gigabit per second access, roughly 100 to 200 times the speed of typical DSL or mobile broadband. More than 1,000 U.S. towns and cities applied for the service when it was first announced, but Google is gauging the results of this initial effort before making any decisions about rolling it out.

AT&T initially launched a similar 1 gigabit service in competition with Google in the Austin area called AT&T FIBER. The firm offers this fast service in dozens of cities across the U.S.

What will widespread use of fast internet access mean to consumers? The opportunities for new or enhanced products and services are endless, and the amount of entertainment, news, commerce and personal services designed to take advantage of broadband will continue to grow rapidly. For example, education support and classes via broadband is rapidly growing into a major industry.

Broadband in the home is essential for everyday activities ranging from children's homework to shopping to managing financial accounts. Online entertainment and information options, already vast, will grow daily. Some online services are becoming indispensable, and always-on is the new accepted standard. The quality of streaming video and audio is becoming clear and reliable, making music and movie downloads extremely fast, and allowing internet telephone users to see their parties on the other end as if they were in the same room. Compression and caching techniques are evolving, and distribution and storage costs are expected to plummet. A very significant portion of today's radio, television and movie entertainment is migrating to the web.

12) Telecommunications Systems Move Online Including Unified Communications, Telepresence

VOIP (internet-based telephony or Voice Over Internet Protocol) is only the first step in the telecommunications revolution taking place online. The next generation is known as "unified communications," a technology pursued by Microsoft, Cisco and many others. Special software operating on a local office server, or in the cloud, enables each office worker to have access, via the desktop PC (and via mobile devices with remote access), to communications tools that include VOIP phone service, e-mail, voice mail, fax, instant messaging (IM), collaborative calendars and schedules, contact information such as address books, audio conferencing and video conferencing.

For example, to make a phone call with unified communications you open your communications screen, click on a name in your address book, and the call is placed via VOIP. To send a fax, you click on a contact's fax number in your address book, click on the item you want to fax in your documents folder, and the fax is gone. When you have an incoming phone call, your screen will tell you what number is calling and match that number to a name in your address book. You might take the call, or you might send it to voice mail with one click. In the same way that you currently save and archive e-mails, you can digitally record, save, review and archive phone calls. Mobility, security and advanced collaboration with coworkers are among the features emphasized.

Network equipment giant Cisco is pushing a next-generation video-conferencing technology it refers to as TelePresence. Cisco provides hardware and software needed to fully equip a state-of-the-art video conferencing room. Such a room may include as many as three 65-inch plasma monitors, advanced high-speed cameras, projectors, microphones, speakers that are set up so that the sound seems to come directly from the participant who is talking, along with appropriate software. There are virtually no delays in the conversation, and images are crisp and life-like. Indeed, the image of a participant, from the chest up, is life-size. High bandwidth and extremely reliable, fast internet connections are prerequisites. Its latest TelePresence iteration, the IX5000 Series, offers triple screen video and takes half the installation time, power and bandwidth of existing products.

Cisco offers a smaller system for use in individual offices and a large telepresence room designed for group training and team meetings. (The company also offers much simpler Unified IP Phone sets with video touchscreens for person-to-person meetings that turns any flat panel display into a telepresence system for small meeting rooms,)

However, the quality of a telepresence meeting relies very heavily on the quality of the equipment involved. For many major corporate users, near life-like quality is required, such as that delivered by Cisco's TelePresence equipment.

The advantages to telepresence systems such as Cisco's are based on making remote meetings more effective and life-like, while enhancing collaboration and communications, as well as greatly reducing the need for expensive, time-consuming business travel. The goal is to have results from teleconference meetings that near the results of F2F (face-to-face) meetings. High-end equipment makers are feeling strong competition from startups and lower-priced products.

Another player in videoconferencing is LifeSize Communications based in Austin, Texas. LifeSize uses high-definition (HD) cameras to send video streams over ordinary internet hardware, and was previously owned by Logitech before spinning off as a standalone company. Another noteworthy entrant to the telepresence market is Vidyo, Inc., a New Jersey firm which offers a mid-priced, three- to 20-screen videoconferencing room. In May 2019, Vidyo was acquired by Enghouse Systems Limited for approximately $40 million.

SPOTLIGHT: Zoom Video Communications, Inc.

Zoom Video Communications, Inc. designs and develops cloud-based video and web conferencing software. Zoom's software unifies cloud video conferencing, online meetings, group messaging and conference room solutions into one simple-to-use platform. The company's products work across multiple room systems. Zoom's software conferencing application program interface (API) runs within a business' conference room or workspace. Cloud video conferencing features include HD video and voice, full-screen with gallery views, dual stream/dual screen, feature-rich mobile apps and various ways to join the conferences (Zoom Room, view-only, voice only and more). The firm also offers a hybrid cloud service with 24/7 online monitoring and instant global service backup. Partner integrations provide content sharing, scheduling/starting meetings, unified log-in, marketing/process automation and room collaboration. Pricing plans range from free to $100 per month, among other options. Zoom is headquartered in San Jose, California, with additional U.S. offices in California, Colorado and Kansas, as well as international offices in Sydney, London, Paris and Amsterdam. The Coronavirus pandemic spurred phenomenal growth in Zoom meetings. The firm reported usage of up to 300 million meeting participants per day in the second quarter of 2022, compared to only 10 million per day in December 2019. For all of 2021, the firm reported over 3.3 trillion meeting minutes facilitated.

Meanwhile, for simpler needs and smaller budgets, startups such as Fuze, formerly FuzeBox, and Blue Jeans Network offer inexpensive subscription software that uses smartphones, tablets and PCs to provide videoconferencing. Both companies have signed big name corporate customers, including Facebook for Blue Jeans Network and Groupon for Fuze. Cisco offers its own subscription-based software. Over the mid-term, watch for more Cloud-based subscription options that offer videoconferencing and telepresence at a fraction of full-blown conference room system prices.

13) Carriers Reinstate Unlimited Access Plans for Smartphones/Face Intense Subscription Price Competition

Starting in 2016, competition among the U.S.'s major cellphone carriers grew to very intense levels. After years of charging users for the amount of data they consumed, carriers one-by-one reinstated unlimited access plans in order to stay competitive.

Carriers must look to the future needs of their subscribers and realizing that massive amounts of capital investment will be needed on a continuous basis in order to provide the fast internet service that customers want. The wireless industry has very clearly evolved from a voice-based business to a data- and video-based business, and pricing plans must support this change while providing the cash that carriers need for capital expenditures. Hundreds of billions of dollars will be invested in a gradual, global rollout of ultra-fast 5G wireless networks over the mid-term. A recent IHS Markit study estimated that 5G will create $3.5 trillion in global economic output and 22 million global jobs by 2035.

The continued growth of mobile video, mobile music, web sites tailored to the mobile screen, as well as social media tools such as Facebook will lead to ever-higher download levels. The Cisco Visual Networking Index (VNI) forecasts mobile (cellular) speeds will reach 44 Mbps by 2023. Cisco further estimated that 82% of all Internet traffic would be video by 2022. In fact, landline internet service companies, both cable and DSL, face the same challenges and opportunities from growing bandwidth use due to online videos, the download of full-length feature films from firms like Netflix, and other popular web trends.

Meanwhile, billions of text messages are sent every day around the world. New mobile messaging tools from firms like WhatsApp and WeChat enable mobile users to send and receive text messages via internet sites, rather than via the cellular carriers' services. For example, WhatsApp (www.whatsapp.com) offers an app that enables users to send each other text messages via a smartphone's internet connection. Facebook acquired WhatsApp for $19 billion in 2014.

Cellular service providers are seeking unique ways to increase their revenues in an environment where subscription rates will remain fiercely competitive between providers. One way to increase total revenue is to increase mobile advertising. This explains Verizon's recent acquisitions of advertising technologies and sites, including its 2016 agreement to purchase the online assets of Yahoo! In 2021, both AT&T and Verizon were offering discounts or incentives generous enough to pay for high-end smartphones such as Apple's iPhone 12.

Cable companies, including Comcast Corp. and Charter Communications, Inc., are now offering mobile phone service, often bundled with cable service. Subscribers enjoy low prices and the ability to adjust their phone plans quickly and easily. Traditional wireless companies remain dominant. However, the cable companies are aggressively seeking mobile business as a way to make up for growing losses of pay TV subscribers.

14) Smaller Satellites (SmallSats and CubeSats) and Low Earth Orbit Revolutionize Telecommunications

Until recently, satellites orbiting the Earth were large, heavy craft. In recent years, however, a growing fleet of smaller satellites has been launched, ranging from the size of kitchen refrigerators down to golf balls. These relatively tiny SmallSats have an exponentially growing number of uses and, thanks to their miniaturization coupled with rocket technology innovation, are far cheaper to launch and maintain than ever before. A class of the smallest SmallSats are called CubeSats, which, according to NASA, refers to a class of nanosatellites that use a standard size and form factor. The standard CubeSat size uses one standard module or unit ("1U") measuring 10x10x10 centimeters, and is extendable to larger sizes; 1.5, 2, 3, 6 and even 12U. They are particularly easy to launch because they have electronic and physical interfaces that allow 1U modules to stack together in a dispenser that increases payload in a rocket, sometimes called a "secondary payload." CubeSats were originally developed in 1999 by California Polytechnic State University at San Luis Obispo (Cal Poly) and Stanford University to provide a platform for education and space exploration.

Satellites of all sizes are being streamlined to rely more on 3-D printing which saves manufacturing costs and increases orbit efficiency. For example, Boeing Co. is incorporating 3-D printing in the development of its Phoenix satellites.

Internet Research Tip: CubeSats

For more on CubeSats, see NASA's web site which includes comprehensive images, videos and media resources:
www.nasa.gov/mission_pages/cubesats/index.html

Some of the uses for these new baby satellites include gathering weather data to aid farmers, transportation and logistics firms, and disaster relief workers. Environmental impact data can be used by government agencies to monitor deforestation, polar icecap and ocean changes over time. Satellites are also bringing internet service to all parts of the globe, including rural areas that other providers cannot reach.

SmallSats promise vast increases in cost-efficiency, utility and flexibility. Today's state-of-the-art satellites feature computer circuitry that can be reprogrammed from the ground, on an as-needed basis. This is vast improvement over traditional satellites. SmallSats are easier and faster to manufacture as well.

Thanks to advanced technologies and miniaturization, today's tiny satellites are much lighter in weight than former generations. This means that a launch rocket can carry a much higher number of satellites in one payload. This is important, because launch costs are one of the major investments required in completing the tasks required to get satellites into service. Entrepreneur Elon Musk, founder of electric car company Tesla, is revolutionizing satellite launches. His SpaceX firm has shown that rocket bodies can be recovered and reused in a highly efficient manner. His methods are slashing the cost of putting satellites into orbit.

Space launch firm SpaceX has its own system, called Starlink, of as many as 12,000 LOE satellites (although the company hopes to add as many as 30,000 more). As of mid-2022, the company had launched about 2,600 Starlink satellites and reported in excess of 400,000 subscribers worldwide for its satellite internet service. Officials at NASA have stated concern over the potential of this huge number of satellites to cause space collisions. This firm, associated with Tesla founder Elon Musk, will have significant advantages, since it operates an extremely cost-effective rocket launch service. Starlink's satellites are deployed at three different altitudes. Its satellites weight only about 200 kilograms each. A grouping of 60 of these units can provide up to 1 terabit per second of bandwidth. In June 2022, the U.S. Federal Communications Commission (FCC) approved the firm's request to service moving vehicles including boats, airplanes and recreational vehicles, as well as commercial trucks and consumer's cars. This will open up a very broad new market.

A firm called OneWeb owns system that may eventually encompass from 648 low-Earth orbit (LOE) satellites. This will drive down costs and improve performance in telecommunications and internet applications. Unfortunately, after raising $1.7 billion and launching only 75 satellites, OneWeb took bankruptcy in the spring of 2020 during the Coronavirus pandemic. It was quickly acquired, in July 2020, by the UK Department for Business,

Energy and Industrial Strategy. As of mid-2022, about 428 OneWeb satellites were in operation.

LEO satellites are typically in orbit at altitudes of 400 to 1,000 miles and are used for telecommunications and internet access. Because of their low altitude, the time required to send and receive Earth signals is shorter. This can be extremely advantageous in many circumstances. For example, passengers on cruise ships are dependent upon satellite systems for internet access, which traditionally has been much slower than land-based access. LEO satellites are speeding up ship-board internet to the extent that movies, gaming and intense business applications are much more practical.

Cruise ships are perfect customers for fast internet access via low-orbit satellites. Guests are encouraged to Tweet, Skype, Stream movies (using accounts with DIRECTV, Netflix or Hulu), connect with friends or play Xbox Live with gamers worldwide.

Jeff Bezos' Blue Origin is investing $10 billion to launch more than 3,200 low-Earth-orbit satellites between 2020 and 2029. The network, called Project Kuiper, received FCC approval in July 2020 and promises to provide reliable, affordable broadband service to underserved communities around the globe. By the end of 2022, Project Kuiper plans to have two prototype satellites launched, and it has scheduled 83 multi-satellite launches over a five-year period.

15) The Internet of Things (IoT) and M2M to Boom, Enhanced by Artificial Intelligence (AI)/Open New Avenues for Hacking

The phrase "Internet of Things" or "IoT" will become increasingly commonplace. It refers to wireless communications known as M2M or machine-to-machine. M2M can be as simple as a refrigerator that lets a smartphone app know when you are running low on milk (via Wi-Fi) to a vast, exceedingly complex network of wireless devices connecting all of the devices in a massive factory. Analysts at network device giant Cisco expect M2M connections to grow dramatically and rapidly, to 14.7 billion by 2023.

A Wireless Sensor Network (WSN) consists of a grouping of remote sensors that transmit data wirelessly to a receiver that is collecting information into a database. Special controls may alert the network's manager to changes in the environment, traffic or hazardous conditions within the vicinity of the sensors. Long-term collection of data from

remote sensors can be used to establish patterns and make predictions, as well as to manage surveillance in real time. Another term that is coming into wide use is M2M2P or machine-to-machine-to-people. The "to-people" part refers to the fact that consumers, workers and professionals will increasingly be actively involved in the gathering of data, its analysis and its usage. For example, M2M2P systems that automatically collect data from patients' bedsides; analyze, chart and store that data; and make the data available to doctors or nurses so that they may take any necessary actions are becoming increasingly powerful. Such systems, part of the growing trend of electronic health records (EHR), can also include bedside comments spoken into tablet computers by physicians that are transcribed automatically by voice recognition software and then stored into EHR.

Connected Devices are a Notorious Channel for Hackers' Entry into Networks and Data

The Internet of Things (IoT) is a vital component of machine-to-machine (M2M) communications and will become even more important with the rollout of fast, urban 5G wireless networks. IoT sensors, monitors and cameras can gather the types of data that can make cities more efficient (in a wide range of areas, from traffic flow to lighting efficiency); make agricultural technology advance (such as better efficiency in irrigation and fertilization) and enhance operations in manufacturing and distribution facilities of all types. However, connected devices such as these are notorious nodes through which hackers have had stunning and costly success at taking over networks and stealing data. There is a massive need, and an accompanying business opportunity, to make M2M networks as cybersafe as possible.

The long-term trend of miniaturization is playing a vital role in M2M. Intel and other firms are working on convergence of MEMS (microelectromechanical systems—tiny devices or switches that can measure changes such as acceleration or vibration), RFID (wireless radio frequency identification devices) and sometimes tiny computer processors (microprocessors embedded with software). In a small but powerful package, such remote sensors can monitor and transmit the stress level or metal fatigue in a highway bridge or an aircraft wing, or monitor manufacturing processes and product quality in a factory. In our age of growing focus on environmental quality, they can be designed to analyze surrounding air for chemicals, pollutants or particles, using lab on a chip technology that already largely exists. Some observers have referred to these wireless sensors as "smart dust," expecting vast quantities of them to be scattered about the Earth as the sensors become smaller and less expensive over the near future. Energy efficiency is going to benefit greatly, particularly in newly-built offices and factories. An important use of advanced sensors will be to monitor and control energy efficiency on a room-by-room, or even square meter-by-square meter, basis in large buildings.

In an almost infinite variety of possible, efficiency-enhancing applications, artificial intelligence (AI) software can use data gathered from smart dust to forecast needed changes, and robotics or microswitches can then act upon that data, making adjustments in processes automatically. For example, such a system of sensors and controls could make adjustments to the amount of an ingredient being added to the assembly line in a paint factory or food processing plant; increase fresh air flow to a factory room; or adjust air conditioning output in one room while leaving a nearby hallway as is. The ability to monitor conditions such as these 24/7, and provide instant analysis and reporting to engineers, means that potential problems can be deterred, manufacturing defects can be avoided and energy efficiency can be enhanced dramatically. Virtually all industry sectors and processes will benefit.

Look for data sensors in homes to proliferate over the near- to mid-term. In the insurance business, live data emanating from sensors in homes could lead to more intelligent policies. Monitoring data via smartphone could be a significant opportunity for companies in the senior care, child care and pet care sectors.

Internet Research Tip: The Internet of Things Connections Counter and Infographic:
Network equipment maker Cisco has posted a "Connections Counter" online, which provides a running count of people and things connected to the internet, newsroom.cisco.com/ioe . This page also provides many useful links to Internet of Things resources. In addition, Cisco posts a highly informative Internet of Things page at www.cisco.com/web/solutions/trends/iot/overview.html which includes a one-minute IoT video.

Meanwhile, French technology firm SigFox offers a simple, inexpensive wireless network, designed specifically for M2M needs. The network transmits data at a rate of 100 bits per second, which is slower by a factor of 1,000 than most smartphone networks but does so cheaply while it fills simple transmission needs such as those from many wireless sensors (such as Whistle, a clip-on collar sensor that tracks dog activity levels). Base stations use a wireless chip that costs only $1 to $2, and customers pay modest service charges per year per device. As of early-2020, SigFox had deployed its technology in about 63 countries.

Intel and other firms have developed methods that enable such remote sensors to bypass the need for internal batteries. Instead, they can run on "power harvesting circuits" that are able to reap power from nearby television signals, FM radio signals, Wi-Fi networks or RFID readers.

Memory chips used in sensors are much smaller than those in smartphones and laptops, opening a major opportunity for manufacturers such as Adesto Technologies. The firm makes chips that store between 32 kilobits and one megabit of data, making them a good fit for small monitors such as fitness data tracking wristbands. Future applications might include location-based beacons in retail stores that alert nearby customers to selected items by cellphone. Smoke detectors with small memory chips could sense battery life, while blood transfusion bags could track their locations, ages and content viabilities.

Internet Research Tip: Internet of Things (IoT) Networks:
For more information on wireless network systems and remote sensors, see:
Analog Devices, Inc. (which acquired Linear Technology Corp.), www.analog.com/en/applications/technology/smartmesh-pavilion-home.html%20#
Moog, Inc., www.moog.com
C3ai, c3iot.com/industries

Chapter 2

TELECOMMUNICATIONS INDUSTRY STATISTICS

CONTENTS:

Telecommunications Industry Statistics and Market Size Overview	**24**
Wired Telecommunications Carriers, Estimated Sources of Revenue, U.S.: 2016-2020	**25**
Wired Telecommunications Carriers, Estimated Breakdown of Revenue by Type of Customer, U.S.: 2017-2020	**26**
ATT & the Bell Companies, Then & Now	**27**
Total Retail Local Telephone Service Connections, U.S.: 2016-2019	**28**
Wireless Telecommunications Carriers (except Satellite): Estimated Sources of Revenue & Expenses, U.S.: 2016-2020	**29**
Top Mobile Operators by Number of Subscribers, Worldwide	**30**
Satellite Telecommunications: Estimated Sources of Revenue & Expenses, U.S.: 2017-2020	**31**
Number of Business & Residential High Speed Internet Lines, U.S.: 2016-2021	**32**
Exports & Imports of Telecommunications Equipment, U.S.: 2016-2021	**33**
Employment in the Telecommunications Industry, U.S.: 1990-May 2022	**34**

Telecommunications Industry Statistics and Market Size Overview

	Number	Unit	Year	Source
U.S. Telecommunications Spending (ICT)[1], 2022**	1.5	Tril. US$	2022	STA
U.S. Telecommunications Spending (ICT)[1], 2020**	1.8	Tril. US$	2020	TIA
Worldwide Telecommunications Spending (ICT)**	7.1	Tril. US$	2020	PRE
Landline				
Total Wireline Retail Local Telephone Service Connections, U.S.	107.0	Mil. Units	Jun-19	FCC
Global Landline Subscriptions	915.0	Mil. Units	2019	ITU
Wireless				
Total Wireless Service Revenues, U.S.	197.5	Bil. US$	2021	PRE
Households with Wireless-only Service, U.S.	73.9	%	Dec-21	NHIS
Portion of Mobile Consumers with a Smartphone, U.S.	85.0	%	Feb-21	Pew
Mobile Advertising Spending, U.S.	113.2	Bil. US$	2020	eMarketer
Mobile Cellular Subscriptions, Worldwide*	110.0	Per 100 inhabitants	2021	ITU
Total Mobile Phone Shipments	1.5	Bil. Units	2022	Gartner
Mobile Application Downloads, Worldwide	218	Bil. Units	2020	App Annie
Mobile App Store Revenue, Worldwide	65.0	Bil. US$	Jun-22	Sensor Tower
Equipment Revenue				
U.S. Exports of Telecommunications Equipment	30.3	Bil. US$	2021	PRE
U.S. Imports of Telecommunications Equipment	99.2	Bil. US$	2021	PRE
TV, Cable & Internet				
Number of High Speed Internet Lines, U.S. (including mobile wireless)	359.4	Mil. Units	Dec-21	PRE
Number of High Speed Internet Lines, U.S. (not including mobile wireless)	104.0	Mil. Units	Dec-21	PRE
Number of Global Internet Users	5.4	Bil. Units	Jun-22	IWS
Monthly Global Internet Protocol (IP) Traffic	278	EB per Month	2021	Cisco VNI
Projected Monthly Global IP Traffic	396	EB per Month	2023	Cisco VNI
Number of TV households	122	Mil. Units	2021	Nielsen
Employment				
Employment in the Telecommunications Industry, U.S.	661.5	Thous.	Jun-22	BLS

TIA = Telecommunications Industry Association; FCC = U.S. Federal Communications Commission; ITU = International Telecommunication Union; PRE = Plunkett Research Estimate; NHIS = National Health Interview Survey; IWS = Internetworldstats.com; Cisco VNI = Cisco Visual Networking Index; BLS = U.S. Bureau of Labor Statistics.

* The actual number of individuals who own cell phones is lower, as some people own more than one "subscription."

** Forecasted or estimated figure.

[1] ICT, or Information and Communication Technologies, is a term used to describe the relationship between the myriad types of goods, services and networks that make up the global information and communications system. Sectors involved in ICT include landlines, data networks, the Internet, wireless communications, (including cellular and remote wireless sensors) and satellites.

Source: Plunkett Research®, Ltd. Copyright © 2022, All Rights Reserved

Plunkett's Telecommunications Industry Almanac 2023

www.plunkettresearch.com

Wired Telecommunications Carriers,
Estimated Sources of Revenue, U.S.: 2016-2020

(In Millions of US$; Latest Year Available)

NAICS Code: 5171[1]	2020	2019	2018	2017	2016	% Chg. 20/19
Total Operating Revenue	**312,008**	**320,681**	**311,465**	**315,390**	**326,119**	**-2.7%**
Fixed local	12,843	13,865	15,165	17,087	27,386	**-7.4%**
Fixed long-distance	17,562	20,000	20,401	S	10,820	**-12.2%**
Fixed all distance	S	3,554	2,761	1,587	1,952	**NA**
Carrier services	14,459	14,493	14,279	13,254	S	**-0.2%**
Private network services	6,976	8,073	7,598	S	17,607	**-13.6%**
Internet access services	103,256	100,371	95,786	88,714	87,250	**2.9%**
Internet telephony	9,495	10,100	9,583	10,244	11,291	**-6.0%**
Telecommunications network installation services	1,686	S	1,293	842	818	**NA**
Reselling services for telecommunications equipment, retail	1,838	1,873	802	1,466	1,476	**-1.9%**
Rental of telecommunications equipment	8,496	7,327	6,923	4,458	1,701	**16.0%**
Repair and maintenance services for telecommunications equipment	S	S	S	858	S	**NA**
Subscriber line charges	2,123	2,312	2,784	2,896	3,222	**-8.2%**
Basic programming package	60,721	61,887	60,321	59,733	63,659	**-1.9%**
Premium programming package	20,047	22,954	22,180	22,978	24,394	**-12.7%**
Pay-per-view	1,399	1,645	1,986	2,782	2,898	**-15.0%**
Air time	5,838	5,919	6,451	5,607	6,345	**-1.4%**
Rental and reselling services for program distribution equipment	10,771	10,679	11,001	12,049	13,862	**0.9%**
Installation services for connections to program distribution networks	S	S	S	S	S	**NA**
Website hosting services	S	S	S	S	S	**NA**
All other operating revenue	28,600	30,709	28,931	33,544	31,271	**-6.9%**

Note: Dollar volume estimates are published in millions of dollars; consequently, results may not be additive.

S = Estimate does not meet publication standards because of high sampling variability (coefficient of variation is greater than 30%) or poor response quality (total quantity response rate is less than 50%) or other concerns about the estimate's quality. Unpublished estimates derived from this table by subtraction are subject to these same limitations and should not be attributed to the U.S. Census Bureau. NA = Not Avaialble.

[1] Includes 2002 NAICS 5171 (Wired Telecommunications Carriers), 2002 NAICS 5175 (Cable and Other Program Distribution), and a portion of 2002 NAICS 518111 (Internet Service Providers). Historical data, 2010 and prior, have been restated to reflect comparable data on the 2007 NAICS basis.

Source: U.S. Census Bureau

Plunkett Research, ® Ltd

www.plunkettresearch.com

Wired Telecommunications Carriers, Estimated Breakdown of Revenue by Type of Customer, U.S.: 2017-2020

(In Millions of US$; Latest Year Available)

NAICS 5171* Kind of Business	2020	2019	2018	2017
Total Operating Revenue	**312,008**	**320,681**	**311,465**	**315,390**
Household consumers & individual users	206,345	209,519	223,043	194,532
Business firms, not-for-profit organizations & Gov't (Federal, state & local)	105,573	111,162	88,422	120,858
Fixed local telephony	12,843	13,865	15,165	17,087
Household consumers & individual users	6,618	7,154	7,787	8,628
Business firms, not-for-profit organizations & Gov't (Federal, state & local)	6,225	6,711	7,378	8,459
Fixed long-distance telephony	17,562	20,000	20,401	S
Household consumers & individual users	10,555	11,839	15,422	S
Business firms, not-for-profit organizations & Gov't (Federal, state & local)	7,007	8,161	4,979	S
Subscriber line charges	2,123	2,312	2,784	2,896
Household consumers & individual users	S	S	1,698	1,760
Business firms, not-for-profit organizations & Gov't (Federal, state & local)	849	906	1,086	1,136

Note: Dollar volume estimates are published in millions of dollars; consequently, results may not be additive.

* Includes 2002 NAICS 5171 (Wired Telecommunications Carriers), 2002 NAICS 5175 (Cable and Other Program Distribution), and a portion of 2002 NAICS 518111 (Internet Service Providers). Historical data, 2010 and prior, have been restated to reflect comparable data on the 2007 NAICS basis.

S = Estimate does not meet publication standards because of high sampling variability (coefficient of variation is greater than 30%) or poor response quality (total quantity response rate is less than 50%). Unpublished estimates derived from this table by subtraction are subject to these same limitations and should not be attributed to the U.S. Census Bureau. NA = Not Available.

Source: U.S. Census Bureau

Plunkett Research, ® Ltd

www.plunkettresearch.com

ATT and the Bell Companies, Then & Now

AT&T's seven Regional Bell Operating Companies (RBOCs, also known as the "Baby Bells") at the time of its breakup on January 1, 1984:

- Ameritech (merged with SBC, SBC is now AT&T, Inc.)
- Bell Atlantic (merged in 2000 with GTE and now called Verizon)
- BellSouth (merged with AT&T in 2006)
- NYNEX (merged in 1997 with Bell Atlantic, now called Verizon)
- Pacific Telesis Group (merged with SBC; SBC is now AT&T, Inc.)
- Southwestern Bell (renamed itself SBC Communications; SBC is now AT&T, Inc.)
- US West (acquired in 2000 by Qwest)

Remaining major consolidated firms:

- Verizon
- AT&T, Inc.
    - Originally Southwestern Bell. Renamed SBC Communications.
    - Acquired AT&T Corp. in November 2005 for $16.5 billion. Renamed AT&T, Inc.
    - Acquired BellSouth and Cingular Wireless in December 2006.
    - Acquired satellite TV provider DIRECTV in July 2015 for $49 billion.
    - Acquired media company Time Warner, Inc. in 2018 for $85 billion.
    - Established a new company in 2021 that now owns the DIRECTV, AT&T TV and U-verse video services businesses. The firm is called DIRECTV and is jointly owned by AT&T and private equity firm TPG Capital.
    - Spun-off its content business to combine it with Discovery, Inc. in 2022.

Source: Plunkett Research, Ltd. Copyright © 2022, All Rights Reserved
www.plunkettresearch.com

Total Retail Local Telephone Service Connections, U.S.: 2016-2019

(Figures in Thousands; Latest Year Available)

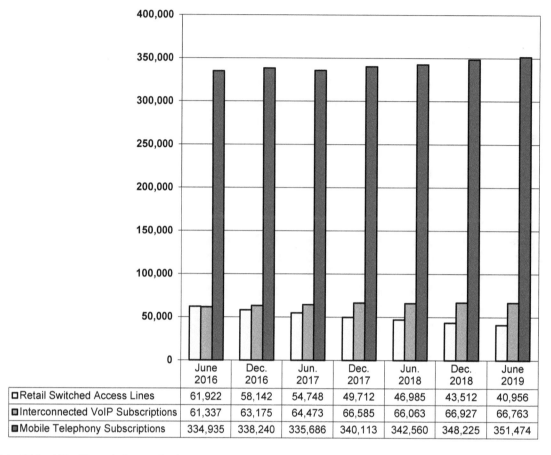

	June 2016	Dec. 2016	Jun. 2017	Dec. 2017	Jun. 2018	Dec. 2018	June 2019
☐Retail Switched Access Lines	61,922	58,142	54,748	49,712	46,985	43,512	40,956
☐Interconnected VoIP Subscriptions	61,337	63,175	64,473	66,585	66,063	66,927	66,763
■Mobile Telephony Subscriptions	334,935	338,240	335,686	340,113	342,560	348,225	351,474

Note: Of the 107 million wireline retail voice telephone service connections (including both switched access lines and interconnected VoIP subscriptions) in June 2019, 54 million (or 50%) were residential connections and 54 million (or 50%) were business connections.

Source: U.S. Federal Communications Commission (FCC)

Plunkett Research, Ltd.

www.plunkettresearch.com

Wireless Telecommunications Carriers (except Satellite): Estimated Sources of Revenue & Expenses, U.S.: 2016-2020

(In Millions of US$; Latest Year Available)

NAICS 5172[1]	2020	2019	2018	2017	2016	% Chg. 20/19
Total Operating Revenue	**278,868**	**276,114**	**272,046**	**257,778**	**259,321**	**1.0%**
Messaging (paging) services	S	S	2,217	S	2,956	NA
Mobile telephony	S	28,579	47,477	S	45,659	NA
Mobile long-distance	1,971	2,585	2,484	2,530	2,441	-23.8%
Mobile all distance	86,778	59,712	40,031	37,722	S	45.3%
Internet access services	96,949	92,179	90,949	96,257	97,296	5.2%
Installation services for telecommunication networks	267	195	S	192	242	36.9%
Reselling services for telecommunications equipment, retail	56,276	58,217	57,730	46,767	44,367	-3.3%
Rental of telecommunications equipment	6,954	6,643	6,768	4,714	S	4.7%
Repair & maintenance svcs. for telecom equipment	1,656	1,694	1,684	1,861	2,338	-2.2%
All other operating revenue	19,364	24,052	22,518	18,092	17,484	-19.5%
Total Operating Expenses	**181,153**	**182,269**	**181,197**	**189,092**	**188,983**	**-0.6%**
Gross annual payroll	18,026	18,306	18,052	18,561	18,899	-1.5%
Health insurance	NA	NA	NA	2,065	2,067	NA
Defined benefit pension plans	NA	NA	NA	S	196	NA
Defined contribution plans	NA	NA	NA	521	425	NA
Payroll taxes, employer paid insurance premiums (except health), and other employer benefits	NA	NA	NA	2,230	2,213	NA
Temporary staff and leased employee expense	S	5,421	5,470	S	7,914	NA
Expensed equipment	D	NA	S	42,766	43,164	NA
Expensed purchases of other materials, parts, and supplies	S	S	S	1,742	1,673	NA
Expensed purchases of software	D	D	S	945	921	NA
Purchased electricity	1,540	NA	NA	S	1,029	NA
Purchased fuels (except motor fuels)	S	NA	NA	3	2	NA
Lease and rental payments for machinery, equipment, and other tangible items	NA	NA	NA	S	887	NA
Lease and rental payments for land, buildings, structures, store spaces, and offices	S	NA	NA	11,059	10,815	NA
Purchased repairs and maintenance to machinery and equipment	1,992	NA	NA	188	184	NA
Purchased repairs and maintenance to buildings, structures, and offices	480	NA	NA	486	310	NA
Purchased advertising and promotional services	7,248	NA	NA	8,337	8,145	NA
Access charges	6,037	4,775	4,811	5,593	5,147	-0.7%
Universal service contributions (USC) and other similar charges	S	S	3,586	S	4,571	NA
Depreciation and amortization charges	S	36,077	37,176	35,634	34,721	-3.0%
Governmental taxes and license fees	2,361	NA	NA	2,278	2,353	NA
Data processing and other purchased computer services	NA	NA	NA	1,953	2,035	NA
Purchased communication services	NA	NA	NA	6,605	7,012	NA
Water, sewer, refuse removal, and other utility payments	NA	NA	NA	13	7	NA
Purchased professional and technical services	S	NA	NA	4,755	4,498	NA
All other operating expenses	40,726	D	60,930	S	29,795	NA

Note: Dollar volume estimates are published in millions of dollars; consequently, results may not be additive.

S = Estimate does not meet publication standards because of high sampling variability (coefficient of variation is greater than 30%) or poor response quality (total quantity response rate is less than 50%). Unpublished estimates derived from this table by subtraction are subject to these same limitations and should not be attributed to the U.S. Census Bureau. NA = Not Available.

D = Estimate in table is withheld to avoid disclosing data of individual companies; data are included in higher level totals.
[1] Includes 2002 NAICS 517211 (Paging), 2002 NAICS 517212 (Cellular and Other Wireless Telecommunications) and a portion of 2002 NAICS 518111 (Internet Service Providers). Historical data, 2010 and prior, have been restated to reflect comparable data on the 2007 NAICS basis.

Source: U.S. Census Bureau
Plunkett Research, ® Ltd.
www.plunkettresearch.com

Top Mobile Operators by Number of Subscribers, Worldwide

(In Millions)

#	Mobile Operator	Major Operating Regions	Mobile Subscribers	Date
1	China Mobile	China	956.9	May-22
2	China Unicom	China	795.5	May-22
3	Airtel	India, South Asia, Africa	458.0	Mar-22
4	Reliance Jio	India	415.7	Dec-21
5	China Telecom	China	372.4	Dec-21
6	Telefónica	Spain, Europe, Latin America	369.1	Dec-21
7	Vodafone	Europe, Africa, Australia	315.0	Dec-21
8	América Móvil	Latin America	286.5	Dec-21
9	MTN Group	Africa, Asia	272.4	Dec-21
10	Orange	Europe, Africa, Caribbean	271.0	Dec-21
11	Vodafone Idea	EMEA, India, APAC	255.7	Dec-21
12	Deutsche Telekom	Europe, USA	248.2	Dec-21
13	Veon	Asia, Africa, Europe	217.0	Dec-21
14	AT&T Mobility	United States	202.0	Dec-21
15	Telkomsel	Indonesia	176.0	Dec-21
16	Telenor	Eastern Europe, SE Asia	172.0	Dec-21
17	Axiata	South Asia, SE Asia	163.0	Dec-21
18	Etisalat	Middle East, Asia, Africa	159.0	Dec-21
19	Verizon Wireless	United States	142.0	Dec-21
20	Ooredoo QSC	Middle East, SE Asia, North Africa	121.4	Dec-21

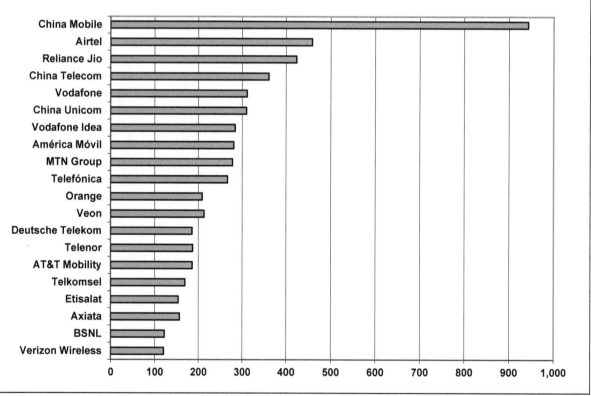

Notes: In some cases, these figures represent proportionate subscribers (adjusted according to percentage of company ownership). Note that some companies listed also have sizeable fixed-line customer bases, in addition to their mobile operations.

Source: Plunkett Research, ® Ltd.

Plunkett Research, ® Ltd. Copyright© 2022, All Rights Reserved, www.plunkettresearch.com

Satellite Telecommunications:
Estimated Sources of Revenue & Expenses, U.S.: 2017-2020

(In Millions of US$; Latest Year Available)

NAICS 5174	2020	2019	2018	2017	% Chg. 20/19
Total Operating Revenue	**5,630**	**5,635**	**5,844**	**6,947**	**-0.1%**
Carrier services	1,840	1,780	1,709	2,190	3.3%
Private network services	S	S	S	S	NA
All other operating revenue	2,615	2,584	2,807	3,120	1.2%
Total Operating Expense	**4,494**	**4,521**	**5,511**	**5,432**	**-0.6%**
Gross annual payroll	1,017	973	1,089	1,165	4.3%
Employer's cost for fringe benefits	185	S	S	265	NA
Temporary staff and leased employee expense	53	46	64	61	13.2%
Expensed equipment, materials, parts, and supplies	148	S	S	S	NA
Expensed purchases of software	S	S	S	16	NA
Access charges	S	13	17	S	NA
Depreciation and amortization charges	903	812	920	767	10.1%
All other operating expenses	1,618	2,092	2,992	2,978	-29.3%

Note: Dollar volume estimates are published in millions of dollars; consequently, results may not be additive. Estimates cover taxable and tax-exempt firms and are not adjusted for price changes.

S = Estimate does not meet publication standards because of high sampling variability (coefficient of variation is greater than 30%) or poor response quality (total quantity response rate is less than 50%). Unpublished estimates derived from this table by subtraction are subject to these same limitations and should not be attributed to the U.S. Census Bureau.

NA = Not Available

Source: U.S. Census Bureau

Plunkett Research, ® Ltd. Copyright© 2022, All Rights Reserved

www.plunkettresearch.com

Number of Business & Residential High Speed Internet Lines, U.S.: 2016-2021

(In Thousands)

Types of Technology	Dec-16	Dec-17	Dec-18	Dec-19*	Dec-20*	Dec-21*
ADSL	26,567	24,193	21,934	18,462	16,353	14,484
SDSL	42	28	20	3	1	1
Other Wireline[1]	599	581	555	530	506	483
Cable Modem	63,325	66,196	68,871	65,415	65,415	66,127
FTTP[2]	11,986	13,890	15,928	16,097	17,760	19,595
Satellite	1,864	1,872	1,971	1,826	1,813	1,800
Fixed Wireless	1,250	1,306	1,477	1,400	1,454	1,510
Total Fixed	**105,593**	**108,065**	**110,756**	**101,283**	**103,303**	**104,001**
Mobile Wireless (Smartphone)	299,256	312,773	330,532	272,096	263,602	255,373
Total Lines	**404,849**	**420,838**	**441,288**	**368,473**	**366,905**	**359,374**

Notes: High-speed lines are connections to end-user locations that deliver services at speeds exceeding 200 kbps in at least one direction. Advanced services lines, which are a subset of high-speed lines, are connections that deliver services at speeds exceeding 200 kbps in both directions. Line counts presented in this report are not adjusted for the number of persons at a single end-user location who have access to, or who use, the Internet-access services that are delivered over the high-speed connection to that location.

[1] Power Line and Other are summarized with Other Wireline to maintain firm confidentiality.

[2] Fiber to the premises.

* Plunkett Research Estimate.

Source: U.S. Federal Communications Bureau (FCC)

Plunkett Research, ® Ltd

www.plunkettresearch.com

Exports & Imports of Telecommunications Equipment, U.S.: 2016-2021

(In Millions of US$)

Exports

Partner	2016	2017	2018	2019	2020	2021
World Total	201,759	208,234	213,593	211,454	199,551	222,996
Mexico	42,795	41,947	44,892	43,844	38,294	41,890
Canada	24,344	25,572	26,148	24,785	22,695	24,802
China	17,108	17,300	18,152	18,945	21,345	24,503
Hong Kong	11,500	13,231	13,100	12,089	10,759	13,455
Netherlands	6,522	7,239	7,700	7,744	7,566	8,872
Taiwan	6,036	5,814	6,158	5,713	6,701	8,357
Germany	7,272	7,787	8,085	7,904	7,486	8,354
Japan	8,285	7,961	8,084	7,822	7,239	7,237
Malaysia	6,523	6,245	5,575	5,035	4,745	6,584
South Korea	6,157	7,508	7,038	6,858	6,695	6,446
Singapore	5,425	5,866	5,985	5,838	5,787	5,903
United Kingdom	5,064	5,022	5,532	5,241	4,479	4,640
United Arab Emirates	2,900	3,320	3,218	3,367	3,596	4,403
Brazil	4,330	4,382	4,444	4,482	3,972	4,355
France	3,083	3,771	3,578	3,263	3,022	3,303

Imports

Partner	2016	2017	2018	2019	2020	2021
World Total	372,664	400,123	413,017	388,491	392,218	464,884
China	161,233	184,054	185,933	145,930	141,387	161,710
Mexico	58,299	58,992	64,364	65,452	62,959	70,824
Vietnam	11,164	11,758	11,303	21,218	26,691	36,499
Taiwan	14,574	15,838	17,066	24,020	29,178	36,038
Malaysia	26,357	25,871	26,575	26,203	26,019	31,042
South Korea	15,516	15,848	16,225	15,419	17,171	22,130
Thailand	12,659	12,985	12,555	11,769	14,287	17,967
Japan	15,332	15,513	16,351	15,482	13,873	15,637
Germany	9,587	9,724	10,632	10,926	9,709	11,207
Switzerland	4,506	4,669	4,876	5,315	5,072	7,309
Philippines	3,689	5,155	5,937	5,788	5,251	6,686
Canada	6,479	6,640	7,027	6,836	5,693	6,312
Singapore	3,594	3,513	3,853	3,427	3,989	4,416
Ireland	4,197	3,829	2,941	3,040	2,636	3,796
United Kingdom	3,626	3,571	3,851	4,035	3,652	3,769

Note: "Telecommunications Equipment" refers to NAICS (North American Industry Classification System) Code 3342.

Source: Foreign Trade Division, U.S. Census Bureau

Plunkett Research, ® Ltd.

www.plunkettresearch.com

Employment in the Telecommunications Industry, U.S.: 1990-May 2022

(Annual Estimates in Thousands of Employed Workers)

Year	Telecom (517)	Wired Telecom Carriers (517311)	Wireless Telecom Carriers (except Satellite) (517312)	Other Telecom (5179)	Telecom Resellers (517911)	Cable & Other Subscription Programming (5152)
1990	1,008.5	759.5	35.8	213.2	179.5	51.7
1991	999.9	749.9	41.8	208.3	178.1	52.5
1992	972.9	725.8	47.8	199.3	172.6	53.6
1993	969.5	716.5	56.3	196.7	170.5	55.9
1994	989.5	719.2	71.7	198.6	171.8	59.3
1995	1,009.3	717.6	90.3	201.4	171.2	63.8
1996	1,038.1	722.2	110.1	205.8	171.6	69.8
1997	1,108.0	755.7	132.1	220.2	181.3	71.5
1998	1,167.4	789.8	144.2	233.4	188.7	75.5
1999	1,270.8	853.2	159.9	257.6	200.2	81.8
2000	1,396.6	921.8	185.6	289.3	213.6	90.7
2001	1,423.9	933.8	201.4	288.8	214.1	95.6
2002	1,280.9	837.1	197.3	246.6	179.5	92.9
2003	1,166.8	761.8	189.9	215.1	154.9	85.9
2004	1,115.1	720.4	189.7	205.1	147.3	85.5
2005	1,071.3	689.6	191.3	190.5	135.1	88.9
2006	1,047.6	669.2	200.2	178.1	125.6	90.0
2007	1,030.6	664.5	203.4	162.7	117.2	88.5
2008	1,019.4	666.4	200.5	152.6	108.6	85.6
2009	965.7	635.0	186.9	143.8	100.7	85.3
2010	902.9	602.9	170.2	129.8	91.5	80.4
2011	873.6	585.1	165.5	123.0	87.1	73.6
2012	856.8	589.5	156.8	110.4	75.1	73.3
2013	853.2	605.2	154.6	93.4	57.6	69.1
2014	838.5	600.7	145.9	91.8	54.4	62.8
2015	810.9	593.1	127.9	89.9	54.0	59.3
2016	801.1	592.3	121.4	87.4	51.8	53.8
2017	780.8	575.1	117.1	88.6	51.3	52.7
2018	749.7	547.5	112.0	90.2	50.3	53.4
2019	714.6	519.6	104.7	90.3	49.8	52.7
2020	687.9	497.5	98.9	91.5	48.6	48.2
2021	666.8	485.0	92.3	89.5	46.3	45.8
May-22*	661.8	480.2	90.7	87.4	43.0	46.6

*Preliminary Estimate

Source: U.S. Bureau of Labor Statistics (BLS)

Plunkett Research, ® Ltd

www.plunkettresearch.com

Chapter 3

IMPORTANT TELECOMMUNICATIONS INDUSTRY CONTACTS

Addresses, Telephone Numbers and Internet Sites

Contents:

1) Apps Industry Associations
2) Broadcasting, Cable, Radio & TV Associations
3) Canadian Government Agencies-General
4) Canadian Government Agencies-Scientific
5) Careers-Computers/Technology
6) Careers-First Time Jobs/New Grads
7) Careers-General Job Listings
8) Careers-Job Reference Tools
9) Chinese Government Agencies-Science & Technology
10) Computer & Electronics Industry Associations
11) Computer & Electronics Industry Resources
12) Corporate Information Resources
13) Ecommerce and Data Interchange Technology Associations
14) Economic Data & Research
15) Engineering Industry Resources
16) Engineering, Research & Scientific Associations
17) Industry Research/Market Research
18) Internet Industry Associations
19) Internet Industry Resources
20) Internet Usage Statistics
21) MBA Resources

22) Outsourcing Industry Resources
23) Patent Organizations
24) Research & Development, Laboratories
25) Robotics & Automation Industry Associations
26) Satellite Industry Associations
27) Satellite-Related Professional Organizations
28) Software Industry Associations
29) Software Industry Resources
30) Technology Law Associations
31) Technology Transfer Associations
32) Telecommunications Industry Associations
33) Telecommunications Resources
34) Telecommunications-VOIP Resources
35) Trade Associations-General
36) Trade Associations-Global
37) Trade Resources
38) U.S. Government Agencies
39) Wireless & Cellular Industry Associations
40) Wireless & Cellular Industry Resources

1) Apps Industry Associations

Association for Competitive Technology (ACT)
1401 K St. NW, Ste. 502
Washington, DC 20005 US
Phone: 202-331-2130
Web Address: www.actonline.org

The Association for Competitive Technology (ACT) is an international advocacy and education organization representing more than 3,000 small and mid-size app developers and information technology firms. It has offices in Washington, DC and in Brussels, Belgium.

2) Broadcasting, Cable, Radio & TV Associations

Cable Center (The)
2000 Buchtel Blvd.
Denver, CO 80210 USA
Phone: 720-502-7500
E-mail Address: info@cablecenter.org
Web Address: www.cablecenter.org
The Cable Center supports communication in the business, technology and programming of cable telecommunications and provides education, training and research for all aspects of the industry.

National Cable and Telecommunications Association (NCTA)
25 Massachusetts Ave. NW, Ste. 100
Washington, DC 20001-1413 USA
Phone: 202-222-2300
Fax: 202-222-2514
E-mail Address: info@ncta.com
Web Address: www.ncta.com
The National Cable and Telecommunications Association (NCTA) is the principal trade association of the cable television industry in the United States. It represents cable operators as well as over 200 cable program networks that produce TV shows.

3) Canadian Government Agencies-General

Canadian Intellectual Property Office (CIPO)
Place du Portage, 50 Victoria St., Rm. C-229
Gatineau, QC K1A 0C9 Canada
Phone: 819-934-0544
Fax: 819-953-2476
Toll Free: 866-997-1936
Web Address: www.cipo.ic.gc.ca
The Canadian Intellectual Property Office (CIPO) is the agency responsible for the administration and processing of intellectual property in Canada, including patents, trademarks, copyrights, industrial designs and integrated circuit topographies.

4) Canadian Government Agencies-Scientific

Institute for Microstructural Sciences (IMS)
1200 Montreal Rd., Bldg. M-58
Ottawa, ON K1A 0R6 Canada
Phone: 613-993-9101
Fax: 613-952-9907
Toll Free: 877-672-2672
E-mail Address: info@nrc-cnrc.gc.ca
Web Address: www.nrc-cnrc.gc.ca
The Institute for Microstructural Sciences (IMS) is a branch of Canada's National Research Council (NRC) that focuses its research on the information and telecommunications technology sector.

5) Careers-Computers/Technology

Dice.com
6465 S. Greenwood Plaza Blvd., Ste. 400
Centennial, CO 80111 USA
Phone: 515-280-1144
Fax: 515-280-1452
Toll Free: 888-321-3423
E-mail Address: techsupport@dice.com
Web Address: www.dice.com
Dice.com provides free employment services for IT jobs. The site includes advanced job searches by geographic location and category, availability announcements and resume postings, as well as employer profiles, a recruiter's page and career links. It is maintained by Dice Holdings, Inc., a publicly traded company.

Institute for Electrical and Electronics Engineers (IEEE) Job Site
445 Hoes Ln.
Piscataway, NJ 08855-1331 USA
Phone: 732-981-0060
Toll Free: 800-678-4333
E-mail Address: candidatejobsite@ieee.org
Web Address: careers.ieee.org
The Institute for Electrical and Electronics Engineers (IEEE) Job Site provides a host of employment services for technical professionals, employers and recruiters. The site offers job listings by geographic area, a resume bank and links to employment services.

Pencom Systems, Inc.
152 Remsen St.
Brooklyn, NY 11201 USA

Phone: 718-923-1111
Fax: 718-923-6065
E-mail Address: tom@pencom.com
Web Address: www.pencom.com
Pencom Systems, Inc., an open system recruiting company, hosts a career web site geared toward high-technology and scientific professionals, featuring an interactive salary survey, career advisor, job listings and technology resources. Its focus is the financial services industry within the New York City area.

6) Careers-First Time Jobs/New Grads

CollegeGrad.com, Inc.
950 Tower Ln., Fl. 6
Foster City, CA 94404 USA
E-mail Address: info@quinstreet.com
Web Address: www.collegegrad.com
CollegeGrad.com, Inc. offers in-depth resources for college students and recent grads seeking entry-level jobs.

National Association of Colleges and Employers (NACE)
62 Highland Ave.
Bethlehem, PA 18017-9085 USA
Phone: 610-868-1421
E-mail Address: customerservice@naceweb.org
Web Address: www.naceweb.org
The National Association of Colleges and Employers (NACE) is a premier U.S. organization representing college placement offices and corporate recruiters who focus on hiring new grads.

7) Careers-General Job Listings

CareerBuilder, Inc.
200 N La Salle Dr., Ste. 1100
Chicago, IL 60601 USA
Phone: 773-527-3600
Fax: 773-353-2452
Toll Free: 800-891-8880
Web Address: www.careerbuilder.com
CareerBuilder, Inc. focuses on the needs of companies and also provides a database of job openings. The site has over 1 million jobs posted by 300,000 employers, and receives an average 23 million unique visitors monthly. The company also operates online career centers for 140 newspapers and 9,000 online partners. Resumes are sent directly to the company, and applicants can set up a special e-mail account for job-seeking purposes. CareerBuilder

is primarily a joint venture between three newspaper giants: The McClatchy Company, Gannett Co., Inc. and Tribune Company.

CareerOneStop
Toll Free: 877-872-5627
E-mail Address: info@careeronestop.org
Web Address: www.careeronestop.org
CareerOneStop is operated by the employment commissions of various state agencies. It contains job listings in both the private and government sectors, as well as a wide variety of useful career resources and workforce information. CareerOneStop is sponsored by the U.S. Department of Labor.

LaborMarketInfo (LMI)
Employment Development Dept.
P.O. Box 826880, MIC 57
Sacramento, CA 94280-0001 USA
Phone: 916-262-2162
Fax: 916-262-2352
Web Address: www.labormarketinfo.edd.ca.gov
LaborMarketInfo (LMI) provides job seekers and employers a wide range of resources, namely the ability to find, access and use labor market information and services. It provides statistics for employment demographics on both a local and regional level, as well as career searching tools for California residents. The web site is sponsored by California's Employment Development Office.

Recruiters Online Network
E-mail Address: rossi.tony@comcast.net
Web Address: www.recruitersonline.com
The Recruiters Online Network provides job postings from thousands of recruiters, Careers Online Magazine, a resume database, as well as other career resources.

USAJOBS
USAJOBS Program Office
1900 E St. NW, Ste. 6500
Washington, DC 20415-0001 USA
Phone: 818-934-6600
Web Address: www.usajobs.gov
USAJOBS, a program of the U.S. Office of Personnel Management, is the official job site for the U.S. Federal Government. It provides a comprehensive list of U.S. government jobs, allowing users to search for employment by location; agency; type of work; or by senior executive positions. It also has special employment sections for individuals with disabilities, veterans and recent college graduates; an information

center, offering resume and interview tips and other information; and allows users to create a profile and post a resume.

8) Careers-Job Reference Tools

Vault.com, Inc.
132 W. 31st St., Fl. 16
New York, NY 10001 USA
Fax: 212-366-6117
Toll Free: 800-535-2074
E-mail Address: customerservice@vault.com
Web Address: www.vault.com
Vault.com, Inc. is a comprehensive career web site for employers and employees, with job postings and valuable information on a wide variety of industries. Its features and content are largely geared toward MBA degree holders.

9) Chinese Government Agencies-Science & Technology

China Ministry of Science and Technology (MOST)
15B Fuxing Rd.
Beijing, 100862 China
Web Address: www.most.gov.cn
The China Ministry of Science and Technology (MOST) is the PRC's official body for science and technology related activities. It drafts laws, policies and regulations regarding science and technology; oversees budgeting and accounting for funds; and supervises research institutes operating in China, among other duties.

10) Computer & Electronics Industry Associations

Asian-Oceanian Computing Industry Organization (ASOCIO)
No. 2, Jalan PJU 8/8A,
c/o PIKOM, Block E1, Empire Damansara E-01-G,
Petaling Jaya, Selangor 47820 Malaysia
Phone: 603-7622-0079
Fax: 603-7622-4879
E-mail Address: secretariat@asocio.org
Web Address: www.asocio.org
The Asian-Oceanian Computing Industry Organization's (ASOCIO) objective is to promote the development of the computing industry in the region.

Association for Information Systems (AIS)
35 Broad St., Ste. 917
Atlanta, GA 30303 USA
Phone: 404-413-7445
Fax: 404-413-7443
E-mail Address: membership@aisnet.org
Web Address: www.aisnet.org
The Association for Information Systems (AIS) is an organization for information system researchers and educators working in colleges and universities worldwide. Its web site offers substantial resources regarding computer systems research and new developments.

Association of Information Technology Professionals (AITP)
3500 Lacey Rd., Ste. 100
Downers Grove, IL 60515 USA
Phone: 630-678-8300
Fax: 630-678-8384
Toll Free: 866-835-8020
E-mail Address: aitp_hq@aitp.org
Web Address: www.aitp.org
The Association of Information Technology Professionals (AITP) is a trade organization that provides training and education through partnerships within the information technology industry.

Association of the Computer and Multimedia Industry of Malaysia (PIKOM)
E1, Empire Damansara, E-01-G
No.2, Jalan PJU 8/8A, Damansara Perdana
Petaling Jaya, Selangor Darul Ehsan 47820 Malaysia
Phone: 603-7622-0079
Fax: 603-7622-4879
E-mail Address: info@pikom.org.my
Web Address: www.pikom.org.my
The Association of the Computer and Multimedia Industry of Malaysia, or, in Malay, Persatuan Industri Komputer dan Multimedia Malaysia (PIKOM), is the national association representing more than 1,000 companies active in the information and communications technology (ICT) industry in Malaysia.

Canadian Advanced Technology Alliances (CATAAlliance)
207 Bank St., Ste. 416
Ottawa, ON K2P 2N2 Canada
Phone: 613-236-6550
E-mail Address: info@cata.ca
Web Address: www.cata.ca

The Canadian Advanced Technology Alliances (CATAAlliance) is one of Canada's leading trade organizations for the research, development and technology sectors.

China Electronic Components Association (CECA)
23 Shijingshan Rd.
ZhongChu Building
Beijing, 100049 China
Phone: 86-10-6887-1587
E-mail Address: icceca@ic-ceca.org.cn
Web Address: www.ic-ceca.org.cn
The China Electronic Components Association (CECA) acts as the representative of the Chinese electronics components industry. Its web site provides consultation services and research reports on components for a wide variety of markets.

China Electronics Chamber of Commerce (CECC)
No. 15 Bldg., Cuiwei Zhongli
Haidian District
Beijing, 100036 China
Phone: 86-10-6825-6762
Fax: 86-10-6825-6764
E-mail Address: ceccinfo@126.com
Web Address: www.cecc.org.cn
China Electronics Chamber of Commerce (CECC), which is led by the Ministry of Information Industry, is the national professional organization for telecommunications and mobile electronics. The group circulates industry information and mediates between its members and the government.

Communications and Information Network Association of Japan (CIAJ)
2-2-12 Hamamatsucho, Minato-ku
Fl. 3, JEI Hamamatsucho Bldg.
Tokyo, 105-0013 Japan
Phone: 81-3-5403-9363
Fax: 81-3-5463-9360
E-mail Address: webmaster@ciaj.or.jp
Web Address: www.ciaj.or.jp/en/
Communications and Information Network Association of Japan (CIAJ) works to help the development of the communication and information network industry in Japan through the promotion of info-communication technologies.

Computer & Communications Industry Association (CCIA)
25 Massachusetts Ave., Ste. 300C
Washington, DC 20001 USA
Phone: 202-783-0070
Fax: 202-783-0534
Web Address: www.ccianet.org
The Computer & Communications Industry Association (CCIA) is a non-profit membership organization for companies and senior executives representing the computer, Internet, information technology (IT) and telecommunications industries.

Computer Technology Industry Association (CompTIA)
3500 Lacey Rd., Ste. 100
Downers Grove, IL 60515 USA
Phone: 630-678-8300
Fax: 630-678-8384
Toll Free: 866-835-8020
Web Address: www.comptia.org
The Computer Technology Industry Association (CompTIA) is the leading association representing the international technology community. Its goal is to provide a unified voice, global advocacy and leadership, and to advance industry growth through standards, professional competence, education and business solutions.

Electronics Technicians Association international (ETA International)
5 Depot St.
Greencastle, IN 46135 USA
Phone: 765-653-8262
Fax: 765-653-4287
Toll Free: 800-288-3824
E-mail Address: eta@eta-i.org
Web Address: www.eta-i.org
The Electronics Technicians Association International (ETA International) is a nonprofit professional association for electronics technicians worldwide. The organization provides recognized professional credentials for electronics technicians.

Federation of Malaysia Manufacturers (FMM)
Wisma FMM, No. 3 Persiaran Dagang, PJU 9
Bandar Sri Damansara
Kuala Lumpur, 52200 Malaysia
Phone: 603-6286-7200
Fax: 603-6274-1266
E-mail Address: webmaster@fmm.org.my
Web Address: www.fmm.org.my
The Federation of Malaysian Manufacturers is an economic organization for the electrical and electronics industry in Malaysia.

Federation of Thai Electrical, Electronics and Allied Industries Club (FTI)
2 Nang Linchi Rd., Fl. 8
Bangkok, 10120 Thailand
Phone: 66-02-345-1000
E-mail Address: information@ft.or.th
Web Address: www.fti.or.th
The Federation of Thai Electrical, Electronics and Allied Industries Club (FTI) represents manufacturers and related firms in these industries within Thailand.

German Association for Information Technology, Telecom and New Media (BITKOM)
Bundersverband Informationswirtschaft Telekommunikation und neue Medien
Albrechtstr. 10A
Berlin-Mitte, 10117 Germany
Phone: 49-30-27576-0
Fax: 49-30-27576-400
E-mail Address: bitkom@bitkom.org
Web Address: www.bitkom.org
German Association for Information Technology, Telecom and New Media (BITKOM) represents information technology and telecommunications specialists and companies.

German Electrical and Electronic Manufacturers' Association (ZVEI)
Zentralverband Elektrotechnik- und Elektronikindustrie e.V.
Lyoner St. 9
Frankfurt am Main, 60528 Germany
Phone: 49-69-6302-0
Fax: 49-69-6302-317
E-mail Address: zvei@zvei.org
Web Address: www.zvei.org
The German Electrical and Electronic Manufacturers' Association (ZVEI) represents its members' interests at the national and international level.

Indian Electrical & Electronics Manufacturers Association (IEEMA)
501 Kakad Chambers
132 Dr. Annie Besant Rd., Worli
Mumbai, 400018 India
Phone: 91-22-2493-0532
Fax: 91-22-2493-2705
E-mail Address: mumbai@ieema.org
Web Address: www.ieema.org
The Indian Electrical & Electronics Manufacturers Association (IEEMA) represents all sectors of the electrical and allied products businesses of the Indian electrical industry.

Information Technology Association of Canada (ITAC)
5090 Explorer Dr., Ste. 801
Mississauga, ON L4W 4T9 Canada
Phone: 905-602-8345
Fax: 905-602-8346
E-mail Address: dwhite@itac.ca
Web Address: www.itac.ca
The Information Technology Association of Canada (ITAC) represents the IT, software, computer and telecommunications industries in Canada.

Information Technology Management Association (ITMA)
Robinson Rd.
P.O. Box 3297
Singapore, 905297 Singapore
Phone: 65-8171-4456
Fax: 65-6410-8008
E-mail Address: secretariat@itma.org.sg
Web Address: www.itma.org.sg
Information Technology Management Association (ITMA) represents professionals working in the field of IT management in Singapore.

Institute for Interconnecting and Packaging Electronic Circuits (IPC)
3000 Lakeside Dr., Ste. 105 N
Bannockburn, IL 60015 USA
Phone: 847-615-7100
Fax: 847-615-7105
E-mail Address: answers@ipc.org
Web Address: www.ipc.org
The Institute for Interconnecting and Packaging Electronic Circuits (IPC) is a trade association for companies in the global printed circuit board and electronics manufacturing services industries.

International Microelectronics Assembly and Packaging Society (IMAPS)
P.O. Box 110127
Research Triangle Park, NC 27709-5127 USA
Phone: 919-293-5000
Fax: 919-287-2339
E-mail Address: info@imaps.org
Web Address: www.imaps.org
The International Microelectronics Assembly and Packaging Society (IMAPS) is dedicated to the advancement and growth of the use of microelectronics and electronic packaging through professional education, workshops and conferences.

Korea Association of Information and Telecommunications (KAIT)
NO. 1678-2, 2nd Fl. Dong-Ah Villat 2 Town
Seocho-dong, Seocho-gu
Seoul, 137-070 Korea
Phone: 82-2-580-0582
E-mail Address: webmaster@kait.or.kr
Web Address: www.kait.or.kr/eng
The Korea Association of Information and
Telecommunications (KAIT) was created to develop
and promote the InfoTech, computer, consumer
electronics, wireless, software and
telecommunications sectors in Korea.

Korea Semiconductor Industry Association (KSIA)
182, Pangyoyeok-ro, Bundang-gu, Seongnam-si
Fl. 9-12, KSIA Bldg.
Gyeonggi-do, Korea
Phone: 82-2-576-3472
Fax: 82-2-570-5269
E-mail Address: admin@ksia.or.kr
Web Address: www.ksia.or.kr
The Korean Semiconductor Industry Association
(KSIA) represents the interests of Korean
semiconductor manufacturers.

National Electrical Manufacturers Association (NEMA)
1300 N. 17th St., Ste. 900
Arlington, VA 22209 USA
Phone: 703-841-3200
E-mail Address: press@nema.org
Web Address: www.nema.org
The National Electrical Manufacturers Association
(NEMA) develops standards for the electrical
manufacturing industry and promotes safety in the
production and use of electrical products.

Network Professional Association (NPA)
3517 Camino Del Rio S., Ste. 215
San Diego, CA 92108-4089 USA
Fax: 888-672-6720
Toll Free: 888-672-6720
Web Address: www.npanet.org
The Network Professionals Association (NPA) is a
self-regulating, nonprofit association of network
computing professionals that sets standards of
technical expertise and professionalism.

Semiconductor Equipment and Materials International (SEMI)
673 S. Milpitas Blvd.

Milpitas, CA 95035 USA
Phone: 408-943-6900
Fax: 408-428-9600
E-mail Address: semihq@semi.org
Web Address: www.semi.org
Semiconductor Equipment and Materials
International (SEMI) is a trade association serving
the global semiconductor equipment, materials and
flat-panel display industries.

Semiconductor Industry Association (SIA)
1101 K St. NW, Ste. 450
Washington, DC 20005 USA
Phone: 202-446-1700
Fax: 202-216-9745
Toll Free: 866-756-0715
Web Address: www.semiconductors.org
The Semiconductor Industry Association (SIA) is a
trade association representing the semiconductor
industry in the U.S. Through its coalition of more
than 60 companies, SIA members represent roughly
80% of semiconductor production in the U.S. The
coalition aims to advance the competitiveness of the
chip industry and shape public policy on issues
particular to the industry.

Singapore Computer Society
53/53A Neil Rd.
Singapore, 088891 Singapore
Phone: 65-6226-2567
Fax: 65-6226-2569
E-mail Address: scs.secretariat@scs.org.sg
Web Address: www.scs.org.sg
The Singapore Computer Society is a membership
society for infocomm professionals in Singapore.

Surface Mount Technology Association (SMTA)
6600 City West Pkwy., Ste. 300
Eden Prairie, MN 55424 USA
Phone: 952-920-7682
Fax: 952-926-1819
E-mail Address: smta@smta.org
Web Address: www.smta.org
The Surface Mount Technology Association (SMTA)
is an international network of professionals whose
careers encompass electronic assembly technologies,
microsystems, emerging technologies and associated
business operations.

Taiwan Electrical and Electronic Manufacturers' Association (TEEMA)
Min Chuan E. Rd.
Fl. 6, No. 109, Sec. 6

Taipei, 11490 Taiwan
Phone: 886-2-8792-6666
Fax: 886-2-8792-6088
E-mail Address: teema@teema.org.tw
Web Address: www.teema.org.tw
The Taiwan Electrical and Electronic Manufacturers'
Association (TEEMA) works as an intermediary
between its members and the government to help the
industry to succeed.

Telecom Equipment Manufacturers Association of India (TEMA)
PHD House, 4th Fl.
Khel Gaon Marg, Hauz Khas
New Delhi, 110 0016 India
Phone: 91-11-49545454
E-mail Address: tema@tematelecom.in
Web Address: www.tematelecom.in
The Telecom Equipment Manufacturers Association
of India (TEMA) is national organization for
companies in the telecommunications industry. The
group disseminates and exchanges information with
the Indian government, foreign agencies, embassies,
trade missions, Indian missions abroad and leading
international trade associations.

World Information Technology and Services Alliance (WITSA)
8300 Boone Blvd., Ste. 450
Vienna, VA 22182 USA
Phone: 571-265-5964
Fax: 703-893-1269
E-mail Address: admin@witsa.org
Web Address: www.witsa.org
The World Information Technology and Services
Alliance (WITSA) is a consortium of over 70
information technology (IT) industry associations
from economies around the world. WITSA members
represent over 90% of the world IT market. Founded
in 1978 and originally known as the World
Computing Services Industry Association, WITSA is
an advocate in international public policy issues
affecting the creation of a robust global information
infrastructure.

11) Computer & Electronics Industry Resources

Cisco Cloud Index
170 W. Tasman Dr.
San Jose, CA 95134 USA
Toll Free: 800-553-6387
Web Address: www.cisco.com/go/cloudindex

The Cisco Cloud Index covers three areas focused on
data center and cloud traffic trends and next-
generation service or application adoption. They
include: Data center and cloud traffic forecast;
Workload transition, which provides projections for
workloads moving from traditional IT to cloud-based
architectures; and Cloud readiness, which provides
regional statistics on broadband adoption as a
precursor for cloud services.

Information Technology and Innovation Foundation (ITIF)
700 K St. NW, Ste. 600
Washington, DC 20001 USA
Phone: 202-449-1351
E-mail Address: mail@itif.org
Web Address: www.itif.org
Information Technology and Innovation Foundation
(ITIF) is a non-partisan research and educational
institute (a think tank) with a mission to formulate
and promote public policies to advance technological
innovation and productivity internationally, in
Washington, and in the States. Recognizing the vital
role of technology in ensuring American prosperity,
ITIF focuses on innovation, productivity, and digital
economy issues.

Ministry of Electronics and Information Technology (India)
Electronics Niketan
6 CGO Complex, Lodhi Rd.
New Delhi, 110003 India
Phone: 91-11-2430-1851
E-mail Address: webmaster@deity.gov.in
Web Address: www.meity.gov.in
Ministry of Communications & Information
Technology (MIT) of the Government of India, is
charged with promoting the information technology
and communications industries.

12) Corporate Information Resources

Business Journals (The)
120 W. Morehead St., Ste. 400
Charlotte, NC 28202 USA
Toll Free: 866-853-3661
E-mail Address: gmurchison@bizjournals.com
Web Address: www.bizjournals.com
Bizjournals.com is the online media division of
American City Business Journals, the publisher of
dozens of leading city business journals nationwide.
It provides access to research into the latest news
regarding companies both small and large. The

organization maintains 42 websites and 64 print publications and sponsors over 700 annual industry events.

Business Wire
101 California St., Fl. 20
San Francisco, CA 94111 USA
Phone: 415-986-4422
Fax: 415-788-5335
Toll Free: 800-227-0845
E-mail Address: info@businesswire.com
Web Address: www.businesswire.com
Business Wire offers news releases, industry- and company-specific news, top headlines, conference calls, IPOs on the Internet, media services and access to tradeshownews.com and BW Connect On-line through its informative and continuously updated web site.

Edgar Online, Inc.
35 W. Wacker Dr.
Chicago, IL 60601 USA
Phone: 301-287-0300
Fax: 301-287-0390
Toll Free: 800-823-5304
Web Address: www.edgar-online.com
Edgar Online, Inc. is a gateway and search tool for viewing corporate documents, such as annual reports on Form 10-K, filed with the U.S. Securities and Exchange Commission.

PR Newswire Association LLC
200 Vesey St., Fl. 19
New York, NY 10281 USA
Fax: 800-793-9313
Toll Free: 800-776-8090
E-mail Address: mediainquiries@cision.com
Web Address: www.prnewswire.com
PR Newswire Association LLC provides comprehensive communications services for public relations and investor relations professionals, ranging from information distribution and market intelligence to the creation of online multimedia content and investor relations web sites. Users can also view recent corporate press releases from companies across the globe. The Association is owned by United Business Media plc.

Silicon Investor
E-mail Address: si.admin@siliconinvestor.com
Web Address: www.siliconinvestor.com
Silicon Investor is focused on providing information about technology companies. Its web site serves as a financial discussion forum and offers quotes, profiles and charts.

13) Ecommerce and Data Interchange Technology Associations

RosettaNet
7877 Washington Village Dr., Ste. 300
Dayton, OH 45459 USA
Phone: 937-435-3870
E-mail Address: info@gs1us.org
Web Address: www.resources.gs1us.org/rosettanet
RosettaNet, a subsidiary of GS1 US, is a nonprofit organization whose mission is to develop e-business process standards that serve as a frame of reference for global trading networks. The organization's standards provide a common language for companies within the global supply chain.

14) Economic Data & Research

Centre for European Economic Research (The, ZEW)
L 7, 1
Mannheim, 68161 Germany
Phone: 49-621-1235-01
Fax: 49-621-1235-224
E-mail Address: empfang@zew.de
Web Address: www.zew.de/en
Zentrum fur Europaische Wirtschaftsforschung, The Centre for European Economic Research (ZEW), distinguishes itself in the analysis of internationally comparative data in a European context and in the creation of databases that serve as a basis for scientific research. The institute maintains a special library relevant to economic research and provides external parties with selected data for the purpose of scientific research. ZEW also offers public events and seminars concentrating on banking, business and other economic-political topics.

Economic and Social Research Council (ESRC)
Polaris House
North Star Ave.
Swindon, SN2 1UJ UK
Phone: 44-01793 413000
E-mail Address: esrcenquiries@esrc.ac.uk
Web Address: www.esrc.ac.uk
The Economic and Social Research Council (ESRC) funds research and training in social and economic issues. It is an independent organization, established by Royal Charter. Current research areas include the

global economy; social diversity; environment and energy; human behavior; and health and well-being.

Eurostat
5 Rue Alphonse Weicker
Joseph Bech Bldg.
Luxembourg, L-2721 Luxembourg
Phone: 352-4301-1
E-mail Address: eurostat-pressoffice@ec.europa.eu
Web Address: ec.europa.eu/eurostat
Eurostat is the European Union's service that publishes a wide variety of comprehensive statistics on European industries, populations, trade, agriculture, technology, environment and other matters.

Federal Statistical Office of Germany
Gustav-Stresemann-Ring 11
Wiesbaden, D-65189 Germany
Phone: 49-611-75-2405
Fax: 49-611-72-4000
Web Address: www.destatis.de
Federal Statistical Office of Germany publishes a wide variety of nation and regional economic data of interest to anyone who is studying Germany, one of the world's leading economies. Data available includes population, consumer prices, labor markets, health care, industries and output.

India Brand Equity Foundation (IBEF)
Fl. 20, Jawahar Vyapar Bhawan
Tolstoy Marg
New Delhi, 110001 India
Phone: 91-11-43845500
Fax: 91-11-23701235
E-mail Address: info.brandindia@ibef.org
Web Address: www.ibef.org
India Brand Equity Foundation (IBEF) is a public-private partnership between the Ministry of Commerce and Industry, the Government of India and the Confederation of Indian Industry. The foundation's primary objective is to build positive economic perceptions of India globally. It aims to effectively present the India business perspective and leverage business partnerships in a globalizing marketplace.

National Bureau of Statistics (China)
57, Yuetan Nanjie, Sanlihe
Xicheng District
Beijing, 100826 China
Fax: 86-10-6878-2000
E-mail Address: info@gj.stats.cn

Web Address: www.stats.gov.cn/english
The National Bureau of Statistics (China) provides statistics and economic data regarding China's economy and society.

Organization for Economic Co-operation and Development (OECD)
2 rue Andre Pascal, Cedex 16
Paris, 75775 France
Phone: 33-1-45-24-82-00
Fax: 33-1-45-24-85-00
E-mail Address: webmaster@oecd.org
Web Address: www.oecd.org
The Organization for Economic Co-operation and Development (OECD) publishes detailed economic, government, population, social and trade statistics on a country-by-country basis for over 30 nations representing the world's largest economies. Sectors covered range from industry, labor, technology and patents, to health care, environment and globalization.

Statistics Bureau, Director-General for Policy Planning (Japan)
19-1 Wakamatsu-cho
Shinjuku-ku
Tokyo, 162-8668 Japan
Phone: 81-3-5273-2020
E-mail Address: toukeisoudan@soumu.go.jp
Web Address: www.stat.go.jp/english
The Statistics Bureau, Director-General for Policy Planning (Japan) and Statistical Research and Training Institute, a part of the Japanese Ministry of Internal Affairs and Communications, plays the central role of producing and disseminating basic official statistics and coordinating statistical work under the Statistics Act and other legislation.

Statistics Canada
150 Tunney's Pasture Driveway
Ottawa, ON K1A 0T6 Canada
Phone: 514-283-8300
Fax: 514-283-9350
Toll Free: 800-263-1136
E-mail Address: STATCAN.infostats-infostats.STATCAN@canada.ca
Web Address: www.statcan.gc.ca
Statistics Canada provides a complete portal to Canadian economic data and statistics. Its conducts Canada's official census every five years, as well as hundreds of surveys covering numerous aspects of Canadian life.

15) Engineering Industry Resources

Cornell Engineering Library (The)
Engineering Library Cornell University
Carpenter Hall, Fl. 1
Ithaca, NY 14853 USA
Phone: 607-254-6261
E-mail Address: engrref@cornell.edu
Web Address: engineering.library.cornell.edu
Cornell University's Engineering Library web site has a number of resources concerning engineering research, as well as links to other engineering industry information sources.

16) Engineering, Research & Scientific Associations

American Association for the Advancement of Science (AAAS)
1200 New York Ave. NW
Washington, DC 20005 USA
Phone: 202-326-6400
Web Address: www.aaas.org
The American Association for the Advancement of Science (AAAS) is the world's largest scientific society and the publisher of Science magazine. It is an international nonprofit organization dedicated to advancing science around the globe.

American National Standards Institute (ANSI)
1899 L St. NW, Fl. 11
Washington, DC 20036 USA
Phone: 202-293-8020
Fax: 202-293-9287
E-mail Address: info@ansi.org
Web Address: www.ansi.org
The American National Standards Institute (ANSI) is a private, nonprofit organization that administers and coordinates the U.S. voluntary standardization and conformity assessment system. Its mission is to enhance both the global competitiveness of U.S. business and the quality of life by promoting and facilitating voluntary consensus standards and conformity assessment systems and safeguarding their integrity.

American Society for Engineering Education (ASEE)
1818 North St. NW, Ste. 600
Washington, DC 20036-2479 USA
Phone: 202-331-3500
Fax: 202-265-8504
E-mail Address: board@asee.org
Web Address: www.asee.org
The American Society for Engineering Education (ASEE) is nonprofit organization dedicated to promoting and improving engineering and technology education.

Association for Electrical, Electronic & Information Technologies (VDE)
Stresemannallee 15
Frankfurt, 60596 Germany
Phone: 49-69-6308-0
Fax: 49-69-6308-9865
E-mail Address: service@vde.com
Web Address: www.vde.com
The Association for Electrical, Electronic & Information Technologies (VDE) is a German organization with roughly 36,000 members, representing one of the largest technical associations in Europe.

Association of Federal Communications Consulting Engineers (AFCCE)
P.O. Box 19333
Washington, DC 20036 USA
Web Address: www.afcce.org
The Association of Federal Communications Consulting Engineers (AFCCE) is a professional organization of individuals who regularly assist clients on technical issues before the Federal Communications Commission (FCC).

Audio Engineering Society, Inc. (AES)
551 Fifth Ave., Ste. 1225
New York, NY 10176 USA
Phone: 212-661-8528
Web Address: www.aes.org
The Audio Engineering Society (AES) provides information on educational and career opportunities in audio technology and engineering.

Center for Innovative Technology (CIT)
2214 Rock Hill Rd., Ste. 600
Herndon, VA 20170-4228 USA
Phone: 703-689-3000
Fax: 703-689-3041
E-mail Address: info@cit.org
Web Address: www.cit.org
The Center for Innovative Technology is a nonprofit organization designed to enhance the research and development capabilities by creating partnerships between innovative technology start-up companies and advanced technology consumers.

China Association for Science and Technology (CAST)
3 Fuxing Rd.
Beijing, 100863 China
Phone: 8610-6857-1898
Fax: 8610-6857-1897
E-mail Address: cast-liasion@cast.org.cn
Web Address: english.cast.org.cn
The China Association for Science and Technology (CAST) is the largest national non-governmental organization of scientific and technological workers in China. The association has nearly 207 member organizations in the fields of engineering, science and technology.

Engineer's Club (The) (TEC)
1737 Silverwood Dr.
San Jose, CA 95124 USA
Phone: 408-316-0488
E-mail Address: INFO@engineers.com
Web Address: www.engineers.com
The Engineer's Club (TEC) provides a variety of resources for engineers and technical professionals.

IEEE Communications Society (ComSoc)
3 Park Ave., Fl. 17
New York, NY 10016 USA
Phone: 212-705-8900
Fax: 212-705-8999
Web Address: www.comsoc.org
The IEEE Communications Society (ComSoc) is composed of industry professionals with a common interest in advancing communications technologies.

Industrial Research Institute (IRI)
P.O. Box 13968
Arlington, VA 22219 USA
Phone: 703-647-2580
Fax: 703-647-2581
E-mail Address: information@iriweb.org
Web Address: www.iriweb.org
The Innovation Research Interchange (IRI) is a nonprofit organization of over 200 leading industrial companies, representing industries such as aerospace, automotive, chemical, computers and electronics, which carry out industrial research efforts in the U.S. manufacturing sector. IRI helps members improve research and development capabilities.

Institute of Electrical and Electronics Engineers (IEEE)
3 Park Ave., Fl. 17
New York, NY 10016-5997 USA

Phone: 212-419-7900
Fax: 212-752-4929
Toll Free: 800-678-4333
E-mail Address: society-info@ieee.org
Web Address: www.ieee.org
The Institute of Electrical and Electronics Engineers (IEEE) is a nonprofit, technical professional association of more than 430,000 individual members in approximately 160 countries. The IEEE sets global technical standards and acts as an authority in technical areas ranging from computer engineering, biomedical technology and telecommunications to electric power, aerospace and consumer electronics.

Institution of Engineering and Technology (The) (IET)
Michael Faraday House
Six Hills Way
Stevenage, Herts SG1 2AY UK
Phone: 44-1438-313-311
Fax: 44-1438-765-526
E-mail Address: postmaster@theiet.org
Web Address: www.theiet.org
The Institution of Engineering and Technology (IET) is an innovative international organization for electronics, electrical, manufacturing and IT professionals.

International Electrotechnical Commission (IEC)
3, rue de Varembe
P.O. Box 131
Geneva 20, CH-1211 Switzerland
Phone: 41-22-919-02-11
Fax: 41-22-919-03-00
E-mail Address: info@iec.ch
Web Address: www.iec.ch
The International Electrotechnical Commission (IEC), based in Switzerland, promotes international cooperation on all questions of standardization and related matters in electrical and electronic engineering.

International Standards Organization (ISO)
Chemin de Blandonnet 8
1214 Vernier
Geneva, CP 401 Switzerland
Phone: 41-22-749-01-11
Fax: 41-22-733-34-30
E-mail Address: central@iso.org
Web Address: www.iso.org
The International Standards Organization (ISO) is a global consortium of national standards institutes from 162 countries. The established International

Standards are designed to make products and services more efficient, safe and clean.

Optical Society of America (OSA)
2010 Massachusetts Ave. NW
Washington, DC 20036-1023 USA
Phone: 202-223-8130
Fax: 202-223-1096
E-mail Address: info@osa.org
Web Address: www.osa.org
The Optical Society of America (OSA) is an interdisciplinary society offering synergy between all components of the optics industry, from basic research to commercial applications such as fiber-optic networks. It has a membership group of over 16,000 individuals from over 100 countries. Members include scientists, engineers, educators, technicians and business leaders.

Research in Germany, German Academic Exchange Service (DAAD)
Kennedyallee 50
Bonn, 53175 Germany
Phone: 49-228-882-743
Web Address: www.research-in-germany.de
The Research in Germany portal, German Academic Exchange Service (DAAD), is an information platform and contact point for those looking to find out more about Germany's research landscape and its latest research achievements. The portal is an initiative of the Federal Ministry of Education and Research.

Royal Society (The)
6-9 Carlton House Ter.
London, SW1Y 5AG UK
Phone: 44-20-7451-2500
E-mail Address: science.policy@royalsociety.org
Web Address: royalsociety.org
The Royal Society, originally founded in 1660, is the UK's leading scientific organization and the oldest scientific community in continuous existence. It operates as a national academy of science, supporting scientists, engineers, technologists and researchers. Its web site contains a wealth of data about the research and development initiatives of its fellows and foreign members.

Society of Cable Telecommunications Engineers (SCTE)
140 Philips Rd.
Exton, PA 19341-1318 USA
Phone: 610-363-6888

Fax: 610-884-7237
Toll Free: 800-542-5040
E-mail Address: info@scte.org
Web Address: www.scte.org
The Society of Cable Telecommunications Engineers (SCTE) is a nonprofit professional association dedicated to advancing the careers and serving the industry of telecommunications professionals by providing technical training, certification and information resources.

World Federation of Engineering Organizations
Maison de l'UNESCO
1, rue Miollis
Paris, 75015 France
Phone: 33-1-45-68-48-47
Fax: 33-1-45-68-48-65
E-mail Address: secretariat@wfeo.net
Web Address: www.wfeo.org
World Federation of Engineering Organizations (WFEO) is an international non-governmental organization that represents major engineering professional societies in over 90 nations. It has several standing committees including engineering and the environment, technology, communications, capacity building, education, energy and women in engineering.

17) Industry Research/Market Research

Forrester Research
60 Acorn Park Dr.
Cambridge, MA 02140 USA
Phone: 617-613-5730
Toll Free: 866-367-7378
E-mail Address: press@forrester.com
Web Address: www.forrester.com
Forrester Research is a publicly traded company that identifies and analyzes emerging trends in technology and their impact on business. Among the firm's specialties are the financial services, retail, health care, entertainment, automotive and information technology industries.

Gartner, Inc.
56 Top Gallant Rd.
Stamford, CT 06902 USA
Phone: 203-964-0096
E-mail Address: info@gartner.com
Web Address: www.gartner.com
Gartner, Inc. is a publicly traded IT company that provides competitive intelligence and strategic

consulting and advisory services to numerous clients worldwide.

MarketResearch.com

6116 Executive Blvd., Ste. 550
Rockville, MD 20852 USA
Phone: 240-747-3093
Fax: 240-747-3004
Toll Free: 800-298-5699
E-mail Address:
customerservice@marketresearch.com
Web Address: www.marketresearch.com
MarketResearch.com is a leading broker for professional market research and industry analysis. Users are able to search the company's database of research publications including data on global industries, companies, products and trends.

Plunkett Research, Ltd.

P.O. Drawer 541737
Houston, TX 77254-1737 USA
Phone: 713-932-0000
Fax: 713-932-7080
E-mail Address:
customersupport@plunkettresearch.com
Web Address: www.plunkettresearch.com
Plunkett Research, Ltd. is a leading provider of market research, industry trends analysis and business statistics. Since 1985, it has served clients worldwide, including corporations, universities, libraries, consultants and government agencies. At the firm's web site, visitors can view product information and pricing and access a large amount of basic market information on industries such as financial services, InfoTech, ecommerce, health care and biotech.

Portio Research Limited

16 Moss Mead
Chippenham, Wilts SN14 OTN UK
Phone: 44-1249-656-964
Fax: 44-1249-656-967
E-mail Address: info@portioresearch.com
Web Address: www.portioresearch.com
Portio Research Limited is a research firm engaged in the study of the mobile and wireless sector. It provides data-centric reports and database products.

Pyramid Research

179 South St., Ste. 200
Boston, MA 02111 USA
Phone: 617-747-4100
Web Address: www.pyramidresearch.com

Pyramid Research provides international market analysis and advisory services to the global communications, media and technology industries. It advises vendors, equipment manufacturers, service providers and the financial community.

18) Internet Industry Associations

Asia & Pacific Internet Association (APIA)

P.O. Box 1908
Milton, 4064 Australia
E-mail Address: apiasec@apia.org
Web Address: www.apia.org
Asia & Pacific Internet Association (APIA) is a nonprofit trade association whose aim is to promote the business interests of the Internet-related service industry in the Asia Pacific region. The site contains a list of organizations, standards, regional Internet registries and related Asia Pacific organizations.

Cooperative Association for Internet Data Analysis (CAIDA)

9500 Gilman Dr.
Mail Stop 0505
La Jolla, CA 92093-0505 USA
Phone: 858-534-5000
E-mail Address: info@caida.org
Web Address: www.caida.org
The Cooperative Association for Internet Data Analysis (CAIDA), representing organizations from the government, commercial and research sectors, works to promote an atmosphere of greater cohesion in the engineering and maintenance of the Internet. CAIDA is located at the San Diego Supercomputer Center (SDSC) on the campus of the University of California, San Diego (UCSD).

Federation of Internet Service Providers of the Americas (FISPA)

c/o Jim Hollis
8200 Raintree Ln., Ste. 100
Charlotte, NC 28277 USA
Phone: 704-844-2540
Fax: 704-844-2728
Toll Free: 813-574-2556
E-mail Address: executive.director@fipsa.org
Web Address: www.fispa.org
The Federation of Internet Service Providers of the Americas (FISPA) encourages discussion, education and collective buying power for organizations involved in providing Internet access, web hosting, web design and other Internet products and services.

Internet and Mobile Association of India (IAMAI)
406 Ready Money Terr.
167 Dr. Annie Besant Rd.
Mumbai, 400 018 India
E-mail Address: gaurav@iamai.in
Web Address: www.iamai.in
The Internet & Mobile Association of India (IAMAI)
is an industry organization representing the interests
of India's online and mobile value-added services
industry.

Internet Law & Policy Forum (ILPF)
2440 Western Ave., Ste. 709
Seattle, WA 98121 USA
Phone: 206-727-0700
Fax: 206-374-2263
E-mail Address: admin@ilpf.org
Web Address: www.ilpf.org
The Internet Law & Policy Forum (ILPF) is
dedicated to the global development of the Internet
through legal and public policy initiatives. It is an
international nonprofit organization whose member
companies develop and deploy the Internet in every
aspect of business today.

Internet Society (ISOC)
11710 Plaza America Dr., Ste. 400
Reston, VA 20190 USA
Phone: 703-439-2120
Fax: 703-326-9881
E-mail Address: isoc@isoc.org
Web Address: www.isoc.org
The Internet Society (ISOC) is a nonprofit
organization that provides leadership in public policy
issues that influence the future of the Internet. The
organization is the home of groups that maintain
infrastructure standards for the Internet, such as the
Internet Engineering Task Force (IETF) and the
Internet Architecture Board (IAB).

Internet Systems Consortium, Inc. (ISC)
P.O. Box 360
Newmarket, NH 03857 USA
Phone: 650-423-1300
Fax: 650-423-1355
E-mail Address: info@isc.org
Web Address: www.isc.org
The Internet Systems Consortium, Inc. (ISC) is a
nonprofit organization with extensive expertise in the
development, management, maintenance and
implementation of Internet technologies.

US Internet Service Provider Association (US ISPA)
700 12th St. NW, Ste. 700E
Washington, DC 20005 USA
Phone: 202-904-2351
E-mail Address: kdean@usispa.org
Web Address: www.usispa.org
US Internet Service Provider Association (US ISPA)
is a leading provider of technical, business, policy
and regulatory support to ISPs (Internet service
providers).

W3C (World Wide Web Consortium)
32 Vassar St., Bldg. 32-G515
Cambridge, MA 02139 USA
Phone: 617-253-2613
Fax: 617-258-5999
E-mail Address: susan@w3.org
Web Address: www.w3.org
The World Wide Web Consortium (W3C) develops
technologies and standards to enhance the
performance and utility of the World Wide Web. The
W3C is hosted by three different organizations: the
European Research Consortium for Informatics and
Mathematics (ERICM) handles inquiries about the
W3C in the EMEA region; Keio University handles
W3C's Japanese and Korean correspondence; and the
Computer Science & Artificial Intelligence Lab
(CSAIL) at MIT handles all other countries, include
Australia and the U.S.

19) Internet Industry Resources

American Registry for Internet Numbers (ARIN)
P.O. Box 232290
Centreville, VA 20120 USA
Phone: 703-227-9840
Fax: 703-263-0417
E-mail Address: info@arin.net
Web Address: www.arin.net
The American Registry for Internet Numbers (ARIN)
is a nonprofit organization that administers and
registers Internet protocol (IP) numbers. The
organization also develops policies and offers
educational outreach services.

Berkman Center for Internet & Society
23 Everett St., Fl. 2
Cambridge, MA 02138 USA
Phone: 617-495-7547
Fax: 617-495-7641
E-mail Address: cyber@law.harvard.edu
Web Address: cyber.law.harvard.edu

The Berkman Center for Internet & Society, housed at Harvard University's law school, focuses on the exploration of the development and inner-workings of laws pertaining to the Internet. The center offers Internet courses, conferences, advising and advocacy.

Congressional Internet Caucus Advisory Committee (CICA)
1440 G St. NW
Washington, DC 20005 USA
Phone: 202-638-4370
E-mail Address: tlordan@netcaucus.org
Web Address: www.netcaucus.org
The Congressional Internet Caucus Advisory Committee (ICAC) works to educate the public, as well as a bipartisan group from the U.S. House and Senate about Internet-related policy issues.

Internet Assigned Numbers Authority (IANA)
12025 Waterfront Dr., Ste. 300
Los Angeles, CA 90094 USA
Phone: 424-254-5300
Fax: 424-254-5033
E-mail Address: iana@iana.org
Web Address: www.iana.org
The Internet Assigned Numbers Authority (IANA) serves as the central coordinator for the assignment of parameter values for Internet protocols. IANA is operated by the Internet Corporation for Assigned Names and Numbers (ICANN).

20) Internet Usage Statistics

comScore, Inc.
11950 Democracy Dr., Ste. 600
Reston, VA 20190 USA
Phone: 703-438-2000
Fax: 703-438-2051
Toll Free: 866-276-6972
Web Address: www.comscore.com
comScore, Inc. provides excellent data on consumer behavior and audiences, particularly in terms of how consumers access and use online sites and digital data and entertainment. They are global leaders in Internet usage data.

eMarketer
11 Times Square
New York, NY 10036 USA
Toll Free: 800-405-0844
Web Address: www.emarketer.com
eMarketer is a comprehensive, objective and easy-to-use resource for any person or business interested in

online marketing and emerging media. The firm offers news articles, market projections and analytical commentaries.

Nielsen
85 Broad St.
New York, NY 10004 USA
Toll Free: 800-864-1224
Web Address: www.nielsen.com
Nielsen offers detailed, real-time Internet, retail and media research and analysis.

Pew Internet & American Life Project
1615 L St. NW, Ste. 800
Washington, DC 20036 USA
Phone: 202-419-4300
Fax: 202-857-8562
E-mail Address: info@pewinternet.org
Web Address: www.pewinternet.org
The Pew Internet & American Life Project, an initiative of the Pew Research Center, produces reports that explore the impact of the Internet on families, communities, work and home, daily life, education, health care and civic and political life.

21) MBA Resources

MBA Depot
Web Address: www.mbadepot.com
MBA Depot is an online community and information portal for MBAs, potential MBA program applicants and business professionals.

22) Outsourcing Industry Resources

CIO Outsourcing Center
492 Old Connecticut Path
P.O. Box 9208
Framingham, MA 01701-9208 USA
Phone: 508-872-0080
E-mail Address: rhein@cio.com
Web Address: www.cio.com/topic/3195/Outsourcing
CIO Outsourcing Center, a feature on CIO.com, provides data for chief information officers about technology outsourcing. CIO.com and the Outsourcing Center are products of CXO Media Inc., which is itself a division of International Data Group.

23) Patent Organizations

European Patent Office
Bob-van-Benthem-Platz 1

Munich, 80469 Germany
Phone: 49 89 2399-0
Toll Free: 08-800-80-20-20-20
E-mail Address: press@epo.org
Web Address: www.epo.org
The European Patent Office (EPO) provides a uniform application procedure for individual inventors and companies seeking patent protection in up to 38 European countries. It is the executive arm of the European Patent Organization and is supervised by the Administrative Council.

24) Research & Development, Laboratories

Advanced Technology Laboratory (ARL)
10000 Burnet Rd.
University of Texas at Austin
Austin, TX 78758 USA
Phone: 512-835-3200
Fax: 512-835-3259
Web Address: www.arlut.utexas.edu
Advanced Technology Laboratory (ARL) at the University of Texas at Austin provides research programs dedicated to improving the military capability of the United States in applications of acoustics, electromagnetic and information technology.

Commonwealth Scientific and Industrial Research Organization (CSRIO)
CSIRO Enquiries
Private Bag 10
Clayton South, Victoria 3169 Australia
Phone: 61-3-9545-2176
Toll Free: 1300-363-400
Web Address: www.csiro.au
The Commonwealth Scientific and Industrial Research Organization (CSRIO) is Australia's national science agency and a leading international research agency. CSRIO performs research in Australia over a broad range of areas including agriculture, minerals and energy, manufacturing, communications, construction, health and the environment.

Electronics and Telecommunications Research Institute (ETRI)
218 Gajeongno
Yuseong-gu
Daejeon, 34129 Korea
Phone: 82-42-860-6114
E-mail Address: k21human@etri.re.kr

Web Address: www.etri.re.kr
Established in 1976, the Electronics and Telecommunications Research Institute (ETRI) is a nonprofit government-funded research organization that promotes technological excellence. The research institute has successfully developed information technologies such as TDX-Exchange, High Density Semiconductor Microchips, Mini-Super Computer (TiCOM), and Digital Mobile Telecommunication System (CDMA). ETRI's focus is on information technologies, robotics, telecommunications, digital broadcasting and future technology strategies.

Fraunhofer-Gesellschaft (FhG) (The)
Fraunhofer-Gesellschaft zur Forderung der angewandten Forschung e.V.
Postfach 20 07 33
Munich, 80007 Germany
Phone: 49-89-1205-0
Fax: 49-89-1205-7531
Web Address: www.fraunhofer.de
The Fraunhofer-Gesellschaft (FhG) institute focuses on research in health, security, energy, communication, the environment and mobility. FhG includes over 80 research units in Germany. Over 70% of its projects are derived from industry contracts.

Helmholtz Association
Anna-Louisa-Karsch-Strasse 2
Berlin, 10178 Germany
Phone: 49-30-206329-0
E-mail Address: info@helmholtz.de
Web Address: www.helmholtz.de/en
The Helmholtz Association is a community of 18 scientific-technical and biological-medical research centers. Helmholtz Centers perform top-class research in strategic programs in several core fields: energy, earth and environment, health, key technologies, structure of matter, aeronautics, space and transport.

Institute for Telecommunication Sciences (ITS)
325 Broadway
Boulder, CO 80305-3337 USA
Phone: 303-497-3571
E-mail Address: info@its.bldrdoc.gov
Web Address: www.its.bldrdoc.gov
The Institute for Telecommunication Sciences (ITS) is the research and engineering branch of the National Telecommunications and Information Administration (NTIA), a division of the U.S. Department of Commerce (DOC). Its research

activities are focused on advanced telecommunications and information infrastructure development.

National Research Council Canada (NRC)
1200 Montreal Rd., Bldg. M-58
Ottawa, ON K1A 0R6 Canada
Phone: 613-993-9101
Fax: 613-952-9907
Toll Free: 877-672-2672
E-mail Address: info@nrc-cnrc.gc.ca
Web Address: www.nrc-cnrc.gc.ca
National Research Council Canada (NRC) is comprised of 12 government organization, research institutes and programs that carry out multidisciplinary research. It maintains partnerships with industries and sectors key to Canada's economic development.

SRI International
1100 Wilson Blvd., Ste. 2800
Arlington, VA 22209 USA
Phone: 650-859-2000
Web Address: www.sri.com
SRI International is a nonprofit research organization that offers contract research services to government agencies, as well as commercial enterprises and other private sector institutions. It is organized around broad divisions including biosciences, global partnerships, education, products and solutions division, advanced technology and systems and information and computing sciences division.

25) Robotics & Automation Industry Associations

Singapore Industrial Automation Association (SIAA)
9, Town Hall Rd., Ste. 02-23
Singapore, 609431 Singapore
Phone: 65-6749-1822
Fax: 65-6841-3986
E-mail Address: secretariat@siaa.org
Web Address: www.siaa.org
The Singapore Industrial Automation Association (SIAA) is a non-profit organization which promotes the application of industrial automation with reference to business, technology & information services.

26) Satellite Industry Associations

Satellite Broadcasting & Communications Association (SBCA)
230 Washington Ave., Ste. 101
Albany, NY 12203 USA
Phone: 202-349-3620
Fax: 202-349-3621
Toll Free: 800-541-5981
E-mail Address: info@sbca.org
Web Address: www.sbca.com
The Satellite Broadcasting & Communications Association (SBCA) is the national trade organization representing all segments of the satellite consumer services industry in America.

World Teleport Association (WTA)
250 Park Ave., Fl. 7
New York, NY 10177 USA
Phone: 212-825-0218
Fax: 212-825-0075
E-mail Address: wta@worldteleport.org
Web Address: www.worldteleport.org
The World Teleport Association (WTA) is a nonprofit trade association representing the key commercial players in satellite communications.

27) Satellite-Related Professional Organizations

Geospatial Information and Technology Association (GITA)
1360 University Ave. W., Ste. 455
St. Paul, MN 55104 USA
Phone: 844-447-4482
Fax: 844-223-8218
E-mail Address: president@gita.org
Web Address: www.gita.org
The Geospatial Information and Technology Association (GITA) is an educational association for geospatial information and technology professionals.

Society of Satellite Professionals International (SSPI)
250 Park Ave., Fl. 7
The New York Information Technology Ctr.
New York, NY 10177 USA
Phone: 212-809-5199
Fax: 212-825-0075
E-mail Address: rbell@sspi.org
Web Address: www.sspi.org

The Society of Satellite Professionals International (SSPI) is a nonprofit member-benefit society that serves satellite professionals worldwide.

28) Software Industry Associations

European Software Institute (ESI)
Parque Tecnologico de Bizkaia
Edificio 202
Zamudio, Bizkaia E-48170 Spain
Phone: 34-946-430-850
Fax: 34-901-706-009
Web Address: www.esi.es
The European Software Institute (ESI) is a nonprofit foundation launched as an initiative of the European Commission, with the support of leading European companies working in the information technology field.

Korea Software Industry Association (KOSA)
IT Venture Tower W., 12F
135 Jung-daero, Songpa-gu
Seoul, 05717 South Korea
Phone: 82-2-2188-6900
Fax: 82-2-2188-6901
E-mail Address: choicy@sw.or.kr
Web Address: www.sw.or.kr
The Korea Software Industry Association (KOSA) is Korea's nonprofit trade organization representing more than 1,200 member companies in the software industry.

National Association of Software and Service Companies of India (NASSCOM)
Plot No. 7-10 NASSCOM Campus, Sector 126
Noida, 201303 India
Phone: 91-120-4990111
Fax: 91-120-4990119
E-mail Address: north@nasscom.in
Web Address: www.nasscom.im
The National Association of Software and Service Companies (NASSCOM) is the trade body and chamber of commerce for the IT and business process outsourcing (BPO) industry in India. The association's 1,400 members consist of corporations located around the world involved in software development, software services, software products, IT-enabled/BPO services and e-commerce.

Singapore Infocomm Technology Federation (SiTF)
79 Ayer Rajah Crescent, Ste. 02-03/04/05
Singapore, 139955 Singapore

Phone: 65-6325-9700
Fax: 65-6325-4993
E-mail Address: info@sitf.org.sg
Web Address: sitf.org.sg
Singapore Infocomm Technology Federation (SiTF) is an infocom industry association that has four chapters: Cloud Computing Chapter, Digital Media Wireless Chapter, Security and Governance Chapter and Singapore Enterprise Chapter.

Software & Information Industry Association (SIIA)
1090 Vermont Ave. NW, Fl. 6
Washington, DC 20005-4095 USA
Phone: 202-289-7442
Fax: 202-289-7097
Web Address: www.siia.net
The Software & Information Industry Association (SIIA) is a principal trade association for the software and digital content industry.

29) Software Industry Resources

Software Engineering Institute (SEI)-Carnegie Mellon
4500 5th Ave.
Pittsburgh, PA 15213-2612 USA
Phone: 412-268-5800
Fax: 412-268-5758
Toll Free: 888-201-4479
E-mail Address: info@sei.cmu.edu
Web Address: www.sei.cmu.edu
The Software Engineering Institute (SEI) is a federally funded research and development center at Carnegie Mellon University, sponsored by the U.S. Department of Defense through the Office of the Under Secretary of Defense for Acquisition, Technology, and Logistics [OUSD (AT&L)]. The SEI's core purpose is to help users make measured improvements in their software engineering capabilities.

30) Technology Law Associations

International Technology Law Association (ITechLaw)
7918 Jones Branch Dr., Ste. 300
McLean, VA 22102 USA
Phone: 703-506-2895
Fax: 703-506-3266
E-mail Address: memberservices@itechlaw.org
Web Address: www.itechlaw.org

The International Technology Law Association (ITechLaw) offers information concerning Internet and converging technology law. It represents lawyers in the field of technology law.

31) Technology Transfer Associations

Association of University Technology Managers (AUTM)
111 W. Jackson Blvd., Ste. 1412
Chicago, IL 60604 USA
Phone: 847-686-2244
Fax: 847-686-2253
E-mail Address: info@autm.net
Web Address: www.autm.net
The Association of University Technology Managers (AUTM) is a nonprofit professional association whose members belong to over 300 research institutions, universities, teaching hospitals, government agencies and corporations. The association's mission is to advance the field of technology transfer and enhance members' ability to bring academic and nonprofit research to people around the world.

Licensing Executives Society (USA and Canada), Inc.
11130 Sunrise Valley Dr., Ste. 350
Reston, VA 20191 USA
Phone: 703-234-4058
Fax: 703-435-4390
E-mail Address: info@les.org
Web Address: www.lesusacanada.org
Licensing Executives Society (USA and Canada), Inc., established in 1965, is a professional association composed of about 3,000 members who work in fields related to the development, use, transfer, manufacture and marketing of intellectual property. Members include executives, lawyers, licensing consultants, engineers, academic researchers, scientists and government officials. The society is part of the larger Licensing Executives Society International, Inc. (same headquarters address), with a worldwide membership of some 12,000 members from approximately 80 countries.

32) Telecommunications Industry Associations

Alliance for Telecommunications Industry Solutions (ATIS)
1200 G St. NW, Ste. 500
Washington, DC 20005 USA
Phone: 202-628-6380
E-mail Address: moran@atis.org
Web Address: www.atis.org
The Alliance for Telecommunications Industry Solutions (ATIS) is a U.S.-based body committed to rapidly developing and promoting technical and operations standards for the communications and related information technologies industry worldwide.

Asia-Pacific Telecommunity (APT)
Chaengwattana Rd., 12/49 Soi 5
Bangkok, 10210 Thailand
Phone: 66-2-573-0044
Fax: 66-2-573-7479
E-mail Address: aptmail@apt.int
Web Address: www.aptsec.org
The Asia-Pacific Telecommunity (APT) is an organization of governments, telecom service providers, manufacturers of communication equipment, research & development organizations and other stakeholders active in the field of communication and information technology. APT serves as the focal organization for communication and information technology in the Asia-Pacific region.

Association of Telecommunications and Value-added Service Providers (VATM)
Neustadtische Kirchstrasse 8
Berlin, 10117 Germany
Phone: 49-30/50-56-15-38
Fax: 49-30/50-56-15-39
E-mail Address: vatm@vatm.de
Web Address: www.vatm.de
The Association of Telecommunications and Value-added Service Providers (VATM) represents over 90 private companies in the German telecommunications industry at the national and international level.

China Communications Standards Association (CCSA)
52 Garden Rd., Haidian District
Beijing, 100083 China
Phone: 86-10-6230-2730
Fax: 86-10-6230-1849
Web Address: www.ccsa.org.cn/english
The China Communications Standards Association (CCSA) is a nonprofit organization that works to standardize the field of communications technology across China. Its membership includes operators, telecom equipment manufacturers and universities and academies from across China.

DigitalEurope
Rue de la Science 14
Brussels, 1040 Belgium
Phone: 32-2-609-5310
Fax: 32-2-609-5339
E-mail Address: info@digitaleurope.org
Web Address: www.digitaleurope.org
DigitalEurope is dedicated to improving the business
environment for the European information and
communications technology and consumer
electronics sector. Its members include 57 leading
corporations and 37 national trade associations from
across Europe.

European Telecommunications Standards Institute (ETSI)
ETSI Secretariat
650, route des Lucioles
Sophia-Antipolis Cedex, 06921 France
Phone: 33-4-92-94-42-00
Fax: 33-4-93-65-47-16
E-mail Address: info@etsi.org
Web Address: www.etsi.org
The European Telecommunications Standards
Institute (ETSI) is a non-profit organization whose
mission is to produce the telecommunications
standards to be implemented throughout Europe.

Fiber to the Home (FTTH) Council
6841 Elm St., Ste. 843
McLean, VA 22101-0843 USA
Phone: 202-524-9550
E-mail Address: heather.b.gold@ftthcouncil.org
Web Address: www.ftthcouncil.org
The Fiber-to-the-Home (FTTH) Council is a
nonprofit organization established in 2001 to educate
the public on the opportunities and benefits of FTTH
solutions. Its website is an excellent resource for
statistics, general reference and trends in the delivery
of fiber optic cable directly to the home and office.

INCOMPAS
1100 G St. NW
Ste. 800
Washington, DC 20005 USA
Phone: 202-296-6650
E-mail Address: gnorris@comptel.org
Web Address: www.incompas.org
CompTel is a trade organization representing voice,
data and video communications service providers and
their supplier partners. Members are supported
through education, networking, policy advocacy and
trade shows.

International Federation for Information Processing (IFIP)
Hofstrasse 3
Laxenburg, A-2361 Austria
Phone: 43-2236-73616
Fax: 43-2236-73616-9
E-mail Address: ifip@ifip.org
Web Address: www.ifip.org
The International Federation for Information
Processing (IFIP) is a multinational, apolitical
organization in information & communications
technologies and sciences recognized by the United
Nations and other world bodies. It represents
information technology societies from 56 countries or
regions, with over 500,000 members in total.

International Multimedia Telecommunications Consortium (IMTC)
Bishop Ranch 6, 2400 Camino Ramon, Ste. 375
San Ramon, CA 94583 USA
Phone: 925-275-6600
Fax: 925-275-6691
Web Address: www.imtc.org
The International Multimedia Telecommunications
Consortium (IMTC) is a non-profit corporation that
promotes interoperable multimedia conferencing and
telecommunications solutions based on international
standards.

International Telecommunications Union (ITU)
Place des Nations
Geneva 20, 1211 Switzerland
Phone: 41-22-730-5111
Fax: 41-22-733-7256
E-mail Address: itumail@itu.int
Web Address: www.itu.int
The International Telecommunications Union (ITU)
is an international organization for the
standardization of the radio and telecommunications
industry. It is an agency of the United Nations (UN).

National Association of Telecommunications Officers and Advisors (NATOA)
3213 Duke St., Ste. 695
Alexandria, VA 22314 USA
Phone: 703-519-8035
Fax: 703-997-7080
E-mail Address: info@natoa.org
Web Address: www.natoa.org
The National Association of Telecommunications
Officers and Advisors (NATOA) works to support
and serve the telecommunications industry's interests
and the needs of local governments.

National Exchange Carrier Association (NECA)
80 S. Jefferson Rd.
Whippany, NJ 07981-1009 USA
Fax: 973-884-8469
Toll Free: 800-228-8597
E-mail Address: necainfo@neca.org
Web Address: www.neca.org
The National Exchange Carrier Association (NECA), formed by the FCC, helps administer the FCC's access charge plan. The association helps ensure that telephone service remains available in all areas of the country.

National Telecommunications Cooperative Association (NTCA)
4121 Wilson Blvd., Ste. 1000
Arlington, VA 22203 USA
Phone: 703-351-2000
Fax: 703-351-2001
E-mail Address: membership@ntca.org
Web Address: www.ntca.org
The National Telecommunications Cooperative Association (NTCA) is an association representing more than 580 small and rural independent local exchange carriers providing telecommunications services throughout rural America.

Telecommunications Carriers Association (TCA)
E-mail Address: enq@tca.or.jp
Web Address: www.tca.or.jp
The Telecommunications Carriers Association (TCA) represents major telecommunications firms within Japan. It publishes a large amount of research regarding the telecom industry in Japan.

Telecommunications Industry Association (TIA)
1310 N. Courthouse Rd., Ste. 890
Arlington, VA 22201 USA
Phone: 703-907-7700
Fax: 703-907-7727
E-mail Address: smontgomery@tiaonline.org
Web Address: www.tiaonline.org
The Telecommunications Industry Association (TIA) is a leading trade association in the information, communications and entertainment technology industry. TIA focuses on market development, trade promotion, trade shows, domestic and international advocacy, standards development and enabling e-business.

TeleManagement Forum (TM Forum)
240 Headquarters Plz., E. Twr., 10th Fl.
Morristown, NJ 07960-6628 USA

Phone: 973-944-5100
Fax: 973-944-5110
E-mail Address: info@tmforum.org
Web Address: www.tmforum.org
The TeleManagement Forum (TM Forum) is a nonprofit global organization that provides leadership, strategic guidance and practical solutions to improve the management and operation of information and communications services.

United States Telecom Association (USTelecom)
601 New Jersey Ave. NW, Ste. 600
Washington, DC 20001 USA
Phone: 202-326-7300
Fax: 202-315-3603
E-mail Address: membership@ustelecom.org
Web Address: www.ustelecom.org
The United States Telecom Association (USTelecom) is a trade association representing service providers and suppliers for the telecom industry.

Voice On the Net (VON) Coalition, Inc.
1200 Seventh St. NW
Pillsbury Winthrop Shaw Pittman LLP
Washington, DC 20036-3006 USA
Phone: 202-663-8215
E-mail Address: glenn.richards@pillsburylaw.com
Web Address: www.von.org
Voice On the Net (VON) Coalition, Inc. is an organization is an advocate for the IP telephony industry. The VON Coalition supports that the IP industry should remain free of governmental regulations. It also serves to educate consumers and the media on Internet communications technologies.

33) Telecommunications Resources

Call Centers India
Phone: 206-384-4669
E-mail Address: sandy@callcentersindia.com
Web Address: www.callcentersindia.com
Call Centers India provides news and information concerning offshore call center services in India.

Center for Democracy and Technology (CDT)
1401 K St. NW, Ste. 200
Washington, DC 20005 USA
Phone: 202-637-9800
Fax: 202-637-0968
Web Address: www.cdt.org
The Center for Democracy and Technology (CDT) works to promote democratic values and constitutional liberties in the digital age.

Department of Telecommunication (Gov. of India)
20 Ashoka Rd.
Sanchar Bhawan
New Delhi, 110001 India
Phone: 91-11-2373-9191
Fax: 91-11-2372-3330
E-mail Address: secy-dot@nic.in
Web Address: www.dot.gov.in
The Government of India's Department of
Telecommunication web site provides information,
directories, guidelines, news and information related
to the telecom, Internet, Wi-Fi and wireless
communication industries. It is a branch of India's
Ministry of Communications & Information
Technology.

Infocomm Development Authority of Singapore (IMDA)
10 Pasir Panjang Rd.
#10-01 Mapletree Business City
Singapore, 117438 Singapore
Phone: 65-6211-0888
Fax: 65-6211-2222
E-mail Address: info@imda.gov.sg
Web Address: www.imda.gov.sg
The goal of the Infocomm Media Development
Authority of Singapore (IMDA) is to actively seek
opportunities to grow infocomm industry in both the
domestic and international markets.

International Communications Project (The)
Unit 2, Marine Action
Birdhill Industrial Estate
Birdhill, Co Tipperary Ireland
Phone: 353-86-108-3932
Fax: 353-61-749-801
E-mail Address: robert.alcock@intercomms.net
Web Address: www.intercomms.net
The International Communications Project
(InterComms) is an authoritative policy, strategy and
reference publication for the international
telecommunications industry.

International Customer Management Institute (ICMI)
121 S. Tejon St., Ste. 1100
Colorado Springs, CO 80903 USA
Phone: 719-955-8149
Fax: 719-955-8146
Toll Free: 800-672-6177
E-mail Address: icmi@informa.com
Web Address: www.icmi.com

International Customer Management Institute (ICMI)
is a leader in call center consulting, training,
publications and membership services. ICMI is a part
of UBM, a global marketing and communications
company.

Ministry of Communications and Information (Gov. of Singapore)
140 Hill St. #01-01A
Old Hill St. Police Station
Singapore, 179369 Singapore
Fax: 65-6837-9480
Toll Free: 800-837-9655
E-mail Address: MCI_Connects@mci.gov.sg
Web Address: www.mci.gov.sg
The Ministry of Communications and Information is
the department of the Singapore government
responsible for the development of the nation as a
global hub for information, communication and arts
related endeavors through the promotion of creativity
in the arts, design, media and infocomm technology
sectors.

Ministry of Communications and Information Technology (Gov. of India)
Electronics Niketan, 6
CGO Complex, Lodhi Rd.
New Delhi, 110 003 India
Phone: 91-11-2371-0445
Fax: 91-11-2337-2428
E-mail Address: mocit@nic.in
Web Address: www.mit.gov.in
The Government of India's Ministry of
Communications and Information Technology is the
department responsible for matters relating to
communications and IT issues in India. It oversees
three branches: the Department of
Telecommunications, the Department of IT and the
Department of Posts (the Indian postal service).

Telecom Regulatory Authority of India (TRAI)
Mahanagar Doorsanchar Bhawan
Jawaharlal Nehru Marg (Old Minto Rd.)
New Delhi, 110 002 India
Phone: 91-11-2323-6308
Fax: 91-11-2321-3294
E-mail Address: ap@trai.gov.in
Web Address: www.trai.gov.in
The TRAI publishes frequently updated data
regarding wireless, landline and broadcast industry
usage within India.

Total Telecom
Wren House
43 Hatton Garden
London, EC1N 8 EL UK
Phone: 44-20-7092-100
Fax: 44-20-7242-1508
E-mail Address: info@totaltele.com
Web Address: www.totaltele.com
Total Telecom offers information, news and articles
on the telecommunications industry in the U.K. and
worldwide through its web site and the Total
Telecom Magazine. Total Telecom is owned by
Terrapinn Ltd.

34) Telecommunications-VOIP Resources

VOIP News
28 E. 28th St.
New York, NY 10016 USA
Toll Free: 877-864-7275
E-mail Address: info@ziffdavis.com
Web Address: www.voip-news.com
VOIP News provides extensive news and information
about the VOIP industry, including FAQs,
comparison guides, buyer's guides, a dictionary of
terms, case studies, vendor directories and resources
for providers, users and technicians.

35) Trade Associations-General

**Associated Chambers of Commerce and Industry
of India (ASSOCHAM)**
5, Sardar Patel Marg
Chanakyapuri
New Delhi, 110 021 India
Phone: 91-11-4655-0555
Fax: 91-11-2301-7008
E-mail Address: assocham@nic.in
Web Address: www.assocham.org
The Associated Chambers of Commerce and Industry
of India (ASSOCHAM) has a membership of more
than 300 chambers and trade associations and serves
members from all over India. It works with domestic
and international government agencies to advocate
for India's industry and trade activities.

BUSINESSEUROPE
168 Ave. de Cortenbergh 168
Brussels, 1000 Belgium
Phone: 32-2-237-65-11
Fax: 32-2-231-14-45
E-mail Address: main@businesseurope.eu

Web Address: www.businesseurope.eu
BUSINESSEUROPE is a major European trade
federation that operates in a manner similar to a
chamber of commerce. Its members are the central
national business federations of the 34 countries
throughout Europe from which they come.
Companies cannot become direct members of
BUSINESSEUROPE, though there is a support group
which offers the opportunity for firms to encourage
BUSINESSEUROPE objectives in various ways.

Federation of Hong Kong Industries
8 Cheung Yue St.
Fl. 31, Billion Plz., Cheung Sha Wan
Kowloon, Hong Kong Hong Kong
Phone: 852-2732-3188
Fax: 852-2721-3494
E-mail Address: fhki@fhki.org.hk
Web Address: www.industryhk.org
The Federation of Hong Kong Industries promotes
the trade, investment advancement and development
of opportunities for the industrial and business
communities of Hong Kong. The web site hosts a
trade enquiry on products and services and publishes
research reports and trade publications.

**United States Council for International Business
(USCIB)**
1212 Ave. of the Americas
New York, NY 10036 USA
Phone: 212-354-4480
Fax: 212-575-0327
E-mail Address: azhang@uscib.org
Web Address: www.uscib.org
The United States Council for International Business
(USCIB) promotes an open system of world trade
and investment through its global network. Standard
USCIB members include corporations, law firms,
consulting firms and industry associations. Limited
membership options are available for chambers of
commerce and sole legal practitioners.

36) Trade Associations-Global

World Trade Organization (WTO)
Centre William Rappard
Rue de Lausanne 154
Geneva 21, CH-1211 Switzerland
Phone: 41-22-739-51-11
Fax: 41-22-731-42-06
E-mail Address: enquiries@wto.og
Web Address: www.wto.org

The World Trade Organization (WTO) is a global organization dealing with the rules of trade between nations. To become a member, nations must agree to abide by certain guidelines. Membership increases a nation's ability to import and export efficiently.

37) Trade Resources

Made-in-China.com - China Manufacturers Directory
Block A, Software Bldg. No. 9, Xinghuo Rd.
Nanjing New & High Technology Industry
Development Zone
Nanjing, Jiangsu 210032 China
Fax: 86-25-6667-0000
Web Address: www.made-in-china.com
Made-in-China.com - China Manufacturers Directory, one of the largest business to business portals in China, helps to connect Chinese manufacturers, suppliers and traders with international buyers. Made-in-China.com contains additional information on trade shows and important laws and regulations about business with China.

38) U.S. Government Agencies

Bureau of Economic Analysis (BEA)
4600 Silver Hill Rd.
Washington, DC 20233 USA
Phone: 301-278-9004
E-mail Address: customerservice@bea.gov
Web Address: www.bea.gov
The Bureau of Economic Analysis (BEA), is an agency of the U.S. Department of Commerce, is the nation's economic accountant, preparing estimates that illuminate key national, international and regional aspects of the U.S. economy.

Bureau of Labor Statistics (BLS)
2 Massachusetts Ave. NE
Washington, DC 20212-0001 USA
Phone: 202-691-5200
Fax: 202-691-7890
Toll Free: 800-877-8339
E-mail Address: blsdata_staff@bls.gov
Web Address: stats.bls.gov
The Bureau of Labor Statistics (BLS) is the principal fact-finding agency for the Federal Government in the field of labor economics and statistics. It is an independent national statistical agency that collects, processes, analyzes and disseminates statistical data to the American public, U.S. Congress, other federal agencies, state and local governments, business and labor. The BLS also serves as a statistical resource to the Department of Labor.

Cybersecurity & Infrastructure Security Agency (CISA)
245 Murray Ln.
Washington, D.C. 20528-0380 USA
Phone: 888-282-0870
E-mail Address: central@cisa.gov
Web Address: www.cisa.gov
The Cybersecurity & Infrastructure Security Agency (CISA) is the U.S. government agency focused on defending against cyber attacks and the development of new cybersecurity tools. The CISA also responds to attacks against the U.S. Government.

FCC-Office of Engineering & Technology (OET)
Office of Engineering and Technology
445 12th St. SW
Washington, DC 20554 USA
Fax: 866-418-0232
Toll Free: 888-225-5322
E-mail Address: oetinfo@fcc.gov
Web Address: www.fcc.gov/engineering-%26-technology
The Office of Engineering & Technology (OET) unit of the Federal Communications Commission (FCC) acts as a consultant to the FCC regarding matters of engineering and technology. It also provides evaluations of emerging technologies and equipment.

FCC-VoIP Division
445 12th St. SW
Washington, DC 20554 USA
Fax: 866-418-0232
Toll Free: 888-225-5322
E-mail Address: FOIA@fcc.gov
Web Address: www.fcc.gov/voip
The FCC-VoIP Division is dedicated to the promotion and regulation of the VoIP (Voice over Internet Protocol) industry. It operates as part of the Federal Communications Commission (FCC). VoIP allows users to call from their computer (or adapters) over the Internet to regular telephone numbers.

FCC-Wireline Competition Bureau (WCB)
445 12th St. SW
Washington, DC 20554 USA
Phone: 202-418-1500
Fax: 202-418-2825
Toll Free: 888-225-5322
E-mail Address: FOIA@fcc.gov

Web Address: www.fcc.gov/wcb
The FCC-Wireline Competition Bureau (WCB), formerly the Common Carrier Bureau, is a unit of the Federal Communications Commission (FCC). It is responsible for administering the FCC's policies concerning companies that provide wireline telecommunications.

Federal Communications Commission (FCC)
445 12th St. SW
Washington, DC 20554 USA
Fax: 866-418-0232
Toll Free: 888-225-5322
E-mail Address: PRA@fcc.gov
Web Address: www.fcc.gov
The Federal Communications Commission (FCC) is an independent U.S. government agency established by the Communications Act of 1934 responsible for regulating interstate and international communications by radio, television, wire, satellite and cable.

Federal Communications Commission (FCC)-International Bureau
445 12th St. SW
Washington, DC 20554 USA
Fax: 866-418-0232
Toll Free: 888-225-5322
E-mail Address: PRA@fcc.gov
Web Address: www.fcc.gov/international
The Federal Communications Commission (FCC)-International Bureau exists to administer the FCC's international telecommunications and satellite policies and obligations, such as licensing and regulatory functions, as well as promotes U.S. interests in international communications arena.

Federal Communications Commission (FCC)-Wireless Telecommunications Bureau
445 12th St. SW
Washington, DC 20554 USA
Phone: 202-418-0600
Fax: 202-418-0787
Toll Free: 888-225-5322
E-mail Address: PRA@fcc.gov
Web Address: www.fcc.gov/wireless-telecommunications#block-menu-block-4
The Federal Communications Commission (FCC)-Wireless Telecommunications Bureau handles nearly all FCC domestic wireless telecommunications programs and policies, including cellular and smarftphones, pagers and two-way radios. The

bureau also regulates the use of radio spectrum for businesses, aircraft/ship operators and individuals.

National Institute of Standards and Technology (NIST)
100 Bureau Dr.
Gaithersburg, MD 20899-1070 USA
Phone: 301-975-6478
Toll Free: 800-877-8339
E-mail Address: inquiries@nist.gov
Web Address: www.nist.gov
The National Institute of Standards and Technology (NIST) is an agency of the U.S. Department of Commerce that works with various industries to develop and apply technology, measurements and standards.

National Science Foundation (NSF)
2415 Eisenhower Ave.
Alexandria, VA 22314 USA
Phone: 703-292-5111
Toll Free: 800-877-8339
E-mail Address: info@nsf.gov
Web Address: www.nsf.gov
The National Science Foundation (NSF) is an independent U.S. government agency responsible for promoting science and engineering. The foundation provides colleges and universities with grants and funding for research into numerous scientific fields.

National Telecommunications and Information Administration (NTIA)
1401 Constitution Ave. NW
Herbert C. Hoover Bldg.
Washington, DC 20230 USA
Phone: 202-482-2000
Web Address: www.ntia.doc.gov
The National Telecommunications and Information Administration (NTIA), an agency of the U.S. Department of Commerce, is the Executive Branch's principal voice on domestic and international telecommunications and information technology issues.

U.S. Census Bureau
4600 Silver Hill Rd.
Washington, DC 20233-8800 USA
Phone: 301-763-4636
Toll Free: 800-923-8282
E-mail Address: pio@census.gov
Web Address: www.census.gov
The U.S. Census Bureau is the official collector of data about the people and economy of the U.S.

Founded in 1790, it provides official social, demographic and economic information. In addition to the Population & Housing Census, which it conducts every 10 years, the U.S. Census Bureau conducts numerous other surveys annually.

U.S. Department of Commerce (DOC)
1401 Constitution Ave. NW
Washington, DC 20230 USA
Phone: 202-482-2000
E-mail Address: publicaffairs@doc.gov
Web Address: www.commerce.gov
The U.S. Department of Commerce (DOC) regulates trade and provides valuable economic analysis of the economy.

U.S. Department of Labor (DOL)
200 Constitution Ave. NW
Washington, DC 20210 USA
Phone: 202-693-4676
Toll Free: 866-487-2365
E-mail Address: m-DOLPublicAffairs@dol.gov
Web Address: www.dol.gov
The U.S. Department of Labor (DOL) is the government agency responsible for labor regulations. The Department of Labor's goal is to foster, promote, and develop the welfare of the wage earners, job seekers, and retirees of the United States; improve working conditions; advance opportunities for profitable employment; and assure work-related benefits and rights.

U.S. Patent and Trademark Office (PTO)
600 Dulany St.
Madison Bldg.
Alexandria, VA 22314 USA
Phone: 571-272-1000
Toll Free: 800-786-9199
E-mail Address: usptoinfo@uspto.gov
Web Address: www.uspto.gov
The U.S. Patent and Trademark Office (PTO) administers patent and trademark laws for the U.S. and enables registration of patents and trademarks.

U.S. Securities and Exchange Commission (SEC)
100 F St. NE
Washington, DC 20549 USA
Phone: 202-942-8088
Fax: 202-772-9295
Toll Free: 800-732-0330
E-mail Address: help@sec.gov
Web Address: www.sec.gov

The U.S. Securities and Exchange Commission (SEC) is a nonpartisan, quasi-judicial regulatory agency responsible for administering federal securities laws. These laws are designed to protect investors in securities markets and ensure that they have access to disclosure of all material information concerning publicly traded securities. Visitors to the web site can access the EDGAR database of corporate financial and business information.

39) Wireless & Cellular Industry Associations

3GPP (The 3rd Generation Partnership Project)
3GPP Mobile Competence Centre, c/o ETSI
650, route des Lucioles
Sophia-Antipolis, Cedex 06921 France
E-mail Address: info@3gpp.org
Web Address: www.3gpp.org
The 3GPP is engaged in study, discussion and information approval through its members, which includes seven telecommunications standard development organizations (ARIB, ETSI, ATIS, CCSA, TTA, TSDSI and TTC) and mainly covers topics, such as cellular telecommunications network technologies, including radio access, the core transport network and service capabilities.

4GAmericas
1750 112th Ave. NE, Ste. B220
Bellevue, WA 98004 USA
Phone: 425-372-8922
Fax: 425-372-8923
Web Address: www.4gamericas.org
4G Americas is an industry trade organization composed of leading telecommunications service providers and manufacturers. The organization's mission is to promote, facilitate and advocate for the deployment and adoption of the 3GPP family of technologies throughout the Americas. The organization aims to develop the expansive wireless ecosystem of networks, devices, and applications enabled by GSM and its evolution to LTE. The organization publishes a significant amount of research and holds important conferences.

Bluetooth Special Interest Group (SIG)
5209 Lake Washington Blvd. NE, Ste. 350
Kirkland, WA 98033 USA
Phone: 425-691-3535
Fax: 425-691-3524
Web Address: www.bluetooth.com

The Bluetooth Special Interest Group (SIG) is a trade association comprised of leaders in the telecommunications, computing, automotive, industrial automation and network industries that is driving the development of Bluetooth wireless technology, a low cost short-range wireless specification for connecting mobile devices and bringing them to market.

Broadband Wireless Association (BWA)
Phone: 44-7765-250610
E-mail Address: Stephen@thebwa.eu
Web Address: www.thebwa.eu
The Broadband Wireless Association (BWA) provides representation, news and information for the European broadband wireless industry.

Canada Wireless Telecommunications Association (CWTA)
80 Elgin St., Ste. 300
Ottawa, ON K1P 6R2 Canada
Phone: 613-233-4888
Fax: 613-233-2032
E-mail Address: info@cwta.ca
Web Address: www.cwta.ca
The Canada Wireless Telecommunications Association (CWTA) seeks to be the pre-eminent source of input to government policy and public opinion on behalf of the wireless communications industry in Canada, in order to establish and maintain a positive economic environment for the wireless industry.

CDMA Development Group (CDG)
P.O. Box 22249
San Diego, CA 92129-2249 USA
Phone: 714-987-2362
Fax: 714-545-4601
Toll Free: 888-800-2362
E-mail Address: info@mobilitydg.org
Web Address: www.cdg.org
The CDMA Development Group (CDG) is composed of the world's leading code division multiple access (CDMA) service providers and manufacturers that have joined together to lead the adoption and evolution of CDMA wireless systems around the world.

Cellular Telecommunications & Internet Association (CTIA)
1400 16th St. NW, Ste. 600
Washington, DC 20036 USA
Phone: 202-785-0081

Web Address: www.ctia.org
The Cellular Telecommunications & Internet Association (CTIA) is an international nonprofit membership organization that represents a variety of wireless communications sectors including cellular service providers, manufacturers, wireless data and Internet companies. CTIA's industry committees study spectrum allocation, homeland security, taxation, safety and emerging technology.

Competitive Carrier Association (CCS)
805 15th St. NW, Ste. 401
Washington, DC 20005 USA
Fax: 866-436-1080
Toll Free: 800-722-1872
E-mail Address: administrator@rca-usa.org
Web Address: www.rca-usa.org
The Competitive Carrier Association (CCS), formerly the Rural Cellular Association (RCA) represents rural telecommunications providers in the United States before state and federal legislators. It primarily focuses on two-way wireless providers with a subscriber base less than 500,000.

Global System for Mobile Communication Association (GSMA)
The Wallbrook Bldg., Fl. 2, 25 Wallbrook
London, EC4N 8AF UK
Phone: 44-207-356-0600
Fax: 44-20-7356-0601
E-mail Address: info@gsma.com
Web Address: www.gsmworld.com
The Global System for Mobile Communications Association (GSMA) is a global trade association representing nearly 800 GSM mobile phone operators from 219 countries.

Hong Kong Wireless Technology Industry Association
16 Cheung Yue St.
Unit B & D, Fl. 11, Gee Hing
Hong Kong, Hong Kong
Phone: 582-2989-9164
E-mail Address: contact@hkwtia.org
Web Address: www.hkwtia.org
The Hong Kong Wireless Technology Industry Association (WTIA), established in 2001, is a nonprofit trade association intended to provide a platform for wireless-related businesses to work together for the development and growth of the wireless industry.

Industrial Internet Consortium
9C Medway Rd., PMB 274
Milford, MA 01757 USA
Phone: 781-444-0404
E-mail Address: info@iiconsortium.org
Web Address: www.iiconsortium.org
The Industrial Internet Consortium was founded in 2014 to further development, adoption and widespread use of interconnected machines, intelligent analytics and people at work. Through an independently run consortium of technology innovators, industrial companies, academia and government, the goal of the IIC is to accelerate the development and availability of intelligent industrial automation for the public good.

Li-Fi Consortium
E-mail Address: info@lificonsortium.org
Web Address: www.lificonsortium.org
The Li-Fi Consortium is a membership group founded to set standards and promote utilization of next-generation optical wireless networks, which are sometimes referred to as Li-Fi. Li-Fi is somewhat like Wi-Fi, except that it utilizes light as a means to transmit data.

Open Handset Alliance
E-mail Address: press@openhandsetalliance.com
Web Address: www.openhandsetalliance.com
The Open Handset Alliance is a group of about mobile-handset and technology makers, cellular carriers, semiconductor companies and software firms (largely led by Google, Inc.) who are collectively committed to an open system platform for cell phones. The Alliance has developed Android, the first open and free mobile platform.

Open Mobile Alliance (OMA)
2907 Shelter Island, Ste. 105-273
San Diego, CA 92106 USA
Phone: 858-623-0742
Fax: 858-623-0743
E-mail Address: snewberry@omaorg.org
Web Address: www.openmobilealliance.org
The Open Mobile Alliance (OMA) facilitates global user adoption of mobile data services by specifying market driven mobile service enablers that ensure service interoperability across devices, geographies, service providers, operators and networks, while allowing businesses to compete through innovation and differentiation.

Personal Communications Industry Association (PCIA)
500 Montgomery St., Ste. 500
Alexandria, VA 22314 USA
Phone: 703-739-0300
Fax: 703-836-1608
Toll Free: 800-759-0300
E-mail Address: jennifer.blasi@pcia.com
Web Address: www.pcia.com
The Personal Communications Industry Association (PCIA) is an association of companies that own and operate tower, rooftop and other kinds of wireless broadcasting and telecommunications equipment.

Small Cell Forum
P.O. Box 23
Dursley, Gloucestershire GL11 5WA UK
E-mail Address: info@smallcellforum.org
Web Address: www.smallcellforum.org
The Small Cell Forum, formerly the Femto Forum is a not-for-profit membership organization founded in 2007 to promote femtocell deployment worldwide. Comprised of mobile operators, telecoms hardware and software vendors, content providers and innovative start-ups, the group's mission is to advance the development and adoption of femtocell products and services within the residential and small to medium business markets.

Wi-Fi Alliance
10900-B Stonelake Blvd., Ste. 126
Austin, TX 78759 USA
Phone: 512-498-9434
Fax: 512-498-9435
Web Address: www.wi-fi.org
The Wi-Fi Alliance is a non-profit group that promotes wireless interoperability via Wi-Fi (802.11 standards). It also provides consumers with current information about Wi-Fi systems. The alliance currently includes over 350 member organizations.

WiMAX Forum
9009 SE Adams St., Ste. 2259
Clackamas, OR 97015 USA
Phone: 858-605-0978
Fax: 858-461-6041
Web Address: www.wimaxforum.org
The WiMAX Forum supports the implementation and standardization of long-range wireless Internet connections. It is a non-profit organization dedicated to the promotion and certification of interoperability and compatibility of broadband wireless products.

WiMedia Alliance, Inc.
2400 Camino Ramon, Ste. 375
San Ramon, CA 94583 USA
Phone: 925-275-6604
Fax: 925-886-3809
E-mail Address: help@wimedia.org
Web Address: www.wimedia.org
WiMedia Alliance, Inc. is an open, nonprofit wireless
industry association that promotes the adoption and
standardization of ultrawideband (UWB) worldwide
for use in the personal computer, consumer
electronics and mobile market segments.

Wireless Communications Alliance (WCA)
1510 Page Mill Rd.
Palo Alto, CA 94304-1125 USA
E-mail Address: promote@wca.org
Web Address: www.wca.org
The Wireless Communications Alliance (WCA) is a
non-profit business association for companies and
organizations working with wireless technologies. It
promotes networking, education and the exchange of
information amongst its members.

**Wireless Communications Association
International (WCAI)**
1333 H St. NW, Ste. 700 W
Washington, DC 20005-4754 USA
Phone: 202-452-7823
Web Address: www.wcainternational.com/
The Wireless Communications Association
International (WCAI) is a nonprofit trade association
representing the wireless broadband industry.

**40) Wireless & Cellular Industry
Resources**

**Hong Kong Wireless Development Centre
(HKWDC)**
31 Wylie Rd.
Room 1814, Fl. 18, Tung Wah College
Homantin, Kowloon Hong Kong
Phone: 852-3190-6630
E-mail Address: po@twc.edu.hk
Web Address: www.hkwdc.org
The Hong Kong Wireless Development Centre
(HKWDC) aims to facilitate mobile and wireless
application development in Hong Kong.

Wi-Fi Planet
Web Address: www.wi-fiplanet.com
Wi-Fi Planet is a web site devoted to wireless
networking protocols. The site features daily news,

reviews, tutorials, forums and event and product
listings related to the Wi-Fi performance.

Wireless Design Online
5340 Fryling Rd., Ste. 300
Erie, PA 16510 USA
Phone: 814-897-7700
Fax: 814-897-9555
E-mail Address: info@wirelessdesignonline.com
Web Address: www.wirelessdesignonline.com
Wireless Design Online is an Internet resource for
technical information covering the wireless industry.

Chapter 4

THE TELECOMMUNICATIONS 350: WHO THEY ARE AND HOW THEY WERE CHOSEN

Includes Indexes by Company Name, Industry & Location

The companies chosen to be listed in PLUNKETT'S TELECOMMUNICATIONS INDUSTRY ALMANAC comprise a unique list. THE TELECOMMUNICATIONS 350 were chosen specifically for their dominance in the many facets of the telecommunications industry in which they operate. Complete information about each firm can be found in the "Individual Profiles," beginning at the end of this chapter. These profiles are in alphabetical order by company name.

THE TELECOMMUNICATIONS 350 companies are from all parts of the United States, Asia, Canada, Europe and beyond. Essentially, THE TELECOMMUNICATIONS 350 includes companies that are deeply involved in the manufacturing, services and technologies that keep the entire industry forging ahead. To be included in our list, the firms had to meet the following criteria:

1) These companies were selected from top firms on a global basis.
2) Prominence, or a significant presence, in telecommunications and supporting fields. (See the following Industry Codes section for a complete list of types of businesses that are covered).
3) The companies in THE TELECOMMUNICATIONS 350 do not have to be exclusively in the telecommunications industry.
4) Financial data and vital statistics must have been available to the editors of this book, either directly from the company being written about or from outside sources deemed reliable and accurate by the editors. A small number of companies that we would like to have included are not listed because of a lack of sufficient, objective data.

INDEXES TO THE TELECOMMUNICATIONS 350, AS FOUND IN THIS CHAPTER AND IN THE BACK OF THE BOOK:

Index of Companies Within Industry Groups	p. 66
Alphabetical Index	p. 76
Index of U.S. Headquarters Location by State	p. 79
Index of Non-U.S. Headquarters Location by Country	p. 82
Index of Firms Noted as "Hot Spots for Advancement" for Women/Minorities	p. 448
Index by Subsidiaries, Brand Names and Selected Affiliations	p. 450

INDEX OF COMPANIES WITHIN INDUSTRY GROUPS

The industry codes shown below are based on the 2012 NAIC code system (NAIC is used by many analysts as a replacement for older SIC codes because NAIC is more specific to today's industry sectors, see www.census.gov/NAICS). Companies are given a primary NAIC code, reflecting the main line of business of each firm.

Industry Group/Company	Industry Code	2021 Sales	2021 Profits
Advertising, Public Relations and Marketing Services			
TTEC Holdings Inc	541800	2,273,061,888	140,970,000
Cloud, Data Processing, Business Process Outsourcing (BPO) and Internet Content Hosting Services			
Cyxtera Technologies Inc	518210	703,699,968	-257,900,000
GoDaddy Inc	518210	3,815,699,968	242,300,000
Neustar Inc	518210	575,000,000	
Newfold Digital Inc	518210	866,580,000	
Sitel Corporation	518210	4,300,000,000	
Computer and Data Systems Design, Consulting and Integration Services			
Kratos Defense & Security Solutions Inc	541512	811,500,032	-2,000,000
Sopra Steria Group SA	541512	4,775,346,176	191,409,520
Sykes Enterprises Incorporated	541512	1,744,466,212	
Computer Manufacturing, Including PCs, Laptops, Mainframes and Tablets			
Fujitsu Limited	334111	26,621,937,664	1,503,263,104
HP Inc	334111	63,487,000,576	6,503,000,064
MiTAC Holdings Corp	334111	1,522,269,328	431,628,846
Computer Networking & Related Equipment Manufacturing (may incl. Internet of Things, IoT)			
Calix Inc	334210A	679,393,984	238,378,000
Cisco Systems Inc	334210A	49,818,001,408	10,590,999,552
D-Link Corporation	334210A	502,509,300	8,631,400
Extreme Networks Inc	334210A	1,009,417,984	1,936,000
Juniper Networks Inc	334210A	4,735,399,936	252,700,000
NETGEAR Inc	334210A	1,168,072,960	49,387,000
Computer Peripherals and Accessories, including Printers, Monitors and Terminals Manufacturing			
Belkin International Inc	334118	1,915,681,950	
Computer Software: Accounting, Banking & Financial			
Calero-MDSL	511210Q		
Computer Software: Business Management & Enterprise Resource Planning (ERP)			
Microsoft Corporation	511210H	168,087,994,368	61,270,999,040
Computer Software: E-Commerce, Web Analytics & Applications Management			
Realtime Corporation	511210M		
Computer Software: Network Management, System Testing, & Storage			
F5 Inc	511210B	2,603,416,064	331,240,992
NetScout Systems Inc	511210B	831,281,984	19,352,000
OPTERNA	511210B	15,456,180	

Industry Group/Company	Industry Code	2021 Sales	2021 Profits
Computer Software: Sales & Customer Relationship Management			
Alvaria Inc	511210K	525,429,450	
Computer Software: Security & Anti-Virus			
AsiaInfo Technologies Limited	511210E		
Axway Inc	511210E		
Certicom Corp	511210E		
Cloudflare Inc	511210E	656,425,984	-260,308,992
Entrust Corporation	511210E	880,000,000	
McAfee Corp	511210E	1,920,000,000	2,688,000,000
OneSpan Inc	511210E	214,480,992	-30,584,000
VeriSign Inc	511210E	1,327,576,064	784,830,016
Computer Software: Telecom, Communications & VOIP, Internet of Things (IoT)			
BlackBerry Limited	511210C	893,000,000	-1,104,000,000
Skype Technologies Sarl	511210C		
Slack Technologies Inc	511210C	655,638,896	
TeamViewer AG	511210C	511,000,160	51,040,156
Vocera Communications Inc	511210C	200,000,000	
XIUS	511210C	468,231,000	3,979,000
Zoom Video Communications Inc	511210C	2,651,367,936	672,316,032
Computers, Peripherals, Software and Accessories Distribution			
Anixter International Inc	423430	9,400,000,000	
Black Box Corporation	423430	909,346,364	
Ingram Micro Mobility	423430		
ScanSource Inc	423430	3,150,806,016	10,795,000
TESSCO Technologies Incorporated	423430	373,340,704	-8,742,900
Connectors for Electronics Manufacturing			
Belden Inc	334417	2,408,100,096	63,925,000
Construction of Telecommunications Lines and Systems & Electric Power Lines and Systems			
Bouygues SA	237130	38,331,871,232	1,147,233,280
China Communications Services Corp Ltd	237130	19,850,858,496	467,774,944
Dycom Industries Inc	237130	3,199,164,928	34,337,000
IHS Holding Ltd	237130	1,579,730,048	-25,832,000
Consumer Electronics Manufacturing, Including Audio and Video Equipment, Stereos, TVs and Radios			
AAC Technologies Holdings Inc	334310	2,617,367,296	195,007,184
InnoMedia Pte Ltd	334310		
Koninklijke Philips NV (Royal Philips)	334310	17,495,054,336	3,384,593,408
Panasonic Corporation	334310	49,679,577,088	1,224,243,456
Samsung Electronics Co Ltd	334310	215,241,228,288	30,210,072,576
TCL Technology Group Corporation	334310	25,682,570,815	2,347,017,202
Contract Electronics Manufacturing Services (CEM) and Printed Circuits Assembly			
Accton Technology Corporation	334418	2,150,623,955	169,779,455
Celestica Inc	334418	5,634,699,776	103,900,000
Flex Ltd	334418	24,124,000,256	613,000,000
Foxconn Technology Co Ltd	334418	3,469,320,192	149,626,720

Industry Group/Company	Industry Code	2021 Sales	2021 Profits
Jabil Inc	334418	29,284,999,168	696,000,000
Sanmina Corporation	334418	6,756,642,816	268,998,016
SMTC Corporation	334418	390,000,000	
Copper Rolling, Drawing, Extruding and Alloying			
Bangkok Cable Co Ltd	331420		
Distributors of Telecommunications Equipment, Telephones, Cellphones and Electronics Components (Wholesale Distribution)			
Likewize Corp	423690	12,039,300,000	
Simply Inc	423690	68,024,000	4,277,000
Electric Signal, Electricity, and Semiconductor Test and Measuring Equipment Manufacturing			
Anritsu Corporation	334515	785,664,512	119,437,856
Rohde & Schwarz GmbH & Co KG	334515	2,714,157,600	
Tektronix Inc	334515		
Electrical Contractors and Other Wiring Installation Contractors			
TKH Group NV	238210		
Electricity Control Panels, Circuit Breakers and Power Switches Equipment (Switchgear) Manufacturing			
Broadcom Inc	335313	27,449,999,360	6,736,000,000
Factory Automation, Robots (Robotics) Industrial Process, Thermostat, Flow Meter and Environmental Quality Monitoring and Control Manufacturing (incl. Artificial Intelligence, AI)			
Siemens AG	334513	63,495,540,736	6,282,759,680
Fiber Optic Cable, Connectors and Related Products Manufacturing			
Amphenol Corporation	335921	10,876,300,288	1,590,800,000
CommScope Holding Company Inc	335921	8,586,699,776	-462,600,000
Energy Focus Inc	335921	9,865,000	-7,886,000
Optical Cable Corporation	335921	59,136,296	6,610,516
Preformed Line Products Company	335921	517,416,992	35,729,000
Iron and Steel Mills and Ferroalloy Manufacturing			
Tata Group	331110	128,000,000,000	
LCD (Liquid-Crystal Display), Radio Frequency (RF, RFID) and Microwave Equipment Manufacturing			
Qisda Corporation	334419		
Mail Order, Catalogs and Other Direct Marketing, and TV Shopping			
Hello Direct Inc	454113		
Management of Businesses & Enterprises			
LG Corporation	551114	5,767,309,391	2,256,784,646
OJSC JFSC Sistema	551114		
Medical Imaging and Electromedical (Medical Devices) Equipment, including MRI, Ultrasound, Pacemakers, EKG and CAT			
Aware Inc	334510	16,854,000	-5,824,000
Outsourced Computer Facilities Management and Operations Services			
International Business Machines Corporation (IBM)	541513	57,351,000,064	5,742,000,128

Industry Group/Company	Industry Code	2021 Sales	2021 Profits
Photographic and Photocopying Equipment Manufacturing			
BenQ Corporation	333316	3,300,000,000	
Kyocera Corporation	333316	11,323,769,856	669,044,800
Port and Harbor Operations			
CK Hutchison Holdings Limited	488310	35,777,732,608	4,265,602,304
Pressed and Blown Glass and Glassware (except Glass Packaging Containers) Manufacturing			
Corning Incorporated	327212	14,081,999,872	1,906,000,000
Radar, Navigation, Sonar, Space Vehicle Guidance, Flight Systems and Marine Instrument Manufacturing			
Garmin Ltd	334511	4,982,794,752	1,082,200,064
TomTom International BV	334511	516,944,384	-96,523,624
Trimble Inc	334511	3,659,099,904	492,700,000
Radio, Television and Other Electronics Stores			
Hikari Tsushin Inc	443142	4,915,723,264	479,895,232
Satellite Telecommunications			
Asia Satellite Telecommunications Holdings Ltd	517410	213,261,838	
Eutelsat Communications SA	517410	1,258,285,568	218,331,264
Gilat Satellite Networks Ltd	517410	214,970,000	-3,033,000
Globalstar Inc	517410	124,297,000	-112,625,000
Inmarsat Global Limited	517410	1,195,000,000	
Intelsat SA	517410	2,000,000,000	
Iridium Communications Inc	517410	614,499,968	-9,319,000
NextPlat Corp	517410	7,739,910	-8,107,662
OneWeb Ltd	517410	3,000,000	370,800,000
Planet Labs PBC	517410	113,168,000	-127,103,000
SES SA	517410	1,784,585,216	461,952,640
Thuraya Telecommunications Company	517410	187,218,435	
Semiconductor and Solar Cell Manufacturing, Including Chips, Memory, LEDs, Transistors and Integrated Circuits, Artificial Intelligence (AI), & Internet of Things (IoT)			
Advanced Micro Devices Inc (AMD)	334413	16,433,999,872	3,161,999,872
DSP Group Inc	334413	118,000,000	
Infinera Corporation	334413	1,425,204,992	-170,778,000
Intel Corporation	334413	79,023,996,928	19,868,000,256
Marvell Technology Group Ltd	334413	2,968,900,096	-277,297,984
Qualcomm Incorporated	334413	33,565,999,104	9,043,000,320
Toshiba Corporation	334413	22,651,846,656	845,305,536
Telecommunications, Telephone and Network Equipment Manufacturing, including PBX, Routers, Switches, Internet of Things (IoT), and Handsets Manufacturing			
ADTRAN Inc	334210	563,004,032	-8,635,000
Avaya Holdings Corp	334210	2,972,999,936	-13,000,000
Bogen Communications LLC	334210		
Calient Technologies Inc	334210		
Channell Commercial Corporation	334210		
Ciena Corporation	334210	3,620,684,032	500,196,000
Datang Telecom Technology Co Ltd	334210	203,495,409	-8,001,747
Dialogic Inc	334210	130,000,000	
DZS Inc	334210	350,206,016	-34,683,000

Industry Group/Company	Industry Code	2021 Sales	2021 Profits
Fujitsu Network Communications Inc	334210		
General DataComm LLC	334210		
Harmonic Inc	334210	507,148,992	13,254,000
Huawei Technologies Co Ltd	334210		
ITI Limited	334210	321,950,000	1,291,760
Lattice Incorporated	334210	86,500,000	
MedTel Services LLC	334210		
Mitel Networks Corporation	334210	1,319,552,000	
Pineapple Energy Inc	334210		
Plantronics Inc	334210	1,727,607,040	-57,331,000
Tellabs Inc	334210	1,640,000,000	
Telvue Corporation	334210		
VTech Holdings Limited	334210	2,372,300,032	230,900,000
Westell Technologies Inc	334210	29,947,000	-2,734,000
ZTE Corporation	334210	16,966,420,480	1,009,339,520
Telemarketing Bureaus and Other Contact Centers			
Intrado Corporation	561422	2,800,000,000	
Telephone, Internet Access, Broadband, Data Networks, Server Facilities and Telecommunications Services Industry			
AAPT Limited	517110		
Akamai Technologies Inc	517110	3,461,222,912	651,641,984
Alaska Communications	517110	250,191,752	
Altice Portugal SA	517110		
Altice USA Inc	517110	10,090,849,280	990,310,976
Arqiva Limited	517110	745,635,544	246,790,962
AT&T Inc	517110	168,864,006,144	20,081,000,448
ATN International Inc	517110	602,707,008	-22,108,000
Axtel SAB de CV	517110	562,705,344	-39,363,556
B Communications Ltd	517110	2,667,937,024	39,016,424
BCE Inc (Bell Canada Enterprises)	517110	18,238,314,496	2,208,913,408
Bell Aliant Inc	517110	3,551,593,500	
Bell MTS Inc	517110	866,829,600	
Bezeq-The Israel Telecommunication Corp Ltd	517110		
Bharat Sanchar Nigam Limited (BSNL)	517110	2,534,384,690	-1,014,171,205
Boingo Wireless Inc	517110	246,896,000	
BT Global Services plc	517110	137,908,418	551,271
BT Group plc	517110	25,777,954,816	1,778,873,472
Ceske Radiokomunikace as	517110	92,702,610	
Charter Communications Inc	517110	51,682,000,896	4,654,000,128
China Telecom Corporation Limited	517110	68,964,390,144	4,102,542,756
China Unicom (Hong Kong) Limited	517110	51,439,309,038	2,276,104,779
Chorus Limited	517110	552,837,120	29,560,120
Chunghwa Telecom Co Ltd	517110	6,981,203,968	1,187,172,224
Cincinnati Bell Inc	517110	1,693,100,000	-108,500,000
Cisco Webex	517110		
Cogent Communications Group Inc	517110	589,796,992	48,185,000
Colt Technology Services Group Limited	517110	168,000,000	
Comcast Corporation	517110	116,384,997,376	14,158,999,552

Industry Group/Company	Industry Code	2021 Sales	2021 Profits
Consensus Cloud Solutions Inc	517110	352,664,000	109,001,000
Consolidated Communications Holdings Inc	517110	1,282,232,960	-107,085,000
Converge Information and Communications Technology Solutions, Inc.	517110		
Cox Communications Inc	517110	13,104,000,000	
DirecTV LLC (DIRECTV)	517110	29,039,677,664	
EarthLink LLC	517110	1,230,989,760	
EchoStar Corporation	517110	1,985,720,064	72,875,000
eircom Limited	517110	1,439,911,200	
Elisa Corporation	517110	2,037,486,464	350,390,560
Embratel Participacoes SA	517110	11,328,981,300	
Emirates Telecommunications Corporation (Etisalat)	517110	14,520,652,556	2,550,954,878
Empresa Nacional de Telecommunicacions SA (Entel)	517110	2,855,795,033	88,213,702
Enterprise Diversified Inc	517110		
enTouch Systems Inc	517110	633,750,000	
Equinix Inc	517110	6,635,536,896	500,191,008
FASTWEB SpA	517110	3,100,000,000	
Frontier Communications Corporation	517110		
Fusion Connect Inc	517110	140,000,000	
GCI Communication Corp	517110	908,189,776	
Glentel Inc	517110		
Hawaiian Telcom Holdco Inc	517110		
Hellenic Telecommunications Organization SA	517110	3,434,867,712	568,619,840
HKT Trust and HKT Limited	517110	4,326,368,768	612,502,016
iBasis Inc	517110	1,070,000,000	
IDT Corporation	517110	1,446,989,952	96,475,000
Iliad SA	517110	8,591,518,800	595,642,400
INNOVATE Corp	517110	1,205,200,000	-227,500,000
Inteliquent Inc	517110		
Internap Corporation	517110	296,088,000	
Internet Initiative Japan Inc	517110	1,579,663,872	72,022,832
KCOM Group Limited	517110	330,000,000	
Koninklijke KPN NV (Royal KPN NV)	517110	5,375,170,560	1,313,454,720
KT Corporation	517110	19,166,613,504	1,044,531,712
Liberty Global plc	517110	10,311,300,096	13,426,799,616
Liberty Latin America Ltd	517110	4,799,000,064	-440,100,000
Lumen Technologies Inc	517110	19,687,000,064	2,032,999,936
Lumos Networks Corp	517110		
Magyar Telekom plc	517110	1,811,717,120	152,667,936
Mahanagar Telephone Nigam Limited	517110	174,331,264	-309,196,128
Maroc Telecom SA	517110	3,484,907,520	585,004,864
Maxis Berhad	517110	2,065,073,536	293,503,872
Mediacom Communications Corporation	517110	2,200,000,000	
Megacable Holdings SAB de CV	517110	1,217,066,496	171,368,736
Momentum Telecom	517110	140,000,000	
Net2Phone Inc	517110	43,897,000	
Nippon Telegraph and Telephone Corporation (NTT)	517110	88,578,809,856	6,794,578,944
Nortel Inversora SA	517110	3,000,000,000	
O2 Czech Republic AS	517110	1,851,494,484	293,312,356

Industry Group/Company	Industry Code	2021 Sales	2021 Profits
Oblong Inc	517110	7,739,000	-9,051,000
Oi SA	517110	1,874,997,463	
Orange	517110	43,362,365,440	237,604,784
Orange Polska SA	517110	2,590,760,448	363,158,208
Pakistan Telecommunication Company Limited	517110	614,262,208	11,494,153
PCCW Limited	517110	4,924,220,416	161,151,200
Perusahaan Perseroan PT Telekomunikasi Indonesia Tbk (Telkom)	517110	9,655,995,392	1,677,342,336
PLDT Inc	517110	3,483,991,040	475,338,016
Proximus Group	517110	5,646,427,648	451,755,008
Q Beyond AG	517110	158,227,440	9,903,938
Rackspace Technology Inc	517110	3,009,499,904	-218,300,000
Rogers Communications Inc	517110	11,398,459,392	1,211,791,232
Rostelecom PJSC	517110		
Shaw Communications Inc	517110	4,284,825,600	766,897,408
Shenandoah Telecommunications Company	517110	245,239,008	998,830,976
Singapore Technologies Telemedia Pte Ltd	517110	3,000,000,000	
Singapore Telecommunications Limited	517110	11,350,625,280	401,741,312
Singtel Optus Pty Limited	517110	5,400,340,112	1,435,417,716
SK Broadband Co Ltd	517110	3,600,000,000	
SoftBank Group Corp	517110	41,739,595,776	36,991,713,280
Spark New Zealand Limited	517110	2,242,166,528	241,512,480
Speedcast International Limited	517110	700,000,000	
Startec Global Communications Corporation	517110		
Swisscom AG	517110	11,711,419,392	1,918,565,760
Syniverse Technologies LLC	517110	733,000,000	-49,387,000
Tata Communications Limited	517110		
Tata Teleservices Limited	517110	365,430,155	-1,302,939,273
TDC A/S	517110	2,193,046,272	54,271,112
TDS Telecommunications LLC	517110	1,006,000,000	90,000,000
Tele2 AB	517110	2,645,542,400	425,238,144
Telecom Argentina SA	517110	3,185,746,176	64,876,484
Telecom Egypt SAE	517110	1,937,406,080	439,691,936
Telecom Italia SpA	517110	15,618,690,048	-8,822,989,824
Telecomunicaciones de Puerto Rico Inc	517110	963,144,000	
Telefonica Brasil SA	517110	8,616,781,824	1,220,986,752
Telefonica Chile SA	517110	17,500,000,000	
Telefonica de Argentina SA	517110		
Telefonica del Peru SAA	517110	2,000,000,000	
Telefonica SA	517110	40,053,231,616	8,297,811,456
Telefonos de Mexico SAB de CV (Telmex)	517110	4,984,049,623	223,563,640
Telekom Austria AG	517110	4,758,196,736	463,439,424
Telekom Malaysia Berhad	517110	2,587,007,744	200,875,136
Telenet Group NV/SA	517110	2,647,100,928	401,786,624
Telephone and Data Systems Inc (TDS)	517110	5,328,999,936	156,000,000
Telia Company AB	517110	8,724,295,680	1,153,456,128
Telia Lietuva AB	517110		
Telkom SA SOC Limited	517110	2,599,757,056	145,680,720
Telstra Corporation Limited	517110	14,672,773,120	1,297,615,872

Industry Group/Company	Industry Code	2021 Sales	2021 Profits
TELUS Corporation	517110	13,096,368,128	1,287,236,608
TOT pcl	517110		
TPG Telecom Limited	517110		
TruConnect Communications Inc	517110	153,972,000	
True Corporation Public Company Limited	517110	4,043,210,496	-40,202,896
Turk Telekomunikasyon AS	517110	1,910,654,336	321,190,688
United Online Inc	517110	196,716,000	
Verizon Communications Inc	517110	133,613,002,752	22,065,000,448
Virgin Media Business Ltd	517110	1,100,000,000	
Vocus Group Limited	517110		
Vonage Holdings Corp	517110	1,409,015,040	-24,497,000
Windstream Holdings Inc	517110	5,000,000,000	
Wireless Communications and Radio and TV Broadcasting Equipment Manufacturing, including Cellphones (Handsets) and Internet of Things (IoT)			
Anaren Inc	334220	250,000,000	
ARC Group Worldwide Inc	334220		
Aviat Networks Inc	334220	274,911,008	110,139,000
Blonder Tongue Laboratories Inc	334220	15,754,000	84,000
CalAmp Corp	334220	308,587,008	-56,309,000
ClearOne Inc	334220	28,967,000	-7,694,000
Comtech Telecommunications Corp	334220	581,694,976	-73,480,000
EF Johnson Technologies Inc	334220	156,510,900	
Filtronic plc	334220	18,799,018	72,508
Hitachi Kokusai Electric Inc	334220	650,000,000	
InterDigital Inc	334220	425,408,992	55,295,000
L3Harris Technologies Inc	334220	17,813,999,616	1,846,000,000
Laird Connectivity	334220	1,500,000,000	
LG Electronics Inc	334220	57,521,094,656	794,215,680
LG Electronics USA Inc	334220	12,371,585,410	212,212,124
LM Ericsson Telephone Company (Ericsson)	334220	22,942,121,984	2,241,141,504
Lumentum Operations LLC	334220	1,700,000,000	
Nokia Corporation	334220	22,640,777,216	1,655,075,456
Phazar Antenna Corp	334220	14,391,000	
Potevio Corporation	334220	13,122,837,000	
Proxim Wireless Corporation	334220		
Socket Mobile Inc	334220	23,199,060	4,466,257
Sonim Technologies Inc	334220	54,570,000	-38,627,000
Sony Corporation	334220	21,225,175,648	1,931,789
TCI International Inc	334220	1,536,311,000	
Technical Communications Corporation	334220		
Telesat Corporation	334220	589,726,976	80,533,560
Telular Corporation	334220	167,310,000	
Ubiquiti Inc	334220	1,898,093,952	616,584,000
Uniden Holdings Corporation	334220	174,850,006	33,191,558
ViaSat Inc	334220	2,256,107,008	3,691,000
Viavi Solutions Inc	334220	1,198,899,968	46,100,000
Wireless Telecommunications Carriers (except Satellite)			
Advanced Info Service plc	517210	5,103,655,424	757,729,984

Industry Group/Company	Industry Code	2021 Sales	2021 Profits
Altice Europe NV	517210	17,536,546,884	
America Movil SAB de CV	517210	42,268,254,208	9,506,791,424
AT&T Mexico SAU	517210	2,747,000,000	-510,000,000
AT&T Mobility LLC	517210	78,254,000,000	23,312,000,000
Axiata Group Berhad	517210	5,811,884,032	183,754,064
Bharti Airtel Limited	517210	13,800,838,370	-1,685,126,652
Cellcom Israel Ltd	517210		
China Mobile Limited	517210	133,089,135,426	18,248,062,482
Comba Telecom Systems Holdings Ltd	517210	747,750,016	-75,488,448
COSMOTE Mobile Telephones SA	517210	1,551,954,200	219,345,880
Cricket Wireless LLC	517210	4,419,657,900	
Data Select Limited	517210	223,187,200	
Deutsche Telekom AG	517210		
Digi.com Bhd	517210	1,421,670,400	260,766,288
Far EasTone Telecommunications Co Ltd	517210	3,078,763,957	333,203,672
First Pacific Company Limited	517210	9,103,200,256	333,300,000
Freenet AG	517210	2,606,840,576	202,083,376
Global Telecom Holding SAE	517210	2,268,084,000	-373,000,000
Globe Telecom Inc	517210	3,024,107,008	426,407,232
GTT Communications Inc	517210	1,820,000,000	
Hi Sun Technology (China) Limited	517210	532,841,568	448,615,168
Hutchison Telecommunications Hong Kong Holdings Limited	517210	686,007,296	509,569
Indosat Ooredoo Hutchison	517210	1,831,030,020	114,303,105
Jio (Reliance Jio Infocomm Limited)	517210	9,599,828,802	1,637,569,936
KDDI Corporation	517210	39,399,280,640	4,831,622,656
LG Uplus Corp	517210	11,646,546,800	608,949,290
M1 Limited	517210	840,129,680	
Millicom International Cellular SA	517210	4,616,999,936	590,000,000
Mobile Telecommunications Company KSCP (Zain Group)	517210	5,033,000,000	616,000,000
Mobile TeleSystems PJSC	517210	7,179,053,568	852,682,496
NTT DOCOMO Inc	517210	42,874,999,988	8,286,093,708
Ooredoo QPSC	517210	8,167,035,500	287,517,890
Orange Belgium	517210	1,390,418,304	40,508,044
ORBCOMM Inc	517210	250,000,000	
Partner Communications Co Ltd	517210	1,017,149,120	34,782,084
Purple Communications Inc	517210	167,346,270	
SK Telecom Co Ltd	517210	12,893,147,136	1,853,323,520
SmarTone Telecommunications Holdings Limited	517210	865,596,480	57,015,758
Spok Inc	517210	142,152,992	-22,180,000
TalkTalk Telecom Group Limited	517210	1,883,389,530	-15,312,110
Telefonica Deutschland Holding AG	517210	7,919,479,296	215,169,984
Telenor ASA	517210	11,343,850,496	157,233,344
TIM SA	517210	3,533,791,488	578,692,032
T-Mobile Polska SA	517210	1,615,934,800	103,048,400
T-Mobile US Inc	517210	80,117,997,568	3,024,000,000
Total Access Communication PCL	517210	2,288,770,048	94,453,496
Trilogy International Partners Inc	517210	653,564,032	-144,688,992
Turkcell Iletisim Hizmetleri AS	517210	2,002,505,600	280,474,656
United States Cellular Corporation	517210	4,121,999,872	155,000,000

Industry Group/Company	Industry Code	2021 Sales	2021 Profits
Veon Ltd	517210	7,788,000,256	674,000,000
Verio Inc	517210		
VMED O2 UK Limited (Virgin Media O2)	517210	13,965,845,608	24,685,602
Vodafone Group plc	517210	44,674,797,568	114,213,456
WIND Hellas Telecommunications SA	517210	602,436,800	

ALPHABETICAL INDEX

AAC Technologies Holdings Inc
AAPT Limited
Accton Technology Corporation
ADTRAN Inc
Advanced Info Service plc
Advanced Micro Devices Inc (AMD)
Akamai Technologies Inc
Alaska Communications
Altice Europe NV
Altice Portugal SA
Altice USA Inc
Alvaria Inc
America Movil SAB de CV
Amphenol Corporation
Anaren Inc
Anixter International Inc
Anritsu Corporation
ARC Group Worldwide Inc
Arqiva Limited
Asia Satellite Telecommunications Holdings Ltd
AsiaInfo Technologies Limited
AT&T Inc
AT&T Mexico SAU
AT&T Mobility LLC
ATN International Inc
Avaya Holdings Corp
Aviat Networks Inc
Aware Inc
Axiata Group Berhad
Axtel SAB de CV
Axway Inc
B Communications Ltd
Bangkok Cable Co Ltd
BCE Inc (Bell Canada Enterprises)
Belden Inc
Belkin International Inc
Bell Aliant Inc
Bell MTS Inc
BenQ Corporation
Bezeq-The Israel Telecommunication Corp Ltd
Bharat Sanchar Nigam Limited (BSNL)
Bharti Airtel Limited
Black Box Corporation
BlackBerry Limited
Blonder Tongue Laboratories Inc
Bogen Communications LLC
Boingo Wireless Inc
Bouygues SA
Broadcom Inc
BT Global Services plc
BT Group plc
CalAmp Corp
Calero-MDSL
Calient Technologies Inc
Calix Inc
Celestica Inc

Cellcom Israel Ltd
Certicom Corp
Ceske Radiokomunikace as
Channell Commercial Corporation
Charter Communications Inc
China Communications Services Corp Ltd
China Mobile Limited
China Telecom Corporation Limited
China Unicom (Hong Kong) Limited
Chorus Limited
Chunghwa Telecom Co Ltd
Ciena Corporation
Cincinnati Bell Inc
Cisco Systems Inc
Cisco Webex
CK Hutchison Holdings Limited
ClearOne Inc
Cloudflare Inc
Cogent Communications Group Inc
Colt Technology Services Group Limited
Comba Telecom Systems Holdings Ltd
Comcast Corporation
CommScope Holding Company Inc
Comtech Telecommunications Corp
Consensus Cloud Solutions Inc
Consolidated Communications Holdings Inc
Converge Information and Communications Technology Solutions, Inc.
Corning Incorporated
COSMOTE Mobile Telephones SA
Cox Communications Inc
Cricket Wireless LLC
Cyxtera Technologies Inc
Data Select Limited
Datang Telecom Technology Co Ltd
Deutsche Telekom AG
Dialogic Inc
Digi.com Bhd
DirecTV LLC (DIRECTV)
D-Link Corporation
DSP Group Inc
Dycom Industries Inc
DZS Inc
EarthLink LLC
EchoStar Corporation
EF Johnson Technologies Inc
eircom Limited
Elisa Corporation
Embratel Participacoes SA
Emirates Telecommunications Corporation (Etisalat)
Empresa Nacional de Telecommunicacions SA (Entel)
Energy Focus Inc
Enterprise Diversified Inc
enTouch Systems Inc
Entrust Corporation
Equinix Inc
Eutelsat Communications SA
Extreme Networks Inc

F5 Inc
Far EasTone Telecommunications Co Ltd
FASTWEB SpA
Filtronic plc
First Pacific Company Limited
Flex Ltd
Foxconn Technology Co Ltd
Freenet AG
Frontier Communications Corporation
Fujitsu Limited
Fujitsu Network Communications Inc
Fusion Connect Inc
Garmin Ltd
GCI Communication Corp
General DataComm LLC
Gilat Satellite Networks Ltd
Glentel Inc
Global Telecom Holding SAE
Globalstar Inc
Globe Telecom Inc
GoDaddy Inc
GTT Communications Inc
Harmonic Inc
Hawaiian Telcom Holdco Inc
Hellenic Telecommunications Organization SA
Hello Direct Inc
Hi Sun Technology (China) Limited
Hikari Tsushin Inc
Hitachi Kokusai Electric Inc
HKT Trust and HKT Limited
HP Inc
Huawei Technologies Co Ltd
Hutchison Telecommunications Hong Kong Holdings
Limited
iBasis Inc
IDT Corporation
IHS Holding Ltd
Iliad SA
Indosat Ooredoo Hutchison
Infinera Corporation
Ingram Micro Mobility
Inmarsat Global Limited
InnoMedia Pte Ltd
INNOVATE Corp
Intel Corporation
Inteliquent Inc
Intelsat SA
InterDigital Inc
Internap Corporation
International Business Machines Corporation (IBM)
Internet Initiative Japan Inc
Intrado Corporation
Iridium Communications Inc
ITI Limited
Jabil Inc
Jio (Reliance Jio Infocomm Limited)
Juniper Networks Inc
KCOM Group Limited

KDDI Corporation
Koninklijke KPN NV (Royal KPN NV)
Koninklijke Philips NV (Royal Philips)
Kratos Defense & Security Solutions Inc
KT Corporation
Kyocera Corporation
L3Harris Technologies Inc
Laird Connectivity
Lattice Incorporated
LG Corporation
LG Electronics Inc
LG Electronics USA Inc
LG Uplus Corp
Liberty Global plc
Liberty Latin America Ltd
Likewize Corp
LM Ericsson Telephone Company (Ericsson)
Lumen Technologies Inc
Lumentum Operations LLC
Lumos Networks Corp
M1 Limited
Magyar Telekom plc
Mahanagar Telephone Nigam Limited
Maroc Telecom SA
Marvell Technology Group Ltd
Maxis Berhad
McAfee Corp
Mediacom Communications Corporation
MedTel Services LLC
Megacable Holdings SAB de CV
Microsoft Corporation
Millicom International Cellular SA
MiTAC Holdings Corp
Mitel Networks Corporation
Mobile Telecommunications Company KSCP (Zain
Group)
Mobile TeleSystems PJSC
Momentum Telecom
Net2Phone Inc
NETGEAR Inc
NetScout Systems Inc
Neustar Inc
Newfold Digital Inc
NextPlat Corp
Nippon Telegraph and Telephone Corporation (NTT)
Nokia Corporation
Nortel Inversora SA
NTT DOCOMO Inc
O2 Czech Republic AS
Oblong Inc
Oi SA
OJSC JFSC Sistema
OneSpan Inc
OneWeb Ltd
Ooredoo QPSC
OPTERNA
Optical Cable Corporation
Orange

Orange Belgium
Orange Polska SA
ORBCOMM Inc
Pakistan Telecommunication Company Limited
Panasonic Corporation
Partner Communications Co Ltd
PCCW Limited
Perusahaan Perseroan PT Telekomunikasi Indonesia Tbk
(Telkom)
Phazar Antenna Corp
Pineapple Energy Inc
Planet Labs PBC
Plantronics Inc
PLDT Inc
Potevio Corporation
Preformed Line Products Company
Proxim Wireless Corporation
Proximus Group
Purple Communications Inc
Q Beyond AG
Qisda Corporation
Qualcomm Incorporated
Rackspace Technology Inc
Realtime Corporation
Rogers Communications Inc
Rohde & Schwarz GmbH & Co KG
Rostelecom PJSC
Samsung Electronics Co Ltd
Sanmina Corporation
ScanSource Inc
SES SA
Shaw Communications Inc
Shenandoah Telecommunications Company
Siemens AG
Simply Inc
Singapore Technologies Telemedia Pte Ltd
Singapore Telecommunications Limited
Singtel Optus Pty Limited
Sitel Corporation
SK Broadband Co Ltd
SK Telecom Co Ltd
Skype Technologies Sarl
Slack Technologies Inc
SmarTone Telecommunications Holdings Limited
SMTC Corporation
Socket Mobile Inc
SoftBank Group Corp
Sonim Technologies Inc
Sony Corporation
Sopra Steria Group SA
Spark New Zealand Limited
Speedcast International Limited
Spok Inc
Startec Global Communications Corporation
Swisscom AG
Sykes Enterprises Incorporated
Syniverse Technologies LLC
TalkTalk Telecom Group Limited

Tata Communications Limited
Tata Group
Tata Teleservices Limited
TCI International Inc
TCL Technology Group Corporation
TDC A/S
TDS Telecommunications LLC
TeamViewer AG
Technical Communications Corporation
Tektronix Inc
Tele2 AB
Telecom Argentina SA
Telecom Egypt SAE
Telecom Italia SpA
Telecomunicaciones de Puerto Rico Inc
Telefonica Brasil SA
Telefonica Chile SA
Telefonica de Argentina SA
Telefonica del Peru SAA
Telefonica Deutschland Holding AG
Telefonica SA
Telefonos de Mexico SAB de CV (Telmex)
Telekom Austria AG
Telekom Malaysia Berhad
Telenet Group NV/SA
Telenor ASA
Telephone and Data Systems Inc (TDS)
Telesat Corporation
Telia Company AB
Telia Lietuva AB
Telkom SA SOC Limited
Tellabs Inc
Telstra Corporation Limited
Telular Corporation
TELUS Corporation
Telvue Corporation
TESSCO Technologies Incorporated
Thuraya Telecommunications Company
TIM SA
TKH Group NV
T-Mobile Polska SA
T-Mobile US Inc
TomTom International BV
Toshiba Corporation
TOT pcl
Total Access Communication PCL
TPG Telecom Limited
Trilogy International Partners Inc
Trimble Inc
TruConnect Communications Inc
True Corporation Public Company Limited
TTEC Holdings Inc
Turk Telekomunikasyon AS
Turkcell Iletisim Hizmetleri AS
Ubiquiti Inc
Uniden Holdings Corporation
United Online Inc
United States Cellular Corporation

Veon Ltd
Verio Inc
VeriSign Inc
Verizon Communications Inc
ViaSat Inc
Viavi Solutions Inc
Virgin Media Business Ltd
VMED O2 UK Limited (Virgin Media O2)
Vocera Communications Inc
Vocus Group Limited
Vodafone Group plc
Vonage Holdings Corp
VTech Holdings Limited
Westell Technologies Inc
WIND Hellas Telecommunications SA
Windstream Holdings Inc
XIUS
Zoom Video Communications Inc
ZTE Corporation

INDEX OF U.S. HEADQUARTERS LOCATION BY STATE

To help you locate the firms geographically, the city and state of the headquarters of each company are in the following index.

ALABAMA
ADTRAN Inc; Huntsville

ALASKA
Alaska Communications; Anchorage
GCI Communication Corp; Anchorage

ARIZONA
Axway Inc; Scottsdale
GoDaddy Inc; Tempe
Viavi Solutions Inc; Scottsdale

ARKANSAS
Windstream Holdings Inc; Little Rock

CALIFORNIA
Advanced Micro Devices Inc (AMD); Santa Clara
Avaya Holdings Corp; Santa Clara
Belkin International Inc; El Segundo
Boingo Wireless Inc; Los Angeles
Broadcom Inc; San Jose
CalAmp Corp; Irvine
Calient Technologies Inc; Goleta
Calix Inc; San Jose
Cisco Systems Inc; San Jose
Cisco Webex; San Jose
Cloudflare Inc; San Francisco
Consensus Cloud Solutions Inc; Los Angeles
DirecTV LLC (DIRECTV); El Segundo
DSP Group Inc; San Jose
DZS Inc; Alameda
Equinix Inc; Redwood City
Globalstar Inc; Covington
Harmonic Inc; San Jose
HP Inc; Palo Alto
Infinera Corporation; San Jose
Intel Corporation; Santa Clara
Juniper Networks Inc; Sunnyvale
Lumentum Operations LLC; San Jose
McAfee Corp; San Jose
NETGEAR Inc; San Jose
Planet Labs PBC; San Francisco
Plantronics Inc; Santa Cruz
Proxim Wireless Corporation; San Jose
Qualcomm Incorporated; San Diego
Sanmina Corporation; San Jose
Slack Technologies Inc; San Francisco
Socket Mobile Inc; Newark
Sony Corporation; San Diego

TCI International Inc; Fremont
Trimble Inc; Sunnyvale
TruConnect Communications Inc; Los Angeles
Ubiquiti Inc; San Jose
United Online Inc; Woodland Hills
ViaSat Inc; Carlsbad
Vocera Communications Inc; San Jose
Zoom Video Communications Inc; San Jose

COLORADO
EchoStar Corporation; Englewood
Oblong Inc; Conifer
TTEC Holdings Inc; Englewood

CONNECTICUT
Amphenol Corporation; Wallingford
Charter Communications Inc; Stamford
Frontier Communications Corporation; Norwalk
General DataComm LLC; Oxford

DELAWARE
InterDigital Inc; Wilmington

DISTRICT OF COLUMBIA
Cogent Communications Group Inc; Washington

FLORIDA
ARC Group Worldwide Inc; Deland
Cyxtera Technologies Inc; Coral Gables
Dycom Industries Inc; Palm Beach Gardens
Hello Direct Inc; Cape Canaveral
Jabil Inc; St. Petersburg
L3Harris Technologies Inc; Melbourne
MedTel Services LLC; Palmetto
Newfold Digital Inc; Jacksonville
NextPlat Corp; Coconut Grove
Simply Inc; Miami
Sitel Corporation; Miami
Sykes Enterprises Incorporated; Tampa
Syniverse Technologies LLC; Tampa

GEORGIA
AT&T Mobility LLC; Atlanta
Cox Communications Inc; Atlanta
Cricket Wireless LLC; Atlanta
EarthLink LLC; Atlanta
Fusion Connect Inc; Atlanta
Momentum Telecom; Atlanta

HAWAII
Hawaiian Telcom Holdco Inc; Honolulu

ILLINOIS
Anixter International Inc; Glenview
Consolidated Communications Holdings Inc; Mattoon
Inteliquent Inc; Chicago
OneSpan Inc; Chicago

Telephone and Data Systems Inc (TDS); Chicago
Telular Corporation; Chicago
United States Cellular Corporation; Chicago
Westell Technologies Inc; Aurora

LOUISIANA
Lumen Technologies Inc; Monroe

MARYLAND
Ciena Corporation; Hanover
TESSCO Technologies Incorporated; Hunt Valley

MASSACHUSETTS
Akamai Technologies Inc; Cambridge
Alvaria Inc; Westford
ATN International Inc; Beverly
Aware Inc; Bedford
iBasis Inc; Lexington
NetScout Systems Inc; Westford
Technical Communications Corporation; Concord
Verio Inc; Burlington
XIUS; North Chelmsford

MINNESOTA
Entrust Corporation; Shakopee
Pineapple Energy Inc; Minnetonka

MISSOURI
Belden Inc; St. Louis

NEBRASKA
Intrado Corporation; Omaha

NEW JERSEY
Blonder Tongue Laboratories Inc; Old Bridge
Bogen Communications LLC; Mahwah
Dialogic Inc; Parsippany
enTouch Systems Inc; Princeton
IDT Corporation; Newark
Lattice Incorporated; Pennsauken
LG Electronics USA Inc; Englewood Cliffs
Net2Phone Inc; Newark
ORBCOMM Inc; Fort Lee
Telvue Corporation; Mount Laurel
Vonage Holdings Corp; Holmdel

NEW YORK
Altice USA Inc; Long Island City
Anaren Inc; East Syracuse
Calero-MDSL; Rochester
Comtech Telecommunications Corp; Melville
Corning Incorporated; Corning
INNOVATE Corp; New York
International Business Machines Corporation (IBM);
Armonk
Mediacom Communications Corporation; Middletown
OPTERNA; East Syracuse

Verizon Communications Inc; New York

NORTH CAROLINA
CommScope Holding Company Inc; Hickory
Extreme Networks Inc; Morrisville

OHIO
Cincinnati Bell Inc; Cincinnati
Energy Focus Inc; Solon
Laird Connectivity; Akron
Preformed Line Products Company; Mayfield Village

OREGON
Tektronix Inc; Beaverton

PENNSYLVANIA
Black Box Corporation; Lawrence
Comcast Corporation; Philadelphia

SOUTH CAROLINA
ScanSource Inc; Greenville

TEXAS
AT&T Inc; Dallas
Aviat Networks Inc; Austin
Channell Commercial Corporation; Rockwall
EF Johnson Technologies Inc; Irving
Fujitsu Network Communications Inc; Richardson
Kratos Defense & Security Solutions Inc; Round Rock
Likewize Corp; Southlake
Phazar Antenna Corp; Mineral Wells
Purple Communications Inc; Austin
Rackspace Technology Inc; San Antonio
Sonim Technologies Inc; Austin
Speedcast International Limited; Houston
Startec Global Communications Corporation; Irving
Tellabs Inc; Carrollton

UTAH
ClearOne Inc; Salt Lake City

VIRGINIA
Enterprise Diversified Inc; Lynchburg
GTT Communications Inc; McLean
Internap Corporation; Renton
Iridium Communications Inc; McLean
Lumos Networks Corp; Waynesboro
Neustar Inc; Reston
Optical Cable Corporation; Roanoke
Shenandoah Telecommunications Company; Edinburg
Spok Inc; Alexandria
VeriSign Inc; Reston

WASHINGTON
F5 Inc; Seattle
Microsoft Corporation; Redmond
T-Mobile US Inc; Bellevue

Trilogy International Partners Inc; Bellevue

WISCONSIN
TDS Telecommunications LLC; Madison

INDEX OF NON-U.S. HEADQUARTERS LOCATION BY COUNTRY

ARGENTINA
Nortel Inversora SA; Buenos Aires
Telecom Argentina SA; Buenos Aires
Telefonica de Argentina SA; Buenos Aires

AUSTRALIA
AAPT Limited; Sydney
Singtel Optus Pty Limited; Macquarie Park
Telstra Corporation Limited; Melbourne
TPG Telecom Limited; Sydney
Vocus Group Limited; Melbourne

AUSTRIA
Telekom Austria AG; Vienna

BELGIUM
Orange Belgium; Brussels
Proximus Group; Brussels
Telenet Group NV/SA; Mechelen

BERMUDA
Liberty Latin America Ltd; Hamilton
Marvell Technology Group Ltd; Hamilton

BRAZIL
Embratel Participacoes SA; Rio de Janeiro
Oi SA; Rio de Janeiro
Realtime Corporation; Sao Paulo
Telefonica Brasil SA; Sao Paulo
TIM SA; Rio de Janeiro

CANADA
BCE Inc (Bell Canada Enterprises); Verdun
Bell Aliant Inc; Halifax
Bell MTS Inc; Winnipeg
BlackBerry Limited; Waterloo
Celestica Inc; Toronto
Certicom Corp; Mississauga
Glentel Inc; Burnaby
Mitel Networks Corporation; Kanata
Rogers Communications Inc; Toronto
Shaw Communications Inc; Calgary
SMTC Corporation; Markham
Telesat Corporation; Ottawa
TELUS Corporation; Vancouver

CHILE
Empresa Nacional de Telecommunicacions SA (Entel); Santiago
Telefonica Chile SA; Santiago

CHINA
AAC Technologies Holdings Inc; Shenzhen

AsiaInfo Technologies Limited; Beijing
China Telecom Corporation Limited; Beijing
Datang Telecom Technology Co Ltd; Beijing
Huawei Technologies Co Ltd; Shenzhen
Potevio Corporation; Beijing
TCL Technology Group Corporation; Huizhou
ZTE Corporation; Shenzhen

CZECH REPUBLIC
Ceske Radiokomunikace as; Prague
O2 Czech Republic AS; Prague

DENMARK
TDC A/S; Copenhagen

EGYPT
Global Telecom Holding SAE; Cairo
Telecom Egypt SAE; Cairo

FINLAND
Elisa Corporation; Elisa
Nokia Corporation; Espoo

FRANCE
Bouygues SA; Paris
Eutelsat Communications SA; Paris
Iliad SA; Paris
Orange; Paris
Sopra Steria Group SA; Paris

GERMANY
Deutsche Telekom AG; Bonn
Freenet AG; Budelsdorf
Q Beyond AG; Cologne
Rohde & Schwarz GmbH & Co KG; Munich
Siemens AG; Munich
TeamViewer AG; Goppingen
Telefonica Deutschland Holding AG; Munich

GREECE
COSMOTE Mobile Telephones SA; Maroussi
Hellenic Telecommunications Organization SA; Athens
WIND Hellas Telecommunications SA; Athens

HONG KONG
Asia Satellite Telecommunications Holdings Ltd; Hong Kong
China Communications Services Corp Ltd; Hong Kong
China Mobile Limited; Hong Kong
China Unicom (Hong Kong) Limited; Hong Kong
CK Hutchison Holdings Limited; Hong Kong
Comba Telecom Systems Holdings Ltd; Hong Kong
First Pacific Company Limited; Hong Kong
Hi Sun Technology (China) Limited; Hong Kong
HKT Trust and HKT Limited; Hong Kong

Hutchison Telecommunications Hong Kong Holdings
Limited; Hong Kong
PCCW Limited; Hong Kong
SmarTone Telecommunications Holdings Limited; Hong
Kong
VTech Holdings Limited; Hong Kong

HUNGARY
Magyar Telekom plc; Budapest

INDIA
Bharat Sanchar Nigam Limited (BSNL); New Delhi
Bharti Airtel Limited; New Delhi
ITI Limited; Bengaluru
Jio (Reliance Jio Infocomm Limited); Ambawadi
Mahanagar Telephone Nigam Limited; New Delhi
Tata Communications Limited; Mumbai
Tata Group; Mumbai
Tata Teleservices Limited; Mumbai

INDONESIA
Indosat Ooredoo Hutchison; Jakarta
Perusahaan Perseroan PT Telekomunikasi Indonesia Tbk
(Telkom); Jakarta

IRELAND
eircom Limited; Dublin

ISRAEL
B Communications Ltd; Tel Aviv
Bezeq-The Israel Telecommunication Corp Ltd; Tel Aviv
Cellcom Israel Ltd; Netanya
Gilat Satellite Networks Ltd; Petah Tikva
Partner Communications Co Ltd; Rosh Ha'ayin

ITALY
FASTWEB SpA; Milan
Telecom Italia SpA; Rome

JAPAN
Anritsu Corporation; Kanagawa
Fujitsu Limited; Tokyo
Hikari Tsushin Inc; Tokyo
Hitachi Kokusai Electric Inc; Tokyo
Internet Initiative Japan Inc; Tokyo
KDDI Corporation; Tokyo
Kyocera Corporation; Kyoto
Nippon Telegraph and Telephone Corporation (NTT);
Tokyo
NTT DOCOMO Inc; Tokyo
Panasonic Corporation; Osaka
SoftBank Group Corp; Tokyo
Toshiba Corporation; Tokyo
Uniden Holdings Corporation; Tokyo

KOREA
KT Corporation; Gyeonggi-do

LG Corporation; Seoul
LG Electronics Inc; Seoul
LG Uplus Corp; Seoul
Samsung Electronics Co Ltd; Suwon-si
SK Broadband Co Ltd; Seoul
SK Telecom Co Ltd; Seoul

KUWAIT
Mobile Telecommunications Company KSCP (Zain
Group); Safat

LITHUANIA
Telia Lietuva AB; Vilnius

LUXEMBOURG
Intelsat SA; Luxembourg
Millicom International Cellular SA; Luxembourg
SES SA; Luxembourg
Skype Technologies Sarl; Luxembourg

MALAYSIA
Axiata Group Berhad; Kuala Lumpur
Digi.com Bhd; Selangor
Maxis Berhad; Kuala Lumpur
Telekom Malaysia Berhad; Jalan Pantai Baharu, Kuala
Lumpur

MEXICO
America Movil SAB de CV; Mexico City
AT&T Mexico SAU; Delegacion Miguel Hidalgo
Axtel SAB de CV; Unidad San Pedro
Megacable Holdings SAB de CV; Guadalajara
Telefonos de Mexico SAB de CV (Telmex); Mexico City

MOROCCO
Maroc Telecom SA; Rabat

NEW ZEALAND
Chorus Limited; Wellington
Spark New Zealand Limited; Auckland

NORWAY
Telenor ASA; Fornebu

PAKISTAN
Pakistan Telecommunication Company Limited;
Islamabad

PERU
Telefonica del Peru SAA; Lima

PHILIPPINES
Converge Information and Communications Technology
Solutions, Inc.; Brgy. Ugong Pasig City
Globe Telecom Inc; Taguig
PLDT Inc; Makati

POLAND
Orange Polska SA; Warszawa
T-Mobile Polska SA; Warsaw

PORTUGAL
Altice Portugal SA; Lisboa

PUERTO RICO
Telecomunicaciones de Puerto Rico Inc; Guaynabo

QATAR
Ooredoo QPSC; Doha

RUSSIA
Mobile TeleSystems PJSC; Moscow
OJSC JFSC Sistema; Moscow
Rostelecom PJSC; Moscow

SINGAPORE
Flex Ltd; Singapore
InnoMedia Pte Ltd; Singapore
M1 Limited; Singapore
Singapore Technologies Telemedia Pte Ltd; Singapore
Singapore Telecommunications Limited; Singapore

SOUTH AFRICA
Telkom SA SOC Limited; Centurion

SPAIN
Telefonica SA; Madrid

SWEDEN
LM Ericsson Telephone Company (Ericsson); Stockholm
Tele2 AB; Stockholm
Telia Company AB; Stockholm

SWITZERLAND
Garmin Ltd; Schaffhausen
Swisscom AG; Bern

TAIWAN
Accton Technology Corporation; Hsinchu
BenQ Corporation; Taipei
Chunghwa Telecom Co Ltd; Taipei City
D-Link Corporation; Taipei
Far EasTone Telecommunications Co Ltd; Taipei
Foxconn Technology Co Ltd; Taipei
MiTAC Holdings Corp; Taoyuan City
Qisda Corporation; Taoyuan

THAILAND
Advanced Info Service plc; Bangkok
Bangkok Cable Co Ltd; Bangkok
TOT pcl; Bangkok
Total Access Communication PCL; Bangkok

True Corporation Public Company Limited; Bangkok

THE NETHERLANDS
Altice Europe NV; Amsterdam
Koninklijke KPN NV (Royal KPN NV); Rotterdam
Koninklijke Philips NV (Royal Philips); Amsterdam
TKH Group NV; Haaksbergen
TomTom International BV; Amsterdam
Veon Ltd; Amsterdam

TURKEY
Turk Telekomunikasyon AS; Ankara
Turkcell Iletisim Hizmetleri AS; Istanbul

UNITED ARAB EMIRATES
Emirates Telecommunications Corporation (Etisalat); Abu Dhabi
Thuraya Telecommunications Company; Dubai

UNITED KINGDOM
Arqiva Limited; Winchester
BT Global Services plc; London
BT Group plc; London
Colt Technology Services Group Limited; London
Data Select Limited; Theale, Reading
Filtronic plc; West Yorkshire
IHS Holding Ltd; London
Ingram Micro Mobility; Surrey
Inmarsat Global Limited; London
KCOM Group Limited; Hull
Liberty Global plc; London
OneWeb Ltd; London
TalkTalk Telecom Group Limited; Salford
Virgin Media Business Ltd; Peterborough
VMED O2 UK Limited (Virgin Media O2); London
Vodafone Group plc; Newbury

Individual Profiles
On Each Of
THE TELECOMMUNICATIONS 350

AAC Technologies Holdings Inc

www.aactechnologies.com

NAIC Code: 334310

TYPES OF BUSINESS:

Audio and Video Equipment Manufacturing
Audio Components
Manufacture
Miniaturized Solutions

GROWTH PLANS/SPECIAL FEATURES:

AAC Technologies is one of the world's largest manufacturers of miniature acoustic components, such as speakers and receivers, for smartphones, which account for 44% of its 2020 revenue. Founded in 1993, the company has since then expanded into the production of other handset components, such as electromagnetic drives (haptics), precision mechanical (metal casing), lenses and camera modules, which combined make up 56% of the firm's sales. The firm is also engaged in the production of MEMS microphones, 3D glass, RF antennas, and headsets. AAC is headquartered in Shenzhen, China and runs factories in China, Vietnam, Czechia and Malaysia, employing over 41,000 people.

BRANDS/DIVISIONS/AFFILIATES:

Ibeo Automotive Systems GmbH

CONTACTS:
Note: Officers with more than one job title may be intentionally listed here more than once.

Pan Benjamin Zhengmin, CEO
Hongjiang Zhang, Chmn.

FINANCIAL DATA:
Note: Data for latest year may not have been available at press time.

In U.S. $	2021	2020	2019	2018	2017	2016
Revenue	2,617,367,000	2,539,329,000	2,649,485,000	2,686,137,000	3,128,723,000	2,297,342,000
R&D Expense	255,739,600	284,486,400	254,411,300	224,027,000	246,472,800	172,694,300
Operating Income	219,656,300	199,994,400	366,169,700	631,792,800	900,913,200	668,689,900
Operating Margin %	.08%	.08%	.14%	.24%	.29%	.29%
SGA Expense	171,270,700	141,822,500	136,021,600	143,169,100	144,474,100	113,076,000
Net Income	195,007,200	223,219,200	329,245,600	562,361,700	788,838,100	596,403,600
Operating Cash Flow	325,349,000	532,245,500	569,414,700	1,005,837,000	783,270,400	712,896,400
Capital Expenditure	562,960,000	701,755,100	432,839,100	612,513,300	758,156,700	598,465,900
EBITDA	661,277,500	661,295,100	737,335,000	930,262,400	1,106,434,000	838,816,700
Return on Assets %	.03%	.04%	.07%	.13%	.19%	.20%
Return on Equity %	.06%	.07%	.12%	.21%	.34%	.32%
Debt to Equity	.34%	0.254	0.354	0.128	0.111	0.056

CONTACT INFORMATION:

Phone: 86 755 33972018   Fax: 86 755 33018531
Toll-Free:
Address: Blk. A, No. 6 Yuexing 3rd Rd., South Hi-Tech Indus, Shenzhen, Guangdong 518057 China

STOCK TICKER/OTHER:

Stock Ticker: AACAY   Exchange: PINX
Employees: 37,591   Fiscal Year Ends: 12/31
Parent Company:

SALARIES/BONUSES:

Top Exec. Salary: $ Bonus: $
Second Exec. Salary: $ Bonus: $

OTHER THOUGHTS:

Estimated Female Officers or Directors:
Hot Spot for Advancement for Women/Minorities:

Sales, profits and employees may be estimates. Financial information, benefits and other data can change quickly and may vary from those stated here.

AAPT Limited

www.aapt.com.au

NAIC Code: 517110

TYPES OF BUSINESS:

Local & Long Distance Service
Internet & Broadband Service
Cloud Services
Voice & Data Services
Media Services

BRANDS/DIVISIONS/AFFILIATES:

TPG Telecom Limited

GROWTH PLANS/SPECIAL FEATURES:

AAPT Limited is the wholesale arm of TPG Telecom Limited, supplying broadband services and backhaul to small, medium, large and international internet service providers (ISPs) as well as government organizations throughout Australia. AAPT owns and operates its own carrier-grade voice, data and internet network. It comprises more than 6,835 miles of fiber optic cables and 410 exchanges, as well as Ethernet capability in over 210 of the exchanges. AAPT also provides cloud services, and broadcast and media services.

CONTACTS: *Note: Officers with more than one job title may be intentionally listed here more than once.*

David Yuile, CEO
Mark Rafferty, Dir.-Sales
Michael Edwards, Gen. Mgr.-Product & Technology
Art Cartwright, Gen. Mgr.-Customer Svc. & Sales

FINANCIAL DATA: *Note: Data for latest year may not have been available at press time.*

In U.S. $	2021	2020	2019	2018	2017	2016
Revenue						
R&D Expense						
Operating Income						
Operating Margin %						
SGA Expense						
Net Income						
Operating Cash Flow						
Capital Expenditure						
EBITDA						
Return on Assets %						
Return on Equity %						
Debt to Equity						

CONTACT INFORMATION:

Phone: 61 2-8314-1700   Fax: 61-2-9009-9999
Toll-Free:
Address: 680 George St., Sydney, NSW 2000 Australia

STOCK TICKER/OTHER:

Stock Ticker: Subsidiary   Exchange:
Employees: 740   Fiscal Year Ends: 06/30
Parent Company: TPG Telecom Limited

SALARIES/BONUSES:

Top Exec. Salary: $ Bonus: $
Second Exec. Salary: $ Bonus: $

OTHER THOUGHTS:

Estimated Female Officers or Directors:
Hot Spot for Advancement for Women/Minorities:

Sales, profits and employees may be estimates. Financial information, benefits and other data can change quickly and may vary from those stated here.

Accton Technology Corporation

www.accton.com

NAIC Code: 334418

TYPES OF BUSINESS:

Contract Electronics Manufacturing
Technical & Communications Outsourcing
IP Network Switches
Semiconductors & Chipsets
Wireless Hardware
Online Portal
VoIP Hardware

BRANDS/DIVISIONS/AFFILIATES:

CONTACTS: Note: Officers with more than one job title may be intentionally listed here more than once.

Edgar Masri, CEO
Fanny Chen, CFO
George Tchaparian, Sr. VP-R&D
Edward Lin, Contact-News & Media Rel.
Fai-Long Kuo, Exec. VP
Kuo-Tai Choiu, Sr. VP
Meen-Ron Lin, Chmn.
Sheng-Shun Liou, VP-Logistics

GROWTH PLANS/SPECIAL FEATURES:

Accton Technology Corporation is a Taiwan-based company that researches, develops, manufactures and markets computer network system products. The company also manufactures network computers and network peripheral equipment. Accton's solutions are divided into five categories: cloud data center, carrier access, campus network, Internet of Things (IoT) integration and SD-WAN. The firm's cloud data center solution comprises high-speed, high-density, standardized OCP-compliant (open compute project) platforms, which provide an open hardware network fabric for cloud data centers. Accton has developed a full range of Top-of-Rack and spine switches that support interface speeds from 1G up to 400G. Carrier access solutions are provided to carrier, service provider and multiple system operator (MS) customers, with products ranging from passive optical networks for residential broadband access to NEBS-compliant (network equipment building system) aggregation switches and routers, as well as 3G/4G mobile and enterprise IP backhaul devices. Campus network solutions consist of flexible wired connectivity and high-density wireless access for campus and enterprise networks. IoT integration solutions cover three main areas: integrating networking protocols and developing service applications; creating a database for big data applications services; and partnership and collaboration with alliances in the IoT ecosystem. By connecting IoT demands from different communities and through virtualized, digitalized solutions, Accton aims to enhance the quality of life with all alliances. Last, SD-WAN solutions include open software-defined network platforms that provide the foundation for next-generation networks, and involve the transformation of WAN networks into SD-WANs. Accton's R&D division is continually engaged in design and development, with more than 1,000 patents related to network product design and manufacturing.

FINANCIAL DATA: Note: Data for latest year may not have been available at press time.

In U.S. $	2021	2020	2019	2018	2017	2016
Revenue	2,150,623,955	1,936,970,000	1,842,250,000	1,433,710,000	1,180,920,704	951,584,448
R&D Expense						
Operating Income						
Operating Margin %						
SGA Expense						
Net Income	169,779,455	179,555,000	164,602,000	96,409,500	82,924,760	61,170,752
Operating Cash Flow						
Capital Expenditure						
EBITDA						
Return on Assets %						
Return on Equity %						
Debt to Equity						

CONTACT INFORMATION:

Phone: 886 35770270   Fax: 886 35780764
Toll-Free:
Address: No.1 Creation 3rd Rd., Hsinchu Science Park, Hsinchu, 30077 Taiwan

STOCK TICKER/OTHER:

Stock Ticker: 2345   Exchange: TWSE
Employees: 5,200   Fiscal Year Ends: 12/31
Parent Company:

SALARIES/BONUSES:

Top Exec. Salary: $ Bonus: $
Second Exec. Salary: $ Bonus: $

OTHER THOUGHTS:

Estimated Female Officers or Directors:
Hot Spot for Advancement for Women/Minorities:

ADTRAN Inc

www.adtran.com

NAIC Code: 334210

TYPES OF BUSINESS:

Carrier Networks
Network Platforms
Communications Platforms
Broadband Access
Internet of Things
Cloud
Gateways
Fiber Optics

BRANDS/DIVISIONS/AFFILIATES:

GROWTH PLANS/SPECIAL FEATURES:

Adtran Inc is a provider of networking and communications platforms, software, and services focused on the broadband access market. It operate under two reportable segments: Network Solutions, which includes hardware and software products, and Services & Support, which includes a portfolio of network implementation services, support services, and cloud-hosted SaaS applications that complement product portfolio and can be utilized to support other platforms as well. These two segments span across their three revenue categories: Access & Aggregation, Subscriber Solutions & Experience, and Traditional & Other Products.

ADTRAN offers its employees comprehensive health benefits, life and disability insurance, flexible spending accounts and a variety of employee assistance programs and incentives.

CONTACTS: *Note: Officers with more than one job title may be intentionally listed here more than once.*

Thomas Stanton, CEO
Michael Foliano, CFO
Raymond Harris, Chief Information Officer
Daniel Whalen, Other Executive Officer
James Wilson, Other Executive Officer
Ronald Centis, Senior VP, Divisional
Marc Kimpe, Senior VP, Divisional

FINANCIAL DATA: *Note: Data for latest year may not have been available at press time.*

In U.S. $	2021	2020	2019	2018	2017	2016
Revenue	563,004,000	506,510,000	530,061,000	529,277,000	666,900,000	636,781,000
R&D Expense	108,663,000	113,287,000	126,200,000	124,547,000	130,666,000	124,909,000
Operating Income	-14,700,000	-9,708,000	-37,321,000	-45,422,000	37,386,000	34,573,000
Operating Margin %	- .03%	- .02%	- .07%	- .09%	.06%	.05%
SGA Expense	124,414,000	113,972,000	130,288,000	124,440,000	135,583,000	131,848,000
Net Income	-8,635,000	2,378,000	-52,982,000	-19,342,000	23,840,000	35,229,000
Operating Cash Flow	3,008,000	-16,518,000	-2,472,000	55,454,000	-42,379,000	42,002,000
Capital Expenditure	5,669,000	6,413,000	9,494,000	8,110,000	14,720,000	21,441,000
EBITDA	9,813,000	10,386,000	-6,495,000	-16,947,000	60,935,000	61,874,000
Return on Assets %	- .02%	.00%	- .09%	- .03%	.04%	.05%
Return on Equity %	- .02%	.01%	- .13%	- .04%	.05%	.07%
Debt to Equity				0.055	0.051	0.056

CONTACT INFORMATION:

Phone: 256 963-8000   Fax: 256 963-8004
Toll-Free: 800-923-8726
Address: 901 Explorer Blvd., Huntsville, AL 35806-2807 United States

STOCK TICKER/OTHER:

Stock Ticker: ADTN   Exchange: NAS
Employees: 1,335   Fiscal Year Ends: 12/31
Parent Company:

SALARIES/BONUSES:

Top Exec. Salary: $709,988   Bonus: $250,000
Second Exec. Salary: Bonus: $100,000
$381,069

OTHER THOUGHTS:

Estimated Female Officers or Directors:
Hot Spot for Advancement for Women/Minorities:

Advanced Info Service plc

NAIC Code: 517210

www.ais.co.th

TYPES OF BUSINESS:

Cell Phone Service
Phone & Telecom Accessories Distribution
Voice & Data Networks
Internet Service
Broadcast Service
Travel Insurance

BRANDS/DIVISIONS/AFFILIATES:

GROWTH PLANS/SPECIAL FEATURES:

Advanced Info Service PCL is a triple-play telecommunications provider. The company generates revenues from the provision of mobile phone services, mobile handset sales, and broadband services, with the vast majority of revenue derived from mobile phone services. Its broadband service includes various fiber technologies for households, leveraging the company's owned fiber infrastructure. In addition to the fiber networks, it also has mobile infrastructure. The company generates the vast majority of its revenue in Thailand.

CONTACTS: *Note: Officers with more than one job title may be intentionally listed here more than once.*

Somchai Lertsutiwong, CEO
Suwimol Kaewkoon, Chief Organization Dev. Officer
Vilasinee Puddhikarant, Chief Customer Officer
Allen Lew Yoong Keong, Chmn.-Exec. Committee
Vithit Leenutaphong, Chmn.

FINANCIAL DATA: *Note: Data for latest year may not have been available at press time.*

In U.S. $	2021	2020	2019	2018	2017	2016
Revenue	5,103,655,000	4,866,036,000	5,091,294,000	4,780,632,000	4,439,116,000	4,282,293,000
R&D Expense						
Operating Income	1,074,564,000	1,061,386,000	1,161,095,000	1,128,773,000	1,136,657,000	1,108,420,000
Operating Margin %	.21%	.22%	.23%	.24%	.26%	.26%
SGA Expense	610,764,100	684,798,300	780,764,100	740,074,400	705,818,600	838,047,500
Net Income	757,730,000	772,146,400	877,837,600	835,411,700	846,532,900	863,116,800
Operating Cash Flow	2,438,335,000	2,410,059,000	2,156,691,000	1,945,727,000	1,844,315,000	1,734,744,000
Capital Expenditure	1,274,924,000	1,544,933,000	761,300,500	1,146,456,000	1,445,381,000	1,565,532,000
EBITDA	2,530,623,000	2,490,499,000	2,187,323,000	2,098,698,000	2,004,645,000	1,731,164,000
Return on Assets %	.08%	.09%	.11%	.10%	.11%	.13%
Return on Equity %	.34%	.38%	.49%	.55%	.65%	.67%
Debt to Equity	1.40%	1.687	0.999	1.60	1.989	2.05

CONTACT INFORMATION:

Phone: 66 2-029-5000   Fax:
Toll-Free:
Address: 414 Paholyothin Rd., Shinawatra Tower 1, Bangkok, 10400 Thailand

STOCK TICKER/OTHER:

Stock Ticker: AVIFY   Exchange: PINX
Employees: 7,914   Fiscal Year Ends: 12/31
Parent Company:

SALARIES/BONUSES:

Top Exec. Salary: $ Bonus: $
Second Exec. Salary: $ Bonus: $

OTHER THOUGHTS:

Estimated Female Officers or Directors: 4
Hot Spot for Advancement for Women/Minorities: Y

Advanced Micro Devices Inc (AMD)

www.amd.com

NAIC Code: 334413

TYPES OF BUSINESS:

Microprocessors
Semiconductors
Chipsets
Wafer Manufacturing
Multimedia Graphics

GROWTH PLANS/SPECIAL FEATURES:

Advanced Micro Devices designs microprocessors for the computer and consumer electronics industries. The majority of the firm's sales are in the personal computer and data center markets via CPUs and GPUs. Additionally, the firm supplies the chips found in prominent game consoles such as the Sony PlayStation and Microsoft Xbox. AMD acquired graphics processor and chipset maker ATI in 2006 in an effort to improve its positioning in the PC food chain. In 2009, the firm spun out its manufacturing operations to form the foundry GlobalFoundries. In 2022, the firm acquired FPGA-leader Xilinx to diversify its business and augment its opportunities in key end markets such as the data center.

BRANDS/DIVISIONS/AFFILIATES:

AMD
ATI
Athlon
EPYC
Radeon
Ryzen
Threadripper

CONTACTS: *Note: Officers with more than one job title may be intentionally listed here more than once.*

Lisa Su, CEO
Devinder Kumar, CFO
Darla Smith, Chief Accounting Officer
Mark Papermaster, Chief Technology Officer
John Caldwell, Director
Richard Bergman, Executive VP, Divisional
Darren Grasby, Executive VP
Forrest Norrod, General Manager, Divisional
Harry Wolin, Senior VP

FINANCIAL DATA: *Note: Data for latest year may not have been available at press time.*

In U.S. $	2021	2020	2019	2018	2017	2016
Revenue	16,434,000,000	9,763,000,000	6,731,000,000	6,475,000,000	5,253,000,000	4,319,000,000
R&D Expense	2,845,000,000	1,983,000,000	1,547,000,000	1,434,000,000	1,196,000,000	1,008,000,000
Operating Income	3,648,000,000	1,369,000,000	631,000,000	451,000,000	127,000,000	-383,000,000
Operating Margin %	.22%	.14%	.09%	.07%	.02%	-.09%
SGA Expense	1,448,000,000	995,000,000	750,000,000	562,000,000	516,000,000	466,000,000
Net Income	3,162,000,000	2,490,000,000	341,000,000	337,000,000	-33,000,000	-498,000,000
Operating Cash Flow	3,521,000,000	1,071,000,000	493,000,000	34,000,000	12,000,000	81,000,000
Capital Expenditure	301,000,000	294,000,000	217,000,000	163,000,000	113,000,000	77,000,000
EBITDA	4,166,000,000	1,676,000,000	724,000,000	621,000,000	262,000,000	-160,000,000
Return on Assets %	.30%	.33%	.06%	.08%	-.01%	-.16%
Return on Equity %	.47%	.57%	.17%	.36%	-.07%	-249.00%
Debt to Equity	.05%	0.091	0.242	0.88	2.223	3.45

CONTACT INFORMATION:

Phone: 408 749-4000   Fax:
Toll-Free:
Address: 2485 Augustine Dr., Santa Clara, CA 95054 United States

STOCK TICKER/OTHER:

Stock Ticker: AMD   Exchange: NAS
Employees: 15,500   Fiscal Year Ends: 12/31
Parent Company:

SALARIES/BONUSES:

Top Exec. Salary: $1,076,317   Bonus: $
Second Exec. Salary: $700,388   Bonus: $

OTHER THOUGHTS:

Estimated Female Officers or Directors: 3
Hot Spot for Advancement for Women/Minorities: Y

Sales, profits and employees may be estimates. Financial information, benefits and other data can change quickly and may vary from those stated here.

Akamai Technologies Inc

www.akamai.com

NAIC Code: 517110

TYPES OF BUSINESS:
Online Information Service-Streaming Content
Content Delivery Protection
Business Content Applications
Internet Protection
Security Solutions
Edge Computing
Cloud Optimization

BRANDS/DIVISIONS/AFFILIATES:
Guardicore Ltd

GROWTH PLANS/SPECIAL FEATURES:
Akamai operates a content delivery network, or CDN, which entails locating servers at the edges of networks so its customers, which store content on Akamai servers, can reach their own customers faster, more securely, and with better quality. Akamai has over 325,000 servers distributed over 4,000 points of presence in more than 1,000 cities worldwide. Its customers generally include media companies, which stream video content or make video games available for download, and other enterprises that run interactive or high-traffic websites, such as e-commerce firms and financial institutions. Akamai also has a significant security business, which is integrated with its core web and media businesses to protect its customers from cyber threats.

Akamai offers its employees health and dental care, time off, fitness/wellness options and more.

CONTACTS:
Note: Officers with more than one job title may be intentionally listed here more than once.

F. Leighton, CEO
Edward McGowan, CFO
Daniel Hesse, Chairman of the Board
Laura Howell, Chief Accounting Officer
Mani Sundaram, Chief Information Officer
Kim Salem-Jackson, Chief Marketing Officer
Robert Blumofe, Chief Technology Officer
Adam Karon, COO
Paul Joseph, Executive VP, Divisional
Aaron Ahola, Executive VP
Anthony Williams, Executive VP
Rick Mcconnell, General Manager, Divisional

FINANCIAL DATA:
Note: Data for latest year may not have been available at press time.

In U.S. $	2021	2020	2019	2018	2017	2016
Revenue	3,461,223,000	3,198,149,000	2,893,617,000	2,714,474,000	2,489,035,000	2,347,988,000
R&D Expense	335,372,000	269,315,000	261,365,000	246,165,000	222,434,000	167,628,000
Operating Income	807,202,000	701,674,000	577,991,000	416,052,000	392,546,000	477,839,000
Operating Margin %	.23%	.22%	.20%	.15%	.16%	.20%
SGA Expense	918,977,000	966,696,000	947,545,000	982,775,000	887,977,000	799,758,000
Net Income	651,642,000	557,054,000	478,035,000	298,373,000	222,766,000	320,727,000
Operating Cash Flow	1,404,563,000	1,215,000,000	1,058,304,000	1,008,327,000	800,983,000	871,812,000
Capital Expenditure	545,230,000	731,872,000	562,077,000	405,741,000	414,778,000	316,289,000
EBITDA	1,351,185,000	1,163,591,000	1,022,519,000	820,811,000	705,344,000	819,302,000
Return on Assets %	.08%	.08%	.08%	.06%	.05%	.07%
Return on Equity %	.15%	.14%	.14%	.09%	.07%	.10%
Debt to Equity	.59%	0.617	0.692	0.274	0.197	0.199

CONTACT INFORMATION:
Phone: 617 444-3000   Fax:
Toll-Free: 877-425-2624
Address: 145 Broadway, Cambridge, MA 02142 United States

STOCK TICKER/OTHER:
Stock Ticker: AKAM   Exchange: NAS
Employees: 8,700   Fiscal Year Ends: 12/31
Parent Company:

SALARIES/BONUSES:
Top Exec. Salary: $566,692   Bonus: $
Second Exec. Salary: Bonus: $
$537,019

OTHER THOUGHTS:
Estimated Female Officers or Directors: 4
Hot Spot for Advancement for Women/Minorities: Y

Alaska Communications

www.alaskacommunications.com

NAIC Code: 517110

TYPES OF BUSINESS:

Telecommunications Services
Telecommunications
Internet Services
TV Streaming Services
Mobility Services
Cloud and IT Services
Data Networking Services
Data Security Solutions

BRANDS/DIVISIONS/AFFILIATES:

ATN International Inc

GROWTH PLANS/SPECIAL FEATURES:

Alaska Communications is a telecommunications firm based in Anchorage, serving individuals, businesses and enterprises. Products and services include internet, TV streaming, voice, mobility, cloud optimization, information technology (IT) and data networking, as well as bundled packages. Business solutions offered include business connectivity, mobility, productivity, collaboration and security. Support services span customer support, account support, troubleshooting, internet support, voice support and more. Alaska Communications operates as a subsidiary of ATN International, Inc.

CONTACTS: *Note: Officers with more than one job title may be intentionally listed here more than once.*

Bill Bishop, CEO
Sandy Knechtel, COO
Laurie Butcher, CFO
Leonard Steinberg, Secretary

FINANCIAL DATA: *Note: Data for latest year may not have been available at press time.*

In U.S. $	2021	2020	2019	2018	2017	2016
Revenue	250,191,752	240,568,992	231,694,000	232,468,000	226,904,992	226,866,000
R&D Expense						
Operating Income						
Operating Margin %						
SGA Expense						
Net Income		-1,073,000	4,928,000	9,080,000	-6,101,000	2,386,000
Operating Cash Flow						
Capital Expenditure						
EBITDA						
Return on Assets %						
Return on Equity %						
Debt to Equity						

CONTACT INFORMATION:

Phone: 907 297-3000   Fax: 907 297-3100
Toll-Free:
Address: 600 Telephone Ave., Anchorage, AK 99503-6091 United States

STOCK TICKER/OTHER:

Stock Ticker: Subsidiary   Exchange:
Employees: 575   Fiscal Year Ends: 12/31
Parent Company: ATN International Inc

SALARIES/BONUSES:

Top Exec. Salary: $ Bonus: $
Second Exec. Salary: $ Bonus: $

OTHER THOUGHTS:

Estimated Female Officers or Directors: 3
Hot Spot for Advancement for Women/Minorities: Y

Sales, profits and employees may be estimates. Financial information, benefits and other data can change quickly and may vary from those stated here.

Altice Europe NV

NAIC Code: 517210

www.altice.net

TYPES OF BUSINESS:

Wireless Telecommunications Carriers
Telecommunications
Media
Data Technology
Live Broadcasting
Advertising Solutions
Digital Solutions
Fiber-to-the-Home

BRANDS/DIVISIONS/AFFILIATES:

Altice

GROWTH PLANS/SPECIAL FEATURES:

Altice Europe NV is a multinational fiber, broadband, telecommunications, contents and media company. Its global presence covers France, Israel, the Dominican Republic and Portugal. The firm's activities are divided into three segments: telecom, media and content, and data and advertising. The telecom segment distributes, invests and services high speed broadband. The media and content segment provides original content, high-quality TV shows and international, national and local news channels. The firm delivers live broadcast premium sports events as well as well-known media and entertainment. The data and advertising segment innovates with technology in its labs across the world; links leading brands to audiences through its premium advertising solutions; and provides enterprise digital solutions to businesses. All of the company's products are marketed under the Altice brand name.

CONTACTS:
Note: Officers with more than one job title may be intentionally listed here more than once.

Armando Pereira, COO
Malo Corbin, CFO
Patrick Drahi, Pres. of the Board

FINANCIAL DATA:
Note: Data for latest year may not have been available at press time.

In U.S. $	2021	2020	2019	2018	2017	2016
Revenue	17,536,546,884	16,862,064,312	17,383,571,456	16,747,968,512	25,401,894,912	24,498,884,608
R&D Expense						
Operating Income						
Operating Margin %						
SGA Expense						
Net Income			288,312,416	-391,113,312	-590,193,728	-1,838,505,216
Operating Cash Flow						
Capital Expenditure						
EBITDA						
Return on Assets %						
Return on Equity %						
Debt to Equity						

CONTACT INFORMATION:

Phone: 31 79946 4931   Fax:
Toll-Free:
Address: Prins Bernhardplein 200, Amsterdam, 1097JB Netherlands

STOCK TICKER/OTHER:

Stock Ticker: Private   Exchange:
Employees: 35,328   Fiscal Year Ends:
Parent Company:

SALARIES/BONUSES:

Top Exec. Salary: $ Bonus: $
Second Exec. Salary: $ Bonus: $

OTHER THOUGHTS:

Estimated Female Officers or Directors:
Hot Spot for Advancement for Women/Minorities:

Altice Portugal SA

NAIC Code: 517110

www.telecom.pt

TYPES OF BUSINESS:
Telephony Services
Mobile Services
Cable & Satellite Television Service
Internet Services
Wireless Services

BRANDS/DIVISIONS/AFFILIATES:
Altice NV
Altice Group
Altice
MEO
Altice Empresas
SAPO
MOCHE

CONTACTS: *Note: Officers with more than one job title may be intentionally listed here more than once.*
Ana Figueiredo, CEO
Alexander Freese, COO
Alexandre Matos, CFO
Joao Teixeira, CTO
Alexandre Filipe Fonseca, Chmn.

GROWTH PLANS/SPECIAL FEATURES:
Altice Portugal SA is a leading provider of telecommunications and multimedia services in Portugal. The firm develops equipment and solutions in markets such as medicine, education, culture and sports. Altice Portugal's brands include: Altice, offering telecommunications, media, content, data and advertising services, solutions and products; MEO offers telephone, interactive televisions, voice and internet services through its technological infrastructure of fiber optics; Altice Empresas offers business solutions in regards to technology and telecommunication, which are designed for small and medium enterprises, as well as for large corporations and public institutions; SAPO gathers news, shopping services, job opportunities, blogs, weather reports and more and offers the information on its digital platform; and MOCHE is a tactical brand for younger persons engaged in areas such as skateboarding, gaming and nightlife. Altice Portugal develops technology used within its branded products, including television channels with their own content, cloud file storage, payment capabilities via mobile devices, online video games, online radios, the delivery of large volumes of data, applications that allow companies to analyze large amounts of data about their consumer's behavior, a remote diagnosis and medical care tool, the monitoring of health and safety of vulnerable people, a learning management system, Gigabit passive optical network technology, extended reach of television networks to remote locations and much more. Altice Portugal, part of the Altice Group, operates as a wholly-owned subsidiary of Altice NV, a multinational telecommunications company.

FINANCIAL DATA: *Note: Data for latest year may not have been available at press time.*

In U.S. $	2021	2020	2019	2018	2017	2016
Revenue						
R&D Expense						
Operating Income						
Operating Margin %						
SGA Expense						
Net Income						
Operating Cash Flow						
Capital Expenditure						
EBITDA						
Return on Assets %						
Return on Equity %						
Debt to Equity						

CONTACT INFORMATION:
Phone: 351 21-500-1701   Fax: 351-21-500-0800
Toll-Free:
Address: Avenida Fontes Pereira de Melo, 40, Lisboa, 1250 -133 Portugal

STOCK TICKER/OTHER:
Stock Ticker: Subsidiary
Employees: 14
Parent Company: Altice NV

Exchange:
Fiscal Year Ends: 12/31

SALARIES/BONUSES:
Top Exec. Salary: $ Bonus: $
Second Exec. Salary: $ Bonus: $

OTHER THOUGHTS:
Estimated Female Officers or Directors:
Hot Spot for Advancement for Women/Minorities:

Sales, profits and employees may be estimates. Financial information, benefits and other data can change quickly and may vary from those stated here.

Altice USA Inc

www.alticeusa.com

NAIC Code: 517110

TYPES OF BUSINESS:
Cable Television Service
Professional Sports Teams
Television Programming
Communications Services
Movie Theatres
Voice Over Internet Protocol
High-Speed Internet

BRANDS/DIVISIONS/AFFILIATES:
Optimum
Suddenlink
Altice Mobile
News 12 Networks
Cheddar
i24NEWS
a4
New York Interconnect

CONTACTS: Note: Officers with more than one job title may be intentionally listed here more than once.
Dexter Goei, CEO
Michael Grau, CFO
Patrick Drahi, Chairman of the Board
Layth Taki, Chief Accounting Officer
Colleen Schmidt, Executive VP, Divisional
Michael Olsen, Executive VP

GROWTH PLANS/SPECIAL FEATURES:
Altice Europe acquired privately held U.S. cable company Suddenlink in 2015 and Cablevision in 2016. Suddenlink's networks provided television, internet access, and phone services to roughly 3.5 million U.S. homes and businesses located primarily in smaller markets, with major clusters in Texas, West Virginia, Idaho, Arizona, and Louisiana. Cablevision provided comparable services to about 5.5 million homes and business in the New York City metro area. Altice Europe spun off Altice USA, which includes both the Suddenlink and Cablevision operations, to shareholders in 2018. Altice USA also owns News 12 Networks, which broadcasts local 24-hour news networks in New York, i24News, a news operation focused on the Middle East and Israel, and Cheddar, a news upstart.

FINANCIAL DATA: Note: Data for latest year may not have been available at press time.

In U.S. $	2021	2020	2019	2018	2017	2016
Revenue	10,090,850,000	9,894,642,000	9,760,859,000	9,566,608,000	9,306,950,000	6,017,212,000
R&D Expense						
Operating Income	2,541,803,000	2,206,362,000	1,896,789,000	1,720,927,000	993,409,000	703,204,000
Operating Margin %	.25%	.22%	.19%	.18%	.11%	.12%
SGA Expense						
Net Income	990,311,000	436,183,000	138,936,000	18,833,000	1,493,177,000	-832,030,000
Operating Cash Flow	2,854,078,000	2,980,164,000	2,554,169,000	2,508,317,000	2,018,247,000	1,184,455,000
Capital Expenditure	1,231,715,000	1,073,955,000	1,355,350,000	1,153,589,000	953,056,000	625,647,000
EBITDA	4,359,810,000	4,019,127,000	3,986,832,000	3,920,560,000	3,166,115,000	2,065,702,000
Return on Assets %	.03%	.01%	.00%	.00%	.04%	-.02%
Return on Equity %		.77%	.05%	.00%	.40%	-.41%
Debt to Equity			10.801	6.171	3.879	11.051

CONTACT INFORMATION:
Phone: 516 803-2300   Fax: 516 803-2273
Toll-Free:
Address: 1 Court Square West, Long Island City, NY 11101 United States

STOCK TICKER/OTHER:
Stock Ticker: ATUS   Exchange: NYS
Employees: 8,900   Fiscal Year Ends: 12/31
Parent Company: Next Alt Sarl

SALARIES/BONUSES:
Top Exec. Salary: $750,000   Bonus: $
Second Exec. Salary: $500,000   Bonus: $

OTHER THOUGHTS:
Estimated Female Officers or Directors: 4
Hot Spot for Advancement for Women/Minorities: Y

Alvaria Inc

www.alvaria.com

NAIC Code: 511210K

TYPES OF BUSINESS:
Software-Customer Service Request Processing
Contact Center Automation Systems
Customer Experience Solutions
Workforce Engagement Software
Cloud Services Technology
Chat Bots
Voice Agents
Fraud Detection Solutions

BRANDS/DIVISIONS/AFFILIATES:
Vector Capital
Alvaria WEM Suite
Alvaria CX Suite

CONTACTS: *Note: Officers with more than one job title may be intentionally listed here more than once.*
Patrick Dennis, CEO
Michael Harris, CMO
Tim Dahltorp, CFO
Steve Seger, Chief Commercial Officer
Sweety Rath, Sr. VP-Global Human Resources
Michael Regan, Sr. VP-R&D
David Funck, CTO
Stephen Beaver, General Counsel
Manish Chandak, VP-Microsoft Professional Svcs.
David Herzog, Sr. VP
Spence Mallder, Gen. Mgr.-Workforce Optimization
Chris Koziol, Pres.
Edward Skowronski, Chief Customer Officer
Bryan Sheppeck, Exec. VP-Worldwide Sales

GROWTH PLANS/SPECIAL FEATURES:
Alvaria, Inc. provides optimized customer experience and workforce engagement software and cloud services technology solutions. The Alvaria WEM Suite is an enterprise-grade workforce engagement management solution, whether in the cloud, on-premise or hybrid. The suite comprises workforce management and employee experience (EX) solutions as well as a mobile app. The Alvaria CX Suite focuses on customer experiences, and encompasses: a voice agent for inbound/outbound interaction; an omnichannel agent across voice, email, web chat, short message service (SMS) and social through a single agent; an automated agent such as digital bots and assistants; dialing options and compliance controls for driving campaigns and targeted collections; and fraud detection via consumer behavior data, device location, voice biometrics and other identity features. Industries served by Alvaria primarily include banking/finance, collections, insurance providers, healthcare, manufacturing, airlines, automotive, retail, utilities and telecommunications. Headquartered in Boston, Massachusetts, Alvaria has global operations spanning North America, Latin America, Europe, Africa and Asia-Pacific. The company is owned by global private equity firm, Vector Capital.

Alvaria offers its employees comprehensive health benefits, life and disability insurance, 401(k(and retirement plans, group auto and home insurance, pet insurance and other plans and programs.

FINANCIAL DATA: *Note: Data for latest year may not have been available at press time.*

In U.S. $	2021	2020	2019	2018	2017	2016
Revenue	525,429,450	505,220,625	518,175,000	493,500,000	470,000,000	468,000,000
R&D Expense						
Operating Income						
Operating Margin %						
SGA Expense						
Net Income						
Operating Cash Flow						
Capital Expenditure						
EBITDA						
Return on Assets %						
Return on Equity %						
Debt to Equity						

CONTACT INFORMATION:
Phone: 978-250-7900   Fax: 978-244-7410
Toll-Free: 888-412-7728
Address: 5 Technology Park Dr., Westford, MA 01886 United States

STOCK TICKER/OTHER:
Stock Ticker: Private   Exchange:
Employees: 1,911   Fiscal Year Ends: 12/31
Parent Company: Vector Capital

SALARIES/BONUSES:
Top Exec. Salary: $ Bonus: $
Second Exec. Salary: $ Bonus: $

OTHER THOUGHTS:
Estimated Female Officers or Directors: 1
Hot Spot for Advancement for Women/Minorities:

Sales, profits and employees may be estimates. Financial information, benefits and other data can change quickly and may vary from those stated here.

America Movil SAB de CV

www.americamovil.com

NAIC Code: 517210

TYPES OF BUSINESS:

Wireless Telecommunications Carriers (except Satellite)
Wireless Internet
Local & Long Distance
Satellite & Cable TV

BRANDS/DIVISIONS/AFFILIATES:

Telcel
Telmex
Claro
TracFone
A1
Straight Talk
KPN

GROWTH PLANS/SPECIAL FEATURES:

America Movil is the largest telecom carrier in Latin America, serving about 290 million wireless customers across the region. It also provides fixed-line phone, internet access, and television services in most of the countries it serves. Mexico is the firm's largest market, providing about 40% of service revenue. Movil dominates the Mexican wireless market with about 63% customer share and also serves about half of fixed-line internet access customers in the country. Brazil, its second most important market, provides about 30% of service revenue. Movil sold its low-margin wireless resale business in the U.S. to Verizon in 2021 and now owns a 1.4% stake in the U.S. telecom giant. The firm also holds a 51% stake in Telekom Austria and a 20% stake in Dutch carrier KPN.

CONTACTS: *Note: Officers with more than one job title may be intentionally listed here more than once.*

Daniel Hajj Aboumrad, CEO
Carlos Garcia Moreno Elizondo, CFO
Patric Slim Domit, Vice Chmn.
Alejandro Cantu Jimenez, General Counsel
Salvador Cortes Gomez, COO-Mexico
Fernando Ocampo Carapia, CFO-Mexico
Juan Antonio Aguilar, CEO-Central America
Enrique Luna Roshard, CFO-Central America
Juan Carlos Archila Cabal, CEO-Colombia
Fernando Gonzalez Apango, CFO-Colombia
Carlos Slim Domit, Chmn.

FINANCIAL DATA: *Note: Data for latest year may not have been available at press time.*

In U.S. $	2021	2020	2019	2018	2017	2016
Revenue	42,268,250,000	41,486,270,000	42,068,090,000	51,293,320,000	50,474,470,000	48,190,890,000
R&D Expense						
Operating Income	8,207,891,000	7,188,688,000	7,104,448,000	6,894,913,000	4,947,645,000	5,415,369,000
Operating Margin %	.19%	.17%	.17%	.13%	.10%	.11%
SGA Expense	8,934,439,000	9,481,038,000	9,659,193,000	11,224,590,000	11,888,700,000	11,269,480,000
Net Income	9,506,791,000	2,314,783,000	3,346,289,000	2,597,067,000	1,448,866,000	427,330,600
Operating Cash Flow	12,755,630,000	13,874,470,000	11,574,680,000	12,268,930,000	10,759,190,000	11,649,760,000
Capital Expenditure	7,809,597,000	6,398,743,000	7,499,635,000	7,500,849,000	6,754,922,000	7,659,057,000
EBITDA	15,538,260,000	12,859,330,000	15,586,130,000	14,672,590,000	12,594,690,000	10,495,180,000
Return on Assets %	.12%	.03%	.05%	.04%	.02%	.01%
Return on Equity %	.60%	.22%	.36%	.27%	.15%	.05%
Debt to Equity	1.26%	2.254	3.315	2.769	3.328	2.993

CONTACT INFORMATION:

Phone: 52-55-2581-4449   Fax: 52-55-2581-4422
Toll-Free:
Address: Lago Zurich 245, Colonia Granada, Ampliacion, Mexico City, DF 11529 Mexico

STOCK TICKER/OTHER:

Stock Ticker: AMOV   Exchange: NYS
Employees: 181,205   Fiscal Year Ends: 12/31
Parent Company:

SALARIES/BONUSES:

Top Exec. Salary: $ Bonus: $
Second Exec. Salary: $ Bonus: $

OTHER THOUGHTS:

Estimated Female Officers or Directors: 2
Hot Spot for Advancement for Women/Minorities:

Sales, profits and employees may be estimates. Financial information, benefits and other data can change quickly and may vary from those stated here.

Amphenol Corporation

www.amphenol.com

NAIC Code: 335921

TYPES OF BUSINESS:

Cables & Connectors
Fiber Optic Cable
Interconnect Systems

GROWTH PLANS/SPECIAL FEATURES:

Amphenol is a global supplier of connectors, sensors, and interconnect systems. Amphenol holds the second-largest connector market share globally and sells into the end markets of automotive, broadband, commercial air, industrial, IT and data communications, military, mobile devices, and mobile networks. Amphenol is diversified geographically, with operations in 40 countries.

BRANDS/DIVISIONS/AFFILIATES:

MTS Systems Corporation

CONTACTS: *Note: Officers with more than one job title may be intentionally listed here more than once.*

Richard Norwitt, CEO
Craig Lampo, CFO
Martin Loeffler, Director
Lance DAmico, General Counsel
Luc Walter, General Manager, Divisional
Jean-Luc Gavelle, General Manager, Divisional
William Doherty, General Manager, Divisional
Martin Booker, General Manager, Divisional
Yaobin Gu, General Manager, Divisional
David Silverman, Senior VP, Divisional
Dieter Ehrmanntraut, Vice President

FINANCIAL DATA: *Note: Data for latest year may not have been available at press time.*

In U.S. $	2021	2020	2019	2018	2017	2016
Revenue	10,876,300,000	8,598,900,000	8,225,400,000	8,202,000,000	7,011,300,000	6,286,400,000
R&D Expense						
Operating Income	2,175,500,000	1,649,900,000	1,644,600,000	1,695,400,000	1,431,600,000	1,241,800,000
Operating Margin %	.20%	.19%	.20%	.21%	.20%	.20%
SGA Expense	1,226,300,000	1,014,200,000	971,400,000	959,500,000	878,300,000	798,200,000
Net Income	1,590,800,000	1,203,400,000	1,155,000,000	1,205,000,000	650,500,000	822,900,000
Operating Cash Flow	1,540,100,000	1,592,000,000	1,502,300,000	1,112,700,000	1,144,200,000	1,077,600,000
Capital Expenditure	360,400,000	276,800,000	295,000,000	310,600,000	226,600,000	190,800,000
EBITDA	2,500,300,000	1,950,100,000	1,925,600,000	1,989,800,000	1,671,500,000	1,430,700,000
Return on Assets %	.12%	.10%	.11%	.12%	.07%	.10%
Return on Equity %	.27%	.24%	.27%	.30%	.17%	.24%
Debt to Equity	.76%	0.675	0.707	0.699	0.888	0.717

CONTACT INFORMATION:

Phone: 203 265-8900   Fax: 203 265-8516
Toll-Free: 877-267-4366
Address: 358 Hall Ave., Wallingford, CT 06492 United States

STOCK TICKER/OTHER:

Stock Ticker: APH   Exchange: NYS
Employees: 90,000   Fiscal Year Ends: 12/31
Parent Company:

SALARIES/BONUSES:

Top Exec. Salary: $1,355,000   Bonus: $
Second Exec. Salary: $656,000   Bonus: $

OTHER THOUGHTS:

Estimated Female Officers or Directors: 1
Hot Spot for Advancement for Women/Minorities:

Anaren Inc

www.anaren.com

NAIC Code: 334220

TYPES OF BUSINESS:

Wireless Telecommunications Components
Radio Frequency
Microwave Microelectronics
Components
Beamforming Networks
Unicircuit
Printed Circuit Board Fabrication
Ceramic Solutions

BRANDS/DIVISIONS/AFFILIATES:

TTM Technologies Inc

CONTACTS: *Note: Officers with more than one job title may be intentionally listed here more than once.*

Thomas T. Edman, CEO
David M. Ferrara, General Counsel
Timothy P. Ross, Pres., Space & Defense
George A. Blanton, Treas.
Gert Thygesen, Sr. VP-Tech.
Mark Burdick, Pres., Wireless Group

GROWTH PLANS/SPECIAL FEATURES:

Anaren, Inc., a subsidiary of TTM Technologies, Inc., is a global designer and manufacturer of high-frequency radio frequency (RF) and microwave microelectronics, components and assemblies. The firm's engineering expertise and products are utilized by major manufacturers around the world, which are primarily engaged in the space, defense and telecommunications sectors. Anaren's capabilities span beamforming networks, integrated RF solutions, complex build-to-print (unicircuit), high-reliability printed circuit board (PCB) fabrication, thick film substrates, low temperature co-fired ceramic (LTCC) solutions and multi-chip modules. The company's standard RF components include couplers, power dividers, baluns, and radio transceivers, as well as build-to-print complex subassemblies including beamformers, RF modules, hybrid microelectronics, ceramic substrates and more. Products that can be customized include beamforming networks, integrated RF solutions, ceramic circuits and packaging, microelectronics and high-reliability PCB. Anaren's diverse portfolio delivers innovations in design and solutions for military, satellite and telecommunications applications.

Anaren offers its employees health and wealth benefits, as well as a variety of employee assistance programs.

FINANCIAL DATA: *Note: Data for latest year may not have been available at press time.*

In U.S. $	2021	2020	2019	2018	2017	2016
Revenue	250,000,000	242,676,000	226,800,000	189,000,000	180,000,000	175,000,000
R&D Expense						
Operating Income						
Operating Margin %						
SGA Expense						
Net Income						
Operating Cash Flow						
Capital Expenditure						
EBITDA						
Return on Assets %						
Return on Equity %						
Debt to Equity						

CONTACT INFORMATION:

Phone: 315 432-8909   Fax: 315 432-9121
Toll-Free: 800-544-2414
Address: 6635 Kirkville Rd., East Syracuse, NY 13057 United States

STOCK TICKER/OTHER:

Stock Ticker: Subsidiary   Exchange:
Employees: 1,100   Fiscal Year Ends: 12/31
Parent Company: TTM Technologies Inc

SALARIES/BONUSES:

Top Exec. Salary: $ Bonus: $
Second Exec. Salary: $ Bonus: $

OTHER THOUGHTS:

Estimated Female Officers or Directors: 2
Hot Spot for Advancement for Women/Minorities:

Anixter International Inc

www.anixter.com

NAIC Code: 423430

TYPES OF BUSINESS:

Wire & Cable Distribution
Network and Security Solutions
Electrical and Electronic Solutions
Utility Power Solutions
Connectivity Solutions
Supply Chain Management
Utility Products and Repair

BRANDS/DIVISIONS/AFFILIATES:

WESCO International Inc
Rapid Fire
Speedpull
Anixter Trakr

CONTACTS: Note: Officers with more than one job title may be intentionally listed here more than once.

William Galvin, CEO
Theodore Dosch, CFO
Rodney Smith, Executive VP, Divisional
Robert Graham, Executive VP, Divisional
William Geary, Executive VP, Divisional
Orlando McGee, Executive VP, Divisional
Justin Choi, Executive VP

GROWTH PLANS/SPECIAL FEATURES:

Anixter International, Inc. is a leading global distributor of network and security solutions, electrical and electronic solutions, and utility power solutions. The company helps build, connect, power and protect assets and critical infrastructures. Anixter's build division consists of electrical engineering staff that offers global support, as well as supply chain solutions to reduce costs and improve performance. It manages multiple engineering specifications, offers inventory management solutions, and provides code and standards guidance. Brands within the build division include Rapid Fire, Speedpull and Anixter Trakr. The connect division works with integrators, end users and contractors to enable connectivity that can handle the data needs of its customers. This division comprises a global distribution network and offers supply chain management to quickly respond to customer demand. Infrastructure offerings include industrial communication and control, network cabling, security application, data center and enterprise cabling, as well as customizable supply chain solutions. The power division offers innovative options for energy professionals facing reliability, regulation, security and safety challenges. It offers an integrated perimeter security solution that enables access control and video surveillance; material management and logistics services; electrical wholesale supply chain services; over 300,000 products for utility customers; a technology platform to streamline processes; smart grid solutions; maintenance/repair/overhaul products; high-voltage transmission line and substation packaging solutions. Last, the protect division provides solutions for security applications including transportation, data centers, residential, commercial, retail, public safety, government, education, healthcare, natural resources, industrial, manufacturing/factory floor and utilities. Solutions range from standalone products to fully integrated IPs. Anixter International operates as a subsidiary of WESCO International, Inc.

FINANCIAL DATA: Note: Data for latest year may not have been available at press time.

In U.S. $	2021	2020	2019	2018	2017	2016
Revenue	9,400,000,000	9,287,879,731	8,845,599,744	8,400,200,192	7,927,399,936	7,622,799,872
R&D Expense						
Operating Income						
Operating Margin %						
SGA Expense						
Net Income			262,900,000	156,300,000	109,000,000	120,500,000
Operating Cash Flow						
Capital Expenditure						
EBITDA						
Return on Assets %						
Return on Equity %						
Debt to Equity						

CONTACT INFORMATION:

Phone: 224-521-8000 Fax: 224-521-8100
Toll-Free: 800-323-8167
Address: 2301 Patriot Blvd., Glenview, IL 60026 United States

SALARIES/BONUSES:

Top Exec. Salary: $ Bonus: $
Second Exec. Salary: $ Bonus: $

STOCK TICKER/OTHER:

Stock Ticker: Subsidiary Exchange:
Employees: 9,400 Fiscal Year Ends: 12/31
Parent Company: WESCO International Inc

OTHER THOUGHTS:

Estimated Female Officers or Directors: 2
Hot Spot for Advancement for Women/Minorities: Y

Sales, profits and employees may be estimates. Financial information, benefits and other data can change quickly and may vary from those stated here.

Anritsu Corporation

NAIC Code: 334515

www.anritsu.com

TYPES OF BUSINESS:
Test & Measurement Equipment
Communications Applications
Optical Measurement Devices
Industrial Equipment
Internet of Things

GROWTH PLANS/SPECIAL FEATURES:
Anritsu Corp is an electronic components manufacturer. The company has three business segments: Measurement, Products Quality Assurance, and Others. The Measurement segment offers measuring devices for mobile phone acceptance testing by mobile phone operators. The Product Quality Assurance Segment offers automatic electronic weighing equipment and contaminant detectors for food, cosmetics, and pharmaceutical industries, and x-ray detectors for contaminants. Anritsu's x-ray machines can detect and remove metal fragments. The Other business segment has a variety of interests, including telecommunication equipment, logistics, and real estate. The company earns the vast majority of its revenue in Japan.

BRANDS/DIVISIONS/AFFILIATES:

CONTACTS: Note: Officers with more than one job title may be intentionally listed here more than once.
Hirokazu Hamada, CEO
Akifumi Kubota, CFO
Yoshiyuki Amano, VP-Global Sales
Takashi Sakamoto, Chief Human Resources Officer
Hanako Noda, CTO
Kenji Tanaka, Sr. Exec. VP
Frank Tiernan, Exec. VP
Fumihiro Tsukasa, Sr. VP
Junkichi Shirono, Sr. VP

FINANCIAL DATA: Note: Data for latest year may not have been available at press time.

In U.S. $	2021	2020	2019	2018	2017	2016
Revenue	785,664,500	793,703,700	739,090,800	637,548,200	649,940,700	
R&D Expense	80,895,880	96,225,150	86,880,750	75,318,900	80,881,050	
Operating Income	145,743,100	129,145,700	83,409,970	36,435,780	31,400,180	
Operating Margin %	.19%	.16%	.11%	.06%	.05%	
SGA Expense	198,702,200	207,920,500	207,238,200	196,996,400	201,705,700	
Net Income	119,437,900	99,043,310	66,426,880	21,492,140	20,275,880	
Operating Cash Flow	151,891,100	109,173,800	90,826,170	58,929,100	68,570,160	
Capital Expenditure	19,956,990	20,987,840	15,677,840	18,125,190	15,143,870	
EBITDA	185,167,600	168,770,400	118,800,100	70,668,940	63,949,860	
Return on Assets %	.11%	.10%	.07%	.02%	.02%	
Return on Equity %	.16%	.15%	.11%	.04%	.04%	
Debt to Equity		0.032	0.128	0.147	0.189	

CONTACT INFORMATION:
Phone: 81 46-223-1111   Fax:
Toll-Free:
Address: 5-1-1 Onna, Atsugi-shi, Kanagawa, 243-8555 Japan

STOCK TICKER/OTHER:
Stock Ticker: AITUF   Exchange: PINX
Employees: 4,336   Fiscal Year Ends: 03/31
Parent Company:

SALARIES/BONUSES:
Top Exec. Salary: $ Bonus: $
Second Exec. Salary: $ Bonus: $

OTHER THOUGHTS:
Estimated Female Officers or Directors:
Hot Spot for Advancement for Women/Minorities:

ARC Group Worldwide Inc   www.arcw.com
NAIC Code: 334220

TYPES OF BUSINESS:
Telecommunications Equipment
Metal Injection Molding
Plastic Injection Molding
Tooling
Wireless

BRANDS/DIVISIONS/AFFILIATES:

GROWTH PLANS/SPECIAL FEATURES:
ARC Group Worldwide Inc provides metal injection molding solutions. Its operating segments include Precision Components Group. The company solutions include metal injection molding, plastic injection molding, clean room plastic injection molding, and rapid and conforming tooling. Geographically, it derives a majority of revenue from the United States. It serves industries, including aerospace, automotive, consumer durables, and defense, and medical, and others.

CONTACTS: Note: Officers with more than one job title may be intentionally listed here more than once.
Jedidiah Rust, CEO
Eli Davidai, Director

FINANCIAL DATA: Note: Data for latest year may not have been available at press time.

In U.S. $	2021	2020	2019	2018	2017	2016
Revenue		48,526,000	60,060,000	80,018,000	99,069,000	94,124,000
R&D Expense						
Operating Income		-1,634,000	-7,105,000	-7,680,000	-9,441,000	462,000
Operating Margin %		- .03%	- .12%	- .10%	- .10%	.00%
SGA Expense		8,436,000	11,104,000	12,949,000	19,263,000	16,401,000
Net Income		-5,287,000	-24,016,000	-13,180,000	-10,199,000	-2,314,000
Operating Cash Flow		3,140,000	-5,004,000	-556,000	2,850,000	6,466,000
Capital Expenditure		1,198,000	1,918,000	5,144,000	6,641,000	2,633,000
EBITDA		4,442,000	2,927,000	2,841,000	489,000	9,991,000
Return on Assets %		- .08%	- .30%	- .14%	- .09%	- .02%
Return on Equity %		- .82%	-1.17%	- .39%	- .26%	- .05%
Debt to Equity		8.977	5.297	1.17	1.29	0.892

CONTACT INFORMATION:
Phone: 386-736-4890   Fax:
Toll-Free:
Address: 810 Flightline Boulevard, Deland, FL 32724 United States

STOCK TICKER/OTHER:
Stock Ticker: ARCW   Exchange: PINX
Employees: 412   Fiscal Year Ends: 06/30
Parent Company:

SALARIES/BONUSES:
Top Exec. Salary: $300,000   Bonus: $
Second Exec. Salary: $138,462   Bonus: $

OTHER THOUGHTS:
Estimated Female Officers or Directors: 1
Hot Spot for Advancement for Women/Minorities:

Sales, profits and employees may be estimates. Financial information, benefits and other data can change quickly and may vary from those stated here.

Arqiva Limited

NAIC Code: 517110

www.arqiva.com

TYPES OF BUSINESS:

Satellite Media Distribution Services
Broadcast Infrastructure
Digital Television
Satellite Television
Radio Distribution
Smart Meter

BRANDS/DIVISIONS/AFFILIATES:

CONTACTS: *Note: Officers with more than one job title may be intentionally listed here more than once.*

Paul Donovan, CEO
Adrian Twyning, Chief of Operations
Sean West, CFO
Shuja Khan, CCO
Vivian Leinster, Chief People Officer
Clive White, CTO
Doug Umbers, Managing Dir.-Bus. Oper.
Wendy McMillan, Dir.-Strategy & Bus. Dev.
Michael Giles, Group Comm. Dir.
Matthew Brearley, Dir.-People & Organization
Charles Constable, Managing Dir.-Digital Platforms
Steve Holebrook, Managing Dir.-Broadcast & Media
Nicolas Ott, Managing Dir.-Gov't, Mobile & Enterprise

GROWTH PLANS/SPECIAL FEATURES:

Arqiva Limited is a broadcast and machine-to-machine (M2M) infrastructure solutions provider based in the U.K. The company's television distribution platforms include over-the-top media services, digital terrestrial television and satellite television channels and content. Its radio distribution platforms include analogue radio (FM/AM) and digital radio (DAB), with related services including coding, multiplexing, distribution, managed transmission, network access and service management. In addition, Arqiva offers smart metering products and services, offering gas, electricity and water smart metering communications across the U.K.; providing gas and electricity smart meter networks in Scotland and northern England for the Data Communications Company; and providing smart water network in partnership with Anglian Water and Thames Water, two leading water companies in the U.K. In July 2020, Arqiva sold its telecommunications infrastructure to Cellnex, including towers, pylons and masts, across the U.K. As a result of the deal, Cellnex acquired more than 7,000 cellular sites that were operated by Arqiva.

Arqiva offers apprentice, intern and graduate recruitment programs.

FINANCIAL DATA: *Note: Data for latest year may not have been available at press time.*

In U.S. $	2021	2020	2019	2018	2017	2016
Revenue	745,635,544	804,399,744	910,873,844	900,220,175	827,143,686	789,093,559
R&D Expense						
Operating Income						
Operating Margin %						
SGA Expense						
Net Income	246,790,962	391,479,471	410,388,132	472,401,880	351,062,709	292,544,616
Operating Cash Flow						
Capital Expenditure						
EBITDA						
Return on Assets %						
Return on Equity %						
Debt to Equity						

CONTACT INFORMATION:

Phone: 44 1962 823434   Fax:
Toll-Free:
Address: Crawley Ct., Winchester, SO21 2QA United Kingdom

STOCK TICKER/OTHER:

Stock Ticker: Subsidiary   Exchange:
Employees: 1,505   Fiscal Year Ends: 06/30
Parent Company: Arqiva Holdings Limited

SALARIES/BONUSES:

Top Exec. Salary: $ Bonus: $
Second Exec. Salary: $ Bonus: $

OTHER THOUGHTS:

Estimated Female Officers or Directors: 1
Hot Spot for Advancement for Women/Minorities:

Asia Satellite Telecommunications Holdings Ltd www.asiasat.com
NAIC Code: 517410

TYPES OF BUSINESS:
Satellites
Satellite Services
Broadcast Services
Telecommunications
Broadband Services
Mobility Services
Hosting Services
Video and Streaming Services

BRANDS/DIVISIONS/AFFILIATES:
Asia Satellite Telecommunications Company Limited
AsiaSat 5
AsiaSat 6
AsiaSat 7
AsiaSat 8
AsiaSat 9

CONTACTS: Note: Officers with more than one job title may be intentionally listed here more than once.
Roger Tong, CEO
William Wade, Pres.
Sue Yeung, CFO
Fred Ho, VP-Technical Oper.
Wayne Bannon, Chmn.

GROWTH PLANS/SPECIAL FEATURES:
Asia Satellite Telecommunications Holdings Ltd. is the parent company of Asia Satellite Telecommunications Company Limited (AsiaSat), a leading provider of satellite transponder capacity in Asia. The firm's satellite services connect countries across the Asia Pacific and bridge communication for over two-thirds of the world's population. The firm operates five in-orbit satellites: AsiaSat 5, AsiaSat 6, AsiaSat 7, AsiaSat 8 and AsiaSat 9. AsiaSat 5 is a Space Systems/Loral 1300 series satellite; its footprint covers more than 53 countries, including New Zealand, Japan and Russia. It also covers the Middle East and parts of Africa. Additionally, the AsiaSat 5 offers two high-power fixed Ku-band beams over East Asia and South Africa. AsiaSat 6 is a Space Systems/Loral 1300 series satellite, and carries 28 C band transponders covering southern Asia, Australia and New Zealand. AsiaSat 7 is also a Space Systems/Loral 1300 series satellite that supports a broad range of applications across its coverage area. The satellite's C-band footprint covers Central Asia, the Middle East and Australia, while its K-beam provides service to South and East Asia. AsiaSat 8 is a Space Systems/Loral 1300 series satellite, and carries 24 Ku band transponders and a Ka band payload, covering southern and south-eastern Asia, China and the Middle East. AsiaSat 9 is a Space Systems Loral 1300 satellite equipped with 28 C-band and 32 Ku-band transponders, and a Ka-band payload, covering Asia and Australia. The satellites are managed by a control center in Hong Kong, and together deliver more than 500 television and radio channels. The firm also offers applications such as broadcast, telecommunications, broadband and mobility, as well as video, streaming and hosting services.

FINANCIAL DATA: Note: Data for latest year may not have been available at press time.

In U.S. $	2021	2020	2019	2018	2017	2016
Revenue	213,261,838	205,059,460	195,294,724	185,994,976	174,686,944	163,707,664
R&D Expense						
Operating Income						
Operating Margin %						
SGA Expense						
Net Income			58,161,154	55,391,576	51,179,724	55,316,192
Operating Cash Flow						
Capital Expenditure						
EBITDA						
Return on Assets %						
Return on Equity %						
Debt to Equity						

CONTACT INFORMATION:
Phone: 852 2500-0888   Fax: 852-2500-0895
Toll-Free:
Address: 15 Dai Kwai St., Tai Po Industrial Estate, Hong Kong, Hong Kong 999077 Hong Kong

STOCK TICKER/OTHER:
Stock Ticker: Private   Exchange:
Employees: 135   Fiscal Year Ends: 12/31
Parent Company:

SALARIES/BONUSES:
Top Exec. Salary: $ Bonus: $
Second Exec. Salary: $ Bonus: $

OTHER THOUGHTS:
Estimated Female Officers or Directors: 2
Hot Spot for Advancement for Women/Minorities: Y

Sales, profits and employees may be estimates. Financial information, benefits and other data can change quickly and may vary from those stated here.

AsiaInfo Technologies Limited

www.asiainfo.com

NAIC Code: 511210E

TYPES OF BUSINESS:

Computer Software: Network Security, Managed Access, Digital ID, Cybersecurity & Anti-Virus
Business Operation Support Systems
Business Intelligence Systems
Service and Data Applications
Network Infrastructure Services
Internet of Things
Artificial Intelligence
Big Data

BRANDS/DIVISIONS/AFFILIATES:

AISWare

GROWTH PLANS/SPECIAL FEATURES:

AsiaInfo Technologies Limited is based in Beijing and is a provider of software products, solutions and services. The company's AISWare product series is an artificial intelligence (AI) platform that enables the digital transformation of large enterprises. The AI-based platform includes solutions such as billing, big data, customer relations management, DevOps, intelligent/smart operation, Internet of Things (IoT), global AI, infrastructure foundations and 5G network intelligence. AsiaInfo's products, solutions and services offer operators and enterprises the ability to acquire new customers, promote customers, retain customers and increase business volume based on AI technology. It provides governments with public services and city management and operation services based on big data. AsiaInfo also offers management consulting services, from corporate strategy to organization, and from marketing and service to customer experience. End-to-end hardware and software integration services are provided.

CONTACTS: *Note: Officers with more than one job title may be intentionally listed here more than once.*

Steve Zhang, Pres.
Ying Huang, CFO
Ye Ouyang, CTO
Alex Hawker, Corp. VP
Jerry Li, VP-Bus. Mgmt. Support
Feng Liu, Gen. Mgr.-Global Delivery Svcs.
Suning Tian, Chmn.
Jason Ong, Managing Dir.-AsiaInfo Int'l Pte Ltd.

FINANCIAL DATA: *Note: Data for latest year may not have been available at press time.*

In U.S. $	2021	2020	2019	2018	2017	2016
Revenue		922,293,000	818,693,000	757,610,000	759,799,000	650,000,000
R&D Expense						
Operating Income						
Operating Margin %						
SGA Expense						
Net Income		101,422,000	58,529,100	29,688,100	51,468,900	
Operating Cash Flow						
Capital Expenditure						
EBITDA						
Return on Assets %						
Return on Equity %						
Debt to Equity						

CONTACT INFORMATION:

Phone: 86 1082166688   Fax: 86 1082166699
Toll-Free:
Address: AsianInfo Plaza East Area, #10, Xibeiwang East Rd., Beijing, Beijing 100193 China

STOCK TICKER/OTHER:

Stock Ticker: 1675
Employees: 13,216
Parent Company: CITIC Capital Partners

Exchange: Hong Kong
Fiscal Year Ends: 12/31

SALARIES/BONUSES:

Top Exec. Salary: $ Bonus: $
Second Exec. Salary: $ Bonus: $

OTHER THOUGHTS:

Estimated Female Officers or Directors:
Hot Spot for Advancement for Women/Minorities:

AT&T Inc

NAIC Code: 517110

www.att.com

TYPES OF BUSINESS:

Local Telephone Service
Telecommunications
Media Services
Technology
Video Services
Broadband
Television and Film Production
Entertainment Services

BRANDS/DIVISIONS/AFFILIATES:

WanerMedia
HBO Max
Home Box Office
Warner Bros
Xandr
Otter Media Holdings
Vrio

GROWTH PLANS/SPECIAL FEATURES:

The wireless business contributes about two thirds of AT&T's revenue following the spinoff of WarnerMedia. The firm is the third- largest U.S. wireless carrier, connecting 67 million postpaid and 17 million prepaid phone customers. Fixed-line enterprise services, which account for about 20% of revenue, include internet access, private networking, security, voice, and wholesale network capacity. Residential fixed-line services, about 10% of revenue, primarily consist of broadband internet access service. AT&T also has a sizable presence in Mexico, serving 20 million customers, but this business only accounts for 2% of revenue. The firm still holds a 70% equity stake in satellite television provider DirecTV but does not consolidate this business in its financial statements.

CONTACTS: *Note: Officers with more than one job title may be intentionally listed here more than once.*

Jason Kilar, CEO, Subsidiary
Jeffery McElfresh, CEO, Subsidiary
Lori Lee, CEO, Subsidiary
John Stankey, CEO
Pascal Desroches, CFO
William Kennard, Chairman of the Board
David Huntley, Chief Compliance Officer
David McAtee, General Counsel
Angela Santone, Senior Executive VP, Divisional
Edward Gillespie, Senior Executive VP, Subsidiary

FINANCIAL DATA: *Note: Data for latest year may not have been available at press time.*

In U.S. $	2021	2020	2019	2018	2017	2016
Revenue	168,864,000,000	171,760,000,000	181,193,000,000	170,756,000,000	160,546,000,000	163,786,000,000
R&D Expense						
Operating Income	28,251,000,000	25,285,000,000	29,413,000,000	26,142,000,000	22,884,000,000	23,904,000,000
Operating Margin %	.17%	.15%	.16%	.15%	.14%	.15%
SGA Expense	37,944,000,000	38,039,000,000	39,422,000,000	36,765,000,000	35,465,000,000	36,845,000,000
Net Income	20,081,000,000	-5,176,000,000	13,903,000,000	19,370,000,000	29,450,000,000	12,976,000,000
Operating Cash Flow	41,957,000,000	43,130,000,000	48,668,000,000	43,602,000,000	38,010,000,000	38,442,000,000
Capital Expenditure	16,527,000,000	15,675,000,000	19,635,000,000	21,251,000,000	20,647,000,000	21,516,000,000
EBITDA	56,693,000,000	33,585,000,000	55,107,000,000	61,260,000,000	45,826,000,000	50,569,000,000
Return on Assets %	.04%	- .01%	.03%	.04%	.07%	.03%
Return on Equity %	.12%	- .03%	.08%	.12%	.22%	.11%
Debt to Equity	1.05%	1.088	0.94	0.903	0.894	0.923

CONTACT INFORMATION:

Phone: 210 821-4105   Fax:
Toll-Free:
Address: 208 S. Akard St., Dallas, TX 75202 United States

STOCK TICKER/OTHER:

Stock Ticker: T   Exchange: NYS
Employees: 203,000   Fiscal Year Ends: 12/31
Parent Company:

SALARIES/BONUSES:

Top Exec. Salary: $2,400,000   Bonus: $
Second Exec. Salary: $1,300,000   Bonus: $

OTHER THOUGHTS:

Estimated Female Officers or Directors: 4
Hot Spot for Advancement for Women/Minorities: Y

AT&T Mexico SAU

NAIC Code: 517210

www.att.com.mx

TYPES OF BUSINESS:

Cell Phone Service
Telephone Service
Wireless Telecommunications
Personal Communication Services
Messaging Services
Mobile Television Services
Wireless Broadband
Business Solutions

BRANDS/DIVISIONS/AFFILIATES:

AT&T Inc

GROWTH PLANS/SPECIAL FEATURES:

AT&T Mexico SAU is a wireless telecommunications and personal communication service (PCS) provider, as well as one of the largest telecom providers in Mexico. The company's mobile network is available in 90% of the country, serving approximately 13% of the Mexican wireless market. AT&T Mexico also offers local and long-distance telephone services, messaging services, mobile television and wireless broadband services. Pre-paid plans are available. The firm serves consumers and businesses, and operates as a wholly-owned subsidiary of AT&T, Inc.

CONTACTS: Note: Officers with more than one job title may be intentionally listed here more than once.

Monica Aspe Bernal, CEO
Ricardo B. Salinas Pliego, Pres.
Diego Foyo, Gen. Dir.-Admin.
Francisco Borrego, Gen. Dir.-Legal
Diego Foyo, Gen. Dir.-Oper.
Joaquin Arrangoiz, Dir.-Strategic Svcs.
Tristan Canales, Gen. Dir.-Corp. Comm.
Rodrigo Pliego, Gen. Dir.-Finance
Pedro Padilla, Gen. Dir.
Gustavo Vega, Gen. Dir.-Systems & Telecommunications
Luis J. Echarte, VP-Finance Strategy & Int'l Rel.

FINANCIAL DATA: Note: Data for latest year may not have been available at press time.

In U.S. $	2021	2020	2019	2018	2017	2016
Revenue	2,747,000,000	2,562,000,000	2,869,000,000	2,868,000,000	2,831,000,000	2,472,000,000
R&D Expense						
Operating Income						
Operating Margin %						
SGA Expense						
Net Income	-510,000,000	-587,000,000	-718,000,000	-1,057,000,000	-788,000,000	
Operating Cash Flow						
Capital Expenditure						
EBITDA						
Return on Assets %						
Return on Equity %						
Debt to Equity						

CONTACT INFORMATION:

Phone: 52-55-5109-4400   Fax: 52-55-5109-5925
Toll-Free: 800-333-4321
Address: Montes Urales 460, Colonia Lomas de Chapultepec, Delegacion Miguel Hidalgo, 11000 Mexico

STOCK TICKER/OTHER:

Stock Ticker: Subsidiary
Employees: 18,000
Parent Company: AT&T Inc

Exchange:
Fiscal Year Ends: 12/31

SALARIES/BONUSES:

Top Exec. Salary: $ Bonus: $
Second Exec. Salary: $ Bonus: $

OTHER THOUGHTS:

Estimated Female Officers or Directors:
Hot Spot for Advancement for Women/Minorities:

AT&T Mobility LLC

www.att.com/wireless

NAIC Code: 517210

TYPES OF BUSINESS:
Mobile Phone and Wireless Services
Wireless Data Services
Cell Phone Services

GROWTH PLANS/SPECIAL FEATURES:
AT&T Mobility, LLC, also referred to as AT&T Wireless, is a leading wireless telecommunications service provider predominantly serving the U.S. The wholly-owned subsidiary of AT&T, Inc. provides wireless voice and data services to consumer and wholesale subscribers. AT&T Mobility offers a comprehensive range of nationwide wireless voice and data communications services in a variety of pricing plans, including postpaid and prepaid service plans. The firm provides 5G services in select locations throughout the U.S., which offers seamless connection and ultra-fast speeds at home, business or on-the-go. AT&T also offers 5G phones and devices for 5G connection and compatibility.

BRANDS/DIVISIONS/AFFILIATES:
AT&T Inc
AT&T Wireless

CONTACTS:
Note: Officers with more than one job title may be intentionally listed here more than once.

John T. Stankey, CEO-Corporate
Ralph de la Vega, Pres.

FINANCIAL DATA:
Note: Data for latest year may not have been available at press time.

In U.S. $	2021	2020	2019	2018	2017	2016
Revenue	78,254,000,000	72,564,000,000	71,056,000,000	70,521,000,000	70,259,000,000	72,587,000,000
R&D Expense						
Operating Income						
Operating Margin %						
SGA Expense						
Net Income	23,312,000,000	22,372,000,000	22,321,000,000	21,568,000,000	20,204,000,000	20,743,000,000
Operating Cash Flow						
Capital Expenditure						
EBITDA						
Return on Assets %						
Return on Equity %						
Debt to Equity						

CONTACT INFORMATION:
Phone: 404-236-7895   Fax:
Toll-Free:
Address: 1025 Lenox Park Blvd., Atlanta, GA 30319 United States

STOCK TICKER/OTHER:
Stock Ticker: Subsidiary   Exchange:
Employees: 40,000   Fiscal Year Ends: 12/31
Parent Company: AT&T Inc

SALARIES/BONUSES:
Top Exec. Salary: $ Bonus: $
Second Exec. Salary: $ Bonus: $

OTHER THOUGHTS:
Estimated Female Officers or Directors:
Hot Spot for Advancement for Women/Minorities:

Sales, profits and employees may be estimates. Financial information, benefits and other data can change quickly and may vary from those stated here.

ATN International Inc

NAIC Code: 517110

www.atni.com

TYPES OF BUSINESS:

Wired & Wireless Telecommunications
Long Distance & International Telecom Services
Local Exchange Telecom Services
Communications Transport Services

GROWTH PLANS/SPECIAL FEATURES:

ATN International Inc is a telecommunications and utilities company. The company derives revenues from wireless and wireline communication services, solar power services, and terrestrial and submarine fiber optic services. Atlantic Tele-Network has operations in North America, Bermuda, and the Caribbean. Wireless services are provided in North America, Bermuda, and the Caribbean, whereas wireline services are only provided within North America and the Caribbean. From a product perspective, the majority of revenue stems from the provision of wireless and wireline services. Atlantic Tele-Network is an owner of telecommunications infrastructure, such as terrestrial and submarine fiber optic cables.

BRANDS/DIVISIONS/AFFILIATES:

Choice
Commnet
Geoverse
Fireminds
GTT+
One
Viya
Vibrant Energy

CONTACTS:
Note: Officers with more than one job title may be intentionally listed here more than once.

Michael Prior, CEO
Justin Benincasa, CFO
Brad Martin, COO
Mary Mabey, General Counsel

FINANCIAL DATA:
Note: Data for latest year may not have been available at press time.

In U.S. $	2021	2020	2019	2018	2017	2016
Revenue	602,707,000	455,444,000	438,722,000	451,207,000	481,193,000	457,003,000
R&D Expense						
Operating Income	18,541,000	32,393,000	19,741,000	38,505,000	61,543,000	73,482,000
Operating Margin %	.03%	.07%	.04%	.09%	.13%	.16%
SGA Expense	188,283,000	139,011,000	139,264,000	139,474,000	137,478,000	120,684,000
Net Income	-22,108,000	-14,122,000	-10,806,000	19,815,000	31,488,000	12,101,000
Operating Cash Flow	80,548,000	86,284,000	87,903,000	115,865,000	145,725,000	111,656,000
Capital Expenditure	106,142,000	75,323,000	72,725,000	185,921,000	179,203,000	135,142,000
EBITDA	97,433,000	93,751,000	100,207,000	147,434,000	143,325,000	126,716,000
Return on Assets %	-.02%	-.01%	-.01%	.02%	.03%	.01%
Return on Equity %	-.04%	-.02%	-.02%	.03%	.05%	.02%
Debt to Equity	.75%	0.186	0.205	0.124	0.21	0.213

CONTACT INFORMATION:

Phone: 978 619-1300   Fax:
Toll-Free:
Address: 500 Cummings Ctr., Beverly, MA 01915 United States

STOCK TICKER/OTHER:

Stock Ticker: ATNI   Exchange: NAS
Employees: 1,700   Fiscal Year Ends: 12/31
Parent Company:

SALARIES/BONUSES:

Top Exec. Salary: $661,000   Bonus: $
Second Exec. Salary: $398,000 Bonus: $

OTHER THOUGHTS:

Estimated Female Officers or Directors: 1
Hot Spot for Advancement for Women/Minorities:

Sales, profits and employees may be estimates. Financial information, benefits and other data can change quickly and may vary from those stated here.

Avaya Holdings Corp

www.avaya.com

NAIC Code: 334210

TYPES OF BUSINESS:

Telecommunications Systems
Telecommunications Software
Consulting Services
Networking Systems & Software
Network Maintenance, Management & Security Services
Systems Planning & Integration
Unified Communications Systems

BRANDS/DIVISIONS/AFFILIATES:

Avaya Inc

GROWTH PLANS/SPECIAL FEATURES:

Avaya Holdings Corp provides digital communications products, solutions, and services for businesses. The company has two operating segments namely Products and Solutions, and Services. Products and Solutions offer Unified Communications and Contact Center platforms, applications and devices. It helps to offer an open, extensible development platform so that customers and third parties can easily create custom applications and automated workflows for their unique needs. Whereas Services consists of three business areas: Global Support Services, Enterprise Cloud and Managed Services and Professional Services. The company generates maximum revenue from the Services segment. Geographically, it derives a majority of revenue from the U.S.

CONTACTS: *Note: Officers with more than one job title may be intentionally listed here more than once.*

James Chirico, CEO
Kieran McGrath, CFO
William Watkins, Chairman of the Board
Kevin Speed, Chief Accounting Officer
Shefali Shah, Chief Administrative Officer
Stephen Spears, Executive VP

FINANCIAL DATA: *Note: Data for latest year may not have been available at press time.*

In U.S. $	2021	2020	2019	2018	2017	2016
Revenue	2,973,000,000	2,873,000,000	2,887,000,000	2,887,000,000	3,272,000,000	3,702,000,000
R&D Expense	228,000,000	207,000,000	204,000,000	204,000,000	225,000,000	273,000,000
Operating Income	239,000,000	221,000,000	215,000,000	208,000,000	318,000,000	358,000,000
Operating Margin %	.08%	.08%	.07%	.07%	.10%	.10%
SGA Expense	1,024,000,000	991,000,000	994,000,000	1,001,000,000	1,261,000,000	1,401,000,000
Net Income	-13,000,000	-680,000,000	-671,000,000	-671,000,000	-182,000,000	-730,000,000
Operating Cash Flow	30,000,000	147,000,000	241,000,000	241,000,000	301,000,000	113,000,000
Capital Expenditure	106,000,000	98,000,000	113,000,000	113,000,000	59,000,000	96,000,000
EBITDA	649,000,000	31,000,000	11,000,000	11,000,000	374,000,000	126,000,000
Return on Assets %	.00%	-.10%	-.09%	-.10%	-.04%	-.12%
Return on Equity %	-.04%	-.89%	-.40%			
Debt to Equity	7.44%	12.775	2.377	1.51		

CONTACT INFORMATION:

Phone: 908-953-6000   Fax:
Toll-Free: 866-462-8292
Address: 4655 Great American Pkwy., Santa Clara, CA 95054 United States

STOCK TICKER/OTHER:

Stock Ticker: AVYA
Employees: 8,266
Parent Company:

Exchange: NYS
Fiscal Year Ends: 09/30

SALARIES/BONUSES:

Top Exec. Salary: $600,000   Bonus: $750,000
Second Exec. Salary: $1,250,000   Bonus: $

OTHER THOUGHTS:

Estimated Female Officers or Directors: 1
Hot Spot for Advancement for Women/Minorities:

Sales, profits and employees may be estimates. Financial information, benefits and other data can change quickly and may vary from those stated here.

Aviat Networks Inc

NAIC Code: 334220

www.aviatnetworks.com

TYPES OF BUSINESS:

Wireless Transmission Systems
Network Management Services
Network Operations Centers

BRANDS/DIVISIONS/AFFILIATES:

GROWTH PLANS/SPECIAL FEATURES:

Aviat Networks Inc is a networking solutions provider. It designs, manufactures and sells wireless networking products, solutions, and services to mobile and fixed operators, private network operators, government agencies, transportation and utility companies, public safety agencies and broadcast network operators around the world. The company's product categories include point-to-point microwave and millimeter wave radios that are licensed (subject to local frequency regulatory requirements), lightly-licensed and license-exempt (operating in license-exempt frequencies), and element and network management software. Primarily the firm's concentrations for most of the Sales and Service resources are in the United States, Western and Southern Africa, the Philippines, and the European Union.

CONTACTS: *Note: Officers with more than one job title may be intentionally listed here more than once.*

Peter Smith, CEO
David Gray, CFO
John Mutch, Director
Bryan Tucker, Senior VP, Geographical

FINANCIAL DATA: *Note: Data for latest year may not have been available at press time.*

In U.S. $	2021	2020	2019	2018	2017	2016
Revenue	274,911,000	238,642,000	243,858,000	242,506,000	241,874,000	268,690,000
R&D Expense	21,810,000	19,284,000	21,111,000	19,750,000	18,684,000	20,806,000
Operating Income	24,481,000	7,427,000	2,104,000	2,596,000	-396,000	-24,991,000
Operating Margin %	.09%	.03%	.01%	.01%	.00%	- .09%
SGA Expense	56,324,000	57,985,000	56,055,000	58,157,000	57,184,000	65,902,000
Net Income	110,139,000	257,000	9,738,000	1,845,000	-823,000	-29,907,000
Operating Cash Flow	17,298,000	17,493,000	2,944,000	8,209,000	9,405,000	356,000
Capital Expenditure	2,847,000	4,608,000	5,246,000	6,563,000	4,021,000	1,574,000
EBITDA	27,823,000	8,150,000	6,120,000	6,494,000	5,285,000	-21,791,000
Return on Assets %	.46%	.00%	.06%	.01%	- .01%	- .15%
Return on Equity %	.87%	.00%	.15%	.03%	- .02%	- .44%
Debt to Equity	.02%	0.034				

CONTACT INFORMATION:

Phone: 408 941-7100   Fax: 512-582-4605
Toll-Free:
Address: 200 Parker Dr., Ste. C100A, Austin, TX 78728 United States

STOCK TICKER/OTHER:

Stock Ticker: AVNW   Exchange: NAS
Employees: 687   Fiscal Year Ends: 07/01
Parent Company:

SALARIES/BONUSES:

Top Exec. Salary: $444,231   Bonus: $
Second Exec. Salary: Bonus: $
$315,000

OTHER THOUGHTS:

Estimated Female Officers or Directors: 1
Hot Spot for Advancement for Women/Minorities:

Aware Inc

NAIC Code: 334510

www.aware.com

TYPES OF BUSINESS:

Biometric Systems
Seismic Data Storage Software
ADSL Diagnostics
Medical Imaging Software

BRANDS/DIVISIONS/AFFILIATES:

Nexa

GROWTH PLANS/SPECIAL FEATURES:

Aware Inc is a provider of software and services to the biometrics industry. The company's software products are used in government and commercial biometrics systems to identify or authenticate people. The government applications of biometrics systems include border control, visitor screening, law enforcement, national defense, intelligence, secure credentialing, access control, and background checks. Its commercial applications include user authentication for login to mobile devices, computers, networks, and software programs, user authentication for financial transactions and purchases. Its geographical segments include the United States, Brazil, the United Kingdom, and the Rest of the world.

The company offers employees health and life insurance as well as a 401(k) plan.

CONTACTS: Note: Officers with more than one job title may be intentionally listed here more than once.

Robert Eckel, CEO
David Barcelo, CFO
Brent Johnstone, Chairman of the Board
Mohamed Lazzouni, Chief Technology Officer
Robert Mungovan, Other Executive Officer

FINANCIAL DATA: Note: Data for latest year may not have been available at press time.

In U.S. $	2021	2020	2019	2018	2017	2016
Revenue	16,854,000	11,309,000	12,204,000	16,131,000	15,465,000	21,566,000
R&D Expense	9,259,000	9,093,000	7,928,000	7,105,000	7,769,000	6,938,000
Operating Income	-6,097,000	-9,424,000	-4,178,000	312,000	-475,000	5,099,000
Operating Margin %	-.36%	-.83%	-.34%	.02%	-.03%	.24%
SGA Expense	12,482,000	10,830,000	7,193,000	7,474,000	7,296,000	7,428,000
Net Income	-5,824,000	-7,614,000	-8,340,000	1,233,000	1,001,000	4,103,000
Operating Cash Flow	-6,234,000	-5,274,000	-2,945,000	660,000	3,669,000	3,814,000
Capital Expenditure	27,000	484,000	111,000	206,000	82,000	87,000
EBITDA	-5,410,000	-8,863,000	-3,737,000	755,000	44,000	5,721,000
Return on Assets %	-.12%	-.14%	-.13%	.02%	.02%	.07%
Return on Equity %	-.13%	-.15%	-.14%	.02%	.02%	.07%
Debt to Equity						

CONTACT INFORMATION:

Phone: 781 276-4000   Fax: 781 276-4001
Toll-Free:
Address: 40 Middlesex Turnpike, Bedford, MA 01730 United States

STOCK TICKER/OTHER:

Stock Ticker: AWRE   Exchange: NAS
Employees: 81   Fiscal Year Ends: 12/31
Parent Company:

SALARIES/BONUSES:

Top Exec. Salary: $574,437   Bonus: $
Second Exec. Salary: $275,000   Bonus: $

OTHER THOUGHTS:

Estimated Female Officers or Directors:
Hot Spot for Advancement for Women/Minorities:

Axiata Group Berhad

NAIC Code: 517210

www.axiata.com

TYPES OF BUSINESS:
Mobile Phone Service
Telecommunications Services

GROWTH PLANS/SPECIAL FEATURES:
Axiata Group Bhd is a telecommunications company. It primarily provides mobile and infrastructure services and operates in six main geographic areas: Malaysia, Indonesia, Bangladesh, Sri Lanka, Nepal, and Cambodia. Mobile services are derived through controlling interests in five mobile operators: Celcom in Malaysia, XL in Indonesia, Dialog in Sri Lanka, Robi in Bangladesh, and Smart in Cambodia. The company generates the vast majority of its revenue in Malaysia and Indonesia. It also owns mobile tower and fiber infrastructure and generates infrastructure revenue through its infrastructure company, Edotco.

BRANDS/DIVISIONS/AFFILIATES:
Celcom
XL Axiata
Dialog
Robi
Boost
Aspirasi
Axiata Digital Labs
Edotco

CONTACTS:
Note: Officers with more than one job title may be intentionally listed here more than once.

Dato Izzaddin Idris, CEO
Vivek Sood, CFO
Norlida Azmi, Chief People Officer
Anthony Rodrigo, CIO
Amandeep Singh, Head-Tech.
Tan Gim Boon, Gen. Counsel
Supun Weerasinghe, Chief Strategy Officer
Nik Hasnan Nik Abd Kadir, Chief Internal Auditor
Datin Badrunnisa Mohd Yasin Khan, Chief Talent Officer
Annis Sheikh Mohamad, Head-Corp. Dev.
Suryani Binti Hussein, Sec.
Mohamad Idham Nawawi, Chief Corp. Officer
Ghazzali Shelkh Abdul Khalid, Chmn.

FINANCIAL DATA:
Note: Data for latest year may not have been available at press time.

In U.S. $	2021	2020	2019	2018	2017	2016
Revenue	5,811,884,000	5,430,982,000	5,516,282,000	5,359,762,000	5,475,687,000	4,839,088,000
R&D Expense						
Operating Income	856,856,800	625,574,800	882,204,400	312,128,400	760,581,800	564,401,200
Operating Margin %	.15%	.12%	.16%	.06%	.14%	.12%
SGA Expense	541,748,900	471,003,300	516,048,500	567,085,400	549,567,200	502,791,900
Net Income	183,754,100	81,937,620	327,061,600	-1,068,557,000	204,079,400	113,150,200
Operating Cash Flow						
Capital Expenditure	1,557,816,000	1,152,964,000	1,624,975,000	1,605,233,000	1,124,969,000	1,473,650,000
EBITDA	2,439,379,000	2,341,560,000	2,439,910,000	870,805,800	2,048,186,000	1,781,575,000
Return on Assets %	.01%	.01%	.02%	-.07%	.01%	.01%
Return on Equity %	.05%	.02%	.09%	-.23%	.04%	.02%
Debt to Equity	1.29%	1.285	1.025	0.991	0.649	0.642

CONTACT INFORMATION:
Phone: 60 3-2263-8888   Fax: 60 3-2263-8822
Toll-Free:
Address: 9 Jalan Stesen Sentral 5, Kuala Lumpur Sentral, Kuala Lumpur, 50470 Malaysia

STOCK TICKER/OTHER:
Stock Ticker: AXXTF				Exchange: PINX
Employees: 12,600				Fiscal Year Ends: 12/31
Parent Company:

SALARIES/BONUSES:
Top Exec. Salary: $			Bonus: $
Second Exec. Salary: $		Bonus: $

OTHER THOUGHTS:
Estimated Female Officers or Directors: 3
Hot Spot for Advancement for Women/Minorities: Y

Sales, profits and employees may be estimates. Financial information, benefits and other data can change quickly and may vary from those stated here.

Axtel SAB de CV

www.axtelcorp.mx

NAIC Code: 517110

TYPES OF BUSINESS:

Wired Telecommunications Carriers
Information, Communication Technology (ICT)
Data Center Solutions

BRANDS/DIVISIONS/AFFILIATES:

GROWTH PLANS/SPECIAL FEATURES:

Axtel SAB de CV is a Mexican based information and communication technology company. The company has two operating segments namely Business operating segment offers communication services such as information, data and Internet technologies, managed through the company's network and infrastructure for both multinational companies, as well as for international and national businesses. Its Government operating segment offers communication services such as information, data and Internet technologies, administered through the company's network and infrastructure, for the federal, state and municipal governments. The company generates maximum revenue from the Business segment. Geographically, it derives majority revenue from the Mexico region.

CONTACTS: *Note: Officers with more than one job title may be intentionally listed here more than once.*

Rolando Zubiran Shetler, CEO
Adrian De Los Santos Escobedo, Exec. Dir.-Finance

FINANCIAL DATA: *Note: Data for latest year may not have been available at press time.*

In U.S. $	2021	2020	2019	2018	2017	2016
Revenue	562,705,300	610,455,300	631,583,700	631,823,400	619,749,500	688,582,300
R&D Expense						
Operating Income	31,896,630	38,172,340	41,366,760	24,137,180	5,154,689	31,043,990
Operating Margin %	.06%	.06%	.07%	.04%	.01%	.05%
SGA Expense	244,961,100	267,386,500	288,623,700	296,876,300	291,749,900	363,866,000
Net Income	-39,363,560	17,848,040	-688,665	54,083,280	3,071,599	-177,824,100
Operating Cash Flow	162,621,400	207,891,000	146,077,900	267,336,400	217,122,300	192,566,100
Capital Expenditure	75,714,750	106,014,900	87,054,230	92,423,200	123,866,200	204,824,100
EBITDA	152,831,400	288,135,200	223,924,700	214,126,600	230,326,200	-11,921,340
Return on Assets %	- .04%	.02%	.00%	.04%	.00%	- .13%
Return on Equity %	- .25%	.10%	.00%	.36%	.03%	-1.15%
Debt to Equity	4.47%	3.824	4.178	4.186	7.641	8.535

CONTACT INFORMATION:

Phone: 52 8181140000   Fax:
Toll-Free:
Address: Blvd. Diaz Ordaz Km. 3.33, L-1, Unidad San Pedro, 66215 Mexico

STOCK TICKER/OTHER:

Stock Ticker: AXTLF   Exchange: PINX
Employees: 4,643   Fiscal Year Ends: 12/31
Parent Company:

SALARIES/BONUSES:

Top Exec. Salary: $ Bonus: $
Second Exec. Salary: $ Bonus: $

OTHER THOUGHTS:

Estimated Female Officers or Directors:
Hot Spot for Advancement for Women/Minorities:

Sales, profits and employees may be estimates. Financial information, benefits and other data can change quickly and may vary from those stated here.

Axway Inc

NAIC Code: 511210E

www.axway.com

TYPES OF BUSINESS:

Computer Software, Network Security, Managed Access, Digital ID,
Cybersecurity & Anti-Virus
Email, Firewall & Anti-Spam Software
Secure File Transfer Software
Professional Services
Security Software
Business Processing Management
Tracking Software
Data Integration and Transformation

BRANDS/DIVISIONS/AFFILIATES:

Axway AMPLIFY

CONTACTS: *Note: Officers with more than one job title may be intentionally listed here more than once.*

Patrick Donovan, CEO
Rahim Bhatia, Exec. VP-Product
Cecile Allmacher, CFO
Paul French, CMO
Dominique Fougerat, Exec. VP-Human Resources
Vince Padua, CTO
Rohit Khanna, Sr. VP-Consulting & Professional Svcs.
Pierre Pasquier, Chmn.

GROWTH PLANS/SPECIAL FEATURES:

Axway, Inc. is a leading provider of multi-enterprise solutions
serving more than 11,000 enterprises in 100 countries. Its
Axway AMPLIFY platform unifies a business' employees,
suppliers, partners and developers in order to create a powerful
customer experience network that meets consumer demands.
The platform's capabilities include analytics, application
program interface (API) lifecycle management, app
development, community management and integration. For IT
and architects, Axway's product offers big data integration,
cloud integration, identity federation, identity validation,
mobility and Internet of Things (IoT). For developers, it offers
cross-platform application, development, API development and
DevOps. Businesses and industries served by the firm include
automotive, banking/financial services, healthcare, life
sciences, manufacturing, consumer packaged goods, retail,
U.S. federal government and many more. Axway offers various
types of support, including downloads, documentation, contact
support, how-to videos, webinars, training courses, resource
library, cloud consulting, API management consulting and
managed file transfer consulting. Corporately headquartered in
Arizona, USA, the firm has global headquarters in France,
Singapore and Brazil, and offices throughout North America,
Asia-Pacific, Europe and Latin America.

FINANCIAL DATA: *Note: Data for latest year may not have been available at press time.*

In U.S. $	2021	2020	2019	2018	2017	2016
Revenue	293,923,350	365,076,000	395,928,698	377,074,950	359,119,000	317,227,000
R&D Expense						
Operating Income						
Operating Margin %						
SGA Expense						
Net Income	20,832,462	10,413,100	11,565,164	12,581,700	5,270,600	33,187,100
Operating Cash Flow						
Capital Expenditure						
EBITDA						
Return on Assets %						
Return on Equity %						
Debt to Equity						

CONTACT INFORMATION:

Phone: 480-627-1800   Fax: 480-627-1801
Toll-Free: 877-564-7700
Address: 16220 N. Scottsdale Rd., Ste. 500, Scottsdale, AZ 85254
United States

STOCK TICKER/OTHER:

Stock Ticker: AXW   Exchange: Amsterdam
Employees: 1,885   Fiscal Year Ends: 12/31
Parent Company:

SALARIES/BONUSES:

Top Exec. Salary: $ Bonus: $
Second Exec. Salary: $ Bonus: $

OTHER THOUGHTS:

Estimated Female Officers or Directors:
Hot Spot for Advancement for Women/Minorities:

B Communications Ltd

www.bcommunications.co.il

NAIC Code: 517110

TYPES OF BUSINESS:
Wired Telecommunications Carriers
Telecommunications
Mobile Voice
Internet
Pay TV
Fiber Optic
Broadband

GROWTH PLANS/SPECIAL FEATURES:

B Communications Ltd is a holding company with a single asset, a controlling interest in Bezeq, an Israeli triple play telecommunications company. Bezeq generates revenue through the provision of mobile, broadband, and data. It operates through four business segments: Bezeq, Pelephone Communications Ltd, Bezeq International Ltd, and DBS Satellite Services. The Bezeq segment generates revenue from fixed line communications. Bezeq's other division, Pelephone, derives revenue from the provision of mobile services. The last two segments, Bezeq International and DBS Satellite Services, produce revenue from the provision of Internet services and satellite TV services, respectively. Bezeq is an owner of telecommunications infrastructure, such as fiber networks.

BRANDS/DIVISIONS/AFFILIATES:
Eurocom Group
Internet Gold-Golden Lines Ltd
Israeli Telecommunications Corp Ltd
Bezeq

CONTACTS: *Note: Officers with more than one job title may be intentionally listed here more than once.*

Tomer Raved, CEO
Itzik Tadmor, CFO
Darren Glatt, Chmn.

FINANCIAL DATA: *Note: Data for latest year may not have been available at press time.*

In U.S. $	2021	2020	2019	2018	2017	2016
Revenue	2,667,937,000	2,638,297,000	2,700,602,000	2,819,163,000	2,960,711,000	3,049,935,000
R&D Expense						
Operating Income	535,644,100	535,644,100	432,205,200	425,853,700	517,799,400	538,366,100
Operating Margin %	.20%	.20%	.16%	.15%	.17%	.18%
SGA Expense	792,124,200	781,235,800	819,344,900	841,726,400	878,928,100	904,636,600
Net Income	39,016,420	47,485,100	-257,992,300	-322,414,800	23,591,330	-71,378,880
Operating Cash Flow	854,731,800	970,571,300	878,625,700	1,054,351,000	1,054,653,000	1,047,092,000
Capital Expenditure	511,447,900	453,376,900	446,722,900	485,739,400	462,752,900	428,273,300
EBITDA	1,119,681,000	1,050,419,000	769,137,700	265,251,200	1,103,953,000	1,198,318,000
Return on Assets %	.01%	.01%	-.05%	-.05%	.00%	-.01%
Return on Equity %				-1.45%	.06%	-.21%
Debt to Equity	661.19%			47.118	9.982	9.783

CONTACT INFORMATION:

Phone: 972 3-6796121   Fax: 972-3-6796111
Toll-Free:
Address: 144 Menachem Begin Rd., Tel Aviv, 6492102 Israel

SALARIES/BONUSES:

Top Exec. Salary: $115,000   Bonus: $15,000
Second Exec. Salary: $ Bonus: $

STOCK TICKER/OTHER:

Stock Ticker: BCOMF   Exchange: PINX
Employees: 10,212   Fiscal Year Ends: 12/31
Parent Company: Eurocom Group

OTHER THOUGHTS:

Estimated Female Officers or Directors:
Hot Spot for Advancement for Women/Minorities:

Sales, profits and employees may be estimates. Financial information, benefits and other data can change quickly and may vary from those stated here.

Bangkok Cable Co Ltd

www.bangkokcable.com

NAIC Code: 331420

TYPES OF BUSINESS:

Telecommunications & Electrical Wire Manufacturing
Solar Panel Manufacturing

BRANDS/DIVISIONS/AFFILIATES:

Pongtipa Co Ltd
Indra Industrial Park Co Ltd
Bangkok Solar Power Co Ltd
Bangkok Magnet Wire Co Ltd
Bangkok Cable Ventures Co Ltd
HBC Telecom Co Ltd
Tawana Container Co Ltd
Thai Copper Rod Co Ltd

CONTACTS: *Note: Officers with more than one job title may be intentionally listed here more than once.*

Sompong Nakornsri, CEO
AthikomTongnumtago, Sr. Exec. VP
Kantika Prechaharn, Exec. Dir.
Titinan Nakornsri, Exec. Dir.

GROWTH PLANS/SPECIAL FEATURES:

Bangkok Cable Co., Ltd. (BCC), established in 1964, is a Thai manufacturer of aluminum and copper cables and wires for various applications, primarily electrical and telecommunications systems. Aluminum cables principally consist of overhead power distribution lines for low and medium voltages. Copper conductors comprise both power transmission lines for low to high voltages as well as wiring for buildings, household appliances, industrial equipment and other applications. The firm also manufactures specialty cables designed to resist threats such as fire and vermin. These specialty cables are general purpose conductors used in everything from fire alarm systems to general building wiring. BCC also manufactures several wires and cables specifically for the electronics and telecommunications markets, including coaxial cables, automobile wires, heavy-duty transmission cables and light-duty end point wires. Additionally, the company produces photovoltaic interconnection systems, which include photovoltaic cables, junction boxes and connector compliances. BCC maintains three factories, in Samut Prakan, Ban Pho and Chachoengsao, Thailand, which manufacture copper and aluminum wires and cables. Currently, BCC sells its cables domestically and exports them to countries such as India, Bangladesh, Australia and the Philippines. Cable subsidiaries include, but are not limited to: Pongtipa Co. Ltd., Indra Industrial Park Co. Ltd., Bangkok Solar Power Co. Ltd., Bangkok Magnet Wire Co. Ltd. and Bangkok Cable Ventures Co. Ltd. Joint ventures include HBC Telecom Co. Ltd., Tawana Container Co. Ltd., Thai Copper Rod Co. Ltd., Thai Refrigeration Components Co. Ltd., and Hayakawa Electronics (Thailand) Co. Ltd.

FINANCIAL DATA: *Note: Data for latest year may not have been available at press time.*

In U.S. $	2021	2020	2019	2018	2017	2016
Revenue						
R&D Expense						
Operating Income						
Operating Margin %						
SGA Expense						
Net Income						
Operating Cash Flow						
Capital Expenditure						
EBITDA						
Return on Assets %						
Return on Equity %						
Debt to Equity						

CONTACT INFORMATION:

Phone: 66-2254-4550   Fax: 66-2253-6028
Toll-Free:
Address: 187/1 Rajdamri Rd., Lumpinee, Pathumwan, Bangkok, 10330 Thailand

STOCK TICKER/OTHER:

Stock Ticker: Private   Exchange:
Employees: Fiscal Year Ends:
Parent Company:

SALARIES/BONUSES:

Top Exec. Salary: $ Bonus: $
Second Exec. Salary: $ Bonus: $

OTHER THOUGHTS:

Estimated Female Officers or Directors: 3
Hot Spot for Advancement for Women/Minorities: Y

BCE Inc (Bell Canada Enterprises)

www.bce.ca

NAIC Code: 517110

TYPES OF BUSINESS:

Telecommunication Service Provider
Telecommunications
Media
Wireless
Internet
Television Services
Retail

BRANDS/DIVISIONS/AFFILIATES:

GROWTH PLANS/SPECIAL FEATURES:

BCE is both a wireless and internet service provider, offering wireless, broadband, television, and landline phone services in Canada. It is one of the big three national wireless carriers, with its roughly 10 million customers constituting about 30% of the market. It is also the ILEC (incumbent local exchange carrier-- the legacy telephone provider) throughout much of the eastern half of Canada, including in the most populous Canadian provinces--Ontario and Quebec. Additionally, BCE has a media segment, which holds television, radio, and digital media assets. BCE licenses the Canadian rights to movie channels including HBO, Showtime, and Starz. In 2021, the wireline segment accounted for 54% of total EBITDA, while wireless composed 39%, and media provided the remainder.

CONTACTS: Note: Officers with more than one job title may be intentionally listed here more than once.

Mirko Bibic, CEO
Claire Gillies, Pres., Subsidiary
Glen LeBlanc, CFO
Gordon Nixon, Chairman of the Board
Robert Malcolmson, Chief Legal Officer
Bernard le Duc, Executive VP, Divisional
Michael Cole, Executive VP, Subsidiary
Stephen Howe, Executive VP, Subsidiary
Devorah Lithwick, Other Executive Officer
Rizwan Jamal, President, Subsidiary
Blaik Kirby, President, Subsidiary
Thomas Little, President, Subsidiary
John Watson, President, Subsidiary
Wade Oosterman, President, Subsidiary

FINANCIAL DATA: Note: Data for latest year may not have been available at press time.

In U.S. $	2021	2020	2019	2018	2017	2016
Revenue	18,238,310,000	17,798,090,000	18,505,870,000	18,253,090,000	17,700,090,000	16,892,740,000
R&D Expense						
Operating Income	4,109,823,000	4,046,823,000	4,403,827,000	4,231,936,000	4,288,714,000	4,106,712,000
Operating Margin %	.23%	.23%	.24%	.23%	.24%	.24%
SGA Expense				62,222,920	-59,111,770	
Net Income	2,208,913,000	2,048,689,000	2,481,917,000	2,278,136,000	2,328,692,000	2,357,471,000
Operating Cash Flow	6,228,514,000	6,030,956,000	6,189,624,000	5,743,175,000	5,722,953,000	5,166,835,000
Capital Expenditure	5,381,505,000	3,335,148,000	3,090,923,000	3,132,146,000	3,137,590,000	2,933,033,000
EBITDA	7,503,305,000	6,863,965,000	7,688,419,000	7,039,745,000	6,992,300,000	6,746,520,000
Return on Assets %	.04%	.04%	.05%	.05%	.05%	.06%
Return on Equity %	.15%	.15%	.18%	.17%	.19%	.21%
Debt to Equity	1.45%	1.407	1.313	1.208	1.118	1.135

CONTACT INFORMATION:

Phone: 514 870-8777   Fax: 514 766-5735
Toll-Free: 800-339-6353
Address: 1, carrefour Alexander-Graham Bell, Bldg. A, 7/Fl, Verdun, QC H3E 3B3 Canada

STOCK TICKER/OTHER:

Stock Ticker: BCE   Exchange: NYS
Employees: 49,781   Fiscal Year Ends: 12/31
Parent Company:

SALARIES/BONUSES:

Top Exec. Salary: $1,300,000   Bonus: $
Second Exec. Salary: $900,000   Bonus: $

OTHER THOUGHTS:

Estimated Female Officers or Directors: 4
Hot Spot for Advancement for Women/Minorities: Y

Sales, profits and employees may be estimates. Financial information, benefits and other data can change quickly and may vary from those stated here.

Belden Inc

www.belden.com

NAIC Code: 334417

TYPES OF BUSINESS:

Cable & Wire Connectors Manufacturing
Electronic Products
Broadcasting Equipment
Aerospace & Automotive Electronics
Enclosures

GROWTH PLANS/SPECIAL FEATURES:

Belden Inc provides signal transmission products to distributors, end-users, installers, and original equipment manufacturers. The firm operates in two segments - Enterprise Solutions and Industrial Solutions. The Enterprise Solutions segment is a provider in network infrastructure solutions, as well as cabling and connectivity solutions for commercial audio/video and security applications. The Industrial Solutions segment is a provider of high-performance networking components and machine connectivity products.

BRANDS/DIVISIONS/AFFILIATES:

OTN Systems NV

CONTACTS: Note: Officers with more than one job title may be intentionally listed here more than once.

Roel Vestjens, CEO
Jeremy Parks, CFO
David Aldrich, Chairman of the Board
Douglas Zink, Chief Accounting Officer
Ashish Chand, Executive VP, Divisional
Brian Anderson, General Counsel
Dean McKenna, Senior VP, Divisional
Anshuman Mehrotra, Senior VP, Divisional

FINANCIAL DATA: Note: Data for latest year may not have been available at press time.

In U.S. $	2021	2020	2019	2018	2017	2016
Revenue	2,408,100,000	1,862,716,000	2,131,278,000	2,165,702,000	2,087,185,000	2,356,672,000
R&D Expense	124,660,000	107,296,000	94,360,000	91,552,000	88,748,000	140,519,000
Operating Income	265,021,000	125,410,000	207,207,000	251,867,000	233,641,000	256,014,000
Operating Margin %	.11%	.07%	.10%	.12%	.11%	.11%
SGA Expense	426,335,000	366,188,000	417,329,000	411,352,000	389,743,000	486,403,000
Net Income	63,925,000	-55,162,000	-377,015,000	160,894,000	93,210,000	128,003,000
Operating Cash Flow	272,055,000	173,364,000	276,893,000	289,220,000	255,300,000	314,794,000
Capital Expenditure	94,632,000	90,215,000	110,002,000	97,847,000	64,261,000	53,974,000
EBITDA	238,345,000	233,702,000	347,483,000	439,551,000	330,289,000	367,104,000
Return on Assets %	.02%	-.02%	-.11%	.03%	.02%	.03%
Return on Equity %	.07%	-.06%	-.34%	.09%	.04%	.10%
Debt to Equity	1.60%	2.158	1.558	1.055	1.088	1.109

CONTACT INFORMATION:

Phone: 314 854-8000   Fax: 314 854-8001
Toll-Free: 800-235-3361
Address: 1 N. Brentwood Blvd., 15/Fl, St. Louis, MO 63105 United States

STOCK TICKER/OTHER:

Stock Ticker: BDC   Exchange: NYS
Employees: 7,000   Fiscal Year Ends: 12/31
Parent Company:

SALARIES/BONUSES:

Top Exec. Salary: $385,608   Bonus: $400,000
Second Exec. Salary: Bonus: $
$778,125

OTHER THOUGHTS:

Estimated Female Officers or Directors: 2
Hot Spot for Advancement for Women/Minorities: Y

Belkin International Inc

www.belkin.com

NAIC Code: 334118

TYPES OF BUSINESS:
Keyboards & Mouse Devices, Computer Peripheral Equipment, Manufacturing
Computer & Networking Accessories
Cables
Adapters
USB Devices
Power Supplies
MP3 Accessories
Energy Conservation

BRANDS/DIVISIONS/AFFILIATES:
Hon Hai Precision Industry Co Ltd
Foxconn Interconnect Technology Limited
Belkin
Wemo

CONTACTS: Note: Officers with more than one job title may be intentionally listed here more than once.
Steve Malony, CEO
Mark Reynoso, Pres.
Romain Cholat, Managing Dir.-Europe Region
Gary Tubb, Managing Dir.-U.K. & Ireland & OEM in EMEA
Chet Pipkin, Chmn.

GROWTH PLANS/SPECIAL FEATURES:
Belkin International, Inc. is a leading designer, manufacturer and supplier of accessories for consumer computers, electronics, phones and mobile products. The company's primary brands include Belkin and Wemo. Products include screen protectors, wireless chargers, speakers, headphones, USB accessories, power banks, cables, cases, armbands, in-car devices and accessories, lightning accessories, mobile accessories, smartwatch accessories, tablet keyboards, tablet cases, smart home, Internet of Things (IoT), networking and more. Belkin's solutions span health, work from home, connect from home, charging smart devices, charging mobile devices and phone protection. Products, solutions and services for businesses include docks, hubs, adapters, cables, keyboards, mice, wrist pads, surge protectors, power strips, multi-device charging, desktop accessories, wireless charging, cyber security, classroom accessories, and more. Based in California, USA, Belkin International has worldwide offices throughout the Americas, Europe, Asia, the Middle East, Australia and New Zealand. Belkin International is privately-owned by Foxconn Interconnect Technology Limited, itself a subsidiary of Hon Hai Precision Industry Co., Ltd.

FINANCIAL DATA: Note: Data for latest year may not have been available at press time.

In U.S. $	2021	2020	2019	2018	2017	2016
Revenue	1,915,681,950	1,842,001,875	1,797,075,000	1,711,500,000	1,630,000,000	1,610,000,000
R&D Expense						
Operating Income						
Operating Margin %						
SGA Expense						
Net Income						
Operating Cash Flow						
Capital Expenditure						
EBITDA						
Return on Assets %						
Return on Equity %						
Debt to Equity						

CONTACT INFORMATION:
Phone: 310 751-5100   Fax:
Toll-Free: 800-223-5546
Address: 555 S. Aviation Blvd., Ste. 180, El Segundo, CA 90245-4852 United States

STOCK TICKER/OTHER:
Stock Ticker: Subsidiary   Exchange:
Employees: 1,450   Fiscal Year Ends: 12/31
Parent Company: Hon Hai Precision Industry Co Ltd

SALARIES/BONUSES:
Top Exec. Salary: $ Bonus: $
Second Exec. Salary: $ Bonus: $

OTHER THOUGHTS:
Estimated Female Officers or Directors:
Hot Spot for Advancement for Women/Minorities:

Bell Aliant Inc

NAIC Code: 517110

TYPES OF BUSINESS:

Telecommunications Services
Local, Long-Distance & Cellular Services
System Integration & Software Engineering
Infrastructure Services
IT Consulting & Services
Fiber Optic Cable & Satellite TV Service
Internet Service

BRANDS/DIVISIONS/AFFILIATES:

BCE Inc (Bell Canada Enterprises)
Bell
Bell Aliant

CONTACTS: Note: Officers with more than one job title may be intentionally listed here more than once.

Stephen Wetmore, CEO
John Watson, Director Nominee
Frederick Crooks, Secretary
Charles Hartlen, Senior VP, Subsidiary
Daniel McKeen, Senior VP, Subsidiary
Mary-Ann Bell, Senior VP, Subsidiary
Rod MacGregor, Vice President, Divisional

GROWTH PLANS/SPECIAL FEATURES:

Bell Aliant, Inc. provides regional telecommunications services to residential and business customers in the Atlantic region of Canada, including Ontario and Quebec. The company's communications services are offered under the Bell and Bell Aliant brands, and consist of fiber-based internet protocol television (IPTV) and high-speed internet services, 4G and 5G LTE wireless, home phone and business network and communications services (data hosting, cloud computing). Bell Aliant also provides license, hosting and maintenance services. The firm sells devices and related equipment, including mobile phones, landline phones, television receivers (including wireless and satellite) and smart home devices. Bell Aliant is wholly-owned and privatized by its parent BCE, Inc. (Bell Canada Enterprises).

Bell Aliant offers its employees performance incentives, unit purchase programs, employee discounts and retirement savings plans

FINANCIAL DATA: Note: Data for latest year may not have been available at press time.

In U.S. $	2021	2020	2019	2018	2017	2016
Revenue	3,551,593,500	3,414,993,750	3,252,375,000	3,097,500,000	2,950,000,000	2,972,000,000
R&D Expense						
Operating Income						
Operating Margin %						
SGA Expense						
Net Income						
Operating Cash Flow						
Capital Expenditure						
EBITDA						
Return on Assets %						
Return on Equity %						
Debt to Equity						

CONTACT INFORMATION:

Phone: 902 487-4609   Fax: 902 425-0708
Toll-Free: 877-248-3113
Address: 6 S. Maritime Centre, 1505 Barrington St., Halifax, NS B3J 2W3 Canada

STOCK TICKER/OTHER:

Stock Ticker: Subsidiary   Exchange:
Employees: 5,800   Fiscal Year Ends: 12/31
Parent Company: BCE Inc

SALARIES/BONUSES:

Top Exec. Salary: $ Bonus: $
Second Exec. Salary: $ Bonus: $

OTHER THOUGHTS:

Estimated Female Officers or Directors: 4
Hot Spot for Advancement for Women/Minorities: Y

Bell MTS Inc

www.bellmts.ca

NAIC Code: 517110

TYPES OF BUSINESS:

Telephone Service
Internet Service
Smart Home
Business Voice & Data Connectivity
Digital Television Services
Security Monitoring Services

BRANDS/DIVISIONS/AFFILIATES:

BCE Inc
Fibe

CONTACTS:
Note: Officers with more than one job title may be intentionally listed here more than once.

Ryan Klaussen, VP
Marvin Boakye, Other Executive Officer
Heather Tulk, Other Executive Officer
Paul Beauregard, Other Executive Officer
Kevin Jessiman, Senior VP, Divisional
Patricia Solman, Senior VP, Divisional
Brenda McInnes, Treasurer

GROWTH PLANS/SPECIAL FEATURES:

Bell MTS, Inc. is a full-service telecommunications company that serves the Manitoba region within Canada. The BCE, Inc. subsidiary provides its services and solutions for residential customers, including mobility, internet, TV, smart home, and connected things. Mobile services encompass long-term evolution (LTE)-A and 5G networks, offering internet, streaming and sharing capabilities. Devices can be purchased through Bell MTS, as well as travel protection insurance and a variety of rate plans and pre-paid options. Bell MTS Fibe TV offers an array of channels and movies and shows on-demand. TV features include 4K picture quality, recording, pausing, rewinding, set-top boxes and Crave and Netflix capabilities. Related accessories and hardware are available, as well as satellite TV and a mobile TV app. Internet services is powered by Bell MTS' fiber optic network, providing whole home smart and wi-fi and internet capability for every room of the home. Business solutions include mobility, internet, phone, TV, security, automation and more.

FINANCIAL DATA:
Note: Data for latest year may not have been available at press time.

In U.S. $	2021	2020	2019	2018	2017	2016
Revenue	866,829,600	833,490,000	793,800,000	756,000,000	720,000,000	725,000,000
R&D Expense						
Operating Income						
Operating Margin %						
SGA Expense						
Net Income						
Operating Cash Flow						
Capital Expenditure						
EBITDA						
Return on Assets %						
Return on Equity %						
Debt to Equity						

CONTACT INFORMATION:

Phone: 204 269-4727   Fax: 204 786-4514
Toll-Free: 888-544-5554
Address: 333 Main St., Winnipeg, MB R3C 3V6 Canada

STOCK TICKER/OTHER:

Stock Ticker: Subsidiary   Exchange:
Employees: 3,000   Fiscal Year Ends: 12/31
Parent Company: BCE Inc

SALARIES/BONUSES:

Top Exec. Salary: $ Bonus: $
Second Exec. Salary: $ Bonus: $

OTHER THOUGHTS:

Estimated Female Officers or Directors: 4
Hot Spot for Advancement for Women/Minorities: Y

Sales, profits and employees may be estimates. Financial information, benefits and other data can change quickly and may vary from those stated here.

BenQ Corporation

www.benq.com

NAIC Code: 333316

TYPES OF BUSINESS:

Photographic and Photocopying Equipment Manufacturing
Consumer Electronics
Communication Devices
Device Components
Manufacturing
Product Design
Network Convergence Technologies
Product Design

BRANDS/DIVISIONS/AFFILIATES:

Qisda Corporation
BenQ America Corporation
BenQ Asia Pacific Corporation
BenQ Latin America Corporation
BenQ China
BenQ Europe BV
AQCOLOR
ZOWIE

CONTACTS: Note: Officers with more than one job title may be intentionally listed here more than once.

Conway Lee, CEO
Tony Yang, Pres., BenQ China
Lars Yoder, Pres., BenQ America
Peter Tan, Pres., BenQ Latin America
Adrian Chang, Pres., BenQ Asia Pacific
Adams Lee, Pres., BenQ Europe

GROWTH PLANS/SPECIAL FEATURES:

BenQ Corporation, a subsidiary of Qisda Corporation, is a Taiwan-based producer of consumer electronics communication devices and components. The firm operates worldwide, maintaining an active presence in more than 100 countries. BenQ continuously invests and utilizes most of its resources and expertise in product design, visual display mobile solutions and network convergence technologies These include digital projectors, professional monitors interactive large-format displays and panels, imaging solutions mobile computing devices, speakers, light emitting diode (LED lighting solutions and accessories/attachments. The firm's products and solutions primarily serve the business, education and e-sports industries. BenQ's AQCOLOR program offers BenQ partners innovative and reliable solutions for professional audio/visual products. All Integrators Choice partners receive access to product assortment, deal registration, preferential pricing, a three-year warranty, access to an exclusive partner portal and more. The company's ZOWIE brand is dedicated to the development and refurbishment of professional eSports equipment, including monitors, mouse products, audio systems and accessories BenQ is headquartered in Taiwan, with offices in China, the Netherlands and the U.S. Subsidiaries of the firm include BenQ America Corporation, BenQ Asia Pacific Corporation, BenQ Latin America Corporation, BenQ China and BenQ Europe BV.

FINANCIAL DATA: Note: Data for latest year may not have been available at press time.

In U.S. $	2021	2020	2019	2018	2017	2016
Revenue	3,300,000,000	3,281,118,750	3,365,250,000	3,202,500,000	3,050,000,000	3,000,000,000
R&D Expense						
Operating Income						
Operating Margin %						
SGA Expense						
Net Income						
Operating Cash Flow						
Capital Expenditure						
EBITDA						
Return on Assets %						
Return on Equity %						
Debt to Equity						

CONTACT INFORMATION:

Phone: 886-2-2727-8899   Fax: 886-2-2797-9288
Toll-Free:
Address: 16 Jihu Rd., Neihu, Taipei, 114 Taiwan

STOCK TICKER/OTHER:

Stock Ticker: Subsidiary
Employees: 1,500
Parent Company: Qisda Corporation

Exchange:
Fiscal Year Ends: 12/31

SALARIES/BONUSES:

Top Exec. Salary: $ Bonus: $
Second Exec. Salary: $ Bonus: $

OTHER THOUGHTS:

Estimated Female Officers or Directors:
Hot Spot for Advancement for Women/Minorities:

Bezeq-The Israel Telecommunication Corp Ltd

ir.bezeq.co.il/phoenix.zhtml?c=159870&p=irol-IRHome
NAIC Code: 517110

TYPES OF BUSINESS:

Telecommunications Provider
Domestic & International Phone Service
Cell Phone Service
Broadband Internet & Data Services
Network Infrastructure Installation
Call Center Services
Internet Portal

BRANDS/DIVISIONS/AFFILIATES:

Eurocom Group
Internet Gold-Golden Lines Ltd
B Communications

GROWTH PLANS/SPECIAL FEATURES:

Bezeq The Israeli Telecommunication Corp Ltd is a triple-play telecommunications company. The company generates revenue through the provision of mobile, broadband, and data. It operates through four business segments: Bezeq, Pelephone, Bezeq International, and DBS Satellite Services. The Bezeq segment generates revenue from fixed-line communications and contributes the majority of overall company revenue. Pelephone derives revenue from the provision of mobile services. Bezeq International and DBS Satellite Services produce revenue from the provision of Internet services and satellite TV services, respectively. The company owns telecommunications infrastructure, such as fiber networks. It generates the vast majority of its revenue in Israel.

CONTACTS: *Note: Officers with more than one job title may be intentionally listed here more than once.*

Dudu Mizrahi, CEO
Tobi Fischbein, CFO
Gil Sharon, CEO-Telephone Communications Ltd
Ron Eilon, CEO-Yes
Gil Sharon, Chmn.
Isaac Benbenisti, CEO-Bezeq International

FINANCIAL DATA: *Note: Data for latest year may not have been available at press time.*

In U.S. $	2021	2020	2019	2018	2017	2016
Revenue		2,638,297,000	2,700,602,000	2,819,163,000	2,960,711,000	3,049,935,000
R&D Expense						
Operating Income		538,366,100	534,131,800	494,812,900	652,390,800	698,666,200
Operating Margin %		.20%	.20%	.18%	.22%	.23%
SGA Expense		139,733,200	147,899,500	167,861,400	372,319,500	401,052,500
Net Income		240,752,500	-361,128,800	-330,278,600	373,529,300	376,251,400
Operating Cash Flow		973,898,300	884,372,300	1,062,215,000	1,066,146,000	1,066,449,000
Capital Expenditure		453,376,900	446,722,900	485,739,400	462,752,900	428,273,300
EBITDA		975,108,100	783,050,500	473,338,800	1,146,296,000	1,199,528,000
Return on Assets %		.06%	-.08%	-.07%	.08%	.08%
Return on Equity %				-.85%	.57%	.54%
Debt to Equity				24.753	4.771	4.143

CONTACT INFORMATION:

Phone: 972 3626-2600   Fax: 972 3626-2609
Toll-Free:
Address: Azrieli Center 2, 132 Menachem Begin Ave., 27/Fl., Tel Aviv, 61620 Israel

STOCK TICKER/OTHER:

Stock Ticker: BZQIF   Exchange: PINX
Employees: 5,256   Fiscal Year Ends: 12/31
Parent Company: Eurocom Group

SALARIES/BONUSES:

Top Exec. Salary: $ Bonus: $
Second Exec. Salary: $ Bonus: $

OTHER THOUGHTS:

Estimated Female Officers or Directors: 3
Hot Spot for Advancement for Women/Minorities: Y

Bharat Sanchar Nigam Limited (BSNL)   www.bsnl.co.in

NAIC Code: 517110

TYPES OF BUSINESS:

Telecommunications Service Provider
Local & Long-Distance Telephone Service
Mobile Service
Telecommunications
Wireline Services
Internet and Broadband
Fiber-to-the-Home
Data Center Services

BRANDS/DIVISIONS/AFFILIATES:

GROWTH PLANS/SPECIAL FEATURES:

Bharat Sanchar Nigam Limited (BSNL) is a government-owned telecommunications company in India. BSNL offers wireline services, GSM mobile services up through 4G and value-added solutions, internet services, broadband, fiber-to-the-home (FTTH), Wi-Fi, data center services, enterprise data services, national long-distance services and international long-distance services. The company is part of a global network, providing access to countries worldwide via voice, data and video platforms. BSNL specializes in the planning, installation, integration and maintenance of switches and transmission networks. In addition, the firm has an ISO 9000 certified telecom training institute. BSNL serves both consumers and businesses.

CONTACTS: *Note: Officers with more than one job title may be intentionally listed here more than once.*

Yojana Das, Dir.-Finance
Rajesh Wadhwa, Dir.-Oper.
Pravin Kumar Purwar, Chmn.

FINANCIAL DATA: *Note: Data for latest year may not have been available at press time.*

In U.S. $	2021	2020	2019	2018	2017	2016
Revenue	2,534,384,690	2,512,930,000	2,777,080,000	3,854,000,000	4,939,320,000	5,116,828,267
R&D Expense						
Operating Income						
Operating Margin %						
SGA Expense						
Net Income	-1,014,171,205	-2,060,100,000	-2,147,140,000	-1,229,130,000	-7,507,960,000	-603,077,989
Operating Cash Flow						
Capital Expenditure						
EBITDA						
Return on Assets %						
Return on Equity %						
Debt to Equity						

CONTACT INFORMATION:

Phone: 91 11-2373-4064   Fax: 91-11-2373-0392
Toll-Free:
Address: Harish Chandra Mathur Ln., New Delhi, 110 001 India

STOCK TICKER/OTHER:

Stock Ticker: Government-Owned   Exchange:
Employees: 64,536   Fiscal Year Ends: 03/31
Parent Company:

SALARIES/BONUSES:

Top Exec. Salary: $ Bonus: $
Second Exec. Salary: $ Bonus: $

OTHER THOUGHTS:

Estimated Female Officers or Directors:
Hot Spot for Advancement for Women/Minorities:

Bharti Airtel Limited

NAIC Code: 517210

www.airtel.in

TYPES OF BUSINESS:
Cell Phone Service
Telecommunications
Mobile Services
Digital TV
Telephone Services
Broadband Services
ICT Services

BRANDS/DIVISIONS/AFFILIATES:
Bharti Enterprises Limited
Singapore Telecommunications Limited

CONTACTS: *Note: Officers with more than one job title may be intentionally listed here more than once.*
Gopal Vittal, CEO
Soumen Ray, CFO
Jyoti Pawar, Dir.-Legal & Regulatory
Raghunath Mandava, Dir.-Customer Experience
Manoj Kohli, CEO-International
Jagbir Singh, Dir.-Network Svcs. Group
Sunil Bharti Mittal, Chmn.

GROWTH PLANS/SPECIAL FEATURES:
Bharti Airtel Limited is an Indian telecommunications service provider. With customers and operations spanning India, as well as countries across south Asia and Africa, Airtel is one of India's largest telecommunications firms. The company is divided into four business units: mobile services, home services, Airtel business and digital TV services. The mobile services unit offers prepaid and postpaid voice, data (2G, 3G, 4G LTE and 4G+) and other services, including news updates, mobile internet and email. The home services business unit offers fixed-line telephone and broadband services across pan-India. Product offerings include voice connectivity over fixed-line and high-speed broadband on copper and fiber with speeds up to 100 megabits per second. The Airtel business unit provides information and communication technology (ICT) services in India, and offers a portfolio of services to enterprises, governments, carriers and small/medium businesses. Services include voice, data and video, as well as network integration, data centers, managed services, enterprise mobility applications, digital media, international toll-free services and short-message-system (SMS) hubbing. The digital TV services business unit offers standard and high-definition (HD) digital TV services with 3D capabilities and Dolby surround sound. This division offers hundreds of channels, including HD channels, international channels and interactive services. The firm is 64%-owned by Bharti Enterprises Limited and 36%-owned by Singapore Telecommunications Limited (SingTel).

Bharti Airtel offers its employees sabbaticals, onsite day care and fitness center, concierge service and maternity leave.

FINANCIAL DATA: *Note: Data for latest year may not have been available at press time.*

In U.S. $	2021	2020	2019	2018	2017	2016
Revenue	13,800,838,370	11,458,722,139	11,652,878,750	12,741,896,310	14,709,500,000	14,791,625,789
R&D Expense						
Operating Income						
Operating Margin %						
SGA Expense						
Net Income	-1,685,126,652	-4,075,670,815	242,554,500	335,658,538	585,452,000	840,076,842
Operating Cash Flow						
Capital Expenditure						
EBITDA						
Return on Assets %						
Return on Equity %						
Debt to Equity						

CONTACT INFORMATION:
Phone: 91 11-4666-6100   Fax: 91-11-4166-6137
Toll-Free:
Address: 1 Nelson Mandela Rd., Vasant Kunj, Phase II, New Delhi, 110 070 India

STOCK TICKER/OTHER:
Stock Ticker: 532454   Exchange: Bombay
Employees: 14,000   Fiscal Year Ends: 03/31
Parent Company: Bharti Enterprises

SALARIES/BONUSES:
Top Exec. Salary: $ Bonus: $
Second Exec. Salary: $ Bonus: $

OTHER THOUGHTS:
Estimated Female Officers or Directors: 4
Hot Spot for Advancement for Women/Minorities: Y

Sales, profits and employees may be estimates. Financial information, benefits and other data can change quickly and may vary from those stated here.

Black Box Corporation

NAIC Code: 423430

www.blackbox.com

TYPES OF BUSINESS:

Networking Products, Distribution
Technology Solutions and Services
Audio Visual Solutions and Services
Custom Networking Products
Cybersecurity Solutions
Digital Workplace Solutions and Services
Infrastructure Integration Solutions and Services
Product Repair and Management

BRANDS/DIVISIONS/AFFILIATES:

Essar Group
AGC Networks Ltd

CONTACTS: *Note: Officers with more than one job title may be intentionally listed here more than once.*

Sanjeev Verma, CEO
Deepak Kumar Bansal, CFO
Bikram Sahoo, CTO
Ronald Basso, Executive VP, Divisional

GROWTH PLANS/SPECIAL FEATURES:

Black Box Corporation is a leading technology solutions provider. The firm helps customers build, manage, optimize and secure their IT infrastructure. As a technology solutions provider, Black Box connects people to devices through a range of products and services. Solutions include 5G, private long-term evolution (LTE), audio visual integrations, connected buildings, customer experience, cybersecurity, data center, digital workplace, enterprise networking, global multi-site deployment, managed unified-communications-as-a-service (UCaaS), reimagined workplace, structured cabling and infrastructure integrations. Black Box services span carrier services, consulting and design, field services, managed services, project management, refurbishment and repair, and related support services. Products by Black Box include audio video, cables, Internet of Things (IoT), keyboard/video/mouse (KVM) and networking, as well as specific products per industry, application and/or technology. Industries Black Box specializes in include broadcast, data center, education, finance, government, healthcare, manufacturing/industrial military/defense, retail, transportation and work spaces. Based in the U.S., Black Box has global locations throughout the Americas, Europe, Asia and Asia Pacific. Black Box is privately-owned by AGC Networks Ltd., part of the Essar Group.

FINANCIAL DATA: *Note: Data for latest year may not have been available at press time.*

In U.S. $	2021	2020	2019	2018	2017	2016
Revenue	909,346,364	874,371,504	832,734,766	774,636,992	855,731,008	912,654,976
R&D Expense						
Operating Income						
Operating Margin %						
SGA Expense						
Net Income			-105,099,750	-100,095,000	-7,051,000	-171,102,000
Operating Cash Flow						
Capital Expenditure						
EBITDA						
Return on Assets %						
Return on Equity %						
Debt to Equity						

CONTACT INFORMATION:

Phone: 724 746-5500   Fax:
Toll-Free: 800-316-7107
Address: 1000 Park Dr., Lawrence, PA 15055 United States

STOCK TICKER/OTHER:

Stock Ticker: Private   Exchange:
Employees: 3,264   Fiscal Year Ends: 03/31
Parent Company: Essar Group

SALARIES/BONUSES:

Top Exec. Salary: $ Bonus: $
Second Exec. Salary: $ Bonus: $

OTHER THOUGHTS:

Estimated Female Officers or Directors: 2
Hot Spot for Advancement for Women/Minorities:

Sales, profits and employees may be estimates. Financial information, benefits and other data can change quickly and may vary from those stated here.

BlackBerry Limited

NAIC Code: 511210C

www.blackberry.com

TYPES OF BUSINESS:

Computer Software: Telecom, Communications & VOIP
Security Software & Services
Internet of Things
Machine Learning
Endpoint Security
Artificial Intelligence
Intelligent Vehicle Solutions

BRANDS/DIVISIONS/AFFILIATES:

BlackBerry Spark
BlackBerry Protect
BlackBerry Persona
BlackBerry QNX
BlackBerry IVY
BlackBerry AtHoc
BlackBerry Alert
Cylance Inc

CONTACTS: Note: Officers with more than one job title may be intentionally listed here more than once.

John Chen, CEO
Steve Rai, CFO
Randall Cook, Chief Legal Officer
Mark Wilson, Chief Marketing Officer
Thomas Eacobacci, COO
Sai Yuen Ho, Executive VP, Divisional
Mattias Eriksson, General Manager, Divisional
Nita White-Ivy, Other Executive Officer
Marjorie Dickman, Other Executive Officer
John Giamatteo, President, Divisional

GROWTH PLANS/SPECIAL FEATURES:

BlackBerry, once known for being the world's largest smartphone manufacturer, is now exclusively a software provider with a stated goal of end-to-end secure communication for enterprises. The firm provides endpoint management and protection to enterprises, specializing in regulated industries like government, as well as embedded software to the automotive, medical, and industrial markets.

BlackBerry's employee benefits vary by location, but can include comprehensive health benefits, reimbursement programs, stock purchase and savings plans, as well as a variety of employee assistance programs.

FINANCIAL DATA: Note: Data for latest year may not have been available at press time.

In U.S. $	2021	2020	2019	2018	2017	2016
Revenue	893,000,000	1,040,000,000	904,000,000	932,000,000	1,309,000,000	
R&D Expense	215,000,000	259,000,000	219,000,000	239,000,000	306,000,000	
Operating Income	-98,000,000	-183,000,000	-66,000,000	-198,000,000	-428,000,000	
Operating Margin %	-.11%	-.18%	-.07%	-.21%	-.33%	
SGA Expense	344,000,000	493,000,000	409,000,000	476,000,000	553,000,000	
Net Income	-1,104,000,000	-152,000,000	93,000,000	405,000,000	-1,206,000,000	
Operating Cash Flow	82,000,000	26,000,000	100,000,000	704,000,000	-224,000,000	
Capital Expenditure	44,000,000	44,000,000	49,000,000	45,000,000	69,000,000	
EBITDA	100,000,000	29,000,000	83,000,000	-21,000,000	-189,000,000	
Return on Assets %	-.33%	-.04%	.02%	.11%	-.27%	
Return on Equity %	-.55%	-.06%	.04%	.18%	-.46%	
Debt to Equity	.54%	0.047	0.252	0.312	0.287	

CONTACT INFORMATION:

Phone: 519 888-7465   Fax: 519 888-7884
Toll-Free:
Address: 2200 University Ave. E., Waterloo, ON N2K 0A7 Canada

STOCK TICKER/OTHER:

Stock Ticker: BB   Exchange: NYS
Employees: 3,325   Fiscal Year Ends: 02/28
Parent Company:

SALARIES/BONUSES:

Top Exec. Salary: $1,000,000   Bonus: $2,000,000
Second Exec. Salary: $568,976   Bonus: $114,993

OTHER THOUGHTS:

Estimated Female Officers or Directors:
Hot Spot for Advancement for Women/Minorities: Y

Sales, profits and employees may be estimates. Financial information, benefits and other data can change quickly and may vary from those stated here.

Blonder Tongue Laboratories Inc

www.blondertongue.com

NAIC Code: 334220

TYPES OF BUSINESS:

Digital Video Electronics Manufacturing
Telephone Products
Networking & Broadband Equipment
Receivers & Antennas
Cable TV Equipment
Microwave Products
Technology Development

GROWTH PLANS/SPECIAL FEATURES:

Blonder Tongue Laboratories Inc is a technology development and manufacturing company that provides television signal encoding, transcoding, digital transport, and broadband product solutions in the United States. The company serves the multi-dwelling unit market, the lodging/hospitality market, and the institutional market, including hospitals, prisons, and schools, primarily throughout the United States and Canada.

BRANDS/DIVISIONS/AFFILIATES:

RL Drake Holdings LLC
EdgeQAM Collection

CONTACTS: *Note: Officers with more than one job title may be intentionally listed here more than once.*

Edward Grauch, CEO
Eric Skolnik, CFO
Steven Shea, Chairman of the Board
Ronald Alterio, Chief Technology Officer
Robert Palle, Director
Allen Horvath, Senior VP, Divisional

FINANCIAL DATA: *Note: Data for latest year may not have been available at press time.*

In U.S. $	2021	2020	2019	2018	2017	2016
Revenue	15,754,000	16,379,000	19,842,000	21,707,000	23,283,000	22,506,000
R&D Expense						
Operating Income						
Operating Margin %						
SGA Expense						
Net Income	84,000	-7,474,000	-742,000	-1,339,000	-384,000	-1,195,000
Operating Cash Flow						
Capital Expenditure						
EBITDA						
Return on Assets %						
Return on Equity %						
Debt to Equity						

CONTACT INFORMATION:

Phone: 732 679-4000   Fax: 732 679-4353
Toll-Free: 800-523-6049
Address: 1 Jake Brown Rd., Old Bridge, NJ 08857 United States

STOCK TICKER/OTHER:

Stock Ticker: BDR   Exchange: ASE
Employees: 85   Fiscal Year Ends: 12/31
Parent Company:

SALARIES/BONUSES:

Top Exec. Salary: $ Bonus: $
Second Exec. Salary: $ Bonus: $

OTHER THOUGHTS:

Estimated Female Officers or Directors: 1
Hot Spot for Advancement for Women/Minorities:

Bogen Communications LLC

www.bogen.com

NAIC Code: 334210

TYPES OF BUSINESS:

Sound Processing Systems

GROWTH PLANS/SPECIAL FEATURES:

Bogen Communications International develops computerized telecommunication peripherals and sound-processing products. These products include audio amplifiers and related equipment for sound-system applications, such as telephone paging, intercommunications and administrative communication systems for schools, correctional facilities and other institutions.

BRANDS/DIVISIONS/AFFILIATES:

CONTACTS: Note: Officers with more than one job title may be intentionally listed here more than once.

Jonathan G. Guss, CEO

FINANCIAL DATA: Note: Data for latest year may not have been available at press time.

In U.S. $	2021	2020	2019	2018	2017	2016
Revenue						
R&D Expense						
Operating Income						
Operating Margin %						
SGA Expense						
Net Income						
Operating Cash Flow						
Capital Expenditure						
EBITDA						
Return on Assets %						
Return on Equity %						
Debt to Equity						

CONTACT INFORMATION:

Phone: 201 934-8500   Fax: 201 934-9832
Toll-Free: 800 999-2809
Address: 1200 MacArthur Blvd., Ste. 304, Mahwah, NJ 07430 United States

STOCK TICKER/OTHER:

Stock Ticker: Private
Employees: 175
Parent Company:

Exchange:
Fiscal Year Ends: 12/31

SALARIES/BONUSES:

Top Exec. Salary: $ Bonus: $
Second Exec. Salary: $ Bonus: $

OTHER THOUGHTS:

Estimated Female Officers or Directors:
Hot Spot for Advancement for Women/Minorities:

Sales, profits and employees may be estimates. Financial information, benefits and other data can change quickly and may vary from those stated here.

Boingo Wireless Inc

NAIC Code: 517110

TYPES OF BUSINESS:

Wi-Fi Internet Access
Next-Generation Wireless Connectivity
Neutral Host Networks
5G
Distributed Antenna Systems
Small Cell Deployment
Private Networks
Digital Services and Solutions

BRANDS/DIVISIONS/AFFILIATES:

DigitalBridge Group Inc

CONTACTS: *Note: Officers with more than one job title may be intentionally listed here more than once.*

Mike Finley, CEO
Peter Hovenier, CFO
Dawn Callahan, CMO
Derek Peterson, CTO
Michael Zeto, Chief Commercial Officer

GROWTH PLANS/SPECIAL FEATURES:

Boingo Wireless, Inc. provides next-generation wireless connectivity through neutral host networks. The firm's solutions connect people and things 24/7 with cellular, Wi-Fi and shared-spectrum technologies. Boingo powers 5G connectivity using a combination of licensed, unlicensed and shared spectrum. Other wireless solutions include distributed antenna system (DAS) deployment, small cell deployment, private networks and more. Products and services include network monetization, customizable digital services, passpoint and roaming solutions, unlimited free Wi-Fi at more than 1 million hotspots worldwide, managed wireless services, a single Wi-Fi network for venues, Edge and cloud computing, Internet of Things (IoT), digital transformation, and advertising solutions. Industries served by Boingo Wireless include airports, transportation, sports, entertainment, military, government, commercial real estate, healthcare, manufacturing and logistics. Boingo operates as a subsidiary of DigitalBridge Group, Inc.

Boingo Wireless offers its employees health benefits, life and AD&D insurance, a 401(k) savings plan, flexible spending accounts, training and development programs, tuition reimbursement and company programs and perks.

FINANCIAL DATA: *Note: Data for latest year may not have been available at press time.*

In U.S. $	2021	2020	2019	2018	2017	2016
Revenue	246,896,000	237,400,000	263,790,000	250,820,992	204,368,992	159,344,000
R&D Expense						
Operating Income						
Operating Margin %						
SGA Expense						
Net Income		-17,100,000	-10,296,000	-1,220,000	-19,366,000	-27,331,000
Operating Cash Flow						
Capital Expenditure						
EBITDA						
Return on Assets %						
Return on Equity %						
Debt to Equity						

CONTACT INFORMATION:

Phone: 310 586-5180   Fax: 310 586-4060
Toll-Free: 800-880-4117
Address: 10960 Wilshire Blvd., Fl. 23, Los Angeles, CA 90024 United States

STOCK TICKER/OTHER:

Stock Ticker: Subsidiary   Exchange:
Employees: 411   Fiscal Year Ends: 12/31
Parent Company: DigitalBridge Group Inc

SALARIES/BONUSES:

Top Exec. Salary: $ Bonus: $
Second Exec. Salary: $ Bonus: $

OTHER THOUGHTS:

Estimated Female Officers or Directors: 1
Hot Spot for Advancement for Women/Minorities:

Bouygues SA

NAIC Code: 237130

www.bouygues.com

TYPES OF BUSINESS:

Construction & Telecommunications
Construction
Road Building
Property Development
Precasting
Cellular Phone Service
Media Operation
Research & Development

BRANDS/DIVISIONS/AFFILIATES:

Bouygues Construction
Bouygues Immobilier
Colas
Buoygues Telecom
Bbox
TF1
TMC

GROWTH PLANS/SPECIAL FEATURES:

Bouygues is a French conglomerate made up of a disparate range of assets: a construction business, a TV business, and a telecom business. It is one of the biggest construction companies in France and Europe with construction sales of around EUR 25 billion-EUR 30 billion and one of the four telecom operators in France, with both mobile and fixed operations and EUR 6 billion in revenue. It is also the owner of TF1, one of the main media and TV companies in France.

CONTACTS: *Note: Officers with more than one job title may be intentionally listed here more than once.*

Olivier Roussat, CEO
Pascal Grange, CFO
Jean-Claude Tostivin, Sr. VP-Admin.
Jean Francois Guillemin, Corp. Sec.
Olivier Bouygues, Deputy CEO
Philippe Marien, Chmn.-Bourgues Telecom
Martin Bouygues, Chmn.

FINANCIAL DATA: *Note: Data for latest year may not have been available at press time.*

In U.S. $	2021	2020	2019	2018	2017	2016
Revenue	38,331,870,000	35,379,660,000	38,761,190,000	36,446,330,000	33,573,660,000	32,395,830,000
R&D Expense						
Operating Income	2,180,253,000	1,715,241,000	1,846,791,000	2,290,388,000	1,885,542,000	1,435,826,000
Operating Margin %	.06%	.05%	.05%	.06%	.06%	.04%
SGA Expense	8,784,238,000	7,741,021,000	8,107,116,000	7,638,025,000	7,471,803,000	6,997,614,000
Net Income	1,147,233,000	709,755,100	1,207,399,000	1,333,850,000	1,103,384,000	746,466,500
Operating Cash Flow	3,646,672,000	3,473,313,000	3,439,661,000	2,634,048,000	2,099,692,000	2,465,787,000
Capital Expenditure	2,586,119,000	2,700,332,000	1,916,135,000	2,389,305,000	1,969,162,000	2,206,767,000
EBITDA	4,469,621,000	3,727,234,000	4,309,518,000	4,342,151,000	3,412,127,000	3,101,099,000
Return on Assets %	.03%	.02%	.03%	.03%	.03%	.02%
Return on Equity %	.10%	.07%	.12%	.14%	.13%	.09%
Debt to Equity	.65%	0.665	0.547	0.661	0.641	0.759

CONTACT INFORMATION:

Phone: 33-1-44-20-10-00   Fax: 33-1-30-60-4861
Toll-Free:
Address: 32 Ave. Hoche, Paris, 75008 France

STOCK TICKER/OTHER:

Stock Ticker: BOUYY   Exchange: PINX
Employees: 130,500   Fiscal Year Ends: 12/31
Parent Company:

SALARIES/BONUSES:

Top Exec. Salary: $ Bonus: $
Second Exec. Salary: $ Bonus: $

OTHER THOUGHTS:

Estimated Female Officers or Directors: 5
Hot Spot for Advancement for Women/Minorities: Y

Broadcom Inc

NAIC Code: 335313

www.broadcom.com

TYPES OF BUSINESS:

Electrical Switches, Sensors, MEMS, Optomechanicals
Semiconductors
Connectivity Technology
Wireless Applications
Optical Products
Mainframe Software
Enterprise Software
Security Software

BRANDS/DIVISIONS/AFFILIATES:

AppNeta Inc

GROWTH PLANS/SPECIAL FEATURES:

Broadcom--the combined entity of Broadcom and Avago--
boasts a highly diverse product portfolio across an array of end
markets. Avago focused primarily on radio frequency filters and
amplifiers used in high-end smartphones, such as the Apple
iPhone and Samsung Galaxy devices, in addition to an
assortment of solutions for wired infrastructure, enterprise
storage, and industrial end markets. Legacy Broadcom
targeted networking semiconductors, such as switch and
physical layer chips, broadband products (such as television
set-top box processors), and connectivity chips that handle
standards such as Wi-Fi and Bluetooth. The company has
acquired Brocade, CA Technologies, Symantec's enterprise
security business, and has a pending deal to acquire VMware
to bolster its offerings in software.

CONTACTS:
*Note: Officers with more than one job title may be
intentionally listed here more than once.*

Hock Tan, CEO
Kirsten Spears, CFO
Henry Samueli, Chairman of the Board
Mark Brazeal, Chief Legal Officer
Charles Kawwas, COO
Thomas Krause, President, Divisional

FINANCIAL DATA:
Note: Data for latest year may not have been available at press time.

In U.S. $	2021	2020	2019	2018	2017	2016
Revenue	27,450,000,000	23,888,000,000	22,597,000,000	20,848,000,000	17,636,000,000	13,240,000,000
R&D Expense	4,854,000,000	4,968,000,000	4,696,000,000	3,768,000,000	3,302,000,000	2,674,000,000
Operating Income	8,667,000,000	4,212,000,000	4,180,000,000	5,368,000,000	2,654,000,000	587,000,000
Operating Margin %	.32%	.18%	.18%	.26%	.15%	.04%
SGA Expense	1,347,000,000	1,935,000,000	1,709,000,000	1,056,000,000	789,000,000	806,000,000
Net Income	6,736,000,000	2,960,000,000	2,724,000,000	12,259,000,000	1,784,000,000	-1,739,000,000
Operating Cash Flow	13,764,000,000	12,061,000,000	9,697,000,000	8,880,000,000	6,551,000,000	3,411,000,000
Capital Expenditure	443,000,000	463,000,000	432,000,000	635,000,000	1,069,000,000	723,000,000
EBITDA	14,691,000,000	11,125,000,000	9,478,000,000	9,254,000,000	7,016,000,000	2,520,000,000
Return on Assets %	.08%	.04%	.05%	.23%	.03%	-.06%
Return on Equity %	.26%	.11%	.10%	.52%	.09%	-.15%
Debt to Equity	1.58%	1.685	1.203	0.656	0.859	0.698

CONTACT INFORMATION:

Phone: 408-433-8000   Fax:
Toll-Free:
Address: 1320 Ridder Park Dr., San Jose, CA 95131-2313 United States

STOCK TICKER/OTHER:

Stock Ticker: AVGO   Exchange: NAS
Employees: 20,000   Fiscal Year Ends: 10/31
Parent Company:

SALARIES/BONUSES:

Top Exec. Salary: $1,200,000   Bonus: $
Second Exec. Salary: Bonus: $
$678,462

OTHER THOUGHTS:

Estimated Female Officers or Directors: 5
Hot Spot for Advancement for Women/Minorities: Y

BT Global Services plc

www.globalservices.bt.com

NAIC Code: 517110

TYPES OF BUSINESS:

Telecommunications Equipment-Networking
Network Software
Managed Network Services
Consulting Services
Broadcast Transmission Products

BRANDS/DIVISIONS/AFFILIATES:

BT Group plc

GROWTH PLANS/SPECIAL FEATURES:

BT Global Services plc is the international business services and information technology division of BT Group plc, one of the leading telecommunications carriers in the U.K. The firm provides global multinational organizations with security, cloud and networking services. BT Global helps clients move to the cloud, integrates collaboration into part of their workforce culture, improves their contact centers, transforms their customer service and simplifies their network. The company's technologies span the areas of security, cloud infrastructure, cloud collaboration, cloud contact center and networking.

Employees of BT Group receive benefits including retirement plans, employee discounts and flexible work hours.

CONTACTS:
Note: Officers with more than one job title may be intentionally listed here more than once.

Bas Burger, CEO

FINANCIAL DATA:
Note: Data for latest year may not have been available at press time.

In U.S. $	2021	2020	2019	2018	2017	2016
Revenue	137,908,418	143,008,000	150,086,000	152,233,000	158,112,000	159,520,000
R&D Expense						
Operating Income						
Operating Margin %						
SGA Expense						
Net Income	551,271	809,236	-359,440	-1,071,990	-473,229	-3,304,560
Operating Cash Flow						
Capital Expenditure						
EBITDA						
Return on Assets %						
Return on Equity %						
Debt to Equity						

CONTACT INFORMATION:

Phone: 44-20-7356-5000   Fax: 44-20-7356-5520
Toll-Free:
Address: 81 Newgate St., BT Centre, London, EC1A 7AJ United Kingdom

STOCK TICKER/OTHER:

Stock Ticker: Subsidiary
Employees: 16,800
Parent Company: BT Group plc

Exchange:
Fiscal Year Ends: 03/31

SALARIES/BONUSES:

Top Exec. Salary: $ Bonus: $
Second Exec. Salary: $ Bonus: $

OTHER THOUGHTS:

Estimated Female Officers or Directors:
Hot Spot for Advancement for Women/Minorities: Y

BT Group plc
NAIC Code: 517110

TYPES OF BUSINESS:
Telecommunications Services
Communications Networks
Telecommunications Equipment Distribution
International Broadband Networks
Consulting Services
Internet Service Provider
Local & Long-Distance Phone Service
Networking Services

BRANDS/DIVISIONS/AFFILIATES:
British Telecommunications plc
BT
EE
Plusnet
Openreach

GROWTH PLANS/SPECIAL FEATURES:
BT Group is the incumbent telecommunications operator in the U.K. It is the owner of Openreach, the largest fixed-line network in the U.K. Openreach operates as a separate entity due to regulatory requirements but is still 100% owned by BT. Since 2020 Openreach has accelerated its fiber investments and expects to have 90% of the U.K. covered with FTTH by 2026. In the consumer division, BT has approximately 30% market share in broadband and mobile services. The enterprise segment serves large corporations while the global services division provides communications consultancy services.

Most BT employees in the UK can receive broadband, BT TV, BT Sport and BT Mobile products either for free or at discounted rates. The firm also offers a retirement savings scheme and optional health benefits.

CONTACTS: Note: Officers with more than one job title may be intentionally listed here more than once.
Philip Jansen, CEO
Simon Lowth, CFO
Dan Fitz, General Counsel
Jan du Plessis, Chmn.
Stephen Yeo, CEO-South East Asia

FINANCIAL DATA: Note: Data for latest year may not have been available at press time.

In U.S. $	2021	2020	2019	2018	2017	2016
Revenue	25,777,950,000	27,680,090,000	28,312,120,000	28,668,620,000	29,078,290,000	
R&D Expense						
Operating Income	3,841,738,000	4,617,578,000	4,610,328,000	4,795,224,000	4,972,870,000	
Operating Margin %	.15%	.17%	.16%	.17%	.17%	
SGA Expense	308,160,800	366,167,600	389,128,600	383,086,200	441,092,900	
Net Income	1,778,873,000	2,095,494,000	2,609,095,000	2,455,619,000	2,305,768,000	
Operating Cash Flow	7,206,129,000	7,578,339,000	5,143,264,000	5,954,150,000	7,461,117,000	
Capital Expenditure	5,925,147,000	4,960,785,000	4,444,767,000	4,062,889,000	3,800,650,000	
EBITDA	8,317,925,000	9,120,351,000	8,410,978,000	8,303,423,000	8,152,364,000	
Return on Assets %	.03%	.03%	.05%	.05%	.04%	
Return on Equity %	.11%	.14%	.22%	.22%	.21%	
Debt to Equity	1.81%	1.506	1.453	1.21	1.209	

CONTACT INFORMATION:
Phone: 44 2073565000   Fax: 44 2073565520
Toll-Free:
Address: 81 Newgate St., London, EC1A 7AJ United Kingdom

STOCK TICKER/OTHER:
Stock Ticker: BTGOF   Exchange: PINX
Employees: 100,475   Fiscal Year Ends: 03/31
Parent Company:

SALARIES/BONUSES:
Top Exec. Salary: $1,329,321   Bonus: $1,595,185
Second Exec. Salary: Bonus: $1,067,082
$888,228

OTHER THOUGHTS:
Estimated Female Officers or Directors: 1
Hot Spot for Advancement for Women/Minorities: Y

CalAmp Corp

www.calamp.com

NAIC Code: 334220

TYPES OF BUSINESS:

Microwave Communications Equipment
Wireless Broadband Access Systems
Satellite Products

GROWTH PLANS/SPECIAL FEATURES:

CalAmp Corp provides wireless communications solutions applications to customers. The company's products, software, and application services are sold into a broad array of market verticals, including transportation, government, construction, and automotive to customers in the United States; Latin America; Europe the Middle East, and Africa; Asia-Pacific; and others. Its business activities are organized into its Telematics Systems and Software & Subscription Services. The Telematics Systems segment offers a series of Mobile Resource Management ('MRM') telematics products and applications. The software & Subscription Services segment offers cloud-based application enablement and telematics service platforms.

BRANDS/DIVISIONS/AFFILIATES:

CONTACTS: Note: Officers with more than one job title may be intentionally listed here more than once.

Jeffery Gardner, CEO
Kurtis Binder, CFO
Amal Johnson, Chairman of the Board
Arym Diamond, Other Executive Officer
Anand Rau, Senior VP, Divisional

FINANCIAL DATA: Note: Data for latest year may not have been available at press time.

In U.S. $	2021	2020	2019	2018	2017	2016
Revenue	308,587,000	321,773,000	311,538,000	365,912,000	351,102,000	
R&D Expense	25,811,000	26,993,000	25,541,000	25,761,000	22,005,000	
Operating Income	-3,466,000	-4,229,000	35,776,000	7,955,000	123,000	
Operating Margin %	- .01%	- .01%	.11%	.02%	.00%	
SGA Expense	95,279,000	96,858,000	60,630,000	102,185,000	106,163,000	
Net Income	-56,309,000	-79,304,000	18,398,000	16,617,000	-7,904,000	
Operating Cash Flow	28,585,000	11,544,000	47,740,000	66,894,000	25,796,000	
Capital Expenditure	11,356,000	21,301,000	10,399,000	8,339,000	7,962,000	
EBITDA	16,893,000	12,840,000	53,359,000	61,946,000	25,182,000	
Return on Assets %	- .12%	- .14%	.03%	.04%	- .02%	
Return on Equity %	- .48%	- .46%	.09%	.09%	- .04%	
Debt to Equity	2.10%	1.46	1.342	0.776	0.899	

CONTACT INFORMATION:

Phone: 949-600-5600   Fax: 805-856-3857
Toll-Free:
Address: 15635 Alton Pkwy, Ste. 250, Irvine, CA 92618 United States

STOCK TICKER/OTHER:

Stock Ticker: CAMP   Exchange: NAS
Employees: 887   Fiscal Year Ends: 02/28
Parent Company:

SALARIES/BONUSES:

Top Exec. Salary: $515,000   Bonus: $
Second Exec. Salary: Bonus: $
$385,654

OTHER THOUGHTS:

Estimated Female Officers or Directors: 1
Hot Spot for Advancement for Women/Minorities:

Calero-MDSL

NAIC Code: 511210Q

TYPES OF BUSINESS:

Financial Software for the Telecommunications Industry
Software
Expense Management
Telecom Solutions
Mobility Solutions
Software-as-a-Service Solutions

BRANDS/DIVISIONS/AFFILIATES:

CONTACTS: *Note: Officers with more than one job title may be intentionally listed here more than once.*

Scott Gilbert, CEO
Andrew Taylor, Pres.
Brian Brady, CFO
David Bliss, Exec. VP-Product Mgmt..
Kris Sleeper, VP-Human Resources
James Jones, Dir.-R&D
Simon Mendoza, CTO
Kristie Shanks, Dir.-Prod. Management & Quality Assurance
Brian Martin, Pres., Telesoft Recovery Corp.
Robert Sullivan, Exec. VP-Telesoft Recovery Corp.
Joan Lara, Dir.-Managed Services
Devin Gentry, Dir.-Implementation
Patrick Mulvehill, Exec. VP-Oper.

GROWTH PLANS/SPECIAL FEATURES:

Calero-MDSL is a software provider, with a focus on expense management for the technology industry. The firm's solutions are designed to offer clarity, control, compliance and cost savings. Solutions are grouped into three categories: telecom, mobility and software-as-a-service (SaaS). The telecom solutions category provides an easy-to-use platform that enables order management, usage analysis, call accounting and expense management for telecom environments, spanning a wide range of applications such as inventory tracking, automating expenses and gaining insight. The mobility solution category offers managed mobility services, which centralizes enterprise mobility across global carriers from a single platform. This division support customer's teams with resources for all mobile hardware, software and carrier-related issues. It prepares and delivers devices for immediate use, including pre-loading of apps, configuration and customized packaging. Calero-MDSL verifies device eligibility and places orders with carriers around the world to keep mobile inventory up-to-date, which also offering a centralized view of everything. The SaaS solution category provides expense management across all unified communications spend, including subscriptions, variable teleconferencing and call usage. Turn-key integrations are provided for enabling end-to-end invoice, subscription (phone, webinar, rooms, etc.), call accounting, contract and renewals management. Based in the U.S., Calero-MDSL has offices worldwide, including North America, Europe and Asia.

FINANCIAL DATA: *Note: Data for latest year may not have been available at press time.*

In U.S. $	2021	2020	2019	2018	2017	2016
Revenue						
R&D Expense						
Operating Income						
Operating Margin %						
SGA Expense						
Net Income						
Operating Cash Flow						
Capital Expenditure						
EBITDA						
Return on Assets %						
Return on Equity %						
Debt to Equity						

CONTACT INFORMATION:

Phone: 866-769-5992   Fax:
Toll-Free:
Address: 1040 University Ave., Ste. 200, Rochester, NY 14607 United States

STOCK TICKER/OTHER:

Stock Ticker: Private
Employees: 140
Parent Company:

Exchange:
Fiscal Year Ends: 11/30

SALARIES/BONUSES:

Top Exec. Salary: $ Bonus: $
Second Exec. Salary: $ Bonus: $

OTHER THOUGHTS:

Estimated Female Officers or Directors: 2
Hot Spot for Advancement for Women/Minorities:

Calient Technologies Inc

www.calient.net

NAIC Code: 334210

TYPES OF BUSINESS:

Fiber-Optic Switches
Photonic Switches & Software
MEMS Technologies
Professional Services

BRANDS/DIVISIONS/AFFILIATES:

Suzhou Chunxing Precision Mechanical Co Ltd
S Series Optical Circuit Switch
LightConnect Fabric Manager
Edge/640 Optical Circuit Switch
Calient Optical Components

CONTACTS: *Note: Officers with more than one job title may be intentionally listed here more than once.*

Arjun Gupta, CEO
Kevin Walsh, Pres.-Oper.
Rick Santos, CFO
Jitender K. Miglani, VP-Eng.
Kevin Welsh, Sr. VP-Oper.
John Reynolds, Head-Bus. Dev. & OEM Partnerships
Shifu Yuan, Co-CTO
Chris Lee, Co-CTO
Daniel Tardent, VP-Mktg.
Erik Leonard, VP-Sales, Svc. Providers & Digital Entertainment

GROWTH PLANS/SPECIAL FEATURES:

Calient Technologies, Inc. manufactures adaptive photonic switching products and software for global service provider, cloud computing, content delivery and government network clients. Products include the S Series Optical Circuit Switch, a family of scalable optical circuit switches with port densities ranging from 160 to 320 ports. These switches are suited for data center, cloud computing, service provider and government applications. Calient's OEM (original equipment manufacturer) subsystem solutions build on the S Series Optical Circuit Switch family, aiming to bring pure-photonoic layer optical circuit switching to new markets. The subsystem solutions provide white label optical switching system solutions for OEM applications, and can also be integrated into partner router, switching, next generation ROADM, optical transport and cloud computing platforms. The company's LightConnect Fabric Manager provides software management of the dynamic optical layer in Optical Circuit Switch-powered LightConnect Fabric networks. Core functions of the fabric manager include a topology manager, which maintains a database of the data center network topology and a cross-connect manager. The Edge/640 Optical Circuit Switch is a variant of the S-Series and built for reconfiguring connectivity between network or device domains. Subsidiary Calient Optical Components manufactures MEMS (microelectromechanical systems) silicon semiconductor systems on micro- to nano-scale wafers for the biotechnology and nanotechnology industries. Its 3D MEMs switches allow data center operators to dramatically improve utilization rates of expensive compute resources. Partners of Calient include Spirent Communications, ADARA Networks, Phoenix Datacom, Cumulus Networks, Signa Solutions Nordic Oy and Plexxi, Inc.

FINANCIAL DATA: *Note: Data for latest year may not have been available at press time.*

In U.S. $	2021	2020	2019	2018	2017	2016
Revenue						
R&D Expense						
Operating Income						
Operating Margin %						
SGA Expense						
Net Income						
Operating Cash Flow						
Capital Expenditure						
EBITDA						
Return on Assets %						
Return on Equity %						
Debt to Equity						

CONTACT INFORMATION:

Phone: 805-695-4800   Fax:
Toll-Free:
Address: 25 Castilian Dr., Goleta, CA 93117 United States

STOCK TICKER/OTHER:

Stock Ticker: Subsidiary   Exchange:
Employees: 80   Fiscal Year Ends: 12/31
Parent Company: Suzhou Chunxing Precision Mechanical Co.Ltd

SALARIES/BONUSES:

Top Exec. Salary: $ Bonus: $
Second Exec. Salary: $ Bonus: $

OTHER THOUGHTS:

Estimated Female Officers or Directors:
Hot Spot for Advancement for Women/Minorities:

Calix Inc

NAIC Code: 334210A

www.calix.com

TYPES OF BUSINESS:

Network Access Equipment
Cloud
Software
Real-Time Data
Insights
Customer Acquisition
Technologies
Professional Services

BRANDS/DIVISIONS/AFFILIATES:

GROWTH PLANS/SPECIAL FEATURES:

Calix cloud and software platforms enable service providers of all types and sizes to innovate and transform. The company's customers utilize the real-time data and insights from Calix platforms to simplify their business and deliver experiences that excite their subscribers. The resulting growth in subscriber acquisition, loyalty and revenue creates more value for their businesses and communities. This is the Calix mission; to enable communications service providers of all sizes to simplify, excite and grow.

CONTACTS: *Note: Officers with more than one job title may be intentionally listed here more than once.*

Carl Russo, CEO
Cory Sindelar, CFO
Donald Listwin, Chairman of the Board
Michael Weening, COO

FINANCIAL DATA: *Note: Data for latest year may not have been available at press time.*

In U.S. $	2021	2020	2019	2018	2017	2016
Revenue	679,394,000	541,239,000	424,330,000	441,320,000	510,367,000	458,787,000
R&D Expense	101,747,000	85,258,000	81,184,000	89,963,000	127,541,000	106,869,000
Operating Income	73,152,000	43,132,000	-12,927,000	-19,513,000	-77,307,000	-32,619,000
Operating Margin %	.11%	.08%	- .03%	- .04%	- .15%	- .07%
SGA Expense	181,688,000	138,629,000	119,668,000	126,932,000	122,656,000	125,267,000
Net Income	238,378,000	33,484,000	-17,694,000	-19,298,000	-83,032,000	-27,402,000
Operating Cash Flow	56,793,000	51,409,000	4,654,000	3,560,000	-62,772,000	24,419,000
Capital Expenditure	10,463,000	7,819,000	13,353,000	10,426,000	8,026,000	9,839,000
EBITDA	88,068,000	49,587,000	-5,258,000	-8,949,000	-66,316,000	-18,495,000
Return on Assets %	.41%	.09%	- .06%	- .06%	- .26%	- .08%
Return on Equity %	.56%	.15%	- .12%	- .13%	- .46%	- .12%
Debt to Equity	.02%	0.046	0.093			

CONTACT INFORMATION:

Phone: 408-514-3000   Fax:
Toll-Free:
Address: 2777 Orchard Pkwy, San Jose, CA 95131 United States

STOCK TICKER/OTHER:

Stock Ticker: CALX   Exchange: NYS
Employees: 763   Fiscal Year Ends: 12/31
Parent Company:

SALARIES/BONUSES:

Top Exec. Salary: $483,179   Bonus: $900,000
Second Exec. Salary: Bonus: $
$500,000

OTHER THOUGHTS:

Estimated Female Officers or Directors: 2
Hot Spot for Advancement for Women/Minorities:

Celestica Inc

NAIC Code: 334418

www.celestica.com

TYPES OF BUSINESS:

Contract Electronics Manufacturing
Product Design
Engineering and Design
Distribution Services

BRANDS/DIVISIONS/AFFILIATES:

GROWTH PLANS/SPECIAL FEATURES:

Celestica Inc offers supply chain solutions. The firm operates in two segments: Advanced Technology Solutions (ATS) and Connectivity & Cloud Solutions (CCS). ATS segment consists of the ATS end market and is comprised of A&D, Industrial, Energy, HealthTech, and Capital Equipment businesses. Capital Equipment business is comprised of our semiconductor, display, and power & signal distribution equipment businesses. CCS segment that derives majority revenue consists of Communications and Enterprise end markets.

CONTACTS: Note: Officers with more than one job title may be intentionally listed here more than once.

Robert Mionis, CEO
Mandeep Chawla, CFO
Michael Wilson, Chairman of the Board
Todd Cooper, COO
John Lawless, President, Divisional
Jason Phillips, President, Divisional

FINANCIAL DATA: Note: Data for latest year may not have been available at press time.

In U.S. $	2021	2020	2019	2018	2017	2016
Revenue	5,634,700,000	5,748,100,000	5,888,300,000	6,633,200,000	6,142,700,000	6,046,600,000
R&D Expense	38,400,000	29,900,000	28,400,000	28,800,000	26,200,000	24,900,000
Operating Income	178,000,000	151,400,000	95,300,000	167,300,000	178,300,000	184,200,000
Operating Margin %	.03%	.03%	.02%	.03%	.03%	.03%
SGA Expense	245,100,000	230,700,000	231,400,000	219,000,000	205,100,000	211,100,000
Net Income	103,900,000	60,600,000	70,300,000	98,900,000	105,500,000	138,300,000
Operating Cash Flow	226,800,000	239,600,000	345,000,000	33,100,000	127,000,000	173,300,000
Capital Expenditure	52,200,000	52,800,000	80,500,000	82,200,000	102,600,000	64,100,000
EBITDA	294,000,000	252,600,000	284,700,000	195,400,000	219,700,000	248,600,000
Return on Assets %	.02%	.02%	.02%	.03%	.04%	.05%
Return on Equity %	.07%	.04%	.05%	.07%	.08%	.12%
Debt to Equity	.51%	0.345	0.412	0.488	0.122	0.15

CONTACT INFORMATION:

Phone: 416 448-2211   Fax: 416 448-4810
Toll-Free: 888-899-9998
Address: 844 Don Mills Rd., Toronto, ON M3C 1V7 Canada

STOCK TICKER/OTHER:

Stock Ticker: CLS   Exchange: NYS
Employees: 20,550   Fiscal Year Ends: 12/31
Parent Company:

SALARIES/BONUSES:

Top Exec. Salary: $950,000   Bonus: $
Second Exec. Salary: $538,356   Bonus: $

OTHER THOUGHTS:

Estimated Female Officers or Directors: 4
Hot Spot for Advancement for Women/Minorities: Y

Sales, profits and employees may be estimates. Financial information, benefits and other data can change quickly and may vary from those stated here.

Cellcom Israel Ltd

NAIC Code: 517210

www.cellcom.co.il

TYPES OF BUSINESS:

Cell Phone Service
Virtual Private Network Provider
Internet Service Provider

GROWTH PLANS/SPECIAL FEATURES:

Cellcom Israel Ltd is the largest wireless provider in Israel and holds about one third of the market. The company offers basic voice services and data services such as Web browsing and music downloads. Cellcom is also launching a next-generation wireless service that supports higher data capacity. The company went public in February 2007, and its largest shareholder, Discount Investment, owns about 47% of its shares.

BRANDS/DIVISIONS/AFFILIATES:

Cellcom Fixed Line Communications LP

CONTACTS: *Note: Officers with more than one job title may be intentionally listed here more than once.*

Avi Gabbay, CEO
Liat Menahemi-Stadler, Gen. Legal Counsel
Eli Nir, Chief Internal Audit
Amos Maor, VP-Sales & Svcs.
Keren Shtevy, VP-Bus. Customers
Itamar Bartov, VP-Exec. & Regulatory Affairs
Galit Tuchterman, Spokeswoman

FINANCIAL DATA: *Note: Data for latest year may not have been available at press time.*

In U.S. $	2021	2020	2019	2018	2017	2016
Revenue		1,111,817,000	1,121,495,000	1,115,446,000	1,170,795,000	1,217,978,000
R&D Expense						
Operating Income		-6,956,417		22,381,510	86,501,530	93,760,400
Operating Margin %		- .01%		.02%	.07%	.08%
SGA Expense		275,232,100	199,014,000	218,371,000	226,537,200	254,967,800
Net Income		-51,416,990	-32,362,460	-18,752,080	33,874,720	44,763,030
Operating Cash Flow		300,335,700	313,341,200	232,888,700	234,098,500	236,215,700
Capital Expenditure		150,924,000	168,466,300	179,354,600	176,330,000	111,302,700
EBITDA		271,300,300	284,910,600	205,063,100	268,275,700	265,251,200
Return on Assets %		- .02%	- .02%	- .01%	.02%	.02%
Return on Equity %		- .09%	- .06%	- .04%	.08%	.12%
Debt to Equity		1.718	1.774	1.937	1.964	2.425

CONTACT INFORMATION:

Phone: 972 529990052   Fax:
Toll-Free:
Address: 10 Hagavish St., Netanya, 4250708 Israel

STOCK TICKER/OTHER:

Stock Ticker: CELJF   Exchange: PINX
Employees: 3,358   Fiscal Year Ends: 12/31
Parent Company:

SALARIES/BONUSES:

Top Exec. Salary: $1,087,261   Bonus: $
Second Exec. Salary: Bonus: $
$589,168

OTHER THOUGHTS:

Estimated Female Officers or Directors: 6
Hot Spot for Advancement for Women/Minorities: Y

Sales, profits and employees may be estimates. Financial information, benefits and other data can change quickly and may vary from those stated here.

Certicom Corp

www.certicom.com

NAIC Code: 511210E

TYPES OF BUSINESS:

Computer Software: Network Security, Managed Access, Digital ID,
Cybersecurity & Anti-Virus
Cryptography Software
Technology Solutions
Elliptic Curve Cryptography
Key Management Solutions
Internet of Things
Product Protection Solutions
Semiconductor and Automotive Chip Solutions

BRANDS/DIVISIONS/AFFILIATES:

BlackBerry Limited
BlackBerry Certicom

CONTACTS: *Note: Officers with more than one job title may be intentionally listed here more than once.*

John Chen, CEO

GROWTH PLANS/SPECIAL FEATURES:

Certicom Corp. is a subsidiary of BlackBerry Limited and thus operates as BlackBerry Certicom. The firm is part of BlackBerry's technology solutions business unit, and has specialized in elliptic curve cryptography (ECC) since 1985. Other specialties of BlackBerry Certicom include: key management and public key infrastructure (PKI) solutions; ZigBee Smart Energy Certification Authority operation services, which secures over 120 million related devices; key management solutions for automotive gateways; vehicle-to-anything communication systems based on ECC technology innovations; and quantum-resistant code signing and key management servers. BlackBerry Certicom's applied cryptography and key management capabilities help customers protect the integrity of their silicon chips and devices from the point of manufacturing through the device lifecycle. The company's solutions are primarily used to prevent product counterfeiting, re-manufacturing and rogue network access; and used to protect next-generation connected vehicles, critical infrastructure and IoT deployments.

FINANCIAL DATA: *Note: Data for latest year may not have been available at press time.*

In U.S. $	2021	2020	2019	2018	2017	2016
Revenue						
R&D Expense						
Operating Income						
Operating Margin %						
SGA Expense						
Net Income						
Operating Cash Flow						
Capital Expenditure						
EBITDA						
Return on Assets %						
Return on Equity %						
Debt to Equity						

CONTACT INFORMATION:

Phone: 905-507-4220   Fax: 905-507-4230
Toll-Free: 800-561-6100
Address: 4701 Tahoe Blvd., Fl. 5, Mississauga, ON L4W 0B5 Canada

STOCK TICKER/OTHER:

Stock Ticker: Subsidiary   Exchange:
Employees: 209   Fiscal Year Ends: 04/30
Parent Company: BlackBerry Limited

SALARIES/BONUSES:

Top Exec. Salary: $ Bonus: $
Second Exec. Salary: $ Bonus: $

OTHER THOUGHTS:

Estimated Female Officers or Directors:
Hot Spot for Advancement for Women/Minorities:

Sales, profits and employees may be estimates. Financial information, benefits and other data can change quickly and may vary from those stated here.

Ceske Radiokomunikace as

NAIC Code: 517110

www.cra.cz

TYPES OF BUSINESS:

Telephone Services
Television Infrastructure
Radio Infrastructure
Internet Infrastructure
Cloud
Streaming
Data Center
Telecommunications

BRANDS/DIVISIONS/AFFILIATES:

Cordiant Digital Infrastructure Limited

GROWTH PLANS/SPECIAL FEATURES:

Ceske Radiokomunikace a.s. (CRA) is a provider of television, radio and internet infrastructure throughout the Czech Republic. CRA's products and services include cloud, streaming, audio, over-the-top (OTT), data centers, telecommunications, infrastructure, Internet of Things (IoT), television broadcasting and radio broadcasting. The company also offers General Data Protection Regulation (GDPR) data security and cybernetic security solutions. CRA operates as a wholly-owned subsidiary of Cordiant Digital Infrastructure Limited.

CONTACTS: *Note: Officers with more than one job title may be intentionally listed here more than once.*

Milos Mastnik, CEO
Michal Omelka, CFO
Pavel Kos, CTO
Pavla Tloustova, Legal & Regulatory Dir.
Marie Fianova, Dir.-Comm.

FINANCIAL DATA: *Note: Data for latest year may not have been available at press time.*

In U.S. $	2021	2020	2019	2018	2017	2016
Revenue	92,702,610	89,137,125	84,892,500	80,850,000	77,000,000	76,000,000
R&D Expense						
Operating Income						
Operating Margin %						
SGA Expense						
Net Income						
Operating Cash Flow						
Capital Expenditure						
EBITDA						
Return on Assets %						
Return on Equity %						
Debt to Equity						

CONTACT INFORMATION:

Phone: 420 2 4241 1111   Fax: 420 2 4241 7595
Toll-Free:
Address: Skokanska 2117/1, Praha 6 SkokanskÃ¡ 2117/1 169 00, Prague, 160 00 Czech Republic

STOCK TICKER/OTHER:

Stock Ticker: Private   Exchange:
Employees: 230   Fiscal Year Ends: 12/31
Parent Company: Cordiant Digital Infrastructure Limited

SALARIES/BONUSES:

Top Exec. Salary: $ Bonus: $
Second Exec. Salary: $ Bonus: $

OTHER THOUGHTS:

Estimated Female Officers or Directors: 3
Hot Spot for Advancement for Women/Minorities: Y

Channell Commercial Corporation

www.channell.com

NAIC Code: 334210

TYPES OF BUSINESS:
Telecommunications Infrastructure Hardware
Thermoplastic & Metal Enclosures
Fiber-Optic Cable Management Systems
Termination & Connection Products
Splice Closures
Terminals
Splitters
Pedestals

BRANDS/DIVISIONS/AFFILIATES:

GROWTH PLANS/SPECIAL FEATURES:

Channell Commercial Corporation, founded in 1922, manufactures and supplies fiber optic solutions, thermoplastic pedestals and grade level vaults for outside plant construction needs. The company specializes in end-to-end, full network solutions. Channell serves the fiber optic, telecommunication, broadband, water, power and utility industries. Channell's products include splice closures, splitters, terminals, pedestals, vaults, grade level boxes, cabinets, covers and self-lock protection systems. Channell's Rockwall, Texas facility is a 280,000-square-foot state-of-the-art plant featuring high-automation robotics and end-to-end material processing. The firm's other manufacturing facilities are located in Canada, the U.K. (which serves Europe, the Middle East and Africa) and Australia (which also services Asia and the Pacific Rim). Channell is headquartered in Texas, USA.

CONTACTS: *Note: Officers with more than one job title may be intentionally listed here more than once.*
William H. Channell, Jr., Pres.
Jacqueline M. Channell, Sec.
Greg Balla, Managing Dir.-Australasia

FINANCIAL DATA: *Note: Data for latest year may not have been available at press time.*

In U.S. $	2021	2020	2019	2018	2017	2016
Revenue						
R&D Expense						
Operating Income						
Operating Margin %						
SGA Expense						
Net Income						
Operating Cash Flow						
Capital Expenditure						
EBITDA						
Return on Assets %						
Return on Equity %						
Debt to Equity						

CONTACT INFORMATION:
Phone: 214-304-7800   Fax:
Toll-Free: 800-423-1863
Address: 1700 Justin Rd., Rockwall, TX 75087 United States

STOCK TICKER/OTHER:
Stock Ticker: Private   Exchange:
Employees: 200   Fiscal Year Ends: 12/31
Parent Company:

SALARIES/BONUSES:
Top Exec. Salary: $ Bonus: $
Second Exec. Salary: $ Bonus: $

OTHER THOUGHTS:
Estimated Female Officers or Directors: 1
Hot Spot for Advancement for Women/Minorities:

Charter Communications Inc

NAIC Code: 517110

corporate.charter.com

TYPES OF BUSINESS:

Cable TV Service
Internet Access
Advanced Broadband Cable Services
Telephony Services
Voice Over Internet Protocol

BRANDS/DIVISIONS/AFFILIATES:

Spectrum
Spectrum TV
Spectrum Internet Gig
Spectrum Voice
Spectrum Mobile
Spectrum Enterprise Solutions
Spectrum Community Solutions
Spectrum Reach

GROWTH PLANS/SPECIAL FEATURES:

Charter is the product of the 2016 merger of three cable companies, each with a decades-long history in the business: Legacy Charter, Time Warner Cable, and Bright House Networks. The firm now holds networks capable of providing television, internet access, and phone services to roughly 54 million U.S. homes and businesses, around 40% of the country. Across this footprint, Charter serves 29 million residential and 2 million commercial customer accounts under the Spectrum brand, making it the second-largest U.S. cable company behind Comcast. The firm also owns, in whole or in part, sports and news networks, including Spectrum SportsNet (long-term local rights to Los Angeles Lakers games), SportsNet LA (Los Angeles Dodgers), SportsNet New York (New York Mets), and Spectrum News NY1.

Charter Communications offers its employees comprehensive benefits and retirement options.

CONTACTS: *Note: Officers with more than one job title may be intentionally listed here more than once.*

Thomas Rutledge, CEO
Catherine Bohigian, Exec. VP, Divisional
Jessica Fischer, CFO
Kevin Howard, Chief Accounting Officer
Jonathan Hargis, Chief Marketing Officer
Stephanie Mitchko-Beale, Chief Technology Officer
Richard DiGeronimo, Chief Technology Officer
Christopher Winfrey, COO
Tom Montemagno, Executive VP, Divisional
Charles Fisher, Executive VP, Divisional
Cliff Hagan, Executive VP, Divisional
Thomas Adams, Executive VP, Divisional
James Nuzzo, Executive VP, Divisional
Adam Ray, Executive VP, Divisional
Magesh Srinivasan, Executive VP, Divisional

FINANCIAL DATA: *Note: Data for latest year may not have been available at press time.*

In U.S. $	2021	2020	2019	2018	2017	2016
Revenue	51,682,000,000	48,097,000,000	45,764,000,000	43,634,000,000	41,581,000,000	29,003,000,000
R&D Expense						
Operating Income	10,855,000,000	8,463,000,000	6,614,000,000	5,456,000,000	4,452,000,000	3,441,000,000
Operating Margin %	.21%	.18%	.14%	.13%	.11%	.12%
SGA Expense	3,071,000,000	3,031,000,000	3,044,000,000	3,042,000,000	3,036,000,000	2,136,000,000
Net Income	4,654,000,000	3,222,000,000	1,668,000,000	1,230,000,000	9,895,000,000	3,522,000,000
Operating Cash Flow	16,239,000,000	14,562,000,000	11,748,000,000	11,767,000,000	11,954,000,000	8,041,000,000
Capital Expenditure	7,555,000,000	7,956,000,000	7,140,000,000	9,595,000,000	7,861,000,000	33,532,000,000
EBITDA	19,770,000,000	17,854,000,000	16,154,000,000	15,544,000,000	14,706,000,000	10,226,000,000
Return on Assets %	.03%	.02%	.01%	.01%	.07%	.04%
Return on Equity %	.25%	.12%	.05%	.03%	.25%	.18%
Debt to Equity	6.30%	3.434	2.403	1.916	1.745	1.488

CONTACT INFORMATION:

Phone: 203-905-7801   Fax:
Toll-Free:
Address: 400 Atlantic St., Stamford, CT 06901 United States

STOCK TICKER/OTHER:

Stock Ticker: CHTR   Exchange: NAS
Employees: 93,700   Fiscal Year Ends: 12/31
Parent Company:

SALARIES/BONUSES:

Top Exec. Salary: $2,500,000   Bonus: $
Second Exec. Salary: Bonus: $
$1,875,000

OTHER THOUGHTS:

Estimated Female Officers or Directors: 1
Hot Spot for Advancement for Women/Minorities:

China Communications Services Corp Ltd   www.chinaccs.com.hk
NAIC Code: 237130

TYPES OF BUSINESS:
Power and Communication Line and Related Structures Construction
Telecommunications
Business Process Outsourcing

GROWTH PLANS/SPECIAL FEATURES:
China Communications Services Corp Ltd is active in the technology sector. The company provides integrated comprehensive solutions for the informatization and digitalization sector. It provides integrated solutions, including telecommunications infrastructure services, business process outsourcing services as well as applications, content and other services to telecommunications operators, government agencies, industrial customers and small-and-medium enterprises.

BRANDS/DIVISIONS/AFFILIATES:
China Telecom Corporation Limited

CONTACTS: Note: Officers with more than one job title may be intentionally listed here more than once.
Xiaoqing Huang, Pres.
Xu Zhang, CFO
Xiaochu Wang, Chmn.

FINANCIAL DATA: Note: Data for latest year may not have been available at press time.

In U.S. $	2021	2020	2019	2018	2017	2016
Revenue	19,850,860,000	18,170,480,000	17,394,790,000	15,730,100,000	14,010,940,000	13,103,800,000
R&D Expense						
Operating Income	414,752,500	420,065,600	432,487,700	423,458,400	427,428,700	410,337,200
Operating Margin %	.02%	.02%	.02%	.03%	.03%	.03%
SGA Expense	1,918,802,000	1,752,096,000	1,702,900,000	1,572,034,000	1,464,529,000	1,407,648,000
Net Income	467,774,900	456,521,600	451,744,300	429,832,100	402,111,600	375,746,200
Operating Cash Flow	667,559,000	807,935,700	712,847,700	631,264,300	1,031,755,000	780,048,300
Capital Expenditure	262,109,500	523,248,800	174,848,400	146,280,500	150,171,100	125,431,300
EBITDA	768,282,300	726,130,800	713,211,600	635,360,500	608,508,400	585,631,600
Return on Assets %	.03%	.03%	.04%	.04%	.04%	.04%
Return on Equity %	.08%	.09%	.09%	.10%	.10%	.10%
Debt to Equity	.03%	0.021	0.02	0.00	0.001	0.001

CONTACT INFORMATION:
Phone: 852 3699 0000   Fax:
Toll-Free:
Address: Great Eagle Ctr., Ste. 3203-3205, 23 Harbour Rd., Wanchai, Hong Kong, Hong Kong 999077 Hong Kong

STOCK TICKER/OTHER:
Stock Ticker: CUCSY   Exchange: PINX
Employees: 90,003   Fiscal Year Ends: 12/31
Parent Company: China Telecom Corporation Limited

SALARIES/BONUSES:
Top Exec. Salary: $ Bonus: $
Second Exec. Salary: $ Bonus: $

OTHER THOUGHTS:
Estimated Female Officers or Directors:
Hot Spot for Advancement for Women/Minorities:

China Mobile Limited

www.chinamobileltd.com

NAIC Code: 517210

TYPES OF BUSINESS:

Mobile Phone Service
Wireless Music Service
News & Information Service
Instant Messaging Service
Data Services

BRANDS/DIVISIONS/AFFILIATES:

China Mobile Communications Group Co Ltd

CONTACTS:
Note: Officers with more than one job title may be intentionally listed here more than once.

Xin Dong, CEO
Ronghua Li, CFO
Huang Wnelin, VP
Sha Yuejia, VP
Liu Aili, VP
Jie Yang, Chmn.

GROWTH PLANS/SPECIAL FEATURES:

China Mobile Limited (CML) is a leading provider of mobile telecommunications services in Mainland China. CML is one of the largest wireless communications providers in the world in terms of total subscribers, with more than 940 million mobile customers and 210 million wireline broadband customers. The firm is a publicly-traded subsidiary of China Mobile Communications Group Co., Ltd., owning a majority 72.72%. CML provides wireless products and services in all 31 Chinese provinces, regions and municipalities, including Hong Kong, through several wholly-owned subsidiaries. The company offers voice services comprising local calls, domestic long-distance calls, international long-distance calls, intra-provincial roaming, inter-provincial roaming and international roaming, as well as providing voice value-added services such as caller identity display, caller restrictions, call waiting, call forwarding, call holding, voice mail, conference calls and more. The company also offers: SMS (short message services) and other data services; and MMS (multimedia message services), which allows users to combine and deliver several types of messages, including graphics, sounds, text, and motion pictures over wireless networks. In addition, CML provides high-speed internet through wireless local area networks (WLAN) throughout Mainland China. The company's 4G and 5G business is based on the FDD (frequency division duplex) mode long-term evolution (LTE) technology.

FINANCIAL DATA: *Note: Data for latest year may not have been available at press time.*

In U.S. $	2021	2020	2019	2018	2017	2016
Revenue	133,089,135,426	117,674,000,000	110,869,217,280	109,516,939,264	104,327,135,232	104,893,775,872
R&D Expense						
Operating Income						
Operating Margin %						
SGA Expense						
Net Income	18,248,062,482	16,567,600,000	15,850,562,560	17,506,353,152	16,100,168,704	16,100,953,088
Operating Cash Flow						
Capital Expenditure						
EBITDA						
Return on Assets %						
Return on Equity %						
Debt to Equity						

CONTACT INFORMATION:

Phone: 85 231218888   Fax: 85 225119092
Toll-Free:
Address: 99 Queen's Rd. Central, The Center, Fl. 60, Hong Kong, Hong Kong 999077 Hong Kong

STOCK TICKER/OTHER:

Stock Ticker: 941   Exchange: Hong Kong
Employees: 449,934   Fiscal Year Ends: 12/31
Parent Company: China Mobile Communications Group Co Ltd

SALARIES/BONUSES:

Top Exec. Salary: $ Bonus: $
Second Exec. Salary: $ Bonus: $

OTHER THOUGHTS:

Estimated Female Officers or Directors: 1
Hot Spot for Advancement for Women/Minorities: Y

Sales, profits and employees may be estimates. Financial information, benefits and other data can change quickly and may vary from those stated here.

China Telecom Corporation Limited

www.chinatelecom-h.com

NAIC Code: 517110

TYPES OF BUSINESS:

Fixed-Line & Wireless Telecommunications Services
Internet & Ecommerce Services
Outsourcing Services
Telecommunications
Broadband
Wireless Networks
Internet of Things
Optical Broadband

BRANDS/DIVISIONS/AFFILIATES:

CONTACTS: *Note: Officers with more than one job title may be intentionally listed here more than once.*

Ruiwen Ke, CEO
Zhengmao Li, Pres.
Yung Shun Loy, Jacky, Sec.
Gao Jinxing, Financial Controller
Li Ping, Exec. VP
Zhang Jiping, Exec. VP
Yang Xiaowei, Exec. VP
Sun Kangmin, Exec. VP

GROWTH PLANS/SPECIAL FEATURES:

China Telecom Corporation Limited is one of the largest telecommunications and broadband service providers in China. With approximately 160 million wireline access lines in service, over 350 million mobile subscribers, 153 million wireline broadband subscribers and 108 million access lines in service. China Telecom operates in 31 provinces, municipalities and autonomous regions. The company focuses on government, financial and large enterprise customers. China Telecom's wireline telephone services consist of local telephone; domestic long distance; and international, Hong Kong, Macau and Taiwan long distance and interconnection. Mobile voice services include local calls, domestic long-distance calls, international long-distance calls, intra-provincial roaming, inter-provincial roaming and international roaming. The firm is focused on further developing high speed services. It is a leader in 4G and 5G wireless within China. The company also provides industry-specific applications for government and enterprise subscribers. Internet access services consist of wireline internet access services, including dial-up and broadband services, and wireless internet access services. Services are offered through integrated and customized plans. China Telecom continually expanded its fiber network coverage and deployed its Gigabit optical broadband network. Value added services are offered for the wireline and mobile services, and include caller ID, color ringtone, mobile MMS (multimedia messaging service) and industry-specific applications for government and enterprises. Additionally, the company provides integrated information application services and managed data and leased line services.

FINANCIAL DATA: *Note: Data for latest year may not have been available at press time.*

In U.S. $	2021	2020	2019	2018	2017	2016
Revenue	68,964,390,144	60,295,513,005	55,847,145,472	56,053,747,712	51,596,083,200	52,161,781,760
R&D Expense						
Operating Income						
Operating Margin %						
SGA Expense						
Net Income	4,102,542,756	3,194,324,250	3,049,539,840	3,152,543,744	2,622,851,584	2,665,798,400
Operating Cash Flow						
Capital Expenditure						
EBITDA						
Return on Assets %						
Return on Equity %						
Debt to Equity						

CONTACT INFORMATION:

Phone: 86 10 5850 1800   Fax: 86 10 6601 0728
Toll-Free:
Address: 31 Jinrong St., Xicheng Dist., Beijing, Beijing 100033 China

STOCK TICKER/OTHER:

Stock Ticker: 728   Exchange: Hong Kong
Employees: 278,922   Fiscal Year Ends: 12/31
Parent Company:

SALARIES/BONUSES:

Top Exec. Salary: $ Bonus: $
Second Exec. Salary: $ Bonus: $

OTHER THOUGHTS:

Estimated Female Officers or Directors: 3
Hot Spot for Advancement for Women/Minorities: Y

China Unicom (Hong Kong) Limited

www.chinaunicom.com.hk/en/global/home.php

NAIC Code: 517110

TYPES OF BUSINESS:

Wireless Phone Service
Fixed Line Services
Mobile Services
Digital Transformation
Omnichannel
Internet of Things
Cloud Computing
Industrial IT

BRANDS/DIVISIONS/AFFILIATES:

China United Network Communications Group Co Ltd

CONTACTS: *Note: Officers with more than one job title may be intentionally listed here more than once.*

Liehong Liu, CEO
Zhongyue Chen, Pres.
Yuzhuo Li, CFO
Chu Ka Yee, Sec.
Tong Jilu, Sr. VP
Liehong Liu, Chmn.

GROWTH PLANS/SPECIAL FEATURES:

China Unicom (Hong Kong) Limited is an integrated telecommunications operator in China. The company focuses on promoting comprehensive digital transformation led by 5G services, upgrading existing customers and expanding scale in new user markets. The firm's 5G strategy helps to drive mobile service growth, including bundled services across internet, cloud and more. China Unicom coordinates network with business, service interaction and information regarding its fixed-line service, ranging from 3G to 5G capabilities. The firm therefore sells broadband and marketing services as well as network deployment. China Unicom offers a cloud management product, which is cloud-network integrated, secure, reliable, private, customizable and multi-cloud collaborative. Big data services span data application, data technology, artificial intelligence and blockchain, with a focus on key industries such as government, finance, cultural, tourism and transportation. Its Internet of Things (IoT) system for business is driven by 4G and 5G capabilities; and its industrial IT services span areas such as smart city, digital government, big data, IT, innovation and more to further promote digital transformation. China Unicom's marketing services promote 5G and smart living solutions via brand strategies across all marketing channels. The firm actively implements comprehensive digital transformation, online-to-offline integration and end-to-end customer experience services. China Unicom operates as a subsidiary of China United Network Communications Group Company Limited, which holds a majority, 52.1% share.

FINANCIAL DATA: *Note: Data for latest year may not have been available at press time.*

In U.S. $	2021	2020	2019	2018	2017	2016
Revenue	51,439,309,038	46,549,500,790	43,180,634,112	43,234,443,264	38,719,217,664	40,599,527,424
R&D Expense						
Operating Income						
Operating Margin %						
SGA Expense						
Net Income	2,276,104,779	1,926,859,285	1,684,032,128	1,515,629,056	257,537,328	92,541,872
Operating Cash Flow						
Capital Expenditure						
EBITDA						
Return on Assets %						
Return on Equity %						
Debt to Equity						

CONTACT INFORMATION:

Phone: 852 21262018　　Fax:
Toll-Free:
Address: 99 Queens Rd. Central, The Center, Fl. 75, Hong Kong, Hong Kong 999077 Hong Kong

STOCK TICKER/OTHER:

Stock Ticker: 762　　　　　　　　Exchange: Hong Kong
Employees: 254,702　　　　　　　Fiscal Year Ends: 12/31
Parent Company: China United Network Communications Group Co Ltd

SALARIES/BONUSES:

Top Exec. Salary: $　　　　Bonus: $
Second Exec. Salary: $　　　Bonus: $

OTHER THOUGHTS:

Estimated Female Officers or Directors:
Hot Spot for Advancement for Women/Minorities:

Sales, profits and employees may be estimates. Financial information, benefits and other data can change quickly and may vary from those stated here.

Chorus Limited

www.chorus.co.nz

NAIC Code: 517110

TYPES OF BUSINESS:

Wired Telecommunications Carriers
Telecommunications
Open Access Internet Network
Broadband
Telephone Services

BRANDS/DIVISIONS/AFFILIATES:

GROWTH PLANS/SPECIAL FEATURES:

Chorus is New Zealand's largest fixed-line communications infrastructure group, and was demerged in 2011 from Telecom New Zealand. Chorus offers wholesale access to its nationwide copper and fiber network to retail service providers. It also provides backhaul services to retail service providers and connects mobile phone towers. Chorus is currently involved in the government's rollout of ultra-fast broadband, a fiber-to-the-premises network to cover 87% of the population by the end of 2023.

CONTACTS: *Note: Officers with more than one job title may be intentionally listed here more than once.*

JB Rousselot, CEO
David Collins, CFO
Ed Hyde, CCO
Shaun Philip, Chief People Officer
Ewen Powell, CTO

FINANCIAL DATA: *Note: Data for latest year may not have been available at press time.*

In U.S. $	2021	2020	2019	2018	2017	2016
Revenue	552,837,100	603,152,300	563,529,100	622,649,300	654,096,300	630,196,600
R&D Expense						
Operating Income	98,114,440	154,718,900	106,290,600	167,297,700	213,839,200	164,153,000
Operating Margin %	.18%	.26%	.19%	.27%	.33%	.26%
SGA Expense	54,088,730	58,491,300	54,717,670	51,572,980	57,233,420	59,120,240
Net Income	29,560,120	32,704,810	33,333,750	53,459,790	71,070,080	57,233,420
Operating Cash Flow						
Capital Expenditure	406,923,400	427,049,400	506,924,600	481,767,100	401,262,900	357,866,100
EBITDA	408,810,200	415,099,600	406,294,400	415,099,600	416,357,400	377,992,200
Return on Assets %	.01%	.01%	.01%	.02%	.03%	.02%
Return on Equity %	.05%	.05%	.05%	.09%	.12%	.11%
Debt to Equity	2.62%	2.318	2.03	2.00	1.873	1.924

CONTACT INFORMATION:

Phone: 649-975-2983   Fax:
Toll-Free:
Address: 1 Willis St., 10/Fl, Wellington, 6140 New Zealand

STOCK TICKER/OTHER:

Stock Ticker: CHRYY   Exchange: PINX
Employees: 817   Fiscal Year Ends: 06/30
Parent Company:

SALARIES/BONUSES:

Top Exec. Salary: $ Bonus: $
Second Exec. Salary: $ Bonus: $

OTHER THOUGHTS:

Estimated Female Officers or Directors:
Hot Spot for Advancement for Women/Minorities:

Sales, profits and employees may be estimates. Financial information, benefits and other data can change quickly and may vary from those stated here.

Chunghwa Telecom Co Ltd

NAIC Code: 517110

www.cht.com.tw/en/home/cht

TYPES OF BUSINESS:

Telephone Service
Cellular Service
Internet & Data Service

GROWTH PLANS/SPECIAL FEATURES:

Chunghwa Telecom Co Ltd is Taiwan's largest integrated telecom operator, providing fixed-line, wireless, and Internet and data services. The company enjoys a dominant position in all three segments. It has a market share of 35% in mobile 85% in broadband, and greater than 95% in the fixed-line segment as measured by the number of local fixed-line subscribers. Chunghwa Telecom was privatized in 2005, with the government retaining 36.3% of the firm's ordinary shares.

BRANDS/DIVISIONS/AFFILIATES:

Ministry of Transportation and Communications
Light Era Development Co Ltd
Chunghwa System Integration Co Ltd
Chunghwa Telecom Global Inc
HiNet

CONTACTS: Note: Officers with more than one job title may be intentionally listed here more than once.

Shui-yi Kuo, Pres.
Chi-Mau Sheih, Chmn.

FINANCIAL DATA: Note: Data for latest year may not have been available at press time.

In U.S. $	2021	2020	2019	2018	2017	2016
Revenue	6,981,204,000	6,912,839,000	6,910,339,000	7,182,600,000	7,583,624,000	7,666,188,000
R&D Expense	122,930,500	128,330,300	131,363,600	124,163,800	129,530,300	126,163,700
Operating Income	1,544,497,000	1,366,135,000	1,365,902,000	1,451,100,000	1,560,230,000	1,619,996,000
Operating Margin %	.22%	.20%	.20%	.20%	.21%	.21%
SGA Expense	874,546,300	863,946,600	899,245,800	925,278,400	999,410,000	1,001,743,000
Net Income	1,187,172,000	1,113,941,000	1,098,208,000	1,251,871,000	1,299,570,000	1,349,469,000
Operating Cash Flow	2,495,208,000	2,481,809,000	2,414,177,000	2,212,148,000	2,364,345,000	2,165,016,000
Capital Expenditure	1,186,272,000	2,370,478,000	817,614,300	968,244,100	1,272,637,000	793,281,500
EBITDA	2,822,801,000	2,646,538,000	2,571,507,000	2,562,340,000	2,665,305,000	2,730,536,000
Return on Assets %	.07%	.07%	.07%	.08%	.09%	.09%
Return on Equity %	.09%	.09%	.09%	.10%	.11%	.11%
Debt to Equity	.09%	0.07	0.022	0.004	0.004	0.004

CONTACT INFORMATION:

Phone: 886 223445488   Fax: 886 233938188
Toll-Free:
Address: No.21-3, Sec. 1, Xinyi Rd., Zhongzheng Dist.,, Taipei City, 100 Taiwan

STOCK TICKER/OTHER:

Stock Ticker: CHT   Exchange: NYS
Employees: 32,218   Fiscal Year Ends: 12/31
Parent Company:

SALARIES/BONUSES:

Top Exec. Salary: $271,690   Bonus: $
Second Exec. Salary: Bonus: $
$211,511

OTHER THOUGHTS:

Estimated Female Officers or Directors: 3
Hot Spot for Advancement for Women/Minorities: Y

Ciena Corporation

www.ciena.com

NAIC Code: 334210

TYPES OF BUSINESS:
Communications Networking Equipment
Software & Support Services
Consulting Services
Switching Platforms
Packet Interworking Products
Access Products
Network & Service Management Tools

BRANDS/DIVISIONS/AFFILIATES:
Blue Planet Automation
Adaptive Network

GROWTH PLANS/SPECIAL FEATURES:
Ciena Corp is a network strategy and technology company. It provides network hardware, software, and services that support the transport, switching, aggregation, service delivery, and management of video, data, and voice traffic on communications networks. It serves various industries such as communication services providers, web-scale providers, cable operators, government, and large enterprises worldwide. The business activities function through Networking Platforms; Platform Software and Services; Blue Planet Automation Software, and Global Services segments. Geographically, its presence is seen in the markets of the United States, Canada, the Caribbean, Latin America, Europe, the Middle East, Africa, the Asia Pacific, Japan, and India.

CONTACTS:
Note: Officers with more than one job title may be intentionally listed here more than once.

Gary Smith, CEO
James Moylan, CFO
Patrick Nettles, Chairman of the Board
Andrew Petrik, Chief Accounting Officer
Mary Yang, Chief Strategy Officer
Stephen Alexander, Chief Technology Officer
David Rothenstein, General Counsel
Scott McFeely, Senior VP, Divisional
Jason Phipps, Senior VP, Divisional
Rick Hamilton, Senior VP, Divisional

FINANCIAL DATA:
Note: Data for latest year may not have been available at press time.

In U.S. $	2021	2020	2019	2018	2017	2016
Revenue	3,620,684,000	3,532,157,000	3,572,131,000	3,094,286,000	2,801,687,000	2,600,573,000
R&D Expense	536,666,000	529,888,000	548,139,000	491,564,000	475,329,000	451,794,000
Operating Income	527,493,000	513,647,000	374,674,000	253,196,000	238,655,000	165,715,000
Operating Margin %	.15%	.15%	.10%	.08%	.09%	.06%
SGA Expense	634,088,000	585,973,000	597,445,000	554,193,000	498,773,000	482,559,000
Net Income	500,196,000	361,291,000	253,434,000	-344,690,000	1,261,953,000	72,584,000
Operating Cash Flow	541,646,000	493,654,000	413,140,000	229,261,000	234,882,000	289,520,000
Capital Expenditure	79,550,000	82,667,000	62,579,000	67,616,000	94,600,000	107,185,000
EBITDA	625,854,000	619,809,000	473,354,000	314,050,000	334,880,000	285,066,000
Return on Assets %	.11%	.09%	.07%	- .09%	.37%	.03%
Return on Equity %	.18%	.15%	.12%	- .17%	.87%	.10%
Debt to Equity	.26%	0.319	0.343	0.391	0.308	1.359

CONTACT INFORMATION:
Phone: 410 694-5700 Fax: 410 694-5750
Toll-Free: 800-921-1144
Address: 7035 Ridge Rd., Hanover, MD 21076 United States

STOCK TICKER/OTHER:
Stock Ticker: CIEN Exchange: NYS
Employees: 7,241 Fiscal Year Ends: 10/31
Parent Company:

SALARIES/BONUSES:
Top Exec. Salary: $1,000,000 Bonus: $
Second Exec. Salary: Bonus: $
$575,000

OTHER THOUGHTS:
Estimated Female Officers or Directors: 2
Hot Spot for Advancement for Women/Minorities:

Sales, profits and employees may be estimates. Financial information, benefits and other data can change quickly and may vary from those stated here.

Cincinnati Bell Inc

NAIC Code: 517110

www.cincinnatibell.com

TYPES OF BUSINESS:
Local Telephone Service
Wireless Local Phone Service
Long-Distance Service
Data Services
Internet Access
Digital Services
Payphone Services
IT Consulting

BRANDS/DIVISIONS/AFFILIATES:
Cincinnati Bell Telephone Company LLC
Cincinnati Bell Extended Territories LLC
Hawaiian Telecom Holdco Inc

CONTACTS: *Note: Officers with more than one job title may be intentionally listed here more than once.*
Leigh R. Fox, CEO
Thomas E. Simpson, COO
Suzanne Maratta, Controller
Joshua T. Duckworth, CFO
Kevin Murray, CIO
Leigh Fox, Director
Christi Cornette, Other Executive Officer
Shannon Mullen, Vice President, Divisional
Joshua Duckworth, Vice President, Divisional
Christopher Wilson, Vice President

GROWTH PLANS/SPECIAL FEATURES:
Cincinnati Bell, Inc. provides integrated communications and information technology (IT) solutions to residential and business customers. The company operates through two business segments: entertainment and communications and IT services and hardware. The entertainment and communications segment provides high-speed data, video and voice solutions over an expanding fiber network, as well as a copper network. Subsidiary Cincinnati Bell Telephone Company, LLC, is the incumbent local exchange carrier for a geography that covers about 25 miles around Cincinnati and includes parts of northern Kentucky and southeastern Indiana. This segment also provides voice and data services in Dayton and Mason, Ohio, through Cincinnati Bell Extended Territories, LLC, a subsidiary of Cincinnati Bell Telephone; and provides full-service communications on all of Hawaii's major islands through subsidiary Hawaiian Telcom Holdco, Inc. The IT services and hardware segment sells and services end-to-end communications and IT systems and solutions, as well as hybrid cloud services. These products and services are categorized into: consulting, offering IT staffing and application services; cloud, offering virtual data centers, storage, backup, network management/monitoring, security, and cloud consulting; communications, offering unified communications-as-a-services, contact center, software-defined wide area networking, networking solutions, multi-protocol label switching and network-as-a-service; and infrastructure solutions, offering hardware, software licenses and maintenance. In mid-2021, Cincinnati Bell was acquired by Macquarie Infrastructure Partners V, a fund managed by Macquarie Infrastructure and Real Assets.

FINANCIAL DATA: *Note: Data for latest year may not have been available at press time.*

In U.S. $	2021	2020	2019	2018	2017	2016
Revenue	1,693,100,000	1,559,800,064	1,536,700,032	1,378,200,064	1,288,499,968	1,185,799,936
R&D Expense						
Operating Income						
Operating Margin %						
SGA Expense						
Net Income	-108,500,000	-55,600,000	-66,600,000	-69,800,000	35,100,000	102,100,000
Operating Cash Flow						
Capital Expenditure						
EBITDA						
Return on Assets %						
Return on Equity %						
Debt to Equity						

CONTACT INFORMATION:
Phone: 513 397-9900   Fax: 513 784-1613
Toll-Free: 800-345-6301
Address: 221 E. Fourth St., Cincinnati, OH 45202 United States

STOCK TICKER/OTHER:
Stock Ticker: Subsidiary   Exchange:
Employees: 5,200   Fiscal Year Ends: 12/31
Parent Company: Macquarie Infrastructure Partners V

SALARIES/BONUSES:
Top Exec. Salary: $ Bonus: $
Second Exec. Salary: $ Bonus: $

OTHER THOUGHTS:
Estimated Female Officers or Directors:
Hot Spot for Advancement for Women/Minorities: Y

Cisco Systems Inc

www.cisco.com

NAIC Code: 334210A

TYPES OF BUSINESS:

Computer Networking Equipment
Routers & Switches
Real-Time Conferencing Technology
Server Virtualization Software
Data Storage Products
Security Products
Teleconference Systems and Technology
Unified Communications Systems

BRANDS/DIVISIONS/AFFILIATES:

AppDynamics Inc
Acacia Communications Inc

GROWTH PLANS/SPECIAL FEATURES:

Cisco Systems, Inc. is the world's largest hardware and software supplier within the networking solutions sector. The secure, agile networks business contains switching, routing, and wireless solutions. The hybrid work division has products for collaboration and contact center needs. The end-to-end security group has products spanning a variety of threat prevention necessities. The internet for the future division has routed optical networks, silicon, and optics. Optimized application experiences offer solutions such as full stack observability. Services are Cisco's technical support and advanced services offerings. In collaboration with Cisco's initiative on growing software and services, its revenue model is focused on increasing subscriptions and recurring sales.

CONTACTS: Note: Officers with more than one job title may be intentionally listed here more than once.

Charles Robbins, CEO
R. Herren, CFO
Prat Bhatt, Chief Accounting Officer
Deborah Stahlkopf, Chief Legal Officer
Maria Martinez, COO
Geraldine Elliott, Executive VP

FINANCIAL DATA: Note: Data for latest year may not have been available at press time.

In U.S. $	2021	2020	2019	2018	2017	2016
Revenue	49,818,000,000	49,301,000,000	51,904,000,000	49,330,000,000	48,005,000,000	49,247,000,000
R&D Expense	6,549,000,000	6,347,000,000	6,577,000,000	6,332,000,000	6,059,000,000	6,296,000,000
Operating Income	13,719,000,000	14,101,000,000	14,541,000,000	12,667,000,000	12,729,000,000	12,928,000,000
Operating Margin %	.28%	.29%	.28%	.26%	.27%	.26%
SGA Expense	11,411,000,000	11,094,000,000	11,398,000,000	11,386,000,000	11,177,000,000	11,433,000,000
Net Income	10,591,000,000	11,214,000,000	11,621,000,000	110,000,000	9,609,000,000	10,739,000,000
Operating Cash Flow	15,454,000,000	15,426,000,000	15,831,000,000	13,666,000,000	13,876,000,000	13,570,000,000
Capital Expenditure	692,000,000	770,000,000	909,000,000	834,000,000	964,000,000	1,146,000,000
EBITDA	15,558,000,000	16,363,000,000	17,327,000,000	16,174,000,000	15,434,000,000	15,746,000,000
Return on Assets %	.11%	.12%	.11%	.00%	.08%	.09%
Return on Equity %	.27%	.31%	.30%	.00%	.15%	.17%
Debt to Equity	.22%	0.305	0.431	0.471	0.389	0.385

CONTACT INFORMATION:

Phone: 408 526-4000   Fax: 408 526-4100
Toll-Free: 800-553-6387
Address: 170 W. Tasman Dr., San Jose, CA 95134-1706 United States

STOCK TICKER/OTHER:

Stock Ticker: CSCO   Exchange: NAS
Employees: 79,500   Fiscal Year Ends: 07/31
Parent Company:

SALARIES/BONUSES:

Top Exec. Salary: $495,385   Bonus: $8,000,000
Second Exec. Salary: $1,416,731   Bonus: $

OTHER THOUGHTS:

Estimated Female Officers or Directors: 10
Hot Spot for Advancement for Women/Minorities: Y

Sales, profits and employees may be estimates. Financial information, benefits and other data can change quickly and may vary from those stated here.

Cisco Webex

NAIC Code: 517110

www.webex.com

TYPES OF BUSINESS:

Videoconferencing Services
Online Conferencing Services & Software
Workforce Collaboration Solutions
Online and Mobile Meetings
Online Staff Training Solutions
Online Presentations
Cloud Calling and Video
Artificial Intelligence

BRANDS/DIVISIONS/AFFILIATES:

Cisco Systems Inc

CONTACTS: Note: Officers with more than one job title may be intentionally listed here more than once.

Chuck Robbins, CEO

GROWTH PLANS/SPECIAL FEATURES:

Cisco Webex, a subsidiary of Cisco Systems, Inc., provides cloud-based collaboration, meeting, training and conferencing applications for teams of all sizes. Webex's services scale to business needs in order to reach and deliver anywhere in the world. Staff meetings, presentations and trainings can be accessed anywhere, any time, on any device, including mobile. The firm's cloud-based products are designed to integrate with the rest of a business' IT investments. Webex products also include cloud calling, a video support platform for providing real-time technical help, an agile contact center focused on customer experience, and artificial intelligence (AI) and machine learning capabilities for enhanced collaboration experiences. Cisco Webex offers devices, including an all-in-one wireless whiteboard, intelligent video conferencing devices for meeting rooms of all sizes and compact desk video conferencing devices, as well as related accessories such as sound bars, headsets, screen stands and other room kits. Price plans range from free to individual plans to business suite plans.

FINANCIAL DATA: Note: Data for latest year may not have been available at press time.

In U.S. $	2021	2020	2019	2018	2017	2016
Revenue						
R&D Expense						
Operating Income						
Operating Margin %						
SGA Expense						
Net Income						
Operating Cash Flow						
Capital Expenditure						
EBITDA						
Return on Assets %						
Return on Equity %						
Debt to Equity						

CONTACT INFORMATION:

Phone: 408-435-7048   Fax:
Toll-Free: 877-509-3239
Address: 855 E. Tasman Dr., San Jose, CA 95035 United States

STOCK TICKER/OTHER:

Stock Ticker: Subsidiary   Exchange:
Employees: 229   Fiscal Year Ends: 12/31
Parent Company: Cisco Systems Inc

SALARIES/BONUSES:

Top Exec. Salary: $ Bonus: $
Second Exec. Salary: $ Bonus: $

OTHER THOUGHTS:

Estimated Female Officers or Directors:
Hot Spot for Advancement for Women/Minorities:

CK Hutchison Holdings Limited
www.ckh.com.hk/en/global/home.php

NAIC Code: 488310

TYPES OF BUSINESS:
Port Operations
Port Operations
Terminal Operator
Retail
Manufacturing
Telecommunications
Investments
Biopharmaceuticals

BRANDS/DIVISIONS/AFFILIATES:
Hutchison Port Holdings Limited
AS Watson Group (HK) Limited
CK Hutchison Group Telecom
Hutchison Telecommunications Hong Kong Holdings
Hutchison Whampoa
Hutchison China MediTech
Cenovus Energy Inc
Hutchison Water

CONTACTS: *Note: Officers with more than one job title may be intentionally listed here more than once.*
Kin-ning Fok, Group Co-Managing Dir.
Tzar Kuoi (Victor) Li, Group Co-Managing Dir.
Frank J. Sixt, Group Dir.-Finance
Tzar Kuoi Li, Deputy Chmn.
Mo Fong Chow Woo, Deputy Group Managing Dir.
Tzar Kuoi (Victor) Li, Chmn.

GROWTH PLANS/SPECIAL FEATURES:
CK Hutchison Holdings is a Hong Kong headquartered conglomerate with key businesses in ports, retail, infrastructure, and telecommunications. The company was created in 2015 to house the merged assets of Cheung Kong Holdings and Hutchison Whampoa as the group sought to flatten out and simplify its original holding structure. CK Hutchison contains most of the businesses previously housed in Hutchison Whampoa, minus the property assets, which have been spun off into their own listing, CK Asset Holdings. Telecommunications activities now make up the largest share of EBITDA at around 32%.

FINANCIAL DATA: *Note: Data for latest year may not have been available at press time.*

In U.S. $	2021	2020	2019	2018	2017	2016
Revenue	35,777,730,000	33,936,790,000	38,092,960,000	35,304,090,000	31,658,890,000	33,101,860,000
R&D Expense						
Operating Income	4,283,055,000	4,727,654,000	6,609,238,000	5,729,212,000	4,067,635,000	4,709,437,000
Operating Margin %	.12%	.14%	.17%	.16%	.13%	.14%
SGA Expense	3,940,498,000	3,863,425,000	2,792,693,000	3,268,121,000	3,156,271,000	2,442,110,000
Net Income	4,265,602,000	3,712,593,000	5,074,034,000	4,968,298,000	4,471,469,000	4,204,964,000
Operating Cash Flow	6,647,838,000	9,321,420,000	8,824,462,000	7,100,081,000	6,827,716,000	5,138,749,000
Capital Expenditure	5,050,211,000	3,767,499,000	4,623,448,000	4,783,070,000	3,046,586,000	3,126,971,000
EBITDA	12,506,350,000	11,962,770,000	13,520,650,000	10,110,870,000	9,278,998,000	9,168,677,000
Return on Assets %	.03%	.02%	.03%	.03%	.03%	.03%
Return on Equity %	.06%	.06%	.09%	.08%	.08%	.08%
Debt to Equity	.63%	0.745	0.799	0.712	0.683	0.555

CONTACT INFORMATION:
Phone: 852 21281188   Fax: 852 21281705
Toll-Free:
Address: 48/F, Cheung Kong Center, 2 Queen's Road Central, Hong Kong, Hong Kong 999077 Hong Kong

STOCK TICKER/OTHER:
Stock Ticker: CKHUY   Exchange: PINX
Employees: 300,000   Fiscal Year Ends: 12/31
Parent Company:

SALARIES/BONUSES:
Top Exec. Salary: $ Bonus: $
Second Exec. Salary: $ Bonus: $

OTHER THOUGHTS:
Estimated Female Officers or Directors: 2
Hot Spot for Advancement for Women/Minorities:

Sales, profits and employees may be estimates. Financial information, benefits and other data can change quickly and may vary from those stated here.

ClearOne Inc

NAIC Code: 334220

www.clearone.com

TYPES OF BUSINESS:

Teleconferencing Products & Services
Tabletop Conferencing Phones
Installed Audio Systems
Microphones
Conference Furniture
Streaming A/V Equipment
Digital Signage

BRANDS/DIVISIONS/AFFILIATES:

GROWTH PLANS/SPECIAL FEATURES:

ClearOne Inc is a communications solutions company, which is engaged in designing, developing, and selling conferencing, collaboration, streaming, and digital signage solutions for audio, video, and visual communication. It sells its commercial products to end-users primarily through a network of independent distributors, who in turn sells its products to dealers, systems integrators, and other resellers. Its products include professional audio, network media streaming, and professional microphones. Geographically the firm has its business presence across the US and international market.

Employees receive flexible spending accounts; medical, dental and vision coverage; insurance protection; and flexible time-off.

CONTACTS: *Note: Officers with more than one job title may be intentionally listed here more than once.*

Zeynep Hakimoglu, CEO
Narsi Narayanan, CFO

FINANCIAL DATA: *Note: Data for latest year may not have been available at press time.*

In U.S. $	2021	2020	2019	2018	2017	2016
Revenue	28,967,000	29,069,000	25,042,000	28,156,000	41,804,000	48,637,000
R&D Expense	5,794,000	5,512,000	5,775,000	7,840,000	9,342,000	8,584,000
Operating Income	-7,495,000	-5,567,000	-8,562,000	-10,327,000	-3,490,000	3,546,000
Operating Margin %	-.26%	-.19%	-.34%	-.37%	-.08%	.07%
SGA Expense	13,617,000	12,614,000	13,980,000	15,858,000	18,157,000	17,357,000
Net Income	-7,694,000	505,000	-8,408,000	-16,687,000	-14,172,000	2,444,000
Operating Cash Flow	-4,394,000	-982,000	-4,656,000	-6,621,000	-9,271,000	7,834,000
Capital Expenditure	8,347,000	7,217,000	5,367,000	5,135,000	3,205,000	891,000
EBITDA	-3,985,000	-2,612,000	-5,851,000	-8,737,000	-1,959,000	5,419,000
Return on Assets %	-.12%	.01%	-.15%	-.26%	-.18%	.03%
Return on Equity %	-.15%	.01%	-.17%	-.31%	-.21%	.03%
Debt to Equity	.05%	0.095	0.095			

CONTACT INFORMATION:

Phone: 801 975-7200   Fax: 801 977-0087
Toll-Free: 800-945-7730
Address: 5225 Wiley Post Way, Ste. 500, Salt Lake City, UT 84116
United States

STOCK TICKER/OTHER:

Stock Ticker: CLRO   Exchange: NAS
Employees: 107   Fiscal Year Ends: 12/31
Parent Company:

SALARIES/BONUSES:

Top Exec. Salary: $355,000   Bonus: $
Second Exec. Salary: Bonus: $
$196,500

OTHER THOUGHTS:

Estimated Female Officers or Directors: 1
Hot Spot for Advancement for Women/Minorities:

Sales, profits and employees may be estimates. Financial information, benefits and other data can change quickly and may vary from those stated here.

Cloudflare Inc

www.cloudflare.com

NAIC Code: 511210E

TYPES OF BUSINESS:

Computer Software, Network Security, Managed Access, Digital ID,
Cybersecurity & Anti-Virus
Web Content Delivery
Cloud Platform
Business Network Services
Software

BRANDS/DIVISIONS/AFFILIATES:

GROWTH PLANS/SPECIAL FEATURES:

Cloudflare is a software company based in San Francisco,
California, that offers security and web performance offerings
by utilizing a distributed, serverless content delivery network,
or CDN. The firm's edge computing platform, Workers,
leverages this network by providing clients the ability to deploy,
and execute code without maintaining servers.

Cloudflare offers health benefits and other employee
assistance programs.

CONTACTS: Note: Officers with more than one job title may be intentionally listed here more than once.

Matthew Prince, CEO
Thomas Seifert, CFO
Paul Underwood, Chief Accounting Officer
Michelle Zatlyn, Co-Founder
Douglas Kramer, General Counsel

FINANCIAL DATA: Note: Data for latest year may not have been available at press time.

In U.S. $	2021	2020	2019	2018	2017	2016
Revenue	656,426,000	431,059,000	287,022,000	192,674,000	134,915,000	84,791,000
R&D Expense	189,408,000	127,144,000	90,669,000	54,463,000	33,650,000	23,663,000
Operating Income	-127,684,000	-106,768,000	-107,946,000	-84,899,000	-9,730,000	-17,029,000
Operating Margin %	-.19%	-.25%	-.38%	-.44%	-.07%	-.20%
SGA Expense	447,568,000	309,628,000	240,876,000	179,573,000	82,207,000	54,195,000
Net Income	-260,309,000	-119,370,000	-105,828,000	-87,164,000	-10,748,000	-17,334,000
Operating Cash Flow	64,648,000	-17,129,000	-38,917,000	-43,281,000	3,167,000	-13,318,000
Capital Expenditure	107,738,000	74,962,000	57,279,000	34,839,000	22,975,000	18,558,000
EBITDA	-132,135,000	-50,622,000	-74,122,000	-66,190,000	3,321,000	-8,256,000
Return on Assets %	-.14%	-.11%	-.19%	-.38%	-.07%	
Return on Equity %	-.32%	-.15%	-.35%			
Debt to Equity	1.54%	0.503	0.014			

CONTACT INFORMATION:

Phone: 888-993-5273   Fax:
Toll-Free:
Address: 101 Townsend St., San Francisco, CA 94107 United States

STOCK TICKER/OTHER:

Stock Ticker: NET   Exchange: NYS
Employees: 2,440   Fiscal Year Ends: 12/31
Parent Company:

SALARIES/BONUSES:

Top Exec. Salary: $650,000   Bonus: $
Second Exec. Salary: Bonus: $
$550,000

OTHER THOUGHTS:

Estimated Female Officers or Directors: 1
Hot Spot for Advancement for Women/Minorities:

Cogent Communications Group Inc

www.cogentco.com

NAIC Code: 517110

TYPES OF BUSINESS:

Facilities-Based Internet Service Provider
VoIP Service

GROWTH PLANS/SPECIAL FEATURES:

Cogent carries over one fifth of the world's Internet traffic over its network and is an Internet service provider for businesses. Cogent's corporate customers are in high-rise office buildings, and the firm provides them with two types of connections: dedicated Internet access, which connects them to the Internet, and virtual private networking, which offers an internal network for employees in different locations. Cogent's corporate customers are exclusively in North America and account for over half of the firm's revenue. Cogent's netcentric customers include Internet service providers and content providers, to which Cogent provides Internet transit. They hand traffic to Cogent in data centers and rely on Cogent to deliver it. About half of netcentric revenue is from outside the U.S.

BRANDS/DIVISIONS/AFFILIATES:

CONTACTS: Note: Officers with more than one job title may be intentionally listed here more than once.

David Schaeffer, CEO
Jean-Michel Slagmuylder, CFO, Geographical
Sean Wallace, CFO
John Chang, Chief Legal Officer
R. Kummer, Chief Technology Officer
James Bubeck, Other Executive Officer
Thaddeus Weed, Senior VP, Divisional
Bryant Banks, Vice President, Divisional
Henry Kilmer, Vice President, Divisional

FINANCIAL DATA: Note: Data for latest year may not have been available at press time.

In U.S. $	2021	2020	2019	2018	2017	2016
Revenue	589,797,000	568,103,000	546,159,000	520,193,000	485,175,000	446,900,000
R&D Expense						
Operating Income	111,840,000	106,993,000	99,198,000	85,576,000	72,056,000	56,890,000
Operating Margin %	.19%	.19%	.18%	.16%	.15%	.13%
SGA Expense	162,380,000	158,476,000	146,913,000	133,858,000	127,915,000	120,709,000
Net Income	48,185,000	6,216,000	37,520,000	28,667,000	5,876,000	14,929,000
Operating Cash Flow	170,257,000	140,320,000	148,809,000	133,921,000	111,702,000	107,967,000
Capital Expenditure	69,916,000	55,952,000	46,958,000	49,937,000	45,801,000	45,234,000
EBITDA	227,734,000	156,275,000	190,374,000	173,671,000	155,511,000	140,298,000
Return on Assets %	.05%	.01%	.04%	.04%	.01%	.02%
Return on Equity %						
Debt to Equity						

CONTACT INFORMATION:

Phone: 202 295-4200   Fax:
Toll-Free: 877-875-4432
Address: 2450 N St. NW, Washington, DC 20037 United States

STOCK TICKER/OTHER:

Stock Ticker: CCOI   Exchange: NAS
Employees: 1,001   Fiscal Year Ends: 12/31
Parent Company:

SALARIES/BONUSES:

Top Exec. Salary: $350,000   Bonus: $
Second Exec. Salary: Bonus: $
$297,744

OTHER THOUGHTS:

Estimated Female Officers or Directors:
Hot Spot for Advancement for Women/Minorities:

Colt Technology Services Group Limited
www.colt.net
NAIC Code: 517110

TYPES OF BUSINESS:
Business Voice & Data Services
Data Center & Managed Services
Fiber Optics
Internet
Ethernet
Cybersecurity
Cloud
Data Center

BRANDS/DIVISIONS/AFFILIATES:
Fidelity Investments Inc
Colt IQ Network

GROWTH PLANS/SPECIAL FEATURES:
Colt Technology Services Group Limited is a business communications solutions and services provider. The company offers its services to midsize and major businesses and wholesale customers across Europe, Asia and North America. Colt enables the digital transformation of businesses through its intelligent, purpose-built, cloud-integrated network, known as the Colt IQ Network. Colt's solutions include optical network services, business internet, virtual private network (VPN), Ethernet, voice, cybersecurity, cloud connectivity and data centers. Colt Group connects over 900 data centers worldwide, with more than 29,000 on-net buildings. The firm is owned by Fidelity Investments, Inc., a multinational financial services corporation based in Massachusetts. Based in the U.K., Colt has offices in Austria, Belgium, China, Denmark, France, Germany, Hong Kong, India, Ireland, Italy, Japan, Netherlands, Portugal, Romania, Singapore, South Korea, Spain, Sweden, Switzerland, and the U.S.

CONTACTS:
Note: Officers with more than one job title may be intentionally listed here more than once.

Keri Gilder, CEO
Rajiv Datta, COO
Gary Carr, CFO
Andrew Edison, Exec. VP-Mktg. & Sales
Tessa Raum, Exec. VP-Human Resources
Ashish Surti, Exec. VP-IT & Security
Simon Walsh, Exec. VP-Enterprise Solutions
Caroline Griffin Pain, Sec.
Richard Oosterom, Exec. VP-Bus. Dev. & Strategy
Francois Eloy, Exec. VP-Colt Comm. Svcs.
Morten Singleton, VP-Investor Rel.
Mark Leonard, Exec. VP-Infrastructure Svcs.
Jurgen Hernichel, Exec. VP-Bus. Svcs. Unit
Bernard Geoghegan, Exec. VP-Data Center Svcs.
Adriaan Oosthoek, Exec. VP-Data Center Svcs.
Robin Farnan, Exec. VP-Oper. & Engineering
Andy Kankan, Managing Dir.-COLT India

FINANCIAL DATA:
Note: Data for latest year may not have been available at press time.

In U.S. $	2021	2020	2019	2018	2017	2016
Revenue	168,000,000	165,830,000	161,000,000	1,598,000,000	1,556,000,000	1,550,000,000
R&D Expense						
Operating Income						
Operating Margin %						
SGA Expense						
Net Income						
Operating Cash Flow						
Capital Expenditure						
EBITDA						
Return on Assets %						
Return on Equity %						
Debt to Equity						

CONTACT INFORMATION:
Phone: 4420-7390-3900 Fax:
Toll-Free:
Address: 20 Great Eastern St., London, EC2A 3EH United Kingdom

STOCK TICKER/OTHER:
Stock Ticker: Subsidiary Exchange:
Employees: 5,500 Fiscal Year Ends: 12/31
Parent Company: Fidelity Investments Inc

SALARIES/BONUSES:
Top Exec. Salary: $ Bonus: $
Second Exec. Salary: $ Bonus: $

OTHER THOUGHTS:
Estimated Female Officers or Directors: 1
Hot Spot for Advancement for Women/Minorities: Y

Comba Telecom Systems Holdings Ltd   www.comba-telecom.com

NAIC Code: 517210

TYPES OF BUSINESS:

Wireless Telecommunications Carriers (except Satellite)
Wireless Communication enhancement

BRANDS/DIVISIONS/AFFILIATES:

GROWTH PLANS/SPECIAL FEATURES:

Comba Telecom Systems Holdings Ltd is an investment holding company. The company operates in two segments: Wireless telecommunications network system equipment & services and Operator telecommunication services. Its solutions include Antenna Site, In-Vehicle Broadband, Indoor Coverage Enhancement, Large Venue Coverage, LTE Upgrade, Railway/Tunnel Coverage, Wireless Backhaul, and Small Cell. It serves various industries such as airports, Education, Government and Public, Hospitality, Stadium, Transportation, and others. The company generates maximum revenue from the Wireless telecommunications network system equipment & services segment. Its geographical segments are Mainland China, Other countries/areas in the Asia Pacific, Americas, European Union, Middle East, and Other countries.

CONTACTS: Note: Officers with more than one job title may be intentionally listed here more than once.

Hui Jun Xu, Pres.
Fie Fu Chang, CFO
Tung Ling Fok, Chmn.

FINANCIAL DATA: Note: Data for latest year may not have been available at press time.

In U.S. $	2021	2020	2019	2018	2017	2016
Revenue	747,750,000	644,253,200	736,316,500	721,461,800	708,775,500	758,535,300
R&D Expense	77,574,250	60,008,760	44,177,720	44,980,930	42,208,620	28,995,500
Operating Income	-38,764,960	-470,842	51,640,490	-673,268	17,293,370	41,811,800
Operating Margin %	- .05%	.00%	.07%	.00%	.02%	.06%
SGA Expense	152,694,800	141,894,600	144,407,400	153,946,900	138,370,400	159,714,000
Net Income	-75,488,450	-24,727,350	19,331,650	-21,833,000	3,487,109	19,396,360
Operating Cash Flow	47,721,910	-42,614,240	90,395,900	105,178,400	41,052,540	52,443,060
Capital Expenditure	42,584,680	45,400,060	63,004,140	49,530,880	35,001,660	16,521,120
EBITDA	-19,144,760	17,136,810	66,041,550	25,880,120	40,598,900	56,226,360
Return on Assets %	- .05%	- .02%	.01%	- .02%	.00%	.02%
Return on Equity %	- .15%	- .05%	.05%	- .05%	.01%	.04%
Debt to Equity	.23%	0.125	0.502	0.115	0.131	0.196

CONTACT INFORMATION:

Phone: 852 2636-6861   Fax: 852 2637-0966
Toll-Free:
Address: 611 East Wing, No. 8 Science Park W. Ave., Hong Kong Science Park, Tai Po, Hong Kong, Hong Kong 999077 Hong Kong

STOCK TICKER/OTHER:

Stock Ticker: COBJF   Exchange: PINX
Employees: 5,500   Fiscal Year Ends: 12/31
Parent Company:

SALARIES/BONUSES:

Top Exec. Salary: $ Bonus: $
Second Exec. Salary: $ Bonus: $

OTHER THOUGHTS:

Estimated Female Officers or Directors:
Hot Spot for Advancement for Women/Minorities:

Sales, profits and employees may be estimates. Financial information, benefits and other data can change quickly and may vary from those stated here.

Comcast Corporation

corporate.comcast.com

NAIC Code: 517110

TYPES OF BUSINESS:

Cable Television
VoIP Service
Cable Network Programming
High-Speed Internet Service
Video-on-Demand
Advertising Services
Streaming TV Programming
Wireless Services

BRANDS/DIVISIONS/AFFILIATES:

Sky Limited
XFINITY
Universal Pictures
Sky News
Sky Sports
Philadelphia Flyers
Universal Studios
Peacock

GROWTH PLANS/SPECIAL FEATURES:

Comcast is made up of three parts. The core cable business owns networks capable of providing television, internet access, and phone services to roughly 61 million U.S. homes and businesses, or nearly half of the country. About 57% of the homes in this territory subscribe to at least one Comcast service. Comcast acquired NBCUniversal from General Electric in 2011. NBCU owns several cable networks, including CNBC, MSNBC, and USA, the NBC broadcast network, several local NBC affiliates, Universal Studios, and several theme parks. Sky, acquired in 2018, is the dominant television provider in the U.K. and has invested heavily in exclusive and proprietary content to build this position. The firm is also the largest pay-television provider in Italy and has a presence in Germany and Austria.

CONTACTS:
Note: Officers with more than one job title may be intentionally listed here more than once.

Jeffrey Shell, CEO, Subsidiary
David Watson, CEO, Subsidiary
Brian Roberts, CEO
Michael Cavanagh, CFO
Daniel Murdock, Chief Accounting Officer
Adam Miller, Chief Administrative Officer
Thomas Reid, Chief Legal Officer
Sheldon Bonovitz, Director Emeritus

FINANCIAL DATA:
Note: Data for latest year may not have been available at press time.

In U.S. $	2021	2020	2019	2018	2017	2016
Revenue	116,385,000,000	103,564,000,000	108,942,000,000	94,507,000,000	85,029,000,000	80,736,000,000
R&D Expense						
Operating Income	20,817,000,000	17,493,000,000	21,125,000,000	19,009,000,000	18,018,000,000	16,831,000,000
Operating Margin %	.18%	.17%	.19%	.20%	.21%	.21%
SGA Expense	7,695,000,000	6,741,000,000	7,617,000,000	7,036,000,000	6,519,000,000	6,291,000,000
Net Income	14,159,000,000	10,534,000,000	13,057,000,000	11,731,000,000	22,735,000,000	8,678,000,000
Operating Cash Flow	29,146,000,000	24,737,000,000	25,697,000,000	24,297,000,000	21,261,000,000	19,691,000,000
Capital Expenditure	12,057,000,000	11,634,000,000	12,428,000,000	11,709,000,000	11,155,000,000	10,687,000,000
EBITDA	37,178,000,000	31,753,000,000	34,516,000,000	29,801,000,000	28,569,000,000	26,694,000,000
Return on Assets %	.05%	.04%	.05%	.05%	.12%	.05%
Return on Equity %	.15%	.12%	.17%	.17%	.37%	.16%
Debt to Equity	.96%	1.114	1.182	1.499	0.866	1.03

CONTACT INFORMATION:

Phone: 215 286-1700   Fax:
Toll-Free: 800-266-2278
Address: One Comcast Center, Philadelphia, PA 19103 United States

STOCK TICKER/OTHER:

Stock Ticker: CMCSA   Exchange: NAS
Employees: 189,000   Fiscal Year Ends: 12/31
Parent Company:

SALARIES/BONUSES:

Top Exec. Salary: $3,249,415   Bonus: $
Second Exec. Salary: $2,500,000   Bonus: $

OTHER THOUGHTS:

Estimated Female Officers or Directors: 16
Hot Spot for Advancement for Women/Minorities: Y

Sales, profits and employees may be estimates. Financial information, benefits and other data can change quickly and may vary from those stated here.

CommScope Holding Company Inc

www.commscope.com

NAIC Code: 335921

TYPES OF BUSINESS:

Cable-Coaxial & Fiber Optic
Broadband Networks
Outdoor Wireless Networks
Venue Networks
Campus Networks
Home Networks

BRANDS/DIVISIONS/AFFILIATES:

CommScope NEXT

GROWTH PLANS/SPECIAL FEATURES:

CommScope Holding Co Inc provides infrastructure services
for communications networks. It helps customers increase
bandwidth, maximize existing capacity, improve network
performance and availability, and simplify technology
migration. Its product portfolio consists of products and
services such as wired and wireless systems, cables
broadband devices, distribution and transmission equipment
and WiFi devices used by network services providers. The
company organizes itself into five segments based on the
product type: connectivity solutions, mobility solutions
customer premises equipment, network & cloud, and Ruckus
Networks. The connectivity and customer premises equipment
segments together generate majority of the revenue, and
roughly half of the revenue is earned in United States.

CONTACTS: *Note: Officers with more than one job title may be intentionally listed here more than once.*

Charles Treadway, CEO
Alexander Pease, CFO
Frank Drendel, Chairman Emeritus
Claudius Watts, Chairman of the Board
Brooke Clark, Chief Accounting Officer
Justin Choi, Chief Legal Officer
Ric Johnsen, Other Corporate Officer
Robyn Mingle, Other Executive Officer
John Carlson, Other Executive Officer
Kyle Lorentzen, Other Executive Officer

FINANCIAL DATA: *Note: Data for latest year may not have been available at press time.*

In U.S. $	2021	2020	2019	2018	2017	2016
Revenue	8,586,700,000	8,435,900,000	8,345,100,000	4,568,500,000	4,560,600,000	4,923,621,000
R&D Expense	683,200,000	703,300,000	578,500,000	185,700,000	185,600,000	201,321,000
Operating Income	154,200,000	243,300,000	-44,700,000	509,000,000	515,800,000	649,066,000
Operating Margin %	.02%	.03%	- .01%	.11%	.11%	.13%
SGA Expense	1,233,900,000	1,170,700,000	1,277,100,000	674,000,000	733,100,000	881,661,000
Net Income	-462,600,000	-573,400,000	-929,500,000	140,200,000	193,800,000	222,838,000
Operating Cash Flow	122,300,000	436,200,000	596,400,000	494,100,000	586,300,000	640,221,000
Capital Expenditure	131,400,000	121,200,000	104,100,000	82,300,000	68,700,000	68,314,000
EBITDA	813,000,000	746,600,000	274,100,000	770,200,000	844,800,000	949,156,000
Return on Assets %	- .04%	- .04%	- .09%	.02%	.03%	.03%
Return on Equity %	-5.24%	-1.06%	- .75%	.08%	.13%	.17%
Debt to Equity		26.728	11.719	2.269	2.652	3.263

CONTACT INFORMATION:

Phone: 828-324-2200   Fax:
Toll-Free: 800-982-1708
Address: 1100 CommScope Place SE, Hickory, NC 28602 United States

STOCK TICKER/OTHER:

Stock Ticker: COMM   Exchange: NAS
Employees: 30,000   Fiscal Year Ends: 12/31
Parent Company:

SALARIES/BONUSES:

Top Exec. Salary: $1,100,000   Bonus: $
Second Exec. Salary: Bonus: $
$587,500

OTHER THOUGHTS:

Estimated Female Officers or Directors: 2
Hot Spot for Advancement for Women/Minorities:

Comtech Telecommunications Corp

www.comtechtel.com

NAIC Code: 334220

TYPES OF BUSINESS:
Communications Equipment-Microwave & RF
Satellite Equipment & Technologies
Communication Solutions
Next-Generation 911 Solutions

GROWTH PLANS/SPECIAL FEATURES:
Comtech Telecommunications Corp is a provider of advanced communications solutions. The company is engaged in designing, developing, producing and marketing products, systems, and services for communications solutions. It is engaged in two business segments, Commercial Solutions, and Government Solutions Segment.

BRANDS/DIVISIONS/AFFILIATES:
UHP Networks Inc

CONTACTS: *Note: Officers with more than one job title may be intentionally listed here more than once.*
Fred Kornberg, CEO
Michael Bondi, CFO
Michael Porcelain, COO
Yelena Simonyuk, Other Corporate Officer
Nancy Stallone, Secretary

FINANCIAL DATA: *Note: Data for latest year may not have been available at press time.*

In U.S. $	2021	2020	2019	2018	2017	2016
Revenue	581,695,000	616,715,000	671,797,000	570,589,000	550,368,000	411,004,000
R&D Expense	49,148,000	52,180,000	56,407,000	53,869,000	54,260,000	42,190,000
Operating Income	31,994,000	35,928,000	44,074,000	35,075,000	25,022,000	20,700,000
Operating Margin %	.06%	.06%	.07%	.06%	.05%	.05%
SGA Expense	111,796,000	117,130,000	128,639,000	113,922,000	116,080,000	94,932,000
Net Income	-73,480,000	7,020,000	25,041,000	29,769,000	15,827,000	-7,738,000
Operating Cash Flow	-40,638,000	52,764,000	68,031,000	50,344,000	66,917,000	15,075,000
Capital Expenditure	16,037,000	7,225,000	8,785,000	8,642,000	8,150,000	5,667,000
EBITDA	-37,760,000	47,520,000	68,437,000	69,805,000	74,287,000	22,803,000
Return on Assets %	- .08%	.01%	.03%	.04%	.02%	- .01%
Return on Equity %	- .14%	.01%	.05%	.06%	.03%	- .02%
Debt to Equity	.48%	0.316	0.308	0.294	0.371	0.519

CONTACT INFORMATION:
Phone: 631 962-7000   Fax: 631 777-8877
Toll-Free:
Address: 68 S. Service Rd., Ste. 230, Melville, NY 11747 United States

STOCK TICKER/OTHER:
Stock Ticker: CMTL   Exchange: NAS
Employees: 2,038   Fiscal Year Ends: 07/31
Parent Company:

SALARIES/BONUSES:
Top Exec. Salary: $840,000   Bonus: $
Second Exec. Salary: $525,000   Bonus: $

OTHER THOUGHTS:
Estimated Female Officers or Directors:
Hot Spot for Advancement for Women/Minorities:

Sales, profits and employees may be estimates. Financial information, benefits and other data can change quickly and may vary from those stated here.

Consensus Cloud Solutions Inc

www.j2global.com

NAIC Code: 517110

TYPES OF BUSINESS:

Unified Messaging & Communication Services
Internet-Based Faxing
Internet Conferencing
Cloud-Based Communications Services
Customer Relationship Management Solutions

GROWTH PLANS/SPECIAL FEATURES:

Consensus Cloud Solutions Inc is a provider of secure information delivery services with a scalable Software-as-a-Service SaaS platform. Its online fax products have both a fixed and variable subscription component with the substantial majority of revenues derived from the fixed portion. Geographically, it derives a maximum revenue from the United States.

BRANDS/DIVISIONS/AFFILIATES:

MyFax
eVoice
LiveDrive
IPVanish
Campaigner
IGN
Mashable
Everyday Health

CONTACTS: *Note: Officers with more than one job title may be intentionally listed here more than once.*

R. Turicchi, CEO
Steve Emberland, Chief Accounting Officer
Jeffrey Sullivan, Chief Technology Officer
John Nebergall, COO

FINANCIAL DATA: *Note: Data for latest year may not have been available at press time.*

In U.S. $	2021	2020	2019	2018	2017	2016
Revenue	352,664,000	331,168,000	322,559,000	597,975,000		
R&D Expense	7,652,000	7,146,000	9,745,000	27,656,000		
Operating Income	175,136,000	196,675,000	189,827,000	230,120,000		
Operating Margin %	.50%	.59%	.59%	.38%		
SGA Expense	111,876,000	73,968,000	72,997,000	218,045,000		
Net Income	109,001,000	152,913,000	212,967,000	152,058,000		
Operating Cash Flow	233,675,000	238,789,000	226,702,000	235,805,000		
Capital Expenditure	34,509,000	35,573,000	22,943,000	14,191,000		
EBITDA	227,167,000	307,998,000	270,173,000	289,982,000		
Return on Assets %	.11%	.10%	.14%			
Return on Equity %	.28%	.19%	.46%			
Debt to Equity		0.023	1.491			

CONTACT INFORMATION:

Phone: 323 860-9200   Fax: 323 860-9201
Toll-Free:
Address: 700 S. Flower St., Fl. 15, Los Angeles, CA 90017 United States

STOCK TICKER/OTHER:

Stock Ticker: CCSI   Exchange: NAS
Employees: 459   Fiscal Year Ends: 12/31
Parent Company:

SALARIES/BONUSES:

Top Exec. Salary: $175,481   Bonus: $
Second Exec. Salary: Bonus: $
$115,769

OTHER THOUGHTS:

Estimated Female Officers or Directors: 2
Hot Spot for Advancement for Women/Minorities:

Consolidated Communications Holdings Inc www.consolidated.com

NAIC Code: 517110

TYPES OF BUSINESS:
Telephone Service Provider
Internet Service Provider
Internet Protocol Digital TV
VoIP Telephony

BRANDS/DIVISIONS/AFFILIATES:

GROWTH PLANS/SPECIAL FEATURES:

Consolidated Communications Holdings Inc provides communication services for business and residential customers across various states in the U.S. Its business product suite includes data and Internet solutions, voice, data center services, security services, managed and IT services, and an expanded suite of cloud services. It provides wholesale solutions to wireless and wireline carriers and other service providers including data, voice, network connections, and custom fiber builds and last-mile connections. It offers residential high-speed Internet, video, phone, and home security services as well as multi-service residential and small business bundles.

CONTACTS: Note: Officers with more than one job title may be intentionally listed here more than once.
C. Udell, CEO
Steven Childers, CFO
Robert Currey, Chairman of the Board

FINANCIAL DATA: Note: Data for latest year may not have been available at press time.

In U.S. $	2021	2020	2019	2018	2017	2016
Revenue	1,282,233,000	1,304,028,000	1,336,542,000	1,399,074,000	1,059,574,000	743,177,000
R&D Expense						
Operating Income	140,882,000	143,159,000	81,281,000	20,929,000	72,562,000	91,235,000
Operating Margin %	.11%	.11%	.06%	.01%	.07%	.12%
SGA Expense	271,125,000	275,361,000	299,088,000	333,605,000	249,141,000	156,520,000
Net Income	-107,085,000	36,977,000	-20,383,000	-50,834,000	64,945,000	14,931,000
Operating Cash Flow	318,867,000	364,980,000	339,096,000	357,321,000	210,027,000	218,233,000
Capital Expenditure	480,346,000	217,563,000	232,203,000	244,816,000	181,185,000	125,192,000
EBITDA	441,479,000	468,023,000	462,518,000	453,597,000	364,435,000	265,245,000
Return on Assets %	-.03%	.01%	-.01%	-.01%	.02%	.01%
Return on Equity %	-.24%	.09%	-.06%	-.11%	.17%	.07%
Debt to Equity	3.92%	5.676	6.603	5.622	4.068	8.053

CONTACT INFORMATION:
Phone: 217 235-3311   Fax: 217 258-7883
Toll-Free:
Address: 2116 S. 17th ST., Mattoon, IL 61938 United States

STOCK TICKER/OTHER:
Stock Ticker: CNSL   Exchange: NAS
Employees: 3,200   Fiscal Year Ends: 12/31
Parent Company:

SALARIES/BONUSES:
Top Exec. Salary: $612,000   Bonus: $
Second Exec. Salary: $400,000   Bonus: $

OTHER THOUGHTS:
Estimated Female Officers or Directors: 1
Hot Spot for Advancement for Women/Minorities:

Converge Information and Communications Technology Solutions, Inc.

www.convergeict.com

NAIC Code: 517110

<table>
<tr><td valign="top">

TYPES OF BUSINESS:

Wired Telecommunications Carriers
Fiber Broadband
Internet
Connectivity Solutions and Services
Network Facilities
Telecommunications
Data Transmission Services

BRANDS/DIVISIONS/AFFILIATES:

</td><td valign="top">

GROWTH PLANS/SPECIAL FEATURES:

Converge Information and Communications Technology Solutions, Inc. (Converge ICT) provides end-to-end fiber internet connection from its network facilities to customers, including residential, and small/medium businesses and enterprises. Converge ICT ensures no loss in data transmission or slowdown in internet connectivity, and therefore partners with telecommunications brands from all over the world to accomplish this. Partners include, but are not limited to: Level 3, Hutchison Global Communications, Verizon, Sprint, Singtel, TATA Communications, Netflix, Google and Facebook. Converge ICT plans range from 100 megabits per second (Mbps) of broadband speed to 300 Mbps to 800 Mbps, as well as bundled plans. Every plan comes with unlimited data. As of March 2022, Converge ICT had more than 1.8 million FiberX subscribers and its fiber backbone spanned nearly 500 cities and municipalities across the Philippines.

</td></tr>
</table>

CONTACTS: Note: Officers with more than one job title may be intentionally listed here more than once.

Dennis Anthony H. Uy, CEO
Grace Y. Uy, Pres.
Jesus C. Romero, COO
Matthias Vukovich, Chief Financial Office Advisor
Albert Santos, Chief Customer Experience Officer
Ulysses C. Naguit, CIO
Ronald Brusola, CTO

FINANCIAL DATA: Note: Data for latest year may not have been available at press time.

In U.S. $	2021	2020	2019	2018	2017	2016
Revenue	33,861,256	325,589,000	179,884,000			
R&D Expense						
Operating Income						
Operating Margin %						
SGA Expense						
Net Income						
Operating Cash Flow						
Capital Expenditure						
EBITDA						
Return on Assets %						
Return on Equity %						
Debt to Equity						

CONTACT INFORMATION:

Phone: 632 8667 0848   Fax:
Toll-Free:
Address: 99 E. Rodriguez Jr. Ave., Brgy. Ugong Pasig City, 2009 Philippines

STOCK TICKER/OTHER:

Stock Ticker: CNVRG   Exchange: Manila
Employees: 2,710   Fiscal Year Ends: 12/31
Parent Company:

SALARIES/BONUSES:

Top Exec. Salary: $ Bonus: $
Second Exec. Salary: $ Bonus: $

OTHER THOUGHTS:

Estimated Female Officers or Directors:
Hot Spot for Advancement for Women/Minorities:

Corning Incorporated

www.corning.com

NAIC Code: 327212

TYPES OF BUSINESS:

Glass & Optical Fiber Manufacturing
Glass Substrates for LCDs
Optical Switching Products
Photonic Modules & Components
Networking Devices
Semiconductor Materials
Laboratory Supplies
Emissions Control Products

BRANDS/DIVISIONS/AFFILIATES:

Eagle XG
Iris
Vascade
LEAF
ClearCurve
InfiniCor
Gorilla

GROWTH PLANS/SPECIAL FEATURES:

Corning Inc is a leader in materials science, specializing in the production of glass, ceramics and optical fiber. The firm supplies its products for a wide range of applications, from flat-panel displays in televisions to gasoline particulate filters in automobiles to optical fiber for broadband access, with a leading share in many of its end markets.

CONTACTS: *Note: Officers with more than one job title may be intentionally listed here more than once.*

Wendell Weeks, CEO
Martin Curran, Executive VP
R. Tripeny, CFO
Edward Schlesinger, Chief Accounting Officer
Lewis Steverson, Chief Administrative Officer
Anne Mullins, Chief Information Officer
Jeffrey Evenson, Chief Strategy Officer
David Morse, Chief Technology Officer
Eric Musser, COO
Clark Kinlin, Executive VP
Ronald Verkleeren, General Manager, Divisional
John Zhang, General Manager, Divisional
Avery Nelson, General Manager, Divisional

FINANCIAL DATA: *Note: Data for latest year may not have been available at press time.*

In U.S. $	2021	2020	2019	2018	2017	2016
Revenue	14,082,000,000	11,303,000,000	11,503,000,000	11,290,000,000	10,116,000,000	9,390,000,000
R&D Expense	995,000,000	1,154,000,000	1,031,000,000	993,000,000	864,000,000	736,000,000
Operating Income	2,112,000,000	509,000,000	1,306,000,000	1,575,000,000	1,608,000,000	1,501,000,000
Operating Margin %	.15%	.05%	.11%	.14%	.16%	.16%
SGA Expense	1,827,000,000	1,747,000,000	1,585,000,000	1,799,000,000	1,473,000,000	1,462,000,000
Net Income	1,906,000,000	512,000,000	960,000,000	1,066,000,000	-497,000,000	3,695,000,000
Operating Cash Flow	3,412,000,000	2,180,000,000	2,031,000,000	2,919,000,000	2,004,000,000	2,537,000,000
Capital Expenditure	1,637,000,000	1,377,000,000	1,978,000,000	2,310,000,000	1,804,000,000	1,130,000,000
EBITDA	4,178,000,000	2,419,000,000	2,940,000,000	2,987,000,000	2,970,000,000	5,046,000,000
Return on Assets %	.04%	.01%	.03%	.04%	- .02%	.13%
Return on Equity %	.09%	.04%	.08%	.08%	- .04%	.22%
Debt to Equity	.62%	0.771	0.771	0.522	0.354	0.234

CONTACT INFORMATION:

Phone: 607 974-9000   Fax: 607 974-8688
Toll-Free:
Address: 1 Riverfront Plaza, Corning, NY 14831 United States

STOCK TICKER/OTHER:

Stock Ticker: GLW   Exchange: NYS
Employees: 61,200   Fiscal Year Ends: 12/31
Parent Company:

SALARIES/BONUSES:

Top Exec. Salary: $1,492,648   Bonus: $
Second Exec. Salary: Bonus: $
$870,837

OTHER THOUGHTS:

Estimated Female Officers or Directors: 1
Hot Spot for Advancement for Women/Minorities: Y

COSMOTE Mobile Telephones SA

www.cosmote.gr

NAIC Code: 517210

TYPES OF BUSINESS:

Mobile Phone Service
Mobile Services
Telecommunication
Technology
Ecommerce
Television Production
Electronic Payment Solutions

BRANDS/DIVISIONS/AFFILIATES:

Deutsche Telekom AG
Hellenic Telecommunications Organization SA
COSMOTE eValue
CsomoONE
COSMOTE Payments

CONTACTS: *Note: Officers with more than one job title may be intentionally listed here more than once.*

Konstantinos Apostolidis, VP
George Tsonis, General Technical Dir.
Irini Nikolaidi, Chief Exec. Legal & Regulatory Affairs Officer
Deppie Tzimea, Group Corporate Communications Officer
Michalis Tsamaz, Chmn.

GROWTH PLANS/SPECIAL FEATURES:

COSMOTE Mobile Telephones SA is a leading mobile telecommunications company in Greece. Together with its subsidiary in Romania, COSMOTE offers mobile services to over 15 million customers. The firm utilizes state-of-the-art technologies, offering 5G, 4G+ and 4G networks and related services. COSMOTE e-Value is the company's 24/7/365 customer service division, offering telephone service solutions, sales development solutions, technical assistance, consulting, service client services and more. CosmoONE offers ecommerce services between businesses (B2B), and specializes in eProcurement solutions, applications and services for electronic bids, electronic tenders, electronic auctions, electronic orders, invoice transfer and more. COSMOTE's television productions division produces and distributes programs and audiovisual content intended for pay-tv and televised broadcasting. COSMOTE Payments is an electronic money services division. COSMOTE operates as a wholly-owned subsidiary of Hellenic Telecommuncations Organization SA (OTE SA), which is majority-owned by Deutsche Telekom AG.

FINANCIAL DATA: *Note: Data for latest year may not have been available at press time.*

In U.S. $	2021	2020	2019	2018	2017	2016
Revenue	1,551,954,200	1,494,530,000	1,272,400,000	1,278,320,000	1,336,820,000	1,181,040,000
R&D Expense						
Operating Income						
Operating Margin %						
SGA Expense						
Net Income	219,345,880	162,988,000	601,579,000	716,463,000	-493,520,000	129,588,000
Operating Cash Flow						
Capital Expenditure						
EBITDA						
Return on Assets %						
Return on Equity %						
Debt to Equity						

CONTACT INFORMATION:

Phone: 30 210-61-77-777   Fax: 30-210-61-77-999
Toll-Free:
Address: 99 Kifissias Ave., Maroussi, 151 24 Greece

STOCK TICKER/OTHER:

Stock Ticker: Subsidiary
Employees: 1,950
Parent Company: Deutsche Telekom AG

Exchange:
Fiscal Year Ends: 12/31

SALARIES/BONUSES:

Top Exec. Salary: $ Bonus: $
Second Exec. Salary: $ Bonus: $

OTHER THOUGHTS:

Estimated Female Officers or Directors: 4
Hot Spot for Advancement for Women/Minorities: Y

Sales, profits and employees may be estimates. Financial information, benefits and other data can change quickly and may vary from those stated here.

Cox Communications Inc

www.cox.com/residential/home.cox

NAIC Code: 517110

TYPES OF BUSINESS:

Cable TV Service and Internet Access
Broadband
Internet
TV
Streaming
Smart Home
Security
Telephone Service

BRANDS/DIVISIONS/AFFILIATES:

Cox Enterprises Inc

GROWTH PLANS/SPECIAL FEATURES:

Cox Communications, Inc., owned by Cox Enterprises, Inc., is a broadband communications and entertainment company, serving millions of customers throughout the U.S. Cox Communications' products and services include internet, TV, streaming, smart home, security, home phone and bundled deals. The company serves both residential and business customers.

CONTACTS: Note: Officers with more than one job title may be intentionally listed here more than once.

Patrick J. Esser, Pres.
Len Barlik, Exec. VP-Prod. Mgmt. & Dev.
Asheesh Saksena, Chief Strategy Officer
Joseph J. Rooney, Sr. VP-Social Media, Advertising & Brand Mktg.
William (Bill) J. Fitzsimmons, Chief Acct. Officer
Philip G. Meeks, Sr. VP-Cox Bus.
Jennifer W. Hightower, Sr. VP-Law & Policy
David Pugliese, Sr. VP-Product Mktg.
Mark A. Kaish, Sr. VP-Tech. Oper.
George Richter, VP-Supply Chain Mgmt.

FINANCIAL DATA: Note: Data for latest year may not have been available at press time.

In U.S. $	2021	2020	2019	2018	2017	2016
Revenue	13,104,000,000	12,600,000,000	12,300,000,000	12,000,000,000	11,550,000,000	11,000,000,000
R&D Expense						
Operating Income						
Operating Margin %						
SGA Expense						
Net Income						
Operating Cash Flow						
Capital Expenditure						
EBITDA						
Return on Assets %						
Return on Equity %						
Debt to Equity						

CONTACT INFORMATION:

Phone: 404-843-5000   Fax: 404-843-5939
Toll-Free: 888-566-7751
Address: 6205-B Peachtree Dunwoody Rd., Atlanta, GA 30328 United States

STOCK TICKER/OTHER:

Stock Ticker: Subsidiary   Exchange:
Employees: 18,000   Fiscal Year Ends: 12/31
Parent Company: Cox Enterprises Inc

SALARIES/BONUSES:

Top Exec. Salary: $ Bonus: $
Second Exec. Salary: $ Bonus: $

OTHER THOUGHTS:

Estimated Female Officers or Directors: 3
Hot Spot for Advancement for Women/Minorities: Y

Sales, profits and employees may be estimates. Financial information, benefits and other data can change quickly and may vary from those stated here.

Cricket Wireless LLC

www.cricketwireless.com

NAIC Code: 517210

TYPES OF BUSINESS:

Mobile Phone and Wireless Services
Retail Sales
Wireless Services
Mobile Phones
Telecommunications
Lease-to-own

BRANDS/DIVISIONS/AFFILIATES:

AT&T Inc
Cricket

GROWTH PLANS/SPECIAL FEATURES:

Cricket Wireless, LLC provides mobile wireless services targeted to meet the needs of younger and lower-income customers underserved by traditional communications companies. Its 5G and 4G LTE (long-term evolution) network covers approximately 99% of Americans (based on overall coverage in U.S.-licensed areas). Cricket Wireless offers smartphones under brands such as LG, Motorola, iPhone, Alcatel, Samsung, Nokia and Cricket. Cricket-branded retail stores are located nationwide and are either dealer-owned or company-owned. Cricket's rate plans offer options for cell services, and include 5GB of nationwide 5G access, 10GB nationwide 5G access, unlimited data with video streaming at 1.5 Mbps, and unlimited premium data options which includes 15GB of mobile hotspot services. Monthly taxes are included with each plan. There are no annual contracts and no overage fees with unlimited talk, text and data access cell phone plans. Cricket Wireless operates as a subsidiary of AT&T, Inc. In early-2022, Cricket Wireless announced the shutdown of its 3G network starting that February.

CONTACTS:
Note: Officers with more than one job title may be intentionally listed here more than once.

John Dwyer, Pres.
Colin E. Holland, Sr. VP - Eng. & Tech. Oper.
Robert J. Irving, Jr., Sr. VP-Gen. Counsel
William D. Ingram, Exec. VP-Strategy
Anne M. Liu, Sr. VP
John H. Casey III, Sr. VP -Customer Experience & Growth
David B. Davis, Sr. VP
Annette M. Jacobs, Sr. VP
Aaron P. Maddox, Sr. VP-Finance
Catherine Shackleford, Sr. VP-Supply Chain & Procurement

FINANCIAL DATA:
Note: Data for latest year may not have been available at press time.

In U.S. $	2021	2020	2019	2018	2017	2016
Revenue	4,419,657,900	4,290,930,000	4,410,000,000	4,200,000,000	4,000,000,000	3,900,000,000
R&D Expense						
Operating Income						
Operating Margin %						
SGA Expense						
Net Income						
Operating Cash Flow						
Capital Expenditure						
EBITDA						
Return on Assets %						
Return on Equity %						
Debt to Equity						

CONTACT INFORMATION:

Phone: Fax:
Toll-Free: 800 274-2538
Address: 1025 Lenox Park Blvd. NE, Atlanta, GA 30319 United States

STOCK TICKER/OTHER:

Stock Ticker: Subsidiary
Employees: 3,800
Parent Company: AT&T Inc

Exchange:
Fiscal Year Ends: 12/31

SALARIES/BONUSES:

Top Exec. Salary: $ Bonus: $
Second Exec. Salary: $ Bonus: $

OTHER THOUGHTS:

Estimated Female Officers or Directors:
Hot Spot for Advancement for Women/Minorities: Y

Cyxtera Technologies Inc

NAIC Code: 518210

www.cyxtera.com

TYPES OF BUSINESS:
Hosting & Collocation Services
Data Centers
Networking
Digital Computing
Artificial Intelligence
Machine Learning

BRANDS/DIVISIONS/AFFILIATES:
Medina Capital
BC Partners LLP

GROWTH PLANS/SPECIAL FEATURES:
Cyxtera Technologies Inc is engaged in data center colocation and interconnection services. The company operates a footprint of more than 60 data centers around the world, providing services to more than 2,300 leading enterprises and U.S. federal government agencies. Cyxtera brings proven operational excellence, global scale, flexibility and customer-focused innovation together to provide a comprehensive portfolio of data center and interconnection services.

CONTACTS: Note: Officers with more than one job title may be intentionally listed here more than once.
Nelson Fonseca, CEO
Carlos Sagasta, CFO
Manuel Medina, Chairman of the Board
Victor Semah, Chief Compliance Officer
Randy Rowland, COO
Leo Taddeo, Other Executive Officer

FINANCIAL DATA: Note: Data for latest year may not have been available at press time.

In U.S. $	2021	2020	2019	2018	2017	2016
Revenue	703,700,000	690,500,000				
R&D Expense						
Operating Income	-40,200,000	50,400,000				
Operating Margin %	-.06%	.07%				
SGA Expense	112,800,000	115,500,000				
Net Income	-257,900,000	-122,800,000				
Operating Cash Flow	25,800,000	116,600,000				
Capital Expenditure	77,500,000	83,200,000				
EBITDA	89,700,000	277,100,000				
Return on Assets %	-.08%	-.04%				
Return on Equity %	-.44%	-.22%				
Debt to Equity	3.03%	4.024				

CONTACT INFORMATION:
Phone: 305-537-9500   Fax:
Toll-Free: 855-699-8372
Address: BAC Colonnade Office Twr., 2333 Ponce De Leon Blvd, Coral Gables, FL 33134 United States

STOCK TICKER/OTHER:
Stock Ticker: CYXT   Exchange: NAS
Employees: 725   Fiscal Year Ends: 12/31
Parent Company: Medina Capital

SALARIES/BONUSES:
Top Exec. Salary: $500,000   Bonus: $400,000
Second Exec. Salary: $400,000   Bonus: $200,000

OTHER THOUGHTS:
Estimated Female Officers or Directors: 1
Hot Spot for Advancement for Women/Minorities:

Sales, profits and employees may be estimates. Financial information, benefits and other data can change quickly and may vary from those stated here.

Data Select Limited

NAIC Code: 517210

TYPES OF BUSINESS:

Mobile Phone Sales & Distribution
Marketing
Business-to-Business Solutions
Retail Solutions
Ecommerce Solutions
Mobile Network Solutions
Telecommunications

BRANDS/DIVISIONS/AFFILIATES:

Westcoast Group Holdings Limited

GROWTH PLANS/SPECIAL FEATURES:

Data Select Limited provides a marketing platform for mobile telecommunication brands to engage with U.K. business-to-business (B2B) resellers, retailers, ecommerce businesses and mobile network operators. The company's portfolio helps customers leverage advanced mobile technologies to bring their products and services to market. Data Select offers mobile B2B resellers access to: a broad smartphone, tablet, drone and accessories portfolio; mobile solutions; and a B2B loyalty program. The firm offers retailers an end-to-end mobile supply chain solution across a variety of shopping channels, including ecommerce, home shopping and stores. Data Select also manages multiple sales channel and network marketing promotions to help deliver product and sales support for retailers and network operators alike. Data Select serves hundreds of active customers, spanning 30+ brands. Data Select is privately-owned by Westcoast Group Holdings Limited.

CONTACTS: *Note: Officers with more than one job title may be intentionally listed here more than once.*

Chris Brookfield, Mngr.-Creative & Digital Mktg.
Joe Hemani, Chmn.-Westcoast

FINANCIAL DATA: *Note: Data for latest year may not have been available at press time.*

In U.S. $	2021	2020	2019	2018	2017	2016
Revenue	223,187,200	144,769,000	209,857,000	215,119,000	201,940,000	194,508,000
R&D Expense						
Operating Income						
Operating Margin %						
SGA Expense						
Net Income		909,840	2,456,622	2,339,640	2,461,560	986,396
Operating Cash Flow						
Capital Expenditure						
EBITDA						
Return on Assets %						
Return on Equity %						
Debt to Equity						

CONTACT INFORMATION:

Phone: 44 844-249-0792   Fax:
Toll-Free:
Address: Arrowhead Rd., Theale, Reading, RG7 4AH United Kingdom

STOCK TICKER/OTHER:

Stock Ticker: Subsidiary   Exchange:
Employees: 60   Fiscal Year Ends: 04/30
Parent Company: Westcoast Group Holdings Limited

SALARIES/BONUSES:

Top Exec. Salary: $ Bonus: $
Second Exec. Salary: $ Bonus: $

OTHER THOUGHTS:

Estimated Female Officers or Directors:
Hot Spot for Advancement for Women/Minorities:

Sales, profits and employees may be estimates. Financial information, benefits and other data can change quickly and may vary from those stated here.

Datang Telecom Technology Co Ltd

www.datang.com

NAIC Code: 334210

TYPES OF BUSINESS:

Telecommunications Equipment
Mobile Communications Products
Optical Communications Products
Digital Microwave Products
Software
Integrated Circuits
Internet of Things
Microelectronics

BRANDS/DIVISIONS/AFFILIATES:

CONTACTS:
Note: Officers with more than one job title may be intentionally listed here more than once.

Xincheng Lei, Managing Dir.
Gui Xue, CFO
Xiubin Qi, Corp. Sec.
Desheng Zhou, Vice Chmn.
Kun Jiang, Deputy Gen. Mgr.
Hongyan Wang, Deputy Gen. Mgr.
Pengfei Wang, Deputy Gen. Mgr.
Bin Cao, Chmn.

GROWTH PLANS/SPECIAL FEATURES:

Datang Telecom Technology Co., Ltd., owned by the Chinese government, is principally engaged in integrated circuit (IC) design, terminal design, software, applications, mobile internet services, communication devices and Internet of Things. Datang's IC design solutions include security chips and mobile terminal security chips. The firm implements hardware system-level terminal security solutions that comply with GP International Standards and UnionPay specifications. They can be used for enterprise applications, industrial applications, financial payments, digital rights, personal privacy and more. Terminal design products and services include smartphones, data terminals and industry terminals. Software solutions include wireless access, operation support systems, information security and platform products. Related applications include smart city, smart business, smart traffic, tourism, energy conservation, government information and more. Mobile internet products and services include custom-designed terminals, custom services and mobile applications. Datang's communication devices offer wired access, wireless access and optical communication. Its Internet of Things (IoT) and smart device products and solutions include smart cards, smart cities, car networking, sensor networks, information security, converged communication systems and video surveillance. In addition, Datang offers a core cloud integrated security+ solution based on loading a domestic cryptographic algorithm security chip, expanding the IoT category to include solutions such as smart door locks.

FINANCIAL DATA: *Note: Data for latest year may not have been available at press time.*

In U.S. $	2021	2020	2019	2018	2017	2016
Revenue	203,495,409	184,950,836	204,780,000	340,788,000	667,468,000	1,052,584,806
R&D Expense						
Operating Income						
Operating Margin %						
SGA Expense						
Net Income	-8,001,747	-266,214,215	-128,650,000	81,998,400	-406,745,000	-258,531,837
Operating Cash Flow						
Capital Expenditure						
EBITDA						
Return on Assets %						
Return on Equity %						
Debt to Equity						

CONTACT INFORMATION:

Phone: 86 10-5891-9000   Fax: 86-10 58919131
Toll-Free:
Address: Yongjia N. Rd., No. 6, Haidian Dist., Beijing, Beijing 100094 China

STOCK TICKER/OTHER:

Stock Ticker: 600198
Employees: 1,243
Parent Company:

Exchange: Shanghai
Fiscal Year Ends: 12/31

SALARIES/BONUSES:

Top Exec. Salary: $ Bonus: $
Second Exec. Salary: $ Bonus: $

OTHER THOUGHTS:

Estimated Female Officers or Directors:
Hot Spot for Advancement for Women/Minorities:

Sales, profits and employees may be estimates. Financial information, benefits and other data can change quickly and may vary from those stated here.

Deutsche Telekom AG

www.telekom.com

NAIC Code: 517210

TYPES OF BUSINESS:

Local & Long-Distance Telephone Service
Telecommunications
Fixed Network and Broadband
Mobile Communications
Internet and IPTV
Information and Communication Technology
Cloud-based Services
Internet of Things

BRANDS/DIVISIONS/AFFILIATES:

Deutsche Telekom (UK) Ltd
T-Mobile

CONTACTS: *Note: Officers with more than one job title may be intentionally listed here more than once.*

Timotheus Hottges, CEO
Thomas Kremer, Dir.-Legal Affairs, Compliance & Data Privacy
Claudia Nemat, Dir.-Europe & Technology
Reinhard Clemens, Dir.-T-Systems
Niek Jan van Damme, Dir.-Germany

GROWTH PLANS/SPECIAL FEATURES:

Deutsche Telekom AG (DT) is an integrated telecommunications company, with roughly 248 million mobile customers, 26 million fixed-network lines and 22 million broadband lines. DT offers its products and services in more than 50 countries worldwide while operating domestically out of Germany. The firm generally does business under the T-Mobile brand. DT provides fixed-network/broadband, mobile communications, internet and IPTV (internet protocol TV) products and services for consumers, and information and communication technology (ICT) solutions for business and corporate customers. The company's core business is the operation and sale of telecommunications services such as networks and connections. A range of integrated solutions is offered for business customers, and include the operation of legacy systems and IT services, the transformation to cloud-based services, and digitalization projects such as data analytics, Internet of Things (IoT) and artificial intelligence. While DT serves companies in all industries, its focus industries are automotive, manufacturing, logistics and transport, as well as healthcare and the public sector.

FINANCIAL DATA: *Note: Data for latest year may not have been available at press time.*

In U.S. $	2021	2020	2019	2018	2017	2016
Revenue	93,747,942,837	90,142,252,728	86,675,243,008	81,449,975,808	81,013,276,672	80,808,132,608
R&D Expense						
Operating Income						
Operating Margin %						
SGA Expense						
Net Income			4,179,998,208	2,341,317,888	3,741,136,128	2,957,271,552
Operating Cash Flow						
Capital Expenditure						
EBITDA						
Return on Assets %						
Return on Equity %						
Debt to Equity						

CONTACT INFORMATION:

Phone: 49 2281814949   Fax: 49 2281819400
Toll-Free:
Address: Friedrich-Ebert-Allee 140, Bonn, NW 53113 Germany

STOCK TICKER/OTHER:

Stock Ticker: DTEGY   Exchange: OTC
Employees: 226,000   Fiscal Year Ends: 12/31
Parent Company:

SALARIES/BONUSES:

Top Exec. Salary: $ Bonus: $
Second Exec. Salary: $ Bonus: $

OTHER THOUGHTS:

Estimated Female Officers or Directors: 6
Hot Spot for Advancement for Women/Minorities: Y

Dialogic Inc

NAIC Code: 334210

TYPES OF BUSINESS:

Internet Protocol Products
Cloud Solutions
Communications
Voice Over Solutions
Internet of Things
Software
Gateways

BRANDS/DIVISIONS/AFFILIATES:

Enghouse Systems Limited

CONTACTS: *Note: Officers with more than one job title may be intentionally listed here more than once.*

Bill Crank, CEO
John Hanson, CFO
Anthony Housefather, Executive VP, Divisional

GROWTH PLANS/SPECIAL FEATURES:

Dialogic, Inc. is a cloud-optimized solutions provider for real-time communications media, applications and infrastructure. The company serves service providers, enterprises and developers throughout the world, and works with world-leading mobile operators. Dialogic's solutions include applications for service providers and enterprises, NFV and cloud enablement, IP multimedia subsystem (IMS) for voice over long-term evolution (VoLTE) and voice over WiFi, web real-time communication (WebRTC), multimedia value-added services, VoLTE mobile data roaming, internetwork packet exchange (IPX) enablement, diameter and SS7 signaling, IP networking transformation, and Internet of Things (IoT). Products by Dialogic include unified communications, real-time communication apps, load balancer software, media server software, fax boards and related software, control switch systems, session border controllers, media gateways, SS7 components, and more. Dialogic has authorized distributors throughout the Americas, Asia Pacific, Europe, Middle East and Africa. Dialogic operates as a subsidiary of Enghouse Systems Limited.

FINANCIAL DATA: *Note: Data for latest year may not have been available at press time.*

In U.S. $	2021	2020	2019	2018	2017	2016
Revenue	130,000,000	128,750,000	125,000,000	115,000,000	110,000,000	105,000,000
R&D Expense						
Operating Income						
Operating Margin %						
SGA Expense						
Net Income						
Operating Cash Flow						
Capital Expenditure						
EBITDA						
Return on Assets %						
Return on Equity %						
Debt to Equity						

CONTACT INFORMATION:

Phone: 973-967-6000   Fax: 973-285-1832
Toll-Free:
Address: 4 Gatehall Dr., Parsippany, NJ 07054 United States

STOCK TICKER/OTHER:

Stock Ticker: Private   Exchange:
Employees: 1,000   Fiscal Year Ends: 12/31
Parent Company: Enghouse Systems Limited

SALARIES/BONUSES:

Top Exec. Salary: $ Bonus: $
Second Exec. Salary: $ Bonus: $

OTHER THOUGHTS:

Estimated Female Officers or Directors: 1
Hot Spot for Advancement for Women/Minorities:

Sales, profits and employees may be estimates. Financial information, benefits and other data can change quickly and may vary from those stated here.

Digi.com Bhd

NAIC Code: 517210

www.digi.com.my

TYPES OF BUSINESS:

Mobile Phone Services
Mobile Voice Services
Internet Services
Digital Services
Broadband Services
Mobile Applications

BRANDS/DIVISIONS/AFFILIATES:

Telenor Group
Digi Telecommunications Sdn Bhd
DiGi
MusicFreedom
VideoFreedom
MyDigi

GROWTH PLANS/SPECIAL FEATURES:

Digi.com is a telecommunications company and part of Telenor Group. The company is primarily involved in the provision of mobile and wireless Internet services. Digi.com's revenue is generated through a mixture of prepaid and postpaid customers, coupled with device sale revenue. The majority of revenue is from prepaid customers. From a regional and country perspective, the company earns the majority of revenue in Malaysia. Digi.com owns fiber and mobile-related infrastructure.

CONTACTS: *Note: Officers with more than one job title may be intentionally listed here more than once.*

Albern Murty, CEO
Eugene The, Chief Business Officer
Otto Risbakk, CFO
Praveen Rajan, CMO
Elisabeth Melander Stene, Chief Human Resources Officer
Kesavan Sivabalan, CTO
Christian Thrane, Head-Strategy & Bus. Transformation
Chan Nam Kiong, Head-Customer & Channels
Haakon Bruaset Kjoel, Chmn.

FINANCIAL DATA: *Note: Data for latest year may not have been available at press time.*

In U.S. $	2021	2020	2019	2018	2017	2016
Revenue	1,421,670,000	1,380,623,000	1,413,073,000	1,464,627,000	1,422,747,000	1,480,333,000
R&D Expense						
Operating Income	555,563,500	573,206,500	632,106,400	665,036,000	607,582,400	675,695,900
Operating Margin %	.39%	.42%	.45%	.45%	.43%	.46%
SGA Expense	84,522,610	90,784,470	101,864,900	109,904,000	125,188,600	128,460,000
Net Income	260,766,300	273,974,800	321,541,300	345,739,500	331,358,200	366,354,300
Operating Cash Flow	584,748,100	545,732,800	462,614,600	488,294,400	578,171,700	343,839,100
Capital Expenditure	180,265,700	161,613,200	168,453,300	183,653,500	301,634,200	174,056,800
EBITDA	670,609,700	694,956,800	748,429,100	676,503,600	651,420,800	665,868,500
Return on Assets %	.15%	.15%	.20%	.26%	.26%	.32%
Return on Equity %	1.88%	1.93%	2.15%	2.59%	2.85%	3.14%
Debt to Equity	6.06%	7.721	6.759	3.733	5.189	3.464

CONTACT INFORMATION:

Phone: 60 357211800   Fax:
Toll-Free:
Address: Jalan Delima 1/1, Subang Industrial Park, Lot 10, Selangor, 40000 Malaysia

STOCK TICKER/OTHER:

Stock Ticker: DIGBF   Exchange: PINX
Employees: 1,400   Fiscal Year Ends:
Parent Company: Telenor Group

SALARIES/BONUSES:

Top Exec. Salary: $ Bonus: $
Second Exec. Salary: $ Bonus: $

OTHER THOUGHTS:

Estimated Female Officers or Directors: 3
Hot Spot for Advancement for Women/Minorities: Y

DirecTV LLC (DIRECTV)

www.directv.com

NAIC Code: 517110

TYPES OF BUSINESS:

Satellite Broadcasting
Satellite TV
Streaming
Bundled Services
Internet
Fiber Network
Wireless

BRANDS/DIVISIONS/AFFILIATES:

AT&T Inc
TPG Capital
NFL Sunday Ticket
AT&T TV
U-verse

CONTACTS: Note: Officers with more than one job title may be intentionally listed here more than once.

Bill Morrow, CEO
Adolfo Alomar, COO
Ray Carpenter, CFO

GROWTH PLANS/SPECIAL FEATURES:

DirecTV, LLC is a leading provider of satellite TV and streaming TV services via televisions, online or mobile app. DirecTV offers four satellite TV packages, including Entertainment (with 160+ channels), Choice (185+ channels), Ultimate (250+ channels) and Premier (330+ channels), which differ in price depending on what the package offers. On the streaming side, the firm offers four packages as well: Entertainment (65+ channels), Choice (90+ channels), Ultimate (130+ channels) and Premier (140+ channels). Packages can be bundled with AT&T internet, AT&T fiber, AT&T wireless plans or U-verse (based on location availability). Channels include HBO, Cinemax, STARZ, SHOWTIME, AMC, Fox News, CNN, ESPN, HGTV and the Food Network, as well as niche channels such as DogTV and MBC Drama HD. The firm also offers international packages for multi-language programming. Premium sports entertainment is also provided including NFL Sunday Ticket. In August 2021, former parent company AT&T spun off DirecTV, AT&T TV and U-verse into a new separate company called DirecTV. AT&T retained a 70% stake in the company, with TPG Capital owning the remaining 30%.

DirecTV employees receive medical, dental and vision coverage; flexible spending accounts; wellness plans; employee assistance programs; and a 401(k) savings plan with a matching contribution opportunity.

FINANCIAL DATA: Note: Data for latest year may not have been available at press time.

In U.S. $	2021	2020	2019	2018	2017	2016
Revenue	29,039,677,664	27,922,766,985	31,025,296,650	30,416,957,500	32,017,850,000	33,703,000,000
R&D Expense						
Operating Income						
Operating Margin %						
SGA Expense						
Net Income						
Operating Cash Flow						
Capital Expenditure						
EBITDA						
Return on Assets %						
Return on Equity %						
Debt to Equity						

CONTACT INFORMATION:

Phone: 310 964-5000   Fax:
Toll-Free:
Address: 2260 E. Imperial Hwy., El Segundo, CA 90245 United States

STOCK TICKER/OTHER:

Stock Ticker: Subsidiary   Exchange:
Employees: 30,000   Fiscal Year Ends: 12/31
Parent Company: AT&T Inc

SALARIES/BONUSES:

Top Exec. Salary: $ Bonus: $
Second Exec. Salary: $ Bonus: $

OTHER THOUGHTS:

Estimated Female Officers or Directors: 3
Hot Spot for Advancement for Women/Minorities: Y

Sales, profits and employees may be estimates. Financial information, benefits and other data can change quickly and may vary from those stated here.

D-Link Corporation

NAIC Code: 334210A

www.dlinktw.com.tw

TYPES OF BUSINESS:

Networking Equipment Manufacturing & Distribution
Broadband Products
Unified Network Solutions
Switching Products
Cloud-base Network Management Solutions
Software
IP Surveillance Products
Network Security Solutions

BRANDS/DIVISIONS/AFFILIATES:

GROWTH PLANS/SPECIAL FEATURES:

D-Link Corporation is a leading global designer, developer, manufacturer and distributor of networking, broadband, digital voice and data communications products for homes and businesses, as well as service providers. D-Link implements and supports unified network solutions that integrate capabilities in switching, wireless, broadband, internet protocol (IP) surveillance and cloud-based network management. Consumer products include smart home ecosystems, Wi-Fi devices and cloud platforms along with related tools, apps and services by D-Link. Business products span edge and cloud solutions, connectivity solutions, surveillance products, network security solutions, network infrastructure products, Wi-Fi products, along with view-cam, network assistant, content management system software and other management solutions. D-Link is a global brand with employees in 60 countries.

CONTACTS: *Note: Officers with more than one job title may be intentionally listed here more than once.*

An-Ping Chen, Pres.
Benedict Lee, VP-Oper.
Jui-Hsu Chen, Exec. VP
John Lee, Chmn.
Bendict Lee, Pres., Latin America

FINANCIAL DATA: *Note: Data for latest year may not have been available at press time.*

In U.S. $	2021	2020	2019	2018	2017	2016
Revenue	502,509,300	536,339,000	549,600,000	632,998,000	650,070,000	758,781,167
R&D Expense						
Operating Income						
Operating Margin %						
SGA Expense						
Net Income	8,631,400	44,097,900	4,892,572	4,750,070	-15,380,000	-31,585,848
Operating Cash Flow						
Capital Expenditure						
EBITDA						
Return on Assets %						
Return on Equity %						
Debt to Equity						

CONTACT INFORMATION:

Phone: 886 2-6600-0123   Fax: 886-2-6600-9898
Toll-Free:
Address: No. 289, Xinhu Third Rd., Neihu Dist., Taipei, 114 Taiwan

STOCK TICKER/OTHER:

Stock Ticker: 2332   Exchange: TWSE
Employees: 555   Fiscal Year Ends: 12/31
Parent Company:

SALARIES/BONUSES:

Top Exec. Salary: $ Bonus: $
Second Exec. Salary: $ Bonus: $

OTHER THOUGHTS:

Estimated Female Officers or Directors: 1
Hot Spot for Advancement for Women/Minorities:

DSP Group Inc

www.dspg.com

NAIC Code: 334413

TYPES OF BUSINESS:

Integrated Circuits-Digital Signal Processors
Wireless Chipsets
Voice Recognition

BRANDS/DIVISIONS/AFFILIATES:

HDClear
DBM10

CONTACTS: *Note: Officers with more than one job title may be intentionally listed here more than once.*

Ofer Elyakim, CEO
Dror Levy, CFO
Kenneth Traub, Chairman of the Board

GROWTH PLANS/SPECIAL FEATURES:

DSP Group, Inc. produces wireless chipsets for a wide range of smart-enabled devices. Founded in 1987 on the principles of experience, insight and continuous advancement enables DSP to consistently deliver next-generation solutions in the areas of voice, audio, video and data connectivity across diverse mobile, consumer and enterprise products. The group delivers semiconductor system solutions with software and hardware reference designs, enabling original equipment manufacturers (OEMs), original design manufacturers (ODMs), consumer electronics manufacturers and service providers to cost-effectively develop new revenue-generating products with fast time-to-market. DSP's broad portfolio of wireless chipsets integrate digital enhanced cordless telecommunications (DECT), cordless advanced technology-internet and quality (CAT-iq), ultra-low energy (ULE), Wi-Fi, public switched telephone network (PSTN), HDClear, video and voice-over-internet-protocol (VoIP) technologies. HDClear is DSP's own technology and is incorporated into its smart voice product line for mobile, wearables and always-one Internet of Things (IoT) devices. HDClear capitalizes on the voice user interface by incorporating voice command, voice activation, proprietary noise cancellation, acoustic echo cancellation and beam-forming algorithms for the purpose of improving user experience and delivering voice and speech recognition. Recently (2021), DSP Group launched its DBM10 low-power Edge artificial intelligence/machine learning (AI/ML) system on chip (SoC), which comprises a digital signal processor (DSP) and the company's nNetLite neural network processor, both optimized for low-power voice and sensor processing in battery-operated devices. Based in California, USA, the firm has international offices in Israel, Germany, the U.K., Hong Kong, China, Japan, South Korea and India. Export sales account for the majority (approximately 95%) of DSP's total annual revenues.

FINANCIAL DATA: *Note: Data for latest year may not have been available at press time.*

In U.S. $	2021	2020	2019	2018	2017	2016
Revenue	118,000,000	114,480,000	117,613,000	117,438,000	124,753,000	137,868,992
R&D Expense						
Operating Income						
Operating Margin %						
SGA Expense						
Net Income		-6,790,000	-1,190,000	-1,957,000	-3,003,000	4,813,000
Operating Cash Flow						
Capital Expenditure						
EBITDA						
Return on Assets %						
Return on Equity %						
Debt to Equity						

CONTACT INFORMATION:

Phone: 408 986-4300   Fax: 408 986-4323
Toll-Free:
Address: 2055 Gateway Pl., Ste. 480, San Jose, CA 95110 United States

STOCK TICKER/OTHER:

Stock Ticker: Subsidiary   Exchange:
Employees: 352   Fiscal Year Ends: 12/31
Parent Company: Synaptics Incorporated

SALARIES/BONUSES:

Top Exec. Salary: $ Bonus: $
Second Exec. Salary: $ Bonus: $

OTHER THOUGHTS:

Estimated Female Officers or Directors: 2
Hot Spot for Advancement for Women/Minorities:

Sales, profits and employees may be estimates. Financial information, benefits and other data can change quickly and may vary from those stated here.

Dycom Industries Inc

www.dycomind.com

NAIC Code: 237130

TYPES OF BUSINESS:

Construction, Maintenance & Installation Services
Engineering Services
Utility Maintenance Services

GROWTH PLANS/SPECIAL FEATURES:

Dycom Industries Inc provides contracting services in the United States and Canada. It offers program management, engineering, construction, maintenance, and installation services for telecommunications providers and utilities. Engineering services include the design of aerial, underground, and buried telecommunications systems that extend from telephone companies to end-users homes or businesses. Dycom Industries utilizes copper, coaxial cables, and other materials, and constructs trenches and structures to place the cables or improve distribution lines to consumers. In addition, the company provides tower construction, antenna installation, and other equipment for wireless carriers and television system operators. The majority of sales derive from the United States.

BRANDS/DIVISIONS/AFFILIATES:

CONTACTS: *Note: Officers with more than one job title may be intentionally listed here more than once.*

Steven Nielsen, CEO
H. Deferrari, CFO
Sharon Villaverde, Chief Accounting Officer
Daniel Peyovich, COO
Ryan Urness, General Counsel
Scott Horton, Other Executive Officer

FINANCIAL DATA: *Note: Data for latest year may not have been available at press time.*

In U.S. $	2021	2020	2019	2018	2017	2016
Revenue	3,199,165,000	3,339,682,000	3,127,700,000	1,411,348,000	3,066,880,000	
R&D Expense						
Operating Income	121,509,000	117,806,000	116,565,000	59,885,000	275,009,000	
Operating Margin %	.04%	.04%	.04%		.09%	
SGA Expense	259,770,000	254,590,000	269,140,000	124,930,000	239,231,000	
Net Income	34,337,000	57,215,000	62,907,000	68,835,000	157,217,000	
Operating Cash Flow	381,777,000	57,999,000	124,447,000	160,533,000	256,443,000	
Capital Expenditure	58,047,000	120,574,000	164,963,000	87,839,000	201,197,000	
EBITDA	264,785,000	316,951,000	312,010,000	151,163,000	435,695,000	
Return on Assets %	.02%	.03%	.03%		.09%	
Return on Equity %	.04%	.07%	.08%		.26%	
Debt to Equity	.67%	1.022	1.079		1.099	

CONTACT INFORMATION:

Phone: 561 627-7171   Fax: 561 627-7709
Toll-Free:
Address: 11780 U.S. Highway 1, Ste. 600, Palm Beach Gardens, FL 33408 United States

STOCK TICKER/OTHER:

Stock Ticker: DY   Exchange: NYS
Employees: 15,024   Fiscal Year Ends: 01/27
Parent Company:

SALARIES/BONUSES:

Top Exec. Salary: $641,731   Bonus: $486,265
Second Exec. Salary: $1,092,000   Bonus: $

OTHER THOUGHTS:

Estimated Female Officers or Directors:
Hot Spot for Advancement for Women/Minorities:

DZS Inc

dasanzhone.com

NAIC Code: 334210

TYPES OF BUSINESS:

Telecommunications Infrastructure Equipment
Broadband Access Solutions
Mobile Transport

GROWTH PLANS/SPECIAL FEATURES:

DZS Inc is a broad-based network access solutions provider. It designs, develops and manufactures communications network equipment for telecommunications operators and enterprises. The firm provides solutions in five major product areas: broadband access, mobile backhaul, Ethernet switching, passive optical LAN and software-defined networks. It also offers customer premise equipment, network management, cabinets, and channel bank. Its primary geographic markets are the United States, Canada, Latin America, Europe, Middle East, Africa, Korea, and Other Asia Pacific.

BRANDS/DIVISIONS/AFFILIATES:

DASAN Zhone Solutions Inc
RIFT

CONTACTS: Note: Officers with more than one job title may be intentionally listed here more than once.

Charlie Vogt, CEO
Misty Kawecki, CFO
Min Woo Nam, Chairman of the Board
Justin Ferguson, Chief Legal Officer
Andrew Bender, Chief Technology Officer
Michael Martin, Executive VP, Divisional
Ken Stumpf, Other Corporate Officer
Meggin Sawyer, Other Corporate Officer
Anna Klosterman, Other Corporate Officer
Bethe Strickland, Other Corporate Officer
Thomas Cancro, Treasurer
Miguel Alonso, Vice President, Divisional
Doron Paz, Vice President, Divisional

FINANCIAL DATA: Note: Data for latest year may not have been available at press time.

In U.S. $	2021	2020	2019	2018	2017	2016
Revenue	350,206,000	300,640,000	306,882,000	282,348,000	247,114,000	150,304,000
R&D Expense	47,052,000	37,957,000	38,516,000	35,306,000	36,080,000	25,778,000
Operating Income	-18,207,000	-6,053,000	-1,118,000	7,180,000	722,000	-13,353,000
Operating Margin %	- .05%	- .02%	.00%	.03%	.00%	- .09%
SGA Expense	90,241,000	63,543,000	61,206,000	48,321,000	44,641,000	27,725,000
Net Income	-34,683,000	-23,082,000	-13,457,000	2,767,000	1,071,000	-15,326,000
Operating Cash Flow	-14,326,000	5,064,000	-22,702,000	-12,218,000	1,902,000	3,496,000
Capital Expenditure	5,585,000	2,270,000	2,314,000	1,182,000	1,162,000	1,372,000
EBITDA	-26,574,000	-12,403,000	-582,000	9,000,000	3,937,000	-9,838,000
Return on Assets %	- .14%	- .09%	- .06%	.02%	.01%	- .13%
Return on Equity %	- .31%	- .23%	- .15%	.04%	.02%	- .28%
Debt to Equity	.09%	0.488	0.349	0.183	0.134	0.102

CONTACT INFORMATION:

Phone: 510 777-7000   Fax: 510 777-7001
Toll-Free: 877-946-6320
Address: 1350 S. Loop Rd., Ste. 130, Alameda, CA 94502 United States

STOCK TICKER/OTHER:

Stock Ticker: DZSI   Exchange: NAS
Employees: 830   Fiscal Year Ends: 12/31
Parent Company: DASAN Networks Inc

SALARIES/BONUSES:

Top Exec. Salary: $550,192   Bonus: $
Second Exec. Salary: $300,000   Bonus: $

OTHER THOUGHTS:

Estimated Female Officers or Directors: 1
Hot Spot for Advancement for Women/Minorities:

Sales, profits and employees may be estimates. Financial information, benefits and other data can change quickly and may vary from those stated here.

EarthLink LLC

www.earthlink.net

NAIC Code: 517110

TYPES OF BUSINESS:

Internet Service Provider (ISP)
Web Hosting Services
Voice Services
Managed IT & Data Hosting
Value-Added Services
Internet Service Provider
Hyperlink
Mobile Services

BRANDS/DIVISIONS/AFFILIATES:

Trive Capital
EarthLink Mobile

CONTACTS: *Note: Officers with more than one job title may be intentionally listed here more than once.*

Glenn Goad, CEO
Michael Toplisek, Pres.
Julie Shimer, Chairman of the Board
Michael Borellis, CFO
Jena Dunham, Sr. VP-Mktg.
Scott Klinger, Sr. VP-Human Resources & Mktg.
Leon Hounshell, CTO
Bradley Ferguson, Executive VP, Divisional
Louis Alterman, Executive VP
Samuel Desimone, Executive VP
John Dobbins, Executive VP
Gerard Brossard, Executive VP
Rick Froehlich, Executive VP
Valerie Benjamin, Senior VP, Divisional
R. Thurston, Vice President
Brian McLaughlin, COO

GROWTH PLANS/SPECIAL FEATURES:

EarthLink LLC is a leading next-generation internet service provider. With access available across the U.S., the firm offers fast and reliable connectivity, customer support and customizable features. Internet access offerings include dial-up, high-speed digital subscriber line (DSL), free-standing DSL hyperlink internet, and a hyperlink and Apple TV option. Digital marketing solutions span business internet, digital marketing, website development, custom branding, reputation management and more. Website tools and services include site design and domain and email hosting. Software and tools include security, Wi-Fi privacy and access solutions. EarthLink is owned by private equity firm Trive Capital. During 2021, EarthLink introduced EarthLink Mobile, offering wireless phone service as well as mobile phones.

EarthLink offers its employees health plan options, 401(k), life and disability insurance, and a variety of assistance plans and company perks.

FINANCIAL DATA: *Note: Data for latest year may not have been available at press time.*

In U.S. $	2021	2020	2019	2018	2017	2016
Revenue	1,230,989,760	1,183,644,000	1,212,750,000	1,155,000,000	1,100,000,000	1,099,800,000
R&D Expense						
Operating Income						
Operating Margin %						
SGA Expense						
Net Income						
Operating Cash Flow						
Capital Expenditure						
EBITDA						
Return on Assets %						
Return on Equity %						
Debt to Equity						

CONTACT INFORMATION:

Phone: 404 815-0770   Fax:
Toll-Free: 888-327-8454
Address: 980 Hammond Dr. NE, Ste. 400, Atlanta, GA 30328 United States

STOCK TICKER/OTHER:

Stock Ticker: Private   Exchange:
Employees: 2,138   Fiscal Year Ends: 12/31
Parent Company: Trive Capital

SALARIES/BONUSES:

Top Exec. Salary: $ Bonus: $
Second Exec. Salary: $ Bonus: $

OTHER THOUGHTS:

Estimated Female Officers or Directors: 5
Hot Spot for Advancement for Women/Minorities: Y

EchoStar Corporation

www.echostar.com

NAIC Code: 517110

TYPES OF BUSINESS:
Digital Set-Top Boxes & Related Products
Fixed Satellite Services

BRANDS/DIVISIONS/AFFILIATES:
Hughes
ESS

GROWTH PLANS/SPECIAL FEATURES:

EchoStar Corporation is a provider of broadband technology and video delivery solutions for the home and office. The firm operates in two segments - Hughes, which provides satellite broadband Internet access to North American customers and EchoStar Satellite Services, which uses owned and leased in-orbit satellites to provide services primarily to DISH Network.

EchoStar offers employees medical, dental, vision, life, AD&D and disability insurance; various assistance programs; 401(k) and other retirement/savings plans; paid vacation/holidays; tuition reimbursement and more.

CONTACTS: Note: Officers with more than one job title may be intentionally listed here more than once.
Michael Dugan, CEO
David Rayner, CFO
Charles Ergen, Chairman of the Board
Anders Johnson, Chief Strategy Officer
Pradman Kaul, Director
Dean Manson, Executive VP

FINANCIAL DATA: Note: Data for latest year may not have been available at press time.

In U.S. $	2021	2020	2019	2018	2017	2016
Revenue	1,985,720,000	1,887,907,000	1,886,081,000	1,762,638,000	1,525,155,000	1,810,466,000
R&D Expense	31,777,000	29,448,000	25,739,000	27,570,000	31,745,000	31,170,000
Operating Income	217,255,000	114,158,000	73,077,000	101,357,000	41,324,000	296,163,000
Operating Margin %	.11%	.06%	.04%	.06%	.03%	.16%
SGA Expense	461,705,000	474,912,000	509,145,000	436,088,000	370,500,000	325,044,000
Net Income	72,875,000	-40,150,000	-62,917,000	-40,475,000	392,561,000	179,930,000
Operating Cash Flow	632,226,000	534,388,000	656,322,000	734,522,000	726,892,000	803,343,000
Capital Expenditure	471,973,000	447,453,000	482,341,000	586,780,000	614,542,000	745,593,000
EBITDA	715,188,000	645,103,000	746,051,000	691,680,000	686,038,000	835,175,000
Return on Assets %	.01%	-.01%	-.01%	.00%	.04%	.02%
Return on Equity %	.02%	-.01%	-.02%	-.01%	.10%	.05%
Debt to Equity	.49%	0.455	0.677	0.576	0.863	0.924

CONTACT INFORMATION:
Phone: 303 706-4000   Fax:
Toll-Free:
Address: 100 Inverness Terrace E., Englewood, CO 80112-5308 United States

STOCK TICKER/OTHER:
Stock Ticker: SATS
Employees: 2,500
Parent Company:

Exchange: NAS
Fiscal Year Ends: 12/31

SALARIES/BONUSES:
Top Exec. Salary: $1,000,002   Bonus: $
Second Exec. Salary: $880,006   Bonus: $

OTHER THOUGHTS:
Estimated Female Officers or Directors: 1
Hot Spot for Advancement for Women/Minorities:

EF Johnson Technologies Inc

www.efjohnson.com

NAIC Code: 334220

TYPES OF BUSINESS:

Voice Communication Security Products
Secure Wireless Communications Systems
Land Mobile Radio Systems

BRANDS/DIVISIONS/AFFILIATES:

JVCKENWOOD Corporation
KENWOOD Viking
Viking P25
ATLAS P25
NEXEDGE
KAIROS

CONTACTS: *Note: Officers with more than one job title may be intentionally listed here more than once.*

Duane Anderson, CEO
Timi Jackson, Sr. VP
Mitch Urbanczyk, Sr. VP-Sales
Karthik Rangarajan, Sr. VP-Product & Strategy
Timi Jackson, VP
Karthik Rangarajan, VP-Mktg.

GROWTH PLANS/SPECIAL FEATURES:

EF Johnson Technologies, Inc., owned by JVCKENWOOD Corporation, is a provider of secure wireless technologies primarily for the homeland security marketplace. The firm designs, develops, markets and supports private wireless communications, including wireless radios and wireless communications infrastructure and systems for digital and analog platforms. EF's products are grouped into two categories-radios and systems-and serve the fire, law enforcement, public works/utilities federal, military, business, industrial and education industries. KENWOOD Viking radios combine Viking P25 software and KENWOOD's hardware, and are used throughout the world. Radio sets come in portables and mobiles. The KENWOOD Viking portable radios encompass advanced technology such as Armada fleet management and programming software, and TrueVoice software-based noise cancellation. The systems category consists of land mobile radio (LMR) communications systems, which are used for critical communication purposes and keeps teams connected 24/7. ATLAS P25 features patented latitude technology to offer increased transport flexibility in comparison with traditional LMR systems in various aspects of its operations. Latitude technology offers auto-discovery and self-healing sites, distributed call control and simulcast control. NEXEDGE NXDN is one of the largest two-way radio networks in the U.S., and deploys NEXEDGE technology, which incorporates frequency division multiple access (FDMA) for maximum bandwidth utilization and reducing the risk of interference. NEXEDGE comes in a variety of configurations, including conventional, conventional IP, conventional simulcast, Gen1 Trunked and Gen2 Trunked. Last, KAIROS is a digital mobile radio that keeps teams safe and connected via simulcast repeaters. KAIROS features multi-protocol switching, IP multisite multicasting/simulcasting, UHF linking, system redundancy, soft diversity reception, and SIP/RTP integration. EF services encompass support services, training, field service bulletins and repair request processing. Headquartered in Texas, EF Johnson has office locations in Nebraska and California.

FINANCIAL DATA: *Note: Data for latest year may not have been available at press time.*

In U.S. $	2021	2020	2019	2018	2017	2016
Revenue	156,510,900	150,491,250	143,325,000	136,500,000	130,000,000	128,800,000
R&D Expense						
Operating Income						
Operating Margin %						
SGA Expense						
Net Income						
Operating Cash Flow						
Capital Expenditure						
EBITDA						
Return on Assets %						
Return on Equity %						
Debt to Equity						

CONTACT INFORMATION:

Phone: 972-819-0700   Fax: 972-819-0639
Toll-Free: 800-328-3911
Address: 1440 Corporate Dr., Irving, TX 75038 United States

STOCK TICKER/OTHER:

Stock Ticker: Private   Exchange:
Employees: 200   Fiscal Year Ends: 03/31
Parent Company: JVCKENWOOD Corporation

SALARIES/BONUSES:

Top Exec. Salary: $ Bonus: $
Second Exec. Salary: $ Bonus: $

OTHER THOUGHTS:

Estimated Female Officers or Directors: 2
Hot Spot for Advancement for Women/Minorities:

Sales, profits and employees may be estimates. Financial information, benefits and other data can change quickly and may vary from those stated here.

eircom Limited

NAIC Code: 517110

TYPES OF BUSINESS:

Local & Long-Distance Services
Cellular Phone Service
Internet Service Provider
Data Storage & Networking Services
Wholesale Telecommunications Services
Payphones
Directory Services

BRANDS/DIVISIONS/AFFILIATES:

NJJ Group
NJJ Telecom Europe
Open eir
eir Mobile

CONTACTS: Note: Officers with more than one job title may be intentionally listed here more than once.

Carolan Lennon, CEO
Brendan Lynch, Managing Dir.-Group Tech.
Ann Marie Kearney, Chief Legal Officer
Pat Galvin, Dir.-Regulation & Public Affairs
Ronan Kneafsey, Managing Dir.-eircom Business
Chris Hutchings, Managing Dir.-eircom Wholesale
Tony Mealy, Dir.-Program Execution
Kevin White, Managing Dir.-Consumer
David McRedmond, Chmn.

GROWTH PLANS/SPECIAL FEATURES:

eircom Limited (eir)is the principal provider of fixed-line telecommunications and mobile telecommunications services in Ireland, serving approximately 2.1 million customers. The firm has the most extensive telecommunications network in Ireland both in terms of capacity and geographic reach. Services provided by eir include a comprehensive range of advanced voice, data, broadband and TV services to the residential, small business, enterprise and government markets. Open eir is a leading wholesale operator in Ireland with approximately 820,000 fiber connections. The company's smart Wi-Fi hub mesh extender enables customers to enjoy ultrafast broadband in every part of the house. The mobile division operates under the eir Mobile brand and serves approximately 20% of Ireland's mobile market, and is the third largest mobile operator in the country. The service offers general packet radio services (GPRS), 3G mobile services and high-speed 4G mobile broadband network services. 5G network is being rolled out, and available in nearly 300 towns and cities throughout the country (as of May 2021), and 5G outdoor population coverage was over 57%. NJJ Telecom Europe, part of NJJ Group, owns a majority stake in eir.

FINANCIAL DATA: Note: Data for latest year may not have been available at press time.

In U.S. $	2021	2020	2019	2018	2017	2016
Revenue	1,439,911,200	1,384,530,000	1,399,830,000	1,479,270,000	1,482,070,000	1,451,810,000
R&D Expense						
Operating Income						
Operating Margin %						
SGA Expense						
Net Income		24,743,900	34,100,100	-68,722,000	-257,850,000	-175,104,000
Operating Cash Flow						
Capital Expenditure						
EBITDA						
Return on Assets %						
Return on Equity %						
Debt to Equity						

CONTACT INFORMATION:

Phone: 353 1-701-5000   Fax: 353-1-671-6916
Toll-Free:
Address: 2022 Bianconi Ave., Dublin, D24 HX03 Ireland

STOCK TICKER/OTHER:

Stock Ticker: Subsidiary   Exchange:
Employees: 3,200   Fiscal Year Ends: 06/30
Parent Company: NJJ Group

SALARIES/BONUSES:

Top Exec. Salary: $ Bonus: $
Second Exec. Salary: $ Bonus: $

OTHER THOUGHTS:

Estimated Female Officers or Directors: 2
Hot Spot for Advancement for Women/Minorities:

Sales, profits and employees may be estimates. Financial information, benefits and other data can change quickly and may vary from those stated here.

Elisa Corporation

www.elisa.com

NAIC Code: 517110

TYPES OF BUSINESS:

Voice & Data Services
Mobile Phone Service
Internet Services
Data Security Software Development
Cable TV

BRANDS/DIVISIONS/AFFILIATES:

Elisa Videra
Elisa Smart Factory
Enia Oy
Elisa Saunalahti
Elisa Estonia
Elisa Santa Monica
Elisa Automate

GROWTH PLANS/SPECIAL FEATURES:

Elisa Oyj is a telecommunications company that operates in two segments, Consumer Customers and Corporate Customers. The Consumer Customers segment generates revenue by providing voice and data services to households and individuals. Most of the overall company revenue comes from the consumer segment. Elisa's other segment, Corporate Customers, derives revenue from voice and data services and information and communications technology to corporations. The company generates most of its revenue in Finland.

Elisa offers employees occupational health services, staff dining, discounts on telecom services, timeshares in Finland and St. Petersburg and private hospital services including coverage for dependents.

CONTACTS: Note: Officers with more than one job title may be intentionally listed here more than once.

Veli-Matti Mattila, CEO
Jari Kinnunen, CFO
Timo Katajisto, Exec. VP-Prod.
Sami Ylikortes, Exec. VP-Admin.
Katiye Vuorela, Exec. VP-Corp. Comm.
Vesa Sahivirta, Dir.-Investor Relations
Pasi Maenpaa, Exec. VP-Corp. Customers
Asko Kansala, Exec. VP-Consumer Customers
Anssi Vanjoki, Chmn.

FINANCIAL DATA: Note: Data for latest year may not have been available at press time.

In U.S. $	2021	2020	2019	2018	2017	2016
Revenue	2,037,486,000	1,932,043,000	1,880,035,000	1,867,798,000	1,822,724,000	1,668,026,000
R&D Expense						
Operating Income	430,340,000	412,800,100	397,095,700	401,888,600	382,717,100	341,518,600
Operating Margin %	.21%	.21%	.21%	.22%	.21%	.20%
SGA Expense						611,858
Net Income	350,390,600	334,482,300	308,988,200	322,041,200	343,252,200	262,181,100
Operating Cash Flow	607,472,800	611,857,800	564,846,800	525,585,900	510,697,300	496,114,700
Capital Expenditure	263,914,700	254,124,900	236,177,100	240,256,200	259,835,600	213,028,500
EBITDA	711,692,600	705,777,900	673,451,500	653,158,100	667,842,800	577,389,800
Return on Assets %	.11%	.11%	.11%	.12%	.13%	.11%
Return on Equity %	.29%	.28%	.27%	.29%	.33%	.27%
Debt to Equity	1.01%	1.028	0.944	0.765	0.904	0.852

CONTACT INFORMATION:

Phone: 358 10804400   Fax: 358 1026060
Toll-Free:
Address: Ratavartijankatu 5, Helsinki PL 1, Elisa, 00061 Finland

STOCK TICKER/OTHER:

Stock Ticker: ELMUF   Exchange: PINX
Employees: 5,391   Fiscal Year Ends: 12/31
Parent Company:

SALARIES/BONUSES:

Top Exec. Salary: $ Bonus: $
Second Exec. Salary: $ Bonus: $

OTHER THOUGHTS:

Estimated Female Officers or Directors: 3
Hot Spot for Advancement for Women/Minorities: Y

Embratel Participacoes SA

www.embratel.com.br

NAIC Code: 517110

TYPES OF BUSINESS:

Local & Long-Distance Telephone Services
Internet & Data Service
Satellite Services
TV & Radio Broadcasting
Free Dial-up Internet Access
Network Outsourcing
Call Centers
Wi-Fi 6

BRANDS/DIVISIONS/AFFILIATES:

America Movil SAB de CV
Empresa Brasileira de Telecomunicacoes SA
Star One SA (Embratel Star One)

GROWTH PLANS/SPECIAL FEATURES:

Embratel Participacoes SA is the holding company for Empresa Brasileira de Telecomunicacoes SA (Embratel), which offers voice, data, internet, television and other services in Brazil and internationally. The firm is owned by America Movil SAB de CV. The company is one of the largest providers of telecommunications services in Brazil, offering the entire Brazilian market solutions including local telephone, data and internet services; domestic and international long distance; and TV and radio broadcasting. In addition, Embratel provides IT integration solutions, cloud computing, data center, customer experience, security solutions, Internet of Things (IoT), video services/solutions and energy efficiency solutions, as well as corporate mobile/fixed-line solutions. Subsidiary Star One SA, also known as Embratel Star One, operates one of the largest satellite networks in South America. It provides satellite capacity covering the entire continent for voice, data, internet, TV and radio signals through several satellites.

CONTACTS:
Note: Officers with more than one job title may be intentionally listed here more than once.

Jose Formoso Martinez, Pres.
Isaac Berensztejn, Dir.-Investor Rel.
Jose Formoso Martinez, VP

FINANCIAL DATA:
Note: Data for latest year may not have been available at press time.

In U.S. $	2021	2020	2019	2018	2017	2016
Revenue	11,328,981,300	10,893,251,250	10,374,525,000	9,880,500,000	9,410,000,000	9,948,200,000
R&D Expense						
Operating Income						
Operating Margin %						
SGA Expense						
Net Income						
Operating Cash Flow						
Capital Expenditure						
EBITDA						
Return on Assets %						
Return on Equity %						
Debt to Equity						

CONTACT INFORMATION:

Phone: 55 21-2121-6474   Fax: 55-21-2121-6388
Toll-Free:
Address: Rua Reg Feijo 166 Sala 1687B Centro, Rio de Janeiro, RJ 20060060 Brazil

STOCK TICKER/OTHER:

Stock Ticker: Subsidiary   Exchange:
Employees: 7,500   Fiscal Year Ends: 12/31
Parent Company: America Movil SAB de CV

SALARIES/BONUSES:

Top Exec. Salary: $ Bonus: $
Second Exec. Salary: $ Bonus: $

OTHER THOUGHTS:

Estimated Female Officers or Directors:
Hot Spot for Advancement for Women/Minorities:

Sales, profits and employees may be estimates. Financial information, benefits and other data can change quickly and may vary from those stated here.

Emirates Telecommunications Corporation (Etisalat)

www.etisalat.ae
NAIC Code: 517110

TYPES OF BUSINESS:

Local Telephone Service
Internet Services
Wireless Services
Digital Cable Service
Underwater Cable Installation
5G Network Services

BRANDS/DIVISIONS/AFFILIATES:

Etisalat
Etisalat Software Solutions (P) Ltd
E-Vision
Thuraya
Etihad Etisalat Company
Mobily
Itissalat Al-Maghrib
Etisalat Sri Lanka

CONTACTS: Note: Officers with more than one job title may be intentionally listed here more than once.

Hatem Dowidar, CEO
Paul Werne, Group General Counsel
Daniel Ritz, Chief Strategy Officer
Nasser bin Obood, Chief Gov't Rel. & Corp. Comm. Officer
Javier Garcia, Chief Internal Auditor
Kamal Shehadi, Chief Legal & Regulatory Officer
Jamal Aljarwan, Chief Regional Officer Asis Cluster
Rainer Rathgeber, Chief Commercial Officer
John Wilkes, Chief Internal Control Officer
Essa Al Haddad, Chief Regional Officer-Africa
Obaid Bokisha, Chief Procurement Officer

GROWTH PLANS/SPECIAL FEATURES:

Emirates Telecommunications Corporation (Etisalat), majority-owned by the government of United Arab Emirates (UAE), provides telecommunications services to more than 155 million subscribers in 16 countries across the Middle East, Asia and Africa. The firm has a wide coverage of 4G and 5G mobile technologies, as well as an extensive fiber-to-the-home (FTTH) network. Etisalat offers fixed-line, wireless and internet services to customers spread across Asia and Africa and the Middle East. The company's 5G network provides mobile broadband speeds up to 1 Gbps. Etisalat offers post-paid, pre-paid and control/visitor mobile lines, as well as data add-on features. Wholly-owned Etisalat Software Solutions (P) Ltd. (ESSPL), based in Bangalore, develops software for IT solutions and offers its services to other telecommunications companies in the region. Devices sold by the company include smartphones, tablets, routers, gaming devices, smart living and related accessories. Wholly-owned E-Vision offers a turnkey end-to-end over-the-top (OTT) platform that includes video on-demand, a content delivery network (CDN), content aggregation, TV channels hosting (more than 600), advertising sales, creative services and more. Thuraya is a regional mobile satellite service provider, operating two geostationary satellites which provide telecommunications coverage in more than 160 countries. Etihad Etisalat Company is a telecommunications services company that offers its products under the Mobily brand name. Itissalat Al-Maghrib offers telecommunications services in Morocco; and Etisalat Sri Lanka is a mobile telecommunications network in Sri Lanka.

FINANCIAL DATA: Note: Data for latest year may not have been available at press time.

In U.S. $	2021	2020	2019	2018	2017	2016
Revenue	14,520,652,556	14,073,700,000	14,205,700,000	14,263,600,000	14,072,900,000	14,266,458,109
R&D Expense						
Operating Income						
Operating Margin %						
SGA Expense						
Net Income	2,550,954,878	2,449,970,000	2,584,520,000	2,843,010,000	2,659,980,000	2,287,044,517
Operating Cash Flow						
Capital Expenditure						
EBITDA						
Return on Assets %						
Return on Equity %						
Debt to Equity						

CONTACT INFORMATION:

Phone: 971 400444101   Fax: 971 400444105
Toll-Free:
Address: Etisalat Bldg. 1, Sheikh Rashid Bin Saeed Al Maktoum, Abu Dhabi, 3838 United Arab Emirates

STOCK TICKER/OTHER:

Stock Ticker: ETISALAT
Employees: 40,000
Parent Company:

Exchange: Abu Dhabi
Fiscal Year Ends: 12/31

SALARIES/BONUSES:

Top Exec. Salary: $ Bonus: $
Second Exec. Salary: $ Bonus: $

OTHER THOUGHTS:

Estimated Female Officers or Directors:
Hot Spot for Advancement for Women/Minorities:

Sales, profits and employees may be estimates. Financial information, benefits and other data can change quickly and may vary from those stated here.

Empresa Nacional de Telecommunicacions SA (Entel)

www.entel.cl
NAIC Code: 517110

TYPES OF BUSINESS:

Telecommunications Services
Mobile Services
Internet Services
Telephony Services

BRANDS/DIVISIONS/AFFILIATES:

Entel Phone Local Telephone
Entel Telefonia Personal
Entel PCS
Entel Movil

CONTACTS: *Note: Officers with more than one job title may be intentionally listed here more than once.*

Antonio Buchi Buc, CEO
Juan Baraqui Anania, Dir.-Admin.
Cristian Maturana Miquel, Dir.-Legal
Alfredo Parot Donoso, VP-Oper.
Sebastian Dominguez Philippi, Dir.-Strategy & Innovation
Carmen Luz De La Cerda C., Investor Rel. Officer
Felipe Ureta Prieto, Mgr.-Corp. Finance
Manuel Araya Arroyo, Dir.-Regulation & Corp. Affairs
Luis Ceron Puelam, Dir.-Internal Audit
Juan Jose Hurtado Vicuna, Chmn.
Eduardo Bobenrieth Giglio, Gen. Mgr.-Peru Americatel SA

GROWTH PLANS/SPECIAL FEATURES:

Empresa Nacional de Telecommunications SA (Entel) is a leading telecommunications company in Chile. The firm provides mobile services and fixed network services, including data, IT (information technology) integration, internet, local telephone, long distance and related services. The firm provides 3.5G, 4G, advanced HSPA+ (high speed packet access evolution) and fiber optic network systems, and has engaged in applying technological trends such as 5G, Internet of Things (IoT), analytics, video recognition, augmented reality, virtual reality and more. Entel additionally has a fixed network and call center in Peru. The company offers its products and services in three areas: residential, business and corporate. Residential solutions offered by Entel include mobile solutions, long distance, value added services, calling cards, broadband Internet, dial-up Internet and broadband telephony services. This division also offers hardware for purchase such as new and refurbished phones and Internet equipment. Entel's business offerings include mobile solutions, traditional and IP (Internet protocol) telephony, long distance, broadband Internet, broadband telephony, data networks and IT solutions. Its long-distance business offerings include international calling, calling cards and ISDN (integrated services digital network) videoconferencing. The firm's broadband Internet solutions for businesses include ADSL (asymmetrical digital subscriber line), wireless and dedicated options. Entel's corporate solutions include mobile services, fixed solutions, IT solutions, cloud computing services and data center solutions. Corporate value-added services include infrastructure, network administration, security and studies. Wholly-owned subsidiaries of Entel include Entel Phone Local Telephone, Entel Telefonia Personal, Entel PCS and Entel Movil.

FINANCIAL DATA: *Note: Data for latest year may not have been available at press time.*

In U.S. $	2021	2020	2019	2018	2017	2016
Revenue	2,855,795,033	2,932,460,000	2,652,470,000	2,768,160,000	3,157,300,000	2,871,786,398
R&D Expense						
Operating Income						
Operating Margin %						
SGA Expense						
Net Income	88,213,702	118,364,000	206,633,000	-33,991,600	70,536,200	51,060,600
Operating Cash Flow						
Capital Expenditure						
EBITDA						
Return on Assets %						
Return on Equity %						
Debt to Equity						

CONTACT INFORMATION:

Phone: 56 2-3600-123   Fax: 56-2-360-3424
Toll-Free:
Address: Ave. Andres Bello 2687, Fl. 14, Las Condes, Santiago, Casilla 4254 Chile

STOCK TICKER/OTHER:

Stock Ticker: ENTEL   Exchange: Santiago
Employees: 9,000   Fiscal Year Ends: 12/31
Parent Company:

SALARIES/BONUSES:

Top Exec. Salary: $ Bonus: $
Second Exec. Salary: $ Bonus: $

OTHER THOUGHTS:

Estimated Female Officers or Directors: 1
Hot Spot for Advancement for Women/Minorities: Y

Sales, profits and employees may be estimates. Financial information, benefits and other data can change quickly and may vary from those stated here.

Energy Focus Inc

www.energyfocus.com

NAIC Code: 335921

TYPES OF BUSINESS:

Fiber Optic Cable Manufacturing
Lighting Systems
Manufacture
Development
Light-Emitting Diode Products

BRANDS/DIVISIONS/AFFILIATES:

Intellitube

GROWTH PLANS/SPECIAL FEATURES:

Energy Focus Inc is a provider of energy-efficient Light Emitting
Diode (LED) lighting products and ultraviolet-C ligh
disinfection (UVCD) products. It serves the military maritime
market and general commercial markets. Its product offerings
include Military maritime LED lighting products such as military
intellitube, globe lights, berth lights, and military fixtures; and
Commercial products such as LED fixtures and panels, LED
down-lights, LED dock lights, and wall-packs, and LED retrofi
kits. Geographically, it operates in the United States and othe
countries. Its products include Commercial products and MMM
products. It generates a majority of its sales from Military
maritime products primarily in the US.

CONTACTS: Note: Officers with more than one job title may be intentionally listed here more than once.

James Tu, CEO
Tod Nestor, CFO
James Warren, General Counsel
John Davenport, Other Executive Officer

FINANCIAL DATA: Note: Data for latest year may not have been available at press time.

In U.S. $	2021	2020	2019	2018	2017	2016
Revenue	9,865,000	16,828,000	12,705,000	18,107,000	19,846,000	30,998,000
R&D Expense	1,891,000	1,415,000	1,284,000	2,597,000	2,940,000	3,537,000
Operating Income	-8,728,000	-4,130,000	-6,759,000	-8,974,000	-9,434,000	-15,973,000
Operating Margin %	-.88%	-.25%	-.53%	-.50%	-.48%	-.52%
SGA Expense	8,535,000	7,900,000	7,449,000	9,789,000	11,315,000	20,113,000
Net Income	-7,886,000	-5,981,000	-7,373,000	-9,111,000	-11,267,000	-16,887,000
Operating Cash Flow	-9,765,000	-2,451,000	-6,624,000	-6,795,000	-5,874,000	-16,553,000
Capital Expenditure	443,000	223,000	132,000	57,000	162,000	1,624,000
EBITDA	-6,907,000	-5,321,000	-6,720,000	-8,570,000	-10,699,000	-16,043,000
Return on Assets %	-.59%	-.50%	-.49%	-.45%	-.39%	-.37%
Return on Equity %	-1.51%	-1.45%	-.98%	-.60%	-.46%	-.45%
Debt to Equity	.00%	0.137	0.297			

CONTACT INFORMATION:

Phone: 440 715-1300   Fax: 440 715-1314
Toll-Free:
Address: 32000 Aurora Rd., Ste. B, Solon, OH 44139 United States

STOCK TICKER/OTHER:

Stock Ticker: EFOI   Exchange: NAS
Employees: 58   Fiscal Year Ends: 12/31
Parent Company:

SALARIES/BONUSES:

Top Exec. Salary: $276,942   Bonus: $
Second Exec. Salary: Bonus: $
$249,086

OTHER THOUGHTS:

Estimated Female Officers or Directors:
Hot Spot for Advancement for Women/Minorities:

Enterprise Diversified Inc

www.enterprisediversified.com

NAIC Code: 517110

TYPES OF BUSINESS:

Internet Service Provider
Asset Management
Real Estate

BRANDS/DIVISIONS/AFFILIATES:

CONTACTS: *Note: Officers with more than one job title may be intentionally listed here more than once.*

Steven Kiel, CEO
Alea Kleinhammer, CFO

GROWTH PLANS/SPECIAL FEATURES:

Enterprise Diversified, Inc. (formerly White Dove Systems, Inc., Interfoods Consolidated, Inc. and then Sitestar Corporation) is an internet service provider (ISP) with additional interests in asset management and real estate. The firm operates through four reportable segments: Asset Management Operations, Real Estate Operations, Internet Operations and Other Operations. The Asset Management Operations segment includes revenue and expenses derived from various joint ventures, service offerings and initiatives undertaken in the asset management industry. The Real Estate Operations segment includes revenue and expenses related to the management of legacy properties held for investment and held for resale through EDI Real Estate located in Roanoke, Virginia. The Internet Operations segment includes revenue and expenses related to the sale of internet access, hosting, storage and other ancillary services. The Other Operations segment includes any revenue and expenses from nonrecurring or one-time strategic funding or similar activity that is not considered to be one of Enterprise Diversified's primary lines of business, and any revenue or expenses derived from corporate office operations, as well as expenses related to public company reporting, the oversight of subsidiaries and other items that affect the overall company.

FINANCIAL DATA: *Note: Data for latest year may not have been available at press time.*

In U.S. $	2021	2020	2019	2018	2017	2016
Revenue	5,902,000	5,247,732	3,589,899	4,410,374	9,094,051	4,975,246
R&D Expense						
Operating Income						
Operating Margin %						
SGA Expense						
Net Income		3,279,404	-5,378,637	-3,840,685	2,141,915	-107,638
Operating Cash Flow						
Capital Expenditure						
EBITDA						
Return on Assets %						
Return on Equity %						
Debt to Equity						

CONTACT INFORMATION:

Phone: 434 336 7737   Fax:
Toll-Free:
Address: 2400 Old Brick Rd., Ste. 115, Glen Allen, VA 23060 United States

STOCK TICKER/OTHER:

Stock Ticker: SYTE
Employees: 7
Parent Company:

Exchange: OTC
Fiscal Year Ends: 12/31

SALARIES/BONUSES:

Top Exec. Salary: $ Bonus: $
Second Exec. Salary: $ Bonus: $

OTHER THOUGHTS:

Estimated Female Officers or Directors:
Hot Spot for Advancement for Women/Minorities:

enTouch Systems Inc

www.astound.com

NAIC Code: 517110

TYPES OF BUSINESS:

Local & Long-Distance Telephone Service
High-Speed Internet Access
Cable Television Service

BRANDS/DIVISIONS/AFFILIATES:

TPG Capital
RCN Telecom Services LLC
Astound Broadband

GROWTH PLANS/SPECIAL FEATURES:

EnTouch Systems, Inc., doing business as Astound Broadband and formerly known as RCN Telecom Services, LLC, provides advanced high-speed internet, digital TV and phone services to residential and small, medium and enterprise businesses. Astound Broadband serves U.S. cities such as Boston, Chicago, Lehigh Valley, New York, Philadelphia and Washington DC. The company's offerings include 100% digital programming, a fiber-optic network and free video on demand. For businesses, Astound Broadband's communications products and services include internet, voice, video and network solutions. Bundled packaged deals are also available. Astound Broadband is privately-owned by TPG Capital.

CONTACTS: *Note: Officers with more than one job title may be intentionally listed here more than once.*

Jim Holanda, CEO
Douglas Bradbury, Exec. VP
James Holanda, CEO-Video High-Speed Internet & Premium Voice Svcs
Felipe Alvarez, Pres.-RCN Metro Optical Networks

FINANCIAL DATA: *Note: Data for latest year may not have been available at press time.*

In U.S. $	2021	2020	2019	2018	2017	2016
Revenue	633,750,000	609,375,000	625,000,000	561,500,000	547,468,623	521,398,689
R&D Expense						
Operating Income						
Operating Margin %						
SGA Expense						
Net Income						
Operating Cash Flow						
Capital Expenditure						
EBITDA						
Return on Assets %						
Return on Equity %						
Debt to Equity						

CONTACT INFORMATION:

Phone: 703-434-8200   Fax:
Toll-Free: 800-746-4726
Address: 650 College Rd., Ste 3100, Princeton, NJ 20170 United States

STOCK TICKER/OTHER:

Stock Ticker: Private   Exchange:
Employees: 1,500   Fiscal Year Ends: 12/31
Parent Company: TPG Capital

SALARIES/BONUSES:

Top Exec. Salary: $ Bonus: $
Second Exec. Salary: $ Bonus: $

OTHER THOUGHTS:

Estimated Female Officers or Directors:
Hot Spot for Advancement for Women/Minorities:

Entrust Corporation

www.entrust.com

NAIC Code: 511210E

TYPES OF BUSINESS:
Computer Software, Network Security, Managed Access, Digital ID, Cybersecurity & Anti-Virus
Digital Identification & Certificates
Digital Security

BRANDS/DIVISIONS/AFFILIATES:
Entrust

CONTACTS: *Note: Officers with more than one job title may be intentionally listed here more than once.*
Todd Wilkinson, CEO
Jeff Smolinksi, Sr. VP-Operations
Kurt Ishaug, CFO
Karen Kaukol, CMO
Beth Klehr, Chief Human Resources Officer
Anudeep Parhar, CIO
Robert (Bob) VanKirk, Sr. VP-American Sales
Mike Baxter, Sr. VP-Product Dev.
Mark Reeves, Sr. VP-Int'l Sales

GROWTH PLANS/SPECIAL FEATURES:
Entrust Corporation is a global provider of security applications that protect and secure digital identities and information. The firm designs, produces and sells security, policy and access management software products and related services. Solutions include financial cards, passports, identification cards, authentication solutions (cloud-based, mobile, hybrid), digital certificates, border control, credential lifecycle management and transaction technologies. Entrust products consist of categories such as secure socket layer (SSL) certificates, qualified certificates, digital signing certificates, passport systems, identification card printers, financial card printers, central card issuance, software, short message service (SMS) passcode, authentication, Internet of Things (IoT) security, public key infrastructure, nCipher hardware security modules, related supplies and related accessories. Supplies include print ribbons, overlays, laminates, tactile impression solutions, cleaning supplies, topping foils, card delivery stickers/labels and more. Accessories include devices such as pin pads, magnetic strip encoders and digital signature pads. Entrust serves financial, corporate, government, education, healthcare, and retail clients located throughout the world. The company is based in Minnesota, USA, and has additional office locations in the U.S., Asia Pacific, Canada, Europe, Latin America, the Caribbean and the Middle East.

FINANCIAL DATA: *Note: Data for latest year may not have been available at press time.*

In U.S. $	2021	2020	2019	2018	2017	2016
Revenue	880,000,000	840,000,000	800,000,000	700,000,000	600,000,000	570,000,000
R&D Expense						
Operating Income						
Operating Margin %						
SGA Expense						
Net Income						
Operating Cash Flow						
Capital Expenditure						
EBITDA						
Return on Assets %						
Return on Equity %						
Debt to Equity						

CONTACT INFORMATION:
Phone: 952-933-1223   Fax:
Toll-Free: 800-621-6972
Address: 1187 Park Pl., Shakopee, MN 55379 United States

STOCK TICKER/OTHER:
Stock Ticker: Private   Exchange:
Employees: 2,500   Fiscal Year Ends: 12/31
Parent Company:

SALARIES/BONUSES:
Top Exec. Salary: $ Bonus: $
Second Exec. Salary: $ Bonus: $

OTHER THOUGHTS:
Estimated Female Officers or Directors:
Hot Spot for Advancement for Women/Minorities:

Sales, profits and employees may be estimates. Financial information, benefits and other data can change quickly and may vary from those stated here.

Equinix Inc

NAIC Code: 517110

www.equinix.com

TYPES OF BUSINESS:

Data Networks
Internet Exchange Services

GROWTH PLANS/SPECIAL FEATURES:

Equinix is a retail provider of data centers, enabling hundreds of enterprise tenants to house their servers and networking equipment in a collocated environment. Tenants can then connect with each other, cloud service providers, and telecom networks. Equinix operates 240 data centers in 66 markets worldwide and owns just less than half of them. The firm has roughly 10,000 customers, including 2,000 networks, that are dispersed over five verticals: Cloud and IT Services, Content Providers, Network and Mobile Services, Financial Services and Enterprise. About 70% of Equinix's revenue comes from renting space to tenants and related services, and more than 15% comes from connecting customers with each other. Equinix operates as a real estate investment trust.

Equinix offers its employees health and financial benefit options, depending on location.

BRANDS/DIVISIONS/AFFILIATES:

International Business Exchange (IBX)
xScale
Equinix Internet Exchange
Equinix Fabric
Equinix SmartKey
Bare Metal

CONTACTS: *Note: Officers with more than one job title may be intentionally listed here more than once.*

Charles Meyers, CEO
Keith Taylor, CFO
Peter Van Camp, Chairman of the Board
Simon Miller, Chief Accounting Officer
Brandi Morandi, Chief Legal Officer
Karl Strohmeyer, Other Executive Officer
Michael Campbell, Other Executive Officer

FINANCIAL DATA: *Note: Data for latest year may not have been available at press time.*

In U.S. $	2021	2020	2019	2018	2017	2016
Revenue	6,635,537,000	5,998,545,000	5,562,140,000	5,071,654,000	4,368,428,000	3,611,989,000
R&D Expense						
Operating Income	1,120,086,000	1,114,868,000	1,165,892,000	1,005,783,000	847,649,000	657,816,000
Operating Margin %	.17%	.19%	.21%	.20%	.19%	.18%
SGA Expense	2,043,029,000	1,809,337,000	1,586,064,000	1,460,396,000	1,327,630,000	1,133,303,000
Net Income	500,191,000	369,777,000	507,450,000	365,359,000	232,982,000	126,800,000
Operating Cash Flow	2,547,206,000	2,309,826,000	1,992,728,000	1,815,426,000	1,439,233,000	1,019,353,000
Capital Expenditure	2,751,512,000	2,282,504,000	2,079,521,000	2,096,174,000	1,378,725,000	1,113,365,000
EBITDA	2,601,324,000	2,346,060,000	2,457,118,000	2,182,021,000	1,808,010,000	1,389,222,000
Return on Assets %	.02%	.01%	.02%	.02%	.01%	.01%
Return on Equity %	.05%	.04%	.06%	.05%	.04%	.04%
Debt to Equity	1.35%	1.26	1.396	1.507	1.451	1.51

CONTACT INFORMATION:

Phone: 650 598-6000   Fax: 650 513-7900
Toll-Free: 800-322-9280
Address: 1 Lagoon Dr., Redwood City, CA 94065 United States

STOCK TICKER/OTHER:

Stock Ticker: EQIX
Employees: 10,944
Parent Company:

Exchange: NAS
Fiscal Year Ends: 12/31

SALARIES/BONUSES:

Top Exec. Salary: $1,050,000   Bonus: $
Second Exec. Salary: $680,000   Bonus: $

OTHER THOUGHTS:

Estimated Female Officers or Directors: 3
Hot Spot for Advancement for Women/Minorities: Y

Eutelsat Communications SA   www.eutelsat-communications.com

NAIC Code: 517410

TYPES OF BUSINESS:
Satellite Telecommunications

GROWTH PLANS/SPECIAL FEATURES:
Eutelsat Communications is a French commercial satellite provider with coverage spanning the European continent, the Middle East, Africa, India, and the Americas. It operates a fleet of over 30 satellites serving broadcasters, telecommunications operators, Internet service providers, and government agencies. These satellites are used for video broadcasting, news gathering, broadband services, and data connectivity solutions. Eutelsat generates revenue in Europe, the Americas, the Middle East and North Africa, and Asia.

BRANDS/DIVISIONS/AFFILIATES:
Eutelsat SA

CONTACTS: Note: Officers with more than one job title may be intentionally listed here more than once.
Rodolphe Belmer, CEO
Sandrine Teran, CFO
Philippe Oliva, CCO
Marie-Sophie Rouzaud, Chief Human Resources Officer
Yohann Leroy, CTO
Dominique d-Hinnin, Chmn.

FINANCIAL DATA: Note: Data for latest year may not have been available at press time.

In U.S. $	2021	2020	2019	2018	2017	2016
Revenue	1,258,286,000	1,303,563,000	1,347,107,000	1,417,980,000	1,507,108,000	1,559,218,000
R&D Expense						
Operating Income	354,061,700	499,887,800	536,497,300	564,744,800	626,848,300	675,083,100
Operating Margin %	.28%	.38%	.40%	.40%	.42%	.43%
SGA Expense	221,594,500	207,623,700	201,913,100	219,045,100	250,249,800	263,200,800
Net Income	218,331,300	303,481,500	347,127,300	297,362,900	358,752,600	355,387,400
Operating Cash Flow	906,569,300	794,395,400	864,962,900	898,207,200	1,002,325,000	913,401,700
Capital Expenditure	187,024,500	224,653,800	214,966,000	304,705,200	400,766,800	397,911,500
EBITDA	841,610,400	1,069,221,000	1,088,087,000	1,087,475,000	1,170,484,000	1,205,768,000
Return on Assets %	.03%	.04%	.04%	.04%	.04%	.04%
Return on Equity %	.08%	.11%	.13%	.11%	.13%	.14%
Debt to Equity	1.34%	1.10	1.261	0.913	1.20	1.244

CONTACT INFORMATION:
Phone: 33 153983700   Fax:
Toll-Free:
Address: 70 rue Balard, Paris, 75015 France

STOCK TICKER/OTHER:
Stock Ticker: ETCMY   Exchange: PINX
Employees: 1,014   Fiscal Year Ends: 06/30
Parent Company:

SALARIES/BONUSES:
Top Exec. Salary: $ Bonus: $
Second Exec. Salary: $ Bonus: $

OTHER THOUGHTS:
Estimated Female Officers or Directors:
Hot Spot for Advancement for Women/Minorities:

Sales, profits and employees may be estimates. Financial information, benefits and other data can change quickly and may vary from those stated here.

Extreme Networks Inc

www.extremenetworks.com

NAIC Code: 334210A

TYPES OF BUSINESS:

Computer Networking & Related Equipment, Manufacturing
Network Solutions
Cloud Solutions
Internet of Things
Wired and Wireless Infrastructure Equipment
Cloud Networking Platform

GROWTH PLANS/SPECIAL FEATURES:

Extreme Networks provides software-driven networking services for enterprise customers. Its products include wired and wireless network infrastructure equipment and software for network management, policy, analytics, and access controls. It offers high-density Wi-Fi, centralized management, cloud-based network management, and application analytics capabilities. Roughly half of the firm's revenue is generated in the Americas, with the rest coming from Europe, the Middle East, Africa, and Asia-Pacific.

BRANDS/DIVISIONS/AFFILIATES:

Ipanema

CONTACTS: Note: Officers with more than one job title may be intentionally listed here more than once.

Edward Meyercord, CEO
Remi Thomas, CFO
John Shoemaker, Chairman of the Board
Joseph Vitalone, Other Executive Officer

FINANCIAL DATA: Note: Data for latest year may not have been available at press time.

In U.S. $	2021	2020	2019	2018	2017	2016
Revenue	1,009,418,000	948,019,000	995,789,000	983,142,000	607,084,000	519,834,000
R&D Expense	196,995,000	209,606,000	210,132,000	183,877,000	93,724,000	78,721,000
Operating Income	38,976,000	-44,815,000	-6,192,000	23,830,000	28,041,000	-17,894,000
Operating Margin %	.04%	- .05%	- .01%	.02%	.05%	- .03%
SGA Expense	343,042,000	344,623,000	340,949,000	318,095,000	200,490,000	187,575,000
Net Income	1,936,000	-126,845,000	-25,853,000	-46,792,000	-1,744,000	-36,363,000
Operating Cash Flow	144,535,000	35,884,000	104,945,000	19,043,000	59,283,000	30,366,000
Capital Expenditure	17,176,000	15,268,000	22,730,000	40,411,000	10,425,000	5,327,000
EBITDA	88,358,000	-32,921,000	39,596,000	16,332,000	33,022,000	14,243,000
Return on Assets %	.00%	- .15%	- .03%	- .08%	.00%	- .09%
Return on Equity %	.06%	-2.09%	- .23%	- .39%	- .02%	- .37%
Debt to Equity	6.40%	82.405	1.463	1.674	0.64	0.415

CONTACT INFORMATION:

Phone: 408 579-2800   Fax: 408 579-3000
Toll-Free: 888-257-3000
Address: 2121 RDU Center Dr., Ste. 300, Morrisville, NC 27560 United States

STOCK TICKER/OTHER:

Stock Ticker: EXTR   Exchange: NAS
Employees: 2,584   Fiscal Year Ends: 06/30
Parent Company:

SALARIES/BONUSES:

Top Exec. Salary: $725,000   Bonus: $
Second Exec. Salary: $400,000   Bonus: $100,000

OTHER THOUGHTS:

Estimated Female Officers or Directors: 1
Hot Spot for Advancement for Women/Minorities: Y

F5 Inc

www.f5.com

NAIC Code: 511210B

TYPES OF BUSINESS:

Computer Software: Network Management (IT), System Testing & Storage
Multi-Cloud
Security
Managed Services
Software
Hardware

BRANDS/DIVISIONS/AFFILIATES:

Volterra

GROWTH PLANS/SPECIAL FEATURES:

F5 is a market leader in the application delivery controller market. The company sells products for networking traffic, security, and policy management. Its products ensure applications are safely routed in efficient manners within on-premises data centers and across cloud environments. More than half of its revenue is based on providing services, and its three customer verticals are enterprises, service providers, and government entities. The Seattle-based firm was incorporated in 1996 and generates sales globally.

F5 Networks offers its employees comprehensive health benefits and a variety of assistance programs.

CONTACTS: Note: Officers with more than one job title may be intentionally listed here more than once.

Francis Pelzer, CFO
Mika Yamamoto, Chief Marketing Officer
Tom Fountain, Chief Strategy Officer
Geng Lin, Chief Technology Officer
Alan Higginson, Director
Francois Locoh-Donou, Director
Chad Whalen, Executive VP, Divisional
Scot Rogers, Executive VP
Kara Sprague, Executive VP
Haiyan Song, Executive VP
Ana White, Executive VP

FINANCIAL DATA: Note: Data for latest year may not have been available at press time.

In U.S. $	2021	2020	2019	2018	2017	2016
Revenue	2,603,416,000	2,350,822,000	2,242,447,000	2,161,407,000	2,090,041,000	1,995,034,000
R&D Expense	512,627,000	441,324,000	408,058,000	366,084,000	350,365,000	334,227,000
Operating Income	394,025,000	400,067,000	518,463,000	609,325,000	577,065,000	556,428,000
Operating Margin %	.15%	.17%	.23%	.28%	.28%	.28%
SGA Expense	1,203,618,000	1,101,544,000	959,349,000	824,517,000	809,126,000	767,174,000
Net Income	331,241,000	307,441,000	427,734,000	453,689,000	420,761,000	365,855,000
Operating Cash Flow	645,196,000	660,898,000	747,841,000	761,068,000	740,281,000	711,535,000
Capital Expenditure	30,651,000	59,940,000	103,542,000	53,465,000	42,681,000	68,238,000
EBITDA	509,449,000	495,924,000	586,970,000	668,816,000	638,213,000	613,204,000
Return on Assets %	.07%	.08%	.14%	.18%	.18%	.16%
Return on Equity %	.14%	.15%	.28%	.36%	.35%	.29%
Debt to Equity	.27%	0.317				

CONTACT INFORMATION:

Phone: 206 272-5555   Fax: 206 272-5556
Toll-Free: 888-882-4447
Address: 801 5th Ave., Seattle, WA 98104 United States

STOCK TICKER/OTHER:

Stock Ticker: FFIV
Employees: 6,461
Parent Company:

Exchange: NAS
Fiscal Year Ends: 09/30

SALARIES/BONUSES:

Top Exec. Salary: $875,000   Bonus: $
Second Exec. Salary: Bonus: $400,000
$338,009

OTHER THOUGHTS:

Estimated Female Officers or Directors: 3
Hot Spot for Advancement for Women/Minorities: Y

Far EasTone Telecommunications Co Ltd   www.fareastone.com.tw

NAIC Code: 517210

TYPES OF BUSINESS:

Wireless Telecommunications Carriers (except Satellite)
Telecommunications
Digital Applications
Internet
5G
Internet of Things
Telemedicine Services
Smart Applications

BRANDS/DIVISIONS/AFFILIATES:

CONTACTS: Note: Officers with more than one job title may be intentionally listed here more than once.

Chee Ching, Pres.
Douglas Hsu, Chmn.

GROWTH PLANS/SPECIAL FEATURES:

Far EasTone Telecommunications Co., Ltd. (FET) provides telecommunications and digital application services in Taiwan to both consumers and businesses. The company offers mobile, voice, internet and Wi-Fi connectivity services through its 2G-through-5G capabilities. FET's dual-band system automatically switches frequency to track the strongest signal for connected devices. Its design of highly-sensitive base stations reduces the dead space of reception, making signal delivery more flexible and less susceptible to interference. And as an all-region service, the firm is a provider of enhanced full-rate communication, in which the processing of digital sound delivers voice and speech clarity. FET is engaged in the development of Internet of Things (IoT) applications, and has applied core 5G technology toward multi-angle viewing video apps, telemedicine, building a 5G smart factory in Taiwan, and developing a 5G communication network. Other services by FET include data circuit, cloud, information security, smart applications, smart cities, smart transportation, smart manufacturing and more.

FINANCIAL DATA: Note: Data for latest year may not have been available at press time.

In U.S. $	2021	2020	2019	2018	2017	2016
Revenue	3,078,763,957	2,184,530,000	2,934,117,120	3,030,996,224	3,080,077,568	3,233,182,464
R&D Expense						
Operating Income						
Operating Margin %						
SGA Expense						
Net Income	333,203,672	297,115,000	305,600,672	328,214,368	363,196,928	390,380,480
Operating Cash Flow						
Capital Expenditure						
EBITDA						
Return on Assets %						
Return on Equity %						
Debt to Equity						

CONTACT INFORMATION:

Phone: 886 277235000   Fax:
Toll-Free:
Address: 468 Ruei Guang Rd., Nei Hu Dist., Taipei, 11492 Taiwan

STOCK TICKER/OTHER:

Stock Ticker: 4904   Exchange: TWSE
Employees: 5,472   Fiscal Year Ends: 12/31
Parent Company:

SALARIES/BONUSES:

Top Exec. Salary: $ Bonus: $
Second Exec. Salary: $ Bonus: $

OTHER THOUGHTS:

Estimated Female Officers or Directors:
Hot Spot for Advancement for Women/Minorities:

Sales, profits and employees may be estimates. Financial information, benefits and other data can change quickly and may vary from those stated here.

FASTWEB SpA

www.fastweb.it

NAIC Code: 517110

TYPES OF BUSINESS:

Internet Telecommunications Services
Telephony Services
Broadband Internet Access
Video Communications Services
Virtual Private Networks
Digital & Interactive TV
Video-on-Demand Services

BRANDS/DIVISIONS/AFFILIATES:

Swisscom AG
Swisscom Italia Srl

CONTACTS: *Note: Officers with more than one job title may be intentionally listed here more than once.*

Alberto Calcagno, CEO
Walter Renna, COO
Peter Gruter, CFO
Roberto Chieppa, CMO
Matteo Melchiorri Roberto Biazzi, Chief Human Resources Officer
Andrea Lasagna, CTO
Giovanni Moglia, Dir.-Legal & Regulatory Affairs
Federico Ciccone, Chief Strategy Officer
Roberto Biazzi, Dir.-Human Capital
Massimo Mancini, Dir.-Enterprise
Sergio Scalpelli, Dir.-Institutional & External Relations
Danilo Vivarelli, Dir.-Consumer & Micro Bus.
Urs Schaeppi, Chmn.

GROWTH PLANS/SPECIAL FEATURES:

FASTWEB SpA is a leading telecommunications operator based in Italy. The firm has 2.7 million wireline customers and 2 million mobile customers. FASTWEB offers a wide range of voice and data services, fixed communication and mobile, to households and businesses. The company has always focused on innovation and network infrastructure to provide quality ultra-broadband services. Ultra broadband connections can be obtained via personal computers, smartphones and tablets. FASTWEB has developed a national fiber optic network infrastructure of approximately 34,800 miles (56,000 kilometers) with over 2.48 million miles (4 million kilometers) of fiber. Connection speeds range up to 1 Gigabit (Gb). The firm offers customers new generation mobile services based on 4G technology and 4G Plus, and 5G in 2020, with plans to reach 20 million families and businesses by 2024. FASTWEB has 20 retail stores nationwide, enabling consumers to purchase products and services and receive commercial, administrative and technical information. The company has 17 offices, 16 throughout Italy and one in Bruxelles. Swisscom AG owns 100% of FASTWEB through its indirect subsidiary, Swisscom Italia Srl.

FINANCIAL DATA: *Note: Data for latest year may not have been available at press time.*

In U.S. $	2021	2020	2019	2018	2017	2016
Revenue	3,100,000,000	2,829,870,000	2,483,850,000	2,406,540,000	2,328,650,000	2,061,820,000
R&D Expense						
Operating Income						
Operating Margin %						
SGA Expense						
Net Income		190,378,000	156,781,000			
Operating Cash Flow						
Capital Expenditure						
EBITDA						
Return on Assets %						
Return on Equity %						
Debt to Equity						

CONTACT INFORMATION:

Phone: 39-02-4545-1   Fax: 39-02-4545-4811
Toll-Free:
Address: Piazza Adriano Olivetti, 1, Milan, 20139 Italy

STOCK TICKER/OTHER:

Stock Ticker: Subsidiary   Exchange:
Employees: 2,700   Fiscal Year Ends: 12/31
Parent Company: Swisscom AG

SALARIES/BONUSES:

Top Exec. Salary: $ Bonus: $
Second Exec. Salary: $ Bonus: $

OTHER THOUGHTS:

Estimated Female Officers or Directors:
Hot Spot for Advancement for Women/Minorities:

Sales, profits and employees may be estimates. Financial information, benefits and other data can change quickly and may vary from those stated here.

Filtronic plc

NAIC Code: 334220

TYPES OF BUSINESS:

Telecommunications Equipment-Cellular & Broadband Components
Wireless Infrastructure Products
Microwave Components
Radio Frequency

BRANDS/DIVISIONS/AFFILIATES:

GROWTH PLANS/SPECIAL FEATURES:

Filtronic PLC designs and manufactures a broad range of customized RF, microwave and millimeter-wave components and subsystems. Its products are used in mobile wireless communication equipment, point-to-point communication systems, and adjacent defense sectors. The company's customers base includes international original equipment manufacturers as well as a wide range of mobile phone network operators. It has operations in the United Kingdom, Europe, the Americas, and the Rest of the world.

CONTACTS: *Note: Officers with more than one job title may be intentionally listed here more than once.*

Richard Gibbs, CEO
Michael Tyerman, CFO
Maura Moynihan, General Counsel
Reg Gott, Chmn.

FINANCIAL DATA: *Note: Data for latest year may not have been available at press time.*

In U.S. $	2021	2020	2019	2018	2017	2016
Revenue	18,799,020	20,762,790	19,253,400	26,141,710	42,747,340	16,411,070
R&D Expense						
Operating Income	652,576	567,983	282,783	3,865,908	2,056,822	-8,172,909
Operating Margin %	.03%	.03%	.01%	.15%	.05%	- .50%
SGA Expense						
Net Income	72,508	-2,362,566	-1,586,726	1,487,631	3,766,813	-6,183,761
Operating Cash Flow	2,968,012	-3,187,954	-24,169	2,116,038	4,671,960	-6,102,793
Capital Expenditure	488,223	1,502,133	1,277,357	1,279,774	980,072	553,481
EBITDA	1,915,431	636,866	1,086,418	4,103,977	3,813,943	-7,625,470
Return on Assets %	.00%	- .11%	- .07%	.06%	.17%	- .31%
Return on Equity %	.01%	- .19%	- .11%	.11%	.33%	- .61%
Debt to Equity	.17%	0.215	0.011	0.026		

CONTACT INFORMATION:

Phone: 44 174 0625163   Fax:
Toll-Free:
Address: 3 Airport W., Lancaster Way, Leeds, West Yorkshire, LS19 7ZA United Kingdom

STOCK TICKER/OTHER:

Stock Ticker: FLTCF   Exchange: PINX
Employees: 130   Fiscal Year Ends: 05/31
Parent Company:

SALARIES/BONUSES:

Top Exec. Salary: $163,144   Bonus: $50,756
Second Exec. Salary: $145,017   Bonus: $45,922

OTHER THOUGHTS:

Estimated Female Officers or Directors: 1
Hot Spot for Advancement for Women/Minorities:

First Pacific Company Limited

www.firstpacific.com

NAIC Code: 517210

TYPES OF BUSINESS:

Wireless Telecommunications Carriers
Infrastructure
Consumer Food Products
Exploration, Development and Management of Mineral and Energy
Resources
Telecommunications

BRANDS/DIVISIONS/AFFILIATES:

PT Indofood Sukses Makmur Tbk
Indofood CBP
Metro Pacific Investments Corporation
Global Business Power Corporation
Philex Mining Corporation
PXP Energy
PLDT Inc
Smart Communications Inc

GROWTH PLANS/SPECIAL FEATURES:

First Pacific Co Ltd is an investment management company. The company aims to achieve capital appreciation and dividend for its investors through investment across various sectors. Its portfolio comprises telecommunications, consumer food products, infrastructure, and natural resources segment. The company generates maximum revenue from the consumer food products segment. Geographically, it derives a majority of revenue from Indonesia.

CONTACTS: Note: Officers with more than one job title may be intentionally listed here more than once.

Manuel V. Pangilinan, Managing Dir.
Christopher H. Young, CFO
Anthoni Salim, Chmn.

FINANCIAL DATA: Note: Data for latest year may not have been available at press time.

In U.S. $	2021	2020	2019	2018	2017	2016
Revenue	9,103,200,000	7,130,500,000	7,585,000,000	7,742,400,000	7,296,800,000	6,779,000,000
R&D Expense						
Operating Income	1,415,200,000	1,065,000,000	1,034,500,000	1,003,400,000	1,000,100,000	909,600,000
Operating Margin %	.16%	.15%	.14%	.13%	.14%	.13%
SGA Expense	1,403,100,000	1,223,900,000	1,241,900,000	1,174,400,000	1,152,400,000	1,095,100,000
Net Income	333,300,000	201,600,000	-253,900,000	131,800,000	120,900,000	103,200,000
Operating Cash Flow	1,245,900,000	1,036,600,000	1,455,500,000	734,100,000	776,100,000	731,400,000
Capital Expenditure	1,104,300,000	1,065,600,000	1,376,500,000	1,236,000,000	1,063,000,000	696,700,000
EBITDA	2,055,600,000	1,935,600,000	1,451,900,000	1,784,700,000	1,691,900,000	1,516,100,000
Return on Assets %	.01%	.01%	-.01%	.01%	.01%	.01%
Return on Equity %	.10%	.07%	-.08%	.04%	.04%	.03%
Debt to Equity	2.90%	2.858	2.277	2.023	2.017	1.551

CONTACT INFORMATION:

Phone: 852 28424388   Fax: 852 28459243
Toll-Free:
Address: Fl. 24, Two Exchange Square, 8 Connaught Pl., Hong Kong, Hong Kong 999077 Hong Kong

STOCK TICKER/OTHER:

Stock Ticker: FPAFF   Exchange: PINX
Employees: 103,127   Fiscal Year Ends: 12/31
Parent Company:

SALARIES/BONUSES:

Top Exec. Salary: $ Bonus: $
Second Exec. Salary: $ Bonus: $

OTHER THOUGHTS:

Estimated Female Officers or Directors:
Hot Spot for Advancement for Women/Minorities:

Sales, profits and employees may be estimates. Financial information, benefits and other data can change quickly and may vary from those stated here.

Flex Ltd
NAIC Code: 334418

www.flextronics.com

TYPES OF BUSINESS:
Printed Circuit Assembly (Electronic Assembly) Manufacturing
Telecommunications Equipment Manufacturing
Engineering, Design & Testing Services
Logistics Services
Camera Modules
Medical Devices
LCD Displays
Original Design Manufacturing (ODM)

BRANDS/DIVISIONS/AFFILIATES:

GROWTH PLANS/SPECIAL FEATURES:

Flex Ltd is a contract manufacturing company providin
comprehensive electronics design, manufacturing, and produc
management services to global electronics and technolog
companies. The company's operating segments include Fle
Agility Solutions (FAS), Flex Reliability Solutions (FRS), an
Nextracker. Flex Agility Solutions segment includes market
such as Communications, Enterprise and Cloud; Lifestyle; an
Consumer Devices. Flex Reliability Solutions segment include
markets such as Automotive, Health Solutions, and Industria
The Nextracker segment provides solar tracker technologie
that optimize and increase energy production. The company'
geographical segments include China, Mexico, the Unite
States, Brazil, Malaysia, Hungary, and Others.

CONTACTS: Note: Officers with more than one job title may be intentionally listed here more than once.
Revathi Advaithi, CEO
Francois Barbier, Pres.
Christopher Collier, CFO
Phil Ulrich, Chief Human Resources Officer
Mark Kemp, Pres., Medical
Gus Shahin, CIO
Erik Volkerink, CTO
Christopher Obey, Pres., Automotive
Christopher Cook, Pres., Power Solutions
Jeannine Sargent, Pres., Energy
Jonathan Hoak, General Counsel
Francois Barbier, Pres., Global Oper. & Components
David Mark, Chief Strategy Officer
Renee Brotherton, VP-Corp. Comm.
Christopher Collier, Chief Accounting Officer
Paul Humphries, Pres., High Reliability Solutions
Caroline Dowling, Pres., Integrated Network Solutions
Doug Britt, Pres., Industrial & Emerging Solutions
Mike Dennison, Pres., High Velocity Solutions
Tom Linton, Chief Procurement & Supply Chain Officer

FINANCIAL DATA: Note: Data for latest year may not have been available at press time.

In U.S. $	2021	2020	2019	2018	2017	2016
Revenue	24,124,000,000	24,210,000,000	26,211,000,000	25,441,130,000	23,862,930,000	
R&D Expense						
Operating Income	808,000,000	441,000,000	491,000,000	497,843,000	502,210,000	
Operating Margin %	.03%	.02%	.02%	.02%	.02%	
SGA Expense	817,000,000	834,000,000	953,000,000	1,019,399,000	937,339,000	
Net Income	613,000,000	88,000,000	93,000,000	428,534,000	319,564,000	
Operating Cash Flow	144,000,000	-1,533,000,000	-2,971,000,000	-3,866,335,000	-3,822,108,000	
Capital Expenditure	351,000,000	462,000,000	726,000,000	561,997,000	525,111,000	
EBITDA	1,431,000,000	959,000,000	1,093,000,000	1,199,355,000	1,088,486,000	
Return on Assets %	.04%	.01%	.01%	.03%	.03%	
Return on Equity %	.20%	.03%	.03%	.15%	.12%	
Debt to Equity	1.19%	1.137	0.815	0.96	1.093	

CONTACT INFORMATION:
Phone: 65 6890-7188   Fax: 65 6543-1888
Toll-Free:
Address: 2 Changi South Ln., Singapore, 486123 Singapore

STOCK TICKER/OTHER:
Stock Ticker: FLEX   Exchange: NAS
Employees: 172,648   Fiscal Year Ends: 03/31
Parent Company:

SALARIES/BONUSES:
Top Exec. Salary: $1,281,250   Bonus: $2,034,667
Second Exec. Salary: Bonus: $504,557
$642,550

OTHER THOUGHTS:
Estimated Female Officers or Directors: 4
Hot Spot for Advancement for Women/Minorities: Y

Foxconn Technology Co Ltd

www.foxconntech.com.tw

NAIC Code: 334418

TYPES OF BUSINESS:

Contract Electronics Manufacturing
Original Design Manufacturer (ODM)
Electronics Manufacturing & Design
Research and Engineering Services
Outsourcing

BRANDS/DIVISIONS/AFFILIATES:

Future Mobility Corporation
Belkin International Inc

GROWTH PLANS/SPECIAL FEATURES:

Foxconn Technology Co Ltd along with its subsidiaries is engaged in manufacturing, processing, and sales of case, heat dissipation modules, and consumer electronics products. The firm's operating segments include Trading services of electronic products and Manufacturing and sales of mechanical components. It generates a majority of its revenue from the Electronic product trading segment. Geographically, the company operates in China, Japan, Taiwan, USA, and others. It generates a majority of its revenue from Japan followed by China.

CONTACTS: *Note: Officers with more than one job title may be intentionally listed here more than once.*

Han-Ming Li, Pres.
Tzu-Hung Li, Dir.-Finance
Guangyao Li, Chmn.

FINANCIAL DATA: *Note: Data for latest year may not have been available at press time.*

In U.S. $	2021	2020	2019	2018	2017	2016
Revenue	3,469,320,000	3,492,905,000	3,326,660,000	4,735,138,000	4,927,072,000	2,670,287,000
R&D Expense	53,749,110	58,069,450	74,110,470	58,605,000	49,112,190	32,519,570
Operating Income	119,797,300	87,278,800	176,472,300	286,872,800	350,863,200	377,464,700
Operating Margin %	.03%	.02%	.05%	.06%	.07%	.14%
SGA Expense	68,399,140	70,299,160	81,089,050	104,269,800	75,323,740	76,771,880
Net Income	149,626,700	157,274,400	237,654,500	304,881,500	332,171,800	357,362,000
Operating Cash Flow	151,779,600	81,100,870	419,815,300	391,281,200	278,478,300	483,668,500
Capital Expenditure	45,681,770	15,168,950	39,040,620	63,879,440	48,067,810	47,293,230
EBITDA	240,959,900	241,405,300	364,578,900	459,802,400	467,909,400	552,415,100
Return on Assets %	.03%	.03%	.04%	.05%	.05%	.08%
Return on Equity %	.04%	.04%	.07%	.08%	.08%	.11%
Debt to Equity	.00%	0.002	0.003			

CONTACT INFORMATION:

Phone: 886 222680970   Fax: 886 222687176
Toll-Free:
Address: No. 3-2, Zhongshan Road, Taipei, 236 Taiwan

STOCK TICKER/OTHER:

Stock Ticker: FXCOF
Employees: 700,000
Parent Company:

Exchange: PINX
Fiscal Year Ends: 12/31

SALARIES/BONUSES:

Top Exec. Salary: $ Bonus: $
Second Exec. Salary: $ Bonus: $

OTHER THOUGHTS:

Estimated Female Officers or Directors:
Hot Spot for Advancement for Women/Minorities:

Sales, profits and employees may be estimates. Financial information, benefits and other data can change quickly and may vary from those stated here.

Freenet AG

www.freenet-group.de

NAIC Code: 517210

TYPES OF BUSINESS:

Mobile Phone Service
Digital Services
Telecommunications
Internet
Mobile Telephone
IPTV
Business Communications

BRANDS/DIVISIONS/AFFILIATES:

Freenet TV
Mobilcom Debitel GmbH
Klarmobil GmbH
Freenet Digital GmbH
GRAVIS-Computervertriebsgesellschaft mbH
Freenet Energy GmbH
waipu.tv

GROWTH PLANS/SPECIAL FEATURES:

Freenet AG is a German mobile communication and mobile Internet company. It operates as an independent service provider without its own network. The company distributes mobile communications tariffs and options throughout Germany, using a subscription agreement and multi-brand strategy. The company has three operating segments: Mobile communications, TV and media and Other/holding. The Mobile communications segment generates almost all of the firm's revenue. This segment offers a product portfolio of voice and data services for mobile communication operators. It also buys mobile communications services from the network operators and sells them to its end customers.

CONTACTS: *Note: Officers with more than one job title may be intentionally listed here more than once.*

Ingo Arnold, CFO
Rickmann Platen, CCO
Stephan Esch, CTO
Christoph Vilanek, Chmn.

FINANCIAL DATA: *Note: Data for latest year may not have been available at press time.*

In U.S. $	2021	2020	2019	2018	2017	2016
Revenue	2,606,841,000	2,627,144,000	2,990,500,000	2,954,729,000	3,576,577,000	3,428,858,000
R&D Expense						
Operating Income	313,048,900	303,369,300	315,569,800	318,153,800	266,201,000	283,307,500
Operating Margin %	.12%	.12%	.11%	.11%	.07%	.08%
SGA Expense			110,386,300	119,700,800	137,903,600	122,888,600
Net Income	202,083,400	577,017,600	194,671,700	227,547,900	292,334,400	232,936,300
Operating Cash Flow	374,469,200	364,113,500	371,430,300	335,369,500	392,971,800	397,349,600
Capital Expenditure	49,179,090	50,866,800	46,047,400	58,323,300	62,325,880	64,032,960
EBITDA	411,228,600	435,793,700	444,619,700	425,431,900	530,039,200	430,409,300
Return on Assets %	.05%	.12%	.04%	.05%	.07%	.07%
Return on Equity %	.11%	.36%	.15%	.17%	.21%	.17%
Debt to Equity	.55%	0.652	1.449	1.536	1.345	1.231

CONTACT INFORMATION:

Phone: 49 43-31-69-1000   Fax:
Toll-Free:
Address: Hollerstrasse 126, Budelsdorf, SH 24782 Germany

STOCK TICKER/OTHER:

Stock Ticker: FRTAY   Exchange: PINX
Employees: 4,004   Fiscal Year Ends: 12/31
Parent Company:

SALARIES/BONUSES:

Top Exec. Salary: $ Bonus: $
Second Exec. Salary: $ Bonus: $

OTHER THOUGHTS:

Estimated Female Officers or Directors:
Hot Spot for Advancement for Women/Minorities:

Frontier Communications Corporation

www.frontier.com

NAIC Code: 517110

TYPES OF BUSINESS:

Telecommunications
Internet Services
Long-Distance Phone Services
Directory Service
Access Services
Wireless Internet Services

GROWTH PLANS/SPECIAL FEATURES:

Frontier Communications Parent Inc offers a variety of services to residential and business customers over its fiber-optic and copper networks, including video, high-speed internet, advanced voice, and Frontier Secure digital protection solutions. It offers communications solutions to small, medium, and enterprise businesses.

BRANDS/DIVISIONS/AFFILIATES:

FiOS
Vantage

CONTACTS: Note: Officers with more than one job title may be intentionally listed here more than once.

Bernard Han, CEO
Scott Beasley, CFO
Donald Daniels, Chief Accounting Officer
Mark Nielsen, Chief Legal Officer

FINANCIAL DATA: Note: Data for latest year may not have been available at press time.

In U.S. $	2021	2020	2019	2018	2017	2016
Revenue		7,155,000,000	8,107,000,000			
R&D Expense						
Operating Income		1,208,000,000	1,466,000,000			
Operating Margin %		.17%	.18%			
SGA Expense		1,648,000,000	1,804,000,000			
Net Income		-402,000,000	-5,911,000,000			
Operating Cash Flow		1,989,000,000	1,508,000,000			
Capital Expenditure		1,181,000,000	1,226,000,000			
EBITDA		1,874,000,000	-3,207,000,000			
Return on Assets %		-.02%				
Return on Equity %						
Debt to Equity						

CONTACT INFORMATION:

Phone: 203 614-5600   Fax: 203 614-4602
Toll-Free:
Address: 401 Merritt 7, Norwalk, CT 06851 United States

STOCK TICKER/OTHER:

Stock Ticker: FYBR   Exchange: NAS
Employees: 15,600   Fiscal Year Ends: 12/31
Parent Company:

SALARIES/BONUSES:

Top Exec. Salary: $1,113,424   Bonus: $3,750,000
Second Exec. Salary: Bonus: $2,540,659
$666,666

OTHER THOUGHTS:

Estimated Female Officers or Directors: 7
Hot Spot for Advancement for Women/Minorities: Y

Sales, profits and employees may be estimates. Financial information, benefits and other data can change quickly and may vary from those stated here.

Fujitsu Limited
NAIC Code: 334111

www.fujitsu.com

TYPES OF BUSINESS:
Computer Manufacturing
Information Technology
Internet of Things
Artificial Intelligence
Cyber Security
Digital Workplace
Hybrid IT
Digital Transformation

BRANDS/DIVISIONS/AFFILIATES:

GROWTH PLANS/SPECIAL FEATURES:
Fujitsu Ltd delivers total solutions in the field of information and communication technology. The company provides solutions/system integration services focused on information system consulting and construction, and infrastructure services centered on outsourcing services. Fujitsu provides services across a wide range of countries and regions, including Europe, the Americas, Asia, and Oceania. It operates in three segments namely, Technology Solutions; Ubiqitous Solutions and Device Solutions. Ubiquitous Solutions consists of PCs, mobile phones, and mobilewear. In PCs, Fujitsu's lineup includes desktop and laptop PCs known for energy efficiency, security, and other enhanced features, as well as water- and dust-resistant tablets.

CONTACTS:
Note: Officers with more than one job title may be intentionally listed here more than once.

Takahito Tokita, Pres.
Masami Fujita, Sr. Exec. VP
Hideyuki Saso, Sr. Exec. VP
Masahiro Koezuka, Vice Chmn.
Kazuhiko Kato, Exec. VP

FINANCIAL DATA:
Note: Data for latest year may not have been available at press time.

In U.S. $	2021	2020	2019	2018	2017	2016
Revenue	26,621,940,000	28,610,180,000	29,312,050,000	30,394,390,000	30,650,940,000	
R&D Expense						
Operating Income	1,975,111,000	1,568,400,000	965,789,100	1,353,374,000	871,069,400	
Operating Margin %	.07%	.05%	.03%	.04%	.03%	
SGA Expense	6,188,957,000	6,412,675,000	6,922,026,000	7,487,304,000	7,517,058,000	
Net Income	1,503,263,000	1,186,903,000	775,452,400	1,255,859,000	656,251,900	
Operating Cash Flow	2,283,796,000	2,575,371,000	737,288,600	1,486,317,000	1,856,504,000	
Capital Expenditure	954,991,100	986,131,700	894,667,800	1,001,224,000	1,472,093,000	
EBITDA	3,498,057,000	3,320,194,000	2,411,132,000	3,127,870,000	2,365,685,000	
Return on Assets %	.06%	.05%	.03%	.05%	.03%	
Return on Equity %	.15%	.13%	.09%	.17%	.11%	
Debt to Equity	.10%	0.166	0.164	0.245	0.402	

CONTACT INFORMATION:
Phone: 81 362522220   Fax:
Toll-Free:
Address: Shiodome City Center, 1-5-2 Higashi-Shimbashi, Tokyo, 105-7123 Japan

STOCK TICKER/OTHER:
Stock Ticker: FJTSY
Employees: 145,845
Parent Company:

Exchange: PINX
Fiscal Year Ends: 03/31

SALARIES/BONUSES:
Top Exec. Salary: $ Bonus: $
Second Exec. Salary: $ Bonus: $

OTHER THOUGHTS:
Estimated Female Officers or Directors: 1
Hot Spot for Advancement for Women/Minorities:

Sales, profits and employees may be estimates. Financial information, benefits and other data can change quickly and may vary from those stated here.

Fujitsu Network Communications Inc

www.fujitsu.com/us/products/network
NAIC Code: 334210

TYPES OF BUSINESS:
Telecommunications Equipment & Software
Digital Transformation Solutions
Optical Network Solutions
5G Network Solutions
Network Automation Services
Network Deployment Solutions
Open RAN Solutions
Consultancy Services

BRANDS/DIVISIONS/AFFILIATES:
Fujitsu Limited

GROWTH PLANS/SPECIAL FEATURES:
Fujitsu Network Communications, Inc. is a subsidiary of Fujitsu Limited and a provider of network products and services that drive digital transformation. The company's network solutions span optical networks, 5G networks, 5G transport, network services, network automation, rural broadband networks, network deployment solutions and open radio access networks (RAN). Fujitsu Networks' products and services help customers achieve goals such as reducing expenses, stretching capital, speeding up deployment time and modernizing legacy networks. Services include cybersecurity, network automation and operational efficiency consultancy; network design, build and integration solutions; supply chain optimization and program management solutions; network operations and maintenance; digital transformation; and managed network services.

CONTACTS: Note: Officers with more than one job title may be intentionally listed here more than once.
Doug Moore, CEO

FINANCIAL DATA: Note: Data for latest year may not have been available at press time.

In U.S. $	2021	2020	2019	2018	2017	2016
Revenue						
R&D Expense						
Operating Income						
Operating Margin %						
SGA Expense						
Net Income						
Operating Cash Flow						
Capital Expenditure						
EBITDA						
Return on Assets %						
Return on Equity %						
Debt to Equity						

CONTACT INFORMATION:
Phone: 972-690-6000   Fax: 972-479-4647
Toll-Free: 800-873-3822
Address: 2801 Telecom Pkwy., Richardson, TX 75082 United States

STOCK TICKER/OTHER:
Stock Ticker: Subsidiary   Exchange:
Employees: 1,358   Fiscal Year Ends: 03/31
Parent Company: Fujitsu Limited

SALARIES/BONUSES:
Top Exec. Salary: $ Bonus: $
Second Exec. Salary: $ Bonus: $

OTHER THOUGHTS:
Estimated Female Officers or Directors:
Hot Spot for Advancement for Women/Minorities:

Sales, profits and employees may be estimates. Financial information, benefits and other data can change quickly and may vary from those stated here.

Fusion Connect Inc

NAIC Code: 517110

TYPES OF BUSINESS:

Wired Telecommunications Carriers
Internet Technology
Business Connection Solutions
Cloud Communication
Network Security Solutions
Network Management Platforms
Artificial Intelligence

BRANDS/DIVISIONS/AFFILIATES:

CONTACTS: *Note: Officers with more than one job title may be intentionally listed here more than once.*

Brian Crotty, CEO
John Dobbins, COO
Keith Soldan, CFO
Brian George, CTO
Michael Miller, CIO
Jan Sarro, Executive VP, Divisional
James Prenetta, Executive VP
Jonathan Kaufman, Other Executive Officer
Russell Markman, President, Divisional
Timothy J. Bernlohr, Chmn.

GROWTH PLANS/SPECIAL FEATURES:

Fusion Connect, Inc. manages, orchestrates and secures critical internet technology that keeps businesses connected. The company tailors its cloud communication, collaboration, security and network management platforms to meet unique business needs. Its management systems are artificial intelligence (AI)-based. Fusion Connect optimizes business communications, costs and collaboration from any place or device, including remote workers. The firm's leading-edge technology seamlessly integrates solutions across virtual and fixed platforms. Fusion Connect's U.S.-based tech support is available 24/7, and its proactive monitoring system is designed to automatically respond to issues before they impact their customer's business. Other services by Fusion Connect include hosted voice, UCaaS, calling services for Microsoft Teams, contact/call center solutions, SIP trunking, dedicated internet access, broadband, wireless access, SD-WAN, MPLS, remote access VPN, unified threat management and a range of managed IT. In early-2021, to focus on core U.S. business operations, Fusion Connect sold wholly-owned subsidiary Primus, in Canada, to Distributel.

FINANCIAL DATA: *Note: Data for latest year may not have been available at press time.*

In U.S. $	2021	2020	2019	2018	2017	2016
Revenue	140,000,000	137,891,638	135,853,831	143,004,032	150,530,560	122,045,320
R&D Expense						
Operating Income						
Operating Margin %						
SGA Expense						
Net Income			-15,451,011	-14,715,249	-14,014,523	-12,716,323
Operating Cash Flow						
Capital Expenditure						
EBITDA						
Return on Assets %						
Return on Equity %						
Debt to Equity						

CONTACT INFORMATION:

Phone: 212 201-2400   Fax:
Toll-Free: 888 301-1721
Address: 210 Interstate N. Pkwy., Ste. 200, Atlanta, GA 30339 United States

STOCK TICKER/OTHER:

Stock Ticker: Private   Exchange:
Employees: 300   Fiscal Year Ends: 12/31
Parent Company:

SALARIES/BONUSES:

Top Exec. Salary: $ Bonus: $
Second Exec. Salary: $ Bonus: $

OTHER THOUGHTS:

Estimated Female Officers or Directors:
Hot Spot for Advancement for Women/Minorities:

Garmin Ltd

www.garmin.com

NAIC Code: 334511

TYPES OF BUSINESS:
Communications Equipment-GPS-Based
Aviation Electronics
Marine Electronics
Automotive Electronics
Recreation & Fitness Electronics
Navigational Equipment

GROWTH PLANS/SPECIAL FEATURES:
Garmin produces GPS-enabled hardware and software for five verticals: fitness, outdoors, auto, aviation, and marine. The company relies on licensing mapping data to enable its hardware specialized for often niche activities like scuba diving or sailing. Garmin operates in 100 countries and sells its products via distributors as well as relationships with original equipment manufacturers.

BRANDS/DIVISIONS/AFFILIATES:
Garmin Connect
Connect IQ
Vesper Marine

CONTACTS: *Note: Officers with more than one job title may be intentionally listed here more than once.*
Clifton Pemble, CEO
Andrew Etkind, VP-Gen. Counsel
Doug Boessen, CFO
Frank McLoughlin, VP-Aviation Eng.
Andrew Etkind, General Counsel
Brian J. Pokorny, VP-Oper.
Dawn Iddings, VP-Bus. Dev. & Customer Care
Jon Cassat, VP-Comm. Affairs
Patrick Desbois, VP-Exec. Office
Matthew Munn, VP
Philip Straub, VP
Michael Wiegers, VP-Consumer Eng.
Min Kao, Exec. Chmn.

FINANCIAL DATA: *Note: Data for latest year may not have been available at press time.*

In U.S. $	2021	2020	2019	2018	2017	2016
Revenue	4,982,795,000	4,186,573,000	3,757,505,000	3,347,444,000	3,121,560,000	3,045,797,000
R&D Expense	840,024,000	705,685,000	605,366,000	567,805,000	511,634,000	467,960,000
Operating Income	1,218,620,000	1,054,240,000	945,586,000	778,343,000	683,637,000	632,864,000
Operating Margin %	.24%	.25%	.25%	.23%	.22%	.21%
SGA Expense	831,815,000	721,411,000	683,024,000	633,571,000	602,670,000	587,701,000
Net Income	1,082,200,000	992,324,000	952,486,000	694,080,000	709,007,000	517,724,000
Operating Cash Flow	1,012,427,000	1,135,267,000	698,549,000	919,520,000	660,842,000	705,682,000
Capital Expenditure	309,587,000	187,466,000	120,408,000	160,355,000	151,928,000	96,675,000
EBITDA	1,373,438,000	1,180,955,000	1,051,761,000	874,537,000	769,889,000	719,204,000
Return on Assets %	.15%	.15%	.16%	.13%	.15%	.11%
Return on Equity %	.19%	.19%	.21%	.17%	.20%	.15%
Debt to Equity	.01%	0.014	0.01			

CONTACT INFORMATION:
Phone: 41 526301600   Fax:
Toll-Free:
Address: Muhlentalstrasse 2, Schaffhausen, 8200 Switzerland

STOCK TICKER/OTHER:
Stock Ticker: GRMN   Exchange: NYS
Employees: 16,000   Fiscal Year Ends: 12/31
Parent Company:

SALARIES/BONUSES:
Top Exec. Salary: $1,200,000   Bonus: $ 307
Second Exec. Salary: Bonus: $ 307
$705,000

OTHER THOUGHTS:
Estimated Female Officers or Directors: 2
Hot Spot for Advancement for Women/Minorities: Y

Sales, profits and employees may be estimates. Financial information, benefits and other data can change quickly and may vary from those stated here.

GCI Communication Corp

www.gci.com

NAIC Code: 517110

TYPES OF BUSINESS:

Local Telephone Service
Data Services
Mobile Services
Video Services
Voice and Telephone Services
Managed Services
Retail Stores
5G Network

BRANDS/DIVISIONS/AFFILIATES:

Liberty Broadband Corporation

GROWTH PLANS/SPECIAL FEATURES:

GCI Communication Corp. provides data, mobile, video, voice and managed services to consumer, business, government and carrier customers throughout Alaska, serving more than 200 communities. GCI offers telemedicine and online education capabilities to communities across the state. The company recently launched 5G new radio (NR) service launched in Anchorage, and is currently the nation's northernmost 5G service area. GCI Communications operates as a subsidiary of Liberty Broadband Corporation. The firm has retail stores throughout Alaska.

CONTACTS: Note: Officers with more than one job title may be intentionally listed here more than once.

Gregory F. Chapados, Pres.
Peter Pounds, CFO
Albert Rosenthaler, Other Executive Officer

FINANCIAL DATA: Note: Data for latest year may not have been available at press time.

In U.S. $	2021	2020	2019	2018	2017	2016
Revenue	908,189,776	873,259,400	894,732,992	739,761,984	23,000,000	428,000,000
R&D Expense						
Operating Income						
Operating Margin %						
SGA Expense						
Net Income				-873,302,976	1,232,999,936	762,000,000
Operating Cash Flow						
Capital Expenditure						
EBITDA						
Return on Assets %						
Return on Equity %						
Debt to Equity						

CONTACT INFORMATION:

Phone: 907-276-6222   Fax:
Toll-Free: 800-390-2782
Address: 2550 Denali St., Ste. 1000, Anchorage, AK 99503 United States

STOCK TICKER/OTHER:

Stock Ticker: Subsidiary   Exchange:
Employees: 2,051   Fiscal Year Ends: 12/31
Parent Company: Liberty Broadband Corporation

SALARIES/BONUSES:

Top Exec. Salary: $ Bonus: $
Second Exec. Salary: $ Bonus: $

OTHER THOUGHTS:

Estimated Female Officers or Directors: 1
Hot Spot for Advancement for Women/Minorities:

General DataComm LLC

www.gdc.com

NAIC Code: 334210

TYPES OF BUSINESS:

Local Area Network (LAN) Communications Equipment (e.g., Bridges, Gateways, Routers) Manufacturing

BRANDS/DIVISIONS/AFFILIATES:

CONTACTS: *Note: Officers with more than one job title may be intentionally listed here more than once.*

Katherine McClerkin, CEO
Mark Johns, COO
Joe Autem, CFO

GROWTH PLANS/SPECIAL FEATURES:

General DataComm LLC (GDC) provides technology solutions and migration from older architectures to newer Internet Protocol and MPLS packet-based networks as well as bandwidth management, multiprotocol label switching, voice over IP, Ethernet and power over Ethernet. GDC's SpectraComm family of products offer a broad range of NEBS Level 3 Certified, carrier-class network access products including integrated access systems for T1/T3, xDSL, DDS and analog services; modems/CSUs/DSUs, Routers, Ethernet Switches, Ethernet extension devices; frame relay probes and monitors; and network management systems. GDCs Xedge hybrid Multiservice Packet Exchange provides solutions for network operators in transportation, utilities, education, and government, as well as revenue generating platforms for regional carriers. By integrating TDM, Frame Relay, ATM, IP and Ethernet services over an MPLS or ATM WAN, Xedge simplifies the commissioning of cost saving solutions. GDC's customers include leading service providers, government and business enterprises in North America and Latin America.

FINANCIAL DATA: *Note: Data for latest year may not have been available at press time.*

In U.S. $	2021	2020	2019	2018	2017	2016
Revenue						
R&D Expense						
Operating Income						
Operating Margin %						
SGA Expense						
Net Income						
Operating Cash Flow						
Capital Expenditure						
EBITDA						
Return on Assets %						
Return on Equity %						
Debt to Equity						

CONTACT INFORMATION:

Phone: 203 729-0271   Fax: 203 723-2883
Toll-Free:
Address: 353 Christian St., Oxford, CT 06478 United States

STOCK TICKER/OTHER:

Stock Ticker: Private   Exchange:
Employees: 60   Fiscal Year Ends: 09/30
Parent Company:

SALARIES/BONUSES:

Top Exec. Salary: $ Bonus: $
Second Exec. Salary: $ Bonus: $

OTHER THOUGHTS:

Estimated Female Officers or Directors:
Hot Spot for Advancement for Women/Minorities:

Gilat Satellite Networks Ltd

www.gilat.com

NAIC Code: 517410

TYPES OF BUSINESS:

Satellite-Based Internet Service
Satellite-Based Communication Services
VSAT Technology
Cloud-Based Satellite Network

BRANDS/DIVISIONS/AFFILIATES:

GROWTH PLANS/SPECIAL FEATURES:

Gilat Satellite Networks Ltd is a provider of satellite-based broadband communications. The company designs and manufactures ground-based satellite communications equipment and provides comprehensive solutions and end-to-end services. Its portfolio includes a cloud-based satellite network platform, very small aperture terminals (VSATs), amplifiers, high-speed modems, on-the-move antennas and high-power solid-state power amplifiers (SSPAs), block up converters (BUCs) and Trancievers. The company's solutions support multiple applications with a full portfolio of products to address key applications including broadband access, cellular backhaul, enterprise, in-flight connectivity, maritime, trains, defense and public safety.

CONTACTS: Note: Officers with more than one job title may be intentionally listed here more than once.

Adi Sfadia, CEO
Yuval Shani, COO
Bosmat Halpern-Levy, CFO
Ron Levin, VP-Global Accounts & Mobility
Lior Moyal, VP-Human Resources
Gai Berkovich, VP-R&D
Alik Shimelmits, CTO
Alon Levy, General Counsel
Yair Shahrabany, VP-Global Oper. & Customer Svcs.
Doron Elinav, VP-Strategic Accounts
Ari Krashin, VP-Finance
Glenn Katz, CEO-Spacenet, Inc.
Assaf Eyal, VP-Commercial Div.
Moshe Tamir, VP-Defense & Homeland Security
Isaac Angel, Chmn.
Danny Fridman, CEO-Gilat Peru & Colombia

FINANCIAL DATA: Note: Data for latest year may not have been available at press time.

In U.S. $	2021	2020	2019	2018	2017	2016
Revenue	214,970,000	166,135,000	257,334,000	266,391,000	282,756,000	279,551,000
R&D Expense	31,336,000	26,303,000	30,184,000	33,023,000	28,014,000	24,853,000
Operating Income	2,832,000	-15,857,000	26,010,000	21,284,000	10,861,000	755,000
Operating Margin %	.01%	-.10%	.10%	.08%	.04%	.00%
SGA Expense	37,099,000	30,934,000	40,003,000	39,730,000	43,620,000	49,882,000
Net Income	-3,033,000	35,076,000	36,858,000	18,409,000	6,801,000	-5,340,000
Operating Cash Flow	18,903,000	43,160,000	34,782,000	32,017,000	-17,223,000	-36,879,000
Capital Expenditure	8,933,000	4,716,000	7,982,000	10,759,000	3,692,000	4,307,000
EBITDA	11,450,000	46,355,000	34,648,000	30,749,000	20,538,000	10,118,000
Return on Assets %	-.01%	.09%	.09%	.05%	.02%	-.01%
Return on Equity %	-.01%	.14%	.15%	.08%	.03%	-.03%
Debt to Equity	.01%	0.012	0.029	0.034	0.058	0.081

CONTACT INFORMATION:

Phone: 972 39252000   Fax:
Toll-Free:
Address: 21 Yegia Kapayim St., Kiryat Arie, Petah Tikva, 49130 Israel

STOCK TICKER/OTHER:

Stock Ticker: GILT
Employees: 796
Parent Company:

Exchange: NAS
Fiscal Year Ends: 12/31

SALARIES/BONUSES:

Top Exec. Salary: $408,875   Bonus: $253,519
Second Exec. Salary: Bonus: $98,582
$262,052

OTHER THOUGHTS:

Estimated Female Officers or Directors: 3
Hot Spot for Advancement for Women/Minorities: Y

Glentel Inc

www.glentel.com

NAIC Code: 517110

TYPES OF BUSINESS:

Wireless Communications
Retail - Wireless Products

BRANDS/DIVISIONS/AFFILIATES:

BCE Inc
Rogers Communication Inc
WIRELESSWAVE
tBooth Wireless
WIRELESS etc
Samsung Canada
Wireless Zone
Diamond

CONTACTS: Note: Officers with more than one job title may be intentionally listed here more than once.

Thomas Skidmore, CEO
Mike Kimball, CFO, Divisional
Cary Skidmore, Chief Marketing Officer
David Hartman, COO, Divisional
Joe Johnson, COO, Divisional
Jas Boparai, Executive VP
Rick Christiaanse, General Manager, Divisional
Damon Jones, General Manager, Divisional
Kevin Sinclair, Managing Director, Divisional
Shaun Colligan, Managing Director, Divisional
Erika Tse, Other Corporate Officer
Troy Crosland, President, Divisional
Danielle Nielsen, President, Divisional
Jeff Nielson, Senior VP, Divisional
Tony Baker, Vice President, Divisional
Daniel Lowndes, Vice President, Divisional

GROWTH PLANS/SPECIAL FEATURES:

Glentel, Inc. provides telecommunications services throughout North America. The company, based in Canada, operates through three segments: Retail Canada, Retail U.S. and Retail Australia. Retail Canada operates more than 350 retail locations under four business brands. WIRELESSWAVE provides expertise to tech-forward consumers inspired by the latest in mobile phones, and has partnerships with Rogers, Bell, Fido, Virgin mobile, chatr mobile and SaskTel. WIRELESSWAVE also provides an extensive selection of wireless devices and accessories. Tbooth Wireless provides wireless phones, plans and accessories, as well as connection choices from six wireless service providers. WIRELESS etc. is a third-party operator of wireless kiosks in Costco Wholesale Canada warehouses. Samsung Canada offers a line of tablets, smartphones, notebook PCs, connected cameras and accessories. Retail U.S. operates more than 730 retail stores under the Wireless Zone and Diamond brands, which are both retailers of Verizon Wireless products and services. Moreover, BJ's Wholesale Club is a membership warehouse club that includes Diamond Wireless kiosks, which offer Verizon Wireless products and services. Retail Australia has two business units that market wireless services in Australia: ALLPHONES and ARMS. ALLPHONES is an in-department multi-carrier telecommunications retailer in Australia, representing major wireless brands such as Vodafone, Boost Mobile, Amaysim and Belong by Telstra. ARMS manages a number of branded stores of behalf of major retailers and carriers across Australia, including Costco, Vodafone and Woolworths. In addition, SAMSUNG Australia Experience stores represent Samsung and its line of products. In 2015, the firm was acquired by Canadian communications provider, BCE Inc., which then sold 50% of its ownership to Rogers Communication Inc., causing Glentel to become a joint venture.

Glentel offers its employees flexible hours and a professional development training program.

FINANCIAL DATA: Note: Data for latest year may not have been available at press time.

In U.S. $	2021	2020	2019	2018	2017	2016
Revenue						
R&D Expense						
Operating Income						
Operating Margin %						
SGA Expense						
Net Income						
Operating Cash Flow						
Capital Expenditure						
EBITDA						
Return on Assets %						
Return on Equity %						
Debt to Equity						

CONTACT INFORMATION:

Phone: 604 415-6500   Fax: 604 415-6565
Toll-Free:
Address: 8501 Commerce Ct., Burnaby, BC V5A 4N3 Canada

STOCK TICKER/OTHER:

Stock Ticker: Joint Venture   Exchange:
Employees: 1,465   Fiscal Year Ends: 12/31
Parent Company:

SALARIES/BONUSES:

Top Exec. Salary: $ Bonus: $
Second Exec. Salary: $ Bonus: $

OTHER THOUGHTS:

Estimated Female Officers or Directors:
Hot Spot for Advancement for Women/Minorities:

Sales, profits and employees may be estimates. Financial information, benefits and other data can change quickly and may vary from those stated here.

Global Telecom Holding SAE

NAIC Code: 517210

TYPES OF BUSINESS:

Mobile Phone Service
Telecommunications Services
Mobile Services
Internet Operations
Wireless Services
Broadband Services

BRANDS/DIVISIONS/AFFILIATES:

VEON Ltd
Orascom Telecom Algeria SpA
Djezzy
Pakistan Mobile Communications Ltd
Mobilink
Sheba Telecom (Pvt) Limited
bangalink

GROWTH PLANS/SPECIAL FEATURES:

Global Telecom Holding SAE (GTH) is an international telecommunications company that provides mobile telecommunication services in Africa and Asia. The company operates networks in: Algeria through Orascom Telecom Algeria SpA, which does business as Djezzy; Pakistan through Pakistan Mobile Communications Ltd., which does business as Mobilink; and Bangladesh through Sheba Telecom (Pvt. Limited, which does business as bangalink. Services offered span mobile, internet, wireless, broadband and other related services. Global Telecom operates as a subsidiary of the VEON Ltd., a multinational telecommunication services company based in Netherlands.

CONTACTS: *Note: Officers with more than one job title may be intentionally listed here more than once.*

Hesham Shoukry, Dir.-Admin.
David Dobbie, Chief Legal & Regulatory Affairs Officer
Murat Kirkgoz, Chmn.

FINANCIAL DATA: *Note: Data for latest year may not have been available at press time.*

In U.S. $	2021	2020	2019	2018	2017	2016
Revenue	2,268,084,000	2,180,850,000	2,077,000,000	2,827,000,000	3,014,700,000	2,955,500,000
R&D Expense						
Operating Income						
Operating Margin %						
SGA Expense						
Net Income	-373,000,000	283,000,000	1,301,000,000	-252,000,000	-79,100,000	193,900,000
Operating Cash Flow						
Capital Expenditure						
EBITDA						
Return on Assets %						
Return on Equity %						
Debt to Equity						

CONTACT INFORMATION:

Phone: 20 224619654   Fax: 20 224615054
Toll-Free:
Address: 2005A Nile City Towers, Cornish El Nile RamletBeau, Cairo, 11221 Egypt

STOCK TICKER/OTHER:

Stock Ticker: Subsidiary
Employees: 12,500
Parent Company: VEON Ltd

Exchange:
Fiscal Year Ends: 12/31

SALARIES/BONUSES:

Top Exec. Salary: $ Bonus: $
Second Exec. Salary: $ Bonus: $

OTHER THOUGHTS:

Estimated Female Officers or Directors:
Hot Spot for Advancement for Women/Minorities:

Sales, profits and employees may be estimates. Financial information, benefits and other data can change quickly and may vary from those stated here.

Globalstar Inc

NAIC Code: 517410

TYPES OF BUSINESS:

Satellite Phone & Data Service
Satellite Network Operations
Satellite Communications Equipment
Logistics & Transportation Data Services
Shipping Container Data Services

BRANDS/DIVISIONS/AFFILIATES:

Thermo Capital Partners LLC
Globalstar System
SPOT
SmartOne
STX-3
STINGR

GROWTH PLANS/SPECIAL FEATURES:

Globalstar Inc is a telecommunications company that derives revenue from the provision of mobile satellite services. Mobile satellite services are typically used by customers where existing terrestrial wireline and wireless communications networks are impaired or do not exist. The company provides communications services such as two-way voice and data transmission. In addition, one-way data transmission is also offered. Both services are offered using mobile or fixed devices. The company is an owner of satellite assets. The company generates the vast majority of its revenue within the United States.

CONTACTS: Note: Officers with more than one job title may be intentionally listed here more than once.

David Kagan, CEO
Rebecca Clary, CFO
James Monroe, Chairman of the Board
L. Ponder, General Counsel

FINANCIAL DATA: Note: Data for latest year may not have been available at press time.

In U.S. $	2021	2020	2019	2018	2017	2016
Revenue	124,297,000	128,487,000	131,718,000	130,113,000	112,660,000	96,861,000
R&D Expense						
Operating Income	-65,261,000	-58,747,000	-62,922,000	-67,857,000	-51,406,000	-62,903,000
Operating Margin %	-.53%	-.46%	-.48%	-.52%	-.46%	-.65%
SGA Expense	41,358,000	41,738,000	45,233,000	55,443,000	38,759,000	40,559,000
Net Income	-112,625,000	-109,639,000	15,324,000	-6,516,000	-89,074,000	-132,646,000
Operating Cash Flow	131,881,000	22,215,000	3,048,000	5,920,000	13,857,000	8,813,000
Capital Expenditure	45,536,000	14,536,000	11,491,000	10,369,000	8,866,000	11,381,000
EBITDA	30,976,000	38,068,000	32,850,000	22,581,000	26,092,000	14,487,000
Return on Assets %	-.13%	-.12%	.02%	-.01%	-.08%	-.11%
Return on Equity %	-.29%	-.26%	.04%	-.02%	-.39%	-.66%
Debt to Equity	.73%	0.804	1.176	1.023	1.493	3.094

CONTACT INFORMATION:

Phone: 408 933-4000   Fax: 409 933-4100
Toll-Free:
Address: 1351 Holiday Sqare Blvd., Covington, CA 70433 United States

STOCK TICKER/OTHER:

Stock Ticker: GSAT
Employees: 329
Parent Company:

Exchange: ASE
Fiscal Year Ends: 12/31

SALARIES/BONUSES:

Top Exec. Salary: $543,225   Bonus: $150,000
Second Exec. Salary: $313,719   Bonus: $100,000

OTHER THOUGHTS:

Estimated Female Officers or Directors: 1
Hot Spot for Advancement for Women/Minorities:

Sales, profits and employees may be estimates. Financial information, benefits and other data can change quickly and may vary from those stated here.

Globe Telecom Inc

NAIC Code: 517210

TYPES OF BUSINESS:
Wireless Telecommunications Carriers (except Satellite)
Telecommunications
Digital
Mobile
Broadband

BRANDS/DIVISIONS/AFFILIATES:

GROWTH PLANS/SPECIAL FEATURES:
Globe Telecom Inc is a telecommunications company that provides mobile, voice, and broadband services. It operates through two segments: mobile and fixed-line/broadband. Mobile involves traditional mobile services and contributes the majority of company revenue. Within the mobile division, the majority of subscribers are considered prepaid customers. The other division, fixed-line and broadband, provides fixed-line voice services, corporate data, and Internet for its customers. The company generates the vast majority of its revenue in the Philippines.

CONTACTS: Note: Officers with more than one job title may be intentionally listed here more than once.
Ernest L. Cu, CEO
Rosemarie Maniego-Eala, CFO
Maria Louisa Guevarra-Cabreira, CCO
Renato M. Jiao, Chief Human Resources Officer
Carlomagno E. Malana, CIO
Jaime Augusto Zobel de Ayala, Chmn.

FINANCIAL DATA: Note: Data for latest year may not have been available at press time.

In U.S. $	2021	2020	2019	2018	2017	2016
Revenue	3,024,107,000	2,893,808,000	3,004,513,000	2,725,307,000	2,438,809,000	2,285,590,000
R&D Expense						
Operating Income	667,394,800	763,182,400	806,250,900	678,736,600	528,812,700	515,542,500
Operating Margin %	.22%	.26%	.27%	.25%	.22%	.23%
SGA Expense	598,405,400	533,335,400	556,333,600	484,673,400	507,031,500	514,547,000
Net Income	426,407,200	334,925,600	401,466,400	336,050,800	271,602,300	286,252,300
Operating Cash Flow	1,174,338,000	1,174,689,000	1,334,815,000	1,042,916,000	906,873,300	675,366,800
Capital Expenditure	1,673,127,000	1,087,101,000	919,844,200	779,858,600	766,923,100	662,431,700
EBITDA	1,390,047,000	1,236,790,000	1,319,286,000	1,149,881,000	970,445,800	886,874,800
Return on Assets %	.06%	.06%	.07%	.06%	.06%	.07%
Return on Equity %	.24%	.22%	.29%	.28%	.26%	.29%
Debt to Equity	1.87%	1.978	1.584	1.844	2.122	1.734

CONTACT INFORMATION:
Phone: 63 27302000   Fax: 63 27390072
Toll-Free:
Address: 32nd Street Corner 7th Avenue, Taguig, 1634 Philippines

STOCK TICKER/OTHER:
Stock Ticker: GTMEY   Exchange: PINX
Employees: 8,285   Fiscal Year Ends: 12/31
Parent Company:

SALARIES/BONUSES:
Top Exec. Salary: $ Bonus: $
Second Exec. Salary: $ Bonus: $

OTHER THOUGHTS:
Estimated Female Officers or Directors:
Hot Spot for Advancement for Women/Minorities:

GoDaddy Inc

www.goDaddy.com

NAIC Code: 518210

TYPES OF BUSINESS:
Domain Name Registration
Domain Name Reselling
Research & Development, Internet Services

BRANDS/DIVISIONS/AFFILIATES:
Poynt

GROWTH PLANS/SPECIAL FEATURES:
GoDaddy is a provider of domain registration and aftermarket services, website hosting, security, design, and business productivity tools, commerce solutions, and domain registry services. The company primarily targets micro- to small businesses, website design professionals, registrar peers, and domain investors. Since acquiring payment processing platform Poynt in 2021, the company has expanded into omni commerce solutions, including offering an online payment gateway and offline point-of-sale devices.

GoDaddy offers its employees 100% paid medical and dental premiums, employee appreciation outings, a 401(k) plan, life and disability insurance, maternity and paternity leave, adoption assistance, subsidized lunches and employee discounts.

CONTACTS: Note: Officers with more than one job title may be intentionally listed here more than once.
Amanpal Bhutani, CEO
Ray Winborne, CFO
Charles Robel, Chairman of the Board
Nick Daddario, Chief Accounting Officer
Nima Kelly, Chief Legal Officer

FINANCIAL DATA: Note: Data for latest year may not have been available at press time.

In U.S. $	2021	2020	2019	2018	2017	2016
Revenue	3,815,700,000	3,316,700,000	2,988,100,000	2,660,100,000	2,231,900,000	1,847,900,000
R&D Expense	706,300,000	560,400,000	492,600,000	434,000,000	355,800,000	287,800,000
Operating Income	381,800,000	-358,900,000	211,300,000	164,500,000	190,100,000	37,600,000
Operating Margin %	.10%	- .11%	.07%	.06%	.09%	.02%
SGA Expense	849,700,000	762,300,000	707,700,000	625,400,000	535,600,000	450,000,000
Net Income	242,300,000	-495,100,000	137,000,000	77,100,000	136,400,000	-16,500,000
Operating Cash Flow	829,300,000	764,600,000	723,400,000	559,800,000	475,600,000	386,500,000
Capital Expenditure	253,200,000	81,500,000	92,300,000	97,000,000	135,200,000	62,800,000
EBITDA	579,200,000	-201,400,000	428,200,000	405,500,000	395,600,000	195,800,000
Return on Assets %	.03%	- .08%	.02%	.01%	.03%	.00%
Return on Equity %	7.04%	-1.30%	.18%	.12%	.26%	- .03%
Debt to Equity	48.97%		3.329	3.02	4.955	1.841

CONTACT INFORMATION:
Phone: 480-505-8800   Fax: 480-505-8844
Toll-Free:
Address: 2155 E. GoDaddy Way, Tempe, AZ 85284 United States

STOCK TICKER/OTHER:
Stock Ticker: GDDY   Exchange: NYS
Employees: 7,024   Fiscal Year Ends: 12/31
Parent Company:

SALARIES/BONUSES:
Top Exec. Salary: $1,000,000   Bonus: $
Second Exec. Salary: Bonus: $250,000
$298,846

OTHER THOUGHTS:
Estimated Female Officers or Directors: 3
Hot Spot for Advancement for Women/Minorities: Y

Sales, profits and employees may be estimates. Financial information, benefits and other data can change quickly and may vary from those stated here.

GTT Communications Inc

NAIC Code: 517210

www.gtt.net/us-en

TYPES OF BUSINESS:

Wireless Telecommunications Carriers (except Satellite)
Network Solutions
Digital Business
Cloud Application
Advanced Solutions
Wide Area Networking
Enterprise Voice
Internet

BRANDS/DIVISIONS/AFFILIATES:

GROWTH PLANS/SPECIAL FEATURES:

GTT Communications, Inc. is a network solutions provider tha
enables digital business and cloud application optimizatior
across six continents, reaching more than 140 countries. The
company serves thousands of multinational and nationa
enterprise, government and carrier customers with a portfolic
of advanced connectivity and security services. GTT's service
are grouped into six categories: transport and infrastructure
consisting of wavelengths, low latency, Ethernet direc
colocation and dark fiber; advanced solutions, such as
advanced security, hybrid cloud and advanced services; wide
area networking (WAN), consisting of SD-WAN, MPLS, VPLS
uCPE, Ethernet, managed network services and clouc
connect; enterprise voice, including SIP trunking; and internet
including IP transit, dedicated internet and broadband. GTT's
industry solutions span financial markets, manufacturing, retai
ecommerce, media, broadcast and event services.

CONTACTS: *Note: Officers with more than one job title may be intentionally listed here more than once.*

Richard Calder, CEO
Michael Bauer, CFO
H. Thompson, Chairman of the Board
Christopher McKee, Executive VP, Divisional

FINANCIAL DATA: *Note: Data for latest year may not have been available at press time.*

In U.S. $	2021	2020	2019	2018	2017	2016
Revenue	1,820,000,000	1,800,000,000	1,727,800,000	1,490,800,000	827,900,000	521,688,000
R&D Expense						
Operating Income						
Operating Margin %						
SGA Expense						
Net Income			-105,900,000	-243,400,000	-71,500,000	5,260,000
Operating Cash Flow						
Capital Expenditure						
EBITDA						
Return on Assets %						
Return on Equity %						
Debt to Equity						

CONTACT INFORMATION:

Phone: 703 442-5500   Fax:
Toll-Free:
Address: 7900 Tysons One Pl., Ste. 1450, McLean, VA 22102 United States

STOCK TICKER/OTHER:

Stock Ticker: Private
Employees: 1,257
Parent Company:

Exchange:
Fiscal Year Ends: 12/31

SALARIES/BONUSES:

Top Exec. Salary: $ Bonus: $
Second Exec. Salary: $ Bonus: $

OTHER THOUGHTS:

Estimated Female Officers or Directors:
Hot Spot for Advancement for Women/Minorities:

Sales, profits and employees may be estimates. Financial information, benefits and other data can change quickly and may vary from those stated here.

Harmonic Inc

NAIC Code: 334210

www.harmonicinc.com

TYPES OF BUSINESS:
Networking Equipment
Video Stream Processing
Cable Edge & Access
Software

GROWTH PLANS/SPECIAL FEATURES:
Harmonic Inc designs and manufactures video infrastructure products and system solutions to deliver video and broadband services to consumer devices. The firm operates in two segments: Video, which sells video processing, production, and playout solutions to cable operators and satellite and telecommunications providers; and Cable Access, which sells cable edge solutions to cable operators. Roughly half of the firm's revenue is generated in the Americas, with the rest coming from Europe, the Middle East, and Africa, as well as the Asia Pacific region.

BRANDS/DIVISIONS/AFFILIATES:

CONTACTS: *Note: Officers with more than one job title may be intentionally listed here more than once.*
Patrick Harshman, CEO
Sanjay Kalra, CFO
Patrick Gallagher, Chairman of the Board
Nimrod Ben-Natan, General Manager, Divisional
Ian Graham, Senior VP, Divisional
Neven Haltmayer, Senior VP, Divisional

FINANCIAL DATA: *Note: Data for latest year may not have been available at press time.*

In U.S. $	2021	2020	2019	2018	2017	2016
Revenue	507,149,000	378,831,000	402,874,000	403,558,000	358,246,000	405,911,000
R&D Expense	102,231,000	82,494,000	84,614,000	89,163,000	95,978,000	98,401,000
Operating Income	18,919,000	-10,127,000	16,224,000	-2,093,000	-65,570,000	-52,434,000
Operating Margin %	.04%	-.03%	.04%	-.01%	-.18%	-.13%
SGA Expense	138,085,000	119,611,000	119,035,000	118,952,000	136,270,000	144,381,000
Net Income	13,254,000	-29,271,000	-5,924,000	-21,035,000	-82,955,000	-72,314,000
Operating Cash Flow	41,017,000	39,163,000	31,295,000	12,284,000	3,064,000	438,000
Capital Expenditure	12,975,000	32,205,000	10,328,000	7,044,000	11,399,000	15,107,000
EBITDA	32,549,000	999,000	24,661,000	15,791,000	-50,708,000	-36,147,000
Return on Assets %	.02%	-.05%	-.01%	-.04%	-.16%	-.13%
Return on Equity %	.05%	-.11%	-.02%	-.09%	-.34%	-.24%
Debt to Equity	.48%	0.641	0.49	0.559	0.568	0.433

CONTACT INFORMATION:
Phone: 408 542-2500   Fax:
Toll-Free: 800-788-1330
Address: 2590 Orchard Pkwy., San Jose, CA 95131 United States

STOCK TICKER/OTHER:
Stock Ticker: HLIT   Exchange: NAS
Employees: 1,267   Fiscal Year Ends: 12/31
Parent Company:

SALARIES/BONUSES:
Top Exec. Salary: $545,592   Bonus: $
Second Exec. Salary: $402,463   Bonus: $

OTHER THOUGHTS:
Estimated Female Officers or Directors: 1
Hot Spot for Advancement for Women/Minorities:

Hawaiian Telcom Holdco Inc
www.hawaiiantel.com

NAIC Code: 517110

TYPES OF BUSINESS:
Telecommunications Service Wired, Satellite or Cable

GROWTH PLANS/SPECIAL FEATURES:
Hawaiian Telcom Holdco, Inc. is the largest full-service provider of communications services and products in Hawaii. The firm provides local telephone service including voice and data transport.

BRANDS/DIVISIONS/AFFILIATES:

CONTACTS: *Note: Officers with more than one job title may be intentionally listed here more than once.*
Scott Barber, CEO
Dan Bessey, CFO
John Komeiji, General Counsel
Kevin Paul, Senior VP, Divisional
Paul Krueger, Vice President, Divisional
Jason Fujita, Vice President, Divisional
Amy Aapala, Vice President, Divisional
Benjamin Morgan, Vice President, Divisional
Sunshine Topping, Vice President, Divisional
Gregory Chamberlain, Vice President, Divisional

FINANCIAL DATA: *Note: Data for latest year may not have been available at press time.*

In U.S. $	2021	2020	2019	2018	2017	2016
Revenue					368,419,008	392,963,008
R&D Expense						
Operating Income						
Operating Margin %						
SGA Expense						
Net Income						
Operating Cash Flow						
Capital Expenditure						
EBITDA						
Return on Assets %						
Return on Equity %						
Debt to Equity						

CONTACT INFORMATION:
Phone: 808 546-4511   Fax:
Toll-Free:
Address: 1177 Bishop Street, Honolulu, HI 96813 United States

STOCK TICKER/OTHER:
Stock Ticker: Subsidiary   Exchange:
Employees: 1,300   Fiscal Year Ends:
Parent Company: Cincinnati Bell Inc

SALARIES/BONUSES:
Top Exec. Salary: $ Bonus: $
Second Exec. Salary: $ Bonus: $

OTHER THOUGHTS:
Estimated Female Officers or Directors:
Hot Spot for Advancement for Women/Minorities:

Sales, profits and employees may be estimates. Financial information, benefits and other data can change quickly and may vary from those stated here.

Hellenic Telecommunications Organization SA   www.cosmote.gr
NAIC Code: 517110

TYPES OF BUSINESS:
Local Telephone Service
Long-Distance Service
Internet Service
Mobile Phone Service
Equipment Sales
Hosting Services
Payphones

BRANDS/DIVISIONS/AFFILIATES:
Deutsche Telekom AG
COSMOTE
CosmoOne
OTE Rural
OTEGlobe
OTE SAT-MARITEL
COSMOTE TV PRODUCTIONS
OTE Estate

GROWTH PLANS/SPECIAL FEATURES:
Hellenic Telecommunication Organization is a telecommunications company that offers Internet access services, TV services, broadband, fixed-line services, and mobile telecommunication. The company is organized into four segments: OTE, Cosmote Group, Telekom Romania, and Telekom Romania Mobile. The company generates most of its revenue from the OTE segment which provides information and communication technology services in Greece. The company earns most of its revenue in Greece and the rest in Romania.

CONTACTS: Note: Officers with more than one job title may be intentionally listed here more than once.
Michael Tsamaz, CEO
Eirini Nikolaidi, Exec. Dir.-Legal & Regulatory Affairs
Ioannis Konstantinidis, Chief Strategic Planning & Transformation Officer
Deppie Tzimea, Dir.-Corp. Comm.
Maria Rontogianni, Chief Internal Audit Officer
Aris Dimitriadis, Chief Compliance, ERM & Insurance Officer
Konstantinos Liamidis, Chief Intl Oper. Officer

FINANCIAL DATA: Note: Data for latest year may not have been available at press time.

In U.S. $	2021	2020	2019	2018	2017	2016
Revenue	3,434,868,000	3,323,305,000	3,368,277,000	3,873,774,000	3,871,938,000	3,985,336,000
R&D Expense						
Operating Income	827,945,600	335,298,100	462,462,500	448,083,900	377,210,300	359,364,500
Operating Margin %	.24%	.10%	.14%	.12%	.10%	.09%
SGA Expense	67,304,360	64,754,950	66,998,430	85,354,160	94,328,080	106,565,200
Net Income	568,619,800	367,012,700	209,153,400	178,458,500	68,528,070	142,766,800
Operating Cash Flow	1,248,394,000	1,272,562,000	1,175,583,000	1,015,174,000	816,422,300	1,045,359,000
Capital Expenditure	597,581,100	680,997,700	557,504,400	734,127,400	922,069,700	665,905,200
EBITDA	1,460,301,000	1,195,672,000	1,332,535,000	1,280,007,000	1,276,437,000	1,308,560,000
Return on Assets %	.10%	.06%	.03%	.03%	.01%	.02%
Return on Equity %	.28%	.18%	.09%	.07%	.03%	.06%
Debt to Equity	.48%	0.634	0.649	0.549	0.544	0.824

CONTACT INFORMATION:
Phone: 30 2106111000   Fax: 30 2106111030
Toll-Free:
Address: 99 Kifissias Ave., Amaroussion, Athens, 15124 Greece

STOCK TICKER/OTHER:
Stock Ticker: HLTOY   Exchange: PINX
Employees: 16,291   Fiscal Year Ends: 12/31
Parent Company: Deutsche Telekom AG

SALARIES/BONUSES:
Top Exec. Salary: $ Bonus: $
Second Exec. Salary: $ Bonus: $

OTHER THOUGHTS:
Estimated Female Officers or Directors: 5
Hot Spot for Advancement for Women/Minorities: Y

Sales, profits and employees may be estimates. Financial information, benefits and other data can change quickly and may vary from those stated here.

Hello Direct Inc

NAIC Code: 454113

www.hellodirect.com

TYPES OF BUSINESS:

Telephone Products & Equipment Interface Solutions Marketing
Telephones & Accessories

BRANDS/DIVISIONS/AFFILIATES:

Synercy Communications Management

GROWTH PLANS/SPECIAL FEATURES:

Hello Direct, Inc. is a provider and direct marketer of innovative hands-free devices for businesses and governments in the U.S. Hands-free devices include headsets, speakerphones, webcams and conferencing solutions. Hello Direct is a business-to-business leader in telecommunications solutions with nearly 2 million customers. The company has also developed special services and pricing exclusively for corporate and public-sector customers, with all of the support necessary for a successful implementation. Hello Direct offers more than 1,500 high-quality solutions, including headsets from Jabra, Plantronics, Sennheiser and VXi, as well as business communication products from manufacturers such as Polycom, Logitech, AT&T and Panasonic. The firm offers an extensive line of unified communication (UC) devices compatible with all the leading UC platforms. Hello Direct is privately-owned by Synergy Communications Management.

CONTACTS: *Note: Officers with more than one job title may be intentionally listed here more than once.*

Lou Desiderio, Pres.-Synergy
E. Alexander Glover, Pres.
Dean Witter III, Secretary

FINANCIAL DATA: *Note: Data for latest year may not have been available at press time.*

In U.S. $	2021	2020	2019	2018	2017	2016
Revenue						
R&D Expense						
Operating Income						
Operating Margin %						
SGA Expense						
Net Income						
Operating Cash Flow						
Capital Expenditure						
EBITDA						
Return on Assets %						
Return on Equity %						
Debt to Equity						

CONTACT INFORMATION:

Phone: Fax: 800 456-2566
Toll-Free: 800-435-5634
Address: 400 Imperial Blvd., Cape Canaveral, FL 32920 United States

STOCK TICKER/OTHER:

Stock Ticker: Private   Exchange:
Employees: Fiscal Year Ends: 12/31
Parent Company: Synergy Communications Management

SALARIES/BONUSES:

Top Exec. Salary: $ Bonus: $
Second Exec. Salary: $ Bonus: $

OTHER THOUGHTS:

Estimated Female Officers or Directors: 1
Hot Spot for Advancement for Women/Minorities:

Hi Sun Technology (China) Limited

www.hisun.com.hk

NAIC Code: 517210

TYPES OF BUSINESS:

Wireless Telecommunications Carriers

BRANDS/DIVISIONS/AFFILIATES:

Beijing Hi Sunsray Information Technology Limited

GROWTH PLANS/SPECIAL FEATURES:

Hi Sun Technology (China) Ltd has five operating segments: payment processing, financial solutions, electronic power meters, Information security chips & solutions, and platform operation solutions. The payment processing solutions segment is engaged in the provision of payment processing services, merchant recruiting, and related products and solutions. Financial solutions provide consulting, integration, and operation services of information technology products to financial institutions and banks. The firm manufactures electronic power meters primarily for the public sector. Platform operation solutions provide operational support for telecom and payment platform solutions. Its geographical segments are Mainland China and Hong Kong.

CONTACTS:
Note: Officers with more than one job title may be intentionally listed here more than once.

Man Chun Kui, CEO
Yuk Fung Cheung, Chmn.

FINANCIAL DATA:
Note: Data for latest year may not have been available at press time.

In U.S. $	2021	2020	2019	2018	2017	2016
Revenue	532,841,600	489,774,600	710,336,800	593,896,900	370,708,700	244,750,700
R&D Expense						
Operating Income	62,391,120	48,959,400	69,675,160	36,847,060	23,664,260	14,032,900
Operating Margin %	.12%	.10%	.10%	.06%	.06%	.06%
SGA Expense	97,851,780	83,013,760	102,300,300	99,980,120	73,247,240	64,391,310
Net Income	448,615,200	83,656,580	72,500,340	35,437,340	44,096,070	36,174,820
Operating Cash Flow	-94,676,530	12,857,960	159,808,800	89,041,330	-19,547,960	115,778,000
Capital Expenditure	21,589,550	29,857,310	28,039,290	61,329,940	54,687,080	21,777,710
EBITDA	521,107,500	153,053,600	150,155,400	93,139,420	79,674,050	55,764,940
Return on Assets %	.34%	.07%	.08%	.05%	.06%	.06%
Return on Equity %	.59%	.13%	.13%	.07%	.09%	.08%
Debt to Equity	.00%	0.005	0.007			

CONTACT INFORMATION:

Phone: 852 2588 8800   Fax: 852 2802 3300
Toll-Free:
Address: Room 2416 24th Floor, Wanchai, Hong Kong, Hong Kong 999077 Hong Kong

STOCK TICKER/OTHER:

Stock Ticker: HISNF
Employees: 2,759
Parent Company:

Exchange: PINX
Fiscal Year Ends: 12/31

SALARIES/BONUSES:

Top Exec. Salary: $ Bonus: $
Second Exec. Salary: $ Bonus: $

OTHER THOUGHTS:

Estimated Female Officers or Directors:
Hot Spot for Advancement for Women/Minorities:

Sales, profits and employees may be estimates. Financial information, benefits and other data can change quickly and may vary from those stated here.

Hikari Tsushin Inc

www.hikari.co.jp

NAIC Code: 443142

TYPES OF BUSINESS:

Cell Phone Retail Sales
Mobile Phone Distribution
Office Automation Equipment Sales
Insurance
Advertising
Venture Capital

BRANDS/DIVISIONS/AFFILIATES:

FT Group Co ltd
Premium Water Holdings Co Ltd
NFC Holdings Inc
CHIC Holdings Inc
IE Group Inc
Members Mobile Inc
Telecom Service Co Ltd
J-Communication Inc

CONTACTS: *Note: Officers with more than one job title may be intentionally listed here more than once.*

Hideaki Wada, Pres.
Koh Gidoh, Chief Dir.-Admin.

GROWTH PLANS/SPECIAL FEATURES:

Hikari Tsushin, Inc. is a Japanese telecommunications company that operates through various subsidiaries that sell and service goods and provides other services. FT Group Co., Ltd. provides information communication services, environment and energy-saving services, and electricity retailing services. Premium Water Holdings Co., Ltd. is a mineral water delivery business that sells its products and services under the Premium Water brand name. NFC Holdings, Inc. offers insurance services, worker dispatch services, IT services, water serving services and alternative power services. CHIC Holdings, Inc. provides emergency rush services that handles various equipment issues, and offers call center services that connect real estate companies with residents. IE Group, Inc. provides office consulting services in areas such as original equipment manufacturing of phones and network equipment, as well as broadband access, light-emitting diode (LED) lighting and commercial air conditioners. Member's Mobile, Inc. offers corporate mobile and landline services, as well as information and communication technology (ICT) lifecycle services. Telecom Service Co., Ltd. provides mobile and telecommunication integration services, with devices including smartphones, tablet terminals, wearable terminals and related accessories. J-Communication, Inc. is a telecommunications company that provides subscription telecommunications services, telecommunications equipment (for sale and for lease), manufacturing devices, importing/exporting and training. Last, EPARK, Inc. provides a hub that matches users with stores and facilities, and offers information such as when best to visit the store, what the store has in stock, where stores are located, etc.

FINANCIAL DATA: *Note: Data for latest year may not have been available at press time.*

In U.S. $	2021	2020	2019	2018	2017	2016
Revenue	4,915,723,264	4,609,416,192	4,256,317,696	3,756,809,728	3,987,440,128	5,468,418,560
R&D Expense						
Operating Income						
Operating Margin %						
SGA Expense						
Net Income	479,895,232	454,026,176	435,371,328	367,842,944	362,884,160	238,154,608
Operating Cash Flow						
Capital Expenditure						
EBITDA						
Return on Assets %						
Return on Equity %						
Debt to Equity						

CONTACT INFORMATION:

Phone: 81 359513718   Fax: 81 359513709
Toll-Free:
Address: Hikari W. Gate Bldg, 1-4-10 Nishi-Ikebukuro Toshim, Tokyo, 171-0021 Japan

STOCK TICKER/OTHER:

Stock Ticker: HKTGF   Exchange: PINX
Employees: 7,572   Fiscal Year Ends:
Parent Company:

SALARIES/BONUSES:

Top Exec. Salary: $ Bonus: $
Second Exec. Salary: $ Bonus: $

OTHER THOUGHTS:

Estimated Female Officers or Directors:
Hot Spot for Advancement for Women/Minorities:

Hitachi Kokusai Electric Inc

www.hitachi-kokusai.co.jp

NAIC Code: 334220

TYPES OF BUSINESS:

Radio and Television Broadcasting and Wireless Communications
Equipment Manufacturing
Information Communication Systems
Wireless Communication Solutions
Film Processing Solutions
CATV Equipment
Manufacturing
Innovation

BRANDS/DIVISIONS/AFFILIATES:

Hitachi Kokusai Electric America Ltd
Hitachi Kokusai Electric Europe GmbH
Goyo Electronics Co Ltd
Hitachi Kokusai Electric Asia (Singapore) Pte Ltd
HYS Engineering Service Inc
Hitachi Kokusai Electric Comark LLC
Hitachi Kokusai Linear Equip. Eletronicos S/A
Hitachi Kokusai Electric Turkey

CONTACTS:
Note: Officers with more than one job title may be intentionally listed here more than once.

Kaichiro Sakuma, CEO

GROWTH PLANS/SPECIAL FEATURES:

Hitachi Kokusai Electric, Inc., part of the Hitachi Group, is a Japanese global provider of information communication systems that offer capabilities through compatibility with global standards on which the next generation of mobile communication systems will be based. The firm is actively engaged in creating new value in the areas of wireless communications, information, broadcasting and video, as well as in the area of eco- and thin-film processing solutions (semiconductor manufacturing equipment). Hitachi Kokusai's products and systems include wireless communications and information systems, financial/stock exchange information systems, broadcasting and video systems, community antenna TV (CATV) equipment and antennas, security and surveillance systems, industrial video cameras and semiconductor manufacturing systems. In addition, Hitachi Kokusai's defense electronics division produces aircraft communication systems, shipboard communication systems and wireless telephone systems for air traffic control purposes. The company has manufacturing plants worldwide, and its subsidiaries include Hitachi Kokusai Electric America Ltd.; Hitachi Kokusai Electric Europe GmbH; Goyo Electronics Co., Ltd.; Hitachi Kokusai Electric Asia (Singapore) Pte. Ltd.; HYS Engineering Service, Inc.; Hitachi Kokusai Electric Comark LLC; Hitachi Kokusai Linear Equipamentos Eletronicos S/A; and Hitachi Kokusai Electric Turkey Elektronik UrUnleri Sanayi ve Ticaret A.S.

FINANCIAL DATA:
Note: Data for latest year may not have been available at press time.

In U.S. $	2021	2020	2019	2018	2017	2016
Revenue	650,000,000	622,570,000	774,944,000	1,532,865,360	1,572,169,600	1,653,432,320
R&D Expense						
Operating Income						
Operating Margin %						
SGA Expense						
Net Income						
Operating Cash Flow						
Capital Expenditure						
EBITDA						
Return on Assets %						
Return on Equity %						
Debt to Equity						

CONTACT INFORMATION:

Phone: 81 3-5510-5931   Fax: 81 3-3502-2502
Toll-Free:
Address: 2-15-12, Nishi-shimbashi, 6/Fl, Hitachi Atago Bldg, Tokyo, 105-8039 Japan

STOCK TICKER/OTHER:

Stock Ticker: Subsidiary   Exchange:
Employees: 1,500   Fiscal Year Ends: 03/31
Parent Company: Hitachi Group

SALARIES/BONUSES:

Top Exec. Salary: $ Bonus: $
Second Exec. Salary: $ Bonus: $

OTHER THOUGHTS:

Estimated Female Officers or Directors:
Hot Spot for Advancement for Women/Minorities:

Sales, profits and employees may be estimates. Financial information, benefits and other data can change quickly and may vary from those stated here.

HKT Trust and HKT Limited

www.hkt.com

NAIC Code: 517110

TYPES OF BUSINESS:

Wired Telecommunications Carriers
Telecommunications Services
Telephone Services
Broadband
Mobile Services

GROWTH PLANS/SPECIAL FEATURES:

HKT Trust and HKT Ltd is a triple-play telecommunications provider that operates through three segments telecommunications services, mobile, and Other businesses Telecommunication services are the larger business segment and generates revenue by providing voice services, broadband services, local and international data services, and the sales o equipment. The company's mobile segment generates revenue by selling mobile services and handsets. The Other Businesses segment primarily comprises new business areas such as Tap & Go mobile payment service and The Club program, and corporate support functions. The company owns fiber and mobile infrastructure. HKT Trust generate the vast majority of its revenue in Hong Kong.

BRANDS/DIVISIONS/AFFILIATES:

PCCW Limited
NETVIGATOR
PCCW Global
PCCW Teleservices

CONTACTS: *Note: Officers with more than one job title may be intentionally listed here more than once.*

Hanqing Xu, Managing Dir.
Tzar Kai (Richard) Li, Chmn.

FINANCIAL DATA: *Note: Data for latest year may not have been available at press time.*

In U.S. $	2021	2020	2019	2018	2017	2016
Revenue	4,326,369,000	4,126,108,000	4,217,066,000	4,482,551,000	4,212,480,000	4,311,846,000
R&D Expense						
Operating Income	925,632,200	952,384,500	1,010,603,000	949,709,300	886,395,400	869,452,200
Operating Margin %	.21%	.23%	.24%	.21%	.21%	.20%
SGA Expense	294,403,500	313,894,500	343,067,400	319,627,200	1,291,375,000	1,595,333,000
Net Income	612,502,000	675,561,200	664,605,400	614,667,600	604,476,300	622,820,700
Operating Cash Flow	1,355,581,000	1,342,714,000	1,412,016,000	1,357,874,000	1,307,172,000	1,562,084,000
Capital Expenditure	670,720,300	607,406,300	653,394,900	595,304,100	489,950,700	903,083,700
EBITDA	1,611,512,000	1,631,895,000	1,635,335,000	1,601,193,000	1,549,727,000	1,598,263,000
Return on Assets %	.05%	.05%	.05%	.05%	.05%	.05%
Return on Equity %	.13%	.14%	.14%	.13%	.12%	.13%
Debt to Equity		1.115	1.109	1.12	1.069	1.029

CONTACT INFORMATION:

Phone: 852 28882888   Fax: 852 28778877
Toll-Free:
Address: 979 Kings Rd., PCCW Tower, Fl. 39, Hong Kong, Hong Kong 999077 Hong Kong

STOCK TICKER/OTHER:

Stock Ticker: HKTTF
Employees: 15,400
Parent Company: PCCW Limited

Exchange: PINX
Fiscal Year Ends: 12/31

SALARIES/BONUSES:

Top Exec. Salary: $ Bonus: $
Second Exec. Salary: $ Bonus: $

OTHER THOUGHTS:

Estimated Female Officers or Directors:
Hot Spot for Advancement for Women/Minorities:

HP Inc

www.hp.com

NAIC Code: 334111

TYPES OF BUSINESS:
Computer Manufacturing
Computer Software
Printers & Supplies
Scanners
Computing Devices

BRANDS/DIVISIONS/AFFILIATES:
HP Labs

GROWTH PLANS/SPECIAL FEATURES:
HP Incorporated is a leading provider of computers, printers, and printer supplies. The company's mains segments are personal systems and printing. Its personal systems segment contains notebooks, desktops, and workstations. Its printing segment contains supplies, consumer hardware, and commercial hardware. In 2015, Hewlett-Packard was separated into HP Incorporated and Hewlett Packard Enterprise and the Palo Alto, California-based HP Incorporated sells on a global scale.

CONTACTS: *Note: Officers with more than one job title may be intentionally listed here more than once.*
Enrique Lores, CEO
Marie Myers, CFO
Charles Bergh, Chairman of the Board
Harvey Anderson, Chief Legal Officer
Sarabjit Singh Baveja, Chief Strategy Officer
Tolga Kurtoglu, Chief Technology Officer
Barb Weiszhaar, Other Corporate Officer
Tracy Keogh, Other Executive Officer
Christoph Schell, Other Executive Officer
Tuan Tran, President, Divisional
Alex Cho, President, Divisional

FINANCIAL DATA: *Note: Data for latest year may not have been available at press time.*

In U.S. $	2021	2020	2019	2018	2017	2016
Revenue	63,487,000,000	56,639,000,000	58,756,000,000	58,472,000,000	52,056,000,000	48,238,000,000
R&D Expense	1,907,000,000	1,478,000,000	1,499,000,000	1,404,000,000	1,190,000,000	1,209,000,000
Operating Income	5,652,000,000	3,727,000,000	3,001,000,000	3,424,000,000	3,902,000,000	4,412,000,000
Operating Margin %	.09%	.07%	.05%	.06%	.07%	.09%
SGA Expense	5,704,000,000	5,120,000,000	5,368,000,000	5,099,000,000	4,532,000,000	3,833,000,000
Net Income	6,503,000,000	2,844,000,000	3,152,000,000	5,327,000,000	2,526,000,000	2,496,000,000
Operating Cash Flow	6,409,000,000	4,316,000,000	4,654,000,000	4,528,000,000	3,677,000,000	3,252,000,000
Capital Expenditure	582,000,000	580,000,000	671,000,000	546,000,000	402,000,000	433,000,000
EBITDA	8,550,000,000	4,259,000,000	3,509,000,000	3,853,000,000	3,939,000,000	4,366,000,000
Return on Assets %	.18%	.08%	.09%	.16%	.08%	.04%
Return on Equity %						.21%
Debt to Equity						

CONTACT INFORMATION:
Phone: 650 857-1501   Fax:
Toll-Free:
Address: 1501 Page Mill Rd., Palo Alto, CA 94304 United States

STOCK TICKER/OTHER:
Stock Ticker: HPQ
Employees: 53,000
Parent Company:

Exchange: NYS
Fiscal Year Ends: 10/31

SALARIES/BONUSES:
Top Exec. Salary: $1,200,000   Bonus: $
Second Exec. Salary: Bonus: $250,000
$664,445

OTHER THOUGHTS:
Estimated Female Officers or Directors: 6
Hot Spot for Advancement for Women/Minorities: Y

Sales, profits and employees may be estimates. Financial information, benefits and other data can change quickly and may vary from those stated here.

Huawei Technologies Co Ltd

www.huawei.com

NAIC Code: 334210

TYPES OF BUSINESS:

Telecommunications Equipment Manufacturing
Network Equipment
Software
Wireless Technology
Smartphones
5G Wireless Technology
Watches

BRANDS/DIVISIONS/AFFILIATES:

Union of Huawei Investment & Holding Co
Huaewi

CONTACTS: Note: Officers with more than one job title may be intentionally listed here more than once.

Ren Zhengfei, Pres.
Ding Yun (Ryan Ding), Chief Prod. & Solutions Officer
Yu Chengdong (Richard Yu), Chief Strategy Officer
Chen Lifang, Corp. Sr. VP-Public Affairs & Comm. Dept.
Guo Ping, Chmn.-Finance Committee
Zhang Ping'an (Alex Zhang), CEO-Huawei Symantec
Hu Houkun (Ken Hu), Chmn.-Huawei USA
Liang Hua, Chmn.
Wan Biao, Pres., Russia

GROWTH PLANS/SPECIAL FEATURES:

Huawei Technologies Co., Ltd., founded in 1987, is a leading global information and communications technology (ICT) solutions provider. Huawei is one of the world's leading manufacturers of smartphones. The company's ICT portfolio of end-to-end solutions in telecom, enterprise networks, and consumers are used in more than 170 countries and regions, serving more than one-third of the world's population. Huawei's consumer products include the Huawei brand of mobile smart phones, the laptops, tablets, watches, ear buds, speakers, Wi-Fi connection devices and more. The company's business products include switches, routers, WLAN (wireless local area network), servers, storage, cloud computing, network energy services and more. Its carrier products include cloud data centers, wireless network, fixed network, cloud core network carrier software, IT infrastructure and network energy global services. Huawei Technologies has rolled out 53 NB-IoT city-aware network using a one network/one platform/N-tier applications model. NB-IoT stands for NarrowBand Internet of Things, and is a low-power, wide-area network radio technology standard that enables a wide range of devices and services to be connected using cellular telecommunications bands. Huawei's smart city solutions senses, processes and delivers informed decisions for improving the environment for citizens, with recent information and communications technology (ICT) offering real-time situation reporting and analysis, empowered by a combination of cloud computing, IoT technologies, big data analytics and artificial intelligence (AI). The company has global joint innovation centers and research and development centers and offices. It has invested very heavily in 5G wireless technologies. Huawei Technologies operates as a subsidiary of the Union of Huawei Investment & Holding Co., Ltd.

FINANCIAL DATA: Note: Data for latest year may not have been available at press time.

In U.S. $	2021	2020	2019	2018	2017	2016
Revenue						
R&D Expense						
Operating Income						
Operating Margin %						
SGA Expense						
Net Income						
Operating Cash Flow						
Capital Expenditure						
EBITDA						
Return on Assets %						
Return on Equity %						
Debt to Equity						

CONTACT INFORMATION:

Phone: 86 755-28780808   Fax: 86-755-28789251
Toll-Free:
Address: Section H, Bantian, Longgang Distr., Shenzhen, Guangdong 518129 China

STOCK TICKER/OTHER:

Stock Ticker: Subsidiary   Exchange:
Employees: 195,000   Fiscal Year Ends: 12/31
Parent Company: Union of Huawei Investment & Holding Co Ltd

SALARIES/BONUSES:

Top Exec. Salary: $ Bonus: $
Second Exec. Salary: $ Bonus: $

OTHER THOUGHTS:

Estimated Female Officers or Directors: 3
Hot Spot for Advancement for Women/Minorities: Y

Hutchison Telecommunications Hong Kong Holdings Limited

www.hthkh.com

NAIC Code: 517210

TYPES OF BUSINESS:

Mobile Telecommunication Service Provider
Fixed-Line Operations
Internet Services

BRANDS/DIVISIONS/AFFILIATES:

CK Hutchison Holdings Limited
Hutchison Telephone Company Limited
Hutchison 3G Hong Kong Holdings Limited
Hutchison Telecom Macau
3

GROWTH PLANS/SPECIAL FEATURES:

Hutchison Telecommunications Hong Kong Holdings Ltd is an investment holding company. It operates in a single segment of the mobile telecommunications business in Hong Kong and Macau. The company offers a wide range of mobile telecommunications services and products with an increasing focus on data-centric services that converge with Internet and PC services. It provides mobile communications services such as local voice, SMS, MMS, IDD, and international roaming, as well as data services and applications including direct carrier billing offerings, mobile device security management, eBooks, music downloads, movies-on-demand, mobile social networking applications, and FinTech.

CONTACTS: Note: Officers with more than one job title may be intentionally listed here more than once.

Wai Sin Cheng, CFO
Edith Shih, Sec.
Kin Ning Fok, Chmn.

FINANCIAL DATA: Note: Data for latest year may not have been available at press time.

In U.S. $	2021	2020	2019	2018	2017	2016
Revenue	686,007,300	578,997,800	711,103,600	1,007,928,000	860,152,600	1,061,432,000
R&D Expense						
Operating Income	15,287,070	49,173,410	49,300,800	40,383,350	-229,178,700	68,027,470
Operating Margin %	.02%	.08%	.07%	.04%	-.27%	.06%
SGA Expense	15,796,640	14,395,330	21,529,290	21,529,290	17,580,130	16,815,780
Net Income	509,569	45,988,610	54,651,280	51,466,470	607,151,600	86,881,520
Operating Cash Flow	430,076,300	165,227,800	189,177,500	65,861,800	247,523,200	312,493,200
Capital Expenditure	371,221,000	101,276,800	89,938,940	65,352,230	129,048,400	369,947,100
EBITDA	182,170,900	216,312,100	224,847,300	163,699,100	228,541,700	260,644,600
Return on Assets %	.00%	.02%	.03%	.02%	.22%	.03%
Return on Equity %	.00%	.03%	.03%	.03%	.35%	.06%
Debt to Equity	.01%	0.016	0.011			0.389

CONTACT INFORMATION:

Phone: 852 2128-1188   Fax: 852-2128-1778
Toll-Free:
Address: Fl. 15, Hutchison Telecom. Twr., 99 Cheung Fai Rd., Tsing Yi, Hong Kong, Hong Kong 999077 Hong Kong

STOCK TICKER/OTHER:

Stock Ticker: HUTCY   Exchange: PINX
Employees: 1,045   Fiscal Year Ends: 12/31
Parent Company: CK Hutchison Holdings Limited

SALARIES/BONUSES:

Top Exec. Salary: $ Bonus: $
Second Exec. Salary: $ Bonus: $

OTHER THOUGHTS:

Estimated Female Officers or Directors: 1
Hot Spot for Advancement for Women/Minorities: Y

Sales, profits and employees may be estimates. Financial information, benefits and other data can change quickly and may vary from those stated here.

iBasis Inc

NAIC Code: 517110

www.ibasis.com

TYPES OF BUSINESS:

International Long Distance Service
Wholesale Voice Services
Prepaid Calling Cards
VoIP Service

BRANDS/DIVISIONS/AFFILIATES:

Tofane Global

CONTACTS: *Note: Officers with more than one job title may be intentionally listed here more than once.*

Alexandre Pebereau, CEO
John Treece, COO
Ardjan Konijnenberg, CFO
Edwin van Ierland, CCO
Ajay Joseph, CTO
Gert-Jan Huizer, Sr. VP-Prod. Mgmt.
Ellen W. Schmidt, General Counsel
Brad Guth, Sr. VP-Oper.
Chris Ward, Sr. Dir.-Corp. Comm.
Edwin van Ierland, Global Sales Officer

GROWTH PLANS/SPECIAL FEATURES:

IBasis, Inc. provides telecommunications services that enable operators and digital players worldwide to perform and transform. The firm offers services such as managed voice, mobile data, Internet of Things (IoT) connectivity, 5G roaming, cloud, eSIM (electronic subscriber identity module) technology, fraud prevention and business intelligence. IBasis has more than 1.5 billion long-term evolution (LTE)-enabled subscribers, over 2,000 fixed and mobile destinations and over 700 LTE destinations, as well as 700+ mobile operator partnerships. The company provides a single point for its customer's local access needs in Asia, the Middle East, Europe, North America and Latin America, as well as select areas within Africa, the Caribbean and French-/Portuguese-speaking markets. The Paris-based telecommunications and digital player specializing in international carrier services firm, Tofane Global, owns iBasis.

IBasis employees are offered health and dental insurance, life insurance and disability coverage.

FINANCIAL DATA: *Note: Data for latest year may not have been available at press time.*

In U.S. $	2021	2020	2019	2018	2017	2016
Revenue	1,070,000,000	1,050,000,000	1,000,000,000	886,718,700	844,494,000	913,437,000
R&D Expense						
Operating Income						
Operating Margin %						
SGA Expense						
Net Income						
Operating Cash Flow						
Capital Expenditure						
EBITDA						
Return on Assets %						
Return on Equity %						
Debt to Equity						

CONTACT INFORMATION:

Phone: 781-430-7500   Fax: 781-505-7300
Toll-Free:
Address: 10 Maguire Rd., Bldg. 3, Lexington, MA 02421 United States

STOCK TICKER/OTHER:

Stock Ticker: Subsidiary   Exchange:
Employees: 300   Fiscal Year Ends: 12/31
Parent Company: Tofane Global

SALARIES/BONUSES:

Top Exec. Salary: $ Bonus: $
Second Exec. Salary: $ Bonus: $

OTHER THOUGHTS:

Estimated Female Officers or Directors: 2
Hot Spot for Advancement for Women/Minorities:

Sales, profits and employees may be estimates. Financial information, benefits and other data can change quickly and may vary from those stated here.

IDT Corporation

www.idt.net

NAIC Code: 517110

TYPES OF BUSINESS:
Local & Long-Distance Service
Telecommunication Services
Payment Services

GROWTH PLANS/SPECIAL FEATURES:
IDT Corp is a multinational holding company. It primarily operates in the telecommunications and payment industries. It has two reportable business segments, Telecom & Payment Services, and net2phone. The Telecom & Payment Services segment that derives majority revenue provides retail telecommunications and payment offerings as well as wholesale international long-distance traffic termination. The net2phone segment provides unified cloud communications and telephony services to business customers.

IDT offers its employee health, medical, dental, life and disability coverage; flexible spending accounts; tuition reimbursement; and 401(k).

BRANDS/DIVISIONS/AFFILIATES:
IDT Carrier Services
BOSS Revolution
net2phone
National Retail Solutions

CONTACTS: Note: Officers with more than one job title may be intentionally listed here more than once.
Samuel Jonas, CEO
Marcelo Fischer, CFO
Howard Jonas, Chairman of the Board
Mitch Silberman, Chief Accounting Officer
David Wartell, Chief Technology Officer
Bill Pereira, COO
Nadine Shea, Executive VP, Divisional
Menachem Ash, Executive VP, Divisional
Joyce Mason, Executive VP

FINANCIAL DATA: Note: Data for latest year may not have been available at press time.

In U.S. $	2021	2020	2019	2018	2017	2016
Revenue	1,446,990,000	1,345,769,000	1,409,172,000	1,547,495,000	1,501,729,000	1,496,261,000
R&D Expense						
Operating Income	57,914,000	22,689,000	725,000	13,008,000	5,549,000	25,237,000
Operating Margin %	.04%	.02%	.00%	.01%	.00%	.02%
SGA Expense	218,242,000	215,377,000	204,040,000	203,251,000	188,293,000	204,655,000
Net Income	96,475,000	21,430,000	134,000	4,208,000	8,177,000	23,514,000
Operating Cash Flow	66,620,000	-29,591,000	85,137,000	20,394,000	36,094,000	49,054,000
Capital Expenditure	16,765,000	16,041,000	18,681,000	20,567,000	22,949,000	18,370,000
EBITDA	75,678,000	43,095,000	23,357,000	35,809,000	27,253,000	45,772,000
Return on Assets %	.21%	.05%	.00%	.01%	.02%	.05%
Return on Equity %	.81%	.33%	.00%	.05%	.06%	.18%
Debt to Equity	.03%	0.098				

CONTACT INFORMATION:
Phone: 973 438-1000   Fax:
Toll-Free:
Address: 520 Broad St., Newark, NJ 07102 United States

STOCK TICKER/OTHER:
Stock Ticker: IDT   Exchange: NYS
Employees: 1,256   Fiscal Year Ends: 07/31
Parent Company:

SALARIES/BONUSES:
Top Exec. Salary: $495,000   Bonus: $532,500
Second Exec. Salary: $500,000   Bonus: $450,000

OTHER THOUGHTS:
Estimated Female Officers or Directors: 1
Hot Spot for Advancement for Women/Minorities:

IHS Holding Ltd

NAIC Code: 237130

www.ihstowers.com

TYPES OF BUSINESS:
Construction of Telecommunications Lines and Systems & Electric Power Lines and Systems

BRANDS/DIVISIONS/AFFILIATES:

GROWTH PLANS/SPECIAL FEATURES:
IHS Holding Ltd is an independent owner, operator, and developer of shared telecommunications infrastructure. The company provides telecommunications infrastructure to its customers, most of who are MNOs, who in turn provide wireless voice and data services to their end-users. Its geographical segments are Nigeria, Sub-Saharan Africa, MENA, and Latam. The majority of its revenue is derived from Nigeria.

CONTACTS:
Note: Officers with more than one job title may be intentionally listed here more than once.

Sam Darwish, CEO
William Saad, COO
Steve Howden, CFO
Bill Bates, Chief Strategy Officer
Ayotade Oyinlola, Chief Human Resources Officer
Sam Darwish, Chmn.

FINANCIAL DATA:
Note: Data for latest year may not have been available at press time.

In U.S. $	2021	2020	2019	2018	2017	2016
Revenue	1,579,730,000	1,403,149,000	1,231,056,000	1,168,087,000		
R&D Expense						
Operating Income	444,952,000	360,029,000	-109,990,000	220,039,000		
Operating Margin %	.28%	.26%	-.09%	.19%		
SGA Expense	241,828,000	180,225,000	495,040,000	125,186,000		
Net Income	-25,832,000	-321,994,000	-423,492,000	-132,770,000		
Operating Cash Flow	750,189,000	635,256,000	641,940,000	462,307,000		
Capital Expenditure	402,475,000	229,199,000	258,278,000	378,025,000		
EBITDA	591,491,000	468,083,000	186,098,000	419,647,000		
Return on Assets %	-.01%	-.07%	-.10%			
Return on Equity %	-.02%	-.24%	-.30%			
Debt to Equity	1.72%	1.904	1.481			

CONTACT INFORMATION:
Phone: 44 20-8106-1600   Fax:
Toll-Free:
Address: 1 Cathedral Piazza, 123 Victoria St., London, SW1E 5BP United Kingdom

STOCK TICKER/OTHER:
Stock Ticker: IHS
Employees: 2,292
Parent Company:

Exchange: NYS
Fiscal Year Ends: 12/31

SALARIES/BONUSES:
Top Exec. Salary: $ Bonus: $
Second Exec. Salary: $ Bonus: $

OTHER THOUGHTS:
Estimated Female Officers or Directors:
Hot Spot for Advancement for Women/Minorities:

Sales, profits and employees may be estimates. Financial information, benefits and other data can change quickly and may vary from those stated here.

Iliad SA

NAIC Code: 517110

TYPES OF BUSINESS:

Fixed-Line Telecommunications & Broadband Internet
Telecommunications Services
Voice
Internet
IPTV
Mobile and Fixed-Line Services
Business-to-Business Services
Managed Hosting Services

BRANDS/DIVISIONS/AFFILIATES:

Free
Freebox
Free Mobile Plan
Iliad
Scaleway
Jaguar Network
Free Pro

CONTACTS: *Note: Officers with more than one job title may be intentionally listed here more than once.*

Thomas Reynaud, CEO
Benjamin Mbiandjeu, COO
Nicolas Jaeger, CFO
Camille Perrin, CMO
Celine Polo, Chief Human Resources Officer
Alexandre Cassen, Chief Network Officer
Xavier Niel, Chief Strategy Officer
Angelique Berge, Dir.-Client Support
Thomas Reynaud, Head-Bus. Dev.
Xavier Niel, Deputy Chmn.
Celine Von der Weid, CCO
Patrick Fouqueriere, Dir.-Supplier Rel.

GROWTH PLANS/SPECIAL FEATURES:

Iliad SA is a French holding company active in the integrated telecommunications sector. The firm's Free subsidiary developed the Freebox, a multiservice box on asymmetric digital subscriber line (ADSL). Other services by Free include voice on internet protocol (VoIP), internet protocol television (IPTV), flat-rate calling plans and more. Free offers mobile phone usage within France's reach with straightforward, no-commitment contracts at affordable prices. The Free Mobile Plan includes roaming communications all year round from more than 35 countries, and includes unlimited calls, texts and multimedia messaging service (MMS). Free also includes unlimited 4G and 5G in a premium plan for Freebox subscribers. Free Pro offers business-to-business (B2B) fiber internet, automatic 4G-5G backup, Wi-Fi, phone usage and data protection. As of mid-2022, Free had approximately 20 million subscribers in France, consisting of fixed and mobile subscribers. Iliad's mobile network in Italy operates under the Iliad brand. Scaleway is a business-to-business (B2B) brand and subsidiary of Iliad that supplies a range of cloud infrastructure services for public and private business customers in over 150 countries. Scaleway relies on seven data centers located in France and one in the Netherlands. Jaguar Network offers managed hosting services in the cloud, offering business-to-business (B2B) support.

FINANCIAL DATA: *Note: Data for latest year may not have been available at press time.*

In U.S. $	2021	2020	2019	2018	2017	2016
Revenue	8,591,518,800	7,173,052,416	6,264,392,192	5,746,275,840	5,391,158,784	5,573,661,696
R&D Expense						
Operating Income						
Operating Margin %						
SGA Expense						
Net Income	595,642,400	521,698,752	2,019,596,928	379,482,144	430,077,152	473,411,584
Operating Cash Flow						
Capital Expenditure						
EBITDA						
Return on Assets %						
Return on Equity %						
Debt to Equity						

CONTACT INFORMATION:

Phone: 33 173502000   Fax:
Toll-Free:
Address: 16 rue de la Ville l Eveque, Paris, 75008 France

STOCK TICKER/OTHER:

Stock Ticker: Private   Exchange:
Employees: 15,100   Fiscal Year Ends: 12/31
Parent Company:

SALARIES/BONUSES:

Top Exec. Salary: $ Bonus: $
Second Exec. Salary: $ Bonus: $

OTHER THOUGHTS:

Estimated Female Officers or Directors: 4
Hot Spot for Advancement for Women/Minorities: Y

Indosat Ooredoo Hutchison
indosatooredoo.com/portal/en/corplanding

NAIC Code: 517210

TYPES OF BUSINESS:

Mobile Telephone Service
Digital Telecommunications
Internet Services
Wireless Services
Mobile Services
International Direct Dialing Services
Fixed Telecommunications
Multimedia

BRANDS/DIVISIONS/AFFILIATES:

Ooredoo QPSC
CK Hutchison Holdings Limited
IM3
Three

GROWTH PLANS/SPECIAL FEATURES:

Indosat Ooredoo Hutchison (formerly Indosat Ooredoo) is a
fully integrated Indonesian digital telecommunications and
internet company in Indonesia. The company offers wireless
services for mobile phones, and to a lesser extent, broadband
internet lines for residences. Wireless services are provided
through the IM3 and Three brands, which differ by payment
type and price plans. Other services by Indosat Ooredoo
Hutchison include international direct dialing (IDD), fixed
telecommunications and multimedia. In early-2022, parent
Ooredoo Q.P.S.C. and CK Hutchison Holdings Limited merged
their telecommunications businesses in Indonesia and formed
PT Indosat Ooredoo Hutchison TbK. The new company is
jointly controlled by Ooredoo Group and CK Hutchison.

CONTACTS: *Note: Officers with more than one job title may be intentionally listed here more than once.*

Vikram Sinha, CEO
Bayu Hanantasena, Chief Business Officer
Nicky Lee Chi Hung, CFO
Ritesh Singh, CCO
Irsyad Sahroni, Chief Human Resources Officer
Desmond Cheung, CTO
Fadzri Santosa, Chief Wholesale & Infrastructure Officer
Sanjeev Rawat, Chief Digital Officer

FINANCIAL DATA: *Note: Data for latest year may not have been available at press time.*

In U.S. $	2021	2020	2019	2018	2017	2016
Revenue	1,831,030,020	1,856,190,000	1,767,420,928	1,565,895,552	2,015,401,984	1,970,258,432
R&D Expense						
Operating Income						
Operating Margin %						
SGA Expense						
Net Income	114,303,105	115,872,000	106,176,472	-162,604,768	76,490,408	74,601,552
Operating Cash Flow						
Capital Expenditure						
EBITDA						
Return on Assets %						
Return on Equity %						
Debt to Equity						

CONTACT INFORMATION:

Phone: 62 213802615   Fax: 62 213458155
Toll-Free:
Address: Jl. Medan Merdeka Barat No. 21, Jakarta, 10110 Indonesia

STOCK TICKER/OTHER:

Stock Ticker: ISAT   Exchange: Jakarta
Employees: 2,891   Fiscal Year Ends: 12/31
Parent Company: Ooredoo QPSC

SALARIES/BONUSES:

Top Exec. Salary: $ Bonus: $
Second Exec. Salary: $ Bonus: $

OTHER THOUGHTS:

Estimated Female Officers or Directors:
Hot Spot for Advancement for Women/Minorities:

Infinera Corporation

www.infinera.com

NAIC Code: 334413

TYPES OF BUSINESS:
Optical-Electronic Conversion Equipment
Optical Network Infrastructure Equipment
Network Design & Engineering

BRANDS/DIVISIONS/AFFILIATES:

GROWTH PLANS/SPECIAL FEATURES:
Infinera Corp is a global supplier of networking solutions comprised of networking equipment, software and services. Its portfolio of solutions includes optical transport platforms, converged packet-optical transport platforms, compact modular platforms, optical line systems, coherent optical subsystems, a suite of automation software offerings, and support and professional services. The company's customers include operators of fixed line and mobile networks, including telecommunications service providers, internet content providers, cable providers, wholesale carriers, research and education institutions, large enterprises, utilities and government entities.

Infinera offers its employees medical, dental, vision and life insurance, stock options and more.

CONTACTS: Note: Officers with more than one job title may be intentionally listed here more than once.
David Heard, CEO
Nancy Erba, CFO
George Riedel, Chairman of the Board
Michael Fernicola, Chief Accounting Officer
David Teichmann, Chief Legal Officer
David Welch, Co-Founder
Nicholas Walden, Senior VP, Divisional

FINANCIAL DATA: Note: Data for latest year may not have been available at press time.

In U.S. $	2021	2020	2019	2018	2017	2016
Revenue	1,425,205,000	1,355,596,000	1,298,865,000	943,379,000	740,739,000	870,135,000
R&D Expense	299,894,000	265,634,000	287,977,000	244,302,000	224,368,000	232,143,000
Operating Income	-74,164,000	-117,480,000	-267,946,000	-157,637,000	-166,659,000	-23,904,000
Operating Margin %	- .05%	- .09%	- .21%	- .17%	- .22%	- .03%
SGA Expense	254,244,000	241,844,000	277,774,000	205,195,000	180,131,000	179,290,000
Net Income	-170,778,000	-206,723,000	-386,618,000	-214,295,000	-194,506,000	-23,927,000
Operating Cash Flow	28,128,000	-112,300,000	-167,350,000	-99,083,000	-21,925,000	38,377,000
Capital Expenditure	41,379,000	39,009,000	30,202,000	37,692,000	58,041,000	43,335,000
EBITDA	-26,108,000	-53,820,000	-232,174,000	-92,407,000	-115,922,000	45,195,000
Return on Assets %	- .10%	- .12%	- .23%	- .15%	- .17%	- .02%
Return on Equity %	- .46%	- .51%	- .71%	- .31%	- .27%	- .03%
Debt to Equity	1.64%	1.225	1.01	0.654		0.175

CONTACT INFORMATION:
Phone: 408 572-5200   Fax: 408 572-5343
Toll-Free:
Address: 6373 San Ignacio Ave., San Jose, CA 95119 United States

STOCK TICKER/OTHER:
Stock Ticker: INFN   Exchange: NAS
Employees: 3,225   Fiscal Year Ends: 12/31
Parent Company:

SALARIES/BONUSES:
Top Exec. Salary: $700,000   Bonus: $200,000
Second Exec. Salary: Bonus: $
$439,615

OTHER THOUGHTS:
Estimated Female Officers or Directors: 1
Hot Spot for Advancement for Women/Minorities:

Sales, profits and employees may be estimates. Financial information, benefits and other data can change quickly and may vary from those stated here.

Ingram Micro Mobility

corp.ingrammicro.com

NAIC Code: 423430

TYPES OF BUSINESS:

Cell Phone Distribution
IT Products Distribution
Supply Chain Solutions and Services
Brand Promotion
Customized Telecommunication Solutions
Tailored Mobility Solutions
Mobile Devices and Accessories

GROWTH PLANS/SPECIAL FEATURES:

Ingram Micro Mobility (IMM) is a division of Ingram Micro, Inc.
that distributes information technology products and services.
IMM brings manufacturer's new products or services to
markets across multiple European countries, and provides
network operators a supply chain solution that helps reduce
costs in increase efficiency. For device manufacturers, IMM
promotes their brand; for network operators and MVNOs, the
company tailors a suite of end-to-end distribution, supply chain
and customization solutions specifically for the
telecommunications market; and for resellers such as retail,
ecommerce and business-to-business (B2B), IMM offers
tailored mobility solutions to leverage continued end-user
demand for smartphones, tablets and accessories. Throughout
Europe, IMM has 22 advanced logistics and distribution
centers, as well as sales offices in 19 countries (which in turn
sells to more than 30 countries across Europe). Ingram Micro
Inc. itself is owned by private equity company Platinum Equity,
LLC.

BRANDS/DIVISIONS/AFFILIATES:

Platinum Equity LLC
Ingram Micro Inc

CONTACTS:
Note: Officers with more than one job title may be intentionally listed here more than once.

Craig M. Carpenter, General Counsel
Bashar Nejdawi, Pres., North American Mobility
R. Bruce Thomlinson, Pres., Asia Pacific Mobility
Alain Monie, Chmn.
Jac Currie, Pres., Europe Mobility

FINANCIAL DATA:
Note: Data for latest year may not have been available at press time.

In U.S. $	2021	2020	2019	2018	2017	2016
Revenue						
R&D Expense						
Operating Income						
Operating Margin %						
SGA Expense						
Net Income						
Operating Cash Flow						
Capital Expenditure						
EBITDA						
Return on Assets %						
Return on Equity %						
Debt to Equity						

CONTACT INFORMATION:

Phone: 44 1784-227000   Fax:
Toll-Free:
Address: Two Pines Trees, Chertsey Ln., Surrey, TW18 3HR United Kingdom

STOCK TICKER/OTHER:

Stock Ticker: Subsidiary
Employees: 35,000
Parent Company: Platinum Equity LLC

Exchange:
Fiscal Year Ends: 12/31

SALARIES/BONUSES:

Top Exec. Salary: $ Bonus: $
Second Exec. Salary: $ Bonus: $

OTHER THOUGHTS:

Estimated Female Officers or Directors:
Hot Spot for Advancement for Women/Minorities: Y

Sales, profits and employees may be estimates. Financial information, benefits and other data can change quickly and may vary from those stated here.

Inmarsat Global Limited

www.inmarsat.com

NAIC Code: 517410

TYPES OF BUSINESS:

Satellite Carrier & Equipment
Mobile Satellite Communications
Broadband
Satellite Launch Operations
Remote Connectivity
Satellites
Internet of Things
Low Earth Orbit Satellite

BRANDS/DIVISIONS/AFFILIATES:

Connect Bidco Limited
Global Xpress
European Aviation Network
ORCHESTRA

CONTACTS: *Note: Officers with more than one job title may be intentionally listed here more than once.*

Rajeev Suri, CEO
Jason Smith, COO
Tony Bates, CFO
Barry French, CMO
Natasha Dillon, Chief People Officer
Peter Hadinger, CTO
Richard Denny, Sr. VP-Eng. & Global Networks
Alison Horrocks, Corp. Sec.
Diane Cornell, VP-Govt Affairs
Jat Brainch, Chief Product Officer & Chief Commercial Officer

GROWTH PLANS/SPECIAL FEATURES:

Inmarsat Global Limited is a world-leading provider of global mobile satellite communications solutions and services. The firm's broadband global area network (BGAN) offers seamless global coverage, and satellite and ground network availability 99.9% of the time. Through its L-band services, Inmarsat connects people and machines in remote locations (on ships and in airplanes), enabling the global Internet of Things (IoT), voice calls and internet access. L-band service capabilities span email and web browsing to high bandwidth applications such as live streaming and telemedicine. The company's Global Xpress (GX) is a mobile high-speed broadband network from a single provider. Its Ka-band service offers content-rich applications from certified developers. Inmarsat's European Aviation Network (EAN) is an integrated satellite and air-to-ground connectivity network, developed to meet the particular requirements of European skies. The EAN combines S-band satellite coverage with a long-term evolution (LTE)-based terrestrial network to deliver cost-effective, scalable capacity across all 28 member states of the European Union, as well as Switzerland and Norway. Developed in partnership with Deutsche Telekom, EAN gives European aviation cutting-edge inflight connectivity. Inmarsat's sixth generation (I-6) of satellites feature dual-payload satellites, each supporting L-band and Ka-band services. The I-6 satellites support advanced global safety services, very low cost mobile services and Internet of Things (IoT) applications. The Ka-band payload will further extend GX capacity in regions of greatest demand. Inmarsat has locations across every continent, offering 24/7/365 customer support. In Inmarsat is a private company owned by Connect Bidco Limited, a consortium consisting of Apax Partners, Warburg Pincus, the CPP Investment Board and the Ontario Teachers' Pension Plan. During 2021, Inmarsat launched ORCHESTRA, a communications network for global mobility and government communications. ORCHESTRA successfully activated a low-earth orbit (LEO) satellite payload in December 2021.

FINANCIAL DATA: *Note: Data for latest year may not have been available at press time.*

In U.S. $	2021	2020	2019	2018	2017	2016
Revenue	1,195,000,000	1,137,300,000	1,091,500,000	1,465,200,000	1,400,199,936	1,328,999,936
R&D Expense						
Operating Income						
Operating Margin %						
SGA Expense						
Net Income		218,400,000	126,200,000	124,200,000	181,700,000	242,800,000
Operating Cash Flow						
Capital Expenditure						
EBITDA						
Return on Assets %						
Return on Equity %						
Debt to Equity						

CONTACT INFORMATION:

Phone: 44 2077281000   Fax: 44 2077281044
Toll-Free:
Address: 99 City Rd., London, EC1Y 1AX United Kingdom

STOCK TICKER/OTHER:

Stock Ticker: Private   Exchange:
Employees: 1,842   Fiscal Year Ends: 12/31
Parent Company: Connect Bidco Limited

SALARIES/BONUSES:

Top Exec. Salary: $ Bonus: $
Second Exec. Salary: $ Bonus: $

OTHER THOUGHTS:

Estimated Female Officers or Directors: 3
Hot Spot for Advancement for Women/Minorities: Y

Sales, profits and employees may be estimates. Financial information, benefits and other data can change quickly and may vary from those stated here.

InnoMedia Pte Ltd

www.innomedia.com

NAIC Code: 334310

TYPES OF BUSINESS:
Videoconferencing & Computer Telephony Products

BRANDS/DIVISIONS/AFFILIATES:

CONTACTS: *Note: Officers with more than one job title may be intentionally listed here more than once.*
Kai-Wa Ng, CEO
Nan-Sheng Lin, Pres.
Harprit Chhatwal, VP-IT
Jeremy Tzeng, VP-Eng.
Kim Huat Toh, Gen. Mgr.-China
Shailesh Patel, Sr. Dir.-Broadband Prod. Mgmt., U.S.
Kai-Wa Ng, Chmn.
Wymond Choy, Corp. VP-US

GROWTH PLANS/SPECIAL FEATURES:
InnoMedia Pte. Ltd. is a Singapore-based technology company formed by alumni of Creative Technology Ltd. and Teknekron Communications Systems. The firm's goal is to use IP and the power of broadband to facilitate efficient and productive communication. The company operates globally, with offices in Singapore, Taiwan, China and the U.S. InnoMedia operates through four business segments. Its residential & SOHO segment provides rapid and large-scale Internet deployment through its MTA product lines that offer network connectivity via Ethernet and USB WAN (universal serial bus, wide area network) options to its residential and SOHO markets. This division also provides IP phone devices that come with multiple business function keys, full duplex speaker phone, device based 3-way calling and digital voice wideband codec. The business voice solutions segment addresses broadband telephony service providers' needs, such as multi-port ATAs that feature T.30, ground/loop starts, calling party control (CPC)/open loop disconnect, 3-way calling and credit card reader support; time-division multiplexing primary rate interface (TDM PRI) with PRI enterprise session border controllers allowing service providers to offer services to TDM-PBX (private branch exchange) customers, with an easy migration path to SIP (session initiation protocol) trunking or hosted services later when the customers transition from TDM to IP adopting IP-PBX or IP Centrix services. The cable segment offers four mini-size DOCSIS 3.0 compliant ECMM products that are commercial and industrial graded for North America and Europe. Its device-initiated DQoS technology enables edge devices to initiate and manage DQoS UGS (unsolicited grant service) flows based on user and signaling events without the need for packet cable multimedia. The element management system (EMS) segment provides the carrier grade EMS with auto-provisioning and device management. InnoMedia partners with several major companies to provide VoIP technology to end users, including Broadsoft, Clearcable Networks, General Bandwidth, MetaSwitch, Net2Phone and Sonus Networks.

FINANCIAL DATA: *Note: Data for latest year may not have been available at press time.*

In U.S. $	2021	2020	2019	2018	2017	2016
Revenue						
R&D Expense						
Operating Income						
Operating Margin %						
SGA Expense						
Net Income						
Operating Cash Flow						
Capital Expenditure						
EBITDA						
Return on Assets %						
Return on Equity %						
Debt to Equity						

CONTACT INFORMATION:
Phone: 65-6872-0828   Fax: 65-6586-9111
Toll-Free:
Address: 15 Jalan Kilang Barat #06-03, Singapore, 159357 Singapore

STOCK TICKER/OTHER:
Stock Ticker: Private   Exchange:
Employees: Fiscal Year Ends: 12/31
Parent Company:

SALARIES/BONUSES:
Top Exec. Salary: $ Bonus: $
Second Exec. Salary: $ Bonus: $

OTHER THOUGHTS:
Estimated Female Officers or Directors:
Hot Spot for Advancement for Women/Minorities:

INNOVATE Corp

NAIC Code: 517110

TYPES OF BUSINESS:
International & Long-Distance Telephone Service
Industrial Construction Services
Structural Steel
Water Pipes
Digital Engineering
Healthcare Product Development
Medical Technologies
Over-the-Air Broadcasting Stations

GROWTH PLANS/SPECIAL FEATURES:
Innovate Corp is a diversified holding company that has a portfolio of subsidiaries in a variety of operating segments which include Infrastructure, Life Sciences and Spectrum.

BRANDS/DIVISIONS/AFFILIATES:
DBM Global Inc
GrayWolf
Panseed Life Sciences LLC
HC2 Broadcasting Holdings Inc
Azteca America
INNOVATE Corp

CONTACTS: Note: Officers with more than one job title may be intentionally listed here more than once.
Philip Falcone, CEO
Michael Sena, CFO
Suzi Herbst, Chief Administrative Officer
Joseph Ferraro, Other Executive Officer

FINANCIAL DATA: Note: Data for latest year may not have been available at press time.

In U.S. $	2021	2020	2019	2018	2017	2016
Revenue	1,205,200,000	716,900,000	1,077,000,000	1,976,700,000	1,634,100,000	1,558,100,000
R&D Expense						
Operating Income	-10,600,000	-28,300,000	75,300,000	-54,800,000	-1,100,000	-1,500,000
Operating Margin %	-.01%	-.04%	.07%	-.03%	.00%	.00%
SGA Expense	168,300,000	145,500,000	177,300,000	218,400,000	182,800,000	152,900,000
Net Income	-227,500,000	-92,000,000	-31,500,000	162,000,000	-46,900,000	-94,500,000
Operating Cash Flow	27,000,000	41,100,000	110,700,000	341,400,000	6,600,000	79,100,000
Capital Expenditure	24,100,000	17,800,000	24,700,000	39,700,000	31,900,000	29,000,000
EBITDA	16,000,000	54,900,000	42,500,000	296,700,000	51,900,000	26,500,000
Return on Assets %	-.06%	-.01%	.00%	.03%	-.02%	-.04%
Return on Equity %	-.97%	-.21%	-.14%	1.72%	-.85%	-1.52%
Debt to Equity		0.285	1.887	8.058	8.115	9.691

CONTACT INFORMATION:
Phone: 212-235-2690   Fax:
Toll-Free:
Address: 295 Madison Ave. Fl. 12, New York, NY 10017 United States

STOCK TICKER/OTHER:
Stock Ticker: VATE
Employees: 3,902
Parent Company:

Exchange: NYS
Fiscal Year Ends: 12/31

SALARIES/BONUSES:
Top Exec. Salary: $480,000   Bonus: $
Second Exec. Salary: $440,000   Bonus: $

OTHER THOUGHTS:
Estimated Female Officers or Directors:
Hot Spot for Advancement for Women/Minorities:

Intel Corporation

NAIC Code: 334413

www.intel.com

TYPES OF BUSINESS:

Microprocessors
Processors
Chipsets
Technologies
Graphics Processing Units
Memory and Storage Products
Programmable Devices
Internet of Things

BRANDS/DIVISIONS/AFFILIATES:

Mobileye

GROWTH PLANS/SPECIAL FEATURES:

Intel is the world's largest logic chipmaker. It designs and manufactures microprocessors for the global personal computer and data center markets. Intel pioneered the x86 architecture for microprocessors. It was the prime proponent of Moore's law for advances in semiconductor manufacturing, though the firm has recently faced manufacturing delays. While Intel's server processor business has benefited from the shift to the cloud, the firm has also been expanding into new adjacencies as the personal computer market has stagnated. These include areas such as the Internet of Things, artificial intelligence, and automotive. Intel has been active on the merger and acquisitions front, acquiring Altera, Mobileye, and Habana Labs in order to bolster these efforts in non-PC arenas.

CONTACTS: Note: Officers with more than one job title may be intentionally listed here more than once.

Patrick Gelsinger, CEO
George Davis, CFO
Omar Ishrak, Chairman of the Board
Kevin McBride, Chief Accounting Officer
Greg Lavender, Chief Technology Officer
Steven Rodgers, Executive VP
Gregory Bryant, Executive VP
Sandra Rivera, Executive VP
Nick McKeown, Managing Director, Divisional
Raja Koduri, Other Corporate Officer

FINANCIAL DATA: Note: Data for latest year may not have been available at press time.

In U.S. $	2021	2020	2019	2018	2017	2016
Revenue	79,024,000,000	77,867,000,000	71,965,000,000	70,848,000,000	62,761,000,000	59,387,000,000
R&D Expense	15,190,000,000	13,556,000,000	13,362,000,000	13,543,000,000	13,035,000,000	12,685,000,000
Operating Income	22,082,000,000	23,876,000,000	22,428,000,000	23,244,000,000	18,434,000,000	14,877,000,000
Operating Margin %	.28%	.31%	.31%	.33%	.29%	.25%
SGA Expense	6,543,000,000	6,180,000,000	6,350,000,000	6,950,000,000	7,452,000,000	8,377,000,000
Net Income	19,868,000,000	20,899,000,000	21,048,000,000	21,053,000,000	9,601,000,000	10,316,000,000
Operating Cash Flow	29,991,000,000	35,384,000,000	33,145,000,000	29,432,000,000	22,110,000,000	21,808,000,000
Capital Expenditure	20,329,000,000	14,453,000,000	16,213,000,000	15,181,000,000	11,778,000,000	9,625,000,000
EBITDA	34,092,000,000	37,946,000,000	35,373,000,000	32,870,000,000	29,127,000,000	21,459,000,000
Return on Assets %	.12%	.14%	.16%	.17%	.08%	.10%
Return on Equity %	.23%	.26%	.28%	.29%	.14%	.16%
Debt to Equity	.35%	0.418	0.327	0.337	0.363	0.312

CONTACT INFORMATION:

Phone: 408 765-8080 Fax: 408 765-2633
Toll-Free: 800-628-8686
Address: 2200 Mission College Blvd., Santa Clara, CA 95054-1549
United States

STOCK TICKER/OTHER:

Stock Ticker: INTC Exchange: NAS
Employees: 121,100 Fiscal Year Ends: 12/31
Parent Company:

SALARIES/BONUSES:

Top Exec. Salary: $1,098,500 Bonus: $1,750,000
Second Exec. Salary: $925,000 Bonus: $1,000,000

OTHER THOUGHTS:

Estimated Female Officers or Directors: 10
Hot Spot for Advancement for Women/Minorities: Y

Inteliquent Inc

www.inteliquent.com

NAIC Code: 517110

TYPES OF BUSINESS:
Telecommunications Service Wired, Satellite or Cable

GROWTH PLANS/SPECIAL FEATURES:
Inteliquent is a full-scale network solutions provider, offering intelligent networking to solve challenging interconnection and interoperability issues on a global scale. With an advanced MPLS network that is highly-interconnected to carriers around the world, Inteliquent provides voice, IP Transit, Ethernet and hosted service solutions to major carriers, service providers and content management firms based in over 80 countries across six continents.

BRANDS/DIVISIONS/AFFILIATES:

CONTACTS: Note: Officers with more than one job title may be intentionally listed here more than once.
Eric Carlson, CFO
James Hynes, Chairman of the Board
Brett Scorza, Chief Information Officer
John Bullock, Chief Technology Officer
Matthew Carter, Director
John Schoder, Executive VP
Richard Monto, General Counsel
John Harrington, Senior VP, Divisional
Michelle Owczarzak, Senior VP, Divisional

FINANCIAL DATA: Note: Data for latest year may not have been available at press time.

In U.S. $	2021	2020	2019	2018	2017	2016
Revenue						
R&D Expense						
Operating Income						
Operating Margin %						
SGA Expense						
Net Income						
Operating Cash Flow						
Capital Expenditure						
EBITDA						
Return on Assets %						
Return on Equity %						
Debt to Equity						

CONTACT INFORMATION:
Phone: 312 384-8000   Fax: 312 386-2601
Toll-Free:
Address: 550 West Adams Street, Chicago, IL 60661 United States

STOCK TICKER/OTHER:
Stock Ticker: Subsidiary   Exchange:
Employees: 177   Fiscal Year Ends: 12/31
Parent Company: Sinch AB

SALARIES/BONUSES:
Top Exec. Salary: $ Bonus: $
Second Exec. Salary: $ Bonus: $

OTHER THOUGHTS:
Estimated Female Officers or Directors:
Hot Spot for Advancement for Women/Minorities:

Sales, profits and employees may be estimates. Financial information, benefits and other data can change quickly and may vary from those stated here.

Intelsat SA
NAIC Code: 517410

TYPES OF BUSINESS:
Satellite Carrier
Global Network
Broadband
Satellites
Terrestrial Infrastructure
Video Broadcasting
Mobility
Communications

BRANDS/DIVISIONS/AFFILIATES:
Intelsat General Communications LLC
IntelsatOne

GROWTH PLANS/SPECIAL FEATURES:
Intelsat SA is a Luxembourg-based company which operates satellite services business. It is engaged in providing diversified communications services to the media companies, fixed and wireless telecommunications operators, data networking service providers for enterprise and mobile applications in the air and on the seas, and a multinational corporation. In addition, it provides commercial satellite communication services to the United States government and other select military organizations and their contractors. Intelsat operates through a single segment being Providing satellite services. The company earns the majority of its revenue from North America.

CONTACTS: Note: Officers with more than one job title may be intentionally listed here more than once.
Stephen Spengler, CEO
David Tolley, CFO
Samer Halawi, CCO
Bruno Fromont, CTO
Michelle Bryan, Chief Admin. Officer
Michelle Bryan, General Counsel
David McGlade, Chmn.

FINANCIAL DATA: Note: Data for latest year may not have been available at press time.

In U.S. $	2021	2020	2019	2018	2017	2016
Revenue	2,000,000,000	1,913,080,064	2,061,464,960	2,161,189,888	2,148,612,096	2,188,047,104
R&D Expense						
Operating Income						
Operating Margin %						
SGA Expense						
Net Income		-911,664,000	-913,595,008	-599,604,992	-178,728,000	990,196,992
Operating Cash Flow						
Capital Expenditure						
EBITDA						
Return on Assets %						
Return on Equity %						
Debt to Equity						

CONTACT INFORMATION:
Phone: 352-27-84-1600   Fax: 352-27-84-1690
Toll-Free:
Address: 4 rue Albert Borschette, Luxembourg, L-1246 Luxembourg

STOCK TICKER/OTHER:
Stock Ticker: Private
Employees: 1,774
Parent Company:

Exchange:
Fiscal Year Ends: 12/31

SALARIES/BONUSES:
Top Exec. Salary: $ Bonus: $
Second Exec. Salary: $ Bonus: $

OTHER THOUGHTS:
Estimated Female Officers or Directors: 1
Hot Spot for Advancement for Women/Minorities:

InterDigital Inc

www.interdigital.com

NAIC Code: 334220

TYPES OF BUSINESS:

Technologies Engineering for Wireless Communications
Research and Development
Connected Technologies
Communications
Wireless
Video Coding
Artificial Intelligence

BRANDS/DIVISIONS/AFFILIATES:

GROWTH PLANS/SPECIAL FEATURES:

InterDigital Inc is a research and development company focused on wireless, visual and related technologies. it designs and develops technologies that enable connected, immersive experiences in a broad range of communications and entertainment products and services. The company derives revenue from patent licensing and sales, with contributions from technology solutions licensing and sales and engineering services. However, the majority of revenue is recurring in nature as it isÂ from current patent royalties and sales as well as technology solutions revenue. Interdigital is focused on two technology areas: cellular wireless technology and "Internet of Things" technology.

CONTACTS: Note: Officers with more than one job title may be intentionally listed here more than once.

Lawrence Chen, CEO
Richard Brezski, CFO
S. Hutcheson, Chairman of the Board
Joshua Schmidt, Chief Legal Officer

FINANCIAL DATA: Note: Data for latest year may not have been available at press time.

In U.S. $	2021	2020	2019	2018	2017	2016
Revenue	425,409,000	358,991,000	318,924,000	307,404,000	532,938,000	665,854,000
R&D Expense	89,368,000	84,646,000	74,860,000	69,698,000	75,724,000	73,118,000
Operating Income	99,083,000	55,168,000	37,835,000	62,595,000	301,495,000	437,306,000
Operating Margin %	.23%	.15%	.12%	.20%	.57%	.66%
SGA Expense	61,217,000	48,999,000	51,289,000	51,030,000	53,068,000	52,067,000
Net Income	55,295,000	44,801,000	20,928,000	65,031,000	176,220,000	309,001,000
Operating Cash Flow	130,392,000	163,467,000	89,433,000	146,792,000	315,800,000	434,159,000
Capital Expenditure	38,277,000	42,408,000	37,990,000	36,895,000	37,004,000	43,440,000
EBITDA	160,974,000	153,133,000	143,991,000	134,122,000	367,288,000	496,150,000
Return on Assets %	.03%	.03%	.01%	.04%	.10%	.19%
Return on Equity %	.07%	.06%	.02%	.07%	.22%	.49%
Debt to Equity	.57%	0.476	0.46	0.339	0.333	0.368

CONTACT INFORMATION:

Phone: 302 281-3600   Fax: 302 281-3763
Toll-Free:
Address: 200 Bellevue Pkwy., Ste. 300, Wilmington, DE 19809 United States

STOCK TICKER/OTHER:

Stock Ticker: IDCC   Exchange: NAS
Employees: 510   Fiscal Year Ends: 12/31
Parent Company:

SALARIES/BONUSES:

Top Exec. Salary: $504,231   Bonus: $1,000,000
Second Exec. Salary: $420,441   Bonus: $

OTHER THOUGHTS:

Estimated Female Officers or Directors: 1
Hot Spot for Advancement for Women/Minorities:

Internap Corporation

www.internap.com

NAIC Code: 517110

<table>
<tr><td valign="top">

TYPES OF BUSINESS:

Internet Access Provider
Hybrid Infrastructure Solutions
Multi-platform Cloud Solutions
Data Centers
Intelligent Managed Services
Network Solutions

BRANDS/DIVISIONS/AFFILIATES:

INAP

</td><td valign="top">

GROWTH PLANS/SPECIAL FEATURES:

Internap Corporation, branded as INAP, is a global provider of performance-driven, secure hybrid infrastructure solutions. INAP's suite of multi-platform cloud, modern data center, optimized network and intelligent managed services solutions help businesses flexibly move workloads, and thus reducing risk and maximizing value. Cloud solutions include bare metal and multi-cloud, as well as cloud backup for data protection. Network solutions include performance internet protocol (IP) and connectivity solutions. Managed services include intelligent monitoring, managed security, managed storage and managed backups. INAP's data centers are located throughout North America, as well as in Amsterdam, Frankfurt, London, Hong Kong, Singapore, Sydney and Tokyo/Osaka.

</td></tr>
</table>

CONTACTS: *Note: Officers with more than one job title may be intentionally listed here more than once.*

Michael T. Sicoli, CEO
Mike Higgins, Sr. VP-Oper.
Lisa Mayr, CFO
Warren Greenberg, Sr. VP-Sales
Jackie Coats, Sr. VP-Human Resources
Jennifer Curry, Sr. VP-Product & Technology
Richard Diegnan, Executive VP
Richard Ramlall, Other Executive Officer
Joseph DuFresne, Vice President, Divisional

FINANCIAL DATA: *Note: Data for latest year may not have been available at press time.*

In U.S. $	2021	2020	2019	2018	2017	2016
Revenue	296,088,000	284,700,000	292,000,000	317,372,992	280,718,016	298,296,992
R&D Expense						
Operating Income						
Operating Margin %						
SGA Expense						
Net Income			-138,000,000	-62,500,000	-45,343,000	-124,742,000
Operating Cash Flow						
Capital Expenditure						
EBITDA						
Return on Assets %						
Return on Equity %						
Debt to Equity						

CONTACT INFORMATION:

Phone: 404 302-9700   Fax: 404 475-0520
Toll-Free: 877-843-7627
Address: 12120 Sunset Hills Rd., Ste. 330, Renton, VA 20190 United States

STOCK TICKER/OTHER:

Stock Ticker: Private   Exchange:
Employees: 530   Fiscal Year Ends: 12/31
Parent Company:

SALARIES/BONUSES:

Top Exec. Salary: $ Bonus: $
Second Exec. Salary: $ Bonus: $

OTHER THOUGHTS:

Estimated Female Officers or Directors: 4
Hot Spot for Advancement for Women/Minorities: Y

Sales, profits and employees may be estimates. Financial information, benefits and other data can change quickly and may vary from those stated here.

International Business Machines Corporation (IBM)

www.ibm.com
NAIC Code: 541513

TYPES OF BUSINESS:

Computer Facilities and Business Process Outsourcing
Computer Facilities Management
Business Process Outsourcing
Software & Hardware
Cloud-Based Computer Services
IT Consulting & Outsourcing
Financial Services
Data Analytics and Health Care Analytics

BRANDS/DIVISIONS/AFFILIATES:

Aspera
Cognos
IBM
Red Hat OpenShift
Watson
Kyndryl Holdings Inc

GROWTH PLANS/SPECIAL FEATURES:

IBM looks to be a part of every aspect of an enterprise's IT needs. The company primarily sells software, IT services, consulting, and hardware. IBM operates in 175 countries and employs approximately 350,000 people. The company has a robust roster of 80,000 business partners to service 5,200 clients--which includes 95% of all Fortune 500. While IBM is a B2B company, IBM's outward impact is substantial. For example, IBM manages 90% of all credit card transactions globally and is responsible for 50% of all wireless connections in the world.

IBM offers employees medical, vision, dental and disability insurance; a flexible spending account; and 401(k) and stock purchase options.

CONTACTS:
Note: Officers with more than one job title may be intentionally listed here more than once.

Arvind Krishna, CEO
James Kavanaugh, CFO, Divisional
Robert Del Bene, Chief Accounting Officer
Nickle Lamoreaux, Other Executive Officer
Michelle Browdy, Senior VP, Divisional
Gary Cohn, Vice Chairman

FINANCIAL DATA:
Note: Data for latest year may not have been available at press time.

In U.S. $	2021	2020	2019	2018	2017	2016
Revenue	57,351,000,000	55,179,000,000	57,714,000,000	79,591,000,000	79,139,000,000	79,920,000,000
R&D Expense	6,488,000,000	6,262,000,000	5,910,000,000	5,379,000,000	5,590,000,000	5,726,000,000
Operating Income	6,832,000,000	4,621,000,000	7,538,000,000	13,217,000,000	13,139,000,000	13,639,000,000
Operating Margin %	.12%	.08%	.13%	.17%	.17%	.17%
SGA Expense	17,699,000,000	19,375,000,000	18,724,000,000	19,366,000,000	19,680,000,000	20,279,000,000
Net Income	5,742,000,000	5,590,000,000	9,431,000,000	8,728,001,000	5,753,000,000	11,872,000,000
Operating Cash Flow	12,796,000,000	18,197,000,000	14,770,000,000	15,247,000,000	16,724,000,000	17,084,000,000
Capital Expenditure	2,768,000,000	3,230,000,000	2,907,000,000	3,964,000,000	3,773,000,000	4,150,000,000
EBITDA	12,409,000,000	10,555,000,000	14,609,000,000	16,545,000,000	16,556,000,000	17,341,000,000
Return on Assets %	.04%	.04%	.07%	.07%	.05%	.10%
Return on Equity %	.29%	.27%	.50%	.51%	.32%	.73%
Debt to Equity	2.51%	2.764	2.782	2.12	2.264	1.899

CONTACT INFORMATION:

Phone: 914-499-1900   Fax: 800-314-1092
Toll-Free: 800-426-4968
Address: 1 New Orchard Rd., Armonk, NY 10504 United States

STOCK TICKER/OTHER:

Stock Ticker: IBM   Exchange: NYS
Employees: 308,000   Fiscal Year Ends: 12/31
Parent Company:

SALARIES/BONUSES:

Top Exec. Salary: $1,170,000   Bonus: $1,000,000
Second Exec. Salary: Bonus: $
$1,500,000

OTHER THOUGHTS:

Estimated Female Officers or Directors: 6
Hot Spot for Advancement for Women/Minorities: Y

Sales, profits and employees may be estimates. Financial information, benefits and other data can change quickly and may vary from those stated here.

Internet Initiative Japan Inc

www.iij.ad.jp

NAIC Code: 517110

TYPES OF BUSINESS:

Internet Service Provider
Internet Access
Wide Area Network
Systems Integration
Outsourcing Services

BRANDS/DIVISIONS/AFFILIATES:

Trust Networks Inc
PTC Systems Pte Ltd

GROWTH PLANS/SPECIAL FEATURES:

Internet Initiative Japan Inc. is a provider of a variety of services including Internet access, outsourcing, and systems integration. Its Internet access services span dial-up, mobile and broadband technologies. Its wide area network services allow secure data sharing over Internet virtual private networks, independently developed routers, and closed wide area networks. Its outsourcing services include cloud, content delivery, data center, and security services. The firm generates nearly all its revenue in Japan.

CONTACTS: *Note: Officers with more than one job title may be intentionally listed here more than once.*

Koichi Suzuki, CEO
Eijiro Katsu, Pres.
Kazuhiro Tokita, Sr. Exec. Officer
Masayoshi Tobita, Exec. Managing Officer
Junichi Shimagami, Exec. Managing Officer
Kiyoshi Ishida, Exec. Managing Officer

FINANCIAL DATA: *Note: Data for latest year may not have been available at press time.*

In U.S. $	2021	2020	2019	2018	2017	2016
Revenue	1,579,664,000	1,516,416,000	1,427,100,000	1,306,981,000	1,170,195,000	
R&D Expense					3,458,314	
Operating Income	105,663,900	60,999,500	44,667,660	50,204,810	38,077,030	
Operating Margin %	.07%	.04%	.03%	.04%	.03%	
SGA Expense	189,043,800	178,550,600	167,992,000	159,251,700	145,699,100	
Net Income	72,022,830	29,715,020	26,109,210	32,801,270	23,483,460	
Operating Cash Flow	300,683,500	247,654,600	186,534,800	108,749,800	54,640,260	
Capital Expenditure	81,632,580	87,799,730	92,559,700	127,657,100	78,789,620	
EBITDA	315,855,300	269,130,000	162,439,300	153,693,400	123,290,600	
Return on Assets %	.05%	.02%	.02%	.03%	.02%	
Return on Equity %	.11%	.05%	.05%	.06%	.05%	
Debt to Equity	.08%	0.154	0.184	0.208	0.125	

CONTACT INFORMATION:

Phone: 81 352056500   Fax: 81 352596311
Toll-Free:
Address: Iidabashi Grand Bloom, 2-10-2 Fujimi, Tokyo, 102-0071 Japan

STOCK TICKER/OTHER:

Stock Ticker: IIJIY   Exchange: PINX
Employees: 3,402   Fiscal Year Ends: 03/31
Parent Company:

SALARIES/BONUSES:

Top Exec. Salary: $ Bonus: $
Second Exec. Salary: $ Bonus: $

OTHER THOUGHTS:

Estimated Female Officers or Directors:
Hot Spot for Advancement for Women/Minorities:

Sales, profits and employees may be estimates. Financial information, benefits and other data can change quickly and may vary from those stated here.

Intrado Corporation

www.intrado.com

NAIC Code: 561422

TYPES OF BUSINESS:

Call Centers
Communication Solutions
Cloud Collaboration
Notification Solutions
Enterprise Communication Solutions
Public Safety Solutions
Software Development

BRANDS/DIVISIONS/AFFILIATES:

Apollo Global Management Inc

CONTACTS: *Note: Officers with more than one job title may be intentionally listed here more than once.*

John Shlonsky, CEO
Steve Cadden, COO
Nancy Disman, CFO
Anup Nair, CIO
David Treinen, Executive VP, Divisional
David Mussman, Executive VP
Nicole Theophilus, Executive VP
Ronald Beaumont, President, Divisional
Jon Hanson, President, Divisional
J. Etzler, President, Divisional

GROWTH PLANS/SPECIAL FEATURES:

Intrado Corporation provides communication and network infrastructure services, helping clients to more efficiently communicate, collaborate and connect with their audiences. The firm does this through its portfolio of services, which include cloud collaboration, notifications, enterprise communications and public safety solutions. Intrado's cloud collaboration services consist of real-time business communication solutions so that teams can communicate and make decisions any time, anywhere. The cloud collaboration platform enables teams to engage with co-workers, customers, prospects and partners. Collaboration solutions span Microsoft Teams, Cisco Webex and Intrado's own solutions, and also include cloud calling, contact center services and artificial intelligent (AI)-based customer service solutions. Notified is Intrado's communication cloud solution for events, public relations and investor relations. Notified solutions are designed to reach, inform, remind and engage customers, investors, employees and the media. Enterprise communication solutions are provided for building real-time connections across people, organizations, partners, devices, supply chain links and more. Its enterprise telecommunication technology applications help improve performance and create services and solutions that meet client objectives. This service connects people and network by delivering interconnection services for all types of providers. Last, Intrado's public safety solutions consist of event management software designed to help keep employees, customers, patients, students and communities safe across the entire 911 continuum. Safety solutions deliver critical information to first responders, notifications to parents, digital workflow communication to healthcare staff, and to patients/healthcare providers. Intrado has sales and/or operations in the Americas, Europe, the Middle East and Asia Pacific. The firm is controlled by affiliates of certain funds managed by Apollo Global Management, Inc.

FINANCIAL DATA: *Note: Data for latest year may not have been available at press time.*

In U.S. $	2021	2020	2019	2018	2017	2016
Revenue	2,800,000,000	2,730,000,000	2,600,000,000	2,156,000,000	2,300,000,000	2,291,962,880
R&D Expense						
Operating Income						
Operating Margin %						
SGA Expense						
Net Income						
Operating Cash Flow						
Capital Expenditure						
EBITDA						
Return on Assets %						
Return on Equity %						
Debt to Equity						

CONTACT INFORMATION:

Phone: 402-571-7700   Fax:
Toll-Free: 800-841-9000
Address: 11808 Miracle Hills Dr., Omaha, NE 68154 United States

STOCK TICKER/OTHER:

Stock Ticker: Private   Exchange:
Employees: 10,000   Fiscal Year Ends: 12/31
Parent Company: Apollo Global Management Inc

SALARIES/BONUSES:

Top Exec. Salary: $ Bonus: $
Second Exec. Salary: $ Bonus: $

OTHER THOUGHTS:

Estimated Female Officers or Directors: 1
Hot Spot for Advancement for Women/Minorities:

Sales, profits and employees may be estimates. Financial information, benefits and other data can change quickly and may vary from those stated here.

Iridium Communications Inc

www.iridium.com

NAIC Code: 517410

<table>
<tr><td valign="top">

TYPES OF BUSINESS:

Satellite Communications Services
Satellites
Mobile Voice
Data Communications
Surveillance
Telecommunications

BRANDS/DIVISIONS/AFFILIATES:

Iridium NEXT

</td><td valign="top">

GROWTH PLANS/SPECIAL FEATURES:

Iridium Communications Inc offers voice and data communications services and products to businesses, U.S. and international government agencies, and other customers on a global basis. It is a provider of mobile voice and data communications services through a constellation of low earth-orbiting satellites. Reaching across land, sea, and air, including the polar regions, Iridium's solutions are ideally suited for industries such as maritime, aviation, government/military, emergency/humanitarian services, mining, forestry, oil and gas, heavy equipment, transportation, and utilities. Iridium also provides service to subscribers from the U.S. Department of Defense, as well as other civil and government agencies around the world.

Employee benefits include medical, dental and vision coverage, and a 401(k).

</td></tr>
</table>

CONTACTS: *Note: Officers with more than one job title may be intentionally listed here more than once.*

Matthew Desch, CEO
Thomas Fitzpatrick, CFO
Robert Niehaus, Chairman of the Board
Timothy Kapalka, Chief Accounting Officer
Thomas Hickey, Chief Legal Officer
Suzanne McBride, COO
Bryan Hartin, Executive VP, Divisional
Scott Scheimreif, Executive VP, Divisional

FINANCIAL DATA: *Note: Data for latest year may not have been available at press time.*

In U.S. $	2021	2020	2019	2018	2017	2016
Revenue	614,500,000	583,439,000	560,444,000	523,008,000	448,046,000	433,640,000
R&D Expense	11,885,000	12,037,000	14,310,000	22,429,000	15,247,000	16,079,000
Operating Income	46,314,000	35,483,000	10,120,000	41,653,000	115,476,000	176,371,000
Operating Margin %	.08%	.06%	.02%	.08%	.26%	.41%
SGA Expense	100,474,000	90,052,000	93,165,000	97,846,000	84,405,000	82,552,000
Net Income	-9,319,000	-56,054,000	-161,999,000	-13,384,000	233,856,000	111,032,000
Operating Cash Flow	302,874,000	249,767,000	198,143,000	263,709,000	259,621,000	225,199,000
Capital Expenditure	42,147,000	38,689,000	117,819,000	391,390,000	400,107,000	405,687,000
EBITDA	350,449,000	308,481,000	194,982,000	252,707,000	237,742,000	225,765,000
Return on Assets %	.00%	- .02%	- .04%	- .01%	.06%	.03%
Return on Equity %	- .01%	- .04%	- .11%	- .01%	.15%	.07%
Debt to Equity	1.23%	1.125	1.21	1.142	1.014	1.233

<table>
<tr><td valign="top">

CONTACT INFORMATION:

Phone: 703 287-7400   Fax: 703 287-7450
Toll-Free:
Address: 1750 Tysons Blvd., Ste. 1400, McLean, VA 22102 United States

</td><td valign="top">

STOCK TICKER/OTHER:

Stock Ticker: IRDM
Employees: 522
Parent Company:

</td><td valign="top">

Exchange: NAS
Fiscal Year Ends: 12/31

</td></tr>
<tr><td valign="top">

SALARIES/BONUSES:

Top Exec. Salary: $956,123   Bonus: $
Second Exec. Salary: $573,674   Bonus: $

</td><td valign="top" colspan="2">

OTHER THOUGHTS:

Estimated Female Officers or Directors:
Hot Spot for Advancement for Women/Minorities:

</td></tr>
</table>

Sales, profits and employees may be estimates. Financial information, benefits and other data can change quickly and may vary from those stated here.

ITI Limited

www.itiltd.in

NAIC Code: 334210

TYPES OF BUSINESS:
Telecommunications Equipment Manufacturing
Telecommunications Infrastructure
Telecommunications Switching
Phone Call Transmission

BRANDS/DIVISIONS/AFFILIATES:
Indian Telephone Industries

CONTACTS: *Note: Officers with more than one job title may be intentionally listed here more than once.*
Rakesh Mohan Agarwal, Managing Dir.
Rajeev Srivastava, CFO
K.K. Gupta, Dir.-Prod.
K.T.Mayuranathan, Sec.
Rakesh Mohan Agarwal, Chmn.

GROWTH PLANS/SPECIAL FEATURES:
ITI Limited is an India-based telecommunications company and a wing of the Indian Government's Department of Telecommunications. The ITI stands for Indian Telephone Industries. The company offers a complete range of telecom products and services covering the spectrum of switching, transmission, access and subscriber premises equipment. ITI manufactures its mobile infrastructure equipment based on GSM (global system for mobile) and CDMA (code division multiple access) technology. The firm also manufactures switching, transmission, broadband equipment, customer premises equipment, IT and convergence equipment, power plant, network management systems, SIM cards, banking automation equipment, Internet of Things (IoT) products and non-conventional energy system products. ITI has a dedicated network systems business unit for carrying out network planning, engineering, implementation and maintenance services as well as consultancy services. The company also carries out in-house research and development activities focused on specialized areas of encryption, network management, satellite, wireless, system engineering and IT and access products to provide customized solutions to the firm's various customers. ITI is involved in the construction of the Army Static Switched Communication Network (ASCON), a strategic communication network that supports the Indian Army. Additionally, the company entered into the information and communication technology (ICT) market by deploying its telecom expertise and vast infrastructure with initiatives such as network management systems, encryption and networking solutions for internet connectivity. ITI has manufacturing facilities in five locations across the country. The company's regional offices are further supported by area located throughout India, including Bangalore, Naini, Rae Bareli, Mankapur, Palakkad and Srinagar. In-house research and development work is focused on the design and development of encryption solutions to Indian Defense forces.

FINANCIAL DATA: *Note: Data for latest year may not have been available at press time.*

In U.S. $	2021	2020	2019	2018	2017	2016
Revenue	321,950,000	298,092,000	288,168,000	278,492,000	293,356,000	252,628,000
R&D Expense						
Operating Income						
Operating Margin %						
SGA Expense						
Net Income	1,291,760	20,051,000	43,350,600	35,443,600	46,992,700	37,858,300
Operating Cash Flow						
Capital Expenditure						
EBITDA						
Return on Assets %						
Return on Equity %						
Debt to Equity						

CONTACT INFORMATION:
Phone: 91-80-2561-4466   Fax: 91-80-2561-7525
Toll-Free:
Address: F29, Ground Floor, Doorvaninagar, Bengaluru, 560 016 India

STOCK TICKER/OTHER:
Stock Ticker: 523610   Exchange: Bombay
Employees: 2,876   Fiscal Year Ends: 03/31
Parent Company:

SALARIES/BONUSES:
Top Exec. Salary: $ Bonus: $
Second Exec. Salary: $ Bonus: $

OTHER THOUGHTS:
Estimated Female Officers or Directors:
Hot Spot for Advancement for Women/Minorities:

Sales, profits and employees may be estimates. Financial information, benefits and other data can change quickly and may vary from those stated here.

Jabil Inc
NAIC Code: 334418

www.jabil.com

TYPES OF BUSINESS:
Contract Electronics Manufacturing
Electronic Manufacturing Services
Engineering Solutions

BRANDS/DIVISIONS/AFFILIATES:

GROWTH PLANS/SPECIAL FEATURES:

Jabil Inc is a United States-based company engaged in
providing manufacturing services and solutions. It provides
comprehensive electronics design, production and product
management services to companies in various industries and
end markets. It operates in two segments. The Electronics
Manufacturing Services (EMS) segment, which is the key
revenue driver, is focused on leveraging IT, supply chain
design and engineering, technologies largely centered on core
electronics. The Diversified Manufacturing Services (DMS)
segment is focused on providing engineering solutions, with an
emphasis on material sciences, technologies, and healthcare.

CONTACTS: *Note: Officers with more than one job title may be intentionally listed here more than once.*
Michael Loparco, CEO, Divisional
Steven Borges, CEO, Divisional
Kenneth Wilson, CEO, Divisional
Mark Mondello, CEO
Meheryar Dastoor, CFO
Timothy Main, Chairman of the Board
Thomas Sansone, Director
Robert Katz, Executive VP
Bruce Johnson, Executive VP
Daryn Smith, Senior VP, Divisional

FINANCIAL DATA: *Note: Data for latest year may not have been available at press time.*

In U.S. $	2021	2020	2019	2018	2017	2016
Revenue	29,285,000,000	27,266,000,000	25,282,000,000	22,095,420,000	19,063,120,000	18,353,090,000
R&D Expense	34,000,000	43,000,000	43,000,000	38,531,000	29,680,000	31,954,000
Operating Income	1,065,000,000	657,000,000	727,000,000	579,055,000	572,737,000	534,202,000
Operating Margin %	.04%	.02%	.03%	.03%	.03%	.03%
SGA Expense	1,213,000,000	1,175,000,000	1,111,000,000	1,050,716,000	907,702,000	924,427,000
Net Income	696,000,000	54,000,000	287,000,000	86,330,000	129,090,000	254,095,000
Operating Cash Flow	1,433,000,000	1,257,000,000	1,193,000,000	-1,105,448,000	-1,464,085,000	916,207,000
Capital Expenditure	1,159,000,000	983,000,000	1,005,000,000	1,036,651,000	716,485,000	924,239,000
EBITDA	1,950,000,000	1,230,000,000	1,411,000,000	1,296,107,000	1,154,712,000	1,220,333,000
Return on Assets %	.04%	.00%	.02%	.01%	.01%	.03%
Return on Equity %	.35%	.03%	.15%	.04%	.05%	.11%
Debt to Equity	1.50%	1.646	1.124	1.279	0.682	0.851

CONTACT INFORMATION:
Phone: 727 577-9749   Fax: 727 579-8529
Toll-Free:
Address: 10560 Dr. Martin Luther King Jr. St. N., St. Petersburg, FL
33716 United States

STOCK TICKER/OTHER:
Stock Ticker: JBL
Employees: 238,000
Parent Company:

Exchange: NYS
Fiscal Year Ends: 08/31

SALARIES/BONUSES:
Top Exec. Salary: $1,133,221   Bonus: $
Second Exec. Salary: Bonus: $
$630,817

OTHER THOUGHTS:
Estimated Female Officers or Directors:
Hot Spot for Advancement for Women/Minorities: Y

Jio (Reliance Jio Infocomm Limited)   www.jio.com
NAIC Code: 517210

TYPES OF BUSINESS:
Wireless Telecommunications Carriers (except Satellite)
LTE Network
Mobile Network Operations
Fiber-to-the-Home
Internet Services
Communication Technologies
Payment Solutions
Mobile Apps

BRANDS/DIVISIONS/AFFILIATES:
Reliance Industries Limited
Jio
JioFi
JioFiber

CONTACTS: *Note: Officers with more than one job title may be intentionally listed here more than once.*
Mukesh D. Ambani, Chmn.

GROWTH PLANS/SPECIAL FEATURES:
Reliance Jio Infocomm Limited (popularly known as Jio), a subsidiary of Reliance Industries Limited, is an Indian mobile network operator and fiber-to-the-home (FTTH) provider serving the India market. Jio operates a national long-term evolution (LTE) network with coverage across all telecommunication circles. The firm has built a world-class, all-IP data network with next-generation LTE technology, which utilizes voice over LTE (VoLTE) to provide voice service on its network. The company's future-ready network can be easily upgraded to support even more data as technologies advance on to 5G, 6G and beyond. Its data routers and phones are marketed under the JioFi and Jio brand names. Jio apps enable users to manage their Jio accounts and related devices, connect to Wi-Fi, chat, watch entertainment, listen to music, read online magazines and news, browse the web, secure data, file into the cloud, make payments, make calls, make online purchases, and connect with doctors and health test results. Devices can be purchased in-store (in the Mumbai area), as well as online at www.jio.com or via mobile app. Jio has the ability to offer free calls and very cheap internet access by relying on advertising and content revenues. JioFiber offers FTTH broadband throughout a home or business, separate Wi-Fi identifications for guests, Norton's mobile security for a select number of devices, high-definition (HD) voice calls throughout India, TV video calling and conferencing, integrated TV, home networking, security and surveillance solutions, a multi-player platform and more.

FINANCIAL DATA: *Note: Data for latest year may not have been available at press time.*

In U.S. $	2021	2020	2019	2018	2017	2016
Revenue	9,599,828,802	8,504,180,000	6,889,980,000	3,645,450,000	17,797,100	5,699,870
R&D Expense						
Operating Income						
Operating Margin %						
SGA Expense						
Net Income	1,637,569,936	734,478,000	426,095,000	111,143,000	-5,235,450	-6,276,040
Operating Cash Flow						
Capital Expenditure						
EBITDA						
Return on Assets %						
Return on Equity %						
Debt to Equity						

CONTACT INFORMATION:
Phone: 91 79-3503-1200   Fax:
Toll-Free:
Address: 101, Saffron, Nr. Cntr. Point Panchwati 5 Rast, Ambawadi, Ahmedabad 380006 India

STOCK TICKER/OTHER:
Stock Ticker: Subsidiary   Exchange:
Employees: Fiscal Year Ends: 03/31
Parent Company: Reliance Industries Limited

SALARIES/BONUSES:
Top Exec. Salary: $ Bonus: $
Second Exec. Salary: $ Bonus: $

OTHER THOUGHTS:
Estimated Female Officers or Directors:
Hot Spot for Advancement for Women/Minorities:

Sales, profits and employees may be estimates. Financial information, benefits and other data can change quickly and may vary from those stated here.

Juniper Networks Inc

NAIC Code: 334210A

www.juniper.net

TYPES OF BUSINESS:

Networking Equipment
Network Product Development
Network Security Solutions
Artificial Intelligence
Machine Learning
Automated WAN Solutions
Cloud-Based Data Center Solutions

BRANDS/DIVISIONS/AFFILIATES:

Mist
EX
128 Technology
SRX
MX
PTX
ACX
Apstra

GROWTH PLANS/SPECIAL FEATURES:

Juniper Networks Inc develops and sells switching, routing
security, related software products, and services for the
networking industry. The company operates as one segmen
and its primary selling verticals are communication service
providers, cloud, and enterprise. The California-based
company was incorporated in 1996, employs over 3025
individuals, and sells worldwide, with over half of its sales from
the Americas region.

Juniper Networks offers medical, dental, prescription and
vision insurance; paid time off; and stock/savings plans.

CONTACTS: *Note: Officers with more than one job title may be intentionally listed here more than once.*

Rami Rahim, CEO
Kenneth Miller, CFO
Scott Kriens, Chairman of the Board
Thomas Austin, Chief Accounting Officer
Manoj Leelanivas, COO
Anand Athreya, Executive VP
Brian Martin, Senior VP

FINANCIAL DATA: *Note: Data for latest year may not have been available at press time.*

In U.S. $	2021	2020	2019	2018	2017	2016
Revenue	4,735,400,000	4,445,100,000	4,445,400,000	4,647,500,000	5,027,200,000	4,990,100,000
R&D Expense	1,007,200,000	958,400,000	955,700,000	1,003,200,000	980,700,000	1,013,700,000
Operating Income	430,400,000	421,100,000	477,500,000	579,500,000	913,700,000	893,000,000
Operating Margin %	.09%	.09%	.11%	.12%	.18%	.18%
SGA Expense	1,302,500,000	1,194,200,000	1,183,600,000	1,158,500,000	1,177,700,000	1,197,800,000
Net Income	252,700,000	257,800,000	345,000,000	566,900,000	306,200,000	592,700,000
Operating Cash Flow	689,700,000	612,000,000	528,900,000	861,100,000	1,259,300,000	1,126,600,000
Capital Expenditure	100,000,000	100,400,000	109,600,000	147,400,000	151,200,000	214,700,000
EBITDA	598,300,000	554,600,000	713,400,000	846,400,000	1,138,600,000	1,131,800,000
Return on Assets %	.03%	.03%	.04%	.06%	.03%	.06%
Return on Equity %	.06%	.06%	.07%	.12%	.06%	.12%
Debt to Equity	.42%	0.411	0.40	0.371	0.456	0.43

CONTACT INFORMATION:

Phone: 408 745-2000   Fax: 408 745-2100
Toll-Free: 888-586-4737
Address: 1133 Innovation Way, Sunnyvale, CA 94089 United States

STOCK TICKER/OTHER:

Stock Ticker: JNPR
Employees: 3,025
Parent Company:

Exchange: NYS
Fiscal Year Ends: 12/31

SALARIES/BONUSES:

Top Exec. Salary: $1,000,000   Bonus: $
Second Exec. Salary: Bonus: $
$612,500

OTHER THOUGHTS:

Estimated Female Officers or Directors: 3
Hot Spot for Advancement for Women/Minorities: Y

KCOM Group Limited

www.kcom.com

NAIC Code: 517110

TYPES OF BUSINESS:

Voice & Data Services
Fiber Broadband Services
Retail Broadband
Wholesale Broadband

BRANDS/DIVISIONS/AFFILIATES:

Macquarie Group Limited
Macquarie Infrastructure and Real Assets
Macquarie European Infrastructure Fund 6

GROWTH PLANS/SPECIAL FEATURES:

KCOM Group Limited is a regional provider of full fiber broadband services to retail and wholesale customers. The company serves customers across Hull, East Yorkshire and North Lincolnshire. KCOM is owned by Macquarie European Infrastructure Fund 6, an investment fund managed by Macquarie Infrastructure and Real Assets, which itself is owned by Macquarie Group Limited. In August 2021, KCOM sold its national information and communication technology (ICT) business to managed services specialist Nasstar, refocusing its business to regional broadband services and developing its wholesale business.

KCOM offers its employees health and pension benefits, life assurance and a variety of employee assistance programs and company perks.

CONTACTS: Note: Officers with more than one job title may be intentionally listed here more than once.

Dale Raneberg, CEO
Sam Booth, CFO
Kathy Smith, Sec.
Nathan Luckey, Chmn.

FINANCIAL DATA: Note: Data for latest year may not have been available at press time.

In U.S. $	2021	2020	2019	2018	2017	2016
Revenue	330,000,000	325,735,000	366,781,000	423,602,000	453,017,280	467,380,480
R&D Expense						
Operating Income						
Operating Margin %						
SGA Expense						
Net Income		-9,615,410	-44,152,600	38,555,200	33,140,166	95,202,024
Operating Cash Flow						
Capital Expenditure						
EBITDA						
Return on Assets %						
Return on Equity %						
Debt to Equity						

CONTACT INFORMATION:

Phone: 44 1482607000   Fax: 44 1482219289
Toll-Free:
Address: 37 Carr Ln., Hull, HU1 3RE United Kingdom

STOCK TICKER/OTHER:

Stock Ticker: Subsidiary   Exchange:
Employees: 1,500   Fiscal Year Ends: 03/31
Parent Company: Macquarie Group Limited

SALARIES/BONUSES:

Top Exec. Salary: $ Bonus: $
Second Exec. Salary: $ Bonus: $

OTHER THOUGHTS:

Estimated Female Officers or Directors: 1
Hot Spot for Advancement for Women/Minorities:

Sales, profits and employees may be estimates. Financial information, benefits and other data can change quickly and may vary from those stated here.

KDDI Corporation

www.kddi.com

NAIC Code: 517210

TYPES OF BUSINESS:

Wireless Telecommunications Carriers (except Satellite)
Telecommunications
Communication Technology
Internet of Things
Business Services
Internet Service

BRANDS/DIVISIONS/AFFILIATES:

GROWTH PLANS/SPECIAL FEATURES:

KDDI is Japan's second-largest wireless operator (31% market share), the largest pay-TV operator (53% market share) and the second-largest provider of fiber-to-the-home broadband (12% market share). It has grown through acquisition and is focusing on increasing the number of customers who subscribe to more than one telecommunication service. It is also looking to grow its Life Design business which includes commerce, energy, and finance and had over 24.5 million IOT connections by the end of March 2022.

CONTACTS: Note: Officers with more than one job title may be intentionally listed here more than once.

Makoto Takahashi, CEO
Makoto Takahashi, Sr. VP
Kanichiro Aritomi, Vice Chmn.
Hirofumi Morozumi, Exec. VP
Hideo Yuasa, Associate Sr. VP

FINANCIAL DATA: Note: Data for latest year may not have been available at press time.

In U.S. $	2021	2020	2019	2018	2017	2016
Revenue	39,399,280,000	38,840,260,000	37,676,900,000	37,392,300,000	35,214,020,000	
R&D Expense						
Operating Income	7,657,305,000	7,579,213,000	7,482,565,000	7,106,200,000	6,750,386,000	
Operating Margin %	.19%	.20%	.20%	.19%	.19%	
SGA Expense	10,117,430,000	9,637,378,000	8,977,084,000	9,427,581,000	8,703,367,000	
Net Income	4,831,623,000	4,744,638,000	4,580,755,000	4,245,980,000	4,054,124,000	
Operating Cash Flow	12,475,270,000	9,814,269,000	7,635,768,000	7,871,589,000	8,610,754,000	
Capital Expenditure	4,632,424,000	4,592,881,000	4,465,574,000	4,159,582,000	3,853,248,000	
EBITDA	13,157,160,000	12,770,720,000	11,737,530,000	11,227,730,000	10,785,850,000	
Return on Assets %	.06%	.08%	.09%	.09%	.09%	
Return on Equity %	.14%	.15%	.16%	.16%	.16%	
Debt to Equity	.30%	0.323	0.249	0.187	0.256	

CONTACT INFORMATION:

Phone: 81 333470077   Fax:
Toll-Free:
Address: Garden Air Tower, 3-10-10, Iidabashi, Chiyoda-ku, Tokyo, 102-8460 Japan

STOCK TICKER/OTHER:

Stock Ticker: KDDIY   Exchange: PINX
Employees: 78,337   Fiscal Year Ends: 03/31
Parent Company:

SALARIES/BONUSES:

Top Exec. Salary: $ Bonus: $
Second Exec. Salary: $ Bonus: $

OTHER THOUGHTS:

Estimated Female Officers or Directors:
Hot Spot for Advancement for Women/Minorities:

Koninklijke KPN NV (Royal KPN NV)

www.kpn.com

NAIC Code: 517110

TYPES OF BUSINESS:

Wired Telecommunications Carriers
Telecommunications
Information Technology
Mobile Networks
Data
Television

GROWTH PLANS/SPECIAL FEATURES:

KPN is the incumbent telecom operator in the Netherlands. It has around 40% share of the broadband market and 20% of the postpaid mobile market, mainly competing with VodafoneZiggo (mobile and fixed) and T-Mobile Netherlands (mainly mobile). KPN is rolling out fiber to the home in the Netherlands and expects to have most of the country covered by 2025.

BRANDS/DIVISIONS/AFFILIATES:

KPN
XS4ALL
Simyo
KPN Security
Ortel Mobile
Cam IT Solutions
Solcon
Inspark

CONTACTS: Note: Officers with more than one job title may be intentionally listed here more than once.

Joost Farwerck, CEO
Chris Figee, CFO
Hilde Garssen, Chief People Officer
Babak Fouladic, CTO

FINANCIAL DATA: Note: Data for latest year may not have been available at press time.

In U.S. $	2021	2020	2019	2018	2017	2016
Revenue	5,375,171,000	5,388,427,000	5,607,676,000	5,744,325,000	5,853,440,000	6,935,408,000
R&D Expense						
Operating Income	1,044,237,000	938,182,000	871,897,300	829,067,300	766,861,800	896,371,600
Operating Margin %	.19%	.17%	.16%	.14%	.13%	.13%
SGA Expense						
Net Income	1,313,455,000	571,067,300	638,371,600	287,573,200	397,707,600	808,672,100
Operating Cash Flow	2,170,056,000	2,083,376,000	2,045,645,000	2,149,660,000	1,988,538,000	1,954,886,000
Capital Expenditure	1,244,111,000	1,594,909,000	1,136,016,000	1,128,878,000	1,154,372,000	1,243,091,000
EBITDA	3,375,416,000	2,491,281,000	2,547,368,000	2,435,194,000	2,305,684,000	2,478,024,000
Return on Assets %	.10%	.05%	.05%	.02%	.02%	.05%
Return on Equity %	.44%	.22%	.28%	.08%	.09%	.17%
Debt to Equity	2.11%	2.521	2.596	4.137	2.475	2.068

CONTACT INFORMATION:

Phone: 31 703434343   Fax:
Toll-Free:
Address: Wilhelminakade 123, Rotterdam, 3072 AP Netherlands

STOCK TICKER/OTHER:

Stock Ticker: KKPNF   Exchange: PINX
Employees: 9,699   Fiscal Year Ends: 12/31
Parent Company:

SALARIES/BONUSES:

Top Exec. Salary: $ Bonus: $
Second Exec. Salary: $ Bonus: $

OTHER THOUGHTS:

Estimated Female Officers or Directors:
Hot Spot for Advancement for Women/Minorities:

Sales, profits and employees may be estimates. Financial information, benefits and other data can change quickly and may vary from those stated here.

Koninklijke Philips NV (Royal Philips)

NAIC Code: 334310

www.philips.com

TYPES OF BUSINESS:

Manufacturing-Electrical & Electronic Equipment
Consumer Electronics & Appliances
Lighting Systems
Medical Imaging Equipment
Semiconductors
Consulting Services
Nanotech Research
MEMS

BRANDS/DIVISIONS/AFFILIATES:

BioTelemetry Inc

GROWTH PLANS/SPECIAL FEATURES:

Philips is a diversified global healthcare company operating in three segments: diagnosis and treatment, connected care, and personal health. About 48% of the company's revenue comes from the diagnosis and treatment segment, which features imaging systems, ultrasound equipment, image-guided therapy solutions and healthcare informatics. The connected care segment (27% of revenue) encompasses monitoring and analytics systems for hospitals and sleep and respiratory care devices, whereas the personal health business (remainder of revenue) includes electric toothbrushes and men's grooming and personal-care products. In 2020, Philips generated EUR 19.5 billion in sales and had 80,000 employees in over 100 countries.

CONTACTS: *Note: Officers with more than one job title may be intentionally listed here more than once.*

Frans van Houten, CEO
Sophie Bechu, COO
Abhijit Bhattacharya, CFO
Daniela Seabrook, Chief Human Resources Officer
Shez Partovi, Exec. VP-Innovation & Strategy
Eric Coutinho, Chief Legal Officer
Jim Andrew, Chief Strategy & Innovation Officer
Deborah DiSanzo, Exec. VP
Pieter Nota, CEO-Phillips Consumer Lifestyle
Eric Rondolat, Exec. VP
Jeroen van der Veer, Chmn.
Patrick Kung, CEO-Greater China

FINANCIAL DATA: *Note: Data for latest year may not have been available at press time.*

In U.S. $	2021	2020	2019	2018	2017	2016
Revenue	17,495,050,000	17,654,140,000	17,485,880,000	18,479,130,000	18,131,390,000	17,765,290,000
R&D Expense	1,841,692,000	1,858,008,000	1,825,376,000	1,793,763,000	1,798,862,000	1,701,984,000
Operating Income	635,312,300	1,431,747,000	1,418,490,000	1,687,708,000	1,449,083,000	1,494,973,000
Operating Margin %	.04%	.08%	.08%	.09%	.08%	.08%
SGA Expense	4,952,989,000	4,776,570,000	4,804,104,000	5,232,404,000	5,073,321,000	4,894,862,000
Net Income	3,384,593,000	1,210,459,000	1,190,063,000	1,111,542,000	1,689,747,000	1,476,617,000
Operating Cash Flow	1,661,194,000	2,560,625,000	1,848,830,000	1,815,178,000	1,906,957,000	1,193,123,000
Capital Expenditure	778,079,200	912,687,900	969,794,600	859,660,200	875,976,400	770,940,800
EBITDA	2,045,645,000	2,924,680,000	2,864,514,000	2,850,238,000	2,698,293,000	2,354,633,000
Return on Assets %	.11%	.04%	.04%	.04%	.06%	.05%
Return on Equity %	.25%	.10%	.09%	.09%	.14%	.12%
Debt to Equity	.40%	0.471	0.387	0.268	0.289	0.321

CONTACT INFORMATION:

Phone: 31 402791111   Fax:
Toll-Free: 877-248-4237
Address: Breitner Ctr., Amstelplein 2, Amsterdam, 1096 BC Netherlands

STOCK TICKER/OTHER:

Stock Ticker: PHG
Employees: 78,189
Parent Company:

Exchange: NYS
Fiscal Year Ends: 12/31

SALARIES/BONUSES:

Top Exec. Salary: $1,351,186   Bonus: $867,732
Second Exec. Salary: $805,613   Bonus: $367,220

OTHER THOUGHTS:

Estimated Female Officers or Directors: 3
Hot Spot for Advancement for Women/Minorities: Y

Kratos Defense & Security Solutions Inc   www.kratosdefense.com
NAIC Code: 541512

TYPES OF BUSINESS:
IT Consulting
System Design & Engineering Services
Consulting Services
Network Management

BRANDS/DIVISIONS/AFFILIATES:

GROWTH PLANS/SPECIAL FEATURES:
Kratos Defense & Security Solutions Inc develops and fields transformative, affordable technology, platforms, and systems. Its segment include The KGS segment is comprised of an aggregation of KGS operating segments, including its microwave electronic products, space, training and cybersecurity, C5ISR/modular systems, turbine technologies and defense and rocket support services operating segments. The US reportable segment consists of its unmanned aerial, unmanned ground and unmanned seaborne system products.

Kratos offers its employees benefits including a 401(k) plan and a stock purchase program.

CONTACTS:
Note: Officers with more than one job title may be intentionally listed here more than once.
Eric Demarco, CEO
Deanna Lund, CFO
William Hoglund, Chairman of the Board
Maria Cervantes de Burgreen, Controller
Marie Mendoza, General Counsel
Steven Fendley, President, Divisional
Stacey Rock, President, Divisional
Phillip Carrai, President, Divisional
David Carter, President, Divisional
Thomas Mills, President, Divisional
Jonah Adelman, President, Divisional
Benjamin Goodwin, Senior VP, Divisional

FINANCIAL DATA:
Note: Data for latest year may not have been available at press time.

In U.S. $	2021	2020	2019	2018	2017	2016
Revenue	811,500,000	747,700,000	717,500,000	618,000,000	603,300,000	541,900,000
R&D Expense	35,200,000	27,000,000	18,000,000	15,600,000	17,800,000	13,900,000
Operating Income	29,700,000	31,700,000	41,200,000	34,300,000	12,500,000	-8,000,000
Operating Margin %	.04%	.04%	.06%	.06%	.02%	-.01%
SGA Expense	160,200,000	144,500,000	130,800,000	119,800,000	127,300,000	114,600,000
Net Income	-2,000,000	79,600,000	12,500,000	-3,500,000	-42,700,000	-60,500,000
Operating Cash Flow	30,800,000	46,600,000	30,000,000	10,400,000	-27,700,000	-13,100,000
Capital Expenditure	46,500,000	35,900,000	26,300,000	22,600,000	26,100,000	9,000,000
EBITDA	62,800,000	64,400,000	74,300,000	48,200,000	-5,800,000	5,400,000
Return on Assets %	.00%	.06%	.01%	.00%	-.04%	-.07%
Return on Equity %	.00%	.11%	.02%	-.01%	-.11%	-.23%
Debt to Equity	.35%	0.367	0.58	0.567	0.574	1.559

CONTACT INFORMATION:
Phone: 512-238-9840   Fax:
Toll-Free:
Address: 1 Chisholm Trail, Ste. 3200, Round Rock, TX 78681 United States

STOCK TICKER/OTHER:
Stock Ticker: KTOS
Employees: 3,300
Parent Company:

Exchange: NAS
Fiscal Year Ends: 12/31

SALARIES/BONUSES:
Top Exec. Salary: $760,000   Bonus: $742,860
Second Exec. Salary: $460,000   Bonus: $337,219

OTHER THOUGHTS:
Estimated Female Officers or Directors: 4
Hot Spot for Advancement for Women/Minorities: Y

Sales, profits and employees may be estimates. Financial information, benefits and other data can change quickly and may vary from those stated here.

KT Corporation

NAIC Code: 517110

TYPES OF BUSINESS:

Local Telephone Service
Mobile Phone Service
Internet Access & Portal Services
Undersea Cable Construction, Maintenance & Repair
IT Consulting & Outsourcing
Online Billing Services
Security Services
e-Commerce Services

BRANDS/DIVISIONS/AFFILIATES:

BC Card Co Ltd

GROWTH PLANS/SPECIAL FEATURES:

KT is South Korea's largest fixed-line telecom operator, with
around 13 million customers. It is the largest broadband firm in
the country, with 9.5 million customers, and the second- largest
wireless operator with 22.8 million subscribers; the company
also has 9.1 million pay-television customers. Additionally,
has about 30 non-telecom businesses. These non-telecom
businesses, including IPTV and IDC/Cloud Services, are the
focus of its growth strategy.

CONTACTS: Note: Officers with more than one job title may be intentionally listed here more than once.

Hyeon-Mo Ky, CEO
Sang-Bong Nam, Exec. VP-Group Legal & Ethics
Young-Soo Woo, Sr. VP-Corp. Center Strategy & Planning Office
Eun-Hye Kim, Exec. VP-Comm. Office
Hyun-Myung Pyo, Pres., Telecom & Convergence Group
Il Yung Kim, Pres., KT Corp. Center
Seong-Mok Oh, Exec. VP-Network Group
Se-Hyun Oh, Exec. VP-New Bus. Unit

FINANCIAL DATA: Note: Data for latest year may not have been available at press time.

In U.S. $	2021	2020	2019	2018	2017	2016
Revenue	19,166,610,000	18,551,840,000	18,967,810,000	17,874,920,000	17,905,310,000	17,550,270,000
R&D Expense	130,073,200	120,813,200	127,039,400	136,069,200	129,816,100	129,235,700
Operating Income	1,353,018,000	1,071,856,000	972,297,000	847,447,700	822,992,500	1,064,721,000
Operating Margin %	.07%	.06%	.05%	.05%	.05%	.06%
SGA Expense	5,327,474,000	5,235,172,000	5,169,928,000	4,988,637,000	4,992,574,000	4,822,736,000
Net Income	1,044,532,000	539,548,000	497,065,500	496,963,900	355,310,500	573,574,100
Operating Cash Flow	4,281,528,000	3,648,743,000	2,883,054,000	3,087,276,000	2,985,111,000	3,672,548,000
Capital Expenditure	3,272,798,000	2,867,127,000	2,925,528,000	2,314,875,000	2,352,355,000	2,478,856,000
EBITDA	4,533,501,000	3,820,404,000	3,815,165,000	3,614,844,000	3,508,112,000	3,791,655,000
Return on Assets %	.04%	.02%	.02%	.02%	.02%	.02%
Return on Equity %	.09%	.05%	.05%	.05%	.04%	.07%
Debt to Equity	.45%	0.421	0.449	0.402	0.433	0.545

CONTACT INFORMATION:

Phone: 82 31-727-0114   Fax:
Toll-Free:
Address: 90, Buljeong-ro, Gundang-gu, Seongnam-si, Gyeonggi-do,
03155 South Korea

STOCK TICKER/OTHER:

Stock Ticker: KT   Exchange: NYS
Employees: 23,372   Fiscal Year Ends:
Parent Company:

SALARIES/BONUSES:

Top Exec. Salary: $42,801   Bonus: $72,824
Second Exec. Salary: $56,966   Bonus: $30,715

OTHER THOUGHTS:

Estimated Female Officers or Directors:
Hot Spot for Advancement for Women/Minorities:

Sales, profits and employees may be estimates. Financial information, benefits and other data can change quickly and may vary from those stated here.

Kyocera Corporation

global.kyocera.com

NAIC Code: 333316

TYPES OF BUSINESS:

Photographic and Photocopying Equipment Manufacturing
Cell Phone Manufacturing
Semiconductor Components
Optoelectronic Products
Consumer Electronics

BRANDS/DIVISIONS/AFFILIATES:

GROWTH PLANS/SPECIAL FEATURES:

Kyocera is a Japanese conglomerate whose original business consisted of manufacturing fine ceramic components; the firm has since expanded into manufacturing handsets, printers, solar cells, and industrial tools. As a result of reorganization, the firm now consists of three major business segments, which are the core components business (28% of 2020 revenue), electronic components business (18% of revenue), and solutions business (55% of revenue).

CONTACTS:
Note: Officers with more than one job title may be intentionally listed here more than once.

Hideo Tanimoto, Pres.
Tatsumi Maeda, Vice Chmn.
Goro Yamaguchi, Chmn.

FINANCIAL DATA:
Note: Data for latest year may not have been available at press time.

In U.S. $	2021	2020	2019	2018	2017	2016
Revenue	11,323,770,000	11,858,890,000	12,041,750,000	11,695,630,000	10,551,420,000	
R&D Expense						
Operating Income	523,909,800	743,051,100	703,226,000	672,641,700	775,304,100	
Operating Margin %	.05%	.06%	.06%	.06%	.07%	
SGA Expense	2,494,089,000	2,528,782,000	2,738,060,000	2,092,324,000	1,993,029,000	
Net Income	669,044,800	798,880,100	765,425,700	586,895,600	770,120,100	
Operating Cash Flow	1,637,652,000	1,591,738,000	1,631,749,000	1,178,471,000	1,217,970,000	
Capital Expenditure	979,872,400	895,387,100	879,227,300	677,758,800	539,239,100	
EBITDA	1,696,908,000	1,803,078,000	1,525,808,000	1,586,488,000	1,603,345,000	
Return on Assets %	.03%	.03%	.03%	.03%	.03%	
Return on Equity %	.04%	.05%	.04%	.03%	.05%	
Debt to Equity	.04%	0.032	0.002	0.003	0.002	

CONTACT INFORMATION:

Phone: 81 75-604-3500   Fax: 81-75-604-3501
Toll-Free:
Address: 6, Takeda Tobadono-cho, Fushimi-ku,, Kyoto, 612-8501 Japan

STOCK TICKER/OTHER:

Stock Ticker: KYOCY   Exchange: PINX
Employees: 70,153   Fiscal Year Ends: 03/31
Parent Company:

SALARIES/BONUSES:

Top Exec. Salary: $459,804   Bonus: $370,810
Second Exec. Salary: $415,307   Bonus: $333,729

OTHER THOUGHTS:

Estimated Female Officers or Directors:
Hot Spot for Advancement for Women/Minorities:

Sales, profits and employees may be estimates. Financial information, benefits and other data can change quickly and may vary from those stated here.

L3Harris Technologies Inc

www.l3harris.com

NAIC Code: 334220

TYPES OF BUSINESS:

Communications Equipment Manufacturing
Aerospace and Defense Technology
Communication Systems
Electronic Systems
Space Systems
Intelligence Systems
Command and Control Systems
Signals Intelligence Systems

BRANDS/DIVISIONS/AFFILIATES:

GROWTH PLANS/SPECIAL FEATURES:

L3Harris Technologies was created in 2019 from the merger of L3 Technologies and Harris, two defense contractors that provide products for the command, control, communications, computers, intelligence, surveillance, and reconnaissance (C4ISR) market. The firm also has smaller operations serving the civil government, particularly the Federal Aviation Administration's communication infrastructure, and produces various avionics for defense and commercial aviation.

CONTACTS: *Note: Officers with more than one job title may be intentionally listed here more than once.*

Christopher Kubasik, CEO
Jesus Malave, CFO
William Brown, Chairman of the Board
Corliss Montesi, Chief Accounting Officer
Scott Mikuen, General Counsel
James Girard, Other Executive Officer
Edward Zoiss, President, Divisional
Todd Gautier, President, Divisional
Sean Stackley, President, Divisional
Dana Mehnert, President, Divisional

FINANCIAL DATA: *Note: Data for latest year may not have been available at press time.*

In U.S. $	2021	2020	2019	2018	2017	2016
Revenue	17,814,000,000	18,194,000,000	9,263,000,000	6,168,000,000	5,897,000,000	
R&D Expense						
Operating Income	2,096,000,000	1,993,000,000	656,000,000	920,000,000	893,000,000	
Operating Margin %	.12%	.11%		.15%	.15%	
SGA Expense	3,280,000,000	3,315,000,000	1,881,000,000	1,182,000,000	1,150,000,000	
Net Income	1,846,000,000	1,119,000,000	822,000,000	699,000,000	543,000,000	
Operating Cash Flow	2,687,000,000	2,790,000,000	939,000,000	751,000,000	569,000,000	
Capital Expenditure	342,000,000	368,000,000	173,000,000	136,000,000	119,000,000	
EBITDA	3,515,000,000	2,608,000,000	1,473,000,000	1,337,000,000	1,372,000,000	
Return on Assets %	.05%	.03%		.07%	.05%	
Return on Equity %	.09%	.05%		.23%	.18%	
Debt to Equity	.41%	0.37		1.04	1.16	

CONTACT INFORMATION:

Phone: 321 727-9100   Fax: 321 724-3973
Toll-Free: 800-442-7747
Address: 1025 West NASA Blvd., Melbourne, FL 32919 United States

STOCK TICKER/OTHER:

Stock Ticker: LHX   Exchange: NYS
Employees: 47,000   Fiscal Year Ends: 06/30
Parent Company:

SALARIES/BONUSES:

Top Exec. Salary: $1,500,000   Bonus: $
Second Exec. Salary: Bonus: $
$1,500,000

OTHER THOUGHTS:

Estimated Female Officers or Directors: 2
Hot Spot for Advancement for Women/Minorities: Y

Sales, profits and employees may be estimates. Financial information, benefits and other data can change quickly and may vary from those stated here.

Laird Connectivity   www.lairdconnect.com

NAIC Code: 334220

TYPES OF BUSINESS:

Antennas, Transmitting and Receiving, Manufacturing
Wireless Modules
Internet of Things Devices
Radio Frequency Antennas
Product Development
Wearable Design
Electromagnetic Compatibility Testing

BRANDS/DIVISIONS/AFFILIATES:

Advent International

CONTACTS: *Note: Officers with more than one job title may be intentionally listed here more than once.*

Bill Steinike, CEO
Jarrod Paulsen, VP-Oper.
Somer Arroyo, CFO
Jen Sarto, VP-Sales
Anna Burtch, VP-Human Resources
Kurt Furlong, VP-Engineering
David England, VP-IT

GROWTH PLANS/SPECIAL FEATURES:

Laird Connectivity manufactures and sells innovative solutions for the integration of wireless connectivity. The firm's products are grouped into three categories: wireless modules, including Wi-Fi and Bluetooth modules, low-power wide area networking protocol (LoRaWAN) solutions, cellular solutions, RAMP ISM (industrial, scientific and medical) modules and related programming kits; Internet of Things (IoT) devices, including IoT gateways, IoT sensors, IoT starter kits and related programming kits; and internal antennas, including Wi-Fi, Bluetooth, cellular IoT, internal, LoRaWAN, global positioning system (GPS) and GNSS, RFID, IoT, machine-to-machine, near field communication (NFC), public safety and custom antennas, along with related accessories. Services offered by Laird Connectivity include product development, antenna design, wearable design and electromagnetic compatibility (EMC) testing. Laird Connectivity has offices and manufacturing facilities worldwide, including the U.S., the U.K. and Taiwan. The company is privately-owned by Advent International, an American private equity firm focused on buyouts of companies in Western and Central Europe, North America, Latin America and Asia.

FINANCIAL DATA: *Note: Data for latest year may not have been available at press time.*

In U.S. $	2021	2020	2019	2018	2017	2016
Revenue	1,500,000,000	1,461,474,000	1,391,880,000	1,325,600,000	1,241,796,224	1,062,805,760
R&D Expense						
Operating Income						
Operating Margin %						
SGA Expense						
Net Income						
Operating Cash Flow						
Capital Expenditure						
EBITDA						
Return on Assets %						
Return on Equity %						
Debt to Equity						

CONTACT INFORMATION:

Phone: 330-434-7929   Fax:
Toll-Free:
Address: 50 S. Main St. #1100, Akron, OH 44308 United States

STOCK TICKER/OTHER:

Stock Ticker: Private   Exchange:
Employees: 10,000   Fiscal Year Ends: 12/31
Parent Company: Advent International

SALARIES/BONUSES:

Top Exec. Salary: $ Bonus: $
Second Exec. Salary: $ Bonus: $

OTHER THOUGHTS:

Estimated Female Officers or Directors:
Hot Spot for Advancement for Women/Minorities:

Sales, profits and employees may be estimates. Financial information, benefits and other data can change quickly and may vary from those stated here.

Lattice Incorporated
NAIC Code: 334210
www.latticeinc.com

TYPES OF BUSINESS:
Call Control Technology
Inmate Call Control Systems
Calling Cards
Secure Network Applications

BRANDS/DIVISIONS/AFFILIATES:
Corrections Operating Platform
Nexus
NetVisit
CellMate
InTouch

CONTACTS: *Note: Officers with more than one job title may be intentionally listed here more than once.*
Paul Burgess, CEO
Joseph Noto, CFO

GROWTH PLANS/SPECIAL FEATURES:
Lattice Incorporated provides innovative inmate managemen and communications solutions to correctional facilities. Th company's Corrections Operating Platform (COP) provide benefits to Lattice clients, their inmates and the family an friends of the inmates through a range of secure, integrate inmate management and communications products an services. COP consists of three primary categories: depos solutions, providing on-site deposit kiosks online/phone/money-order deposits, commissary integration Java message service (JMS) integration, payment processin and billing; products and solutions, such as the Nexus inmat phone system, the NetVisit video visitation and vide arraignment system, and the CellMate mobile device; an support services and solutions, including the InTouch inmat hotline, billing and technical support, and handling accoun funding transactions. Lattice's related services span seamles installation, corrections staff training, and system maintenanc and technical support.

FINANCIAL DATA: *Note: Data for latest year may not have been available at press time.*

In U.S. $	2021	2020	2019	2018	2017	2016
Revenue	86,500,000	8,364,579	7,966,266	7,771,967	7,582,407	7,397,471
R&D Expense						
Operating Income						
Operating Margin %						
SGA Expense						
Net Income						
Operating Cash Flow						
Capital Expenditure						
EBITDA						
Return on Assets %						
Return on Equity %						
Debt to Equity						

CONTACT INFORMATION:
Phone: 856 910-1166   Fax: 856 910-1811
Toll-Free: 800-910-1316
Address: 7150 N. Park Dr., Ste. 500, Pennsauken, NJ 08109 United States

STOCK TICKER/OTHER:
Stock Ticker: Private
Employees: 25
Parent Company:

Exchange:
Fiscal Year Ends: 12/31

SALARIES/BONUSES:
Top Exec. Salary: $ Bonus: $
Second Exec. Salary: $ Bonus: $

OTHER THOUGHTS:
Estimated Female Officers or Directors:
Hot Spot for Advancement for Women/Minorities:

Sales, profits and employees may be estimates. Financial information, benefits and other data can change quickly and may vary from those stated here.

LG Corporation

www.lgcorp.com

NAIC Code: 551114

TYPES OF BUSINESS:

Corporate, Subsidiary, and Regional Managing Offices
Electronics
Technology
Mobile Communications
Vehicle Components
Commercial Displays
Energy Storage
Network Materials

BRANDS/DIVISIONS/AFFILIATES:

LG Electronics
LG Display
LG Innotek
LG Chem
LG Household & Health Care
LG Energy Solution
LG U+
LX Holdings Corp

CONTACTS: *Note: Officers with more than one job title may be intentionally listed here more than once.*

Kwang-mo Koo, Chmn.

GROWTH PLANS/SPECIAL FEATURES:

LG Corporation is a South Korean firm that operates in the electronics, chemicals and communication & services segments. The electronics division consists of three subsidiaries: LG Electronics, which manufactures and markets home appliances, TVs and mobile communications for consumers, and manufactures and markets commercial air conditioners and car parts for businesses; LG Display, a manufacturer of innovative displays, utilizing technologies such as organic light-emitting diodes (OLEDs) and in-plane switching (IPS); and LG Innotek, a developer and producer of materials parts in the fields of vehicles, mobile devices, Internet of Things (IoT), displays, semiconductors and LEDs. The chemicals division consists of three subsidiaries: LG Chem, a chemical, business-to-business company that produces basic materials and chemicals, battery solutions, IT materials, electronics materials, advanced materials, pharmaceuticals and fine chemicals; LG Household & Health Care, a consumer goods company that manages cosmetics, household goods and beverages businesses; and LG Energy Solution, a battery company whose business covers advanced automotive batteries, mobility and IT batteries, energy storage system batteries and next-generation battery technology. Last, the communication & services segment comprises seven subsidiaries: LG U+, a telecommunications service company; GIIR, an advertising and marketing holding company; LG HelloVision Corp., a cable TV company that provides cable TV, internet services, and operates a mobile virtual network; LG CNS, a global IT service company; S&I Corporation, a space solution company for construction, leisure and other industry clients; LG Management Development Institute, which specializes in management consulting and research for future LG businesses; and LG Sports, a professional sports management company in Korea. In addition, LX Holdings Corp. is a 2021 spinoff company that provides global network materials, components and services through subsidiaries: LX International, LX Pantos, LX Hausys, LX Semicon and LX MMA.

FINANCIAL DATA: *Note: Data for latest year may not have been available at press time.*

In U.S. $	2021	2020	2019	2018	2017	2016
Revenue	5,767,309,391	6,206,060,000	5,494,190,000	6,862,430,000	11,088,100,000	8,018,680,000
R&D Expense						
Operating Income						
Operating Margin %						
SGA Expense						
Net Income	2,256,784,646	1,524,570,000	955,782,000	1,685,530,000	2,280,680,000	904,544,000
Operating Cash Flow						
Capital Expenditure						
EBITDA						
Return on Assets %						
Return on Equity %						
Debt to Equity						

CONTACT INFORMATION:

Phone: 822-3773-1114   Fax: 822-3773-2292
Toll-Free:
Address: LG Twin Towers, 128 Yeoui-daero, Yeongdeungpo-gu, Seoul, 150-721 South Korea

STOCK TICKER/OTHER:

Stock Ticker: 3550   Exchange: Seoul
Employees: 222,000   Fiscal Year Ends: 12/31
Parent Company:

SALARIES/BONUSES:

Top Exec. Salary: $ Bonus: $
Second Exec. Salary: $ Bonus: $

OTHER THOUGHTS:

Estimated Female Officers or Directors:
Hot Spot for Advancement for Women/Minorities:

Sales, profits and employees may be estimates. Financial information, benefits and other data can change quickly and may vary from those stated here.

LG Electronics Inc

www.lg.com

NAIC Code: 334220

TYPES OF BUSINESS:

Manufacturing-Electronics
Technology
Home Appliances
Home Entertainment
Mobile Communications
Vehicle Component Solutions
Commercial Displays
Energy Management

BRANDS/DIVISIONS/AFFILIATES:

LG Corporation
LG Signature
LG ThinQ

GROWTH PLANS/SPECIAL FEATURES:

LG Electronics Inc is a South Korea-based company that produces a broad range of electronic products. Its businesses are the home entertainment segment, which produces and sells TVs and digital media products; the mobile communications segment, which produces and sells mobile communications equipment; the home appliance and air solutions segment, which produces and sells washing machines, refrigerators, and other products; the vehicle components segment, which designs and produces vehicle parts; the business solutions segment, which manufactures and sells PCs, solar panels, and other products; and Innotek which sells substrates, sensors, and other items. The company generates the majority of total revenue from the home entertainment and home appliance and air solutions segments

CONTACTS:
Note: Officers with more than one job title may be intentionally listed here more than once.

William Cho, CEO
Hyun-Hoi Ha, Pres.

FINANCIAL DATA:
Note: Data for latest year may not have been available at press time.

In U.S. $	2021	2020	2019	2018	2017	2016
Revenue	57,521,090,000	44,693,280,000	47,963,620,000	47,221,130,000	47,263,180,000	42,621,830,000
R&D Expense	248,172,900	224,034,900	312,374,600	1,789,044,000	209,978,200	272,825,100
Operating Income	2,974,353,000	3,006,172,000	1,875,352,000	2,081,007,000	1,900,302,000	1,029,817,000
Operating Margin %	.05%	.07%	.04%	.04%	.04%	.02%
SGA Expense	7,481,114,000	5,398,898,000	5,757,784,000	7,739,414,000	5,530,386,000	5,757,356,000
Net Income	794,215,700	1,515,232,000	24,083,350	954,665,400	1,328,510,000	59,181,850
Operating Cash Flow	2,061,063,000	3,563,102,000	2,839,938,000	3,496,121,000	1,667,606,000	2,431,037,000
Capital Expenditure	2,508,865,000	2,373,342,000	1,958,845,000	2,959,417,000	2,478,244,000	1,906,108,000
EBITDA	5,115,916,000	4,688,205,000	2,649,827,000	3,405,010,000	3,612,660,000	2,219,357,000
Return on Assets %	.02%	.04%	.00%	.03%	.04%	.00%
Return on Equity %	.06%	.13%	.00%	.09%	.14%	.01%
Debt to Equity	.51%	0.59	0.656	0.67	0.616	0.589

CONTACT INFORMATION:

Phone: 82 237771114   Fax:
Toll-Free: 800-243-0000
Address: LG Twin Towers, 20 Yoido Dong, Seoul, 150-721 South Korea

STOCK TICKER/OTHER:

Stock Ticker: LGEJY   Exchange: PINX
Employees: 100,000   Fiscal Year Ends: 12/31
Parent Company: LG Corporation

SALARIES/BONUSES:

Top Exec. Salary: $ Bonus: $
Second Exec. Salary: $ Bonus: $

OTHER THOUGHTS:

Estimated Female Officers or Directors:
Hot Spot for Advancement for Women/Minorities:

LG Electronics USA Inc
www.lg.com/us

NAIC Code: 334220

TYPES OF BUSINESS:
Consumer Electronics
Mobile Phones & Accessories
Home Appliances
Computer Products

BRANDS/DIVISIONS/AFFILIATES:
LG Corporation
LG Electronics Inc

CONTACTS: Note: Officers with more than one job title may be intentionally listed here more than once.
Thomas Yoon, CEO

GROWTH PLANS/SPECIAL FEATURES:
LG Electronics USA, Inc., a subsidiary of South Korea-based LG Electronics, Inc., sells consumer electronics, home appliances, mobile phones and digital applications in the U.S., Canada and Mexico. Products are grouped into five categories: mobile phones; television, audio and video; appliances; computer products; and solar. The mobile phone segment works in conjunction with service providers, developing a wide range of phones including Android powered smartphones, quad core processors, 4G, 5G, touch screens and more. Additionally, the company sells an array of mobile phone accessories including batteries, cables, chargers, hands-free headsets and Bluetooth products. The television, audio and video category includes organic light-emitting diode (OLED) and nano cell TVs, sound bars, projectors, Blu-ray players, speakers and more. The appliances segment sells washers, dryers, refrigerators, built-in ovens, cooktops, dishwashers, microwave ovens, air conditioners, vacuums, dehumidifiers and accessories. The computer products segment offers consumer and commercial laptops, monitors, burners/drives and related accessories. Last, the solar segment manufactures and installs solar cells and solar panels, which come with a 25-year guarantee. Headquartered in Englewood Cliffs, New Jersey, the firm operates additional training facilities in California and Georgia and a R&D center in Illinois. LG Electronics is itself a subsidiary of holding company LG Corporation.

FINANCIAL DATA: Note: Data for latest year may not have been available at press time.

In U.S. $	2021	2020	2019	2018	2017	2016
Revenue	12,371,585,410	12,429,400,000	8,941,270,432	8,515,495,650	8,473,130,000	6,984,197,082
R&D Expense						
Operating Income						
Operating Margin %						
SGA Expense						
Net Income	212,212,124	152,231,000	-74,658,681	-78,588,085	-82,724,300	-105,207,004
Operating Cash Flow						
Capital Expenditure						
EBITDA						
Return on Assets %						
Return on Equity %						
Debt to Equity						

CONTACT INFORMATION:
Phone: 201 816-2000   Fax: 201-816-0636
Toll-Free: 800-243-0000
Address: 1000 Sylvan Ave., Englewood Cliffs, NJ 60654 United States

STOCK TICKER/OTHER:
Stock Ticker: Subsidiary
Employees: 3,000
Parent Company: LG Corporation
Exchange:
Fiscal Year Ends: 12/31

SALARIES/BONUSES:
Top Exec. Salary: $ Bonus: $
Second Exec. Salary: $ Bonus: $

OTHER THOUGHTS:
Estimated Female Officers or Directors:
Hot Spot for Advancement for Women/Minorities:

Sales, profits and employees may be estimates. Financial information, benefits and other data can change quickly and may vary from those stated here.

LG Uplus Corp

NAIC Code: 517210

www.uplus.co.kr

TYPES OF BUSINESS:

Mobile Telephone Services
Wireless Internet Service
Mobile Commerce Services
Consulting Services
Internet Data Center Services

BRANDS/DIVISIONS/AFFILIATES:

LG Corporation
LG U+
U+ homeBoy

GROWTH PLANS/SPECIAL FEATURES:

LG Uplus Corp is a telecommunications company that sell
cellular-related products. The firm is a subsidiary of the L(
Corporation. The company operates four business segment
that include fixed-line communications, Internet services
mobile voice, and mobile data. The company's termina
distribution segment sells both wireless and wired terminals
LG Uplus also offers Uplus TV and develops both hardwar
and software. The company generates the vast majority of it
revenue in South Korea. The firm sells its products to botl
individuals and businesses.

CONTACTS:
Note: Officers with more than one job title may be intentionally listed here more than once.

Hyun-sik Hwang, CEO

FINANCIAL DATA:
Note: Data for latest year may not have been available at press time.

In U.S. $	2021	2020	2019	2018	2017	2016
Revenue	11,646,546,800	11,268,308,992	10,398,550,016	10,182,787,072	11,498,500,000	10,202,652,290
R&D Expense						
Operating Income						
Operating Margin %						
SGA Expense						
Net Income	608,949,290	391,994,912	368,556,544	404,485,440	512,216,000	438,126,013
Operating Cash Flow						
Capital Expenditure						
EBITDA						
Return on Assets %						
Return on Equity %						
Debt to Equity						

CONTACT INFORMATION:

Phone: 82 070-4080-1114   Fax:
Toll-Free:
Address: LG Uplus Tower, 82, Seoul, 100-095 South Korea

STOCK TICKER/OTHER:

Stock Ticker: 32640   Exchange: Seoul
Employees: 10,221   Fiscal Year Ends: 12/31
Parent Company: LG Corporation

SALARIES/BONUSES:

Top Exec. Salary: $ Bonus: $
Second Exec. Salary: $ Bonus: $

OTHER THOUGHTS:

Estimated Female Officers or Directors:
Hot Spot for Advancement for Women/Minorities:

Liberty Global plc

www.libertyglobal.com

NAIC Code: 517110

TYPES OF BUSINESS:

Video, Voice & Broadband Internet Access Services
Telephony Services
VoIP Services
Mobile Telephony Services
Video on Demand Services

BRANDS/DIVISIONS/AFFILIATES:

Virgin Media
Telenet
UPC
Vodafone Ziggo
ITV
All3Media
ITI Neovision
Sunrise Communications AG

GROWTH PLANS/SPECIAL FEATURES:

Liberty Global is a holding company with interests in cable and telecom companies in the U.K., Netherlands, Belgium, Switzerland, Ireland, and Slovakia. Liberty is the owner of the main cable network in each of these geographies and has pursued a strategy since 2016 to merge or partner with mobile-network-operators to be able to offer converged services. Liberty also owns minority stakes in other media, entertainment, and cloud companies.

CONTACTS: Note: Officers with more than one job title may be intentionally listed here more than once.

Michael Fries, CEO
John Malone, Chairman of the Board
Charlie Bracken, CFO
Amy Blair, Sr. VP
Enrique Rodriguez, Exec. VP
Bryan Hall, Executive VP
Diederik Karsten, Executive VP
Leonard Stegman, Managing Director
John C. Malone, Chmn.

FINANCIAL DATA: Note: Data for latest year may not have been available at press time.

In U.S. $	2021	2020	2019	2018	2017	2016
Revenue	10,311,300,000	11,545,400,000	11,115,800,000	11,957,900,000	11,276,400,000	13,731,100,000
R&D Expense						
Operating Income	1,301,300,000	2,128,300,000	815,300,000	1,087,300,000	872,300,000	1,694,600,000
Operating Margin %	.13%	.18%	.07%	.09%	.08%	.12%
SGA Expense	2,167,800,000	2,157,600,000	2,044,200,000	2,049,100,000	1,980,400,000	2,494,600,000
Net Income	13,426,800,000	-1,628,000,000	11,521,400,000	725,300,000	-2,778,100,000	1,705,300,000
Operating Cash Flow	3,549,000,000	4,185,800,000	4,585,400,000	5,963,100,000	5,708,000,000	5,940,900,000
Capital Expenditure	1,408,000,000	1,292,800,000	1,168,200,000	1,453,000,000	1,250,000,000	1,539,900,000
EBITDA	17,236,600,000	1,613,000,000	3,688,600,000	5,498,700,000	3,095,600,000	6,227,100,000
Return on Assets %	.25%	-.03%	.23%	.01%	-.04%	.03%
Return on Equity %	.68%	-.12%	1.26%	.13%	-.27%	.14%
Debt to Equity	.59%	1.106	1.786	5.594	4.258	2.535

CONTACT INFORMATION:

Phone: 44-208-483-6300   Fax:
Toll-Free:
Address: Griffin House, 161 Hammersmith Rd., London, W6 8BS United Kingdom

STOCK TICKER/OTHER:

Stock Ticker: LBTYA   Exchange: NAS
Employees: 11,200   Fiscal Year Ends: 12/31
Parent Company:

SALARIES/BONUSES:

Top Exec. Salary: $2,563,000   Bonus: $
Second Exec. Salary: $1,205,462   Bonus: $

OTHER THOUGHTS:

Estimated Female Officers or Directors: 3
Hot Spot for Advancement for Women/Minorities: Y

Sales, profits and employees may be estimates. Financial information, benefits and other data can change quickly and may vary from those stated here.

Liberty Latin America Ltd

www.lla.com

NAIC Code: 517110

<table>
<tr><td colspan="2">

TYPES OF BUSINESS:
Telephone, Internet Access, Broadband, Data Networks, Server Facilities and Telecommunications Services Industry

</td></tr>
</table>

GROWTH PLANS/SPECIAL FEATURES:

Liberty Latin America Ltd is a telecommunications company. It is a provider of video, broadband internet, fixed-line telephony and mobile services to residential and business customers. The company's reportable segments include C&W Caribbean and Networks, C&W Panama, VTR, Liberty Puerto Rico and Costa Rica.

BRANDS/DIVISIONS/AFFILIATES:

CONTACTS: *Note: Officers with more than one job title may be intentionally listed here more than once.*
Balan Nair, CEO
Ray Collins, Chief Strategy Officer
Chirs Noyes, CFO
Rocio Lorenzo, Chief Customer Officer
Kerry Scott, Chief People Officer
Aamir Hussain, Chief Technology & Product Officer

FINANCIAL DATA: *Note: Data for latest year may not have been available at press time.*

In U.S. $	2021	2020	2019	2018	2017	2016
Revenue	4,799,000,000	3,764,600,000	3,867,000,000	3,705,700,000	3,590,000,000	2,723,800,000
R&D Expense						
Operating Income	723,400,000	398,400,000	576,300,000	557,400,000	544,700,000	469,000,000
Operating Margin %	.15%	.11%	.15%	.15%	.15%	.17%
SGA Expense	1,119,100,000	910,900,000	880,400,000	826,300,000	710,700,000	523,000,000
Net Income	-440,100,000	-682,200,000	-106,100,000	-345,200,000	-778,100,000	-432,300,000
Operating Cash Flow	1,016,200,000	640,100,000	918,200,000	816,800,000	573,200,000	468,200,000
Capital Expenditure	736,300,000	565,800,000	589,100,000	776,400,000	639,300,000	490,400,000
EBITDA	1,191,500,000	620,400,000	1,080,500,000	688,800,000	524,300,000	803,600,000
Return on Assets %	-.03%	-.05%	-.01%	-.03%	-.06%	-.05%
Return on Equity %	-.18%	-.24%	-.03%	-.11%	-.21%	-.20%
Debt to Equity	3.34%	3.138	2.634	2.05	1.835	1.411

CONTACT INFORMATION:
Phone: 441 295-5950   Fax:
Toll-Free:
Address: 2 Church Street, Hamilton, HM 11 Bermuda

STOCK TICKER/OTHER:
Stock Ticker: LILA   Exchange: NAS
Employees: 11,900   Fiscal Year Ends: 12/31
Parent Company:

SALARIES/BONUSES:
Top Exec. Salary: $1,250,000   Bonus: $
Second Exec. Salary: Bonus: $
$743,886

OTHER THOUGHTS:
Estimated Female Officers or Directors:
Hot Spot for Advancement for Women/Minorities:

Sales, profits and employees may be estimates. Financial information, benefits and other data can change quickly and may vary from those stated here.

Likewize Corp

likewize.com

NAIC Code: 423690

TYPES OF BUSINESS:

Telecommunication Supply Chain & Distribution Services
Device Protection Solutions and Services
Device Malfunction Solutions and Services
Device Insurance and Warranty Solutions
Device Repair
Device Trade-In Solutions
Tech Support

BRANDS/DIVISIONS/AFFILIATES:

Brightstar Corporation

GROWTH PLANS/SPECIAL FEATURES:

Likewize Corp. (formerly Brightstar Corporation) provides protection products and solutions against technology disruption. Device solutions address situations such as lost, stolen, damaged, malfunction or the need to upgrade, among others. Likewize offers insurance, warranty, repair, trade-in, recycling and tech support services and solutions. Headquartered in Texas, USA, Likewize operates in over 30 countries across North America, Latin America, Europe and Asia Pacific. In late-2021, Brightstar Corporation announced the rebranding of its company to reflect a device services direction and away from wireless technology.

CONTACTS: Note: Officers with more than one job title may be intentionally listed here more than once.

Rod Millar, CEO
Dennis J. Strand, Pres., Brightstar Financial Svcs.
Jack Negro, CFO
Ray Roman, CCO
Catherine Smith, Sr. VP
Bela Lainck, Pres., Buy-Back & Trade-In Solutions
Rafael M. de Guzman, III, VP-Strategy
Oscar J. Fumagali, Chief Treas. Officer
Oscar J. Rojas, Pres., Brightstar Latin America
Arturo A. Osorio, Pres., Asia Pacific, Middle East & Africa
Jeff Gower, Pres., Brightstar U.S. & Canada
David Leach, CEO-eSecuritel
Michael Singer, Sr. VP-Global Strategy & New Bus. Dev.
Ramon Colomina, Pres., Supply Chain Solutions

FINANCIAL DATA: Note: Data for latest year may not have been available at press time.

In U.S. $	2021	2020	2019	2018	2017	2016
Revenue	12,039,300,000	11,576,250,000	11,025,000,000	10,500,000,000	10,000,000,000	7,750,000,000
R&D Expense						
Operating Income						
Operating Margin %						
SGA Expense						
Net Income						
Operating Cash Flow						
Capital Expenditure						
EBITDA						
Return on Assets %						
Return on Equity %						
Debt to Equity						

CONTACT INFORMATION:

Phone: 682-348-0354 Fax:
Toll-Free:
Address: 1900 W. Kirkwood Blvd., Ste. 1600C, Southlake, TX 76092 United States

STOCK TICKER/OTHER:

Stock Ticker: Private Exchange:
Employees: 9,000 Fiscal Year Ends: 12/31
Parent Company:

SALARIES/BONUSES:

Top Exec. Salary: $ Bonus: $
Second Exec. Salary: $ Bonus: $

OTHER THOUGHTS:

Estimated Female Officers or Directors: 1
Hot Spot for Advancement for Women/Minorities:

Sales, profits and employees may be estimates. Financial information, benefits and other data can change quickly and may vary from those stated here.

LM Ericsson Telephone Company (Ericsson)   www.ericsson.com

NAIC Code: 334220

TYPES OF BUSINESS:

Wireless Telecommunications Equipment
Information Technology
Communications
Networks
Digital
Internet of Things
Telecommunications
Artificial Intelligence

BRANDS/DIVISIONS/AFFILIATES:

GROWTH PLANS/SPECIAL FEATURES:

Ericsson is primary supplier of telecommunications equipmen
The company's three major operating segments are network
digital services, and managed services. Ericsson sel
hardware, software, and services primarily to communication
service providers while licensing patents to handse
manufacturers. The Stockholm-based company derives sale
worldwide and had 101,000 employees as of June.

CONTACTS: Note: Officers with more than one job title may be intentionally listed here more than once.

Borje Ekholm, CEO
Carl Mellander, Sr. VP
Stella Medlicott, Sr. VP
MajBritt Arfert, Chief People Officer
Erik Ekudden, Sr. VP
Nina Macpherson, General Counsel
Douglas L. Gilstrap, Sr. VP-Function Strategy
Helena Norrman, Sr. VP-Comm..
Jan Frykhammar, Head-Group Function Finance
Angel Ruiz, Head-North America
Magnus Mandersson, Head-Global Svcs.
Rima Qureshi, Sr. VP-Strategic Projects
Johan Wibergh, Exec. VP-Networks
Ronnie Leten, Chmn.
Mats H. Olsson, Head-Asia Pacific

FINANCIAL DATA: Note: Data for latest year may not have been available at press time.

In U.S. $	2021	2020	2019	2018	2017	2016
Revenue	22,942,120,000	22,949,630,000	22,438,670,000	20,821,260,000	20,282,060,000	21,757,260,000
R&D Expense	4,155,010,000	3,921,948,000	3,833,167,000	3,842,451,000	3,741,523,000	3,123,713,000
Operating Income	3,083,224,000	2,771,554,000	923,357,400	206,496,300	-1,809,089,000	560,137,200
Operating Margin %	.13%	.12%	.04%	.01%	-.09%	.03%
SGA Expense	2,662,133,000	2,635,173,000	2,581,154,000	2,717,634,000	2,866,556,000	2,796,440,000
Net Income	2,241,142,000	1,726,530,000	219,531,900	-644,868,900	-3,217,036,000	82,262,750
Operating Cash Flow	3,857,857,000	2,857,273,000	1,666,290,000	922,567,400	948,144,900	1,383,555,000
Capital Expenditure	456,741,000	524,388,000	658,003,300	483,898,500	525,474,300	1,047,986,000
EBITDA	3,868,226,000	3,609,784,000	1,889,574,000	709,849,500	-2,603,473,000	1,297,144,000
Return on Assets %	.08%	.06%	.01%	-.02%	-.12%	.00%
Return on Equity %	.23%	.21%	.03%	-.07%	-.28%	.01%
Debt to Equity	.27%	0.338	0.434	0.355	0.315	0.139

CONTACT INFORMATION:

Phone: 46 87190000   Fax: 46 87191976
Toll-Free:
Address: Torshamnsgatan 21, Kista, Stockholm, 164 83 Sweden

STOCK TICKER/OTHER:

Stock Ticker: ERIC   Exchange: NAS
Employees: 100,824   Fiscal Year Ends: 12/31
Parent Company:

SALARIES/BONUSES:

Top Exec. Salary: $1,517,130   Bonus: $1,812,275
Second Exec. Salary: $ Bonus: $

OTHER THOUGHTS:

Estimated Female Officers or Directors: 10
Hot Spot for Advancement for Women/Minorities: Y

Sales, profits and employees may be estimates. Financial information, benefits and other data can change quickly and may vary from those stated here.

Lumen Technologies Inc

www.lumen.com

NAIC Code: 517110

TYPES OF BUSINESS:

Local Telephone Service
Enterprise Technology
Industry 4.0
Advanced Architecture
Machine Learning
Artificial Intelligence
Internet of Things

BRANDS/DIVISIONS/AFFILIATES:

Lumen
CenturyLink Inc

GROWTH PLANS/SPECIAL FEATURES:

With 450,000 route miles of fiber, including over 35,000 route miles of subsea fiber connecting Europe, Asia, and Latin America, Lumen Technologies is one of the United States' largest telecommunications carriers serving global enterprises. Its merger with Level 3 further shifted the company's operations toward businesses (over 70% of revenue) and away from its legacy consumer business. Lumen offers businesses a full menu of communications services, providing colocation and data center services, data transportation, and end-user phone and internet service. On the consumer side, Lumen provides broadband and phone service across 37 states, where it has 4.5 million broadband customers.

Lumen offers its employees health coverage, life and disability insurance, 401(k) and many other benefits.

CONTACTS: Note: Officers with more than one job title may be intentionally listed here more than once.

Andrea Genschaw, Assistant Controller
Jeffrey Storey, CEO
Indraneel Dev, CFO
T. Glenn, Chairman of the Board
Eric Mortensen, Chief Accounting Officer
Shaun Andrews, Chief Marketing Officer
W. Hanks, Director
Scott Trezise, Executive VP, Divisional
Stacey Goff, Executive VP

FINANCIAL DATA: Note: Data for latest year may not have been available at press time.

In U.S. $	2021	2020	2019	2018	2017	2016
Revenue	19,687,000,000	20,712,000,000	21,458,000,000	22,580,000,000	17,656,000,000	17,470,000,000
R&D Expense						
Operating Income	4,285,000,000	3,604,000,000	3,780,000,000	3,296,000,000	2,009,000,000	2,333,000,000
Operating Margin %	.22%	.17%	.18%	.15%	.11%	.13%
SGA Expense	2,895,000,000	3,464,000,000	3,715,000,000	4,165,000,000	3,508,000,000	3,447,000,000
Net Income	2,033,000,000	-1,232,000,000	-5,269,000,000	-1,733,000,000	1,389,000,000	626,000,000
Operating Cash Flow	6,501,000,000	6,524,000,000	6,680,000,000	7,032,000,000	3,878,000,000	4,608,000,000
Capital Expenditure	2,900,000,000	3,729,000,000	3,628,000,000	3,175,000,000	3,106,000,000	2,981,000,000
EBITDA	8,242,000,000	5,596,000,000	2,084,000,000	5,734,000,000	5,957,000,000	6,254,000,000
Return on Assets %	.03%	-.02%	-.08%	-.02%	.02%	.01%
Return on Equity %	.18%	-.10%	-.32%	-.08%	.08%	.05%
Debt to Equity	2.32%	2.635	2.405	1.786	1.587	1.357

CONTACT INFORMATION:

Phone: 318 388-9000   Fax: 318 789-8656
Toll-Free:
Address: 100 CenturyLink Dr., Monroe, LA 71203 United States

STOCK TICKER/OTHER:

Stock Ticker: LUMN   Exchange: NYS
Employees: 43,000   Fiscal Year Ends: 12/31
Parent Company:

SALARIES/BONUSES:

Top Exec. Salary: $1,800,011   Bonus: $
Second Exec. Salary: Bonus: $
$750,000

OTHER THOUGHTS:

Estimated Female Officers or Directors: 5
Hot Spot for Advancement for Women/Minorities: Y

Lumentum Operations LLC

www.lumentum.com

NAIC Code: 334220

TYPES OF BUSINESS:

Optical Components & Modules
Telecommunications Equipment
Lasers

BRANDS/DIVISIONS/AFFILIATES:

Lumentum Holdings Inc

CONTACTS: *Note: Officers with more than one job title may be intentionally listed here more than once.*

Alan Lowe, CEO
Wajid Ali, CFO
Jason Reinhardt, Exec. VP-Global Sales
Sharon Parker, Sr. VP-Human Resources
Ralph Loura, CIO
Harold Covert, Chmn.

GROWTH PLANS/SPECIAL FEATURES:

Lumentum Operations, LLC designs and manufacture innovative photonics for the purpose of accelerating the speed and scale of cloud, networking, advanced manufacturing and 3D sensing applications. The company's products are categorized into three groups: optical communications commercial lasers and diode lasers. Optical communications products include dense wavelength division multiplexing (DWDM) and coherent optical transceivers, DWDM transmission components, reconfigurable optical add-drop multiplexers (ROADMs) and wavelength management software-defined networking (SDN) elements, optical amplifiers, passive components and modules, pump lasers source lasers and photodiodes, and submarine components Commercial lasers include kW fiber and direct-diode lasers ultra-fast lasers, Q-switched lasers, and low power continuous wave lasers. Diode lasers include edge-emitting diode laser and fiber-coupled diode lasers. Based in California, USA, the firm has additional offices in North America, Europe and Asia Lumentum operates as a subsidiary of Lumentum Holdings Inc.

Lumentum offers its employees benefits programs and wellness initiatives, as well as performance incentives.

FINANCIAL DATA: *Note: Data for latest year may not have been available at press time.*

In U.S. $	2021	2020	2019	2018	2017	2016
Revenue	1,700,000,000	1,678,600,000	1,565,300,000	1,247,700,000	1,001,600,000	903,000,000
R&D Expense						
Operating Income						
Operating Margin %						
SGA Expense						
Net Income		135,500,000	-36,400,000	248,100,000	-102,500,000	9,300,000
Operating Cash Flow						
Capital Expenditure						
EBITDA						
Return on Assets %						
Return on Equity %						
Debt to Equity						

CONTACT INFORMATION:

Phone: 408-546-5483   Fax: 408-546-4300
Toll-Free:
Address: 1001 Ridder Park Dr., San Jose, CA 95131 United States

STOCK TICKER/OTHER:

Stock Ticker: Subsidiary   Exchange:
Employees: 5,618   Fiscal Year Ends: 06/30
Parent Company: Lumentum Holdings Inc

SALARIES/BONUSES:

Top Exec. Salary: $ Bonus: $
Second Exec. Salary: $ Bonus: $

OTHER THOUGHTS:

Estimated Female Officers or Directors:
Hot Spot for Advancement for Women/Minorities:

Sales, profits and employees may be estimates. Financial information, benefits and other data can change quickly and may vary from those stated here.

Lumos Networks Corp

www.lumosnetworks.com

NAIC Code: 517110

TYPES OF BUSINESS:

Wired Telecommunications Carriers

BRANDS/DIVISIONS/AFFILIATES:

CONTACTS: *Note: Officers with more than one job title may be intentionally listed here more than once.*

Timothy Biltz, CEO
Johan Broekhuysen, CFO
Robert Guth, Chairman of the Board
Thomas Ferry, Chief Technology Officer
Joseph McCourt, Executive VP
Diego Anderson, General Manager, Divisional
William Davis, Other Executive Officer
Mary McDermott, Secretary
Jeffrey Miller, Senior VP, Divisional

GROWTH PLANS/SPECIAL FEATURES:

Lumos Networks Corp is incorporated in the state of Delaware. The company is a fiber-based service provider in the Mid-Atlantic region. It is engaged in providing data, broadband, voice and IP services over an expanding fiber optic network. Lumos has diversified communication services which includes Lumos Networks Business DSL which provides broadband services which is faster in both directions, providing up to 6 Mbps downstream and 1 Mbps upstream; Lumos's Integrated Access is a newest business solution which can integrate local voice, long distance, voicemail, and broadband Internet access into a cost-effective package; Its business voice service is a robust which converged voice and data solutions; Its Metro Ethernet (metropolitan-area networking) is one of Lumos Networks' highest speed data connections and has communications technology for businesses of all types and sizes. The MPLS-based Metro Ethernet not only carries multiple network services but can also have traditional data T-1 segments mapped to one's Ethernet span, reducing the number of expensive interfaces. Lumos Networks offers a complete suite of data and voice products supported by approximately 5,800 fiber-route miles in Virginia, West Virginia, and portions of Pennsylvania, Maryland, Ohio and Kentucky.

FINANCIAL DATA: *Note: Data for latest year may not have been available at press time.*

In U.S. $	2021	2020	2019	2018	2017	2016
Revenue						206,899,008
R&D Expense						
Operating Income						
Operating Margin %						
SGA Expense						
Net Income						
Operating Cash Flow						
Capital Expenditure						
EBITDA						
Return on Assets %						
Return on Equity %						
Debt to Equity						

CONTACT INFORMATION:

Phone: 540 946-2000   Fax:
Toll-Free:
Address: One Lumos Plaza, Waynesboro, VA 22980 United States

STOCK TICKER/OTHER:

Stock Ticker: Private   Exchange:
Employees: 566   Fiscal Year Ends:
Parent Company:

SALARIES/BONUSES:

Top Exec. Salary: $ Bonus: $
Second Exec. Salary: $ Bonus: $

OTHER THOUGHTS:

Estimated Female Officers or Directors:
Hot Spot for Advancement for Women/Minorities:

Sales, profits and employees may be estimates. Financial information, benefits and other data can change quickly and may vary from those stated here.

M1 Limited
NAIC Code: 517210

www.m1.com.sg

TYPES OF BUSINESS:
Cell Phone Service
Telecommunications
Mobile Products and Services
Broadband Products and Services
Digital Network

BRANDS/DIVISIONS/AFFILIATES:
Keppel Corporation

CONTACTS: *Note: Officers with more than one job title may be intentionally listed here more than once.*
Manjot Singh Mann, CEO
Nathan Bell, Chief Digital Officer
Lee Kok Chew, CFO
Mustafa Kapasi, CCO
Chan Sock Leng, Dir.-Human Resources
Denis Seek, CTO
Alex Tan, Dir.-Enterprise Svcs. & Prod.
Anil Sachdev, Dir.-Legal Svcs.
Lim Sock Leng, Dir.-Corp. Dev.
Ivan Lim, Dir.-Corp. Comm.
Ivan Lim, Dir.-Investor Rel.
Lee Kok Chew, Chief Commercial Officer
Terence Teo, Dir.-Customer Svcs.
Chan Weng Keong, Dir.-Mgmt. Assurance Svcs.
Loh Chin Hua, Chmn.

GROWTH PLANS/SPECIAL FEATURES:
M1 Limited, a subsidiary of Keppel Corporation, is a digita
network operator in Singapore that provides a suite c
communications services, including mobile, fixed line and fibe
to over 2 million customers. The firm uses 4G and 5G networks
as well as ultra-high-speed fixed broadband, fixed voice an
other services on its next-generation nationwide broadban
network. M1 offers a wide range of mobile and fixe
communication services to consumers, and delivers a
extensive suite of services and solutions to corporat
customers. These services include symmetrical connectivit
solutions of up to 10 gigabits per second (Gbps), manage
services, cloud solutions, cybersecurity solutions, Internet c
Things (IoT) and data center services. Mobile plans includ
SIM-only and flexible phone and payment options. The hom
broadband division offers devices, fixed voice, special benefit
and a variety of plans. The travel division offers pre- and post
paid roaming services as well as tourist SIM services. Digita
services include entertainment, security, finance/payment
utilities and direct carrier billing.

FINANCIAL DATA: *Note: Data for latest year may not have been available at press time.*

In U.S. $	2021	2020	2019	2018	2017	2016
Revenue	840,129,680	807,817,000	815,253,000	800,596,000	790,200,128	782,705,088
R&D Expense						
Operating Income						
Operating Margin %						
SGA Expense						
Net Income		58,434,600			97,769,712	110,408,512
Operating Cash Flow						
Capital Expenditure						
EBITDA						
Return on Assets %						
Return on Equity %						
Debt to Equity						

CONTACT INFORMATION:
Phone: 65 6655-1111   Fax: 65 6655-1977
Toll-Free:
Address: 10 International Business Park, Singapore, 609928 Singapore

STOCK TICKER/OTHER:
Stock Ticker: Subsidiary
Employees: 1,400
Parent Company: Keppel Corporation

Exchange:
Fiscal Year Ends: 12/31

SALARIES/BONUSES:
Top Exec. Salary: $ Bonus: $
Second Exec. Salary: $ Bonus: $

OTHER THOUGHTS:
Estimated Female Officers or Directors: 1
Hot Spot for Advancement for Women/Minorities: Y

Magyar Telekom plc

www.telekom.hu/lakossagi/english

NAIC Code: 517110

TYPES OF BUSINESS:

Local & Long-Distance Telephone Services
Internet & Data Services
Wireless Telephone & Internet Services
Network Infrastructure
Wholesale Services
Telecommunications Equipment

BRANDS/DIVISIONS/AFFILIATES:

Deutsche Telekom AG
T-Mobile
Telekom Hotel Balatonkenese

GROWTH PLANS/SPECIAL FEATURES:

Magyar Telekom PLC is a telecommunications company that operates in two segments: MT-Hungary and North Macedonia. MT Hungary supplies mobile, information communication, television distribution, and system integration services to both business and residential consumers in Hungary. The North Macedonia segment expands the company's mobile and fixed-line telecommunication services across North Macedonia. Magyar controls the largest share of the Hungarian telecom market and has a footprint in Macedonia, Bulgaria, and Romania. The company receives most of its revenue from Hungary.

CONTACTS: Note: Officers with more than one job title may be intentionally listed here more than once.

Tibor Rekasi, CEO
Daria Dodonova, CFO
Melinda Szabo, CCO
Zsuzsanna Friedl, Chief People Officer
Zabor Zatko, CTO
Balazs Mathe, Chief Legal & Corp. Affairs Officer
Robert Pataki, Chief Bus. Dev. Officer
Robert Budafoki, Chief Commercial Officer-Enterprise
Attila Keszeg, Chief Commercial Officer-Residential
Peter Lakatos, Chief Commercial Officer-SMB

FINANCIAL DATA: Note: Data for latest year may not have been available at press time.

In U.S. $	2021	2020	2019	2018	2017	2016
Revenue	1,811,717,000	1,741,662,000	1,725,114,000			
R&D Expense						
Operating Income	235,889,100	211,903,500	203,335,600			
Operating Margin %	.13%	.12%	.12%			
SGA Expense	23,426,660	22,158,680	27,453,160			
Net Income	152,667,900	109,626,300	106,503,000			
Operating Cash Flow	504,010,900	481,200,200	420,163,500			
Capital Expenditure	286,994,100	396,214,100	252,709,300			
EBITDA	623,232,600	553,972,200	545,538,800			
Return on Assets %	.04%	.03%	.03%			
Return on Equity %	.09%	.07%	.07%			
Debt to Equity	.28%	0.29	0.159			

CONTACT INFORMATION:

Phone: 36 1-458-0332   Fax:
Toll-Free:
Address: Konyves Kalman Korut. 36, Budapest, 1097 Hungary

STOCK TICKER/OTHER:

Stock Ticker: MYTAY   Exchange: PINX
Employees: 7,132   Fiscal Year Ends: 12/31
Parent Company: Deutsche Telekom AG

SALARIES/BONUSES:

Top Exec. Salary: $ Bonus: $
Second Exec. Salary: $ Bonus: $

OTHER THOUGHTS:

Estimated Female Officers or Directors: 1
Hot Spot for Advancement for Women/Minorities:

Mahanagar Telephone Nigam Limited

www.mtnl.net.in

NAIC Code: 517110

TYPES OF BUSINESS:
Telephone Service
Local & Long-Distance Service
Mobile Phone Service
Internet Services

GROWTH PLANS/SPECIAL FEATURES:
Mahanagar Telephone Nigam Ltd is an India based company
engaged in providing telecommunications services. The
company's operating segment includes Basic and other
services and Cellular segment. It generates maximum revenue
from the Basic and other services segment. The company is
providing a host of telecom services that include fixed
telephone service, GSM (including 3G services), Internet
Broadband, ISDN and Leased Line services.

BRANDS/DIVISIONS/AFFILIATES:
United Telecom Limited
Mahanagar Telephone Mauritius Limited

CONTACTS: *Note: Officers with more than one job title may be intentionally listed here more than once.*
Pravin Kumar Purwar, Managing Dir.
Suresh Kumar Gupta, Dir.-Finance
Ashok Kumar Garg, Exec. Dir.-Tech.
S.R. Sayal, Corp. Sec.
Peeyush Aggarwal, Exec. Dir.-MTNL Mumbai
B.K.Mittal, Exec. Dir.-MTNL Delhi

FINANCIAL DATA: *Note: Data for latest year may not have been available at press time.*

In U.S. $	2021	2020	2019	2018	2017	2016
Revenue	174,331,300	203,958,700	261,978,700	310,527,800	360,503,900	
R&D Expense						
Operating Income	-105,263,800	-283,605,000	-273,421,900	-250,377,200	-263,927,100	
Operating Margin %	-.60%	-1.39%	-1.04%	-.81%	-.73%	
SGA Expense	84,589,690	2,562,753	2,854,203	3,150,679	2,624,309	
Net Income	-309,196,100	-464,024,100	-425,626,800	-373,223,500	-369,473,600	
Operating Cash Flow	-28,204,100	-151,458,700	-92,076,950	14,870,250	-131,957,900	
Capital Expenditure	12,529,850	17,907,860	39,001,580	61,719,630	50,356,840	
EBITDA	68,007,170	-98,714,980	-89,369,730	-55,434,610	-55,169,540	
Return on Assets %	-.16%	-.24%	-.22%	-.17%	-.16%	
Return on Equity %						
Debt to Equity						

CONTACT INFORMATION:
Phone: 011 24310212   Fax:
Toll-Free:
Address: Fl. 5, Mahanagar Doorsanchar Sadan, 9, CGO Complex,, New
Delhi, Delhi 110003 India

STOCK TICKER/OTHER:
Stock Ticker: MTENY
Employees: 2,713
Parent Company:

Exchange: PINX
Fiscal Year Ends: 03/31

SALARIES/BONUSES:
Top Exec. Salary: $ Bonus: $
Second Exec. Salary: $ Bonus: $

OTHER THOUGHTS:
Estimated Female Officers or Directors:
Hot Spot for Advancement for Women/Minorities:

Sales, profits and employees may be estimates. Financial information, benefits and other data can change quickly and may vary from those stated here.

Maroc Telecom SA

www.iam.ma

NAIC Code: 517110

TYPES OF BUSINESS:
Telephony & Internet Service Provider
Mobile Service Provider
Business Support Services

GROWTH PLANS/SPECIAL FEATURES:
Maroc Telecom SA telecommunications operator in the Kingdom of Morocco. The company provides a wide range of services covering Fixed-Line and Mobile communications, data transfer, and other value-added services. It operates in the Fixed-Line telephony, Mobile telephony and Internet segments. The Fixed-Line and Mobile operating segments are combined within the Services Division (DGS) and the Networks and Systems Division. The company derives revenues from the sale of Mobile, Fixed-line, and Internet telecommunication services and from the sale of equipment.

BRANDS/DIVISIONS/AFFILIATES:
Mauritel SA
Sotelma
Gabon Telecom
Casanet
Onatel
Moov Benin
Moov Ivory Coast
Moov Togo

CONTACTS: Note: Officers with more than one job title may be intentionally listed here more than once.
Laurent Mairot, Dir.-Gen. Admin
Janie Letrot, Gen. Dir.-Regulatory & Legal Affairs
Larbi Guedira, Dir.-General Svcs.
Rachid Mechahouri, Gen. Dir.-Networks & Systems
Abdeslam Ahizoune, Chmn.

FINANCIAL DATA: Note: Data for latest year may not have been available at press time.

In U.S. $	2021	2020	2019	2018	2017	2016
Revenue	3,484,908,000	3,580,234,000	3,555,696,000	3,508,471,000	3,404,382,000	3,432,522,000
R&D Expense						
Operating Income	1,126,972,000	1,170,399,000	801,558,000	1,091,237,000	1,021,227,000	1,028,919,000
Operating Margin %	.32%	.33%	.23%	.31%	.30%	.30%
SGA Expense	82,375,850	76,533,590	91,139,240	94,060,370	94,547,220	248,685,500
Net Income	585,004,900	528,042,800	265,433,300	585,199,600	555,598,800	545,082,800
Operating Cash Flow	1,319,766,000	1,079,649,000	1,487,926,000	1,413,827,000	1,451,899,000	1,312,853,000
Capital Expenditure	514,995,100	403,213,200	774,002,000	786,270,700	814,995,100	608,666,000
EBITDA	1,842,551,000	1,758,617,000	1,840,896,000	1,797,663,000	1,644,985,000	1,641,188,000
Return on Assets %	.10%	.08%	.04%	.10%	.09%	.09%
Return on Equity %	.43%	.44%	.20%	.38%	.36%	.36%
Debt to Equity	.25%	0.373	0.346	0.222	0.265	0.301

CONTACT INFORMATION:
Phone: 212 537 719000   Fax: 212 537 710600
Toll-Free:
Address: Ave. Annakhil Hay Riad, Rabat, 10100 Morocco

STOCK TICKER/OTHER:
Stock Ticker: MAOTF   Exchange: PINX
Employees: 10,123   Fiscal Year Ends:
Parent Company:

SALARIES/BONUSES:
Top Exec. Salary: $ Bonus: $
Second Exec. Salary: $ Bonus: $

OTHER THOUGHTS:
Estimated Female Officers or Directors: 1
Hot Spot for Advancement for Women/Minorities:

Sales, profits and employees may be estimates. Financial information, benefits and other data can change quickly and may vary from those stated here.

Marvell Technology Group Ltd

NAIC Code: 334413

www.marvell.com

TYPES OF BUSINESS:

Semiconductor Manufacturing
Storage Technology
Broadband Technology
Wireless Technology
Power Management Technology
Switching Technology

GROWTH PLANS/SPECIAL FEATURES:

Marvell Technology is a leading fabless chipmaker focused o
networking and storage applications. Marvell serves the dat
center, carrier, enterprise, automotive, and consumer en
markets with processors, optical interconnections, application
specific integrated circuits (ASICs), and merchant silicon fc
Ethernet applications. The firm is an active acquirer, with fiv
large acquisitions since 2017 helping it pivot out of legac
consumer applications to focus on the cloud and 5G markets.

BRANDS/DIVISIONS/AFFILIATES:

OCTEON
OCTEON Fusion-M
NITROX
LiquidIO Server Adapter
ThunderX
Inphi Corporation
Innovium Inc

CONTACTS: *Note: Officers with more than one job title may be intentionally listed here more than once.*

Matt Murphy, CEO
Andy Micallef, COO
Jean Hu, CFO
Dean Jarnac, Sr. VP-Global Sales
Janice Hall, Sr. VP-Human Resources
Pantas Sutardja, Chief R&D Officer
Chris Koopmans, Exec. VP-Mktg.& Bus. Oper.
Gani Jusuf, VP-Prod. Dev., Comm., & Consumer Bus. Group
Yosef Meyouhas, VP-Enterprise Bus. Unit Eng., Marvell Israel
Tom Savage, VP-Worldwide Legal Affairs
James Laufman, General Counsel
Albert Wu, VP-Oper.
Sukhi Nagesh, VP-Investor Rel.
Chris Chang, VP-Greater China Bus.
Bouchung Lin, VP
Renu Bhatia, VP-Sales, Strategic Partnerships
Gaurav Shah, Gen. Mgr.-Digital Entertainment Bus.
Hoo Kuong, VP

FINANCIAL DATA: *Note: Data for latest year may not have been available at press time.*

In U.S. $	2021	2020	2019	2018	2017	2016
Revenue	2,968,900,000	2,699,161,000	2,865,791,000	2,409,170,000	2,300,992,000	
R&D Expense	1,072,740,000	1,080,391,000	914,009,000	714,444,000	805,029,000	
Operating Income	-51,630,000	-188,030,000	120,023,000	509,330,000	227,208,000	
Operating Margin %	-.02%	-.07%	.04%	.21%	.10%	
SGA Expense	467,240,000	464,580,000	424,360,000	238,166,000	251,191,000	
Net Income	-277,298,000	1,584,391,000	-179,094,000	520,831,000	21,151,000	
Operating Cash Flow	817,287,000	360,297,000	596,744,000	571,113,000	-358,435,000	
Capital Expenditure	119,506,000	86,633,000	87,461,000	45,138,000	54,819,000	
EBITDA	388,624,000	1,408,753,000	363,016,000	538,946,000	266,289,000	
Return on Assets %	-.03%	.15%	-.02%	.11%	.00%	
Return on Equity %	-.03%	.20%	-.03%	.13%	.01%	
Debt to Equity	.13%	0.179	0.237			

CONTACT INFORMATION:

Phone: 441 2966395   Fax: 441 2924720
Toll-Free:
Address: Victoria Place, 5/Fl, 31 Victoria St., Hamilton, HM 10 Bermuda

STOCK TICKER/OTHER:

Stock Ticker: MRVL   Exchange: NAS
Employees: 5,340   Fiscal Year Ends: 01/31
Parent Company:

SALARIES/BONUSES:

Top Exec. Salary: $991,731   Bonus: $
Second Exec. Salary: Bonus: $
$595,865

OTHER THOUGHTS:

Estimated Female Officers or Directors: 2
Hot Spot for Advancement for Women/Minorities:

Maxis Berhad

www.maxis.com.my

NAIC Code: 517110

TYPES OF BUSINESS:

Wired Telecommunications Carriers
Telecommunications
Digital Services
Internet
Cloud
Security Solutions

BRANDS/DIVISIONS/AFFILIATES:

ONERetail
MaxisONE

GROWTH PLANS/SPECIAL FEATURES:

Maxis Bhd is a telecommunications provider. Its primary services include mobile, wireless, fixed, and enterprise services, with the majority of revenue generated from mobile and wireless. The company generates mobile revenue from prepaid and postpaid subscribers, with roughly an even split between both. Within enterprise services, Maxis provides traditional managed services in addition to mobile and fixed services. Additionally, the company owns fiber backhaul infrastructure. The company generates the vast majority of its revenue in Malaysia.

CONTACTS: Note: Officers with more than one job title may be intentionally listed here more than once.

Gokhan Ogut, CEO
Mark Dioguardi, Exec. VP-Tech. & Network
Stephen John Mead, General Counsel
Tan Lay Han, VP-Bus. Transformation
Mariam Bevi Batcha, VP-Corp. Affairs
Chow Chee Yan, Sr. VP-Internal Audit
Mohamed Fitri bin Abdullah, Sr. VP-Bus. Svcs.
Maurice Tan, Sr. VP-Personal Svcs.
Harold Quek, Sr. VP-Home Svcs.
Kala Kularajah Sundram, Chief Talent Officer

FINANCIAL DATA: Note: Data for latest year may not have been available at press time.

In U.S. $	2021	2020	2019	2018	2017	2016
Revenue	2,065,074,000	2,011,893,000	2,089,757,000	2,062,605,000	2,113,607,000	1,932,412,000
R&D Expense						
Operating Income	489,173,100	564,119,800	580,276,000	634,129,900	752,013,200	723,807,000
Operating Margin %	.24%	.28%	.28%	.31%	.36%	.37%
SGA Expense	40,390,440	31,863,570	38,819,700	47,346,570	35,996,640	44,087,290
Net Income	293,503,900	310,108,800	339,279,700	399,416,600	489,182,600	451,735,900
Operating Cash Flow	876,472,600	816,560,100	787,838,000	747,447,600	755,643,900	695,634,000
Capital Expenditure	366,431,000	313,250,300	307,640,500	317,065,000	334,890,200	416,926,500
EBITDA	868,394,500	853,584,600	878,492,100	856,950,500	991,708,100	972,836,100
Return on Assets %	.06%	.06%	.07%	.09%	.11%	.10%
Return on Equity %	.19%	.20%	.21%	.25%	.38%	.46%
Debt to Equity	1.20%	1.416	1.252	1.041	1.071	1.89

CONTACT INFORMATION:

Phone: 60 323307000   Fax: 60 323300008
Toll-Free:
Address: Level 18, Menara Maxis, Kuala Lumpur City Centre, Kuala Lumpur, 50088 Malaysia

STOCK TICKER/OTHER:

Stock Ticker: MAXSF   Exchange: PINX
Employees: 3,862   Fiscal Year Ends:
Parent Company:

SALARIES/BONUSES:

Top Exec. Salary: $ Bonus: $
Second Exec. Salary: $ Bonus: $

OTHER THOUGHTS:

Estimated Female Officers or Directors: 4
Hot Spot for Advancement for Women/Minorities: Y

Sales, profits and employees may be estimates. Financial information, benefits and other data can change quickly and may vary from those stated here.

McAfee Corp

NAIC Code: 511210E

TYPES OF BUSINESS:

Computer Software: Network Security, Managed Access, Digital ID,
Cybersecurity & Anti-Virus
Virus Protection Software
Network Management Software
Cybersecurity
Malware Protection

BRANDS/DIVISIONS/AFFILIATES:

Foundation Technology Worldwide LLC
McAfee LLC
McAfee Global Threat Intelligence
MVISION Device

GROWTH PLANS/SPECIAL FEATURES:

McAfee Corp is a device-to-cloud cybersecurity company. It is
engaged in protecting consumers, enterprises, and
governments from cyberattacks with integrated security
privacy, and trust solutions. The company's Persona
Protection Service provides holistic digital protection for ar
individual or family at home, on the go, and on the web. Its
platform includes device security, privacy and safe Wi-Fi
online protection, and identity protection, creating a seamless
and integrated digital moat. It operating segment include
Consumer and Enterprise of which consumer derives a
majority revenue to the company.

CONTACTS: Note: Officers with more than one job title may be intentionally listed here more than once.

Peter Leav, CEO
Venkat Bhamidipati, CFO
Christine Kornegay, Chief Accounting Officer
Lynne Doherty McDonald, Executive VP, Divisional
Gagan Singh, Executive VP
Ashish Agarwal, Senior VP, Divisional

FINANCIAL DATA: Note: Data for latest year may not have been available at press time.

In U.S. $	2021	2020	2019	2018	2017	2016
Revenue	1,920,000,000	1,558,000,000	1,303,000,000	2,408,999,936	2,076,000,000	2,387,000,000
R&D Expense						
Operating Income						
Operating Margin %						
SGA Expense						
Net Income	2,688,000,000	-289,000,000	-236,000,000	-512,000,000	-686,000,000	-95,000,000
Operating Cash Flow						
Capital Expenditure						
EBITDA						
Return on Assets %						
Return on Equity %						
Debt to Equity						

CONTACT INFORMATION:

Phone: 866-622-3911   Fax:
Toll-Free:
Address: 6220 America Center Dr., San Jose, CA 95002 United States

STOCK TICKER/OTHER:

Stock Ticker: Private   Exchange:
Employees: 2,262   Fiscal Year Ends: 12/31
Parent Company: Advent International Corp

SALARIES/BONUSES:

Top Exec. Salary: $ Bonus: $
Second Exec. Salary: $ Bonus: $

OTHER THOUGHTS:

Estimated Female Officers or Directors: 3
Hot Spot for Advancement for Women/Minorities: Y

Mediacom Communications Corporation   www.mediacomcc.com

NAIC Code: 517110

TYPES OF BUSINESS:

Cable TV Service
Internet Service
Digital Cable
Telephone Service

BRANDS/DIVISIONS/AFFILIATES:

Xtream
Home Controller

CONTACTS: *Note: Officers with more than one job title may be intentionally listed here more than once.*

Rocco B. Commisso, CEO
John G. Pascarelli, VP-Operations
Mark E. Stephan, CFO
David M. McNaughton, Sr. VP-Mktg.
Italia Commisso Weinand, Sr. VP-Human Resources & Programming
Peter Lyons, Sr. VP-IT
Joseph E. Young, General Counsel
John G. Pascarelli, Exec. VP-Oper.
Jack Griffin, Dir.-Corp. Finance
Edward S. Pardini, Sr. VP-Divisional Oper.-North Central Division
Brian M. Walsh, Sr. VP
Tapan Dandnaik, Sr. VP-Customer Service & Financial Oper.
Steve Litwer, Sr. VP-Advertising Sales, OnMedia Div.
Rocco B. Commisso, Chmn.

GROWTH PLANS/SPECIAL FEATURES:

Mediacom Communications Corporation, a leading cable company, supplies an array of broadband products and services to more than 1,500 communities in 22 U.S. states, reaching approximately 1.4 million homes. The firm offers customers a full array of traditional video services such as basic service, digital video service, pay-per-view service, high definition television, digital video recorders and video-on-demand. Xtream is the company's all-in-one integrated platform where TV and internet work together to deliver TV, Wi-Fi, caller ID (without interrupting the show), TiVo, On-Demand and apps. In addition, the company offers three types of high-speed internet access: Internet 60, allows for download speeds up to 60 Mbps (megabits per second); Internet 100, speeds up to 100 Mbps; and Xtreme 1 GIG, which allows for up to 1,000 gigabits of download speeds and upload speeds to 50 Mbps. The firm's phone package offers customers unlimited local, regional and long-distance calling within the U.S., Puerto Rico, U.S. Virgin Islands and Canada. It is delivered over voice over internet protocol (VoIP) that digitizes voice signals and routes them as data packets through Mediacom's controlled broadband cable systems. It includes features such as Caller ID with name and number, call waiting, three-way calling and enhanced Emergency 911 dialing. Mediacom also offers video, HSD (high speed data), phone, network and transport services to commercial and large enterprise customers. It offers large enterprise customers who require high-bandwidth connections solutions such as the point-to-point circuits required by wireless communications providers. Additionally, Mediacom offers a home security product, Home Controller, which provides 24-hour-a-day monitoring from a UL-approved facility.

FINANCIAL DATA: *Note: Data for latest year may not have been available at press time.*

In U.S. $	2021	2020	2019	2018	2017	2016
Revenue	2,200,000,000	2,131,224,000	1,971,428,550	1,877,551,000	1,810,255,000	
R&D Expense						
Operating Income						
Operating Margin %						
SGA Expense						
Net Income						
Operating Cash Flow						
Capital Expenditure						
EBITDA						
Return on Assets %						
Return on Equity %						
Debt to Equity						

CONTACT INFORMATION:

Phone: 845-695-2600   Fax: 845-698-4069
Toll-Free: 800-479-2082
Address: 100 Crystal Run Rd., Middletown, NY 10941 United States

STOCK TICKER/OTHER:

Stock Ticker: Private   Exchange:
Employees: 7,000   Fiscal Year Ends: 12/31
Parent Company:

SALARIES/BONUSES:

Top Exec. Salary: $ Bonus: $
Second Exec. Salary: $ Bonus: $

OTHER THOUGHTS:

Estimated Female Officers or Directors: 1
Hot Spot for Advancement for Women/Minorities:

Sales, profits and employees may be estimates. Financial information, benefits and other data can change quickly and may vary from those stated here.

MedTel Services LLC

www.medtelservices.com

NAIC Code: 334210

TYPES OF BUSINESS:

Telecommunications Hardware & Software
Telecommunications Software Applications
Contract Manufacturing
Digital & VOIP Switches

BRANDS/DIVISIONS/AFFILIATES:

Cerato

CONTACTS: *Note: Officers with more than one job title may be intentionally listed here more than once.*

Ewen R. Cameron, Pres.
Robert B. Ramey, Sr. VP-Mfg. Oper.
Angela L. Marvin, VP-Finance
Richard W. Begando, Sr. VP-Int'l Sales

GROWTH PLANS/SPECIAL FEATURES:

MedTel Services, LLC designs, installs, develops manufactures and markets electronic hardware and applicatio software products primarily for more than 20,000 businesses including correctional facilities and government agencies. Th firm's offerings include business phone systems, unifie communication systems, contact center solutions and clou hosted solutions. MedTel's communication server supports i SIP (session initiation protocol) phones and endpoints; it customer interface management software connects customer to the right resource, manages the interaction and provide reports needed for optimizing contact center operations; it services alarms management solutions provides intelliger monitoring and management through a single source fc visibility and control of all related monitored systems; and it voice cyber call recording solution allows businesses to recor multimedia interactions, including chat and email, across fixe and mobile devices over a wide-range of networ configurations from an unlimited number of locations. Product by the company include the Cerato brand of voic communication servers, voice over internet protocol (VoIP phones, hosted cloud systems, cloud softphones, unifie communication systems, monitoring solutions and remot agent appliances. MedTel offers training and support services Its partners include AT&T, Verizon, Telus, Marriott, FixAFone Waterford Crystal, Black Box Network Devices, Carousel Industries, St. Alexius Hospital and many more.

FINANCIAL DATA: *Note: Data for latest year may not have been available at press time.*

In U.S. $	2021	2020	2019	2018	2017	2016
Revenue						
R&D Expense						
Operating Income						
Operating Margin %						
SGA Expense						
Net Income						
Operating Cash Flow						
Capital Expenditure						
EBITDA						
Return on Assets %						
Return on Equity %						
Debt to Equity						

CONTACT INFORMATION:

Phone: 941-753-5000　　　　Fax: 941-758-8469
Toll-Free:
Address: 2511 Corporate Way, Palmetto, FL 34221 United States

STOCK TICKER/OTHER:

Stock Ticker: Private　　　　　　Exchange:
Employees: 200　　　　　　　　Fiscal Year Ends: 12/31
Parent Company:

SALARIES/BONUSES:

Top Exec. Salary: $　　　　Bonus: $
Second Exec. Salary: $　　　Bonus: $

OTHER THOUGHTS:

Estimated Female Officers or Directors: 1
Hot Spot for Advancement for Women/Minorities:

Megacable Holdings SAB de CV

www.megacable.com.mx

NAIC Code: 517110

TYPES OF BUSINESS:

Wired Telecommunications Carriers
Cable Television
Internet Services
Telephone Services
Digital Services

BRANDS/DIVISIONS/AFFILIATES:

GROWTH PLANS/SPECIAL FEATURES:

Megacable Holdings SAB de CV is a diversified media company with interests in cable television services. The company has several business segments, including Cable Network, Internet, Telephone, Business, and other. The Cable Network segment distributes television broadcasting programs. The Internet segment provides Internet subscription services to corporate clients and individual customers. The Telephone segment provides fixed-line communications services. The Business segment provides IT services to enterprise clients, and the Other business segment provides advertising services. Megacable Holdings SAB de CV generates the majority of its revenue in Mexico.

CONTACTS: Note: Officers with more than one job title may be intentionally listed here more than once.

Enrique Yamuni Robles, Managing Dir.

FINANCIAL DATA: Note: Data for latest year may not have been available at press time.

In U.S. $	2021	2020	2019	2018	2017	2016
Revenue	1,217,066,000	1,105,848,000	1,067,914,000	965,100,500	851,659,300	840,015,900
R&D Expense						
Operating Income	346,740,100	323,020,100	308,394,300	309,104,400	263,017,800	247,275,400
Operating Margin %	.28%	.29%	.29%	.32%	.31%	.29%
SGA Expense	67,957,620	63,241,160	64,318,990	64,373,880	52,240,540	49,890,960
Net Income	171,368,700	215,080,800	211,960,700	224,397,600	187,759,300	191,381,300
Operating Cash Flow	567,586,800	661,368,800	352,106,000	404,908,800	375,673,100	247,020,000
Capital Expenditure	497,076,200	262,958,400	385,824,500	264,267,500	211,012,300	266,586,800
EBITDA	564,019,000	549,438,100	522,302,600	483,875,500	399,359,200	349,682,000
Return on Assets %	.07%	.09%	.10%	.12%	.11%	.13%
Return on Equity %	.11%	.14%	.15%	.17%	.16%	.19%
Debt to Equity	.11%	0.203	0.226	0.005	0.159	0.094

CONTACT INFORMATION:

Phone: 52 3337500010   Fax:
Toll-Free:
Address: Lazaro Cardenas 1694, Guadalajara, 44900 Mexico

STOCK TICKER/OTHER:

Stock Ticker: MHSDF   Exchange: PINX
Employees: 18,513   Fiscal Year Ends: 12/31
Parent Company:

SALARIES/BONUSES:

Top Exec. Salary: $ Bonus: $
Second Exec. Salary: $ Bonus: $

OTHER THOUGHTS:

Estimated Female Officers or Directors:
Hot Spot for Advancement for Women/Minorities:

Sales, profits and employees may be estimates. Financial information, benefits and other data can change quickly and may vary from those stated here.

Microsoft Corporation

NAIC Code: 511210H

TYPES OF BUSINESS:

Computer Software, Operating Systems, Languages & Development Tools
Enterprise Software
Game Consoles
Operating Systems
Software as a Service (SAAS)
Search Engine and Advertising
E-Mail Services
Instant Messaging

BRANDS/DIVISIONS/AFFILIATES:

Office 365
Exchange
SharePoint
Microsoft Teams
Skype for Business
Outlook.com
OneDrive
LinkedIn

CONTACTS: Note: Officers with more than one job title may be intentionally listed here more than once.

Satya Nadella, CEO
Amy Hood, CFO
Alice Jolla, Chief Accounting Officer
Bradford Smith, Chief Legal Officer
Christopher Capossela, Chief Marketing Officer
Christopher Young, Executive VP, Divisional
Kathleen Hogan, Executive VP, Divisional
Judson Althoff, Executive VP
Jean-Philippe Courtois, Executive VP

GROWTH PLANS/SPECIAL FEATURES:

Microsoft develops and licenses consumer and enterprise software. It is known for its Windows operating systems and Office productivity suite. The company is organized into three equally sized broad segments: productivity and business processes (legacy Microsoft Office, cloud-based Office 365 Exchange, SharePoint, Skype, LinkedIn, Dynamics) intelligence cloud (infrastructure- and platform-as-a-service offerings Azure, Windows Server OS, SQL Server), and more personal computing (Windows Client, Xbox, Bing search display advertising, and Surface laptops, tablets, and desktops).

Microsoft offers its employees comprehensive benefits, a 401(k) and employee stock purchase plans; and employee assistance programs.

FINANCIAL DATA: Note: Data for latest year may not have been available at press time.

In U.S. $	2021	2020	2019	2018	2017	2016
Revenue	168,088,000,000	143,015,000,000	125,843,000,000	110,360,000,000	96,571,000,000	
R&D Expense	20,716,000,000	19,269,000,000	16,876,000,000	14,726,000,000	13,037,000,000	
Operating Income	69,916,000,000	52,959,000,000	42,959,000,000	35,058,000,000	29,331,000,000	
Operating Margin %	.42%	.37%	.34%	.32%	.30%	
SGA Expense	25,224,000,000	24,709,000,000	23,098,000,000	22,223,000,000	19,942,000,000	
Net Income	61,271,000,000	44,281,000,000	39,240,000,000	16,571,000,000	25,489,000,000	
Operating Cash Flow	76,740,000,000	60,675,000,000	52,185,000,000	43,884,000,000	39,507,000,000	
Capital Expenditure	20,622,000,000	15,441,000,000	13,925,000,000	11,632,000,000	8,129,000,000	
EBITDA	85,134,000,000	68,423,000,000	58,056,000,000	49,468,000,000	40,901,000,000	
Return on Assets %	.19%	.15%	.14%	.07%	.11%	
Return on Equity %	.47%	.40%	.42%	.19%	.32%	
Debt to Equity	.42%	0.568	0.712	0.941	0.929	

CONTACT INFORMATION:

Phone: 425 882-8080   Fax: 425 936-7329
Toll-Free: 800-642-7676
Address: One Microsoft Way, Redmond, WA 98052 United States

STOCK TICKER/OTHER:

Stock Ticker: MSFT   Exchange: NAS
Employees: 163,000   Fiscal Year Ends: 06/30
Parent Company:

SALARIES/BONUSES:

Top Exec. Salary: $541,875   Bonus: $3,500,000
Second Exec. Salary: Bonus: $
$2,500,000

OTHER THOUGHTS:

Estimated Female Officers or Directors: 4
Hot Spot for Advancement for Women/Minorities: Y

Millicom International Cellular SA
www.millicom.com

NAIC Code: 517210

TYPES OF BUSINESS:

Cell Phone Service
Telecommunications
Cable TV
Satellite TV
Broadband
Cloud
Financial Services

BRANDS/DIVISIONS/AFFILIATES:

Tigo
Tigo Business

GROWTH PLANS/SPECIAL FEATURES:

Millicom offers wireless and fixed-line telecom services primarily in smaller, less congested markets or in less developed countries in Latin America. Countries served include Bolivia (100% owned), Honduras (67%), Nicaragua (100%), Panama (80%), El Salvador (100%), Guatemala (100% following the buyout of minority partners in 2021), Paraguay (100%), Colombia (50%), and Costa Rica (100%). The firm's fixed-line networks reach nearly 13 million homes and businesses while its wireless networks cover about 120 million people. Increasingly, Millicom offers a converged package that may include fixed-line phone, broadband, and pay television in conjunction with wireless services. The firm hopes to spin off portions of its tower business and mobile payments operation over the next couple years.

CONTACTS: Note: Officers with more than one job title may be intentionally listed here more than once.

Mauricio Ramos, CEO
Tim Pennington, CFO
Susy Bobenrieth, Chief Human Resources Officer
Xavier Rocoplan, CIO
Mario Zanotti, Sr. Exec. VP-Oper.
Martin Weiss, Exec. VP-Strategy & Corp. Dev.
Justine Dimovic, Head-Investor Rel.
Arthur Bastings, Exec. VP-Africa
Martin Lewerth, Exec. VP-Home & Digital Media
Anders Nilsson, Exec. VP-Commerce & Svcs.
Marc Zagar, Exec. VP-Controlling & Analytics
Jose Antonio Rios Garcia, Chmn.

FINANCIAL DATA: Note: Data for latest year may not have been available at press time.

In U.S. $	2021	2020	2019	2018	2017	2016
Revenue	4,617,000,000	4,172,000,000	4,335,000,000	3,946,000,000	3,936,000,000	4,042,000,000
R&D Expense						
Operating Income	453,000,000	294,000,000	436,000,000	410,000,000	425,000,000	389,000,000
Operating Margin %	.10%	.07%	.10%	.10%	.11%	.10%
SGA Expense	998,000,000	874,000,000	899,000,000	1,043,000,000	1,033,000,000	1,052,000,000
Net Income	590,000,000	-344,000,000	149,000,000	-10,000,000	87,000,000	-32,000,000
Operating Cash Flow	956,000,000	821,000,000	801,000,000	792,000,000	820,000,000	878,000,000
Capital Expenditure	875,000,000	824,000,000	907,000,000	780,000,000	783,000,000	862,000,000
EBITDA	2,404,000,000	1,479,000,000	1,834,000,000	1,274,000,000	1,362,000,000	1,351,000,000
Return on Assets %	.04%	-.03%	.01%	.00%	.01%	.00%
Return on Equity %	.25%	-.15%	.06%	.00%	.03%	-.01%
Debt to Equity	2.67%	3.145	2.811	1.622	1.163	1.207

CONTACT INFORMATION:

Phone: 352 27759021   Fax: 352 27759359
Toll-Free:
Address: 2, Rue du Fort Bourbon, Luxembourg, L-1249 Luxembourg

STOCK TICKER/OTHER:

Stock Ticker: TIGO   Exchange: NAS
Employees: 20,687   Fiscal Year Ends: 12/31
Parent Company:

SALARIES/BONUSES:

Top Exec. Salary: $1,173,000   Bonus: $1,301,131
Second Exec. Salary: Bonus: $508,896
$669,757

OTHER THOUGHTS:

Estimated Female Officers or Directors: 1
Hot Spot for Advancement for Women/Minorities: Y

MiTAC Holdings Corp

NAIC Code: 334111

TYPES OF BUSINESS:

Computer Manufacturing
Server Products
Mobile Communications Products
Storage Products
Tablet PCs & All-in-One PCs

BRANDS/DIVISIONS/AFFILIATES:

MiTAC Computing Technology Corp
MiTAC Digital Technology Corp
MiTAC International Corp
Mio
Magellan
Navman
TYAN

CONTACTS: Note: Officers with more than one job title may be intentionally listed here more than once.

Billy Ho, Pres.
Ting Hui-Yuan, Head-Acct.
C.J. Lin, Sr. Vice Gen. Mgr.
Michael Lin, Sr. Vice Gen. Mgr.
Alice Fang, Vice Gen. Mgr.
Doris Huang, VP-Financial Center
Matthew Miau, Chmn.

GROWTH PLANS/SPECIAL FEATURES:

MiTAC Holdings Corp. is a leading information technology (IT)
and service group. The firm has developed into a multi-national
organization of manufacturers (Japanese domestic market
(JDM), original design manufacturers (ODM), original
equipment manufacturers (OEM) and operational performance
manufacturing (OPM)), designers, research and developers,
testers, assemblers, marketers and servicers in its years of
business. With headquarters in Taiwan, and manufacturing
and logistics centers in China, Taiwan and USA, MiTAC sells
products in more than 30 countries worldwide with its leading
brands include Mio, Magellan and Navman for auto electronics
and TYAN for servers. Subsidiary MiTAC Computing
Technology Corp. offers hyper-scale data centers, high
performance computing (HPC), graphics processing unit
(GPU) and embedded inter-process communication (IPC)
products and services. MiTAC Digital Technology Corp. offers
auto electronics, auto-related Internet of Things (IoT) and
more. MiTAC International Corp. offers smart environment
control products and services, smart service and Smart Factory
4.0.

FINANCIAL DATA: Note: Data for latest year may not have been available at press time.

In U.S. $	2021	2020	2019	2018	2017	2016
Revenue	1,522,269,328	1,463,350,000	1,191,520,000	1,004,260,000	1,640,990,000	1,602,922,379
R&D Expense						
Operating Income						
Operating Margin %						
SGA Expense						
Net Income	431,628,846	101,388,000	92,236,800	107,646,000	86,861,900	90,144,905
Operating Cash Flow						
Capital Expenditure						
EBITDA						
Return on Assets %						
Return on Equity %						
Debt to Equity						

CONTACT INFORMATION:

Phone: 886 3-3289000   Fax:
Toll-Free:
Address: No. 202, Wnehua 2nd Rd., Guishan Dist., Taoyuan City,
33383 Taiwan

STOCK TICKER/OTHER:

Stock Ticker: 3706
Employees: 7,126
Parent Company:

Exchange: TWSE
Fiscal Year Ends: 12/31

SALARIES/BONUSES:

Top Exec. Salary: $ Bonus: $
Second Exec. Salary: $ Bonus: $

OTHER THOUGHTS:

Estimated Female Officers or Directors:
Hot Spot for Advancement for Women/Minorities: Y

Mitel Networks Corporation

www.mitel.com

NAIC Code: 334210

TYPES OF BUSINESS:

Telephony Products & Services
Security Monitoring Products
Network Management Products
Messaging Products
Conferencing Products
Call Recording Products
Wireless Platforms & Phones
Emergency Response Software

BRANDS/DIVISIONS/AFFILIATES:

Searchlight Capital Partners LP
MiCloud Connect CX
MiContact Center
Mitel Teamwork

CONTACTS:
Note: Officers with more than one job title may be intentionally listed here more than once.

Mary T. McDowell, CEO
James Yersh, CFO
Dave Silke, CMO
Billie Hartless, Chief Human Resources Officer
Jamshid Rezaei, CIO
Todd Abbott, Executive VP, Divisional
Graham Bevington, Executive VP, Divisional
Robert Dale, Executive VP
Tarun Loomba, Chief Product Officer

GROWTH PLANS/SPECIAL FEATURES:

Mitel Networks Corporation provides communication solutions. The firm's products are grouped into four categories: contact center, collaboration, business phone systems and phones & accessories. Mitel's contact center portfolio includes the MiCloud Connect CX (customer experience) solution that delivers personalized customer experiences via cloud architecture; and the MiContact Center, an enterprise-grade, omnichannel CX management platform created for customer-centric organizations. The collaboration product category encompasses software that enables employees to remain fluid and connected by sharing information via voice, video and instant messaging. This division's Mitel Teamwork collaborative web and mobile application is designed for the MiCloud Connect user, offering integrated tools ranging from real-time messaging to audio/video conferencing to desktop sharing. The business phone systems product category offers a range of business phone systems and solutions, including a targeted call center, unified communications and cloud communications solutions. This division's business phone systems can be arranged on-site, in the cloud and integrated with third-party applications. The phones & accessories product category offers a range of internet protocol (IP) and digital phones, consoles, conference phones and peripherals tailored for executives and/or employees. Mitel's solutions for business needs include cloud migration, on-site strategies and next-generation applications. The firm serves businesses of all sizes, whether small, medium or an enterprise. Industries served by Mitel primarily include healthcare, hospitality, government, education, field services, manufacturing, retail and sports/entertainment. Mitel is owned by Searchlight Capital Partners LP. The firm has global operations throughout North and South America, Europe, the Middle East, Africa, Asia and Asia Pacific.

FINANCIAL DATA:
Note: Data for latest year may not have been available at press time.

In U.S. $	2021	2020	2019	2018	2017	2016
Revenue	1,319,552,000	1,268,800,000	1,300,000,000	1,112,055,033	1,059,100,032	987,600,000
R&D Expense						
Operating Income						
Operating Margin %						
SGA Expense						
Net Income						
Operating Cash Flow						
Capital Expenditure						
EBITDA						
Return on Assets %						
Return on Equity %						
Debt to Equity						

CONTACT INFORMATION:

Phone: 613-592-2122   Fax: 613-592-4784
Toll-Free:
Address: 4000 Innovation Dr., Kanata, ON L2K 3K1 Canada

STOCK TICKER/OTHER:

Stock Ticker: Private   Exchange:
Employees: 3,300   Fiscal Year Ends: 04/30
Parent Company: Searchlight Capital Partners LP

SALARIES/BONUSES:

Top Exec. Salary: $ Bonus: $
Second Exec. Salary: $ Bonus: $

OTHER THOUGHTS:

Estimated Female Officers or Directors:
Hot Spot for Advancement for Women/Minorities:

Mobile Telecommunications Company KSCP (Zain Group)

www.zain.com
NAIC Code: 517210

TYPES OF BUSINESS:

Mobile Telecommunications Services
Mobile Services
Data Telecommunications Services
Investment

BRANDS/DIVISIONS/AFFILIATES:

Kuwait Investment Authority
Omantel
MADA Bahrain
Zain Ventures

CONTACTS: Note: Officers with more than one job title may be intentionally listed here more than once.

Bader AlpKharafi, CEO
Saud Al-Zaid, Chief Corp. Affairs Officer
Omar Al Omar, CEO-Zain Kuwait
Wael Ghanayem, COO
Fraser Curley, CEO-Zain Saudi Arabia

GROWTH PLANS/SPECIAL FEATURES:

Mobile Telecommunications Company KSCP, operating as Zain Group, provides mobile and data telecommunications services in seven Middle Eastern and African countries. The company provides mobile voice and data services to more than 48.3 million individual and business customers through its 4G LTE and 5G network (as of June 30, 2021). Zain Group also purchases, delivers, installs, manages and maintains mobile telephones and paging systems. Wholly-owned MADA Bahrain offers data and voice technology services. Within the Middle East and North Africa region (MENA), Zain Group has established a corporate entity focused on the delivery of drone-powered solutions. Approximately 24.6% of Zain Group is owned by Kuwait Investment Authority, and 21.9% is owned by Omantel, each of which are the company's largest shareholders. In August 2021, Zain Group created Zain Ventures, which invested in Pipe.com and swvl.com to accelerate company growth across the Middle East and beyond.

FINANCIAL DATA: Note: Data for latest year may not have been available at press time.

In U.S. $	2021	2020	2019	2018	2017	2016
Revenue	5,033,000,000	5,311,000,000	5,471,000,000	4,362,000,000	3,398,000,000	3,549,480,000
R&D Expense						
Operating Income						
Operating Margin %						
SGA Expense						
Net Income	616,000,000	605,000,000	715,000,000	649,000,000	527,000,000	517,963,000
Operating Cash Flow						
Capital Expenditure						
EBITDA						
Return on Assets %						
Return on Equity %						
Debt to Equity						

CONTACT INFORMATION:

Phone: 965 2-464-4444   Fax: 965-2-464-1111
Toll-Free:
Address: Airport Rd., Shuwaikh, Safat, 13083 Kuwait

STOCK TICKER/OTHER:

Stock Ticker: ZAIN
Employees: 7,200
Parent Company:

Exchange: Kuwait
Fiscal Year Ends: 12/31

SALARIES/BONUSES:

Top Exec. Salary: $ Bonus: $
Second Exec. Salary: $ Bonus: $

OTHER THOUGHTS:

Estimated Female Officers or Directors: 1
Hot Spot for Advancement for Women/Minorities:

Sales, profits and employees may be estimates. Financial information, benefits and other data can change quickly and may vary from those stated here.

Mobile TeleSystems PJSC

www.mtsgsm.com

NAIC Code: 517210

TYPES OF BUSINESS:

Mobile Telephone Service
Broadband Internet Services
Pay TV Services
Fixed-Line Telephone Services
Value-Added Services

GROWTH PLANS/SPECIAL FEATURES:

Mobile TeleSystems PJSC is a wireless telephone operator in Russia. The company is a large provider of fixed-line broadband and pay-TV services in the country, with a strong fixed-line presence in Moscow. The company's segment includes Telecom; Fintech and others. It generates maximum revenue from the Telecom segment.

BRANDS/DIVISIONS/AFFILIATES:

MTS Bank
MTS
JUST AI Limited
SWIPGLOBAL Limited

CONTACTS: *Note: Officers with more than one job title may be intentionally listed here more than once.*

Vyacheslav Nikolaev, CEO
Ilya V. Filatov, VP-Financial Svcs.
Olga Ziborova, VP-Mktg.
Pavel Voronin, VP-Technology
Ruslan Ibragimov, VP-Corp. & Legal Matters
Michael Hecker, VP-Strategy & Corp. Dev.
Vadim Savchenko, VP-Sales & Customer Svcs.
Konstantin Markov, Dir.
Felix Evtushenkov, Chmn.
Ivan Zolochevskiy, CEO
Valery Shorzhin, Dir.-Procurement Mgmt.

FINANCIAL DATA: *Note: Data for latest year may not have been available at press time.*

In U.S. $	2021	2020	2019	2018	2017	2016
Revenue	7,179,053,568	6,648,728,064	6,395,904,512	6,452,151,808	6,016,063,488	5,820,083,200
R&D Expense						
Operating Income						
Operating Margin %						
SGA Expense						
Net Income	852,682,496	824,995,456	728,661,824	91,994,544	761,218,880	647,527,872
Operating Cash Flow						
Capital Expenditure						
EBITDA						
Return on Assets %						
Return on Equity %						
Debt to Equity						

CONTACT INFORMATION:

Phone: 7 4952232025   Fax: 7 4959116567
Toll-Free:
Address: 4, Marksistskaya St., Moscow, 109147 Russia

STOCK TICKER/OTHER:

Stock Ticker: MBT   Exchange: NYS
Employees: 58,415   Fiscal Year Ends: 12/31
Parent Company: Sistema JSGC

SALARIES/BONUSES:

Top Exec. Salary: $ Bonus: $
Second Exec. Salary: $ Bonus: $

OTHER THOUGHTS:

Estimated Female Officers or Directors: 1
Hot Spot for Advancement for Women/Minorities:

Sales, profits and employees may be estimates. Financial information, benefits and other data can change quickly and may vary from those stated here.

Momentum Telecom

www.momentumtelecom.com

NAIC Code: 517110

TYPES OF BUSINESS:

Local Exchange Carrier
Cloud Voice
Cloud-based Applications
Business Communication Services
Wholesale Services
Telecommunication Security Services

BRANDS/DIVISIONS/AFFILIATES:

GROWTH PLANS/SPECIAL FEATURES:

Momentum Telecom develops and integrates cloud voice and cloud-based applications for enhanced business communication purposes. The company's solutions span unified communications, Microsoft Teams, managed network, mobility, collaboration, contact center and more. Momentum voice solution architecture interconnects with more than 15 major carriers. Wholesale services include white label cloud voice, broadband management, customer service, real-time subscriber provisioning and management portal, cable diagnostic expansion, fiber expansion, WiFi expansion, network maintenance, bandwidth management and more. Momentum operates one of the most advanced telecommunication platforms in the world and uses state-of-the-art technology to detect and prevent attacks at the network, host and service levels.

CONTACTS: *Note: Officers with more than one job title may be intentionally listed here more than once.*

Todd Zittrouer, CEO
William Birnie, Chief Marketing Officer
Robert Hagan, CFO
Chuck Piazza, Exec. VP-Sales & Mktg.
Heather Dromgoole, VP-Human Resources
Brian Kelley, Director
Mark Marquez, Exec. VP-IT
Andrea McHugh, Vice President

FINANCIAL DATA: *Note: Data for latest year may not have been available at press time.*

In U.S. $	2021	2020	2019	2018	2017	2016
Revenue	140,000,000	137,550,000	131,000,000	124,000,000	99,300,000	30,510,000
R&D Expense						
Operating Income						
Operating Margin %						
SGA Expense						
Net Income						
Operating Cash Flow						
Capital Expenditure						
EBITDA						
Return on Assets %						
Return on Equity %						
Debt to Equity						

CONTACT INFORMATION:

Phone: 877-251-5554   Fax:
Toll-Free:
Address: 1 Concourse Pkwy. NE, Ste. 600, Atlanta, GA 30328 United States

STOCK TICKER/OTHER:

Stock Ticker: Private   Exchange:
Employees: 400   Fiscal Year Ends: 12/31
Parent Company: MBS Holdings Inc

SALARIES/BONUSES:

Top Exec. Salary: $ Bonus: $
Second Exec. Salary: $ Bonus: $

OTHER THOUGHTS:

Estimated Female Officers or Directors: 2
Hot Spot for Advancement for Women/Minorities: Y

Sales, profits and employees may be estimates. Financial information, benefits and other data can change quickly and may vary from those stated here.

Net2Phone Inc

www.net2phone.com

NAIC Code: 517110

TYPES OF BUSINESS:

VoIP Service Providers
Outsourced Telecommunications Services
Calling Card Services
ISP Solutions

BRANDS/DIVISIONS/AFFILIATES:

IDT Corporation
Huddle
N2P Remote

CONTACTS: *Note: Officers with more than one job title may be intentionally listed here more than once.*

Jonah Fink, Pres.
Zali Ritholtz, COO
Dovey Forman, Dir.-Operations
George Longyear, Dir.-Sales
Denise D'Arienzo, Dir.-Mktg.
Jeffrey Skelton, CTO
Andreea Cojocariu, Dir.-Digital Mktg.

GROWTH PLANS/SPECIAL FEATURES:

Net2Phone, Inc. is a provider of retail voice over internet protocol (VoIP) cloud telephone products and services. The firm enables businesses to handle 450,000 simultaneous calls, but serves businesses of any size, whether small or large. Net2Phone's onboarding and implementation team sets up each customer's phone system as needed/required, and provides ongoing support services. The company offers products such as hosted PBX cloud phone systems, Huddle, N2P Remote, Microsoft Teams and SIP (session initiation protocol) trunking for business and call centers. Hosted business PBX (private branch exchange) is a phone system that is completely hosted, managed and maintained by Net2Phone. It features VoIP solutions for a fully-featured cloud PBX, and has a single platform that enables businesses to transfer calls, set up auto-attendant to answer and route calls, manage multiple offices and keep businesses connected even while on-the-go. The hosted PBX phone service includes unlimited calls, both domestic and international, for one low monthly price. Huddle enables virtual face-to-face communications globally. N2P Remote enables work-from-home, work-from-anywhere (remote) connectivity. Microsoft Teams offers a direct routing phone application for productivity, collaboration and content sharing. SIP trunking is a VoIP technology and streaming media service by which internet telephone service providers deliver telephone services and unified communications to customers equipped with SIP-based PBX facilities. This allows businesses and call centers to use their existing internet connection as a phone line, instead of physical leased line, providing cost-savings. Net2Phone's plans offer unlimited calling to/from over 30 countries and virtual phone numbers in over 50 countries. Net2Phone is a subsidiary of IDT Corporation, a holding company with strong interest in telecommunications.

FINANCIAL DATA: *Note: Data for latest year may not have been available at press time.*

In U.S. $	2021	2020	2019	2018	2017	2016
Revenue	43,897,000	31,781,000	47,264,000	34,857,000	29,450,000	
R&D Expense						
Operating Income						
Operating Margin %						
SGA Expense						
Net Income						
Operating Cash Flow						
Capital Expenditure						
EBITDA						
Return on Assets %						
Return on Equity %						
Debt to Equity						

CONTACT INFORMATION:

Phone: 973-438-3111   Fax:
Toll-Free: 866-978-8260
Address: 520 Broad St., Newark, NJ 07102 United States

STOCK TICKER/OTHER:

Stock Ticker: Subsidiary   Exchange:
Employees: 800   Fiscal Year Ends: 07/31
Parent Company: IDT Corporation

SALARIES/BONUSES:

Top Exec. Salary: $ Bonus: $
Second Exec. Salary: $ Bonus: $

OTHER THOUGHTS:

Estimated Female Officers or Directors:
Hot Spot for Advancement for Women/Minorities:

NETGEAR Inc

www.netgear.com

NAIC Code: 334210A

TYPES OF BUSINESS:

Networking Equipment
Wireless Networking Products
Broadband Products
Entertainment Management Software
Security Products
Wi-Fi Phones
Gaming Router

BRANDS/DIVISIONS/AFFILIATES:

Orbi Voice
Meural
Nighthawk

GROWTH PLANS/SPECIAL FEATURES:

Netgear Inc is a provider of networking solutions. The reportable segments of the company are connected home, and Small and Medium Business (SMB). The Connected Home segment focuses on consumers and consists of high performance, dependable and easy-to-use 4G/5G mobile, Wi-Fi internet networking solutions and smart devices such as Orbi Voice smart speakers and Meural digital canvas; and SMB focused on small and medium-sized businesses and consists of business networking, storage, wireless LAN and security solutions that bring enterprise-class functionality to small and medium-sized businesses at an affordable price.

CONTACTS: *Note: Officers with more than one job title may be intentionally listed here more than once.*

Patrick Lo, CEO
Bryan Murray, CFO
Andrew Kim, Chief Legal Officer
Mark Merrill, Co-Founder
Michael Falcon, COO
David Henry, Director
Martin Westhead, Other Executive Officer
Vikram Mehta, Senior VP, Divisional
Michael Werdann, Senior VP, Divisional
Tamesa Rogers, Senior VP, Divisional
Heidi Cormack, Senior VP, Divisional

FINANCIAL DATA: *Note: Data for latest year may not have been available at press time.*

In U.S. $	2021	2020	2019	2018	2017	2016
Revenue	1,168,073,000	1,255,202,000	998,763,000	1,058,816,000	1,039,169,000	1,143,445,000
R&D Expense	92,967,000	88,788,000	77,982,000	82,416,000	71,893,000	70,904,000
Operating Income	66,597,000	75,544,000	26,188,000	38,714,000	42,553,000	109,411,000
Operating Margin %	.06%	.06%	.03%	.04%	.04%	.10%
SGA Expense	205,620,000	209,002,000	187,582,000	217,426,000	193,025,000	193,587,000
Net Income	49,387,000	58,293,000	25,791,000	-9,162,000	19,436,000	75,851,000
Operating Cash Flow	-4,579,000	181,150,000	13,525,000	-103,211,000	87,524,000	118,181,000
Capital Expenditure	9,864,000	10,296,000	14,230,000	12,251,000	10,140,000	10,231,000
EBITDA	80,503,000	94,475,000	45,594,000	57,565,000	65,082,000	139,343,000
Return on Assets %	.05%	.06%	.03%	- .01%	.02%	.07%
Return on Equity %	.07%	.09%	.04%	- .01%	.03%	.10%
Debt to Equity	.03%	0.037	0.042			

CONTACT INFORMATION:

Phone: 408 907-8000   Fax: 408 907-8097
Toll-Free: 888-638-4327
Address: 350 E. Plumeria Dr., San Jose, CA 95134 United States

STOCK TICKER/OTHER:

Stock Ticker: NTGR   Exchange: NAS
Employees: 809   Fiscal Year Ends: 12/31
Parent Company:

SALARIES/BONUSES:

Top Exec. Salary: $935,791   Bonus: $
Second Exec. Salary: Bonus: $
$535,932

OTHER THOUGHTS:

Estimated Female Officers or Directors: 4
Hot Spot for Advancement for Women/Minorities: Y

Sales, profits and employees may be estimates. Financial information, benefits and other data can change quickly and may vary from those stated here.

NetScout Systems Inc

www.netscout.com

NAIC Code: 511210B

TYPES OF BUSINESS:

Computer Software, Network Management (IT), System Testing & Storage
Digital Cyber Security
Digital Business Network Security Solutions
Cyber Threat Solutions
Technologies
Business Transformation Solutions
Digital Transformation Solutions
Software Security Solutions

BRANDS/DIVISIONS/AFFILIATES:

Ngenius
Omnis
Arbor
Spectra

GROWTH PLANS/SPECIAL FEATURES:

NetScout Systems Inc is a provider of service assurance and cybersecurity solutions to enterprise and government networks. It bases its solutions on proprietary adaptive service intelligence technology, which helps customers monitor and identify performance issues and provides insight into network-based security threats. These solutions also deliver real-time and historical information, which provides insight to restore service and understand the quality of user experience. The company derives revenue primarily from the sale of network management tools and security solutions. Its geographical regions include USA, Europe, Asia, and Rest of the World.

CONTACTS: Note: Officers with more than one job title may be intentionally listed here more than once.

Anil Singhal, CEO
Jean Bua, CFO
Michael Szabados, COO
John Downing, Executive VP, Divisional

FINANCIAL DATA: Note: Data for latest year may not have been available at press time.

In U.S. $	2021	2020	2019	2018	2017	2016
Revenue	831,282,000	891,820,000	909,918,000	986,787,000	1,162,112,000	
R&D Expense	179,163,000	188,294,000	203,588,000	215,076,000	232,701,000	
Operating Income	37,192,000	20,312,000	-7,544,000	1,151,000	66,065,000	
Operating Margin %	.04%	.02%	-.01%	.00%	.06%	
SGA Expense	331,699,000	376,517,000	385,442,000	422,015,000	447,066,000	
Net Income	19,352,000	-2,754,000	-73,324,000	79,812,000	33,291,000	
Operating Cash Flow	213,921,000	225,023,000	149,838,000	222,454,000	226,764,000	
Capital Expenditure	16,523,000	19,922,000	23,526,000	16,594,000	32,148,000	
EBITDA	139,011,000	138,625,000	71,109,000	147,477,000	222,232,000	
Return on Assets %	.01%	.00%	-.02%	.02%	.01%	
Return on Equity %	.01%	.00%	-.04%	.04%	.01%	
Debt to Equity	.21%	0.269	0.266	0.29	0.123	

CONTACT INFORMATION:

Phone: 978 614-4000   Fax: 978 614-4004
Toll-Free: 800-357-7666
Address: 310 Littleton Rd., Westford, MA 01886 United States

STOCK TICKER/OTHER:

Stock Ticker: NTCT   Exchange: NAS
Employees: 2,331   Fiscal Year Ends: 03/31
Parent Company:

SALARIES/BONUSES:

Top Exec. Salary: $577,500   Bonus: $
Second Exec. Salary: $423,500   Bonus: $

OTHER THOUGHTS:

Estimated Female Officers or Directors: 2
Hot Spot for Advancement for Women/Minorities:

Sales, profits and employees may be estimates. Financial information, benefits and other data can change quickly and may vary from those stated here.

Neustar Inc

NAIC Code: 518210

TYPES OF BUSINESS:

Clearinghouse Services
Marketing Services
Software
Risk Solutions
Communications Solutions
Security Solutions
Domain Name Services

BRANDS/DIVISIONS/AFFILIATES:

TransUnion

CONTACTS: *Note: Officers with more than one job title may be intentionally listed here more than once.*

Charles E. Gottdiener, CEO
Dorean Kass, Chief Sales Officer
Carey Pellock, Chief Human Resources Officer
Venkat Achanta, CTO
Leonard Kennedy, General Counsel
Brian Foster, Senior VP, Divisional
Venkat Achanta, Senior VP, Divisional
Steve Edwards, Senior VP, Divisional
Henry (Hank) Skorny, Senior VP, Divisional

GROWTH PLANS/SPECIAL FEATURES:

Neustar, Inc., owned by TransUnion, provides data analytic and modeling software that help marketers send timely an relevant messages to target audiences. The firm's solutions ar grouped into four categories: marketing, risk, communication and security. Marketing solutions encompass custome analytics, customer experience and customer intelligence. Ris solutions address compliance intelligence, fraud intelligenc and operational intelligence. Communications solutions includ branded content management, caller intelligence, globa numbering insights, number management and orde management. Security solutions include application security domain name service (DNS), security intelligence and websit performance management. Neustar helps organization across a diverse set of industries understand their customers personalize their marketing and protect themselves from cybe crime. Industries served by the firm primarily include retai consumer packaged goods, quick service restaurants, financia services, insurance, automotive, travel, hospitality communications, technology and government. Its solutions ca also be incorporated into the pharmaceutical/biosciences media/advertising and healthcare industries, among others Headquartered in Virginia, Neustar has additional office throughout the U.S., as well as in India, Costa Rica, Germany India, and the U.K. In late-2020, Neustar completed th divestiture of its registry business to GoDaddy, Inc. December 2021, TransUnion completed its $3.1 billic acquisition of Neustar.

FINANCIAL DATA: *Note: Data for latest year may not have been available at press time.*

In U.S. $	2021	2020	2019	2018	2017	2016
Revenue	575,000,000	420,000,000	400,000,000			
R&D Expense						
Operating Income						
Operating Margin %						
SGA Expense						
Net Income						
Operating Cash Flow						
Capital Expenditure						
EBITDA						
Return on Assets %						
Return on Equity %						
Debt to Equity						

CONTACT INFORMATION:

Phone: 571 434-5400   Fax: 571 434-5401
Toll-Free: 855-683-2677
Address: 1906 Reston Metro Plz., Ste. 500, Reston, VA 20190 United States

STOCK TICKER/OTHER:

Stock Ticker: Subsidiary   Exchange:
Employees: 1,988   Fiscal Year Ends: 12/31
Parent Company: TransUnion

SALARIES/BONUSES:

Top Exec. Salary: $ Bonus: $
Second Exec. Salary: $ Bonus: $

OTHER THOUGHTS:

Estimated Female Officers or Directors: 3
Hot Spot for Advancement for Women/Minorities: Y

Newfold Digital Inc

newfold.com

NAIC Code: 518210

TYPES OF BUSINESS:

Web Hosting Products & Services
Web Design Services
Search Engine
eCommerce
Advertising
Domain Name

BRANDS/DIVISIONS/AFFILIATES:

Siris Capital Group LLC
Clearlake Capital Group LP
Web.com Online Marketing
Web.com Contractor Services
Network Solutions
Register.com
Name Jet
bluehost inc.

CONTACTS: *Note: Officers with more than one job title may be intentionally listed here more than once.*

Sharon Rowlands, CEO
Christine Barry, COO
Christina Clohecy, CFO
Paula Drum, CMO
Deb Myers, Chief People Officer
Michael Bouchet, CIO
Roseann Duran, Executive VP

GROWTH PLANS/SPECIAL FEATURES:

Newfold Digital Inc, a result of the combination of Web.com Group, Inc. and Endurance Web Presence, is a web technology company servicing millions of customers worldwide. The firm's portfolio of brands include Web.com, Network Solutions, Register.com, Name Jet, bluehost inc. Web.com Online Marketing Agency offers search engine optimization and placement solutions. Web.com Contractor Services is a online networking platform that displays over 3,000 remodeling contractors who carry out bathroom remodeling, kitchen remodeling, attic remodeling and basement remodeling projects, from which homeowners can locate and hire. Network Solutions helps small businesses to start and market their businesses on the web, offering a full range of web-related services. Register.com is a leading provider of global domain name registration, website design and management services. The platform is also a business web hosting provider. NameJet is the premier aftermarket domain name service. Bluehost inc. is a web hosting solutions company. In February 2021, Siris Capital Group, LLC and Clearlake Capital Group, L.P. announced the formation of Newfold Digital through the combination of Endurance Web Presence and Web.com Group, Inc.

FINANCIAL DATA: *Note: Data for latest year may not have been available at press time.*

In U.S. $	2021	2020	2019	2018	2017	2016
Revenue	866,580,000	833,250,000	825,000,000	798,000,000	749,260,992	710,505,024
R&D Expense						
Operating Income						
Operating Margin %						
SGA Expense						
Net Income			55,795,611	54,701,580	53,629,000	3,990,000
Operating Cash Flow						
Capital Expenditure						
EBITDA						
Return on Assets %						
Return on Equity %						
Debt to Equity						

CONTACT INFORMATION:

Phone: 904 680-6600 Fax: 904 880-0350
Toll-Free:
Address: 5335 Gate Pkwy., Jacksonville, FL 32256 United States

STOCK TICKER/OTHER:

Stock Ticker: Private Exchange:
Employees: 3,600 Fiscal Year Ends: 12/31
Parent Company: Siris Capital Group LLC

SALARIES/BONUSES:

Top Exec. Salary: $ Bonus: $
Second Exec. Salary: $ Bonus: $

OTHER THOUGHTS:

Estimated Female Officers or Directors: 3
Hot Spot for Advancement for Women/Minorities: Y

Sales, profits and employees may be estimates. Financial information, benefits and other data can change quickly and may vary from those stated here.

NextPlat Corp

NAIC Code: 517410

nextplat.com

TYPES OF BUSINESS:

Satellite Telecommunications Resellers
Satellite Internet
Communication Services
Asset and Personnel Tracking
Mobile Satellite Communication
Ecommerce
Internet of Things
Satellite Constellations

BRANDS/DIVISIONS/AFFILIATES:

Global Telesat Communications Ltd
Orbital Satcom Corp
GTCTrack
Orbsat Corp

GROWTH PLANS/SPECIAL FEATURES:

NextPlat Corp is a global e-commerce platform company created to capitalize on multiple sectors and markets for physical and digital assets. The company intends to collaborate with businesses, optimizing their ability to sell their goods online, domestically, and internationally, and enabling customers and partners to maximize their e-commerce presence and revenue.

CONTACTS: *Note: Officers with more than one job title may be intentionally listed here more than once.*

Charles M. Fernandez, CEO
David Phipps, Pres
Charles Fernandez, CEO
Paul R. Thomson, CFO
Andrew Cohen, Sr. VP-Oper.
Douglas S. Ellenoff, Chief Business Development Strategist
Louis Wise, Chief Technology Advisor
Douglas Ellenoff, Director
Andrew Cohen, Senior VP, Divisional
Charles M. Fernandez, Chmn.

FINANCIAL DATA: *Note: Data for latest year may not have been available at press time.*

In U.S. $	2021	2020	2019	2018	2017	2016
Revenue	7,739,910	5,689,796	5,869,558	5,726,572	6,004,955	4,698,638
R&D Expense						
Operating Income	-6,622,333	-2,033,880	-1,111,328	-1,129,636	-1,557,606	-1,630,524
Operating Margin %	- .86%	- .36%	- .19%	- .20%	- .26%	- .35%
SGA Expense	8,164,954	2,964,274	2,059,378	1,875,596	2,423,959	2,413,843
Net Income	-8,107,662	-2,763,375	-1,379,756	-1,194,706	-3,939,309	-2,589,923
Operating Cash Flow	-4,092,090	-836,980	-659,203	-590,185	-315,245	-1,044,036
Capital Expenditure	229,307	34,903	70,194	30,331	33,193	26,448
EBITDA	-6,290,297	-1,418,352	-800,634	-897,198	-3,630,633	-1,694,960
Return on Assets %	- .70%	-1.09%	- .57%	- .43%	-1.29%	- .71%
Return on Equity %	- .90%	-4.59%	-1.37%	- .74%	-1.89%	-1.00%
Debt to Equity	.01%	0.978	0.536			

CONTACT INFORMATION:

Phone: 305-560-5355   Fax:
Toll-Free:
Address: 3260 Mary St., Ste. 410, Coconut Grove, FL 33133 United States

STOCK TICKER/OTHER:

Stock Ticker: NXPL
Employees: 12
Parent Company:

Exchange: NAS
Fiscal Year Ends: 12/31

SALARIES/BONUSES:

Top Exec. Salary: $251,133   Bonus: $
Second Exec. Salary: $142,923   Bonus: $

OTHER THOUGHTS:

Estimated Female Officers or Directors:
Hot Spot for Advancement for Women/Minorities:

Sales, profits and employees may be estimates. Financial information, benefits and other data can change quickly and may vary from those stated here.

Nippon Telegraph and Telephone Corporation (NTT) group.ntt
NAIC Code: 517110

TYPES OF BUSINESS:
Local Telephone Service
Long-Distance Service
Internet Service
Information Technology Services
Cellular Phone Service
Wireless Internet

GROWTH PLANS/SPECIAL FEATURES:

NTT owns NTT DoCoMo, the largest wireless operator in Japan, with 80.3 million subscribers. It also owns NTT East and NTT West, the two regional incumbent fixed-line operators in Japan, with about 13 million traditional fixed-line and 24 million broadband lines (around 72% of which are wholesaled). The firm also provides IT and communications systems integration via NTT Communications and 52.4%-owned NTT Data.

BRANDS/DIVISIONS/AFFILIATES:
NTT East
NTT West
NTT Communications Corporation
NTT Ltd
NTT Data Corporation
NTT DoCoMo Inc
NTT Urban Solutions Inc
NTT Anode Energy Corporation

CONTACTS: Note: Officers with more than one job title may be intentionally listed here more than once.
Jun Sawada, CEO
Akira Shimada, CFO
Hiromichi Shinohara, Dir.-R&D Planning
Hiroki Watanabe, Chief Compliance Officer
Mitsuyoshi Kobayashi, Sr. VP-Strategic Bus. Dev. Div.
Yoshikiyo Sakai, Sr. VP-Finance & Acct.
Akira Shimada, Sr. VP-Gen. Affairs & Internal Control
Hiroshi Tsujigami, Sr. VP-Corp. Strategy
Hiromichi Shinohara, Chmn.
Tsunehisa Okuno, Sr. VP-Global Bus. Office

FINANCIAL DATA: Note: Data for latest year may not have been available at press time.

In U.S. $	2021	2020	2019	2018	2017	2016
Revenue	88,578,810,000	88,248,400,000	88,103,250,000	87,378,730,000	84,478,020,000	
R&D Expense						
Operating Income	12,585,950,000	11,708,740,000	13,499,190,000	13,489,730,000	11,967,080,000	
Operating Margin %	.14%	.13%	.15%	.15%	.14%	
SGA Expense					20,667,610,000	
Net Income	6,794,579,000	6,343,118,000	6,337,593,000	6,658,907,000	5,933,915,000	
Operating Cash Flow	22,315,810,000	22,213,070,000	17,844,540,000	18,846,560,000	21,635,690,000	
Capital Expenditure	13,315,260,000	13,767,060,000	12,402,480,000	12,964,350,000	12,620,940,000	
EBITDA	23,810,130,000	22,884,190,000	22,559,110,000	23,617,570,000	22,454,500,000	
Return on Assets %	.04%	.04%	.04%	.04%	.04%	
Return on Equity %	.11%	.09%	.09%	.10%	.09%	
Debt to Equity	.66%	0.281	0.309	0.326	0.369	

CONTACT INFORMATION:
Phone: 81 352055581   Fax: 81 352055589
Toll-Free:
Address: 5-1 Otemachi 1-Chome, Chiyoda-ku, Tokyo, 100-8116 Japan

STOCK TICKER/OTHER:
Stock Ticker: NPPXF   Exchange: PINX
Employees: 274,850   Fiscal Year Ends: 03/31
Parent Company:

SALARIES/BONUSES:
Top Exec. Salary: $ Bonus: $
Second Exec. Salary: $ Bonus: $

OTHER THOUGHTS:
Estimated Female Officers or Directors:
Hot Spot for Advancement for Women/Minorities:

Sales, profits and employees may be estimates. Financial information, benefits and other data can change quickly and may vary from those stated here.

Nokia Corporation
NAIC Code: 334220

www.nokia.com

TYPES OF BUSINESS:
Smartphones and Cellphones
Network Systems & Services
Internet Software & Services
Multimedia Equipment
Brand Licensing
Collaboration Devices
5G
Innovation

BRANDS/DIVISIONS/AFFILIATES:
Nokia Bell Labs

GROWTH PLANS/SPECIAL FEATURES:
Nokia is a primary vendor in the telecommunication equipment industry. The company's network business derive revenue from selling wireless and fixed-line hardware software, and services. Nokia's main operating segments ar mobile networks, network infrastructure, cloud and networ services, and Nokia technologies. The company headquartered in Espoo, Finland, operates on a global scale with most of its revenue from communication service providers

CONTACTS: Note: Officers with more than one job title may be intentionally listed here more than once.
Pekka Lundmark, CEO
Marco Wiren, CFO
Stephanie Werner-Dietz, Chief People Officer
Nishant Batra, CTO
Louise Pentland, Chief Legal Officer
Juha Rutkiranta, Exec. VP-Oper.
Kai Oistamo, Chief Dev. Officer
Stephen Elop, Exec. VP-Devices & Svcs.
Timo Toikkanen, Exec. VP-Mobile Phones
Jo Harlow, Exec. VP-Smart Devices

FINANCIAL DATA: Note: Data for latest year may not have been available at press time.

In U.S. $	2021	2020	2019	2018	2017	2016
Revenue	22,640,780,000	22,283,860,000	23,775,770,000	23,008,910,000	23,604,450,000	24,108,220,000
R&D Expense	4,297,282,000	4,167,771,000	4,621,565,000	4,871,408,000	5,013,155,000	5,095,756,000
Operating Income	1,825,376,000	1,041,178,000	514,980,300	-98,917,010	643,470,500	-150,924,900
Operating Margin %	.08%	.05%	.02%	.00%	.03%	- .01%
SGA Expense	2,847,178,000	2,955,273,000	3,282,617,000	3,619,139,000	3,686,443,000	3,841,447,000
Net Income	1,655,075,000	-2,572,862,000	7,138,341	-341,620,600	-1,523,526,000	-781,138,400
Operating Cash Flow	2,676,878,000	1,793,763,000	397,707,600	367,114,700	1,846,791,000	-1,482,735,000
Capital Expenditure	571,067,300	488,466,500	703,636,500	685,280,700	612,877,600	486,426,900
EBITDA	3,290,775,000	2,160,878,000	2,170,056,000	1,404,214,000	1,621,423,000	552,711,600
Return on Assets %	.04%	- .07%	.00%	- .01%	- .03%	- .02%
Return on Equity %	.11%	- .18%	.00%	- .02%	- .08%	- .05%
Debt to Equity	.31%	0.46	0.31	0.185	0.214	0.182

CONTACT INFORMATION:
Phone: 358 10-44-88-000   Fax: 358-10-44-81-002
Toll-Free:
Address: Karakaari 7A, Espoo, FI-02610 Finland

STOCK TICKER/OTHER:
Stock Ticker: NOK
Employees: 87,900
Parent Company:

Exchange: NYS
Fiscal Year Ends: 12/31

SALARIES/BONUSES:
Top Exec. Salary: $1,325,692   Bonus: $
Second Exec. Salary: Bonus: $
$774,372

OTHER THOUGHTS:
Estimated Female Officers or Directors: 4
Hot Spot for Advancement for Women/Minorities: Y

Sales, profits and employees may be estimates. Financial information, benefits and other data can change quickly and may vary from those stated here.

Nortel Inversora SA

www.nortelsa.com.ar

NAIC Code: 517110

TYPES OF BUSINESS:

Wired Telecommunications Carriers
Telecommunications Services
Telephone Services

BRANDS/DIVISIONS/AFFILIATES:

Fintech Telecom LLC
Telecom Argentina SA
Telecom Personal SA
Personal Envios SA
Micro Sistemas SA
Telecom Argentina USA Inc
Nucleo SA

CONTACTS: Note: Officers with more than one job title may be intentionally listed here more than once.

Baruki Luis Alberto Gonzalez, Chmn.

GROWTH PLANS/SPECIAL FEATURES:

Nortel Inversora SA is a stock corporation organized by a consortium of Argentinian and international investors to acquire a controlling interest in the common stock of Telecom Argentina SA. Nortel owns a 54.74% stake in Telecom Argentina, which provides fixed-link public telecommunications services, as well as basic telephone services to Iran (via a roaming mobile services agreement with Mobile Company of Iran (MCI). Telecom's international telecommunications services agreements with international carriers (fixed services) covers delivery of traffic to Iran through non-Iranian carriers. MCI allows its mobile customers to use their mobile device on a network outside their subscriber's home network. Additional subsidiaries owned by the firm include Telecom Personal SA, Personal Envios SA and Micro Sistemas SA, each of which provide mobile telecommunication services in Argentina; Telecom Argentina USA, Inc., which provides fixed line services in the U.S.; and Nucleo SA, which provides mobile telecommunication services in Paraguay. Nortel operates as a majority-owned subsidiary of Fintech Telecom, LLC.

FINANCIAL DATA: Note: Data for latest year may not have been available at press time.

In U.S. $	2021	2020	2019	2018	2017	2016
Revenue	3,000,000,000	2,975,813,366	3,048,989,105	3,123,964,247	3,200,783,040	3,048,364,800
R&D Expense						
Operating Income						
Operating Margin %						
SGA Expense						
Net Income						
Operating Cash Flow						
Capital Expenditure						
EBITDA						
Return on Assets %						
Return on Equity %						
Debt to Equity						

CONTACT INFORMATION:

Phone: 541 49683631   Fax: 54 1143131298
Toll-Free:
Address: Alicia Moreau de Justo 50, Buenos Aires, AG 1107 Argentina

STOCK TICKER/OTHER:

Stock Ticker: Private   Exchange:
Employees: 15,500   Fiscal Year Ends: 12/31
Parent Company: Fintech Telecom LLC

SALARIES/BONUSES:

Top Exec. Salary: $ Bonus: $
Second Exec. Salary: $ Bonus: $

OTHER THOUGHTS:

Estimated Female Officers or Directors:
Hot Spot for Advancement for Women/Minorities:

Sales, profits and employees may be estimates. Financial information, benefits and other data can change quickly and may vary from those stated here.

NTT DOCOMO Inc

www.nttdocomo.co.jp/english

NAIC Code: 517210

TYPES OF BUSINESS:

Mobile Telephone Service
Mobile
5G
Telecommunications
Tablets
Data Communication
Advanced Wireless Networks
LTE Networks

BRANDS/DIVISIONS/AFFILIATES:

Nippon Telegraph and Telephone Corporation (NTT)
docomo

GROWTH PLANS/SPECIAL FEATURES:

NTT DOCOMO, Inc. is a leading telecommunications company in Japan. The firm serves more than 73 million customers in Japan via advanced wireless networks, including long-term evolution (LTE) and LTE-advanced networks. NTT DOCOMO is a world-leading developer of 5G networks, deploying the network and network function virtualization (NFV) and other related technologies during the 2020s. Outside Japan, NTT DOCOMO provides technical and operational expertise to mobile operators and other partner companies, and contributes to the global standardization of new mobile technologies. Products by NTT DOCOMO include iPhones, iPads, docomo 4G and 5G smartphones, docomo tablets and other phones, watches, data communications products, drivers support products and more. The parent company of the firm is Nippon Telegraph and Telephone Corporation (NTT).

CONTACTS: *Note: Officers with more than one job title may be intentionally listed here more than once.*

Motoyuki Ii, CEO
Seizo Onoe, CTO
Fumio Iwasaki, Sr. Exec. VP
Tsutomu Shindou, Exec. VP
Takashi Tanaka, Exec. VP
Kazuhiro Yoshizawa, Exec. VP

FINANCIAL DATA: *Note: Data for latest year may not have been available at press time.*

In U.S. $	2021	2020	2019	2018	2017	2016
Revenue	42,874,999,988	43,134,016,602	46,076,108,800	45,396,140,032	42,620,825,600	43,089,641,472
R&D Expense						
Operating Income						
Operating Margin %						
SGA Expense						
Net Income	8,286,093,708	5,485,300,000	6,339,808,768	7,086,691,328	6,066,396,160	5,219,565,568
Operating Cash Flow						
Capital Expenditure						
EBITDA						
Return on Assets %						
Return on Equity %						
Debt to Equity						

CONTACT INFORMATION:

Phone: 81 351561111   Fax: 81 351560271
Toll-Free:
Address: 2-11-1, Nagata-cho, Chiyoda-ku, Tokyo, 100-6150 Japan

STOCK TICKER/OTHER:

Stock Ticker: Subsidiary   Exchange:
Employees: 38,000   Fiscal Year Ends: 03/31
Parent Company: Nippon Telegraph and Telephone Corporation (NTT)

SALARIES/BONUSES:

Top Exec. Salary: $ Bonus: $
Second Exec. Salary: $ Bonus: $

OTHER THOUGHTS:

Estimated Female Officers or Directors:
Hot Spot for Advancement for Women/Minorities:

O2 Czech Republic AS

www.o2.cz/osobni

NAIC Code: 517110

TYPES OF BUSINESS:

Local & Long-Distance Telephone Service
Telecommunications
Voice Services
Internet Services
Data Services
Smart Box Modems
Product Development and Innovation
Financial Services and Insurance

BRANDS/DIVISIONS/AFFILIATES:

PPF Group NV

GROWTH PLANS/SPECIAL FEATURES:

O2 Czech Republic AS is a leading provider of telecommunications services to the Czech market. The firm provides voice, internet and data services to consumers, businesses and enterprises. O2 continues to build and roll out its fifth generation (5G) network. The company develops many things, including the O2 smart box modem for fixed connectivity and a range of financial services such as equipment insurance and mobile travel insurance. Its O2 TV services include internet television broadcasting, which offers a number of exclusive sports distribution services. O2 is also a leading player in hosting and cloud services, managed services and information and communications technology (ICT). O2 operates as a subsidiary of PPF Group NV.

CONTACTS:
Note: Officers with more than one job title may be intentionally listed here more than once.

Jindrich Fremuth, CEO
Tomas Kouril, CFO
Richard Siebenstich, CCO
Pavel Milec, Dir.-Human Resources
Jan Hruska, CTO
Jakub Chytil, Dir.-Legal Affairs
Petr Slovacek, Dir.-Oper. Div.
Felix Geyr, Dir.-Strategy & Bus. Dev.
Dana Dvorakova, Dir.-Corp. Comm.
Martin Bek, Dir.-Support Units
Frantisek Schneider, Dir.-Bus. Div.
Jindrich Fremuth, Dir.-Consumer Div.
Jindrich Fremuth, Chmn.

FINANCIAL DATA:
Note: Data for latest year may not have been available at press time.

In U.S. $	2021	2020	2019	2018	2017	2016
Revenue	1,851,494,484	1,440,770,000	1,339,520,000	1,685,650,000	1,765,900,000	1,462,680,000
R&D Expense						
Operating Income						
Operating Margin %						
SGA Expense						
Net Income	293,312,356	271,190,000	235,510,000	241,694,000	261,637,000	205,006,000
Operating Cash Flow						
Capital Expenditure						
EBITDA						
Return on Assets %						
Return on Equity %						
Debt to Equity						

CONTACT INFORMATION:

Phone: 420-840-114-114   Fax:
Toll-Free:
Address: Praha 4, Michle, Za Brumlovkou 266/2, Prague, 140 22 Czech Republic

STOCK TICKER/OTHER:

Stock Ticker: Private
Employees: 5,000
Parent Company: PPF Group NV

Exchange:
Fiscal Year Ends: 12/31

SALARIES/BONUSES:

Top Exec. Salary: $ Bonus: $
Second Exec. Salary: $ Bonus: $

OTHER THOUGHTS:

Estimated Female Officers or Directors: 1
Hot Spot for Advancement for Women/Minorities:

Oblong Inc
NAIC Code: 517110

www.oblong.com

TYPES OF BUSINESS:
Videoconferencing Services
Collaboration Technologies
Managed Services
Video Collaboration
Network Applications
Cloud-based Services

BRANDS/DIVISIONS/AFFILIATES:
Mezzanine

GROWTH PLANS/SPECIAL FEATURES:
Oblong Inc delivers visual solutions for collaboration an
computing environments. The company's flagship produ
Mezzanine is the technology platform that defines computing
simultaneous multi-user, multi-screen, multi-device, mult
location for dynamic and immersive visual collaboratior
Mezzanine transforms routine meetings and workflows int
agile, engaging experiences by making data visible an
accessible in a collaborative setting. Mezz-In is a conten
sharing experience. Mezzanine enables multiple, concurrer
pieces of content to be shared, manipulated, created an
captured across distance. It serves various industries, digita
business consulting, federal agency, financial services
healthcare, pharmaceutical, marketing and advertising and o
and gas.

CONTACTS: Note: Officers with more than one job title may be intentionally listed here more than once.
Peter Holst, CEO
David Clark, CFO
Pete Hawkes, Senior VP, Divisional

FINANCIAL DATA: Note: Data for latest year may not have been available at press time.

In U.S. $	2021	2020	2019	2018	2017	2016
Revenue	7,739,000	15,333,000	12,827,000	12,557,000	14,799,000	19,218,000
R&D Expense	2,913,000	3,711,000	2,023,000	921,000	1,148,000	1,117,000
Operating Income	-11,489,000	-8,914,000	-5,257,000	-1,647,000	-654,000	-1,410,000
Operating Margin %	-1.48%	- .58%	- .41%	- .13%	- .04%	- .07%
SGA Expense	8,558,000	10,116,000	7,313,000	4,930,000	4,078,000	5,870,000
Net Income	-9,051,000	-7,421,000	-7,761,000	-7,168,000	5,785,000	-3,533,000
Operating Cash Flow	-7,732,000	-6,566,000	-3,253,000	-1,155,000	1,609,000	183,000
Capital Expenditure	50,000	38,000	45,000	335,000	133,000	382,000
EBITDA	-5,888,000	-2,825,000	-6,253,000	-5,820,000	8,299,000	-126,000
Return on Assets %	- .34%	- .26%	- .36%	- .61%	.36%	- .19%
Return on Equity %	- .41%	- .38%	- .54%	- .75%	.71%	- .69%
Debt to Equity	.01%	0.046	0.223		0.03	

CONTACT INFORMATION:
Phone: 303-640-3838   Fax:
Toll-Free: 866-456-9764
Address: 25587 Conifer Road, Conifer, CO 80433 United States

STOCK TICKER/OTHER:
Stock Ticker: OBLG   Exchange: NAS
Employees: 49   Fiscal Year Ends: 12/31
Parent Company:

SALARIES/BONUSES:
Top Exec. Salary: $246,340   Bonus: $200,000
Second Exec. Salary: Bonus: $100,000
$242,164

OTHER THOUGHTS:
Estimated Female Officers or Directors: 2
Hot Spot for Advancement for Women/Minorities:

Oi SA

NAIC Code: 517110

TYPES OF BUSINESS:

Telecommunications
Fixed & Mobile Telephony Services
Data Transmission Services
Voice Services
Remote Access
Television Services
Internet Portal
Fiber Optics

BRANDS/DIVISIONS/AFFILIATES:

Oi TV

CONTACTS: Note: Officers with more than one job title may be intentionally listed here more than once.

Rodrigo Modesto de Abreu, CEO
Eurico Telles, Exec. Officer-Legal
Fabiano Castello, Internal Audit

GROWTH PLANS/SPECIAL FEATURES:

Oi SA is a leading integrated telecommunications service provider in Brazil. The firm offers a range of services for residential and business-to-business (B2B) consumers. Residential services include local and long-distance fixed-line voice, broadband and Pay-tv. This division serves more than 10.5 million fixed lines, and owns more than 248,548 miles (400,000 km) of installed fiber optic cable. Oi offers a variety of high-speed broadband services, serving over 4 million digital subscriber lines. Pay-tv services are offered under the Oi TV brand and delivered throughout residential areas using direct-to-home (DTH) satellite technology. The B2B division provides voice, broadband, Pay-tv, data transmission and other telecommunications services to small- and medium-sized enterprises, corporations and government agencies throughout Brazil. It also provides wholesale interconnection, network usage services and traffic transportation services to other telecommunications providers. In April 2022, Oi sold its unified payment interface (UPI) mobile assets in order to focus on providing services related to fiber optics.

FINANCIAL DATA: Note: Data for latest year may not have been available at press time.

In U.S. $	2021	2020	2019	2018	2017	2016
Revenue	1,874,997,463	1,802,882,176	3,505,053,696	3,839,929,600	4,089,607,168	4,964,858,368
R&D Expense						
Operating Income						
Operating Margin %						
SGA Expense						
Net Income		-2,044,772,096	-1,566,682,240	4,764,124,672	-642,333,440	-1,326,203,264
Operating Cash Flow						
Capital Expenditure						
EBITDA						
Return on Assets %						
Return on Equity %						
Debt to Equity						

CONTACT INFORMATION:

Phone: 55 2131311211   Fax: 55 614151315
Toll-Free:
Address: Rua General Polidoro, No. 99, Fl. 5, Rio de Janeiro, RJ 22280-001 Brazil

STOCK TICKER/OTHER:

Stock Ticker: Private   Exchange:
Employees: 46,624   Fiscal Year Ends: 12/31
Parent Company:

SALARIES/BONUSES:

Top Exec. Salary: $ Bonus: $
Second Exec. Salary: $ Bonus: $

OTHER THOUGHTS:

Estimated Female Officers or Directors:
Hot Spot for Advancement for Women/Minorities:

Sales, profits and employees may be estimates. Financial information, benefits and other data can change quickly and may vary from those stated here.

OJSC JFSC Sistema

NAIC Code: 551114

www.sistema.com

TYPES OF BUSINESS:

Corporate, Subsidiary, and Regional Managing Offices
Investments
Telecommunications
Ecommerce
Real Estate
Industrial Production
Information Technology
Biotechnology

BRANDS/DIVISIONS/AFFILIATES:

GROWTH PLANS/SPECIAL FEATURES:

OJSC JFSC Sistema is a joint stock financial corporation that invests in and manages a range of companies that have value growth potential exceeding $1 billion. The corporation's diversified portfolio of assets mainly consists of Russian companies representing various industries and operating throughout Russia and abroad. These industries include telecommunications, ecommerce, real estate, industrial production, medicine, agriculture, pharmaceutical, hospitality, retail, energy, microelectronics, information technology, biotechnology and financial services. Sistema manages its own and third-party capital to develop its assets.

CONTACTS: *Note: Officers with more than one job title may be intentionally listed here more than once.*

Vladimir Chirakhov, CEO
Vladimir Travkov, CFO
Vladimir Evtushenkov, Chmn.

FINANCIAL DATA: *Note: Data for latest year may not have been available at press time.*

In U.S. $	2021	2020	2019	2018	2017	2016
Revenue	9,662,799,544	9,291,153,408	8,824,127,488	10,443,489,280	9,569,921,024	9,320,118,272
R&D Expense						
Operating Income						
Operating Margin %						
SGA Expense						
Net Income		137,239,520	384,165,888	-616,556,864	-1,284,993,280	-157,066,320
Operating Cash Flow						
Capital Expenditure						
EBITDA						
Return on Assets %						
Return on Equity %						
Debt to Equity						

CONTACT INFORMATION:

Phone: 7-495-737-0101   Fax: 7-495-730-0330
Toll-Free:
Address: 13/1 Mokhovaya St., Moscow, 125009 Russia

STOCK TICKER/OTHER:

Stock Ticker: AFKS   Exchange: Moscow
Employees: 158,000   Fiscal Year Ends: 12/31
Parent Company:

SALARIES/BONUSES:

Top Exec. Salary: $ Bonus: $
Second Exec. Salary: $ Bonus: $

OTHER THOUGHTS:

Estimated Female Officers or Directors:
Hot Spot for Advancement for Women/Minorities:

OneSpan Inc

www.onespan.com

NAIC Code: 511210E

TYPES OF BUSINESS:

Computer Software: Network Security, Managed Access, Digital ID,
Cybersecurity & Anti-Virus
Identity Security
Authentication
Anti-Fraud Services
Agreement Automation

BRANDS/DIVISIONS/AFFILIATES:

GROWTH PLANS/SPECIAL FEATURES:

OneSpan Inc is a provider of information technology security solutions for banking and financial services and application security markets. Its solutions secure and manage access to digital assets and protect online transactions, via mobile devices and in-person. Authentication and anti-fraud solutions are the organization's primary product offerings and include multifactor authentication and virtual private network access capabilities. The company derives revenues from hardware and license fees, maintenance and support fees, and subscription fees. A large majority of the firm's revenue is generated in Europe, Middle East and Africa, and the rest in the United States and Asia-Pacific region.

CONTACTS: Note: Officers with more than one job title may be intentionally listed here more than once.

Steven Worth, CEO
Jan van Gaalen, CFO
Alfred Nietzel, Chairman of the Board
John Bosshart, Chief Accounting Officer

FINANCIAL DATA: Note: Data for latest year may not have been available at press time.

In U.S. $	2021	2020	2019	2018	2017	2016
Revenue	214,481,000	215,691,000	253,484,000	211,336,000	193,291,000	192,304,000
R&D Expense	47,414,000	41,194,000	42,463,000	32,197,000	23,119,000	23,214,000
Operating Income	-26,128,000	-5,258,000	14,189,000	-920,000	6,192,000	9,599,000
Operating Margin %	-.12%	-.02%	.06%	.00%	.03%	.05%
SGA Expense	115,761,000	103,001,000	101,716,000	105,394,000	96,394,000	88,995,000
Net Income	-30,584,000	-5,455,000	7,864,000	3,044,000	-22,399,000	10,514,000
Operating Cash Flow	-2,745,000	14,922,000	18,244,000	1,226,000	17,627,000	28,415,000
Capital Expenditure	2,204,000	3,234,000	7,453,000	3,685,000	3,088,000	2,043,000
EBITDA	-17,202,000	6,745,000	25,734,000	11,218,000	16,793,000	20,376,000
Return on Assets %	-.09%	-.01%	.02%	.01%	-.07%	.03%
Return on Equity %	-.13%	-.02%	.03%	.01%	-.09%	.04%
Debt to Equity	.05%	0.048	0.043			

CONTACT INFORMATION:

Phone: 312-766-4001   Fax:
Toll-Free:
Address: 121 West Wacker Dr., Ste. 2050, Chicago, IL 60601 United States

STOCK TICKER/OTHER:

Stock Ticker: OSPN   Exchange: NAS
Employees: 879   Fiscal Year Ends: 12/31
Parent Company:

SALARIES/BONUSES:

Top Exec. Salary: $480,000   Bonus: $
Second Exec. Salary: Bonus: $
$380,000

OTHER THOUGHTS:

Estimated Female Officers or Directors:
Hot Spot for Advancement for Women/Minorities:

Sales, profits and employees may be estimates. Financial information, benefits and other data can change quickly and may vary from those stated here.

OneWeb Ltd

NAIC Code: 517410

www.oneweb.world

TYPES OF BUSINESS:

Satellite Internet Access Services
Satellites
Global Communications Network
Internet Connectivity Services
User Terminals
5G

BRANDS/DIVISIONS/AFFILIATES:

UK Department for Business Energy and Industrial
Bharthi Group
Eutelsat

CONTACTS: *Note: Officers with more than one job title may be intentionally listed here more than once.*

Neil Masterson, CEO
Michele Franci, COO
Srikanth Balackandran, CFO
Kate Roddy, Chief People Officer
Massimiliano Ladovaz, CTO

GROWTH PLANS/SPECIAL FEATURES:

OneWeb Ltd. is building a global communications network i
space for the delivery of high-throughput, high-speed service
capable of connecting everywhere and to everyone. The firr
has designed an advanced, low-Earth-orbit satelli
constellation referred to as OneWeb. OneWeb's 5G-read
network enables connectivity solutions for maritime, aviatior
enterprise and government entities, and will cover cell sit
backhaul. The company powers digital transformation an
offers tailored networking solutions for any need, at any leve
OneWeb satellites orbit relatively close to the Earth, allowir
for better internet access speeds. As they orbit, they interloc
with each other electronically to create coverage over the entir
planet. Small, low-cost user terminals communicate with th
satellite network and provide wireless internet access. Th
terminals provide connectivity with no change in latenc
(speed) during satellite handovers in order to ensur
continuous quality of voice, gaming and web surfir
experience. User terminals consist of a satellite antenna,
receiver and a customer network exchange unit (CNX). Th
CNX connects the user terminal to the customer's networ
which in turn connects to end-user devices such as laptop
smartphones, sensors and more. Compared to tradition
satellites, OneWeb units have fewer components, are lighter i
weight, easier to manufacture and cheaper to launch. The
contain on-board propulsion and state-of-the-art positionir
GPS sensors that ground-track their placement within meter
The propulsion systems perform maneuvers for steering cle
of space debris. When a OneWeb satellite nears the end of i
service life, it will de-orbit automatically. Headquartered i
London, UK, OneWeb has an international office in Virgini
USA. The firm is owned by the UK government, Bharti Grou
and Eutelsat, with additional investments from SoftBank an
Hughes Network Group. In January 2022, OneWeb an
Hughes Network Systems LLC announced a strategic six-yea
distribution partner agreement to provide low-Earth orb
connectivity services across India.

FINANCIAL DATA: *Note: Data for latest year may not have been available at press time.*

In U.S. $	2021	2020	2019	2018	2017	2016
Revenue	3,000,000	26,520,760	75,773,600	216,496,000	81,254,000	
R&D Expense						
Operating Income						
Operating Margin %						
SGA Expense						
Net Income	370,800,000			-213,184,000	-73,857,000	
Operating Cash Flow						
Capital Expenditure						
EBITDA						
Return on Assets %						
Return on Equity %						
Debt to Equity						

CONTACT INFORMATION:

Phone: 44 20 3727 1160　　Fax:
Toll-Free:
Address: 195 Wood Ln., W., Works Bldg., Fl.3, London, W12 7FQ
United Kingdom

STOCK TICKER/OTHER:

Stock Ticker: Subsidiary　　　　　　　Exchange:
Employees: 400　　　　　　　　　　Fiscal Year Ends: 03/31
Parent Company: UK Department for Business Energy and Industrial

SALARIES/BONUSES:

Top Exec. Salary: $　　　　Bonus: $
Second Exec. Salary: $　　　Bonus: $

OTHER THOUGHTS:

Estimated Female Officers or Directors:
Hot Spot for Advancement for Women/Minorities:

Ooredoo QPSC

NAIC Code: 517210

www.ooredoo.qa

TYPES OF BUSINESS:

Wireless Telecommunications Carriers (except Satellite)
Communications
Mobile Networks
Fixed Networks
Digital Money Wallet
Rewards Card

BRANDS/DIVISIONS/AFFILIATES:

Ooredoo Money
Nojoom

CONTACTS: Note: Officers with more than one job title may be intentionally listed here more than once.

Saud bin Nasser Al Thani, CEO
Yousef Abdulla Al Kubaisi, COO
Faisal bin Thani Al Thani, Chmn.

GROWTH PLANS/SPECIAL FEATURES:

Ooredoo QPSC is an international communications company with a customer base of more than 100 million across the Middle East, North Africa and Southeast Asia. In Qatar, the firm is the leading communications company, delivering services for consumers, businesses, residences and organizations. Services offered include pre-paid and post-paid calling plans, data plans, international calling, internet via fiber cable lines and broadband. Ooredoo's 4G and 5G networks cover all municipalities of Qatar, and extends to the desert. The company launched its live 5G network on the 3.5GHz spectrum band in 2018, claiming to be the first in the world to do so. Ooredoo's ecommerce platform offers monthly plans, and sells smartphones, tablets, internet devices and related accessories. The firm also offers Ooredoo Money, a digital money wallet platform that enables users to send money or make payments via mobile phone. Nojoom is Ooredoo's rewards card in which users can earn points on money spent for Ooredoo services or for partner services.

FINANCIAL DATA: Note: Data for latest year may not have been available at press time.

In U.S. $	2021	2020	2019	2018	2017	2016
Revenue	8,167,035,500	7,865,600,000	8,200,620,000	8,216,400,000	8,902,300,000	8,929,466,368
R&D Expense						
Operating Income						
Operating Margin %						
SGA Expense						
Net Income	287,517,890	387,443,000	609,659,000	492,195,000	517,297,000	602,349,952
Operating Cash Flow						
Capital Expenditure						
EBITDA						
Return on Assets %						
Return on Equity %						
Debt to Equity						

CONTACT INFORMATION:

Phone: 974 440002133   Fax: 974 44830011
Toll-Free:
Address: Ooredoo Tower, W. Bay Area, Doha, 2211 Qatar

STOCK TICKER/OTHER:

Stock Ticker: ORDS   Exchange: Qatar
Employees: 15,167   Fiscal Year Ends: 12/31
Parent Company:

SALARIES/BONUSES:

Top Exec. Salary: $ Bonus: $
Second Exec. Salary: $ Bonus: $

OTHER THOUGHTS:

Estimated Female Officers or Directors:
Hot Spot for Advancement for Women/Minorities:

Sales, profits and employees may be estimates. Financial information, benefits and other data can change quickly and may vary from those stated here.

OPTERNA

www.opterna.com

NAIC Code: 511210B

TYPES OF BUSINESS:

Computer Software: Network Management (IT), System Testing & Storage
Fiber-Optic Infrastructure Products

GROWTH PLANS/SPECIAL FEATURES:

OPTERNA provides enterprise broadband connectivity
solutions through its fiber optic portfolio. Products include
copper products, fiber optics, related equipment cabinets and
accessories, fiber management, distribution hubs and boxes,
splice closures, access and subscriber boxes, panels,
connectors, modules and more. OPTERNA's fiber-to-the-
everywhere (FTTX) solutions span global on- and off-site
premises, including outdoor connectivity. OPTERNA is a
business brand of Belden, Inc., with manufacturing facilities in
Europe, Asia, Africa, the Middle East and the U.S.

BRANDS/DIVISIONS/AFFILIATES:

Belden Inc

CONTACTS: *Note: Officers with more than one job title may be intentionally listed here more than once.*

Abraham Chandy, Dir.-Oper.
Bret A. Matz, Gen. Mgr.
Javad K. Hassan, Chmn.-NeST Group
David Aldrich, Chmn.-Belden

FINANCIAL DATA: *Note: Data for latest year may not have been available at press time.*

In U.S. $	2021	2020	2019	2018	2017	2016
Revenue	15,456,180	15,006,000	15,375,000			
R&D Expense						
Operating Income						
Operating Margin %						
SGA Expense						
Net Income						
Operating Cash Flow						
Capital Expenditure						
EBITDA						
Return on Assets %						
Return on Equity %						
Debt to Equity						

CONTACT INFORMATION:

Phone: 315-431-7200   Fax:
Toll-Free:
Address: 6176 E. Molloy Rd., East Syracuse, NY 13057 United States

STOCK TICKER/OTHER:

Stock Ticker: Subsidiary   Exchange:
Employees: 62   Fiscal Year Ends: 03/31
Parent Company: Belden Inc

SALARIES/BONUSES:

Top Exec. Salary: $ Bonus: $
Second Exec. Salary: $ Bonus: $

OTHER THOUGHTS:

Estimated Female Officers or Directors:
Hot Spot for Advancement for Women/Minorities:

Sales, profits and employees may be estimates. Financial information, benefits and other data can change quickly and may vary from those stated here.

Optical Cable Corporation

www.occfiber.com

NAIC Code: 335921

TYPES OF BUSINESS:

Fiber Optic Cable Manufacturing
Manufacturer
Fiber Optic Cable
Copper Cable
Connectivity Solutions
Communications

BRANDS/DIVISIONS/AFFILIATES:

Applied Optical Systems Inc
Centric Solutions LLC

GROWTH PLANS/SPECIAL FEATURES:

Optical Cable Corp manufactures tight-buffered fiber optic & copper data communication cables, data communication connectivity solutions for enterprise markets, and customized solutions for specialty use & harsh environments. Its products and services include designs and customized products for specialty applications and harsh environments, cabling, connectors, patch cords, assemblies, racks, cabinets, datacom enclosures, patch panels, faceplates, and multi-media boxes. The company supplies its products and services to industries such as the military, industrial, mining, petrochemical, broadcast, and oil & gas.

CONTACTS: *Note: Officers with more than one job title may be intentionally listed here more than once.*

Neil Wilkin, CEO
Tracy Smith, CFO

FINANCIAL DATA: *Note: Data for latest year may not have been available at press time.*

In U.S. $	2021	2020	2019	2018	2017	2016
Revenue	59,136,300	55,277,400	71,324,450	87,828,590	64,092,850	64,616,000
R&D Expense						
Operating Income	-1,973,323	-5,533,064	-5,161,701	1,740,374	-1,317,012	-1,216,965
Operating Margin %	-.03%	-.10%	-.07%	.02%	-.02%	-.02%
SGA Expense	18,239,150	19,245,500	23,434,360	26,130,960	21,968,760	20,760,740
Net Income	6,610,516	-6,121,224	-5,669,321	1,068,753	-1,738,771	-1,778,822
Operating Cash Flow	2,116,588	-3,553,137	-283,797	3,206,248	-687,071	3,155,480
Capital Expenditure	192,867	168,458	550,397	734,395	583,867	703,401
EBITDA	8,487,708	-4,106,304	-3,433,422	3,478,116	454,345	880,692
Return on Assets %	.18%	-.16%	-.14%	.03%	-.04%	-.04%
Return on Equity %	.35%	-.34%	-.24%	.04%	-.07%	-.07%
Debt to Equity	.36%	0.866	0.243	0.342	0.512	0.47

CONTACT INFORMATION:

Phone: 540 265-0690   Fax: 540 265-0724
Toll-Free:
Address: 5290 Concourse Dr., Roanoke, VA 24019 United States

STOCK TICKER/OTHER:

Stock Ticker: OCC   Exchange: NAS
Employees: 317   Fiscal Year Ends: 10/31
Parent Company:

SALARIES/BONUSES:

Top Exec. Salary: $455,000   Bonus: $
Second Exec. Salary: Bonus: $
$290,000

OTHER THOUGHTS:

Estimated Female Officers or Directors:
Hot Spot for Advancement for Women/Minorities:

Sales, profits and employees may be estimates. Financial information, benefits and other data can change quickly and may vary from those stated here.

Orange
NAIC Code: 517110

www.orange.com

TYPES OF BUSINESS:
Telephone, Internet Access, Broadband, Data Networks, Server Facilities and Telecommunications Services Industry
Telecommunications Solutions and Services
Mobile Solutions
Fixed Broadband Solutions
Business Services
Artificial Intelligence
Telecommunications Networks
Cybersecurity

BRANDS/DIVISIONS/AFFILIATES:
Orange Business Services
Code School
Solidarity FabLab
Orang Fab
Orange Ventures

GROWTH PLANS/SPECIAL FEATURES:
Orange is the incumbent telecom operator in France, formerl known as France Telecom. The company operates fixed an wireless businesses in France, where it is the market leade ahead of Iliad, Bouygues and SFR. Orange also has fixed an wireless (convergent) operations in Spain, Poland, Belgiun Luxembourg and Central Europe (Romania, Slovakia Moldova). Around 15% of revenue comes from emergin African markets, where the company only operates wireles networks and 20% comes from the enterprise segment, whic serves companies with more than 50 employees in France an internationally.

CONTACTS: *Note: Officers with more than one job title may be intentionally listed here more than once.*
Stephane Richard, CEO
Pierre Louette, Deputy CEO
Benoit Scheen, Sr. Exec. VP-Oper., Europe
Elie Girard, Sr.-Strategy & Dev.
Beatrice Mandine, Sr. Exec. VP-Comm. & Brand
Delphine Ernotte Cunci, Deputy CEO-Orange France
Christine Albanel, Sr. Exec. VP-Corp. Social Responsibility & Events
Thierry Bonhomme, Sr. Exec. VP-Orange Bus. Svcs.
Marc Rennard, Sr. Exec. VP-Oper., Africa, Middle East & Asia
Pierre Louette, Deputy CEO-Group Sourcing & Supply Chain

FINANCIAL DATA: *Note: Data for latest year may not have been available at press time.*

In U.S. $	2021	2020	2019	2018	2017	2016
Revenue	43,362,370,000	43,105,380,000	43,072,750,000	42,198,810,000	41,666,500,000	41,512,510,000
R&D Expense						
Operating Income	4,486,957,000	5,758,602,000	5,988,048,000	5,036,609,000	5,537,313,000	5,412,902,000
Operating Margin %	.10%	.13%	.14%	.12%	.13%	.13%
SGA Expense	10,112,990,000	8,657,788,000	8,661,866,000	9,253,329,000	8,743,448,000	9,041,219,000
Net Income	237,604,800	4,917,297,000	3,063,368,000	1,992,617,000	1,879,423,000	2,868,593,000
Operating Cash Flow	11,458,060,000	12,947,930,000	10,391,390,000	9,693,867,000	10,375,070,000	8,922,926,000
Capital Expenditure	8,995,330,000	8,714,894,000	8,588,444,000	7,793,029,000	7,675,756,000	8,659,828,000
EBITDA	11,465,200,000	14,223,650,000	14,575,470,000	12,089,290,000	11,404,010,000	10,151,740,000
Return on Assets %	.00%	.04%	.03%	.02%	.02%	.03%
Return on Equity %	.00%	.14%	.09%	.05%	.05%	.08%
Debt to Equity	1.19%	1.041	1.215	0.872	0.849	0.925

CONTACT INFORMATION:
Phone: 33 144442222   Fax: 33 144448034
Toll-Free:
Address: 78 rue Olivier de Serres, Paris, 75015 France

STOCK TICKER/OTHER:
Stock Ticker: ORAN   Exchange: NYS
Employees: 139,698   Fiscal Year Ends: 12/31
Parent Company:

SALARIES/BONUSES:
Top Exec. Salary: $968,775   Bonus: $668,067
Second Exec. Salary: Bonus: $283,780
$611,858

OTHER THOUGHTS:
Estimated Female Officers or Directors: 5
Hot Spot for Advancement for Women/Minorities: Y

Orange Belgium

www.orange.be

NAIC Code: 517210

TYPES OF BUSINESS:

Cell Phone Service
Fixed-Line Services
Mobile Data Services
Fiber Network

GROWTH PLANS/SPECIAL FEATURES:

Orange Belgium SA is a triple-play telecommunications company that provides mobile, fixed telephone, digital TV, and broadband Internet services to households and companies. In terms of products, the majority of revenue comes from mobile services. most of the group's mobile customers are postpaid subscribers. The company also operates in Luxembourg, but overall company revenue comes mostly from Belgium. Orange Belgium owns telecommunications infrastructure, including fiber networks. The company also acts as a wholesaler, allowing mobile virtual network operators access to Orange Belgium's infrastructure.

BRANDS/DIVISIONS/AFFILIATES:

Orange SA
Orange Communications Luxembourg

CONTACTS: Note: Officers with more than one job title may be intentionally listed here more than once.

Xavier Pichon, CEO
Antoine Chouc, CFO
Paul-Marie Dessart, Sec.
Olivier Ysewijn, Chief Strategy Officer
Gabriel Flichy, Chief Network Officer
Stephane Beauduin, Chief Bus. Unit B2B Officer
Sven Bols, Chief Sales & Dist. Officer
Cristina Zanchi, Chief Customer Relationship Officer

FINANCIAL DATA: Note: Data for latest year may not have been available at press time.

In U.S. $	2021	2020	2019	2018	2017	2016
Revenue	1,390,418,000	1,340,854,000	1,367,326,000	1,305,096,000	1,270,983,000	1,266,183,000
R&D Expense						
Operating Income	77,565,220	77,002,300	60,141,540	51,391,980	67,055,540	124,899,600
Operating Margin %	.06%	.06%	.04%	.04%	.05%	.10%
SGA Expense	64,843,670	62,738,880	59,075,890	67,421,630	68,418,960	59,835,610
Net Income	40,508,040	55,046,810	34,062,120	33,071,930	39,765,660	78,125,060
Operating Cash Flow	375,121,900	351,241,000	346,209,500	266,561,000	272,109,500	304,794,900
Capital Expenditure	230,345,100	181,246,600	183,727,600	182,948,500	192,099,900	170,956,100
EBITDA	396,708,200	379,451,700	346,221,800	284,404,800	293,765,200	323,738,000
Return on Assets %	.02%	.03%	.02%	.02%	.03%	.05%
Return on Equity %	.06%	.09%	.06%	.06%	.07%	.15%
Debt to Equity	.60%	0.429	0.828	0.46	0.549	0.677

CONTACT INFORMATION:

Phone: 32 2745-7111   Fax:
Toll-Free:
Address: Ave. du Bourget 3 Avenue du Bourget 3, Brussels, 1140 Belgium

STOCK TICKER/OTHER:

Stock Ticker: MBSRY
Employees: 1,400
Parent Company: Orange SA

Exchange: PINX
Fiscal Year Ends: 12/31

SALARIES/BONUSES:

Top Exec. Salary: $ Bonus: $
Second Exec. Salary: $ Bonus: $

OTHER THOUGHTS:

Estimated Female Officers or Directors: 2
Hot Spot for Advancement for Women/Minorities:

Sales, profits and employees may be estimates. Financial information, benefits and other data can change quickly and may vary from those stated here.

Orange Polska SA

NAIC Code: 517110

www.orange.pl/start.phtml

TYPES OF BUSINESS:

Local & Long-Distance Telephone Service
Mobile Phone Service
Mobile Internet Services
Equipment Sales
Radio Communications
Fiber-Optic Cable
Internet Portal
e-Commerce Services

BRANDS/DIVISIONS/AFFILIATES:

Orange
Orange Customer Service Sp z o o
Orange Foundation

GROWTH PLANS/SPECIAL FEATURES:

Orange Polska SA is a triple-play telecommunications company. It derives revenue from the provision of Internet, mobile, and television services. The company operates through various divisions which include mobile services, mobile equipment sales, and fixed services. The majority of revenue is derived in both mobile and fixed services. Within mobile services, the majority of revenue is derived from voice services. Fixed services revenue is composed of narrow band services, fixed broadband, TV, and VOIP services and enterprise solutions. The company is an owner of telecommunications infrastructure. The company generates the vast majority of its revenue in Poland.

CONTACTS: *Note: Officers with more than one job title may be intentionally listed here more than once.*

Maciej Witucki, Pres.
Piotr Muszynski, VP-Oper.
Vincent Lobry, VP-Strategy
Jacek Kunicki, Dir.-Investor Rel.
Maciej Krzysztof Witucki, Chmn.

FINANCIAL DATA: *Note: Data for latest year may not have been available at press time.*

In U.S. $	2021	2020	2019	2018	2017	2016
Revenue	2,590,760,000	2,499,536,000	2,477,382,000			
R&D Expense						
Operating Income	168,981,500	104,473,100	102,735,600			
Operating Margin %	.07%	.04%	.04%			
SGA Expense						
Net Income	363,158,200	9,991,195	17,810,390			
Operating Cash Flow	673,536,800	652,685,600	620,757,200			
Capital Expenditure	477,622,600	485,658,900	526,709,700			
EBITDA	1,082,307,000	722,406,800	718,062,800			
Return on Assets %	.07%	.00%	.00%			
Return on Equity %	.14%	.00%	.01%			
Debt to Equity	.57%	0.436	0.811			

CONTACT INFORMATION:

Phone: 48 225272323   Fax: 48 225272341
Toll-Free:
Address: Al. Jerozolimskie 160, Warszawa, 02-326 Poland

STOCK TICKER/OTHER:

Stock Ticker: PTTWF   Exchange: PINX
Employees: 10,924   Fiscal Year Ends: 12/31
Parent Company: Orange SA

SALARIES/BONUSES:

Top Exec. Salary: $ Bonus: $
Second Exec. Salary: $ Bonus: $

OTHER THOUGHTS:

Estimated Female Officers or Directors: 1
Hot Spot for Advancement for Women/Minorities:

ORBCOMM Inc

www.orbcomm.com

NAIC Code: 517210

TYPES OF BUSINESS:

Satellite Telecommunications

BRANDS/DIVISIONS/AFFILIATES:

CONTACTS: Note: Officers with more than one job title may be intentionally listed here more than once.

Marc Eisenberg, CEO
Constantine Milcos, CFO
Jerome Eisenberg, Chairman of the Board
Craig Malone, Executive VP, Divisional
John Stolte, Executive VP, Divisional
Christian Le Brun, Executive VP

GROWTH PLANS/SPECIAL FEATURES:

ORBCOMM Inc. is a satellite-based data communications company that operates a two-way global wireless data messaging system optimized for narrowband data communication. Its system consists of a global network of 27 low-Earth orbit, or LEO, satellites and accompanying ground infrastructure. The company's main products and services are satellite-based data communications services and product sales from subscriber communicators. The company offers terrestrial-based cellular communications services, which consist of reselling airtime using cellular providers' wireless technology networks and product sales from cellular wireless SIMS for use with devices or equipment that enable the use of the cellular providers' wireless networks for data communications. The company's communications services are used by businesses and government agencies that are engaged in tracking, monitoring, controlling, or communicating with fixed or mobile assets globally. Its low cost, industrially-rated subscriber communicators are embedded into many different assets for use with its system. Its products and services are combined with industry or customer specific applications developed by its VARs, which are sold to their end-user customers. For the company's satellite-based data and terrestrial-based cellular communications services, it utilizes a cost-effective sales and marketing strategy of partnering with resellers such as VARs, IVARs and country representatives. These resellers, which are its direct customers, market to end users. The company's products and services enable its customers and end-users to enhance productivity, reduce costs and improve security through a number of commercial, government and emerging homeland security applications. It enables its customers and end-users to achieve these benefits using a single global satellite technology standard for machine-to-machine and telematic, or M2M, data communications.

FINANCIAL DATA: Note: Data for latest year may not have been available at press time.

In U.S. $	2021	2020	2019	2018	2017	2016
Revenue	250,000,000	248,466,000	272,012,992	276,140,000	254,220,000	186,744,000
R&D Expense						
Operating Income						
Operating Margin %						
SGA Expense						
Net Income		-33,940,000	-18,423,000	-26,244,000	-61,284,000	-23,511,000
Operating Cash Flow						
Capital Expenditure						
EBITDA						
Return on Assets %						
Return on Equity %						
Debt to Equity						

CONTACT INFORMATION:

Phone: 201 363-4900   Fax: 201 433-6400
Toll-Free:
Address: 2115 Linwood Avenue, Fort Lee, NJ 07024 United States

STOCK TICKER/OTHER:

Stock Ticker: Private   Exchange:
Employees: 800   Fiscal Year Ends: 12/31
Parent Company: GI Partners LP

SALARIES/BONUSES:

Top Exec. Salary: $ Bonus: $
Second Exec. Salary: $ Bonus: $

OTHER THOUGHTS:

Estimated Female Officers or Directors:
Hot Spot for Advancement for Women/Minorities:

Sales, profits and employees may be estimates. Financial information, benefits and other data can change quickly and may vary from those stated here.

Pakistan Telecommunication Company Limited www.ptcl.com.pk

NAIC Code: 517110

TYPES OF BUSINESS:
Wired Telecommunications Carriers
Information Technology
Communication Technology
Telecommunications Services
Digital Services
Broadband Services
Cloud Applications
Smart TV Applications

BRANDS/DIVISIONS/AFFILIATES:
Pak Telecom Mobile Ltd
Ufone
U Microfinance Bank Limited
Rozgar Microfinance Bank Limited
DVCOM Data Private Limited
Smart Sky Private Limited

GROWTH PLANS/SPECIAL FEATURES:
Pakistan Telecommunication Co Ltd is a telecom company.
is mainly engaged in providing telecommunication service
such as wireline and wireless services, interconnect, dat
services and equipment sales. The services provided by th
company include high-speed internet, smart tv, mobil
services, landline services, digital services such as manage
wi-fi, smart radios, point-to-multipoint and cloud services in th
realm of IaaS (Infrastructure as a Service) and hostin
services. The company operates in three operating segment
i.e. fixed-line communications (Wireline), wireles
communications (Wireless) and banking.

CONTACTS: *Note: Officers with more than one job title may be intentionally listed here more than once.*
Muhammad Sohail Khan Rajput, Chmn.

FINANCIAL DATA: *Note: Data for latest year may not have been available at press time.*

In U.S. $	2021	2020	2019	2018	2017	2016
Revenue	614,262,200	577,650,000	578,185,700	566,223,400	522,290,800	523,108,100
R&D Expense					5,614,117	1,392,140
Operating Income	50,288,400	52,521,400	51,944,900	59,060,030	20,426,230	31,828,120
Operating Margin %	.08%	.09%	.09%	.10%	.04%	.06%
SGA Expense	52,368,560	46,171,770	49,517,820	48,225,400	52,242,250	58,829,240
Net Income	11,494,150	14,606,860	10,609,690	25,485,400	19,270,930	7,243,236
Operating Cash Flow	206,757,700	251,249,200	200,601,000	137,786,300	128,743,000	158,977,800
Capital Expenditure	281,679,200	142,432,000	181,627,200	166,886,300	133,127,500	138,514,900
EBITDA	213,765,700	207,610,400	204,592,800	193,628,000	199,930,200	175,944,900
Return on Assets %	.01%	.01%	.01%	.02%	.01%	.01%
Return on Equity %	.03%	.04%	.03%	.07%	.05%	.02%
Debt to Equity	.77%	0.533	0.451	0.301	0.303	0.296

CONTACT INFORMATION:
Phone: 92 512263732   Fax: 92 512263733
Toll-Free:
Address: Block E, Sector G8/4, Islamabad, 44000 Pakistan

STOCK TICKER/OTHER:
Stock Ticker: PKTLY   Exchange: GREY
Employees: 21,852   Fiscal Year Ends: 06/30
Parent Company:

SALARIES/BONUSES:
Top Exec. Salary: $ Bonus: $
Second Exec. Salary: $ Bonus: $

OTHER THOUGHTS:
Estimated Female Officers or Directors:
Hot Spot for Advancement for Women/Minorities:

Panasonic Corporation

www.panasonic.com/global/home.html

NAIC Code: 334310

TYPES OF BUSINESS:

Audio & Video Equipment, Manufacturing
Appliances
Automotive Systems
Digital Cameras
Housing Construction Systems
Industrial Connected Systems
Batteries
Business-to-Business Solutions

BRANDS/DIVISIONS/AFFILIATES:

Blue Yonder

GROWTH PLANS/SPECIAL FEATURES:

Panasonic Holdings Corp is a conglomerate that has diversified from its consumer electronics roots. It has five main business units: appliances (air conditioners, refrigerators, laundry machines, and TVs); life solutions (LED lighting, housing systems, and solar panels; connected solutions (PCs, factory automations, and in-flight entertainment systems); automotive (infotainment systems and rechargeable batteries); and industrial solutions (electronic devices). After the crisis in 2012, former president Kazuhiro Tsuga has focused on shifting the business portfolio to increase the proportion of B2B businesses to mitigate the tough competition in consumer electronics products.

CONTACTS: *Note: Officers with more than one job title may be intentionally listed here more than once.*

Yuki Kusumi, CEO
Kazuhiro Tsuga, Chmn.

FINANCIAL DATA: *Note: Data for latest year may not have been available at press time.*

In U.S. $	2021	2020	2019	2018	2017	2016
Revenue	49,679,580,000	55,551,770,000	59,349,840,000	59,197,300,000	54,462,380,000	
R&D Expense						
Operating Income	2,071,737,000	2,139,224,000	2,971,262,000	2,747,442,000	1,990,552,000	
Operating Margin %	.04%	.04%	.05%	.05%	.04%	
SGA Expense	12,367,960,000	13,826,620,000	14,383,470,000	14,372,660,000	13,667,520,000	
Net Income	1,224,243,000	1,673,888,000	2,107,305,000	1,750,519,000	1,107,683,000	
Operating Cash Flow	3,738,045,000	3,191,212,000	1,510,509,000	3,138,401,000	2,858,277,000	
Capital Expenditure	2,198,361,000	2,560,724,000	2,958,047,000	3,527,166,000	2,534,960,000	
EBITDA	4,427,603,000	5,177,143,000	5,437,126,000	5,125,074,000	4,222,657,000	
Return on Assets %	.03%	.04%	.05%	.04%	.03%	
Return on Equity %	.07%	.12%	.16%	.14%	.10%	
Debt to Equity	.42%	0.579	0.318	0.506	0.602	

CONTACT INFORMATION:

Phone: 81 669081121   Fax:
Toll-Free:
Address: 1006 Oaza Kadoma, Kadoma City, Osaka, 571-8501 Japan

STOCK TICKER/OTHER:

Stock Ticker: PCRFF   Exchange: PINX
Employees: 271,869   Fiscal Year Ends: 03/31
Parent Company:

SALARIES/BONUSES:

Top Exec. Salary: $ Bonus: $
Second Exec. Salary: $ Bonus: $

OTHER THOUGHTS:

Estimated Female Officers or Directors:
Hot Spot for Advancement for Women/Minorities:

Sales, profits and employees may be estimates. Financial information, benefits and other data can change quickly and may vary from those stated here.

Partner Communications Co Ltd

www.partner.co.il

NAIC Code: 517210

TYPES OF BUSINESS:

Cell Phone Service
Fixed-Line Telephony
Internet Service Provider
Voice over Broadband
Web Video on Demand (VOD)
International Long Distance Telephony

BRANDS/DIVISIONS/AFFILIATES:

S B Israel Telecom Ltd
Partner

GROWTH PLANS/SPECIAL FEATURES:

Partner Communications Co Ltd is an Israeli-based telecommunications company. It provides a wide integrated and customized range of cellular and fixed-line telecommunication services, including infrastructure, international long distance (ILD), internet services provider (ISP), television, and other services. The cellular business segment represents the largest portion of total revenues, and it offers services such as airtime calls, international roaming services, text messaging, internet browsing, value-added and content services, and services provided to other operators that use the company's cellular network.

CONTACTS: *Note: Officers with more than one job title may be intentionally listed here more than once.*

Tamir Amar, CFO
Terry Yaskil, VP-Mktg.
Einat Rom, VP-Human Resources
Yaron Eisenstein, VP-IT
Roly Klinger, VP-Legal & Regulatory Affairs
Roly Klinger, VP-Bus. Dev.
Amalia Glaser, VP-Comm. & Corp. Governance Div.
Ori Watermann, VP-Fixed Line Div.
Avi Cohen, VP-Customer Div.
Guy Emodi, VP-Economics, Planning & Corp. Strategy
Guy Emodi, VP-Operator Rel.
Osnat Ronen, Chmn.

FINANCIAL DATA: *Note: Data for latest year may not have been available at press time.*

In U.S. $	2021	2020	2019	2018	2017	2016
Revenue	1,017,149,000	964,522,200	978,132,600	985,694,000	988,416,100	1,071,893,000
R&D Expense						
Operating Income	43,553,220	26,918,310	23,288,870	35,689,440	101,624,200	69,564,170
Operating Margin %	.04%	.03%	.02%	.04%	.10%	.06%
SGA Expense	98,599,640	86,501,530	95,877,570	102,531,500	101,624,200	159,997,600
Net Income	34,782,080	5,141,699	5,746,605	17,239,810	34,479,630	15,727,550
Operating Cash Flow	234,098,500	237,728,000	253,153,100	189,033,100	294,286,700	285,818,000
Capital Expenditure	203,248,300	173,305,500	190,242,900	151,831,300	113,722,300	59,280,770
EBITDA	266,158,500	238,030,400	247,104,000	212,321,900	255,875,200	229,259,300
Return on Assets %	.02%	.00%	.00%	.01%	.02%	.01%
Return on Equity %	.06%	.01%	.01%	.04%	.09%	.05%
Debt to Equity	1.08%	1.095	1.34	0.856	0.849	1.977

CONTACT INFORMATION:

Phone: 972 547814888   Fax: 972 547814999
Toll-Free:
Address: 8 Amal St., Afeq Industrial Park, Rosh Ha'ayin, 48103 Israel

STOCK TICKER/OTHER:

Stock Ticker: PTNR
Employees: 2,655
Parent Company:

Exchange: NAS
Fiscal Year Ends: 12/31

SALARIES/BONUSES:

Top Exec. Salary: $485,134   Bonus: $292,774
Second Exec. Salary: $323,927   Bonus: $388,954

OTHER THOUGHTS:

Estimated Female Officers or Directors: 5
Hot Spot for Advancement for Women/Minorities: Y

PCCW Limited

www.pccw.com

NAIC Code: 517110

TYPES OF BUSINESS:

Local & Long-Distance Telephone Services
Telecommunications
Media
Information Technology
Property Development
Property Investment

BRANDS/DIVISIONS/AFFILIATES:

PCCW Media
HKT
Pacific Century Premium Developments
PCCW Solutions
Now TV

GROWTH PLANS/SPECIAL FEATURES:

PCCW Ltd is a Hong Kong-based company engaged in the businesses of telecommunications, media, information technology solutions, property development and investment, and others. Its operating segments are HKT Limited (HKT), Media Business, and Solutions Business. The entity derives key revenue from the HKT segment, which includes the provision of telecommunications and related services such as local telephony, local data and broadband, international telecommunications, mobile, and other telecommunications businesses such as customer premises equipment sales, outsourcing, consulting, and contact centers. The company operates in Hong Kong and other countries, of which maximum revenue is derived from the operations in Hong Kong.

CONTACTS:
Note: Officers with more than one job title may be intentionally listed here more than once.

Bangalore Gangaiah Srinivas, Group Managing Dir.
Hon Hing (Susanna) Hui, CFO
Janice Le, Managing Dir.-TV & News Media
George Fok, Managing Dir.-PCCW Solutions
Tzar (Richard) Kai Li, Chmn.

FINANCIAL DATA:
Note: Data for latest year may not have been available at press time.

In U.S. $	2021	2020	2019	2018	2017	2016
Revenue	4,924,220,000	4,584,083,000	4,779,885,000	4,949,189,000	4,692,112,000	4,889,825,000
R&D Expense						
Operating Income	631,865,600	585,749,600	694,670,000	701,421,800	717,091,100	704,097,000
Operating Margin %	.13%	.13%	.15%	.14%	.15%	.14%
SGA Expense	551,226,300	594,030,100	668,299,800	590,335,700	1,638,137,000	1,925,407,000
Net Income	161,151,200	-129,940,100	86,754,130	114,270,900	259,625,400	261,281,500
Operating Cash Flow	1,236,214,000	1,617,245,000	1,418,258,000	833,782,300	1,314,179,000	1,267,935,000
Capital Expenditure	1,008,819,000	907,797,200	1,200,290,000	944,741,000	752,506,000	1,039,393,000
EBITDA	1,516,732,000	1,266,152,000	1,456,985,000	1,468,960,000	1,421,315,000	1,555,969,000
Return on Assets %	.01%	-.01%	.01%	.01%	.02%	.03%
Return on Equity %	.09%	-.08%	.04%	.05%	.13%	.18%
Debt to Equity	3.18%	5.706	3.618	3.052	2.534	4.153

CONTACT INFORMATION:

Phone: 852 2883 7747   Fax: 852 2962 5634
Toll-Free:
Address: 979 King's Rd., Fl. 39, PCCW Tower, Hong Kong, Hong Kong 999077 Hong Kong

STOCK TICKER/OTHER:

Stock Ticker: PCWLF   Exchange: PINX
Employees: 20,600   Fiscal Year Ends: 12/31
Parent Company: Pacific Century Group Holdings Limited

SALARIES/BONUSES:

Top Exec. Salary: $ Bonus: $
Second Exec. Salary: $ Bonus: $

OTHER THOUGHTS:

Estimated Female Officers or Directors: 2
Hot Spot for Advancement for Women/Minorities: Y

Sales, profits and employees may be estimates. Financial information, benefits and other data can change quickly and may vary from those stated here.

Perusahaan Perseroan PT Telekomunikasi Indonesia Tbk (Telkom)

www.telkom.co.id

NAIC Code: 517110

TYPES OF BUSINESS:

Local Telephone Service
Broadband
Mobile Digital Services
Internet of Things
Voice Services
Data Center
Business Process Outsourcing
Wholesale Telecommunications

BRANDS/DIVISIONS/AFFILIATES:

GROWTH PLANS/SPECIAL FEATURES:

Perusahaan Perseroan PT Telekomunikasi Indonesia Tbk
the largest integrated telecommunications provider
Indonesia. It is the principal provider of fixed-line services
Indonesia, and its 65%-owned subsidiary, Telkom, is th
largest wireless carrier in the country, with about 48% marke
share. It also provides a wide range of other communicatio
services, including telephone network, interconnectio
services, multimedia, data and internet communication-relate
services, satellite transponder leasing, leased line, intelliger
network and related services, cable television and Vol
services.

CONTACTS:
Note: Officers with more than one job title may be intentionally listed here more than once.

Ririek Adriansyah, Pres.
Heri Supriadi, Dir.-Finance
Afriwandi, Dir.-Human Capital
Herlan Wijanarko, Dir.-IT & Network
Muhammad Awaluddin, Dir.-Enterprise & Wholesale
Sukardi Silalahi, Dir.-Consumer
Priyantono Rudito, Dir.-Human Capital & Gen. Affairs
Rizkan Chandra, Dir.-Network & Solution
Ririek Adriansyah, Dir.-Compliance & Risk Mgmt.

FINANCIAL DATA:
Note: Data for latest year may not have been available at press time.

In U.S. $	2021	2020	2019	2018	2017	2016
Revenue	9,655,995,000	9,199,997,000	9,139,988,000	8,818,437,000	8,647,716,000	7,843,802,000
R&D Expense	5,528,885	3,506,122	3,034,144			
Operating Income	2,967,325,000	2,906,373,000	2,911,970,000	2,521,509,000	2,975,619,000	2,760,532,000
Operating Margin %	.31%	.32%	.32%	.29%	.34%	.35%
SGA Expense	1,592,184,000	1,486,056,000	1,395,100,000	1,454,029,000	1,521,320,000	1,457,131,000
Net Income	1,677,342,000	1,419,440,000	1,285,668,000	1,200,307,000	1,491,450,000	1,303,536,000
Operating Cash Flow						
Capital Expenditure	2,209,396,000	2,153,636,000	2,515,643,000	2,328,470,000	2,211,689,000	1,970,373,000
EBITDA	5,383,718,000	4,900,076,000	4,784,171,000	4,115,783,000	4,441,583,000	4,013,970,000
Return on Assets %	.10%	.09%	.09%	.09%	.12%	.11%
Return on Equity %	.22%	.21%	.19%	.19%	.25%	.24%
Debt to Equity	.38%	0.397	0.449	0.342	0.303	0.313

CONTACT INFORMATION:

Phone: 6221 808 63539   Fax:
Toll-Free:
Address: Telkom Tower, Fl. 39, Jl. Jendral Gatot Subroto Kav 52 Rt.6/RW.1, Jakarta, 12710 Indonesia

STOCK TICKER/OTHER:

Stock Ticker: TLK   Exchange: NYS
Employees: 25,348   Fiscal Year Ends: 12/31
Parent Company:

SALARIES/BONUSES:

Top Exec. Salary: $ 252   Bonus: $
Second Exec. Salary: $ 252   Bonus: $

OTHER THOUGHTS:

Estimated Female Officers or Directors:
Hot Spot for Advancement for Women/Minorities:

Sales, profits and employees may be estimates. Financial information, benefits and other data can change quickly and may vary from those stated here.

Phazar Antenna Corp

phazar.com

NAIC Code: 334220

TYPES OF BUSINESS:

Antenna Systems & Towers
Wireless Equipment
Contract Manufacturing

BRANDS/DIVISIONS/AFFILIATES:

Antenna Products Corporation
TACSAT-214-SATCOM

CONTACTS:
Note: Officers with more than one job title may be intentionally listed here more than once.

Robert Fitzgerald, Pres.

GROWTH PLANS/SPECIAL FEATURES:

Phazar Antenna Corp. is the wireless infrastructure antenna division of Antenna Products Corporation (ACP). The firm offers a complete line of distributed antenna system (DAS) and small cell antenna solutions built to meet the needs of wireless service, WiMAX, Wi-Fi and broadband internet system suppliers for the purpose of increasing network data capacity and speed via densification. The DAS antennas for cellular, sensorimotor rhythm (SMR), advanced wireless service (AWS) and personal communication service (PCS) frequencies are installed on utility poles, streetlights, rooftops and lamp posts in urban and remote areas to increase wireless carrier services. These product lines complement APC's existing product lines of cellular, PCS, paging, industrial, scientific, medical, automated meter reading (AMR), omni-directional and sector wireless antennas. The firm's TACSAT-214-SATCOM antenna is custom engineered for special operations soldiers. The antenna is a holster-worn, lightweight mobile system that allows for quick deployment and reliable performance during mission-critical operations. All of Phazar's DAS, oDAS and small cell antennas are designed in-house using state-of-the-art design tools and software. They are available in single, dual, triple, quad and multi-band configurations and in combinations of a plethora of frequencies. Phazar provides fully integrated antenna services, with research and development, engineering, testing, manufacturing, production and tech support housed within its 100,000+ square foot facility located in Texas. Phazar Antenna is utilized by every major U. S. wireless carrier, including Verizon, AT&T, Sprint and T-Mobile.

FINANCIAL DATA:
Note: Data for latest year may not have been available at press time.

In U.S. $	2021	2020	2019	2018	2017	2016
Revenue	14,391,000	13,837,500	13,500,000	13,125,000	12,500,000	12,000,000
R&D Expense						
Operating Income						
Operating Margin %						
SGA Expense						
Net Income						
Operating Cash Flow						
Capital Expenditure						
EBITDA						
Return on Assets %						
Return on Equity %						
Debt to Equity						

CONTACT INFORMATION:

Phone: 940 325-3301   Fax: 940 325-0716
Toll-Free:
Address: 6300 Columbia Rd., Mineral Wells, TX 76067 United States

STOCK TICKER/OTHER:

Stock Ticker: Subsidiary   Exchange:
Employees: 66   Fiscal Year Ends: 06/30
Parent Company: Antenna Products Corporation

SALARIES/BONUSES:

Top Exec. Salary: $ Bonus: $
Second Exec. Salary: $ Bonus: $

OTHER THOUGHTS:

Estimated Female Officers or Directors: 1
Hot Spot for Advancement for Women/Minorities:

Sales, profits and employees may be estimates. Financial information, benefits and other data can change quickly and may vary from those stated here.

Pineapple Energy Inc

pineapple-holdings.com

NAIC Code: 334210

TYPES OF BUSINESS:
Telecommunications Equipment-Voice, Data & Video Communications
Renewable Energy
Solar Power
Solar Storage
Battery Storage
Grid Services

BRANDS/DIVISIONS/AFFILIATES:
Pineapple Energy LLC

GROWTH PLANS/SPECIAL FEATURES:
Pineapple Energy Inc provides households with sustainable solar energy, back-up power and security, control and predictability, and cost savings. It provides a full range of installation services, including the design, engineering procurement, permitting, construction, grid connection warranty, monitoring and maintenance of residential solar energy systems. The company also offers battery storage to customers in select markets and serves a limited number of enterprise customers.

CONTACTS: Note: Officers with more than one job title may be intentionally listed here more than once.
Roger Lacey, CEO
Mark Fandrich, CFO
Kristin Hlavka, Chief Accounting Officer
Michael Siegler, Senior VP, Divisional
Scott Fluegge, Vice President, Divisional

FINANCIAL DATA: Note: Data for latest year may not have been available at press time.

In U.S. $	2021	2020	2019	2018	2017	2016
Revenue		42,575,544	50,906,180	65,762,944	82,322,616	99,352,936
R&D Expense						
Operating Income						
Operating Margin %						
SGA Expense						
Net Income		-171,658	6,469,049	-6,791,735	-11,825,632	-8,113,548
Operating Cash Flow						
Capital Expenditure						
EBITDA						
Return on Assets %						
Return on Equity %						
Debt to Equity						

CONTACT INFORMATION:
Phone: 952 996-1674   Fax: 320 848-2702
Toll-Free:
Address: 10900 Red Circle Dr., Minnetonka, MN 55343 United States

STOCK TICKER/OTHER:
Stock Ticker: PEGY   Exchange: NAS
Employees: 150   Fiscal Year Ends: 12/31
Parent Company:

SALARIES/BONUSES:
Top Exec. Salary: $ Bonus: $
Second Exec. Salary: $ Bonus: $

OTHER THOUGHTS:
Estimated Female Officers or Directors: 2
Hot Spot for Advancement for Women/Minorities:

Planet Labs PBC

www.planet.com

NAIC Code: 517410

TYPES OF BUSINESS:

Satellite Telecommunications Services
Satellite Design and Manufacture
Satellite Imagery
Mapping
Whole-Earth Imaging Datasets
Analytics
Machine Learning
Cloud-based Platform

GROWTH PLANS/SPECIAL FEATURES:

Planet Labs PBC is an Earth imaging company. The company provides daily satellite data that helps businesses, governments, researchers, and journalists understand the physical world and take action.

Planet Labs offers its employees comprehensive health benefits, learning/tuition reimbursement and company perks.

BRANDS/DIVISIONS/AFFILIATES:

CONTACTS: Note: Officers with more than one job title may be intentionally listed here more than once.

Will Marshall, CEO
Kevin Weil, Pres.-Product & Business
Ashley Fieglein Johnson, CFO
Rosanne Saccone, CMO
Kristi Erickson, Chief People Officer
Robbie Schingler, Chief Strategy Officer
Brian Hernacki, Sr. VP-Software

FINANCIAL DATA: Note: Data for latest year may not have been available at press time.

In U.S. $	2021	2020	2019	2018	2017	2016
Revenue	113,168,000	95,736,000				
R&D Expense	43,825,000	37,871,000				
Operating Income	-87,442,000	-106,460,000				
Operating Margin %	-.77%	-1.11%				
SGA Expense	69,402,000	61,932,000				
Net Income	-127,103,000	-123,714,000				
Operating Cash Flow	-4,027,000	-33,687,000				
Capital Expenditure	30,126,000	24,101,000				
EBITDA	-54,371,000	-39,009,000				
Return on Assets %	-.33%	-.34%				
Return on Equity %	-.78%	-.57%				
Debt to Equity	1.45%	0.214				

CONTACT INFORMATION:

Phone: 415-829-3313   Fax:
Toll-Free:
Address: 645 Harrison St., Fl. 4, San Francisco, CA 94107 United States

STOCK TICKER/OTHER:

Stock Ticker: PL   Exchange: NYS
Employees: 800   Fiscal Year Ends: 01/31
Parent Company:

SALARIES/BONUSES:

Top Exec. Salary: $367,500   Bonus: $26,057
Second Exec. Salary: Bonus: $
$275,000

OTHER THOUGHTS:

Estimated Female Officers or Directors:
Hot Spot for Advancement for Women/Minorities:

Sales, profits and employees may be estimates. Financial information, benefits and other data can change quickly and may vary from those stated here.

Plantronics Inc

NAIC Code: 334210

www.poly.com

TYPES OF BUSINESS:
Communications Headsets
Communications Accessories
Specialty Telephone Products
Wireless Headsets

BRANDS/DIVISIONS/AFFILIATES:
Poly
RealPresence

GROWTH PLANS/SPECIAL FEATURES:

Plantronics Inc designs and manufactures lightweigl
communications headsets, telephone headset systems, an
other communications endpoints. The firm's headsets are use
for unified communications applications in contact centers, wi
mobile devices and Internet telephony, for gaming, and f
other applications. Its products are shipped through a netwo
of distributors, retailers, wireless carriers, original equipme
manufacturers, and other service providers. More than half
the firm's revenue is generated in the United States, with th
rest coming from Europe, Africa, Asia Pacific, and oth
regions.

CONTACTS: *Note: Officers with more than one job title may be intentionally listed here more than once.*
David Shull, CEO
Charles Boynton, CFO
Robert Hagerty, Chairman of the Board
Kristine Diamond, Chief Accounting Officer
Lisa Bodensteiner, Chief Compliance Officer
Marvin Tseu, Director
Tom Puorro, Executive VP
Carl Wiese, Executive VP

FINANCIAL DATA: *Note: Data for latest year may not have been available at press time.*

In U.S. $	2021	2020	2019	2018	2017	2016
Revenue	1,727,607,000	1,696,990,000	1,674,535,000	856,903,000	881,176,000	
R&D Expense	209,290,000	218,277,000	201,886,000	84,193,000	88,318,000	
Operating Income	78,883,000	-261,505,000	-75,626,000	125,532,000	129,222,000	
Operating Margin %	.05%	-.15%	-.05%	.15%	.15%	
SGA Expense	488,378,000	595,463,000	567,879,000	229,390,000	223,830,000	
Net Income	-57,331,000	-827,182,000	-135,561,000	-869,000	82,599,000	
Operating Cash Flow	145,180,000	78,019,000	116,047,000	121,148,000	139,387,000	
Capital Expenditure	22,715,000	22,880,000	26,797,000	12,468,000	23,176,000	
EBITDA	182,593,000	-573,681,000	98,677,000	150,702,000	150,199,000	
Return on Assets %	-.02%	-.31%	-.06%	.00%	.08%	
Return on Equity %		-2.59%	-.25%	.00%	.24%	
Debt to Equity			2.274	1.395	1.285	

CONTACT INFORMATION:
Phone: 831 426-5858   Fax: 831 426-6098
Toll-Free: 800-544-4660
Address: 345 Encinal St., Santa Cruz, CA 95060 United States

STOCK TICKER/OTHER:
Stock Ticker: POLY
Employees: 6,500
Parent Company:

Exchange: NYS
Fiscal Year Ends: 03/31

SALARIES/BONUSES:
Top Exec. Salary: $500,000   Bonus: $
Second Exec. Salary: Bonus: $
$493,200

OTHER THOUGHTS:
Estimated Female Officers or Directors: 4
Hot Spot for Advancement for Women/Minorities: Y

PLDT Inc

www.pldt.com.ph

NAIC Code: 517110

TYPES OF BUSINESS:

Diversified Telecommunications Services
Telecommunications
Digital Services
Fixed Line
Broadband
Wireless Services

GROWTH PLANS/SPECIAL FEATURES:

PLDT Inc is the telecommunications carrier in the Philippines. The company operates the nation's most extensive fixed-line network and holds more than half of the market share in the domestic wireless space. First Pacific, a Hong Kong-based investment and management company specializing in consumer and telecommunication businesses, owns about 26% of PLDT, while two subsidiaries of Nippon Telegraph and Telephone jointly own about 21%.

BRANDS/DIVISIONS/AFFILIATES:

Voyager Innovations Holdings Pte Ltd
Multisys Technologies Corporation

CONTACTS:
Note: Officers with more than one job title may be intentionally listed here more than once.

Maria Lourdes C. Rausa-Chan, General Counsel
Anabelle L. Chua, Sr. VP-Corp. Finance
Ray C. Espinosa, Head-Regulatory Affairs & Policies
Ernesto R. Alberto, Exec. VP-Enterprise & Intl Carrier Bus.
Claro Carmelo P. Ramirez, Sr. VP
Jun R. Florencio, Sr. VP-Internal Audit & Fraud Risk Mgmt.
Manuel V. Pangilinan, Chmn.
Rene G. Banez, Sr. VP-Supply Chain, Asset Protection & Mgmt.

FINANCIAL DATA:
Note: Data for latest year may not have been available at press time.

In U.S. $	2021	2020	2019	2018	2017	2016
Revenue	3,483,991,000	3,263,097,000	3,050,063,000	2,936,975,000	2,865,206,000	2,979,304,000
R&D Expense						
Operating Income	871,462,000	838,417,200	815,449,800	416,225,000	320,335,300	644,402,400
Operating Margin %	.25%	.26%	.27%	.14%	.11%	.22%
SGA Expense	838,236,900	824,031,000	773,300,900	968,433,300	927,366,100	840,418,200
Net Income	475,338,000	437,786,200	406,003,200	341,013,200	241,049,200	360,663,400
Operating Cash Flow	1,658,013,000	1,533,730,000	1,250,982,000	1,101,785,000	1,011,610,000	882,927,700
Capital Expenditure	1,845,953,000	1,379,178,000	1,593,492,000	852,136,300	662,574,300	764,701,600
EBITDA	1,792,933,000	1,676,654,000	1,463,998,000	1,403,912,000	1,344,006,000	1,165,333,000
Return on Assets %	.04%	.04%	.04%	.04%	.03%	.04%
Return on Equity %	.22%	.21%	.20%	.17%	.12%	.18%
Debt to Equity	2.10%	1.925	1.668	1.393	1.483	1.41

CONTACT INFORMATION:

Phone: 632-816-8556   Fax: 632-840-1864
Toll-Free:
Address: Makati Ave., Ramon Cojuangco Bldg., Makati, 1200 Philippines

STOCK TICKER/OTHER:

Stock Ticker: PHI
Employees: 18,822
Parent Company:

Exchange: NYS
Fiscal Year Ends: 12/31

SALARIES/BONUSES:

Top Exec. Salary: $ Bonus: $
Second Exec. Salary: $ Bonus: $

OTHER THOUGHTS:

Estimated Female Officers or Directors: 10
Hot Spot for Advancement for Women/Minorities: Y

Sales, profits and employees may be estimates. Financial information, benefits and other data can change quickly and may vary from those stated here.

Potevio Corporation

www.cccme.org.cn/shop/cccme3055/index.aspx

NAIC Code: 334220

TYPES OF BUSINESS:

Communications Equipment-Telecom & Mobile
Information Technology
Telecommunications
Equipment
Electronics
Automotive Parts
Terminal Products
Storage

BRANDS/DIVISIONS/AFFILIATES:

Potevio Capitel
Potevio Eastcom
Ningbo Electronics
Beijing Ericsson Potevio
Nanjing Ericsson Panda
Potevio Taili
Potevio Designing & Planning Institute

CONTACTS: *Note: Officers with more than one job title may be intentionally listed here more than once.*

Liang Sun, Gen. Mgr.-Broadcasting Dept.
Xu Mingwen, VP
Lv Weiping, Chmn.

GROWTH PLANS/SPECIAL FEATURES:

Potevio Corporation is an information technology (IT) products and service provider for the telecommunication industry. Based in China, the company consists of seven subsidiaries: Potevio Capitel, Potevio Eastcom, Ningbo Electronics, Beijing Ericsson Potevio, Nanjing Ericsson Panda, Potevio Taili, and Potevio Designing & Planning Institute for Engineering. Together, these businesses cover both fixed and mobile communications. The firm's product categories consist of communication devices, computer products, electronic apparatus, office supplies, machinery equipment, automotive and motorcycle parts, electrical wire and other electrical parts and components. Products include communication equipment and terminal products include mobile communication network equipment and handsets, optical transmission equipment, communication cables, power connectors, power supplies, microwave communication equipment, telecommunication network operation support systems (OSS), stored program control (SPC) switches, videophones, integrated circuit card payphones, personal handphone system (PHS) handsets, logistics information systems, ITS series products, office information equipment and much more.

FINANCIAL DATA: *Note: Data for latest year may not have been available at press time.*

In U.S. $	2021	2020	2019	2018	2017	2016
Revenue	13,122,837,000	12,618,112,500	12,017,250,000	11,445,000,000	10,900,000,000	10,732,293,749
R&D Expense						
Operating Income						
Operating Margin %						
SGA Expense						
Net Income						
Operating Cash Flow						
Capital Expenditure						
EBITDA						
Return on Assets %						
Return on Equity %						
Debt to Equity						

CONTACT INFORMATION:

Phone: 86 10-6268-3863   Fax: 86-10-62683898
Toll-Free:
Address: 6 Beier St., Haidian Distr., Beijing, Beijing 100080 China

STOCK TICKER/OTHER:

Stock Ticker: Government-Owned
Employees: 22,000
Parent Company: SASAC

Exchange:
Fiscal Year Ends: 12/31

SALARIES/BONUSES:

Top Exec. Salary: $ Bonus: $
Second Exec. Salary: $ Bonus: $

OTHER THOUGHTS:

Estimated Female Officers or Directors: 1
Hot Spot for Advancement for Women/Minorities:

Sales, profits and employees may be estimates. Financial information, benefits and other data can change quickly and may vary from those stated here.

Preformed Line Products Company
www.preformed.com

NAIC Code: 335921

TYPES OF BUSINESS:
Electrical Equipment/Appliances/Tools, Manufacturing
Network Products and Systems
Material & Product Design

BRANDS/DIVISIONS/AFFILIATES:

GROWTH PLANS/SPECIAL FEATURES:
Preformed Line Products Co is a designer and manufacturer of products and systems for construction and maintenance of overhead and underground networks for the energy, telecommunication, cable operators, data communication and other industries. In addition, it provides solar hardware systems and mounting hardware for a variety of solar power applications. Its products consist of Energy Products, Communications Products, and Special Industries Products. The company's majority of the revenue is derived from the sale of products in the United States with operations also in The Americas, Europe, the Middle East and Africa, and Asia-Pacific. Most of its revenue gets driven by Energy products which comprise protecting transmission conductors, spacers, spacer-dampers, Stockbridge dampers.

CONTACTS: Note: Officers with more than one job title may be intentionally listed here more than once.
Robert Ruhlman, CEO
Andrew Klaus, CFO
Dennis Mckenna, COO
Barbara Ruhlman, Director Emeritus
Jon Ruhlman, Director
John Hofstetter, Executive VP, Geographical
Caroline Vaccariello, General Counsel
Tim O'Shaughnessy, Vice President, Divisional
John Olenik, Vice President, Divisional
William Haag, Vice President, Geographical

FINANCIAL DATA: Note: Data for latest year may not have been available at press time.

In U.S. $	2021	2020	2019	2018	2017	2016
Revenue	517,417,000	466,449,000	444,861,000	420,878,000	378,212,000	336,634,000
R&D Expense	19,188,000	17,625,000	17,187,000	15,107,000	14,327,000	14,025,000
Operating Income	47,549,000	40,207,000	32,627,000	35,368,000	27,093,000	21,533,000
Operating Margin %	.09%	.09%	.07%	.08%	.07%	.06%
SGA Expense	95,796,000	91,972,000	88,415,000	81,756,000	77,208,000	73,856,000
Net Income	35,729,000	29,803,000	23,303,000	26,581,000	12,654,000	15,255,000
Operating Cash Flow	33,598,000	41,642,000	27,217,000	22,976,000	33,830,000	25,974,000
Capital Expenditure	18,384,000	24,569,000	29,467,000	9,528,000	11,233,000	24,725,000
EBITDA	66,483,000	56,805,000	47,423,000	46,322,000	39,657,000	33,793,000
Return on Assets %	.08%	.07%	.06%	.07%	.04%	.05%
Return on Equity %	.12%	.11%	.09%	.11%	.05%	.07%
Debt to Equity	.15%	0.144	0.231	0.10	0.145	0.192

CONTACT INFORMATION:
Phone: 440 461-5200   Fax: 440 442-8816
Toll-Free:
Address: 660 Beta Dr., Mayfield Village, OH 44143 United States

STOCK TICKER/OTHER:
Stock Ticker: PLPC   Exchange: NAS
Employees: 2,927   Fiscal Year Ends: 12/31
Parent Company:

SALARIES/BONUSES:
Top Exec. Salary: $925,008   Bonus: $
Second Exec. Salary: Bonus: $
$500,004

OTHER THOUGHTS:
Estimated Female Officers or Directors:
Hot Spot for Advancement for Women/Minorities:

Sales, profits and employees may be estimates. Financial information, benefits and other data can change quickly and may vary from those stated here.

Proxim Wireless Corporation

www.proxim.com

NAIC Code: 334220

TYPES OF BUSINESS:

Wireless Networking Equipment
Home & Office Networking Equipment
Millimeter Wave Products
Wireless Systems
Product Manufacturing
Multi-Point Wireless Systems

BRANDS/DIVISIONS/AFFILIATES:

SRA Holdings Inc
Proxim SmartConnect
Proxim ClearConnect
WORP
Proxim FastConnect

CONTACTS: Note: Officers with more than one job title may be intentionally listed here more than once.

Fred Huey, CEO
Lee Gopadze, Pres.
David L. Renauld, General Counsel
David L. Renauld, VP-Corp. Affairs
Yoram Rubin, VP-Customer Support & Quality Mgmt.

GROWTH PLANS/SPECIAL FEATURES:

Proxim Wireless Corporation produces and sells advanced W
Fi, point-to-point and point-to-multipoint outdoor wireles
systems. These systems are primarily built for mission-critica
and high-availability communications environments. Th
company's products serve a wide variety of markets, includin
enterprises, service providers, carriers, government
municipalities, Wi-Fi operators, hot spot operators and othe
organizations that need high-performance, secure scalabl
wireless solutions. Proxim's solutions include smart cit
mobility, video surveillance, backhaul, wireless broadban
internet service provider (ISP) and Wi-Fi. Technology-wise
Proxim's proprietary products include: Proxim SmartConnec
an intelligent interface avoidance technology designed 1
deliver high-performance in high-density environments an
under challenging radio frequency interference condition
Proxim ClearConnect, a suite of interference mitigatio
technologies ensuring robust and reliable communications i
high-density wireless deployments via continuous analysis an
automatic link tuning; WORP, a wireless outdoor route
protocol that optimizes the performance of multi-stream voic
video and data over wireless networks; and Proxir
FastConnect, offering high-speed mobility technology for us
on public transportation systems such as buses, trains, metro
and ferries. Based in California, USA, the firm serve
customers all over the world, with offices in over 15 countrie
and a partner network in 105 countries. Proxim is majority-hel
by SRA Holdings, Inc., a Japan-based IT holding company.

Proxim offers its employees medical, dental and visio
insurance; short- and long-term disability coverage; flexibl
spending accounts; and a 401(k) plan.

FINANCIAL DATA: Note: Data for latest year may not have been available at press time.

In U.S. $	2021	2020	2019	2018	2017	2016
Revenue						
R&D Expense						
Operating Income						
Operating Margin %						
SGA Expense						
Net Income						
Operating Cash Flow						
Capital Expenditure						
EBITDA						
Return on Assets %						
Return on Equity %						
Debt to Equity						

CONTACT INFORMATION:

Phone: 408 383-7600   Fax: 408 383-7680
Toll-Free: 800-229-1630
Address: 2114 Ringwood Ave., San Jose, CA 95131 United States

STOCK TICKER/OTHER:

Stock Ticker: Subsidiary   Exchange:
Employees: 353   Fiscal Year Ends: 12/31
Parent Company: SRA Holdings Inc

SALARIES/BONUSES:

Top Exec. Salary: $ Bonus: $
Second Exec. Salary: $ Bonus: $

OTHER THOUGHTS:

Estimated Female Officers or Directors:
Hot Spot for Advancement for Women/Minorities:

Sales, profits and employees may be estimates. Financial information, benefits and other data can change quickly and may vary from those stated here.

Proximus Group

www.proximus.com/en

NAIC Code: 517110

TYPES OF BUSINESS:

Local & Long-Distance Telephone Services
Cell Phone Services
Internet Services
Television Broadcasting
Network IT Consulting & Sourcing
Media Activities

BRANDS/DIVISIONS/AFFILIATES:

Proximus
Scarlet
Tango
Telindus
BICS
Mobile Vikings

GROWTH PLANS/SPECIAL FEATURES:

Proximus is the incumbent telecom operator in Belgium. The firm has around 45% market share of the broadband market and 30% of the postpaid mobile market, mainly competing with Telenet (mobile and fixed), Orange (mobile and fixed) and Voo (fixed). Proximus is rolling out fiber to the home in Belgium and expects to have 70% of the country covered by 2028. Its international carrier services division is one of the four largest in the world, serving more than 250 operators, which was strengthened in 2017 with the acquisition of TeleSign.

CONTACTS: *Note: Officers with more than one job title may be intentionally listed here more than once.*

Guillaume Boutin, CEO
Renaud Tilmans, Chief Customer Operations Officer
Didier Bellens, Pres.
Katleen Vandeweyer, Interim CFO
Jim Casteele, Chief Consumer Market Officer
Jan Van Acoleyen, Chief Human Resources Officer
Geert Standaert, CTO
Bruno Chauvat, Exec. VP-Strategy & Content
Bart Van Den Meersche, Exec. VP-Enterprise Bus. Unit
Geert Standaert, Exec. VP-Svc. Delivery Engine & Wholesale
Dominique Leroy, Exec. VP-Consumer Bus. Unit
Stefaan De Clerck, Chmn.

FINANCIAL DATA: *Note: Data for latest year may not have been available at press time.*

In U.S. $	2021	2020	2019	2018	2017	2016
Revenue	5,646,428,000	5,550,570,000	5,749,424,000	5,877,914,000	5,852,420,000	5,944,199,000
R&D Expense						
Operating Income	655,707,600	817,849,900	552,711,600	770,940,800	800,513,900	805,612,700
Operating Margin %	.12%	.15%	.10%	.13%	.14%	.14%
SGA Expense			229,446,700	268,197,700	250,861,700	256,980,300
Net Income	451,755,000	575,146,300	380,371,600	518,039,600	532,316,300	533,336,000
Operating Cash Flow	1,653,036,000	1,544,941,000	1,687,708,000	1,588,791,000	1,499,052,000	1,551,059,000
Capital Expenditure	1,159,470,000	1,110,522,000	1,112,562,000	1,120,720,000	1,008,546,000	981,012,000
EBITDA	1,848,830,000	1,958,965,000	1,733,597,000	1,814,158,000	1,795,803,000	1,751,953,000
Return on Assets %	.05%	.06%	.04%	.06%	.06%	.06%
Return on Equity %	.15%	.20%	.13%	.17%	.18%	.19%
Debt to Equity	.99%	0.938	0.91	0.822	0.65	0.623

CONTACT INFORMATION:

Phone: 32 22024111   Fax: 32 22036593
Toll-Free:
Address: 27 Blvd. du Roi Albert II, Brussels, 1030 Belgium

STOCK TICKER/OTHER:

Stock Ticker: BGAOF   Exchange: PINX
Employees: 11,532   Fiscal Year Ends: 12/31
Parent Company:

SALARIES/BONUSES:

Top Exec. Salary: $ Bonus: $
Second Exec. Salary: $ Bonus: $

OTHER THOUGHTS:

Estimated Female Officers or Directors: 6
Hot Spot for Advancement for Women/Minorities: Y

Sales, profits and employees may be estimates. Financial information, benefits and other data can change quickly and may vary from those stated here.

Purple Communications Inc

NAIC Code: 517210

TYPES OF BUSINESS:

Wireless Data Products
Deaf & Speech-Impaired Communications Systems
Video Relay Services
Telecommunications Services
Internet-based Communication Services

BRANDS/DIVISIONS/AFFILIATES:

ZP Better Together LLC
POP

CONTACTS: Note: Officers with more than one job title may be intentionally listed here more than once.

Sherri Turpin, CEO
Chris Wagner, COO
John Ferron, Pres.
Zarko Roganovic, CTO
Bill Billbrough, CIO
John Goodman, Chief Legal Officer
Francine Cummings, VP-Oper.
Rita Beier Braman, Sr. Dir.-Oper. & Quality, IP-Relay & ClearCaptions
Gordon Ellis, Sr. VP

GROWTH PLANS/SPECIAL FEATURES:

Purple Communications, Inc. is a provider of versatile video relay services (VRS) and professional interpreting for individuals with hearing or speech disabilities. The company offers both major categories of relay services: traditional telecommunications and internet-based services. VRS provides deaf and hard-of-hearing (HOH) individuals and business owners a means to communicate over video desktop and mobile devices (both iOS and Android) with hearing individuals in real-time using American Sign Language (ASL) interpreters. This service is free to deaf and HOH customers. On-site interpreting consists of a team of national-qualified interpreters that can provide services for ASL and English as well as Spanish ASL interpreters. Interpreters provide services for any in-person interpretation needs, including job interviews; special events such as concerts, parties and plays; business meetings; conferences; and employee training. Video remote interpreting (VRI) is an on-demand sign language interpreting service delivered over a live internet video connection. VRI is beneficial for schools, business meetings, medical appointments/hospitals, conferences and more. VRS services are offered 24/7/365 under the Purple brand name. POP is a fully customizable light-emitting diode (LED) light bulb capable of producing millions of colors. When someone calls a Purple customer, Purple sends a signal to the POP to flash, informing the customer that they have an incoming call. POP comprises a management tool built into the mobile app. Purple Communications operates as a subsidiary of ZP Better Together, LLC.

FINANCIAL DATA: Note: Data for latest year may not have been available at press time.

In U.S. $	2021	2020	2019	2018	2017	2016
Revenue	167,346,270	160,909,875	153,247,500	145,950,000	139,000,000	137,700,000
R&D Expense						
Operating Income						
Operating Margin %						
SGA Expense						
Net Income						
Operating Cash Flow						
Capital Expenditure						
EBITDA						
Return on Assets %						
Return on Equity %						
Debt to Equity						

CONTACT INFORMATION:

Phone: 727 260-6838   Fax:
Toll-Free: 800 900-9478
Address: 11900 N. Jollyville Rd. #204209, Austin, TX 78759 United States

STOCK TICKER/OTHER:

Stock Ticker: Subsidiary   Exchange:
Employees: 1,500   Fiscal Year Ends: 12/31
Parent Company: ZP Better Together LLC

SALARIES/BONUSES:

Top Exec. Salary: $ Bonus: $
Second Exec. Salary: $ Bonus: $

OTHER THOUGHTS:

Estimated Female Officers or Directors: 3
Hot Spot for Advancement for Women/Minorities: Y

Q Beyond AG

www.qbeyond.de/en

NAIC Code: 517110

TYPES OF BUSINESS:
Wired Telecommunications Carriers

GROWTH PLANS/SPECIAL FEATURES:
Q.Beyond AG helps its customers find the best digital solutions for their business and then put them into practice. The company is an expert in Cloud, SAP and IoT. It focuses on a digital branch, tailored cloud, SAP S / 4HANA, and cyber security.

BRANDS/DIVISIONS/AFFILIATES:
Next Generation Network
INFO AG
QSC AG

CONTACTS: Note: Officers with more than one job title may be intentionally listed here more than once.
Jurgen Hermann, CEO
Thies Rixen, COO
Christoph Reif, CFO

FINANCIAL DATA: Note: Data for latest year may not have been available at press time.

In U.S. $	2021	2020	2019	2018	2017	2016
Revenue	158,227,400	146,250,300	242,751,500	370,567,600	364,939,500	393,607,100
R&D Expense						
Operating Income	-17,504,230	-21,294,690	-31,184,350	8,575,187	8,384,492	851,502
Operating Margin %	-.11%	-.15%	-.13%	.02%	.02%	.00%
SGA Expense	35,823,260	33,234,080	62,724,600	25,419,630	24,562,010	31,116,030
Net Income	9,903,938	-20,290,220	75,073,930	3,522,262	5,441,456	-25,329,890
Operating Cash Flow	-7,813,424	-5,064,143	-18,060,000	34,799,410	40,073,630	41,087,270
Capital Expenditure	7,418,776	5,692,317	12,680,750	18,144,640	21,830,070	26,601,540
EBITDA	31,720,750	-2,024,230	124,652,800	35,830,390	38,683,690	26,639,270
Return on Assets %	.05%	-.10%	.29%	.01%	.02%	-.08%
Return on Equity %	.07%	-.13%	.59%	.04%	.06%	-.25%
Debt to Equity	.04%	0.091	0.104	1.101	1.502	1.682

CONTACT INFORMATION:
Phone: 49 2216698000   Fax: 49 2216698009
Toll-Free:
Address: Mathias-Brueggen-Street 55, Cologne, NW 50829 Germany

STOCK TICKER/OTHER:
Stock Ticker: QSCGF   Exchange: PINX
Employees: 1,139   Fiscal Year Ends: 12/31
Parent Company:

SALARIES/BONUSES:
Top Exec. Salary: $ Bonus: $
Second Exec. Salary: $ Bonus: $

OTHER THOUGHTS:
Estimated Female Officers or Directors:
Hot Spot for Advancement for Women/Minorities:

Sales, profits and employees may be estimates. Financial information, benefits and other data can change quickly and may vary from those stated here.

Qisda Corporation

NAIC Code: 334419

www.qisda.com

TYPES OF BUSINESS:

LCD (Liquid Crystal Display) Unit Screens Manufacturing
LCD Flat-Panel Displays
LCD TVs
Projectors
Multifunctional Printers
Mobile Phones
Portable Display Devices

BRANDS/DIVISIONS/AFFILIATES:

CONTACTS: *Note: Officers with more than one job title may be intentionally listed here more than once.*

Peter Chen, CEO
James T. Wong, VP-Mfg. Oper.
Mark Hsiao, Sr. VP-Mfg. Oper.
David Wang, Sr. VP-Admin.
Joe Huang, Sr. VP-Display System Products
Jason Tyan, VP-Medical Devices & Products
April Huang, VP-Precise Optical Products
Chinglung Chen, VP-Mobile Communications Products
K.Y. Lee, Chmn.
Mark Hsiao, Managing Dir.-Qisda (Suzhou)
C.M. Wu, Sr. VP-Supply Chain Mgmt.

GROWTH PLANS/SPECIAL FEATURES:

Qisda Corporation is an original design manufacturer (ODM) and an original equipment manufacturer (OEM) of electronic products for consumer, commercial, medical and industrial applications. Qisda's products include LCD (liquid crystal display) monitors, commercial LCD TVs, all-in-one PCs, projectors, multifunctional printers, mobile phones, car infotainment displays, medical electronics, portable display devices and smart business solutions. The firm has four research and development centers in Taiwan and China, as well as manufacturing locations in China, Mexico and Taiwan. Qisda's in-house manufacturing capabilities include surface mount technology (SMT), metal stamping, plastic injection and LCD module assembling. Main product categories include displays, imaging devices, smart business solutions, healthcare solutions, opto-mechatronic products, mobile communication devices, industrial solutions and automobile infotainment devices. LCD monitors include wide format models from 15-27 inches, traditional format models from 15-20 inches, monitor-TVs and pen/touch displays. Its LCD TVs include sizes ranging from 15 to more than 45 inches. Qisda's projectors include LCD projectors, home theater projectors and short-throw projectors for home and office settings. Printer offerings include color inkjet printers, color inkjet multifunctional printers, color laser printers (CLPs), color laser multifunctional printers (MFPs), high-speed inkjet and multifunctional printers, image scanners, mobile scanners and medical electronics. Its mobile communications devices include mobile internet devices (MIDs) and tablets, dual mode phones, wireless modules and USB modem modules. Qisda's infotainment products include car electronics, such as navigation systems, TFT-LCD displays and vehicle rear seat entertainment systems; portable display devices, such as mobile digital TVs, e-readers, digital photo frames and GPS units; and general displays for professional, public, industrial and e-signage use. Smart business solutions include light software-hardware integration and innovative Internet of Things (IoT) applications for six major business domains: retail, manufacturing, healthcare, school, enterprise and energy.

FINANCIAL DATA: *Note: Data for latest year may not have been available at press time.*

In U.S. $	2021	2020	2019	2018	2017	2016
Revenue	709,058,400	6,817,870,000	5,341,780,500	5,150,260,000	4,469,530,000	3,988,530,000
R&D Expense						
Operating Income						
Operating Margin %						
SGA Expense						
Net Income	225,961,000	226,427,000	125,182,450	131,771,000	190,361,000	125,747,000
Operating Cash Flow						
Capital Expenditure						
EBITDA						
Return on Assets %						
Return on Equity %						
Debt to Equity						

CONTACT INFORMATION:

Phone: 886-3-3595-000   Fax: 886-3-3599-000
Toll-Free:
Address: 157 Shan-ying Rd., Gueishan, Taoyuan, 333 Taiwan

STOCK TICKER/OTHER:

Stock Ticker: 2352   Exchange: TWSE
Employees: 9,985   Fiscal Year Ends: 12/31
Parent Company:

SALARIES/BONUSES:

Top Exec. Salary: $ Bonus: $
Second Exec. Salary: $ Bonus: $

OTHER THOUGHTS:

Estimated Female Officers or Directors: 1
Hot Spot for Advancement for Women/Minorities:

Qualcomm Incorporated

NAIC Code: 334413

www.qualcomm.com

TYPES OF BUSINESS:

Telecommunications Equipment
Digital Wireless Communications Products
Integrated Circuits
Mobile Communications Systems
Wireless Software & Services
E-Mail Software
Code Division Multiple Access

BRANDS/DIVISIONS/AFFILIATES:

GROWTH PLANS/SPECIAL FEATURES:

Qualcomm develops and licenses wireless technology and designs chips for smartphones. The company's key patents revolve around CDMA and OFDMA technologies, which are standards in wireless communications that are the backbone of all 3G and 4G networks. The firm is a leader in 5G network technology as well. Qualcomm's IP is licensed by virtually all wireless device makers. The firm is also the world's largest wireless chip vendor, supplying nearly every premier handset maker with leading-edge processors. Qualcomm also sells RF-front end modules into smartphones and chips into automotive and Internet of Things markets.

U.S. employees of the company receive medical, dental and vision insurance; dependent/health care reimbursement accounts; tuition reimbursement; a 401(k); and an employee stock purchase plan.

CONTACTS: Note: Officers with more than one job title may be intentionally listed here more than once.

Cristiano Amon, CEO
Akash Palkhiwala, CFO
Mark McLaughlin, Chairman of the Board
Erin Polek, Chief Accounting Officer
James Thompson, Chief Technology Officer
Ann Cathcart Chaplin, General Counsel
Heather Ace, Other Executive Officer
Alexander Rogers, President, Subsidiary
Rogerio Amon, Vice President, Subsidiary

FINANCIAL DATA: Note: Data for latest year may not have been available at press time.

In U.S. $	2021	2020	2019	2018	2017	2016
Revenue	33,566,000,000	23,531,000,000	24,273,000,000	22,611,000,000	22,258,000,000	23,554,000,000
R&D Expense	7,176,000,000	5,975,000,000	5,398,000,000	5,625,000,000	5,485,000,000	5,151,000,000
Operating Income	9,789,000,000	6,255,000,000	7,667,000,000	621,000,000	2,581,000,000	6,495,000,000
Operating Margin %	.29%	.27%	.32%	.03%	.12%	.28%
SGA Expense	2,339,000,000	2,074,000,000	2,195,000,000	2,986,000,000	2,658,000,000	2,385,000,000
Net Income	9,043,000,000	5,198,000,000	4,386,000,000	-4,964,000,000	2,445,000,000	5,705,000,000
Operating Cash Flow	10,536,000,000	5,814,000,000	7,286,000,000	3,908,000,000	5,001,000,000	7,632,000,000
Capital Expenditure	1,888,000,000	1,407,000,000	887,000,000	784,000,000	690,000,000	539,000,000
EBITDA	12,415,000,000	7,714,000,000	9,509,000,000	2,721,000,000	4,942,000,000	8,558,000,000
Return on Assets %	.24%	.15%	.13%	-.10%	.04%	.11%
Return on Equity %	1.13%	.95%	1.53%	-.31%	.08%	.18%
Debt to Equity	1.38%	2.506	2.737	19.04	0.631	0.315

CONTACT INFORMATION:

Phone: 858 587-1121   Fax: 858 658-2100
Toll-Free:
Address: 5775 Morehouse Dr., San Diego, CA 92121-1714 United States

STOCK TICKER/OTHER:

Stock Ticker: QCOM   Exchange: NAS
Employees: 41,000   Fiscal Year Ends: 09/30
Parent Company:

SALARIES/BONUSES:

Top Exec. Salary: $575,016   Bonus: $1,500,000
Second Exec. Salary: $1,064,590   Bonus: $

OTHER THOUGHTS:

Estimated Female Officers or Directors: 2
Hot Spot for Advancement for Women/Minorities: Y

Sales, profits and employees may be estimates. Financial information, benefits and other data can change quickly and may vary from those stated here.

Rackspace Technology Inc

www.rackspace.com

NAIC Code: 517110

TYPES OF BUSINESS:

Web Hosting Services
Data Centers
Cloud Computing Services
Server Farms

BRANDS/DIVISIONS/AFFILIATES:

Apollo Global Management LLC
Onica

GROWTH PLANS/SPECIAL FEATURES:

Rackspace Technology Inc is an end-to-end multi cloud technology services company. It designs, builds and operates its customers' cloud environments across all technology platforms, irrespective of technology stack or deployment model. The company's solutions include Application Services Data; Colocation; Managed Cloud; Managed Hosting Professional Services; and Security & Compliance. It operates in three reportable segments Multicloud Services; Apps & Cross-Platform; and OpenStack Public Cloud. It generates revenue through the sale of consumption-based contracts for its services offerings and from the sale of professional services related to designing and building custom solutions.

CONTACTS:
Note: Officers with more than one job title may be intentionally listed here more than once.

Kevin Jones, CEO
Amar Maletira, CFO
David Sambur, Chairman of the Board
Mark Marino, Chief Accounting Officer
Holly Windham, Chief Legal Officer
Zarina Stanford, Chief Marketing Officer
Tolga Tarhan, Chief Technology Officer
Subroto Mukerji, COO
John McCabe, Executive VP, Geographical
Sandeep Bhargava, Managing Director, Geographical
Martin Blackburn, Managing Director, Geographical
Neil Emerson, Other Executive Officer
Thomas Wolf, Senior VP, Divisional

FINANCIAL DATA:
Note: Data for latest year may not have been available at press time.

In U.S. $	2021	2020	2019	2018	2017	2016
Revenue	3,009,500,000	2,707,100,000	2,438,100,000	2,452,800,000	2,144,700,000	
R&D Expense						
Operating Income	30,000,000	24,700,000	99,500,000	57,800,000	-122,800,000	
Operating Margin %	.01%	.01%	.04%	.02%	-.06%	
SGA Expense	906,800,000	959,700,000	911,700,000	949,300,000	942,200,000	
Net Income	-218,300,000	-245,800,000	-102,300,000	-470,600,000	-59,900,000	
Operating Cash Flow	370,800,000	116,700,000	292,900,000	429,800,000	291,700,000	
Capital Expenditure	108,400,000	116,500,000	198,000,000	294,300,000	189,500,000	
EBITDA	446,700,000	493,300,000	774,100,000	391,600,000	620,600,000	
Return on Assets %	-.03%	-.04%	-.02%	-.08%		
Return on Equity %	-.16%	-.22%	-.11%	-.52%		
Debt to Equity	2.89%	2.797	4.757	4.651		

CONTACT INFORMATION:

Phone: 210 312-4000   Fax: 210 312-4300
Toll-Free: 800-961-2888
Address: 1 Fanatical Pl., City of Windcrest, San Antonio, TX 78218
United States

STOCK TICKER/OTHER:

Stock Ticker: RXT					Exchange: NAS
Employees: 6,600				Fiscal Year Ends: 12/31
Parent Company: Apollo Global Management LLC

SALARIES/BONUSES:

Top Exec. Salary: $856,731   Bonus: $5,977,625
Second Exec. Salary: $586,822   Bonus: $899,176

OTHER THOUGHTS:

Estimated Female Officers or Directors: 1
Hot Spot for Advancement for Women/Minorities:

Realtime Corporation
www.realtimecorp.com.br

IAIC Code: 511210M

TYPES OF BUSINESS:
Cloud-based Messaging
Web Page Tracking Tools

BRANDS/DIVISIONS/AFFILIATES:
Internet Business Technologies
PowerMarketing
Realtime DMC
iG
HIS
Mobbit
WebSpectator

CONTACTS: Note: Officers with more than one job title may be intentionally listed here more than once.
Andre Tomas Parreira, CEO
Alexandre Botelho, Exec. VP-Sales
David Facter, Dir.-Client Acquisition, U.S.
Gilberto Martins, Head-Brazil Oper.

GROWTH PLANS/SPECIAL FEATURES:
Realtime Corporation is a technology company that offers a variety of cloud-based messaging tools to ensure constant accurate content updates to websites, mobile apps and social channels. Realtime offers its customized services for businesses, individual users and developers. The company's tools allow clients to receive up-to-date site traffic data, including length of visit, page views and number of visitors in real-time. The firm's solutions include PowerMarketing, Realtime DCM, iG, HIS, Mobbit, and WebSpectator. PowerMarketing is a platform for monitoring and interacting in real-time with website visitors (of the company's own website), delivering live and unlimited interaction. Realtime DCM (dynamic cloud manager) is a real-time infrastructure manager for public and private clouds. The iG brand represents the company's online products, including iG email, iG online courses (administrative, management, language, financial and more), iG antivirus security, iG Esoteric psychic services (via telephone, email or online chat), and iG online games (cards, crosswords, roulette and more). HIS is the company's subsidiary and brand of e-Health innovation systems specifically designed for the information systems and technologies sector which caters to healthcare markets, primarily in Brazil. Mobbit boosts Internet of Things (IoT). It develops innovative solutions for live stream management, interaction and dynamic communication, leveraged by mobile devices and the IoT. WebSpectator is a partnership vertical, offering an online advertising solution that combines real-time analytics and ad serving. It measures and monetizes the actual time spent viewing ads or videos. Its technology enables web publishers to earn up to four times the revenue by monetizing TV-type impressions of 20 seconds or more, not simply the number of clicks. Realtime is privately-owned by Internet Business Technologies.

FINANCIAL DATA: Note: Data for latest year may not have been available at press time.

In U.S. $	2021	2020	2019	2018	2017	2016
Revenue						
R&D Expense						
Operating Income						
Operating Margin %						
SGA Expense						
Net Income						
Operating Cash Flow						
Capital Expenditure						
EBITDA						
Return on Assets %						
Return on Equity %						
Debt to Equity						

CONTACT INFORMATION:
Phone: 5.5119752553e+12   Fax:
Toll-Free:
Address: Rua Guararapes, Sao Paulo, SP 2064 Brazil

STOCK TICKER/OTHER:
Stock Ticker: Private   Exchange:
Employees: 100   Fiscal Year Ends:
Parent Company: Internet Business Technologies

SALARIES/BONUSES:
Top Exec. Salary: $ Bonus: $
Second Exec. Salary: $ Bonus: $

OTHER THOUGHTS:
Estimated Female Officers or Directors: 1
Hot Spot for Advancement for Women/Minorities:

Sales, profits and employees may be estimates. Financial information, benefits and other data can change quickly and may vary from those stated here.

Rogers Communications Inc

www.rogers.com

NAIC Code: 517110

TYPES OF BUSINESS:

Cable TV Service
Internet Services
Wireless Phone Service
5G Technology
Wireless Services
Connected Home
Internet of Things
Media Services

BRANDS/DIVISIONS/AFFILIATES:

Rogers
Fido
chatr

GROWTH PLANS/SPECIAL FEATURES:

Rogers is the largest wireless service provider in Canada, wit
its more than 10 million subscribers equating to one third of th
total Canadian market. Rogers' wireless business accounte
for 60% of the company's total sales in 2021 and ha
increasingly provided a bigger portion of total company sale
over the last several years. Rogers' cable segment, whic
provides about one fourth of total sales, offers home interne
television, and landline phone service to consumers an
businesses. Remaining sales come from Rogers' media uni
which owns and operates various television and radio statior
and the Toronto Blue Jays. Rogers' significant exposure t
sports also includes ownership stakes in the Toronto Mapl
Leafs, Raptors, FC, and Argonauts.

CONTACTS: *Note: Officers with more than one job title may be intentionally listed here more than once.*

Alan Horn, CEO, Subsidiary
Jordan Banks, Pres., Divisional
Joseph Natale, CEO
Anthony Staffieri, CFO
Edward Rogers, Chairman of the Board
Jorge Fernandes, Chief Information Officer
Graeme McPhail, Chief Legal Officer
Melinda Rogers-Hixon, Deputy Chairman
Philip Lind, Director
Lisa Durocher, Executive VP, Divisional
James Reid, Other Executive Officer
Eric Agius, Other Executive Officer
Sevaun Palvetzian, Other Executive Officer
Dean Prevost, President, Divisional
Brent Johnston, President, Divisional

FINANCIAL DATA: *Note: Data for latest year may not have been available at press time.*

In U.S. $	2021	2020	2019	2018	2017	2016
Revenue	11,398,460,000	10,823,680,000	11,723,580,000	11,741,460,000	11,176,010,000	10,657,230,000
R&D Expense						
Operating Income	2,568,251,000	2,519,250,000	2,896,477,000	2,933,810,000	2,613,362,000	2,142,802,000
Operating Margin %	.23%	.23%	.25%	.25%	.23%	.20%
SGA Expense	1,696,352,000	1,436,572,000	1,559,462,000	1,624,796,000	1,641,907,000	1,612,351,000
Net Income	1,211,791,000	1,238,236,000	1,589,018,000	1,601,462,000	1,435,016,000	649,451,600
Operating Cash Flow	3,236,369,000	3,360,815,000	3,520,261,000	3,335,148,000	3,062,923,000	3,077,701,000
Capital Expenditure	2,210,469,000	1,842,576,000	2,229,914,000	2,212,025,000	1,940,577,000	1,865,132,000
EBITDA	4,352,493,000	4,441,938,000	4,759,275,000	4,502,606,000	4,254,492,000	3,309,481,000
Return on Assets %	.04%	.04%	.06%	.07%	.06%	.03%
Return on Equity %	.15%	.17%	.23%	.26%	.27%	.14%
Debt to Equity	1.78%	1.912	1.855	1.637	1.693	2.44

CONTACT INFORMATION:

Phone: 416 935-2303   Fax: 416 935-3548
Toll-Free:
Address: 333 Bloor St. E., Fl. 10, Toronto, ON M4W 1G9 Canada

STOCK TICKER/OTHER:

Stock Ticker: RCI   Exchange: NYS
Employees: 25,300   Fiscal Year Ends: 12/31
Parent Company:

SALARIES/BONUSES:

Top Exec. Salary: $1,293,048   Bonus: $
Second Exec. Salary: Bonus: $
$756,731

OTHER THOUGHTS:

Estimated Female Officers or Directors: 6
Hot Spot for Advancement for Women/Minorities: Y

Rohde & Schwarz GmbH & Co KG   www.rohde-schwarz.com

NAIC Code: 334515

TYPES OF BUSINESS:

Electronic Test & Measurement Equipment
Electronic Security Products
Test and Measurement Technology
Broadcast Technology
Internet Protocol Network
Secure Network Infrastructure

BRANDS/DIVISIONS/AFFILIATES:

Rohde & Schwarz USA Inc

CONTACTS: Note: Officers with more than one job title may be intentionally listed here more than once.

Christian Leicher, CEO
Peter Riedel, COO

GROWTH PLANS/SPECIAL FEATURES:

Rohde & Schwarz GmbH & Co. KG is a technology firm that develops, produces and sells a wide range of electronic capital goods, with a focus on solutions that contribute to a secure and networked society. The company operates through four divisions: measuring technology, broadcast and media technology, aerospace/defense/security, and networks and cybersecurity. The measuring technology division offers measuring devices and systems for cellular and wireless applications, including industry-specific applications. Market segments within this division include aerospace and defense, automotive, electronics development, EMC testing, mobile network testing, radio frequency (RF) and microwave components/devices, spectrum monitoring and wireless technologies. The broadcast and media technology division serves network operators, content providers, studios and device manufacturers with solutions for the entire transmission chain of audiovisual content, from camera output to broadcast via terrestrial transmitters, satellites or internet protocol (IP) networks. The aerospace, defense and security division offers communication, reconnaissance and security products for the armed forces, authorities, and organizations with security tasks, and operators of critical infrastructures. Last, the networks and cybersecurity division provides businesses and authorities with secure WAN, LAN and WLAN network infrastructures, and offers products that protect data transmission, end devices and applications. Rohde & Schwarz has branch offices in approximately 60 counties. Subsidiary Rohde & Schwarz USA, Inc. is based in Maryland and serves the U.S. market.

FINANCIAL DATA: Note: Data for latest year may not have been available at press time.

In U.S. $	2021	2020	2019	2018	2017	2016
Revenue	2,714,157,600	2,901,780,000	2,396,500,000	2,333,340,000	2,167,770,000	2,105,680,000
R&D Expense						
Operating Income						
Operating Margin %						
SGA Expense						
Net Income						
Operating Cash Flow						
Capital Expenditure						
EBITDA						
Return on Assets %						
Return on Equity %						
Debt to Equity						

CONTACT INFORMATION:

Phone: 49 89-41-29-0   Fax: 49-89-41-29-12-164
Toll-Free:
Address: Muhldorfstrasse 15, Munich, BY 81671 Germany

STOCK TICKER/OTHER:

Stock Ticker: Private   Exchange:
Employees: 13,000   Fiscal Year Ends: 06/30
Parent Company:

SALARIES/BONUSES:

Top Exec. Salary: $ Bonus: $
Second Exec. Salary: $ Bonus: $

OTHER THOUGHTS:

Estimated Female Officers or Directors:
Hot Spot for Advancement for Women/Minorities:

Sales, profits and employees may be estimates. Financial information, benefits and other data can change quickly and may vary from those stated here.

Rostelecom PJSC

NAIC Code: 517110

company.rt.ru/en

TYPES OF BUSINESS:

Telecommunications Services
Long-Distance & International Phone Service
Pre-Paid Services
Virtual Private Networks

BRANDS/DIVISIONS/AFFILIATES:

Federal Agency for State Property Management

CONTACTS: *Note: Officers with more than one job title may be intentionally listed here more than once.*

Mikhail Oseevskiy, Pres.
Sergey Anokhin, CFO
Inna Pokhodnya, VP-Mktg.
Galina Rysakova, Sr. VP-Human Resources
Kirill Menshov, Sr. VP-IT
Ivan Zima, CTO
Alexander Rogovoy, VP-Legal Affairs
Alexander Rogovoy, VP-Corp. Dev.
Roman Frolov, Chief Accountant
Olga Rumyantseva, Dir.-Corp. Clients Sales Dept.
Anton Khozyainov, Sr. VP
Mikhail Magrilov, VP-Strategy & Operational Efficiency
Vladimir Mironov, VP
Sergey Ivanov, Chmn.

GROWTH PLANS/SPECIAL FEATURES:

Rostelecom PJSC is a telecommunications provider, making use of a digital trunk network to interconnect local public operators in a single national network that supports services throughout Russia. The company offers fixed-line domestic and international long-distance services, as well as internet services, directly to corporate and residential customers. Rostelecom has 13.5 million fixed-line broadband subscribers and 10.8 million pay-TV subscribers, of which over 6.3 million also subscribe to the IPTV service. At the network's core are fiber-optic backbone lines linking Moscow to Novorossiysk, Khabarovsk and St. Petersburg. In addition to fixed line domestic and international long-distance telephony, the firm has expanded its services to include interconnection and voice traffic transit services; data-based services for voice transmission, which use a high quality IP/MPLS-based network; interconnection and IP traffic transit services; virtual private network (VPN) services, which join offices and branches of corporate clients or operators into a private secured telecommunications network; data center services; digital circuits between Europe and Asia; domestic long-distance and international rent of channels; and intelligent network services, such as televoting, freephone (toll free numbers) and mass-calling services. State-controlled Federal Agency for State Property Management (FA SPM) is Rostelecom's majority shareholder.

FINANCIAL DATA: *Note: Data for latest year may not have been available at press time.*

In U.S. $	2021	2020	2019	2018	2017	2016
Revenue	7,736,671,887	7,439,107,584	4,268,342,528	4,050,991,872	4,147,286,272	3,973,358,592
R&D Expense						
Operating Income						
Operating Margin %						
SGA Expense						
Net Income		317,089,984	186,927,584	179,046,704	186,046,464	156,972,816
Operating Cash Flow						
Capital Expenditure						
EBITDA						
Return on Assets %						
Return on Equity %						
Debt to Equity						

CONTACT INFORMATION:

Phone: 7 499-999-82-83   Fax: 7 499-999-82-22
Toll-Free:
Address: Goncharnaya Street, 30 (Building 1), Moscow, 115172 Russia

STOCK TICKER/OTHER:

Stock Ticker: RTKM   Exchange: Moscow
Employees: 15,171   Fiscal Year Ends: 12/31
Parent Company:

SALARIES/BONUSES:

Top Exec. Salary: $ Bonus: $
Second Exec. Salary: $ Bonus: $

OTHER THOUGHTS:

Estimated Female Officers or Directors: 3
Hot Spot for Advancement for Women/Minorities: Y

Sales, profits and employees may be estimates. Financial information, benefits and other data can change quickly and may vary from those stated here.

Samsung Electronics Co Ltd

www.samsung.com

NAIC Code: 334310

TYPES OF BUSINESS:

Consumer Electronics
Semiconductors and Memory Products
Smartphones
Computers & Accessories
Digital Cameras
Fuel-Cell Technology
LCD Displays
Solar Energy Panels

BRANDS/DIVISIONS/AFFILIATES:

Samsung Group

GROWTH PLANS/SPECIAL FEATURES:

Samsung Electronics is a diversified electronics conglomerate that manufactures and sells a wide range of products, including smartphones, semiconductor chips, printers, home appliances, medical equipment, and telecom network equipment. About half of its profit is generated from semiconductor business, and a further 30%-35% is generated from its mobile handset business, although these percentages vary with the fortunes of each of these businesses. It is the largest smartphone and television manufacturer in the world, which helps provide a base demand for its component businesses, such as memory chips and displays, and is also the largest manufacturer of these globally.

CONTACTS: *Note: Officers with more than one job title may be intentionally listed here more than once.*

Ki Nam Kim, CEO
Oh-Hyun Kwon, Vice Chmn.

FINANCIAL DATA: *Note: Data for latest year may not have been available at press time.*

In U.S. $	2021	2020	2019	2018	2017	2016
Revenue	215,241,200,000	182,295,200,000	177,363,800,000	187,656,500,000	184,426,400,000	155,398,100,000
R&D Expense	17,244,960,000	16,251,730,000	15,324,690,000	14,129,060,000	12,590,630,000	10,863,010,000
Operating Income	39,748,010,000	27,708,270,000	21,376,340,000	45,331,260,000	41,296,230,000	22,509,620,000
Operating Margin %	.18%	.15%	.12%	.24%	.22%	.14%
SGA Expense	22,730,820,000	20,080,210,000	20,801,810,000	20,197,510,000	24,749,450,000	24,063,260,000
Net Income	30,210,070,000	20,084,870,000	16,554,700,000	33,787,420,000	31,827,260,000	17,255,690,000
Operating Cash Flow	50,118,510,000	50,258,280,000	34,936,000,000	51,601,480,000	47,852,660,000	36,477,710,000
Capital Expenditure	38,358,630,000	31,001,450,000	22,030,030,000	23,538,270,000	33,698,970,000	19,391,890,000
EBITDA	67,766,510,000	51,779,980,000	46,739,630,000	67,986,590,000	60,790,560,000	40,040,990,000
Return on Assets %	.10%	.07%	.06%	.14%	.15%	.09%
Return on Equity %	.14%	.10%	.09%	.20%	.21%	.12%
Debt to Equity	.01%	0.011	0.012	0.004	0.013	0.007

CONTACT INFORMATION:

Phone: 82 31-200-1114   Fax: 82-31-200-7538
Toll-Free:
Address: 129, Samsung-ro, Suwon-si, 443-742 South Korea

STOCK TICKER/OTHER:

Stock Ticker: SSNHZ   Exchange: PINX
Employees: 95,798   Fiscal Year Ends: 12/31
Parent Company: Samsung Group

SALARIES/BONUSES:

Top Exec. Salary: $ Bonus: $
Second Exec. Salary: $ Bonus: $

OTHER THOUGHTS:

Estimated Female Officers or Directors:
Hot Spot for Advancement for Women/Minorities:

Sanmina Corporation

NAIC Code: 334418

TYPES OF BUSINESS:

Printed Circuit Assembly (Electronic Assembly) Manufacturing
Assembly & Testing
Logistics Services
Support Services
Product Design & Engineering
Repair & Maintenance Services
Printed Circuit Boards

BRANDS/DIVISIONS/AFFILIATES:

GROWTH PLANS/SPECIAL FEATURES:

Sanmina Corp is a provider of integrated manufacturing
solutions, components, and after-market services to original
equipment manufacturers in the communications networks,
storage, industrial, defense and aerospace end markets. The
company operates in two business segments: Integrated
Manufacturing Solutions, which consists of printed circuit board
assembly and represents a majority of the firm's revenue; and
Components, Products, and Services, which include
interconnect systems and mechanical systems. The firm
generates revenue primarily in the United States, China, and
Mexico, but has a presence around the world.

CONTACTS: Note: Officers with more than one job title may be intentionally listed here more than once.

Jure Sola, CEO
Kurt Adzema, CFO
Brent Billinger, Chief Accounting Officer
Dennis Young, Executive VP, Divisional
Alan Reid, Executive VP, Divisional

FINANCIAL DATA: Note: Data for latest year may not have been available at press time.

In U.S. $	2021	2020	2019	2018	2017	2016
Revenue	6,756,643,000	6,960,370,000	8,233,859,000	7,110,130,000	6,868,619,000	6,481,181,000
R&D Expense	20,911,000	22,564,000	27,552,000	30,754,000	33,716,000	37,746,000
Operating Income	296,357,000	262,212,000	304,354,000	182,105,000	227,806,000	228,486,000
Operating Margin %	.04%	.04%	.04%	.03%	.03%	.04%
SGA Expense	234,537,000	240,931,000	260,032,000	250,924,000	251,568,000	244,604,000
Net Income	268,998,000	139,713,000	141,515,000	-95,533,000	138,833,000	187,838,000
Operating Cash Flow	338,342,000	300,555,000	382,965,000	156,424,000	250,961,000	390,116,000
Capital Expenditure	73,296,000	65,982,000	134,674,000	118,881,000	111,833,000	120,400,000
EBITDA	436,212,000	343,879,000	393,331,000	244,093,000	354,165,000	341,438,000
Return on Assets %	.07%	.04%	.04%	-.02%	.04%	.05%
Return on Equity %	.15%	.09%	.09%	-.06%	.09%	.12%
Debt to Equity	.17%	0.202	0.211	0.01	0.238	0.27

CONTACT INFORMATION:

Phone: 408-964-3500   Fax: 408-964-3636
Toll-Free:
Address: 2700 N. First St., San Jose, CA 95134 United States

STOCK TICKER/OTHER:

Stock Ticker: SANM   Exchange: NAS
Employees: 37,000   Fiscal Year Ends: 09/30
Parent Company:

SALARIES/BONUSES:

Top Exec. Salary: $1,125,000   Bonus: $
Second Exec. Salary: Bonus: $
$500,000

OTHER THOUGHTS:

Estimated Female Officers or Directors:
Hot Spot for Advancement for Women/Minorities:

ScanSource Inc

www.scansource.com

NAIC Code: 423430

TYPES OF BUSINESS:
Data Capture Products, Distribution
Bar Code
Point of Sale
Payments
Security
Unified Communications
Cloud
Telecommunications

BRANDS/DIVISIONS/AFFILIATES:

GROWTH PLANS/SPECIAL FEATURES:

ScanSource Inc provides value-added services for technology manufacturers and sells to resellers in specialty technology markets. The firm's operations are organized in two segments: Worldwide Barcode, Networking and Security, which focuses on automatic identification and data capture, point-of-sale, networking, electronic physical security, and 3-D printing technologies; and Worldwide communications and services, which focuses on communications technologies for vertical markets including education, healthcare, and government. The company generates a majority of its revenue from the Worldwide Barcode, Networking, and Security segment.

ScanSource offers its employees comprehensive health benefits, life and disability insurance, retirement and savings options, and a variety of employee assistance programs.

CONTACTS:
Note: Officers with more than one job title may be intentionally listed here more than once.
Michael Baur, CEO
Stephen Jones, CFO
Matthew Dean, General Counsel
John Eldh, Other Executive Officer

FINANCIAL DATA:
Note: Data for latest year may not have been available at press time.

In U.S. $	2021	2020	2019	2018	2017	2016
Revenue	3,150,806,000	3,047,734,000	3,249,799,000	3,164,709,000	3,568,186,000	3,540,226,000
R&D Expense						
Operating Income	71,257,000	63,048,000	118,588,000	106,049,000	93,450,000	98,171,000
Operating Margin %	.02%	.02%	.04%	.03%	.03%	.03%
SGA Expense	247,438,000	259,535,000	244,294,000	232,291,000	265,178,000	240,115,000
Net Income	10,795,000	-192,654,000	57,597,000	33,153,000	69,246,000	63,619,000
Operating Cash Flow	140,940,000	226,271,000	-27,127,000	24,806,000	94,876,000	52,211,000
Capital Expenditure	2,363,000	6,387,000	5,797,000	6,998,000	12,432,000	12,081,000
EBITDA	97,971,000	-24,224,000	130,451,000	106,663,000	129,678,000	115,288,000
Return on Assets %	.01%	- .10%	.03%	.02%	.04%	.04%
Return on Equity %	.02%	- .24%	.06%	.04%	.09%	.08%
Debt to Equity	.21%	0.342	0.354	0.287	0.116	0.099

CONTACT INFORMATION:
Phone: 864 288-2432   Fax: 864 288-5515
Toll-Free: 800-944-2439
Address: 6 Logue Ct., Greenville, SC 29615 United States

STOCK TICKER/OTHER:
Stock Ticker: SCSC   Exchange: NAS
Employees: 2,200   Fiscal Year Ends: 06/30
Parent Company:

SALARIES/BONUSES:
Top Exec. Salary: $765,625   Bonus: $
Second Exec. Salary: Bonus: $
$475,000

OTHER THOUGHTS:
Estimated Female Officers or Directors: 1
Hot Spot for Advancement for Women/Minorities:

Sales, profits and employees may be estimates. Financial information, benefits and other data can change quickly and may vary from those stated here.

SES SA

NAIC Code: 517410

TYPES OF BUSINESS:

Satellite Carrier
Satellite Operations
Broadband Internet Service
Secure Communications
Broadcasting Services
Video on Demand
Direct to Home TV

BRANDS/DIVISIONS/AFFILIATES:

QuetzSat
Ciel
GovSat
YahLive
SES Government Solutions
SES Techcom
HD Plus GmbH
Redu Space Services

CONTACTS: *Note: Officers with more than one job title may be intentionally listed here more than once.*

Steve Collar, CEO
Sandeep Jalan, CFO
Evie Ross, Chief Human Resources Officer
Ruy Pinto, CTO
Gerson Souto, Chief Dev. Officer
Patrick Biewer, Managing Dir.-SES Broadband Svcs.
Wilfried Urner, CEO-SES Platform Svcs.
Frank Esser, Chmn.
Tip Osterthaler, CEO-SES Gov't Solutions

GROWTH PLANS/SPECIAL FEATURES:

SES SA is a communications satellite owner and operator and functions in the television broadcasting space. The company operates in one operating segment, namely the provision of satellite-based data transmission capacity, and ancillary services, to customers around the world. The largest video division delivers television and radio satellite signals to homes through various media providers such as Comcast, Viacom, M7, and QVC. The mobility revenue is derived by offering high bandwidth Internet connections to maritime users, and SES SA's government division offers secure communication links to international institutions and governments. Fixed Data works to bring satellite networks and connectivity to remote areas.

FINANCIAL DATA: *Note: Data for latest year may not have been available at press time.*

In U.S. $	2021	2020	2019	2018	2017	2016
Revenue	1,784,585,000	1,872,285,000	1,979,360,000	1,971,712,000	1,993,331,000	2,109,686,000
R&D Expense						
Operating Income	299,810,300	333,462,500	435,438,800	320,511,500	540,780,300	836,511,600
Operating Margin %	.17%	.18%	.22%	.16%	.27%	.40%
SGA Expense						
Net Income	461,952,600	-87,699,620	301,849,900	298,178,700	607,880,700	981,725,900
Operating Cash Flow	1,319,573,000	1,069,731,000	1,156,411,000	1,214,844,000	1,275,927,000	1,299,280,000
Capital Expenditure	285,533,600	214,150,200	311,027,700	334,686,200	490,710,000	632,253,100
EBITDA	1,923,273,000	1,038,119,000	1,218,617,000	1,261,651,000	1,317,432,000	1,952,540,000
Return on Assets %	.04%	-.01%	.02%	.02%	.05%	.08%
Return on Equity %	.08%	-.01%	.05%	.05%	.09%	.18%
Debt to Equity	.63%	0.623	0.61	0.64	0.57	0.62

CONTACT INFORMATION:

Phone: 352 7107251   Fax:
Toll-Free:
Address: Chateau de Betzdorf, Rue Pierre Werner, Luxembourg, L-6815 Luxembourg

STOCK TICKER/OTHER:

Stock Ticker: SGBAF
Employees: 2,037
Parent Company:

Exchange: PINX
Fiscal Year Ends: 12/31

SALARIES/BONUSES:

Top Exec. Salary: $ Bonus: $
Second Exec. Salary: $ Bonus: $

OTHER THOUGHTS:

Estimated Female Officers or Directors:
Hot Spot for Advancement for Women/Minorities:

Shaw Communications Inc

www.shaw.ca

NAIC Code: 517110

TYPES OF BUSINESS:

Cable TV Service
Internet Service Provider
Satellite Services
Digital and Phone Services
Television Connectivity Services
Business Connectivity Solutions
Wireless Solutions and Services

BRANDS/DIVISIONS/AFFILIATES:

Shaw
Shaw Direct
Shaw Business
Freedom

GROWTH PLANS/SPECIAL FEATURES:

Shaw Communications is a Canadian cable company that is one of the biggest providers of internet, television, and landline telephone services in British Columbia, Alberta, Saskatchewan, Manitoba, and northern Ontario. In fiscal 2021, more than 75% of Shaw's total revenue resulted from this wireline business. Shaw is also now a national wireless service provider after acquiring Wind Mobile in 2016. Shaw has upgraded its wireless network, undertaken an aggressive pricing strategy, and significantly enhanced its spectrum holdings. As a smaller carrier, Shaw has favored bidding status in spectrum auctions, giving it a further boost in enhancing its wireless network. At the 2019 auction, Shaw added significant amounts of 600 MHz spectrum to the 700 MHz spectrum it is currently deploying.

CONTACTS: *Note: Officers with more than one job title may be intentionally listed here more than once.*

Bradley Shaw, CEO
Trevor English, CFO
Peter Johnson, Chief Legal Officer
Zoran Stakic, Chief Technology Officer
Dan Markou, Executive VP
Paul Mcaleese, President
Paul Deverell, President, Divisional
Katherine Emberly, President, Divisional

FINANCIAL DATA: *Note: Data for latest year may not have been available at press time.*

In U.S. $	2021	2020	2019	2018	2017	2016
Revenue	4,284,826,000	4,205,491,000	4,153,380,000	4,035,934,000	3,797,153,000	3,514,039,000
R&D Expense						
Operating Income	985,455,500	902,232,300	875,009,700	455,782,800	777,008,600	867,231,900
Operating Margin %	.23%	.21%	.21%	.11%	.20%	.25%
SGA Expense	516,450,200	511,005,700	515,672,400	900,676,700	668,118,500	603,562,300
Net Income	766,897,400	535,117,100	568,561,900	30,333,670	661,896,300	948,899,500
Operating Cash Flow	1,495,683,000	1,493,350,000	1,219,569,000	1,052,345,000	1,168,235,000	1,293,459,000
Capital Expenditure	791,008,800	895,232,300	1,009,567,000	1,011,900,000	1,254,569,000	819,786,900
EBITDA	1,961,577,000	1,841,798,000	1,694,797,000	1,127,012,000	1,519,795,000	1,382,904,000
Return on Assets %	.06%	.04%	.05%	.00%	.06%	.08%
Return on Equity %	.16%	.11%	.12%	.00%	.14%	.23%
Debt to Equity	.94%	0.96	0.677	0.722	0.694	0.913

CONTACT INFORMATION:

Phone: 403 750-4500   Fax: 403 750-4501
Toll-Free: 888-472-2222
Address: 630 - 3rd Avenue S.W., Calgary, AB T2P 4L4 Canada

STOCK TICKER/OTHER:

Stock Ticker: SJR   Exchange: NYS
Employees: 9,500   Fiscal Year Ends: 08/31
Parent Company:

SALARIES/BONUSES:

Top Exec. Salary: $2,000,000   Bonus: $
Second Exec. Salary: Bonus: $
$1,320,000

OTHER THOUGHTS:

Estimated Female Officers or Directors: 3
Hot Spot for Advancement for Women/Minorities: Y

Sales, profits and employees may be estimates. Financial information, benefits and other data can change quickly and may vary from those stated here.

Shenandoah Telecommunications Company
www.shentel.com

NAIC Code: 517110

TYPES OF BUSINESS:
Local Exchange Carrier
Broadband Services
Data
Video
Telephone
Fiber Optic Services
Tower Leasing

BRANDS/DIVISIONS/AFFILIATES:

GROWTH PLANS/SPECIAL FEATURES:

Shenandoah Telecommunications Company, with subsidiaries, provides various broadband communication products and services via its wireless, cable, fiber optic, and fixed wireless networks to customers in the Mid-Atlantic United States. The company operates through two business unit: tower and broadband. The tower segment leases compan owned cell tower spaces to other wireless communicatic providers, while the broadband segment provides broadbar Internet, video, and voice services to residential an commercial customers. The broadband segment generates th vast majority of the company's revenue, with the bulk of sale flowing from residential and small, and medium businesse within the broadband unit.

STC offers its employees comprehensive health benefit 401(k), tuition assistance, and a variety of employe training/development and assistance programs.

CONTACTS: *Note: Officers with more than one job title may be intentionally listed here more than once.*
Christopher French, CEO
James Volk, CFO
Dennis Romps, Chief Accounting Officer
Elaine Cheng, Chief Information Officer
David Heimbach, COO
Raymond Ostroski, General Counsel
Heather Banks, Other Executive Officer
William Pirtle, Senior VP, Divisional
Edward McKay, Senior VP, Divisional

FINANCIAL DATA: *Note: Data for latest year may not have been available at press time.*

In U.S. $	2021	2020	2019	2018	2017	2016
Revenue	245,239,000	220,775,000	206,862,000	192,683,000	611,991,000	535,288,000
R&D Expense						
Operating Income	5,283,000	-2,601,000	-1,342,000	-2,969,000	57,540,000	64,758,000
Operating Margin %	.02%	-.01%	-.01%	-.02%	.09%	.12%
SGA Expense	82,451,000	85,016,000	77,846,000	70,844,000	165,937,000	133,325,000
Net Income	998,831,000	125,673,000	55,500,000	46,595,000	66,390,000	-895,000
Operating Cash Flow	-250,934,000	302,867,000	259,145,000	265,647,000	222,930,000	161,526,000
Capital Expenditure	160,101,000	120,450,000	67,048,000	56,631,000	146,489,000	173,231,000
EBITDA	60,489,000	46,102,000	45,444,000	41,421,000	228,501,000	170,732,000
Return on Assets %	.69%	.06%	.03%	.03%	.05%	.00%
Return on Equity %	1.64%	.24%	.12%	.12%	.21%	.00%
Debt to Equity	.08%	0.082	1.551	1.695	2.163	2.694

CONTACT INFORMATION:
Phone: 540 984-4141   Fax: 540 984-3438
Toll-Free: 800-743-6835
Address: 500 Shentel Way, Edinburg, VA 22824 United States

STOCK TICKER/OTHER:
Stock Ticker: SHEN   Exchange: NAS
Employees: 860   Fiscal Year Ends: 12/31
Parent Company:

SALARIES/BONUSES:
Top Exec. Salary: $688,585   Bonus: $
Second Exec. Salary: Bonus: $137,000
$390,000

OTHER THOUGHTS:
Estimated Female Officers or Directors: 3
Hot Spot for Advancement for Women/Minorities: Y

Sales, profits and employees may be estimates. Financial information, benefits and other data can change quickly and may vary from those stated here.

Siemens AG

www.siemens.com

NAIC Code: 334513

TYPES OF BUSINESS:

Industrial Control Manufacturing
Digitalization
Smart Infrastructure
Mobility
Advanced Technologies
Artificial Intelligence
Internet of Things
Robotics

BRANDS/DIVISIONS/AFFILIATES:

Siemens Advanta
Siemens Healthineers AG
Siemens Financial Services
Siemens Real Estate
Next47

GROWTH PLANS/SPECIAL FEATURES:

Siemens AG is an industrial conglomerate, with businesses selling components and equipment for factory automation, railway equipment, electrical distribution equipment, and medical equipment. Its separately listed business units include Siemens Healthineers, Siemens Energy, and Siemens Gamesa, which supply medical imaging equipment, power generation, and wind turbines, respectively.

CONTACTS:
Note: Officers with more than one job title may be intentionally listed here more than once.

Joe Kaeser, CEO
Ralf P. Thomas, CFO
Peter Y. Solmssen, Head-Corp. Legal & Compliance
Joe Kaeser, Head-Controlling
Roland Busch, CEO-Infrastructure & Cities Sector
Hermann Requardt, CEO-Health Care Sector
Michael Suess, CEO-Energy Sector
Siegfried Russwurm, CEO-Industry Sector
Jim Hagemenn Snabe, Chmn.
Barbara Kux, Chief Sustainability Officer

FINANCIAL DATA:
Note: Data for latest year may not have been available at press time.

In U.S. $	2021	2020	2019	2018	2017	2016
Revenue	63,495,540,000	56,345,990,000	59,638,800,000	84,685,200,000	84,500,620,000	81,218,000,000
R&D Expense	4,955,028,000	4,659,297,000	4,761,274,000	5,667,843,000	5,266,056,000	4,825,519,000
Operating Income	6,548,918,000	4,472,680,000	6,224,633,000	6,028,839,000	7,365,748,000	7,281,107,000
Operating Margin %	.10%	.08%	.10%	.07%	.09%	.09%
SGA Expense	11,410,130,000	10,893,110,000	10,899,230,000	13,196,750,000	12,604,270,000	11,899,610,000
Net Income	6,282,760,000	4,109,645,000	5,276,254,000	5,921,764,000	6,078,807,000	5,557,708,000
Operating Cash Flow	10,193,550,000	9,037,140,000	8,623,116,000	8,591,503,000	7,317,819,000	7,761,416,000
Capital Expenditure	1,764,190,000	1,527,605,000	1,815,178,000	2,653,423,000	2,453,550,000	2,177,194,000
EBITDA	11,436,640,000	9,600,048,000	10,379,150,000	12,806,180,000	12,697,070,000	11,377,500,000
Return on Assets %	.05%	.03%	.04%	.04%	.05%	.04%
Return on Equity %	.15%	.10%	.11%	.13%	.15%	.16%
Debt to Equity	.92%	1.044	0.632	0.596	0.62	0.72

CONTACT INFORMATION:

Phone: 49 8963633032   Fax: 49 8932825
Toll-Free:
Address: Werner-von-Siemens-Strabe 1, Munich, BY 80333 Germany

STOCK TICKER/OTHER:

Stock Ticker: SMAWF   Exchange: PINX
Employees: 293,000   Fiscal Year Ends: 09/30
Parent Company:

SALARIES/BONUSES:

Top Exec. Salary: $2,039,526   Bonus: $1,386,776
Second Exec. Salary: Bonus: $561,645
$1,386,836

OTHER THOUGHTS:

Estimated Female Officers or Directors: 5
Hot Spot for Advancement for Women/Minorities: Y

Sales, profits and employees may be estimates. Financial information, benefits and other data can change quickly and may vary from those stated here.

Simply Inc

NAIC Code: 423690

TYPES OF BUSINESS:

Wireless Handset Distribution
Wireless Handset Design

GROWTH PLANS/SPECIAL FEATURES:

Simply Inc through its subsidiary, operates a chain of reta
electronics stores and is an authorized reseller of Apple
products and other high-profile consumer electronic brands.
operates business in a single segment in the United State
through its Simply Mac retail stores.

BRANDS/DIVISIONS/AFFILIATES:

Simply Mac
www.simplymac.com

CONTACTS: *Note: Officers with more than one job title may be intentionally listed here more than once.*

Reinier Voigt, CEO
Vernon Loforti, CFO
Kevin Taylor, Chairman of the Board

FINANCIAL DATA: *Note: Data for latest year may not have been available at press time.*

In U.S. $	2021	2020	2019	2018	2017	2016
Revenue	68,024,000	5,285,000	30,385,000	11,615,000	13,615,000	
R&D Expense						
Operating Income	-8,845,000	-850,000	-13,249,000	-8,172,000	-6,714,000	
Operating Margin %	-.13%		-.44%	-.70%	-.49%	
SGA Expense	27,197,000	2,358,000	20,293,000	9,730,000	8,094,000	
Net Income	4,277,000	-2,123,000	-21,016,000	-27,271,000	-7,540,000	
Operating Cash Flow	-2,209,000	-580,000	-2,265,000	-7,685,000	-4,192,000	
Capital Expenditure	1,035,000		61,000	467,000	2,000	
EBITDA	6,175,000	-1,595,000	-11,282,000	-9,430,000	-6,452,000	
Return on Assets %	.16%		-1.01%	-1.51%	-.39%	
Return on Equity %					-1.83%	
Debt to Equity						

CONTACT INFORMATION:

Phone: 786-254-6709   Fax:
Toll-Free:
Address: 2001 NW 84th Avenue, Miami, FL 33122 United States

STOCK TICKER/OTHER:

Stock Ticker: SIMPQ   Exchange: PINX
Employees: 352   Fiscal Year Ends: 12/31
Parent Company:

SALARIES/BONUSES:

Top Exec. Salary: $193,315   Bonus: $54,000
Second Exec. Salary: Bonus: $54,000
$193,315

OTHER THOUGHTS:

Estimated Female Officers or Directors:
Hot Spot for Advancement for Women/Minorities:

Singapore Technologies Telemedia Pte Ltd www.sttelemedia.com

NAIC Code: 517110

TYPES OF BUSINESS:

Telecommunications Services
Cellular Service
Internet Services
Satellite & Broadcasting Services
Wireless Telecommunications Equipment
Data Centers & Internet Exchange
Network Design & Integration Services
Cloud

BRANDS/DIVISIONS/AFFILIATES:

Temasek Holdings Pvt Limited

CONTACTS: *Note: Officers with more than one job title may be intentionally listed here more than once.*

Stephen Miller, CEO
Johnny Ong, CFO
Chan Jen Keet, VP
Nicholas Tan, Sr. VP-Corp. Planning
Melinda Tan, Media Contact
Richard Lim, Exec. VP-Corp. Svcs.
Kek Soon Eng, Sr. VP-Investee Companies Mgmt.
Alvin Oei, Sr. VP-Intl Bus. Dev.
Melinda Tan, VP-Strategic Rel.
Ek Tor Teo, Chmn.
Steven Terrell Clontz, Sr. Exec. VP-North America & Europe

GROWTH PLANS/SPECIAL FEATURES:

Singapore Technologies Telemedia Pte., Ltd. (ST Telemedia) is an active investor in communications, media, data centers and infrastructure technology businesses worldwide. ST Telemedia's team of industry professionals has deep and broad experience in these sectors worldwide, in both developed and emerging markets. The firm has a long-term view on value creation and works with its portfolio companies and business partners to achieve mutual success. ST Telemedia's investments in communications and media companies are primarily located in the Asia-Pacific region. They enable the company to combine data, mobility, voice and content services with state-of-the-art delivery platforms. ST Telemedia's investments in data centers spans a footprint in key global economic hubs, particularly in China, India, Singapore, Thailand and the U.K. This division's agile and multi-tenant business model enables its global data center assets to be differentiated and diversified. Last, ST Telemedia's infrastructure technology investments are in cutting-edge technology companies that are ready to scale up in areas such as cloud computing, system performance, cybersecurity and advanced analytics. They are at the forefront of technology innovation critical for the digital transformation of governments, enterprises and telecommunications companies. ST Telemedia operates as a subsidiary of Temasek Holdings Pvt. Limited.

FINANCIAL DATA: *Note: Data for latest year may not have been available at press time.*

In U.S. $	2021	2020	2019	2018	2017	2016
Revenue	3,000,000,000	2,970,803,015	3,062,386,348	4,909,244,928	4,669,999,616	4,917,730,816
R&D Expense						
Operating Income						
Operating Margin %						
SGA Expense						
Net Income		68,702,179	-158,603,746	362,253,824	361,128,768	356,493,568
Operating Cash Flow						
Capital Expenditure						
EBITDA						
Return on Assets %						
Return on Equity %						
Debt to Equity						

CONTACT INFORMATION:

Phone: 65-6723-8777   Fax: 65-6720-7266
Toll-Free:
Address: 1 Temasek Ave., #33-01 Millenia Twr., Singapore, 567710 Singapore

STOCK TICKER/OTHER:

Stock Ticker: Subsidiary   Exchange:
Employees: 5,000   Fiscal Year Ends: 12/31
Parent Company: Temasek Holdings Pvt Limited

SALARIES/BONUSES:

Top Exec. Salary: $ Bonus: $
Second Exec. Salary: $ Bonus: $

OTHER THOUGHTS:

Estimated Female Officers or Directors: 1
Hot Spot for Advancement for Women/Minorities:

Singapore Telecommunications Limited   www.singtel.com

NAIC Code: 517110

TYPES OF BUSINESS:

Telecommunications Services
Local & Long-Distance Services
Cell Phone & Paging Services
Internet Service Provider
IT & Communications Engineering Services
Satellite Services
Virtual Private Networks
Equipment Sales

BRANDS/DIVISIONS/AFFILIATES:

Optus
NCS Pte Ltd
Singtel Digital Media
Singtel Innov8
Amobee
Trustwave

GROWTH PLANS/SPECIAL FEATURES:

Singapore Telecommunications is Singapore's leadin
telecoms company. It owns extensive wired and wireles
networks offering data and voice services to a broad custome
base. Singtel's diverse investment portfolio spreads across th
region. The firm wholly owns Optus in Australia and minorit
equity stakes in Airtel (33%) in India; Telkomsel (35%) i
Indonesia; Globe Telecom (47%) in the Philippines; an
Advanced Information Services (23%) and Intouch (21%) i
Thailand. Singtel is majority-owned by the Singapor
government.

CONTACTS: Note: Officers with more than one job title may be intentionally listed here more than once.

Chua Sock Koong, Group CEO
Lim Cheng Cheng, Group CFO
Aileen Tan, Group Human Resources Officer
William Woo, Group CIO
Allen Lew, CEO-Digital Life Group
Bill Chang, CEO-Group Enterprise
Paul OSullivan, CEO-Consumer Group
Mark Chong, Group CTO

FINANCIAL DATA: Note: Data for latest year may not have been available at press time.

In U.S. $	2021	2020	2019	2018	2017	2016
Revenue	11,350,630,000	12,002,390,000	12,604,170,000	12,528,930,000	12,125,090,000	
R&D Expense						
Operating Income	729,403,200	1,293,162,000	1,621,041,000	1,801,632,000	1,843,207,000	
Operating Margin %	.06%	.11%	.13%	.14%	.15%	
SGA Expense	2,679,630,000	2,670,270,000	2,948,231,000	3,012,806,000	3,263,196,000	
Net Income	401,741,300	779,684,300	2,245,239,000	3,970,978,000	2,795,356,000	
Operating Cash Flow	4,069,653,000	4,220,787,000	3,894,504,000	4,320,842,000	3,856,122,000	
Capital Expenditure	1,761,945,000	1,731,616,000	1,493,125,000	2,520,152,000	1,827,172,000	
EBITDA	2,777,943,000	3,328,932,000	4,600,907,000	6,362,561,000	5,153,782,000	
Return on Assets %	.01%	.02%	.06%	.11%	.08%	
Return on Equity %	.02%	.04%	.10%	.19%	.14%	
Debt to Equity	.41%	0.381	0.294	0.292	0.287	

CONTACT INFORMATION:

Phone: 65-6838-3388   Fax: 65-6732-8428
Toll-Free:
Address: 31 Exeter Rd., 19-00 Comcentre, Singapore, 239732
Singapore

STOCK TICKER/OTHER:

Stock Ticker: SGAPY   Exchange: PINX
Employees: 22,892   Fiscal Year Ends: 03/31
Parent Company:

SALARIES/BONUSES:

Top Exec. Salary: $ Bonus: $
Second Exec. Salary: $ Bonus: $

OTHER THOUGHTS:

Estimated Female Officers or Directors: 5
Hot Spot for Advancement for Women/Minorities: Y

Singtel Optus Pty Limited

www.optus.com.au

NAIC Code: 517110

TYPES OF BUSINESS:

Diversified Telecommunications Services
Mobile Phone Service
Digital Television Service
Local & Long-Distance Phone Service
Business Network Services
Satellite Services
Multimedia Services
5G

BRANDS/DIVISIONS/AFFILIATES:

Singapore Telecommunications Limited

CONTACTS: *Note: Officers with more than one job title may be intentionally listed here more than once.*

Kelly Bayer Rosmarin, CEO
Michael Venter, CFO
Kate Aitken, VP-People & Culture
Mark Potter, CIO
Martin Mercer, Managing Dir.-Strategy & Fixed
David Epstein, VP-Corp. & Regulatory Affairs
Vicki Brady, Managing Dir.-Customer
John Paitaridis, Managing Dir.-Optus Bus.
Gunther Ottendorfer, Managing Dir.-Optus Networks
Austin R. Bryan, VP-Transformation, Consumer Australia
Paul O'Sullivan, Chmn.

GROWTH PLANS/SPECIAL FEATURES:

Singtel Optus Pty. Limited (Optus), a subsidiary of Singapore Telecommunications Limited (Singtel), provides telecommunications services in Australia. Optus' mobile network reaches 98.5% of the Aussie population through its mobile, satellite and broadband networks. The firm's home and mobile network offers 5G coverage in select locations, which becomes faster, more responsive and has higher capacity than Optus' 3G and 4G networks. The company's 4G+ coverage reaches all of the country's capital cities as well as hundreds of regional towns. Optus owns and operates its own satellite fleet, which provides satellite-based communications services throughout Australia and Asia. The satellites in orbit are tracked by its ground network of domestic and international Earth stations. Over 1 million Australians utilize Optus for internet services, including digital subscriber line (DSL) connectivity. Optus' regional network invests in and builds the company's regional coverage. In addition, Optus has retail stores throughout Australia, which offer mobile phones, office phones, tablets and other communication devices, as well as related accessories and services. For businesses and enterprises, Optus provides digital transformation services and solutions through its managed cloud strategy, which enables firms to build a simple, secure and cost-effective way for driving business outcomes. The company also provides managed security, security technology, cyber security, digital business solutions, mobility solutions and related consultancy services to businesses, enterprises and all tiers of government entities.

FINANCIAL DATA: *Note: Data for latest year may not have been available at press time.*

In U.S. $	2021	2020	2019	2018	2017	2016
Revenue	5,400,340,112	4,768,040,000	5,434,520,000	5,478,460,000	5,285,450,000	5,764,390,000
R&D Expense						
Operating Income						
Operating Margin %						
SGA Expense						
Net Income	1,435,417,716	1,570,080,000	1,761,130,000	1,898,400,000	1,851,480,000	1,884,220,000
Operating Cash Flow						
Capital Expenditure						
EBITDA						
Return on Assets %						
Return on Equity %						
Debt to Equity						

CONTACT INFORMATION:

Phone: 61 2-8082-7800   Fax: 61-2-8082-7100
Toll-Free:
Address: 1 Lyonpark Rd., Macquarie Park, NSW 2113 Australia

STOCK TICKER/OTHER:

Stock Ticker: Subsidiary   Exchange:
Employees: 8,400   Fiscal Year Ends: 03/31
Parent Company: Singapore Telecommunications Limited

SALARIES/BONUSES:

Top Exec. Salary: $ Bonus: $
Second Exec. Salary: $ Bonus: $

OTHER THOUGHTS:

Estimated Female Officers or Directors: 1
Hot Spot for Advancement for Women/Minorities:

Sales, profits and employees may be estimates. Financial information, benefits and other data can change quickly and may vary from those stated here.

Sitel Corporation

NAIC Code: 518210

TYPES OF BUSINESS:

Communications Consulting-Call Centers
Contact Center Technology
Contact Center Solutions
Customer Experience Products and Services
Customer Experience Technology
Customer Experience Analytics
Chat Technology
Customer Experience Digital Solutions

BRANDS/DIVISIONS/AFFILIATES:

CX Learning
CX Technology
CX Operations
CX Analytics
CX Digital

CONTACTS: *Note: Officers with more than one job title may be intentionally listed here more than once.*

Laurent Uberti, CEO
Olivier Camino, COO
Elisabeth Destailleur, CFO
Martin Wilkinson-Brown, CMO
chris Knauer, Chief Security Officer
David Slaviero, CTO
David Beckman, Chief Legal Officer
Sean Erickson, Exec. VP-Corp. Dev.
Neal Miller, Chief Tax Officer
Pedro Lozano, Gen. Mgr.-EMEA
Jason Skaria, CIO
Steve Barker, Exec. VP-Global Oper. Support

GROWTH PLANS/SPECIAL FEATURES:

Sitel Corporation provides contact center technology solutions to enhance brand and consumer relationships. The firm's products and services are customer experience related, and are grouped into five categories: CX Learning, CX Technology, CX Operations, CX Analytics and CX Digital. CX Learning products and services span learning needs analysis, learning ecosystem design, custom blended and digital learning, learning catalogue, learning experience platform technology and chat simulator technology. CX Technology products and services include virtual communication platform, associate performance management tools, unified agent desktop, conversational messaging, case management and orchestration, omnichannel telephone platform, self-service solutions, intelligent automations, digital assistant, robotic process automation and more. CX Operations products and services include customer care, multi-lingual customer support, sales and retention, technology support as a service, first-party collections, back office support, trust and safety, social media management, chat services and messaging services, as well as customer authentication solutions and secure credit card processing solutions. CX Analytics products and services cover customer analytics, customer's voice, next-best action for increased sales, performance analytics, real-time agent assistant, clickstream analytics, quality assurance automation and social fraud engineering. Last, CX Digital products and services include self-service content, knowledge management, social engagement, social intelligence, boosting social media experience, conversational artificial intelligence (AI), call bots, automated image analysis and other solutions. Industries served by Sitel include banking and financial services, government, healthcare, insurance, manufacturing, media and entertainment, retail, technology, telecommunications, travel, hospitality and utilities. Sitel has over 160 CX hub locations supporting more than 50 languages, and a presence in 40 countries across North America, Europe and Asia.

FINANCIAL DATA: *Note: Data for latest year may not have been available at press time.*

In U.S. $	2021	2020	2019	2018	2017	2016
Revenue	4,300,000,000	2,100,000,000	1,785,000,000	1,700,000,000	1,615,000,000	1,400,000,000
R&D Expense						
Operating Income						
Operating Margin %						
SGA Expense						
Net Income						
Operating Cash Flow						
Capital Expenditure						
EBITDA						
Return on Assets %						
Return on Equity %						
Debt to Equity						

CONTACT INFORMATION:

Phone: 866-957-4835   Fax:
Toll-Free:
Address: 600 Brickell Ave., Ste. 3200, Miami, FL 33131 United States

STOCK TICKER/OTHER:

Stock Ticker: Private
Employees: 160,000
Parent Company:

Exchange:
Fiscal Year Ends: 12/31

SALARIES/BONUSES:

Top Exec. Salary: $ Bonus: $
Second Exec. Salary: $ Bonus: $

OTHER THOUGHTS:

Estimated Female Officers or Directors: 1
Hot Spot for Advancement for Women/Minorities:

SK Broadband Co Ltd

www.skbroadband.com/eng/main.do

NAIC Code: 517110

TYPES OF BUSINESS:

Internet Service Provider
Local, Long Distance & International Phone Service
Leased-Line Business Connectivity
VoIP Service
Internet Data Centers

BRANDS/DIVISIONS/AFFILIATES:

SK Telecom Co Ltd

CONTACTS: Note: Officers with more than one job title may be intentionally listed here more than once.

Jin-Hwan Choi, CEO

GROWTH PLANS/SPECIAL FEATURES:

SK Broadband Co., Ltd. is a Korean provider of telecommunications, high-speed broadband internet, voice, internet data center (IDC) and leased line services. The firm's offerings can be divided into two categories: residential and business services. In addition to basic telephone services, its residential services include fiber-to-the-home (FTTH) internet access, voice over internet protocol (VoIP) service and video-on-demand, cable TV and internet protocol television (IPTV). The company's services for business customers include internet phone, digital fiber-optic, intelligent network, internet leased lines, domestic leased line, virtual private network (VPN), internet data center and content delivery network (CDN), information and communication technology (ICT) and fixed-mobile convergence, data and other services for buildings. A variety of enterprise telephone services are also offered under this category, including general phone services; direct inbound and outbound calling; international calls; and Primary Rate Interface, which automatically dials and receives calls without the need of an operator. IDC services are geared toward content providers and small to medium-sized businesses. It provides direct connections to IXs and ISPs. Additionally, SK Broadband provides managed service solutions to its enterprise customers that provide for 24-hour remote monitoring of data and voice communications infrastructure, LAN/WAN lines and traffic conditions; and carries out preventive examinations and onsite support. SK Broadband is wholly-owned by SK Telecom Co. Ltd.

FINANCIAL DATA: Note: Data for latest year may not have been available at press time.

In U.S. $	2021	2020	2019	2018	2017	2016
Revenue	3,600,000,000	3,411,250,000	2,872,370,280	2,730,390,000	2,500,000,000	2,439,430,000
R&D Expense						
Operating Income						
Operating Margin %						
SGA Expense						
Net Income		138,446,000	30,078,038	120,312,150	114,583,000	17,842,900
Operating Cash Flow						
Capital Expenditure						
EBITDA						
Return on Assets %						
Return on Equity %						
Debt to Equity						

CONTACT INFORMATION:

Phone: 82-2-626-64534   Fax: 82-2-626-64609
Toll-Free:
Address: SK Namsan Bldg., 24, Toegye-ro, Jung-gu, Seoul, 100-711 South Korea

STOCK TICKER/OTHER:

Stock Ticker: Subsidiary
Employees: 1,600
Parent Company: SK Telecom Co Ltd

Exchange:
Fiscal Year Ends: 12/31

SALARIES/BONUSES:

Top Exec. Salary: $ Bonus: $
Second Exec. Salary: $ Bonus: $

OTHER THOUGHTS:

Estimated Female Officers or Directors:
Hot Spot for Advancement for Women/Minorities:

Sales, profits and employees may be estimates. Financial information, benefits and other data can change quickly and may vary from those stated here.

SK Telecom Co Ltd

www.sktelecom.com

NAIC Code: 517210

TYPES OF BUSINESS:

Wireless Telecommunications Services
Multimedia Broadcasting
Online Shopping
Internet of Things
Telecommunications

GROWTH PLANS/SPECIAL FEATURES:

SK Telecom is South Korea's largest wireless teleco
operator, with 29 million mobile customers. The firm also ow
SK Broadband (formerly Hanaro Telecom), which has 6
million broadband customers and 8.7 million broadband T
customers. While the firm also purchased stakes in businesse
in security and semiconductor memory production as well a
developing e-commerce and internet platform businesse
these were all spun off into the separate, SK Square busines
in November 2021.

BRANDS/DIVISIONS/AFFILIATES:

Eleven Street
SK Hynix Inc
KEB HanaCard Co Ltd
Content Wavve

CONTACTS: *Note: Officers with more than one job title may be intentionally listed here more than once.*

Jung-Ho Park, CEO
Young Sang Ryu, Dir.-Mobile Network Oper.
Dong Seob Jee, Head-Corp. Vision Dept.
Young Tae Kim, Exec. Dir.
Daesik Cho, Chmn.

FINANCIAL DATA: *Note: Data for latest year may not have been available at press time.*

In U.S. $	2021	2020	2019	2018	2017	2016
Revenue	12,893,150,000	12,384,430,000	11,867,650,000	12,989,660,000	13,487,000,000	13,157,370,000
R&D Expense	267,669,700	271,893,700	249,456,900	298,434,200	304,285,500	265,418,800
Operating Income	1,132,165,000	996,455,900	840,392,400	959,263,400	1,145,183,000	1,219,072,000
Operating Margin %	.09%	.08%	.07%	.07%	.08%	.09%
SGA Expense	6,639,613,000	6,290,600,000	5,917,752,000	6,786,035,000	6,921,573,000	6,684,554,000
Net Income	1,853,324,000	1,158,058,000	684,124,300	2,407,864,000	2,001,362,000	1,290,168,000
Operating Cash Flow	3,873,104,000	4,481,710,000	3,106,151,000	3,335,242,000	2,968,230,000	3,266,413,000
Capital Expenditure	2,546,853,000	2,838,869,000	2,707,322,000	2,536,985,000	2,202,874,000	2,406,289,000
EBITDA	4,705,297,000	4,129,378,000	3,952,729,000	5,825,596,000	5,350,044,000	4,199,404,000
Return on Assets %	.06%	.03%	.02%	.08%	.08%	.06%
Return on Equity %	.14%	.06%	.04%	.15%	.15%	.11%
Debt to Equity	.74%	0.453	0.442	0.382	0.326	0.406

CONTACT INFORMATION:

Phone: 82 2-6100-2114   Fax: 82-2-6110-7830
Toll-Free:
Address: SK T-Tower, 65, Eulji-ro, Jung-gu, Seoul, 100-999 South Korea

STOCK TICKER/OTHER:

Stock Ticker: SKM
Employees: 41,097
Parent Company:

Exchange: NYS
Fiscal Year Ends: 12/31

SALARIES/BONUSES:

Top Exec. Salary: $ Bonus: $
Second Exec. Salary: $ Bonus: $

OTHER THOUGHTS:

Estimated Female Officers or Directors:
Hot Spot for Advancement for Women/Minorities:

Sales, profits and employees may be estimates. Financial information, benefits and other data can change quickly and may vary from those stated here.

Skype Technologies Sarl

www.skype.com

NAIC Code: 511210C

TYPES OF BUSINESS:

Computer Software: Telecom, Communications & VOIP
Voice Communication Services
Video Call Services
Software
Mobile Applications
Group Communication Solutions

BRANDS/DIVISIONS/AFFILIATES:

Microsoft Corporation

CONTACTS: *Note: Officers with more than one job title may be intentionally listed here more than once.*

Satya Nadella, CEO-Microsoft

GROWTH PLANS/SPECIAL FEATURES:

Skype Technologies Sarl, a subsidiary of Microsoft Corporation, offers software capabilities that enable users to make free voice and video calls over the internet. The firm's proprietary software and mobile applications help people stay connected not only through telephone calls, but also via instant messaging and photo- and file-sharing in real-time. Skype Technologies' additional capabilities include purchasing tickets, surfing the internet for recipes and finding and dropping information into conversations. Skype in the Classroom offers live educational experiences for teachers and their students from over 235 countries, including virtual field trips, talks with guest speakers, collaborative lessons and projects with other classrooms around the world. For businesses and friends who are not online, calling their mobile and landline numbers through Skype-to-Phone is offered at affordable rates. Calls can be made by phone, desktop, tablet, Xbox, Amazon Alexa devices and even connected wearables. Skype numbers are available for purchase in several countries and regions, and can be used across devices. Those with Skype numbers pay a flat fee for unlimited incoming calls.

FINANCIAL DATA: *Note: Data for latest year may not have been available at press time.*

In U.S. $	2021	2020	2019	2018	2017	2016
Revenue						
R&D Expense						
Operating Income						
Operating Margin %						
SGA Expense						
Net Income						
Operating Cash Flow						
Capital Expenditure						
EBITDA						
Return on Assets %						
Return on Equity %						
Debt to Equity						

CONTACT INFORMATION:

Phone: 352-26-20-15-82   Fax: 352-26-27-05-88
Toll-Free:
Address: 23-29 Rives de Clausen, Luxembourg, L-2165 Luxembourg

STOCK TICKER/OTHER:

Stock Ticker: Subsidiary   Exchange:
Employees: 1,400   Fiscal Year Ends: 12/31
Parent Company: Microsoft Corporation

SALARIES/BONUSES:

Top Exec. Salary: $ Bonus: $
Second Exec. Salary: $ Bonus: $

OTHER THOUGHTS:

Estimated Female Officers or Directors: 1
Hot Spot for Advancement for Women/Minorities: Y

Sales, profits and employees may be estimates. Financial information, benefits and other data can change quickly and may vary from those stated here.

Slack Technologies Inc

NAIC Code: 511210C

TYPES OF BUSINESS:

Computer Software, Telecom, Communications & VOIP

BRANDS/DIVISIONS/AFFILIATES:

Slack
Slack Enterprise Grid
salesforce.com Inc

CONTACTS: Note: Officers with more than one job title may be intentionally listed here more than once.

Stewart Butterfield, CEO
Allen Shim, CFO
Allen Shim, Chief Accounting Officer
Cal Henderson, Chief Technology Officer
Nadia Rawlinson, Chief People Officer
David Schellhase, General Counsel
Tamar Yehoshua, Other Executive Officer
Robert Frati, Senior VP, Divisional

GROWTH PLANS/SPECIAL FEATURES:

Slack Technologies, Inc. is a computer software firm that operates Slack, a platform that enables team communication through a single hub. The platform provides real-time messaging, archiving and searching services primarily for small-to-medium-sized companies or teams. Slack's solutions are divided into two groups: channels and direct messages. Channels provides a way to organize team conversations in open channels and are displayed per subject categories that relate to the project (such as feedback, product, customer service, app, issues, etc.). Direct messages reach colleagues directly and are completely private and secure. Private group features are available for sensitive information, with the ability to invite select team members that no one else can see or join. The direct messages are also categorized by subject for team members to easily see and access. Prices range from free introductory access to $12.50 per month. Free access includes search and browse of 10,000 most recent messages, 10 service integrations, free native apps for iOS/Android/Mac & Windows Desktop and 1:1 voice and video calls, 5GB file storage and two-factor identification. Standard access includes everything in Free, as well as searchable archive with unlimited messages, unlimited service integrations, custom retention policies, guest access, Google Authentication/Apps for Domains sign-on, configurable email ingestion and group voice/video calls, screen sharing, 10GB file storage and custom profiles. Plus access includes everything in Standard as well as SAML-based single sign-on, compliance exports of all message history, support for external message and archival solutions, 99.99% guaranteed uptime SLA, user provisioning and deprovisioning, real-time active directory sync with OneLogic, Okta and Ping, and 20GB file storage. For enterprise collaboration, Slack offers a communication hub with channels for communication, file-sharing and decision making purposes, as well as the Slack Enterprise Grid for managing large, complex teams. In July 2021, Slack Technologies was acquired by salesforce.com inc.

FINANCIAL DATA: Note: Data for latest year may not have been available at press time.

In U.S. $	2021	2020	2019	2018	2017	2016
Revenue	655,638,896	630,422,016	400,552,000	220,544,000	105,153,000	88,000,000
R&D Expense						
Operating Income						
Operating Margin %						
SGA Expense						
Net Income		-571,057,984	-140,683,008	-140,084,992	-146,864,000	
Operating Cash Flow						
Capital Expenditure						
EBITDA						
Return on Assets %						
Return on Equity %						
Debt to Equity						

CONTACT INFORMATION:

Phone: 415-630-7943   Fax:
Toll-Free:
Address: 500 Howard St., San Francisco, CA 94105 United States

STOCK TICKER/OTHER:

Stock Ticker: Subsidiary   Exchange:
Employees: 2,545   Fiscal Year Ends: 01/31
Parent Company: salesforce.com Inc

SALARIES/BONUSES:

Top Exec. Salary: $ Bonus: $
Second Exec. Salary: $ Bonus: $

OTHER THOUGHTS:

Estimated Female Officers or Directors:
Hot Spot for Advancement for Women/Minorities:

SmarTone Telecommunications Holdings Limited
www.smartoneholdings.com
NAIC Code: 517210

TYPES OF BUSINESS:
Mobile Phone Service
Wireless Data Services
Internet Service
Messaging Services
Telecommunications Services
Broadband Services
Fiber Optics
Ecommerce

BRANDS/DIVISIONS/AFFILIATES:
Sun Hung Kai Properties Limited
SmarTone Solutions
Birdie
Sahabat Setia SmarTone
Barkadahan sa SmarTone

CONTACTS: *Note: Officers with more than one job title may be intentionally listed here more than once.*
Stephen Chau, CTO
Angus Yiu, Sr. Mgr.-Corp. Planning
Chris Lau, Dir.-Future Svcs.
Patrick Chan, Exec. Dir.
Ping-luen Kwok, Chmn.

GROWTH PLANS/SPECIAL FEATURES:
SmarTone Telecommunications Holdings Limited is a leading telecommunications company with operating subsidiaries in Hong Kong and Macau. The firm provides voice, multimedia and mobile broadband services, as well as fixed fiber broadband services for the consumer and corporate markets. Network development by SmarTone encompass 2G (launched), 3G (launched), 4G (launched) and 5G (launched in 2020). SmarTone has more than 30 stores in Hong Kong. The company provides an omnichannel strategy, offering customers a seamless online-to-offline experience. SmarTone Solutions offers a range of information and communication technology (ICT) space to businesses and public sector organizations in Hong Kong and Macau, helping them to apply conventional and emerging technologies in areas such as digital transformation, customer engagement, cybersecurity and more. The firm's Birdie brand is a digital only mobile operator in Hong Kong. Targeting millennials, Birdie offers products accessible through its mobile app, doing away with contracts and administration fees and emphasizing community engagement via rewards program Birdie Friday and game activities such as Birdie Farm. Sahabat Setia SmarTone and Barkadahan sa SmarTone provide network services and customer care to customers in the Philippines and Indonesia. SmarTone's online store, shop.smartone.com, offers internationally-branded smartphones, smart watches, tablets and accessories, as well as service plans. SmarTone Telecommunications Holdings is a publicly-traded subsidiary of Sun Hung Kai Properties Limited, a real estate developer and investment company based in Hong Kong.

FINANCIAL DATA: *Note: Data for latest year may not have been available at press time.*

In U.S. $	2021	2020	2019	2018	2017	2016
Revenue	865,596,480	901,407,000	1,081,500,000	1,272,484,352	1,110,300,288	2,338,413,824
R&D Expense						
Operating Income						
Operating Margin %						
SGA Expense						
Net Income	57,015,758	43,818,100	78,970,800	78,378,904	85,622,456	101,552,952
Operating Cash Flow						
Capital Expenditure						
EBITDA						
Return on Assets %						
Return on Equity %						
Debt to Equity						

CONTACT INFORMATION:
Phone: 852 3128-2828   Fax: 852-2168-3021
Toll-Free:
Address: Fl. 31, Millennium City 2, 378 Kwun Tong Rd., Hong Kong, Hong Kong 999077 Hong Kong

STOCK TICKER/OTHER:
Stock Ticker: 315   Exchange: Hong Kong
Employees: 1,665   Fiscal Year Ends: 06/30
Parent Company: Sun Hung Kai Properties Limited

SALARIES/BONUSES:
Top Exec. Salary: $ Bonus: $
Second Exec. Salary: $ Bonus: $

OTHER THOUGHTS:
Estimated Female Officers or Directors: 2
Hot Spot for Advancement for Women/Minorities:

Sales, profits and employees may be estimates. Financial information, benefits and other data can change quickly and may vary from those stated here.

SMTC Corporation

NAIC Code: 334418

www.smtc.com

TYPES OF BUSINESS:

Printed Circuit Assembly (Electronic Assembly) Manufacturing
Electronics Manufacturing Services
Printed Circuit Board Assembly
New Product Introduction
Post-Production Services
Wireless Technology
Global Supply Chain Management
Specialty Services

BRANDS/DIVISIONS/AFFILIATES:

HIG Capital LLC

CONTACTS: *Note: Officers with more than one job title may be intentionally listed here more than once.*

Edward Smith, CEO
Steven M. Waszak, CFO
Terry Wegman, Sr. VP-Sales & Mktg.
Brian Kingston, VP-Global Human Resources
Baron Thrower, VP-Global IT
Josh Chien, CCO

GROWTH PLANS/SPECIAL FEATURES:

SMTC Corporation, based in Toronto, provides end-to-end electronics manufacturing services. The company's services include design and engineering solutions, printed circuit board assembly, new product introduction, post-production services, radio frequency wireless technology, electronic manufacturing and integration, global supply chain management and specialty services. SMTC has facilities in the U.S., Mexico, Canada and China, and its services extend over the entire electronic product lifecycle, from development and introduction of new products to growth, maturity and end-of-life phases. The firm's fully-integrated contract manufacturing services are offered utilized by global original equipment manufacturers and emerging technology companies across a variety of markets including medical and safety, semiconductor, retail and payment systems, test and measurement, industrial, power and clean technology, telecommunications, networks, and defense and aerospace. SMTC Corporation is privately owned by H.I.G. Capital LLC, a global alternative investment firm.

FINANCIAL DATA: *Note: Data for latest year may not have been available at press time.*

In U.S. $	2021	2020	2019	2018	2017	2016
Revenue	390,000,000	386,500,000	372,511,008	216,131,008	139,231,008	167,868,000
R&D Expense						
Operating Income						
Operating Margin %						
SGA Expense						
Net Income		-600,000	-5,995,000	-448,000	-7,845,000	-232,000
Operating Cash Flow						
Capital Expenditure						
EBITDA						
Return on Assets %						
Return on Equity %						
Debt to Equity						

CONTACT INFORMATION:

Phone: 905 479-1810   Fax:
Toll-Free:
Address: 7050 Woodbine Ave., Ste. 300, Markham, ON L3R 4G8 Canada

STOCK TICKER/OTHER:

Stock Ticker: Private
Employees: 3,215
Parent Company: HIG Capital LLC

Exchange:
Fiscal Year Ends: 12/31

SALARIES/BONUSES:

Top Exec. Salary: $ Bonus: $
Second Exec. Salary: $ Bonus: $

OTHER THOUGHTS:

Estimated Female Officers or Directors: 1
Hot Spot for Advancement for Women/Minorities: Y

Socket Mobile Inc

www.socketmobile.com

NAIC Code: 334220

TYPES OF BUSINESS:
Wireless Handheld Computers
Data Collection Devices
Barcode Scanners
RFID Devices
Bluetooth Devices

BRANDS/DIVISIONS/AFFILIATES:
DuraScan
SocketScan
DuraCase
S700
800 Series
D600
S550

CONTACTS: *Note: Officers with more than one job title may be intentionally listed here more than once.*
Kevin Mills, CEO
Lynn Zhao, CFO
Charlie Bass, Chairman of the Board
Leonard Ott, Chief Technology Officer
David Holmes, Other Executive Officer

GROWTH PLANS/SPECIAL FEATURES:
Socket Mobile Inc is a producer of data capture products. The company's products are integrated into mobile applications used in mobile point of sale (mPOS), enterprise mobility, asset tracking, control systems, logistics, event management, medical and education. Its cordless barcode scanners connect over Bluetooth and work with applications running on smartphones, mobile computers and tablets using various operating systems. The company offers barcode scanning products for both one-dimensional, including imager and laser, and two-dimensional bar code scanning in standard and durable cases. The company's geographical segments are the United States, Europe, and Asia and the rest of the world, out of which the majority of the revenue is generated from the United States.

Socket Mobile offers its employees health, dental, vision, group life and long-term disability coverage; and 401(k) and stock options.

FINANCIAL DATA: *Note: Data for latest year may not have been available at press time.*

In U.S. $	2021	2020	2019	2018	2017	2016
Revenue	23,199,060	15,700,040	19,253,100	16,454,430	21,285,800	20,787,590
R&D Expense	3,964,599	3,140,104	3,893,563	3,640,296	3,473,610	2,889,168
Operating Income	2,697,381	76,429	606,370	-585,988	2,417,715	2,563,016
Operating Margin %	.12%	.00%	.03%	-.04%	.11%	.12%
SGA Expense	5,774,464	5,118,368	5,600,710	5,402,100	5,498,684	4,982,034
Net Income	4,466,257	-3,278,601	286,586	-571,141	-1,430,731	12,147,090
Operating Cash Flow	2,144,270	804,445	873,534	750,145	2,376,303	879,816
Capital Expenditure	691,771	536,481	602,954	423,700	620,575	304,470
EBITDA	3,521,621	-2,634,971	1,069,300	-153,946	2,731,368	2,842,408
Return on Assets %	.20%	-.18%	.02%	-.03%	-.07%	.78%
Return on Equity %	.26%	-.25%	.02%	-.04%	-.09%	1.24%
Debt to Equity	.01%	0.023	0.054	0.119	0.016	0.02

CONTACT INFORMATION:
Phone: 510 933-3000   Fax: 510 933-3030
Toll-Free:
Address: 39700 Eureka Dr., Newark, CA 94560 United States

STOCK TICKER/OTHER:
Stock Ticker: SCKT   Exchange: NAS
Employees: 53   Fiscal Year Ends: 12/31
Parent Company:

SALARIES/BONUSES:
Top Exec. Salary: $277,500   Bonus: $
Second Exec. Salary: Bonus: $
$204,000

OTHER THOUGHTS:
Estimated Female Officers or Directors:
Hot Spot for Advancement for Women/Minorities:

Sales, profits and employees may be estimates. Financial information, benefits and other data can change quickly and may vary from those stated here.

SoftBank Group Corp

NAIC Code: 517110

www.softbank.co.jp

TYPES OF BUSINESS:

Telecommunications Services
Investment
Managed Funds
Technology
Artificial Intelligence
Robotics
Smartphone Payment
Electricity

GROWTH PLANS/SPECIAL FEATURES:

SoftBank is a Japan-based telecom and e-commerce conglomerate that has expanded mainly through acquisitions and its key assets include a 28% stake in Chinese e-commerce giant Alibaba and a 40% owned mobile and fixed broadband telecom operator business in Japan. It also owns 75% of semiconductor chip designer ARM Holdings and plans an IPO of this business, and has a vast portfolio of mainly internet- and e-commerce-focused early stage investments. It is also general partner of the $100 billion SoftBank Vision Fund 1 and sole investor in Softbank Vision Fund 2, both of which primarily invest in pre-IPO internet companies.

BRANDS/DIVISIONS/AFFILIATES:

SoftBank Group Capital Limited
SB Northstar LP
SoftBank Investment Advisers
SoftBank Vision Fund LP
Z Holdings
Arm Limited
Fortress Investment Group LLC
PayPay Corporation

CONTACTS: *Note: Officers with more than one job title may be intentionally listed here more than once.*

Masayoshi Shon, CEO
Marcelo Claure, COO
Ken Miyauchi, COO-SOFTBANK MOBILE Corp.
Ronald D. Fisher, Pres., SOFTBANK Holdings, Inc.
Masayoshi Son, Chmn.

FINANCIAL DATA: *Note: Data for latest year may not have been available at press time.*

In U.S. $	2021	2020	2019	2018	2017	2016
Revenue	41,739,600,000	38,853,000,000	45,190,950,000	67,923,210,000	66,011,600,000	
R&D Expense						
Operating Income	4,475,171,000	4,409,560,000	14,071,310,000	9,669,245,000	7,474,132,000	
Operating Margin %	.11%	.11%	.31%	.14%	.11%	
SGA Expense	16,845,870,000	15,277,960,000	13,840,960,000	18,931,060,000	16,888,540,000	
Net Income	36,991,710,000	-7,131,237,000	10,465,730,000	7,705,258,000	10,577,780,000	
Operating Cash Flow	4,132,676,000	8,290,411,000	8,690,775,000	8,073,443,000	11,129,690,000	
Capital Expenditure	4,797,449,000	9,140,841,000	10,122,770,000	7,897,027,000	6,848,873,000	
EBITDA	50,645,370,000	17,764,810,000	27,579,330,000	18,441,380,000	19,671,510,000	
Return on Assets %	.12%	-.03%	.04%	.04%	.06%	
Return on Equity %	.62%	-.14%	.22%	.24%	.46%	
Debt to Equity	1.16%	1.763	1.601	2.667	3.392	

CONTACT INFORMATION:

Phone: 81 3-6889-2000   Fax:
Toll-Free:
Address: 1-9-1 Higashi-shimbashi, Tokyo, 105-7303 Japan

STOCK TICKER/OTHER:

Stock Ticker: SFTBY   Exchange: PINX
Employees: 92,069   Fiscal Year Ends: 03/31
Parent Company:

SALARIES/BONUSES:

Top Exec. Salary: $ Bonus: $
Second Exec. Salary: $ Bonus: $

OTHER THOUGHTS:

Estimated Female Officers or Directors:
Hot Spot for Advancement for Women/Minorities:

Sonim Technologies Inc

www.sonimtech.com

NAIC Code: 334220

TYPES OF BUSINESS:

Radio and Television Broadcasting and Wireless Communications
Equipment Manufacturing
Rugged Mobile Devices
Rugged Scanners
Related Accessories
Software
Manufacture
Distribution

BRANDS/DIVISIONS/AFFILIATES:

GROWTH PLANS/SPECIAL FEATURES:

Sonim Technologies Inc is a provider of ultra-rugged mobile devices, including phones and accessories designed specifically for task workers physically engaged in their work environments, often in mission-critical roles. The company sells its mobile phones and accessories in the United States to AT&T, T-Mobile, Verizon and in Canada to Bell, Rogers and Telus Mobility. Its solutions include ultra-rugged mobile phones that are capable of attaching to both public and private wireless networks, industrial-grade accessories that meet the requirements of specific applications and software applications and cloud-based tools that provide management and deployment services to its customers. The company generates revenue from discounts, price protection and customer incentives.

CONTACTS: Note: Officers with more than one job title may be intentionally listed here more than once.

Robert Tirva, CFO
John Kneuer, Chairman of the Board

FINANCIAL DATA: Note: Data for latest year may not have been available at press time.

In U.S. $	2021	2020	2019	2018	2017	2016
Revenue	54,570,000	63,992,000	116,251,000	135,665,000	59,031,000	
R&D Expense	17,696,000	16,218,000	26,064,000	23,247,000	13,008,000	
Operating Income	-38,001,000	-27,714,000	-21,645,000	5,394,000	-6,770,000	
Operating Margin %	-.70%	-.43%	-.19%	.04%	-.11%	
SGA Expense	26,719,000	26,707,000	30,090,000	19,448,000	14,073,000	
Net Income	-38,627,000	-29,932,000	-25,834,000	1,277,000	-8,519,000	
Operating Cash Flow	-38,476,000	-10,560,000	-33,523,000	3,861,000	-8,906,000	
Capital Expenditure	46,000	11,000	1,356,000	2,545,000	1,174,000	
EBITDA	-36,331,000	-26,966,000	-19,399,000	5,709,000	-6,249,000	
Return on Assets %	-.85%	-.57%	-.42%	-.18%	-.51%	
Return on Equity %	-1.64%	-1.11%	-1.69%			
Debt to Equity	.00%	0.007	0.014	2.575		

CONTACT INFORMATION:

Phone: 650-378-8100   Fax:
Toll-Free:
Address: 6500 River Place Blvd., Bldg. 7, Ste. 250, Austin, TX 78730 United States

STOCK TICKER/OTHER:

Stock Ticker: SONM   Exchange: NAS
Employees: 102   Fiscal Year Ends: 12/31
Parent Company:

SALARIES/BONUSES:

Top Exec. Salary: $400,000   Bonus: $160,000
Second Exec. Salary: $300,000   Bonus: $90,000

OTHER THOUGHTS:

Estimated Female Officers or Directors:
Hot Spot for Advancement for Women/Minorities:

Sony Corporation

NAIC Code: 334220

TYPES OF BUSINESS:

Smartphones
Consumer Electronics
Mobile Phones
Personal Computers
Monitors
Cameras
Entertainment Products and Digital Technology
Robotics and Artificial Intelligence

BRANDS/DIVISIONS/AFFILIATES:

Sony Group Corporation
Sony Corporation of America
Columbia
RCA
Epic
Sony Music Nashville
Sony Classical
Xperia

CONTACTS: *Note: Officers with more than one job title may be intentionally listed here more than once.*

Neal Manowitz, CEO
Bob Ishida, Deputy CEO

GROWTH PLANS/SPECIAL FEATURES:

Sony Corporation, a product of the merger of Sony Electronic Inc.; Sony Imaging Products & Solutions, Inc.; Sony Hom Entertainment & Sound Products, Inc.; and Sony Mobi Communications, Inc., is a company with a broad portfol encompassing electronics, music, motion pictures, mobil gaming and robotics. The firm's operations include researc and development, engineering, sales, marketing, distributic and customer service. Consumer electronic products spa TVs, monitors, projectors, cameras, camcorders, audio/vide vehicle entertainment, marine entertainment, person computers, digital readers, mobile phones and tablets. Th music division specializes in recorded music across a range music artists and types of music, with labels includir Columbia, RCA, Epic, Sony Music Nashville and Sor Classical. The motion picture division produces, acquires ar distributes motion picture and television content, as well a digital content and related products and technologies. Mobi products include mobile phones and smartphones offere under the Xperia brand name, as well as video monitors ar related accessories. The gaming division produces ar markets PlayStation game products, including video and digit games, gaming hardware and software, and relate connectivity accessories and repair services. Last, the robotic division is engaged in artificial intelligence (AI) and robotic designing and developing technologies and systems for a array of purposes and industries, including industri entertainment, healthcare, consumer, professional and mor Sony Corporation is a subsidiary of Sony Corporation America and an affiliate of Sony Group Corporation.

FINANCIAL DATA: *Note: Data for latest year may not have been available at press time.*

In U.S. $	2021	2020	2019	2018	2017	2016
Revenue	21,225,175,648	19,178,608,074	4,958,156,878	6,719,170,000	6,764,010,000	9,974,120,000
R&D Expense						
Operating Income						
Operating Margin %						
SGA Expense						
Net Income	1,931,789	1,186,085,766	787,064,568	-260,102,000	91,338,500	-546,168,000
Operating Cash Flow						
Capital Expenditure						
EBITDA						
Return on Assets %						
Return on Equity %						
Debt to Equity						

CONTACT INFORMATION:

Phone: 858-942-2400   Fax:
Toll-Free:
Address: 16535 Via Esprillo, San Diego, CA 92127 United States

STOCK TICKER/OTHER:

Stock Ticker: Subsidiary   Exchange:
Employees: 8,500   Fiscal Year Ends: 03/31
Parent Company: Sony Group Corporation

SALARIES/BONUSES:

Top Exec. Salary: $ Bonus: $
Second Exec. Salary: $ Bonus: $

OTHER THOUGHTS:

Estimated Female Officers or Directors: 1
Hot Spot for Advancement for Women/Minorities:

Sopra Steria Group SA

www.soprasteria.com/en

NAIC Code: 541512

TYPES OF BUSINESS:

IT Consulting
Business Strategy Consulting
Business Process Outsourcing

BRANDS/DIVISIONS/AFFILIATES:

GROWTH PLANS/SPECIAL FEATURES:

Sopra Steria Group SA is a digital transformation company. It guides businesses through conversion projects, from the development phase to implementation phase. The firm provides end to end services in the fields of consulting, systems integration, software development, infrastructure management and business process services. It's target markets are financial services, insurance and social welfare, public sector, telecom operators, aerospace and defense, retail and energy sectors. Its geographic segments are France, United Kingdom, Spain, Italy, and Belgium. The company derives the majority of its revenues from France.

CONTACTS: *Note: Officers with more than one job title may be intentionally listed here more than once.*

Vincent Paris, CEO
Christian Levi, Head-Sopra Consulting
Pierre Pasquier, Chmn.

FINANCIAL DATA: *Note: Data for latest year may not have been available at press time.*

In U.S. $	2021	2020	2019	2018	2017	2016
Revenue	4,775,346,000	4,347,148,000	4,521,629,000	4,176,236,000	3,906,814,000	3,815,239,000
R&D Expense						
Operating Income	358,446,700	258,611,900	313,883,000	259,631,700	284,411,900	258,306,000
Operating Margin %	.08%	.06%	.07%	.06%	.07%	.07%
SGA Expense						
Net Income	191,409,500	108,910,700	163,468,000	127,572,400	175,909,100	153,372,400
Operating Cash Flow	472,456,200	409,638,800	431,257,800	233,729,700	242,295,700	228,936,800
Capital Expenditure	55,679,060	54,659,300	50,784,200	63,123,330	63,531,240	47,826,880
EBITDA	520,385,100	444,616,700	496,624,600	312,353,400	325,610,300	323,978,700
Return on Assets %	.04%	.02%	.04%	.03%	.04%	.04%
Return on Equity %	.12%	.08%	.12%	.10%	.15%	.13%
Debt to Equity	.45%	0.566	0.548	0.261	0.331	0.376

CONTACT INFORMATION:

Phone: 33-1-10-67-29-29   Fax: 33-1-40-67-29-30
Toll-Free:
Address: 9 bis, Paris, 75116 France

STOCK TICKER/OTHER:

Stock Ticker: SPPSY   Exchange: GREY
Employees: 44,114   Fiscal Year Ends: 12/31
Parent Company:

SALARIES/BONUSES:

Top Exec. Salary: $ Bonus: $
Second Exec. Salary: $ Bonus: $

OTHER THOUGHTS:

Estimated Female Officers or Directors: 1
Hot Spot for Advancement for Women/Minorities:

Sales, profits and employees may be estimates. Financial information, benefits and other data can change quickly and may vary from those stated here.

Spark New Zealand Limited

NAIC Code: 517110

www.sparknz.co.nz

TYPES OF BUSINESS:

Telecommunications Service
Mobile Phones
Broadband
Landline
Ecommerce
Virtual Store

BRANDS/DIVISIONS/AFFILIATES:

SparkLab.co.nz

GROWTH PLANS/SPECIAL FEATURES:

Spark is one of only two large integrated telecommunications companies in New Zealand. It is the dominant provider of fixed line services in the country and effectively equal-number-one player in the mobile telephony market. It also boasts a commanding presence in the New Zealand corporate and wholesale telecommunications services provision space. Spark's operations are split into mobile, voice, broadband, and digital-related services.

CONTACTS: *Note: Officers with more than one job title may be intentionally listed here more than once.*

Jolie Hodson, CEO
Tessa Tierney, Dir.-Product
Stefan Knight, Dir.-Finance
Matt Bain, Dir.-Mktg.
Heather Polglase, Dir.-Human Resources
Mark Beder, Dir.-Technology
Rod Snodgrass, Head-Telecom Digital Ventures
Chris Quin, Chief Exec.-Telecom Retail
Tim Miles, Chief Exec.-Gne-i
David Yuile, Chief Exec.-AAPT
Justine Smyth, Chmn.

FINANCIAL DATA: *Note: Data for latest year may not have been available at press time.*

In U.S. $	2021	2020	2019	2018	2017	2016
Revenue	2,242,167,000	2,256,632,000	2,212,606,000	2,257,261,000	2,159,775,000	2,104,429,000
R&D Expense						
Operating Income	361,639,800	394,973,500	377,992,200	336,482,200	259,122,800	253,462,300
Operating Margin %	.16%	.18%	.17%	.15%	.12%	.12%
SGA Expense	396,231,400	410,068,100	395,602,500	441,515,000	451,578,000	418,873,200
Net Income	241,512,500	268,556,800	257,235,900	242,141,400	262,896,400	232,707,300
Operating Cash Flow						
Capital Expenditure	242,770,400	247,172,900	261,009,600	260,380,600	246,544,000	262,896,400
EBITDA	706,298,200	683,027,500	696,235,200	632,083,500	649,064,800	631,454,500
Return on Assets %	.09%	.10%	.11%	.11%	.13%	.11%
Return on Equity %	.26%	.29%	.27%	.24%	.25%	.21%
Debt to Equity	.96%	1.189	0.97	0.615	0.419	0.407

CONTACT INFORMATION:

Phone: 64 9-215-7564   Fax: 64-800-278-320
Toll-Free:
Address: Telecom House, 167 Victoria St., Auckland, 1010 New Zealand

STOCK TICKER/OTHER:

Stock Ticker: NZTCF   Exchange: PINX
Employees: 5,083   Fiscal Year Ends: 06/30
Parent Company:

SALARIES/BONUSES:

Top Exec. Salary: $ Bonus: $
Second Exec. Salary: $ Bonus: $

OTHER THOUGHTS:

Estimated Female Officers or Directors: 2
Hot Spot for Advancement for Women/Minorities:

Sales, profits and employees may be estimates. Financial information, benefits and other data can change quickly and may vary from those stated here.

Speedcast International Limited

www.speedcast.com

NAIC Code: 517110

TYPES OF BUSINESS:

Satellite Network Service Provider.
Communications Technology
Satellite Network

BRANDS/DIVISIONS/AFFILIATES:

Centerbridge Partners LP

CONTACTS: *Note: Officers with more than one job title may be intentionally listed here more than once.*

Joe Spytek, CEO
Mike Neugebauer, COO
Lee Eckert, CFO
Olga Pirogova, Chief Human Resources Officer
Chris Hill, CTO

GROWTH PLANS/SPECIAL FEATURES:

Speedcast International Limited is a communications company. The firm's customers are primarily engaged in the maritime, energy, telecommunication, mining, broadcast media, non-government organization (NGO) and government industries. Speedcast's global satellite network and support infrastructure spans across every region in the world. The company's solutions are categorized into: customer experience management, connectivity, professional services, network management and applications and solutions. The customer experience management division works alongside businesses to deliver communication services and support, from implementation to delivery to field support and solution management. The connectivity division unites its diverse connectivity resources into a single, seamless network experience, from both very small aperture terminal (VSAT) two-way satellite ground station and mobile two-way voice and data communication satellite services (MSS). This division has a satellite network of more than 95 satellites and 35+ Speedcast and partner teleports. The professional services division offers problem-solving solutions, enterprise solutions and 24/7/365 support in any time zone and multiple languages. The network management division utilizes a multi-protocol label switching (MPLS) routing technique that directs data from one node to the next based on short path labels rather than long network addresses. It gives a complete view of the customer's network operations, whether it's microwave, fiber, 4G/LTE, terrestrial or satellite services. Last, the applications and solutions division designs solutions to suit its customer's needs, utilizing solutions for Internet of Things (IoT), automation and more. Its solutions are built to satisfy any market, industry, initiative and challenge. Speedcast is privately owned by Centerbridge Partners LP.

FINANCIAL DATA: *Note: Data for latest year may not have been available at press time.*

In U.S. $	2021	2020	2019	2018	2017	2016
Revenue	700,000,000	685,300,000	623,000,000	603,832,256	450,877,312	206,055,248
R&D Expense						
Operating Income						
Operating Margin %						
SGA Expense						
Net Income			1,897,396	1,789,901	4,863,276	5,574,120
Operating Cash Flow						
Capital Expenditure						
EBITDA						
Return on Assets %						
Return on Equity %						
Debt to Equity						

CONTACT INFORMATION:

Phone: 832-668-2300   Fax:
Toll-Free:
Address: 4400 S. Sam Houston Pkwy. E., Houston, TX 77048 United States

STOCK TICKER/OTHER:

Stock Ticker: Subsidiary   Exchange:
Employees: Fiscal Year Ends: 12/31
Parent Company: Centerbridge Partners LP

SALARIES/BONUSES:

Top Exec. Salary: $ Bonus: $
Second Exec. Salary: $ Bonus: $

OTHER THOUGHTS:

Estimated Female Officers or Directors:
Hot Spot for Advancement for Women/Minorities:

Sales, profits and employees may be estimates. Financial information, benefits and other data can change quickly and may vary from those stated here.

Spok Inc
NAIC Code: 517210

www.spok.com

TYPES OF BUSINESS:
Paging Services
Wireless Messaging Services
Instant Text Messaging

GROWTH PLANS/SPECIAL FEATURES:
Spok Holdings Inc is a provider of healthcare communication
It reports three market segments namely Healthcar
Government, and Large enterprise. The company provide
paging services and software solutions in the United States ar
abroad.

Spok offers its employees comprehensive health benefit
retirement and savings plans, training and developme
opportunities and education assistance.

BRANDS/DIVISIONS/AFFILIATES:
Spok Holdings Inc
Spok Go

CONTACTS: *Note: Officers with more than one job title may be intentionally listed here more than once.*
Vincent Kelly, CEO
Michael Wallace, CFO
Royce Yudkoff, Chairman of the Board
Calvin Rice, Chief Accounting Officer
Sharon Woods Keisling, Treasurer

FINANCIAL DATA: *Note: Data for latest year may not have been available at press time.*

In U.S. $	2021	2020	2019	2018	2017	2016
Revenue	142,153,000	148,180,000	160,289,000	169,474,000	171,175,000	179,561,000
R&D Expense	17,920,000	15,828,000	27,543,000	24,464,000	18,702,000	13,467,000
Operating Income	-11,395,000	3,393,000	-6,291,000	-1,549,000	11,234,000	22,596,000
Operating Margin %	-.08%	.02%	-.04%	-.01%	.07%	.13%
SGA Expense	88,613,000	85,550,000	93,300,000	97,371,000	65,474,000	62,898,000
Net Income	-22,180,000	-44,225,000	-10,765,000	-1,479,000	-15,306,000	13,979,000
Operating Cash Flow	7,968,000	26,163,000	11,693,000	10,315,000	15,515,000	37,551,000
Capital Expenditure	15,235,000	14,707,000	4,837,000	5,915,000	9,214,000	6,254,000
EBITDA	-949,000	12,449,000	2,958,000	9,220,000	22,858,000	35,559,000
Return on Assets %	-.08%	-.15%	-.03%	.00%	-.04%	.04%
Return on Equity %	-.12%	-.20%	-.04%	-.01%	-.05%	.04%
Debt to Equity	.07%	0.047	0.046			

CONTACT INFORMATION:
Phone: 800 611-8488   Fax: 508 836-3626
Toll-Free: 800-611-8488
Address: 5911 Kingstown Village Pkwy., Fl. 6, Alexandria, VA 22315
United States

STOCK TICKER/OTHER:
Stock Ticker: SPOK
Employees: 563
Parent Company: Spok Holdings Inc

Exchange: NAS
Fiscal Year Ends: 12/31

SALARIES/BONUSES:
Top Exec. Salary: $566,592   Bonus: $
Second Exec. Salary: Bonus: $
$385,420

OTHER THOUGHTS:
Estimated Female Officers or Directors: 5
Hot Spot for Advancement for Women/Minorities: Y

Startec Global Communications Corporation   www.startec.com

NAIC Code: 517110

TYPES OF BUSINESS:

IP Communications Services
Voice Services
Data Services
Internet Access

BRANDS/DIVISIONS/AFFILIATES:

Lingo Communications LLC
CellConnect
MyCountry Number
Direct Dial
10-10-719

CONTACTS: *Note: Officers with more than one job title may be intentionally listed here more than once.*

Chuck Griffin, CEO-Lingo

GROWTH PLANS/SPECIAL FEATURES:

Startec Global Communications Corporation provides telephone, internet and communications services. Startec also works with international long-distance carriers and internet service providers transacting with the world's emerging economies. The company's products include CellConnect long distance, CellConnect Prepaid long distance, Direct Dial long distance, 10-10-719 long distance and MyCountry Number local. CellConnect is an international calling service that allows the user to register up to four phone numbers, including land and cell phone lines, to dial internationally without using PIN numbers. CellConnect Prepaid offers international calls from any cell phone or landline at low rates. Direct Dial provides international long-distance service for home phones and cell phones. Startec's 10-10-719 service, which does not require a calling plan, bills the calls on the company's rates, even if the user is currently under contract with another service provider. MyCountry Number offers one low rate for same-country local calls and unlimited international incoming calls. Startec also offers online account management, online billing for certain services and a referral credit program for users of Direct Dial long distance. The company's services connect worldwide communities including the Pacific Rim, the Middle East, North Africa, Russia, Central Europe and North and South America. It provides services through a network of owned and leased facilities, operating and termination agreements and resale arrangements. Startec technology includes an extensive network of internet protocol (IP) gateways, international gateways, domestic switches and ownership in undersea fiber optic cables. Startec is a subsidiary and brand of Lingo Communications LLC.

FINANCIAL DATA: *Note: Data for latest year may not have been available at press time.*

In U.S. $	2021	2020	2019	2018	2017	2016
Revenue						
R&D Expense						
Operating Income						
Operating Margin %						
SGA Expense						
Net Income						
Operating Cash Flow						
Capital Expenditure						
EBITDA						
Return on Assets %						
Return on Equity %						
Debt to Equity						

CONTACT INFORMATION:

Phone: Fax: 301-610-4301
Toll-Free: 800-827-3374
Address: 400 E. Las Colinas Blvd., Ste. 500, Irving, TX 75039 United States

STOCK TICKER/OTHER:

Stock Ticker: Subsidiary   Exchange:
Employees: Fiscal Year Ends: 12/31
Parent Company: Lingo Communications LLC

SALARIES/BONUSES:

Top Exec. Salary: $ Bonus: $
Second Exec. Salary: $ Bonus: $

OTHER THOUGHTS:

Estimated Female Officers or Directors:
Hot Spot for Advancement for Women/Minorities:

Sales, profits and employees may be estimates. Financial information, benefits and other data can change quickly and may vary from those stated here.

Swisscom AG
NAIC Code: 517110

TYPES OF BUSINESS:
Integrated Telecommunications Services
Internet Service Provider
Value-added Services
Mobile Phone Service
Local & Long-Distance Service
IT & Outsourcing Services
5G

BRANDS/DIVISIONS/AFFILIATES:
Swisscom Switzerland
Fastweb
www.bluewin.ch
CT Cinetrade AG

GROWTH PLANS/SPECIAL FEATURES:
Swisscom AG is the incumbent telephone operator in Switzerland. It dominates both the broadband and wireless market segments, with a 58% wireless postpaid share and 53% fixed broadband share, a level that dwarfs other developed countries' incumbent operators. Swisscom has also 37% market share in pay television. The firm owns Fastweb, an alternative telecom operator in Italy, offering fixed-line telephone, broadband, and wireless services. It holds a 16% market share in fixed broadband and 2% mobile market share

CONTACTS: Note: Officers with more than one job title may be intentionally listed here more than once.
Urs Schaeppi, CEO
Eugen Stermetz, CFO
Klementina Pejic, Human Resources
Christoph Aeschlimann, IT
Jurgen Galler, Head-Strategy & Innovation
Bart Morselt, Head-Investor Rel.
Mario Rossi, Head-Group Bus. Steering
Michael Rechsteiner, Chmn.
Urs Schaeppi, Head-Switzerland

FINANCIAL DATA: Note: Data for latest year may not have been available at press time.

In U.S. $	2021	2020	2019	2018	2017	2016
Revenue	11,711,420,000	11,624,500,000	11,994,180,000	12,267,510,000	12,213,050,000	12,193,160,000
R&D Expense						
Operating Income	2,130,111,000	1,998,157,000	1,983,495,000	2,158,386,000	2,225,411,000	2,237,978,000
Operating Margin %	.18%	.17%	.17%	.18%	.18%	.18%
SGA Expense	405,286,600	390,625,000	495,350,200	529,909,500	558,185,300	553,996,400
Net Income	1,918,566,000	1,602,296,000	1,751,005,000	1,599,154,000	1,644,186,000	1,679,792,000
Operating Cash Flow	4,235,087,000	4,365,994,000	4,169,110,000	3,895,778,000	4,284,308,000	3,897,872,000
Capital Expenditure	2,377,262,000	2,291,387,000	2,502,932,000	2,517,594,000	2,490,365,000	2,530,161,000
EBITDA	4,892,762,000	4,561,830,000	4,497,948,000	4,402,648,000	4,497,948,000	4,532,507,000
Return on Assets %	.07%	.06%	.07%	.07%	.07%	.08%
Return on Equity %	.18%	.17%	.20%	.19%	.22%	.27%
Debt to Equity	.17%	0.186	0.202	0.044		

CONTACT INFORMATION:
Phone: 41 582219911   Fax:
Toll-Free:
Address: Alte Tiefenaustrasse 6, Bern, CH-3050 Switzerland

STOCK TICKER/OTHER:
Stock Ticker: SWZCF
Employees: 18,905
Parent Company:

Exchange: PINX
Fiscal Year Ends: 12/31

SALARIES/BONUSES:
Top Exec. Salary: $ Bonus: $
Second Exec. Salary: $ Bonus: $

OTHER THOUGHTS:
Estimated Female Officers or Directors: 2
Hot Spot for Advancement for Women/Minorities: Y

Sykes Enterprises Incorporated

www.sykes.com

NAIC Code: 541512

TYPES OF BUSINESS:

Computer Integrated Systems Design
Outsourcing Services
Customer Engagement Solutions
Management Services
Consulting
Sales Services
Brand Marketing Solutions
Digital Transformation Solutions

BRANDS/DIVISIONS/AFFILIATES:

Sitel Group
clearlink
Sykes Digital Services
Assistance Services Group

CONTACTS: Note: Officers with more than one job title may be intentionally listed here more than once.

Charles Sykes, CEO
John Chapman, CFO
James Macleod, Chairman of the Board
David Pearson, Chief Information Officer
William Rocktoff, Controller
Jenna Nelson, Executive VP, Divisional
James Holder, Executive VP
Lawrence Zingale, Executive VP

GROWTH PLANS/SPECIAL FEATURES:

Sykes Enterprises, Incorporated provides multichannel demand generation customer engagement solutions and services primarily to Global 2000 companies and their end customers. These companies and customers are engaged in the communications, education, emerging brands, energy/utilities, financial services, healthcare, insurance services, retail, small/medium businesses, technology and travel industries. Sykes' wide range of services are grouped into five categories: outsourcing, holistic customer experience (CX), sales services, brand marketing, and digital transformation. Outsourcing solutions include talent acquisition and development, transaction handling, coaching, engagement, delivery models, work at home, risk management and technology platform. Holistic CX services include intelligent CX innovation, insight analytics, multi-channel solutions, social media care and chat support. Sales services include sales through service, customer selection, customer acquisition and customer retention. Brand marketing solutions span customer growth, digital marketing, digital inbound sales, assistance services and social media care. Last, digital transformation services cover workforce, CX for driving growth and operating cost reduction. Headquartered in the U.S., Sykes has office locations worldwide, including North America, Latin America, Europe, the Middle East, Africa and Asia-Pacific. Sykes' family of companies include clearlink, Sykes Digital Services, and Assistance Services Group. Sykes Enterprises operates as a subsidiary of Sitel Group, a provider of customer experience products and solutions.

Depending on the country and location, Sykes offers employees comprehensive health benefits, 401(k) and education programs.

FINANCIAL DATA: Note: Data for latest year may not have been available at press time.

In U.S. $	2021	2020	2019	2018	2017	2016
Revenue	1,744,466,212	1,710,260,992	1,614,761,984	1,625,687,040	1,586,008,064	1,460,036,992
R&D Expense						
Operating Income						
Operating Margin %						
SGA Expense						
Net Income		56,432,000	64,081,000	48,926,000	32,216,000	62,390,000
Operating Cash Flow						
Capital Expenditure						
EBITDA						
Return on Assets %						
Return on Equity %						
Debt to Equity						

CONTACT INFORMATION:

Phone: 813 274-1000   Fax: 813 273-0148
Toll-Free:
Address: 400 N. Ashley Dr., Ste. 2800, Tampa, FL 33602 United States

STOCK TICKER/OTHER:

Stock Ticker: Subsidiary   Exchange:
Employees: 61,100   Fiscal Year Ends: 12/31
Parent Company: Sitel Group

SALARIES/BONUSES:

Top Exec. Salary: $ Bonus: $
Second Exec. Salary: $ Bonus: $

OTHER THOUGHTS:

Estimated Female Officers or Directors: 3
Hot Spot for Advancement for Women/Minorities: Y

Sales, profits and employees may be estimates. Financial information, benefits and other data can change quickly and may vary from those stated here.

Syniverse Technologies LLC

www.syniverse.com

NAIC Code: 517110

TYPES OF BUSINESS:

Telecommunications & Network Services
Communications Technology Services
Business Management Services
5G Services
Voice Over Long-Term Evolution Solutions
Internet of Things Solutions
Private Wireless Network Solutions
Application Solutions

BRANDS/DIVISIONS/AFFILIATES:

Carlyle Group (The)

CONTACTS: *Note: Officers with more than one job title may be intentionally listed here more than once.*

Andrew Davies, CEO
Simeon Irvine, CFO
Sara DeBella, Chief Human Resources Officer
John Wick, CTO
Gary Weisenborn, Sr. VP-Prod. Dev.
David W. Hitchcock, Chief Admin. Officer
Laura Binion, General Counsel
John McRae, Sr. VP-Global Customer Oper.
Ed Lewis, Chief Strategy Officer
Amy Moini, Sr. VP-Global Sales Dev.
Pablo Milikota, Sr. VP-Global Bus. Svcs. & Solutions
David Wasserman, Sr. VP-North America Sales
Michael O'Brien, Sr. VP-Strategic Market Initiatives
James A. Attwood, Chmn.
Mahesh Prasad, Pres., Americas

GROWTH PLANS/SPECIAL FEATURES:

Syniverse Technologies, LLC is a provider of wireless voice and data services for telecommunications companies worldwide, including North America, Asia Pacific, the Caribbean, Latin America, Europe, the Middle East and Africa. The company's product line is used to implement transformational technologies, including 3G to voice over long term evolution (VoLTE) roaming, 5G roaming, 5G messaging, private wireless networks and Internet of Things (IoT) connectivity. Customer companies span mobile operators, cable and internet providers and enterprises as well as machine-to-machine (M2M), voice over internet protocol (VoIP), VoLTE and application service providers. Syniverse makes seamless communications between operators, technologies, devices and networks possible for mobile users. The firm's services can be divided into three primary divisions: messaging solutions, which includes advanced messaging, short message service (SMS) interoperability, multimedia messaging service (MMS) interoperability, mobile video broadcast service, application-to-person (A2P)/enterprise services and value-added messaging services; roaming solutions, which includes roaming hub services, data clearing, financial clearing and settlement, fraud management, roaming interoperability and business intelligence solutions; and network and database solutions, which includes network signaling and transport, interworking gateway, database, data management, number portability and virtual network enablement. Syniverse is owned by The Carlyle Group.

FINANCIAL DATA: *Note: Data for latest year may not have been available at press time.*

In U.S. $	2021	2020	2019	2018	2017	2016
Revenue	733,000,000	652,000,000	749,868,840	833,187,600	793,512,000	781,892,000
R&D Expense						
Operating Income						
Operating Margin %						
SGA Expense						
Net Income	-49,387,000	-143,511,000			-20,799,000	-65,170,000
Operating Cash Flow						
Capital Expenditure						
EBITDA						
Return on Assets %						
Return on Equity %						
Debt to Equity						

CONTACT INFORMATION:

Phone: 813-637-5000   Fax:
Toll-Free:
Address: 8125 Highwoods Palm Way, Tampa, FL 33647 United States

STOCK TICKER/OTHER:

Stock Ticker: Private
Employees: 2,200
Parent Company: Carlyle Group (The)

Exchange:
Fiscal Year Ends: 12/31

SALARIES/BONUSES:

Top Exec. Salary: $ Bonus: $
Second Exec. Salary: $ Bonus: $

OTHER THOUGHTS:

Estimated Female Officers or Directors: 5
Hot Spot for Advancement for Women/Minorities: Y

TalkTalk Telecom Group Limited

www.talktalkgroup.com

NAIC Code: 517210

TYPES OF BUSINESS:

Wireless Telecommunications Carriers (except Satellite)
Telecommunications Services
Landline Services
Broadband Services
Television Services and Solutions
Mobile Services and Solutions
Fiber Network

BRANDS/DIVISIONS/AFFILIATES:

Toscafund Asset Management LLP
TalkTalk
TalkTalk Business
FibreNation
Virtual1

GROWTH PLANS/SPECIAL FEATURES:

TalkTalk Telecom Group Limited provides landline, broadband, fiber, TV and mobile services to more than 4 million customers, including 2.4 million fiber customers. The firm operates in the U.K. and comprises a major broadband network, covering approximately 96% of the population. TalkTalk supplies services to consumers through the TalkTalk brand, to businesses through TalkTalk Business, and by wholesaling to resellers. Subsidiary FibreNation has been rolling out fiber-to-the-premises (FTTP) to consumers and businesses since its 2018 inception. TalkTalk Telecom Group is privately owned by Toscafund Asset Management LLP. In May 2022, TalkTalk acquired Virtual1, a U.K. wholesaler of high-bandwidth services. Virtual1 will continue to operate as a standalone business as part of the TalkTalk group.

CONTACTS: Note: Officers with more than one job title may be intentionally listed here more than once.

Tristia Harrison, CEO
Jonathan Thackray, CFO
Daniel Kasmir, Chief People Officer
Phil Haslam, CTO

FINANCIAL DATA: Note: Data for latest year may not have been available at press time.

In U.S. $	2021	2020	2019	2018	2017	2016
Revenue	1,883,389,530	1,944,398,940	2,119,948,544	2,218,671,616	2,199,062,784	2,378,160,384
R&D Expense						
Operating Income						
Operating Margin %						
SGA Expense						
Net Income	-15,312,110	189,606,780	41,567,620	-102,620,064	71,534,288	2,592,000
Operating Cash Flow						
Capital Expenditure						
EBITDA						
Return on Assets %						
Return on Equity %						
Debt to Equity						

CONTACT INFORMATION:

Phone: 44 20 3417 1000   Fax:
Toll-Free:
Address: Soapworks, Ordsall Lane, Salford, M5 3TT United Kingdom

STOCK TICKER/OTHER:

Stock Ticker: Private   Exchange:
Employees: 2,019   Fiscal Year Ends: 02/28
Parent Company: Toscafund Asset Management LLP

SALARIES/BONUSES:

Top Exec. Salary: $ Bonus: $
Second Exec. Salary: $ Bonus: $

OTHER THOUGHTS:

Estimated Female Officers or Directors:
Hot Spot for Advancement for Women/Minorities:

Sales, profits and employees may be estimates. Financial information, benefits and other data can change quickly and may vary from those stated here.

Tata Communications Limited

www.tatacommunications.com

NAIC Code: 517110

TYPES OF BUSINESS:

Telecommunications Services
Internet Service Provider
VoIP Service
Satellite Communications
Managed Data Network Services
Managed Network Security Services
Outsourced Contact Center Services

BRANDS/DIVISIONS/AFFILIATES:

Tata Group
Oasis Smart SIM Europe SAS

CONTACTS: *Note: Officers with more than one job title may be intentionally listed here more than once.*

Amur Swaminathan Lakshminarayanan, CEO
Kabir Ahmed Shakir, CFO
Sumeet Walia, Chief Sales & Mktg. Officer
Aadesh Goyal, Chief Human Resources Officer
Genius Wong, CTO
John Hayduk, Pres., Prod. Mgmt. & Service Dev.
John Freeman, General Counsel
Madhusudhan Mysore, Head-Customer Svcs. & Oper.
Srinivasa Addepalli, Chief Strategy Officer
Natalie Chak, Sr. Mgr.-Corp. Comm.
Michel Guyot, Pres., Global Voice Solutions
Sumeet Walia, Head-Global Enterprise Bus.
Allan Chan, Exec. VP-Global Carrier Solutions
Rangu Saigame, CEO-Growth Ventures
Srinivasan CR, Chief Digital Officer
Sunil Joshi, CEO

GROWTH PLANS/SPECIAL FEATURES:

Tata Communications Limited, a division of Indian conglomerate Tata Group, is a provider of global digital infrastructure services. Tata Communications focuses on helping businesses understand and navigate through the vast potential offered by technologies such as network, mobility, collaboration and security services. It also shapes the future by investing in emerging technologies of Internet of Things (IoT), artificial intelligence (AI), automation and analytics. All of this is delivered seamlessly through the cloud via Tata Communication's wholly-owned advanced subsea fiber optic network. This type of cloud and subsea architecture enables an infinite amount of global connections. Tata Communications is present in more than 200 countries and territories worldwide, serving over 7,000 customers that represent more than 300 of the Fortune 500. Tata connects four out of five mobile subscribers worldwide, and connects businesses to 60% of the world's cloud giants. In December 2020, Tata Communications acquired a majority equity stake of 58.1% in Oasis Smart SIM Europe SAS, a France-based embedded subscriber identity module (SIM) technology provider.

FINANCIAL DATA: *Note: Data for latest year may not have been available at press time.*

In U.S. $	2021	2020	2019	2018	2017	2016
Revenue		2,277,820,000	2,322,650,000	2,559,650,000	2,770,260,000	2,733,878,230
R&D Expense						
Operating Income						
Operating Margin %						
SGA Expense						
Net Income		-11,275,000	-1,126,370	-50,514,200	-118,095,000	191,309,413
Operating Cash Flow						
Capital Expenditure						
EBITDA						
Return on Assets %						
Return on Equity %						
Debt to Equity						

CONTACT INFORMATION:

Phone: 9122 66592000   Fax:
Toll-Free: 800-266-0660
Address: Tower 4, Fl. 4-8, Equinox Bus. Pk, LBS Marg, Kurla, Mumbai, 400070 India

STOCK TICKER/OTHER:

Stock Ticker: 500483   Exchange: Bombay
Employees: 5,700   Fiscal Year Ends: 03/31
Parent Company: Tata Group

SALARIES/BONUSES:

Top Exec. Salary: $ Bonus: $
Second Exec. Salary: $ Bonus: $

OTHER THOUGHTS:

Estimated Female Officers or Directors: 2
Hot Spot for Advancement for Women/Minorities:

ata Group

www.tata.com

NAIC Code: 331110

TYPES OF BUSINESS:

Steel Production
Communication & Information Systems
Engineered Products
Energy Utilities
Solar Power
Automobiles
Hotels & Resorts
Pharmaceuticals & Chemicals

BRANDS/DIVISIONS/AFFILIATES:

Tata Sons Private Limited
Tata Consultancy Services
Tata Steel
Tata Motors
Tata Oil Mills Company (TOMCO)
Tata Capital
Tata Communications
Tata Investment Corporation

CONTACTS: *Note: Officers with more than one job title may be intentionally listed here more than once.*

Madhu Kannan, Head-Bus. Dev.
Mukund G. Rajan, Chief Ethics Officer
Nirmalya Kumar, Head-Strategy
N. Chandrasekaran, Chmn.-Tata Sons

GROWTH PLANS/SPECIAL FEATURES:

Tata Group, owned by Tata Sons Private Limited, is an Indian conglomerate founded in 1868 and today oversees 30 companies across 10 verticals. Operating in more than 100 countries worldwide, the verticals include information technology, steel, automotive, consumer and retail, infrastructure, financial services, aerospace and defense, tourism and travel, telecommunications and media, and trading and investments. Information technology is led by Tata Consultancy Services. Tata Steel has an annual crude steel capacity of 36.3 million tons per year. Tata Motors designs, manufactures and markets automobiles across all market segments. Its two iconic brands include Jaguar and Land Rover. Tata Motors has a global network of nearly 100 subsidiaries. The consumer and retail vertical is led by Tata Oil Mills Company (TOMCO), a producer of soaps, detergents, cooling oils, consumer durables, tea, packaged water and more. The infrastructure vertical is engaged in the energy sector, which supplies energy to residences, large cities and industries. Financial services is led by Tata Capital, which offers a range of financial services that cater to individuals and small businesses. This division's Tata Asset Management Company offers mutual funds, portfolio management services, alternative investment funds and offshore funds. The aerospace and defense vertical is a global, single-source supplier for fixed wing and rotary wing programs; and provides products and services for the Ministry of Defense, armed forces and Defense Research and Development Organization within India. Tourism and travel encompasses Tata's Taj Mahal hotel, and its airline operations. Telecom and media includes Tata Communications, which owns and operates mobile and internet networks, and Tata Sky, a joint venture direct-to-home entertainment provider. Last, trading and investments offers private trading strategies through Tata International, Tata Industries and Tata Investment Corporation. In January 2022, Tata Group reacquired Air India Ltd. for $2.4 billion.

FINANCIAL DATA: *Note: Data for latest year may not have been available at press time.*

In U.S. $	2021	2020	2019	2018	2017	2016
Revenue	128,000,000,000	106,000,000,000	113,000,000,000	109,670,000,000	94,620,050,000	90,672,600,000
R&D Expense						
Operating Income						
Operating Margin %						
SGA Expense						
Net Income			1,450,911,000	1,381,820,000	3,702,500,000	6,947,020,000
Operating Cash Flow						
Capital Expenditure						
EBITDA						
Return on Assets %						
Return on Equity %						
Debt to Equity						

CONTACT INFORMATION:

Phone: 91 22-6665-8282   Fax: 91-22-6665-8160
Toll-Free:
Address: 24 Homi Mody St., Mumbai, 400 001 India

STOCK TICKER/OTHER:

Stock Ticker: Private   Exchange:
Employees: 935,000   Fiscal Year Ends: 03/31
Parent Company: Tata Sons Private Limited

SALARIES/BONUSES:

Top Exec. Salary: $ Bonus: $
Second Exec. Salary: $ Bonus: $

OTHER THOUGHTS:

Estimated Female Officers or Directors:
Hot Spot for Advancement for Women/Minorities:

Sales, profits and employees may be estimates. Financial information, benefits and other data can change quickly and may vary from those stated here.

Tata Teleservices Limited

NAIC Code: 517110

corporate.tatateleservices.com/en-in

TYPES OF BUSINESS:

Telecommunications Services
Telecommunications
Mobile Services
Wireless Services
Wireline Services
Digital Solutions
Machine-to-Machine Solutions
Fiber Optic Network

BRANDS/DIVISIONS/AFFILIATES:

Tata Sons Private Limited
Tata Group
Tata Teleservices (Maharashtra) Limited

GROWTH PLANS/SPECIAL FEATURES:

Tata Teleservices Limited is the telecommunication arm of Ta
Group. The firm is a leading mobile telecom service provider
India, delivering mobile connectivity and services to consume
and businesses. Tata Teleservices (Maharashtra) Limite
serves the Maharashtra region. The company's integrate
solutions include wireline and wireless networks on glob
system for mobile communications (GSM), code divisi
multiple access (CDMA) for transmitting digital signa
simultaneously over the same carrier frequency, and ne
generation platforms. Other services include cloud, softwar
as-a-service (SaaS), collaboration services, data service
marketing solutions, cybersecurity solutions and more. No
voice services include eGovernance, machine-to-machin
(M2M) and m-Remittance services. For businesses, Tata offe
voice, data and managed solutions to small-to-larg
enterprises through its fiber optic network, channel partn
network, and sales and service teams. Tata Teleservices' fib
optic network spans more than 82,020 miles (132,000kn
across India.

CONTACTS: *Note: Officers with more than one job title may be intentionally listed here more than once.*

Ratan Naval Tata, Interim Chmn.

FINANCIAL DATA: *Note: Data for latest year may not have been available at press time.*

In U.S. $	2021	2020	2019	2018	2017	2016
Revenue	365,430,155	384,256,800	189,096,000	292,693,000	414,229,000	452,351,000
R&D Expense						
Operating Income						
Operating Margin %						
SGA Expense						
Net Income	-1,302,939,273	-1,735,778,653	-96,109,100	-1,512,960,000	-363,072,000	-54,048,400
Operating Cash Flow						
Capital Expenditure						
EBITDA						
Return on Assets %						
Return on Equity %						
Debt to Equity						

CONTACT INFORMATION:

Phone: 91 22-6661 5111   Fax: 91-22-6660-5517
Toll-Free:
Address: D-26, TTC Industria Area, MIDC Sanpada, P.o., Turbhe, Mumbai, 400 703 India

STOCK TICKER/OTHER:

Stock Ticker: 532371   Exchange: Bombay
Employees: 1,200   Fiscal Year Ends: 03/31
Parent Company: Tata Sons Private Limited

SALARIES/BONUSES:

Top Exec. Salary: $ Bonus: $
Second Exec. Salary: $ Bonus: $

OTHER THOUGHTS:

Estimated Female Officers or Directors:
Hot Spot for Advancement for Women/Minorities:

Sales, profits and employees may be estimates. Financial information, benefits and other data can change quickly and may vary from those stated here.

TCI International Inc

www.tcibr.com

NAIC Code: 334220

TYPES OF BUSINESS:

Transmission, Receiving & Test Equipment
Spectrum Monitoring & Management Systems
Broadcast & Communication Antennas
Communication Intelligence
Direction Finding Systems
System Software

BRANDS/DIVISIONS/AFFILIATES:

SPX Corporation
Blackbird

CONTACTS: Note: Officers with more than one job title may be intentionally listed here more than once.

Gene Lowe, CEO-SPX Corp

GROWTH PLANS/SPECIAL FEATURES:

TCI International, Inc. is a systems engineering and manufacturing company that specializes in spectrum monitoring, signal collection, radio frequency (RF) solutions and radio direction finding solutions, primarily targeting the government, military, civilian and intelligence markets. The company, a subsidiary of SPX Corporation, designs, manufactures, sells and installs high frequency (HF) and medium frequency (MF) broadcast antennas, HF communications antennas and monitoring antennas. The antennas are designed to capture signals from 9 kilohertz (kHz) to 43 gigahertz (GHz). The firm builds and optimizes its antennas for various applications including spectrum monitoring, direction finding (DF), drone detection, signal interception, radio broadcasting and radio communications. TCI also makes computer-controlled radio frequency (RF) distribution systems that interface between multiple-element antenna arrays and communications receivers; specialized receivers for communications, DF/monitoring and COMINT/SIGINT applications; high-speed 32-bit digital signal processors, which deliver signal analysis by simultaneously scanning, detecting, measuring and analyzing RF signals at rates up to 4 GHz per second; and application-specific software to provide real-time control, data processing, digital signal processing, database structures, graphical user interfaces and network management services. TCI's next-generation Blackbird system combines TCI's modular, scalable RF hardware platform and DF/geolocation technology with second-generation Blackbird software, all designed for NextGen automation, smart recording and ease of use. The firm offers installation, training and services in over 105 countries.

FINANCIAL DATA: Note: Data for latest year may not have been available at press time.

In U.S. $	2021	2020	2019	2018	2017	2016
Revenue	1,536,311,000	1,449,350,000	1,414,000,000	1,400,000,000		
R&D Expense						
Operating Income						
Operating Margin %						
SGA Expense						
Net Income						
Operating Cash Flow						
Capital Expenditure						
EBITDA						
Return on Assets %						
Return on Equity %						
Debt to Equity						

CONTACT INFORMATION:

Phone: 510-687-6100   Fax: 510-687-6101
Toll-Free:
Address: 3541 Gateway Blvd., Fremont, CA 94538 United States

STOCK TICKER/OTHER:

Stock Ticker: Subsidiary
Employees: 72
Parent Company: SPX Corporation

Exchange:
Fiscal Year Ends: 09/30

SALARIES/BONUSES:

Top Exec. Salary: $ Bonus: $
Second Exec. Salary: $ Bonus: $

OTHER THOUGHTS:

Estimated Female Officers or Directors:
Hot Spot for Advancement for Women/Minorities:

Sales, profits and employees may be estimates. Financial information, benefits and other data can change quickly and may vary from those stated here.

TCL Technology Group Corporation   www.tcl.com

NAIC Code: 334310

TYPES OF BUSINESS:

Consumer Electronics
Electronic Technology
Electronic Development
Phones
UHD and AI Televisions
Soundbars
Air Conditioners
Headphones

BRANDS/DIVISIONS/AFFILIATES:

CONTACTS: Note: Officers with more than one job title may be intentionally listed here more than once.

Bo Liangming, Pres.
Zhao Zhongyao, Sr. VP
Shi Wanwen, Sr. VP
Guo Aiping, Sr. VP
Huang Wei, VP-TCL Group
Dongshen Li, Chmn.

GROWTH PLANS/SPECIAL FEATURES:

TCL Technology Group Corporation designs, develops manufactures and sells electronic products. The firm invests in state-of-the-art facilities where it creates innovative display technologies and develops new production techniques for phones and televisions. TCL introduced its big-screen QLED TV in 2014, pioneering dot color technology; introduced its Roku TV in 2014; introduced its mini-LED TV in 2019 introduced its TV with THX Certified Game Mode setting for big-screen gaming; developed RAY-DANZ technology, which offers an ultra-wide soundstage from a slim soundbar; and developed a rollable AMOLED smartphone display concept TCL also develops, manufactures and sells ultra-high-definition (UHD) TVs, and artificial intelligent (AI)-based televisions featuring Google assistant Alexa for smart home integration Other products by TCL include headphones, window and portable air conditioners, air purifiers, and dehumidifiers. TCL comprises 22 manufacturing facilities, more than 25 research and development centers throughout the world, and over 10 joint laboratories.

FINANCIAL DATA: Note: Data for latest year may not have been available at press time.

In U.S. $	2021	2020	2019	2018	2017	2016
Revenue	25,682,570,815	11,770,800,000	10,743,900,000	16,493,800,000	17,155,400,000	15,365,990,000
R&D Expense						
Operating Income						
Operating Margin %						
SGA Expense						
Net Income	2,347,017,202	776,013,000	523,433,000	591,028,000	544,278,000	237,647,000
Operating Cash Flow						
Capital Expenditure						
EBITDA						
Return on Assets %						
Return on Equity %						
Debt to Equity						

CONTACT INFORMATION:

Phone: 86 755-33311666   Fax:
Toll-Free:
Address: No. 17 Huifeng San Rd., Zhongkai High-Tech Dev. Zone, Huizhou, Guangdong 516001 China

STOCK TICKER/OTHER:

Stock Ticker: 100
Employees: 50,000
Parent Company:

Exchange: Shenzhen
Fiscal Year Ends: 12/31

SALARIES/BONUSES:

Top Exec. Salary: $ Bonus: $
Second Exec. Salary: $ Bonus: $

OTHER THOUGHTS:

Estimated Female Officers or Directors:
Hot Spot for Advancement for Women/Minorities:

Sales, profits and employees may be estimates. Financial information, benefits and other data can change quickly and may vary from those stated here.

TDC A/S

tdc.com

NAIC Code: 517110

TYPES OF BUSINESS:

Internet Service Provider (ISP)
Telephone Service
Cellular Service
Cable Service
Broadband
Media

BRANDS/DIVISIONS/AFFILIATES:

Macquarie Group Limited
Tele Danmark Commmunications
TDC NET
Nuuday
YouSee
TDC Erhverv
Telmore
Fullrate

GROWTH PLANS/SPECIAL FEATURES:

TDC A/S is a triple-play telecommunications provider. It generates revenue from the provision of mobile, broadband, and fixed-line voice services. The company operates through following segments Consumer, dedicated to residential households in Denmark; Business, dedicated to the business market in Denmark; and Wholesale, delivering services to service providers in Denmark. Other operations consists of the three operating segments Operations, Digital and Headquarters and includes shared Danish functions such as, IT, procurement, and installation.

CONTACTS: *Note: Officers with more than one job title may be intentionally listed here more than once.*

Henrik Clausen, CEO
Lasse Pilgaard, CFO
Jens Aalose, Chief People Officer
Martin Lippert, Sr. Exec. VP
Eva Berneke, Sr. Exec. VP-TDC Bus.
Jacob Konrad Jensen, Chief Press Officer
Miriam Igelso Hvidt, Sr. Exec. VP-Stakeholder Rel.
Niels Breining, CEO-YouSee A
Jens Munch-Hansen, Sr. Exec. VP-TDC Nordic

FINANCIAL DATA: *Note: Data for latest year may not have been available at press time.*

In U.S. $	2021	2020	2019	2018	2017	2016
Revenue	2,193,046,000	2,204,969,000	2,335,850,000	2,378,609,000	2,382,721,000	2,882,262,000
R&D Expense						
Operating Income	313,429,400	209,135,600	198,034,700	366,055,900	374,827,000	494,196,000
Operating Margin %	.14%	.09%	.08%	.15%	.16%	.17%
SGA Expense	242,164,300	260,802,800	292,323,900	319,870,700	336,727,600	439,376,700
Net Income	54,271,110	20,694,290	24,531,640	784,464,300	211,191,400	417,037,900
Operating Cash Flow	738,964,200	745,679,600	715,117,800	734,167,600	988,529,200	994,696,300
Capital Expenditure	648,375,300	648,923,500	688,119,400	499,678,000	521,879,800	610,413,000
EBITDA	834,623,900	870,256,400	881,631,400	753,354,300	919,867,900	1,117,629,000
Return on Assets %	.01%	.00%	.00%	.10%	.02%	.04%
Return on Equity %	.02%	.01%	.01%	.27%	.05%	.13%
Debt to Equity	1.35%	1.738	1.814	1.403	0.676	0.974

CONTACT INFORMATION:

Phone: 45 66637680   Fax: 45-33157579
Toll-Free:
Address: Teglholmsgade 1, Copenhagen, 0900 Denmark

STOCK TICKER/OTHER:

Stock Ticker: TDCAF   Exchange: PINX
Employees: 7,498   Fiscal Year Ends: 12/31
Parent Company: Macquarie Group Limited

SALARIES/BONUSES:

Top Exec. Salary: $ Bonus: $
Second Exec. Salary: $ Bonus: $

OTHER THOUGHTS:

Estimated Female Officers or Directors: 5
Hot Spot for Advancement for Women/Minorities: Y

TDS Telecommunications LLC

www.tdstelecom.com

NAIC Code: 517110

TYPES OF BUSINESS:

Mobile Phone and Wireless Services
Local Telephone Service
Long-Distance Service
Internet Access

BRANDS/DIVISIONS/AFFILIATES:

Telephone and Data Systems Inc (TDS)
TDS Broadband Service LLC
BendBroadband

CONTACTS: *Note: Officers with more than one job title may be intentionally listed here more than once.*

Jim Butman, CEO
Michelle Brukwicki, CFO
Mike Pandow, Sr. VP-Admin.
Mark Barber, VP-Cable Oper.
Vicki L. Villacrez, VP-Finance
Phil LaForge, Pres., TDS Hosted & Managed Svcs.
Kevin Hess, Sr. VP-Gov't & Regulatory Affairs

GROWTH PLANS/SPECIAL FEATURES:

TDS Telecommunications, LLC (TDS Telecom), a wholly owned subsidiary of Telephone and Data Systems, Inc., is telecommunications service company that provides voice internet and entertainment services to rural and suburba communities nationwide. TDS Telecom provides high-spee internet, phone and TV entertainment services to customers i nearly 1,000 rural, suburban and metropolitan communitie The company deploys up to 1Gig internet access, IPT services, cable TV options and traditional wireline services. Fe businesses, TDS Telecom offers advanced communication solutions such as voice over internet protocol (VoIP), high speed internet, fiber optics, data networking and hosted managed services. Subsidiary TDS Broadband Service, LL serves the cable industry within the U.S. states of Arizona Colorado, Nevada, New Mexico, Utah and Texas. Its cabl operations offer businesses metro Ethernet, passive optic network technology, hosted private branch exchange (PBX carrier backhaul and transport solutions. Subsidiar BendBroadband offers internet, cable TV and telephon services in the state of Oregon.

TDS Telecom offers its employees medical, dental, vision an life insurance; a 401(k) plan; a pension plan; an employe stock purchase plan; flexible spending accounts; educatio assistance; telephone and Internet discounts; and training an development p

FINANCIAL DATA: *Note: Data for latest year may not have been available at press time.*

In U.S. $	2021	2020	2019	2018	2017	2016
Revenue	1,006,000,000	976,000,000	930,000,000	927,000,000	919,000,000	1,151,000,000
R&D Expense						
Operating Income						
Operating Margin %						
SGA Expense						
Net Income	90,000,000	100,000,000	92,000,000	89,000,000	138,000,000	42,000,000
Operating Cash Flow						
Capital Expenditure						
EBITDA						
Return on Assets %						
Return on Equity %						
Debt to Equity						

CONTACT INFORMATION:

Phone: 608-664-4000   Fax: 608-664-4035
Toll-Free: 866-571-6662
Address: 525 Junction Rd., Madison, WI 53717 United States

STOCK TICKER/OTHER:

Stock Ticker: Subsidiary   Exchange:
Employees: 2,700   Fiscal Year Ends: 12/31
Parent Company: Telephone and Data Systems Inc (TDS)

SALARIES/BONUSES:

Top Exec. Salary: $ Bonus: $
Second Exec. Salary: $ Bonus: $

OTHER THOUGHTS:

Estimated Female Officers or Directors:
Hot Spot for Advancement for Women/Minorities:

Sales, profits and employees may be estimates. Financial information, benefits and other data can change quickly and may vary from those stated here.

TeamViewer AG

NAIC Code: 511210C

TYPES OF BUSINESS:
Computer Software: Telecom, Communications & VOIP
Screen-Sharing & Conferencing Software

BRANDS/DIVISIONS/AFFILIATES:
TeamViewer
TeamViewer Remote Management
Monitis
TeamViewer Tensor
TeamViewer Pilot
Blizz
TeamViewer Frontline
Ubimax GmbH

CONTACTS: Note: Officers with more than one job title may be intentionally listed here more than once.
Oliver Steil, CEO
Karl Markgraf, COO
Stefan Gaiser, CFO
Gautam Goswami, CMO
Rebecca Keating, Sr. VP-Human Resources
Dr. Mike Eissele, CTO
Oliver Steil, Chmn.

GROWTH PLANS/SPECIAL FEATURES:
TeamViewer AG is a global remote connectivity platform. It empowers users to connect with anyone, anything, anywhere, anytime. The TeamViewer offers secure remote access, support, control, and collaboration capabilities for online endpoints of any kind and supports businesses of all sizes to tap into their full digital potential. Some of its products are TeamViewer Tensor, TeamViewer IoT, TeamViewer Pilot, TeamViewer Remote Management, and Teamviewer Meeting. The company's geographical segments are EMEA(Europe, Middle East, and Africa), AMERICAS(North, Central, and South America), and APAC(Asia-Pacific).

TeamViewer offers employees in-house doctor visits and eye examinations, work development and training programs and other programs and perks.

FINANCIAL DATA: Note: Data for latest year may not have been available at press time.

In U.S. $	2021	2020	2019	2018	2017	2016
Revenue	511,000,200	464,618,300	397,902,300	263,258,900	141,203,500	93,481,670
R&D Expense	63,365,010	47,548,490	38,683,690	23,494,320	16,868,920	13,277,310
Operating Income	136,055,800	182,151,100	171,866,800	117,690,800	26,711,670	-7,416,736
Operating Margin %	.27%	.39%	.43%	.45%	.19%	-.08%
SGA Expense	241,446,200	174,486,500	51,994,660	43,891,620	57,560,520	48,695,720
Net Income	51,040,160	105,063,100	105,911,600	-12,658,320	-70,519,670	-59,985,520
Operating Cash Flow	197,806,500	228,973,500	146,448,200	114,780,400	99,539,060	79,585,360
Capital Expenditure	15,532,010	26,661,700	16,878,100	11,710,960	8,005,140	5,612,776
EBITDA	151,905,900	234,800,400	187,902,500	97,189,530	15,772,670	20,339,170
Return on Assets %	.04%	.10%	.11%	-.01%	-.07%	-.06%
Return on Equity %	.18%	.62%				
Debt to Equity	2.63%	1.829	6.342			

CONTACT INFORMATION:
Phone: 49 7161 60692 50   Fax:
Toll-Free: 800 638 0235
Address: Bahnhofsplatz 2, Goppingen, BW 73033 Germany

STOCK TICKER/OTHER:
Stock Ticker: TMVWY
Employees: 1,477
Parent Company:

Exchange: PINX
Fiscal Year Ends: 12/31

SALARIES/BONUSES:
Top Exec. Salary: $ Bonus: $
Second Exec. Salary: $ Bonus: $

OTHER THOUGHTS:
Estimated Female Officers or Directors:
Hot Spot for Advancement for Women/Minorities:

Sales, profits and employees may be estimates. Financial information, benefits and other data can change quickly and may vary from those stated here.

Technical Communications Corporation

www.tccsecure.com

NAIC Code: 334220

TYPES OF BUSINESS:

Radio and Television Broadcasting and Wireless Communications
Equipment Manufacturing
Communications Security Devices
Communications Security Systems
Communications Services
Product Development
Product Manufacturing

BRANDS/DIVISIONS/AFFILIATES:

Crypto Learning Center

GROWTH PLANS/SPECIAL FEATURES:

Technical Communications Corporation (TCC) designs, develops, manufactures, distributes, markets and sells communications security devices, systems and services. The firm's devices and systems include encryption solutions in relation to data networks, radios, mobile phones and telephones. The company also provides custom encryption solutions to meet the specific needs of its clients. TCC's products protect vital information transmitted over this wide range of data, video, fax and voice networks; and have been sold in more than 115 countries to governments, military agencies, telecommunications carriers, financial institutions and multi-national corporation. Other value-added services include systems integration; and a Crypto Learning Center which offers a blend of in-person instructor-led and online learning (live and on-demand) to provide customers the practical knowledge needed to enhance their communications security infrastructure and facilitate secure operations.

CONTACTS:
Note: Officers with more than one job title may be intentionally listed here more than once.

Michael Malone, Assistant Secretary
Carl Guild, CEO

FINANCIAL DATA:
Note: Data for latest year may not have been available at press time.

In U.S. $	2021	2020	2019	2018	2017	2016
Revenue			7,024,123	3,684,939	4,209,127	2,522,934
R&D Expense						
Operating Income						
Operating Margin %						
SGA Expense						
Net Income			631,425	-1,479,599	-1,429,006	-2,472,288
Operating Cash Flow						
Capital Expenditure						
EBITDA						
Return on Assets %						
Return on Equity %						
Debt to Equity						

CONTACT INFORMATION:

Phone: 978 287-5100   Fax: 978 371-1280
Toll-Free:
Address: 100 Domino Dr., Concord, MA 01742-2892 United States

STOCK TICKER/OTHER:

Stock Ticker: TCCO   Exchange: OTC
Employees: 25   Fiscal Year Ends: 09/30
Parent Company:

SALARIES/BONUSES:

Top Exec. Salary: $ Bonus: $
Second Exec. Salary: $ Bonus: $

OTHER THOUGHTS:

Estimated Female Officers or Directors:
Hot Spot for Advancement for Women/Minorities:

Tektronix Inc

www.tek.com

NAIC Code: 334515

TYPES OF BUSINESS:

Test & Measurement Equipment
Support Services
Oscilloscopes
Logic analyzers
Video test equipment
Communications test equipment

BRANDS/DIVISIONS/AFFILIATES:

Fortive Corporation
Keithley

CONTACTS: *Note: Officers with more than one job title may be intentionally listed here more than once.*

Tami Newcombe, Pres.
Fuki Yoneyama, Pres., Japan Region

GROWTH PLANS/SPECIAL FEATURES:

Tektronix, Inc. develops, manufactures and markets test, measurement and monitoring products to a wide variety of customers. The firm's applications and industry solutions include 3D sensing technology, advanced research, automotive, electromagnetic interference (EMI) and electromagnetic compatibility (EMC) testing, education, teaching labs, high-speed serial communications, materials science, engineering, media production and delivery, medical technologies/devices/systems, military, government, power efficiency, wired communications, wireless, radio frequency (RF) testing, semiconductor design and manufacturing. Tektronix's products are categorized into oscilloscopes and probes, analyzers, Keithley (a Tektronix company) products, signal generators, power/measure sources and related supplies, meters, video test equipment, switching and data acquisition systems, semiconductor test systems, components and accessories, software, and refurbished test equipment. The company's calibration and related services are categorized into: calibration services, asset management services, repair services, factory service plans and testing services. Tektronix maintains its headquarters location in Beaverton, Oregon and has offices in 20+ countries. The firm has received more than 700 patents since the year 2000. Tektronix operates as a wholly-owned subsidiary of Fortive Corporation, a diversified industrial growth company encompassing several businesses.

Employee benefits include medical, dental and disability coverage; life insurance; and retirement plans.

FINANCIAL DATA: *Note: Data for latest year may not have been available at press time.*

In U.S. $	2021	2020	2019	2018	2017	2016
Revenue						
R&D Expense						
Operating Income						
Operating Margin %						
SGA Expense						
Net Income						
Operating Cash Flow						
Capital Expenditure						
EBITDA						
Return on Assets %						
Return on Equity %						
Debt to Equity						

CONTACT INFORMATION:

Phone: 503-627-7111   Fax: 503-627-6108
Toll-Free: 800-833-9200
Address: 14150 SW Karl Braun Dr., Beaverton, OR 97077 United States

STOCK TICKER/OTHER:

Stock Ticker: Subsidiary   Exchange:
Employees: 4,500   Fiscal Year Ends: 12/31
Parent Company: Fortive Corporation

SALARIES/BONUSES:

Top Exec. Salary: $ Bonus: $
Second Exec. Salary: $ Bonus: $

OTHER THOUGHTS:

Estimated Female Officers or Directors:
Hot Spot for Advancement for Women/Minorities:

Sales, profits and employees may be estimates. Financial information, benefits and other data can change quickly and may vary from those stated here.

Tele2 AB
NAIC Code: 517110

TYPES OF BUSINESS:
Fixed & Mobile Telephony
Broadband Services
Data Network Services
Cable TV
Internet of Things

GROWTH PLANS/SPECIAL FEATURES:
Tele2 is a Swedish telecom operator, generating most of th
business in its domestic country. The company also operate
in the Baltics. It has undergone a significant restructuring b
exiting Kazakhstan and acquiring Com Hem, the largest cab
TV operator in Sweden. It is now a strong converged operato
-offering wireless as well as fixed-line broadband, TV, an
voice services in its home market.

BRANDS/DIVISIONS/AFFILIATES:

CONTACTS: *Note: Officers with more than one job title may be intentionally listed here more than once.*
Kjell Johnsen, CEO
Roxanna Zea, Exec. VP-Group Strategy
Lars Torstensson, Dir.-Corp. Comm.
Thomas Ekman, CEO-Tele2 Sweden
Gunther Vogelpoel, CEO-Tele2 Netherlands
Arild Hustad, CEO-Tele2 Norway
Carla Smits-Nusteling, Chmn.
Niklas Sonkin, Exec. VP-Central Europe & Eurasia

FINANCIAL DATA: *Note: Data for latest year may not have been available at press time.*

In U.S. $	2021	2020	2019	2018	2017	2016
Revenue	2,645,542,000	2,622,335,000	2,686,427,000	2,150,386,000	2,119,870,000	2,092,614,000
R&D Expense						
Operating Income	449,531,800	492,687,700	390,772,700	354,529,800	337,741,400	247,084,500
Operating Margin %	.17%	.19%	.15%	.16%	.16%	.12%
SGA Expense	636,375,900	647,239,000	679,235,500	577,024,300	608,329,600	585,221,000
Net Income	425,238,100	731,575,600	494,169,100	84,336,600	18,960,920	-193,756,900
Operating Cash Flow	1,016,878,000	870,622,300	959,501,600	509,574,800	566,062,500	495,452,800
Capital Expenditure	328,656,000	271,575,700	356,504,800	338,136,400	318,484,200	377,638,400
EBITDA	1,058,948,000	1,250,236,000	993,275,800	704,615,500	832,996,700	602,503,000
Return on Assets %	.06%	.10%	.06%	.01%	.00%	-.05%
Return on Equity %	.13%	.22%	.14%	.03%	.01%	-.11%
Debt to Equity	.86%	0.786	0.753	0.599	0.613	0.414

CONTACT INFORMATION:
Phone: 46 856200060   Fax:
Toll-Free:
Address: Skeppsbron 18, Stockholm, 103 13 Sweden

STOCK TICKER/OTHER:
Stock Ticker: TLTZY   Exchange: PINX
Employees: 4,295   Fiscal Year Ends: 12/31
Parent Company:

SALARIES/BONUSES:
Top Exec. Salary: $ Bonus: $
Second Exec. Salary: $ Bonus: $

OTHER THOUGHTS:
Estimated Female Officers or Directors: 2
Hot Spot for Advancement for Women/Minorities: Y

Telecom Argentina SA

NAIC Code: 517110

TYPES OF BUSINESS:
Local & Long-Distance Telephone Services
Telecommunications
Cable TV
Data Transmission
Internet

BRANDS/DIVISIONS/AFFILIATES:
Nucleo SAE
PEM SAU
Adesol SA
AVC Continente Audiovisual SA
Telecom Argentina USA Inc

GROWTH PLANS/SPECIAL FEATURES:
Telecom Argentina SA offers its customers quadruple play services, combining mobile telephony services, cable television services, Internet services and fixed telephony services. It also provides other telephone-related services such as international long-distance and wholesale services, data transmission and IT solutions outsourcing and install, operate and develop cable television and data transmission services. The company provides services in Argentina (mobile, cable television, Internet and fixed and data services), Paraguay (mobile, Internet and satellite TV services), Uruguay (cable television services) and the United States (fixed wholesale services).

CONTACTS: Note: Officers with more than one job title may be intentionally listed here more than once.
Roberto D. Nobile, CEO
Gonzalo Hita, COO
Gabriel P. Blasi, CFO
Pablo Esses, CIO
Sergio D. Faraudo, Human Capital
Miguel A. Fernandez, CTO
Alejandro D. Quiroga Lopez, Dir.-Legal Affairs
Mariano Cornejo, Dir.-Comm. & Media
Ricardo Luttini, Dir.-Internal Audit
Gonzalo A. Martinez, Dir.-Regulatory Affairs
Guillermo O. Rivaben, Dir.-Mobile Telephony
Maximo D. Lema, Dir.-Wholesale
Paolo Perfetti, Dir.-Network
Carlos Alberto Moltini, Chmn.
Estefano M. Esposizione, Dir.-Procurement

FINANCIAL DATA: Note: Data for latest year may not have been available at press time.

In U.S. $	2021	2020	2019	2018	2017	2016
Revenue	3,185,746,000	3,408,423,000	3,646,762,000	2,635,101,000	767,668,900	452,263,600
R&D Expense						
Operating Income	16,486,790	324,854,100	346,417,200	395,615,500	166,612,400	92,272,100
Operating Margin %	.01%	.10%	.09%	.15%	.22%	.20%
SGA Expense	890,720,800	878,449,300	971,380,300	694,421,700	189,785,200	111,776,200
Net Income	64,876,480	-64,584,480	-67,639,260	83,025,430	112,075,700	78,293,530
Operating Cash Flow	973,596,500	1,129,585,000	1,260,551,000	526,970,800	262,770,300	194,547,100
Capital Expenditure	569,969,700	614,338,800	781,550,100	529,508,900	229,474,800	145,461,100
EBITDA	1,460,795,000	1,135,320,000	1,227,015,000	628,347,300	302,400,100	196,591,100
Return on Assets %	.01%	-.01%	-.01%	.03%	.21%	.24%
Return on Equity %	.02%	-.02%	-.02%	.06%	.41%	.57%
Debt to Equity	.40%	0.433	0.395	0.263	0.27	0.447

CONTACT INFORMATION:
Phone: 54 1149684019 Fax:
Toll-Free:
Address: Alicia Moreau de Justo 50, Buenos Aires, C1107AAB Argentina

STOCK TICKER/OTHER:
Stock Ticker: TCMFF
Employees: 22,587
Parent Company:

Exchange: PINX
Fiscal Year Ends: 12/31

SALARIES/BONUSES:
Top Exec. Salary: $ Bonus: $
Second Exec. Salary: $ Bonus: $

OTHER THOUGHTS:
Estimated Female Officers or Directors: 2
Hot Spot for Advancement for Women/Minorities:

Sales, profits and employees may be estimates. Financial information, benefits and other data can change quickly and may vary from those stated here.

Telecom Egypt SAE

ir.te.eg

NAIC Code: 517110

TYPES OF BUSINESS:
Wired Telecommunications Carriers
Telecommunication Solutions
Voice Over Long-Term Evolution Service

GROWTH PLANS/SPECIAL FEATURES:
Telecom Egypt SAE is the country's sole provider of fixed-line
voice telecommunication services, where it generates the vas
majority of its revenue. The company provides voice services
Internet, and data services in four segments: Communication
marine cables and infrastructure, Internet, Outsourcing and A
Other. Additionally, because of Telecom Egypt's ownership o
telecommunication infrastructure, there is wholesale revenue
This constitutes the majority of overall company revenue.

BRANDS/DIVISIONS/AFFILIATES:
Vodafone Egypt

CONTACTS:
Note: Officers with more than one job title may be intentionally listed here more than once.

Adel Hamed, CEO
Mohamed Shamroukh, CFO
Mohamed Abo-Taleb, CCO
Essam Abdeldayem, VP-Human Resources
Antar Kandil, CIO
Magued Osman, Chmn.

FINANCIAL DATA:
Note: Data for latest year may not have been available at press time.

In U.S. $	2021	2020	2019	2018	2017	2016
Revenue	1,937,406,000	1,667,043,000	1,348,010,000	1,189,493,000	969,920,300	728,722,900
R&D Expense						
Operating Income	432,801,800	304,703,100	129,310,900	170,774,900	101,824,100	100,524,700
Operating Margin %	.22%	.18%	.10%	.14%	.10%	.14%
SGA Expense	131,186,600	126,458,400	109,663,200	105,898,100	56,702,930	37,115,360
Net Income	439,691,900	253,360,800	229,802,100	172,214,200	159,426,600	139,490,300
Operating Cash Flow						
Capital Expenditure	583,098,300	576,592,400	504,876,000	419,345,600	318,026,200	172,998,100
EBITDA	644,511,400	408,032,100	348,461,800	283,838,200	215,369,100	179,790,200
Return on Assets %	.10%	.06%	.06%	.06%	.06%	.07%
Return on Equity %	.20%	.13%	.13%	.11%	.10%	.09%
Debt to Equity	.27%	0.085	0.136	0.017	0.02	0.021

CONTACT INFORMATION:
Phone: 20 231316011   Fax: 20 231316015
Toll-Free:
Address: B7 Building, Smart Village, Cairo, 12577 Egypt

STOCK TICKER/OTHER:
Stock Ticker: TEGPY   Exchange: PINX
Employees: Fiscal Year Ends: 12/31
Parent Company:

SALARIES/BONUSES:
Top Exec. Salary: $ Bonus: $
Second Exec. Salary: $ Bonus: $

OTHER THOUGHTS:
Estimated Female Officers or Directors:
Hot Spot for Advancement for Women/Minorities:

Sales, profits and employees may be estimates. Financial information, benefits and other data can change quickly and may vary from those stated here.

Telecom Italia SpA

www.gruppotim.it/en.html

NAIC Code: 517110

TYPES OF BUSINESS:

Local Telephone Service
Mobile Communications Services
Internet Access
Data Communications Services
IT Products & Services
Media & TV Broadcasting

BRANDS/DIVISIONS/AFFILIATES:

Vivendi SA
TIM SpA
Gruppo TIM
Telecom Italia Sparkle SpA
Olivetti
INWIT SpA

GROWTH PLANS/SPECIAL FEATURES:

Telecom Italia is the incumbent telephone operator in Italy with more than 30% market share in the Italian mobile market and 45% market share in broadband. In the mobile market it competes with Vodafone, Wind Tre and Iliad. On the broadband side, its main competitor is Open Fiber, which operates a wholesale network giving equal access to several operators. TIM also has a 20% wireless market share in Brazil and has started building a broadband network in the country, although its reach is very small compared with its competitors Telefonica and America Movil.

CONTACTS: *Note: Officers with more than one job title may be intentionally listed here more than once.*

Luigi Gubitosi, CEO
Piergiorgio Peluso, Head-Admin.
Antonino Cusimano, Head-Legal Affairs
Oscar Cicchetti, Head-Strategy
Franco Brescia, Head-Public & Regulatory Affairs
Piergiorgio Peluso, Head-Finance & Control
Simone Battiferri, Head-Bus.
Stefano De Angelis, CEO-Telecom Argentina
Alessandro Talotta, Head-National Wholesale Svcs.
Paolo Vantellini, Head-Bus. Support Office
Salvatore Rossi, Chmn.
Rodrigo Abreu, CEO-TIM Participacoes

FINANCIAL DATA: *Note: Data for latest year may not have been available at press time.*

In U.S. $	2021	2020	2019	2018	2017	2016
Revenue	15,618,690,000	16,117,350,000	18,329,220,000	19,314,310,000	20,219,860,000	19,400,990,000
R&D Expense						
Operating Income	1,357,305,000	2,732,965,000	4,202,443,000	3,783,321,000	3,783,321,000	4,146,356,000
Operating Margin %	.09%	.17%	.23%	.20%	.19%	.21%
SGA Expense						
Net Income	-8,822,990,000	7,366,768,000	934,102,900	-1,438,886,000	1,143,154,000	1,843,731,000
Operating Cash Flow	4,421,692,000	6,680,467,000	6,051,274,000	4,682,751,000	5,505,700,000	5,818,767,000
Capital Expenditure	4,092,309,000	4,789,827,000	5,021,313,000	4,620,546,000	5,419,020,000	4,750,056,000
EBITDA	1,182,925,000	7,396,341,000	8,238,665,000	4,860,191,000	7,865,432,000	8,785,258,000
Return on Assets %	-.12%	.10%	.01%	-.02%	.02%	.03%
Return on Equity %	-.40%	.31%	.04%	-.07%	.05%	.09%
Debt to Equity	1.48%	0.985	1.398	1.202	1.207	1.333

CONTACT INFORMATION:

Phone: 39 06-36-881   Fax:
Toll-Free:
Address: Corso D'Italia, 41, Rome, 00198 Italy

STOCK TICKER/OTHER:

Stock Ticker: TIIAY   Exchange: PINX
Employees: 52,347   Fiscal Year Ends: 12/31
Parent Company:

SALARIES/BONUSES:

Top Exec. Salary: $927,984   Bonus: $
Second Exec. Salary: Bonus: $
$525,000

OTHER THOUGHTS:

Estimated Female Officers or Directors: 1
Hot Spot for Advancement for Women/Minorities:

Sales, profits and employees may be estimates. Financial information, benefits and other data can change quickly and may vary from those stated here.

Telecomunicaciones de Puerto Rico Inc   www.claropr.com
NAIC Code: 517110

TYPES OF BUSINESS:

Local Exchange Carrier
Telecommunications Equipment Rental, Sales & Billing
Wireless Internet Services
Cellular Service
Mobile Network
Cloud
5G

BRANDS/DIVISIONS/AFFILIATES:

America Movil SAB de CV
Claro Puerto Rico
Claro

GROWTH PLANS/SPECIAL FEATURES:

Telecomunicaciones de Puerto Rico, Inc. (TELPRI),
subsidiary of America Movil SAB de CV and operating throu
Claro Puerto Rico, is a leading telecommunication service
company in Puerto Rico. TELPRI offers wireline and fixed lir
services to residential and business customers, all markete
under the Claro name. Products and services for residenti
customers include mobile, internet, TV and fixed telephony, a
well as bundled packages. Post- and pre-paid mobile service
are offered, as well as services for the deaf community. T
business customers, TELPRI provides mobile, internet, dat
fixed telephony and cloud services. Data services features IB
trunking, virtual Ethernet link and IP-VPN (virtual priva
network) services. Finally, to the corporate client, whic
includes large-scale companies, the firm additionally offe
data center services including collocation and virtual server
TELPRI offers mobile network and 5G coverage througho
Puerto Rico.

CONTACTS: Note: Officers with more than one job title may be intentionally listed here more than once.

Enrique Ortiz de Montellano, CEO

FINANCIAL DATA: Note: Data for latest year may not have been available at press time.

In U.S. $	2021	2020	2019	2018	2017	2016
Revenue	963,144,000	926,100,000	882,000,000	840,000,000	800,000,000	1,020,000,000
R&D Expense						
Operating Income						
Operating Margin %						
SGA Expense						
Net Income						
Operating Cash Flow						
Capital Expenditure						
EBITDA						
Return on Assets %						
Return on Equity %						
Debt to Equity						

CONTACT INFORMATION:

Phone: 787 792-6052   Fax: 787-282-0958
Toll-Free: 800-781-1314
Address: 1515 FD Roosevelt Ave., Guaynabo, 00968 Puerto Rico

STOCK TICKER/OTHER:

Stock Ticker: Private   Exchange:
Employees: 4,900   Fiscal Year Ends: 12/31
Parent Company: America Movil SAB de CV

SALARIES/BONUSES:

Top Exec. Salary: $ Bonus: $
Second Exec. Salary: $ Bonus: $

OTHER THOUGHTS:

Estimated Female Officers or Directors: 1
Hot Spot for Advancement for Women/Minorities:

Telefonica Brasil SA

www.telefonica.com.br

NAIC Code: 517110

TYPES OF BUSINESS:

Local Telephone Services
Long-Distance Service
Mobile Phone Service
Data Transmission
Pay TV

BRANDS/DIVISIONS/AFFILIATES:

Telefonica SA

GROWTH PLANS/SPECIAL FEATURES:

Telefonica Brasil, known as Vivo, is the largest wireless carrier in Brazil with nearly 85 million customers, equal to about 33% market share. The firm is strongest in the postpaid business, where it has 50 million customers, about 37% share of this market. It is the incumbent fixed-line telephone operator in Sao Paulo state and, following the acquisition of GVT, the owner of an extensive fiber network across the country. The firm provides internet access to 6 million households on this network. Following its parent Telefonica's footsteps, Vivo is cross-selling fixed-line and wireless services as a converged offering. The firm also sells pay-tv services to its fixed-line customers.

CONTACTS:

Note: Officers with more than one job title may be intentionally listed here more than once.

Breno Rodrigo Pacheco de Oliveira, Sec. & Legal Officer
Gilmar Roberto Pereira Camurra, Investor Relations Officer
Cristiane Barretto Sales, Comptroller
Mariano Sebastian de Beer, Gen. Mgr.-Fixed Telephony
Paulo Cesar Pereira Teixeira, Gen. & Exec. Officer
Eduardo Navarro, Chmn.

FINANCIAL DATA:

Note: Data for latest year may not have been available at press time.

In U.S. $	2021	2020	2019	2018	2017	2016
Revenue	8,616,782,000	8,439,458,000	8,662,878,000	8,505,263,000	8,455,184,000	8,318,519,000
R&D Expense						
Operating Income	1,282,220,000	1,287,828,000	1,411,699,000	1,850,489,000	1,276,584,000	1,419,275,000
Operating Margin %	.15%	.15%	.16%	.22%	.15%	.17%
SGA Expense	2,605,529,000	2,638,443,000	2,812,807,000	2,909,437,000	2,952,489,000	2,342,306,000
Net Income	1,220,987,000	933,548,700	978,652,900	1,747,179,000	901,898,200	799,444,700
Operating Cash Flow	3,536,644,000	3,785,005,000	3,467,877,000	2,336,829,000	2,473,740,000	2,238,865,000
Capital Expenditure	1,819,042,000	1,622,134,000	1,729,642,000	1,666,789,000	1,637,475,000	1,461,981,000
EBITDA	3,740,886,000	3,525,093,000	3,579,146,000	3,980,891,000	2,867,588,000	2,764,799,000
Return on Assets %	.06%	.04%	.05%	.09%	.05%	.04%
Return on Equity %	.09%	.07%	.07%	.13%	.07%	.06%
Debt to Equity	.14%	0.137	0.138	0.065	0.078	0.066

CONTACT INFORMATION:

Phone: 55 1134303687   Fax: 55 1174202240
Toll-Free:
Address: Avenida Engenheiro Luis Carlos Berrini, 1376, 32 d, Sao Paulo, SP 04571-936 Brazil

STOCK TICKER/OTHER:

Stock Ticker: VIV
Employees: 32,759
Parent Company: Telefonica SA

Exchange: NYS
Fiscal Year Ends: 12/31

SALARIES/BONUSES:

Top Exec. Salary: $ Bonus: $
Second Exec. Salary: $ Bonus: $

OTHER THOUGHTS:

Estimated Female Officers or Directors:
Hot Spot for Advancement for Women/Minorities:

Telefonica Chile SA

NAIC Code: 517110

www.telefonicachile.cl

TYPES OF BUSINESS:

Local Telephone Service
Long-Distance Service
Public Telephones
Automated Teller Machines
Managed Network Services
Internet & Broadband Service
Satellite TV Service
IPTV Service

BRANDS/DIVISIONS/AFFILIATES:

Telefonica SA
Telfonica Moviles Chile SA
Telefonica Empresas Chile SA
Telefonica Multimedia
Telefonica Moviles Chile Larga Distancia SA
Telefonica Gestion de Servicios Compartidos
Movistar

CONTACTS: *Note: Officers with more than one job title may be intentionally listed here more than once.*

Roberto Munoz Laporte, CEO
Jose Andres Wallis Garces, Dir.-Corp. Affairs
Victor G. Page, General Counsel
Juan Parra Hidalgo, Controller
Claudio Munoz Zuniga, Chmn.

GROWTH PLANS/SPECIAL FEATURES:

Telefonica Chile SA is the one of the larges telecommunications companies in Chile. A subsidiary c Telefonica SA, the firm offers its products and services unde the Movistar brand name. Its services include local telephon service, mobile communications (through Telefonica Movile Chile SA), broadband service, pay TV services, domestic an international long distance, public telephone service, dat transmission and terminal equipment sales and leasing as we as value-added services. The firm provides all of its fixe telephone services through its own digital telecommunication network. The company also maintains public telephones an ATMs. Telefonica Chile offers duo or trio bundled service plan for its pay TV, broadband and voice services. The compan operates through a number of majority- and minority-owne subsidiaries. Telefonica Empresas Chile SA provide corporate communications services, such as data an connectivity solutions, IT solutions, telephone and interne protocol (IP) solutions to government agencies, publi institutions, businesses and other large national an international organizations. Telefonica Multimedia offer television services, including cable, satellite, broadband an pay TV as well as on-demand video services. Telefonic Moviles Chile Larga Distancia SA is the firm's long-distanc carrier, and it provides substantially all of its domestic an international long-distance services with its own equipment an long-distance network. Telefonica Gestion de Servicio Compartidos Chile SA (t-gestiona) provides logistics, treasury insurance, collection, payment, payroll, personnel and othe services to all parts of the company.

FINANCIAL DATA: *Note: Data for latest year may not have been available at press time.*

In U.S. $	2021	2020	2019	2018	2017	2016
Revenue	17,500,000,000	18,561,080,000	19,017,500,000	1,116,810,000	1,311,960,000	1,142,576,692
R&D Expense						
Operating Income						
Operating Margin %						
SGA Expense						
Net Income			4,002,380,000	28,209,600	17,701,100	30,034,049
Operating Cash Flow						
Capital Expenditure						
EBITDA						
Return on Assets %						
Return on Equity %						
Debt to Equity						

CONTACT INFORMATION:

Phone: 562-691-2020   Fax:
Toll-Free:
Address: Ave. Providencia 111, Santiago, 7500775 Chile

STOCK TICKER/OTHER:

Stock Ticker: Subsidiary
Employees: 4,513
Parent Company: Telefonica SA

Exchange:
Fiscal Year Ends: 12/31

SALARIES/BONUSES:

Top Exec. Salary: $ Bonus: $
Second Exec. Salary: $ Bonus: $

OTHER THOUGHTS:

Estimated Female Officers or Directors: 2
Hot Spot for Advancement for Women/Minorities:

Sales, profits and employees may be estimates. Financial information, benefits and other data can change quickly and may vary from those stated here.

Telefonica de Argentina SA

NAIC Code: 517110

TYPES OF BUSINESS:
Fixed-Line Telecommunication Services
Internet Service Provider
Data Services
Voice over Internet Protocol (VoIP) Services

BRANDS/DIVISIONS/AFFILIATES:
Telefonica SA
Movistar

CONTACTS: *Note: Officers with more than one job title may be intentionally listed here more than once.*
Frederico Rava, Pres.

GROWTH PLANS/SPECIAL FEATURES:
Telefonica de Argentina SA is a leading telecommunications service provider in Argentina, operating under the brand name Movistar. The company is owned by companies that are all directly or indirectly owned by Telefonica SA. Telefonica de Argentina's offerings include international long-distance, data transmission, voice over internet protocol (VoIP), mobile, television and internet services. The firm's supplementary service offerings include call waiting and call forwarding through telephones and digital switches. Additionally, the company offers special services, including providing digital links between customers and digital trunk access. Telefonica de Argentina has millions of telephone lines in service. The firm primarily provides services to residential customers, but also caters to professional, commercial and governmental clientele. Besides serving its own customers, Telefonica de Argentina operates a wholesale business that provides network access and the use of its facilities for third-party telecommunications operators, such as cell phone companies. Moreover, the company engages in the sale of telephone booth terminals, batteries, computers, cellular handsets and more.

FINANCIAL DATA: *Note: Data for latest year may not have been available at press time.*

In U.S. $	2021	2020	2019	2018	2017	2016
Revenue						
R&D Expense						
Operating Income						
Operating Margin %						
SGA Expense						
Net Income						
Operating Cash Flow						
Capital Expenditure						
EBITDA						
Return on Assets %						
Return on Equity %						
Debt to Equity						

CONTACT INFORMATION:
Phone: 54-11-4332-2051   Fax: 54-11-4303-5586
Toll-Free:
Address: Avenida Ingeniero Huergo 723, Buenos Aires, C1107AOH Argentina

STOCK TICKER/OTHER:
Stock Ticker: Subsidiary
Employees: 11,000
Parent Company: Telefonica SA
Exchange:
Fiscal Year Ends: 12/31

SALARIES/BONUSES:
Top Exec. Salary: $ Bonus: $
Second Exec. Salary: $ Bonus: $

OTHER THOUGHTS:
Estimated Female Officers or Directors:
Hot Spot for Advancement for Women/Minorities:

Sales, profits and employees may be estimates. Financial information, benefits and other data can change quickly and may vary from those stated here.

Telefonica del Peru SAA

NAIC Code: 517110

www.telefonica.com.pe/home

TYPES OF BUSINESS:

Local Phone Service
Long-Distance Phone Service
Public Telephones
Cable Television Service
Cellular Service
Internet Service
Calling Cards

BRANDS/DIVISIONS/AFFILIATES:

Telefonica SA
Movistar
Telefonica Moviles Peru SAC
Terra Networks Perus SA
Telefonica Ingenieria de Seguridad
Tgestiona
Wayra Peru Acelerador de Proyectos SAC
Telxius Torres Peru

CONTACTS: Note: Officers with more than one job title may be intentionally listed here more than once.

Pedro Cortez Rojas, CEO
Denis Fernandez Armas, Dir.-Network Oper. & Maintenance, Movistar

GROWTH PLANS/SPECIAL FEATURES:

Telefonica del Peru SAA, 98%-owned by the Spanis telecommunications conglomerate Telefonica SA, is one of th largest landline and cellular service providers in Peru. The fir offers both domestic and international long-distance fixe telephone services, public telephone services, dat transmission, information technology and internet throughou Peru under the Movistar brand name. Telefonica del Per operates through several subsidiaries. Telefonica Moviles Per SAC provides mobile telephone, internet and televisio services to individual and business customers. Terra Network Peru SA is a provider of cable internet access. Telefonic Ingenieria de Seguridad offers security engineering service including installing safety equipment such as alarm system fire detection and suppression systems, closed circuit vide surveillance cameras and perimeter protection system Subsidiary Tgestiona offers outsourced administrativ functions to corporate clients, with services in areas such a human resources, property management, collections, logistic management, treasury and accounting. Wayra Per Aceleradora de Proyectos SAC provides technology innovatio and project development services. Telxius Torres Peru is th company's global telecommunications infrastructure subsidiary, with telecommunications towers distribute throughout Peru.

FINANCIAL DATA: Note: Data for latest year may not have been available at press time.

In U.S. $	2021	2020	2019	2018	2017	2016
Revenue	2,000,000,000	1,803,670,000	2,616,568,500	2,491,970,000	2,593,250,000	2,882,688,692
R&D Expense						
Operating Income						
Operating Margin %						
SGA Expense						
Net Income		-190,572,000	-127,777,650	-121,693,000	-73,004,700	283,789,472
Operating Cash Flow						
Capital Expenditure						
EBITDA						
Return on Assets %						
Return on Equity %						
Debt to Equity						

CONTACT INFORMATION:

Phone: 51-1-265-7555   Fax: 51-1-470-7484
Toll-Free:
Address: Dean Valdivia 148 Dpto. 201 LIMA, Lima, 15046 Peru

STOCK TICKER/OTHER:

Stock Ticker: Subsidiary
Employees: 4,374
Parent Company: Telefonica SA

Exchange:
Fiscal Year Ends: 12/31

SALARIES/BONUSES:

Top Exec. Salary: $ Bonus: $
Second Exec. Salary: $ Bonus: $

OTHER THOUGHTS:

Estimated Female Officers or Directors:
Hot Spot for Advancement for Women/Minorities:

Sales, profits and employees may be estimates. Financial information, benefits and other data can change quickly and may vary from those stated here.

Telefonica Deutschland Holding AG

www.telefonica.de

NAIC Code: 517210

TYPES OF BUSINESS:
Mobile Telephone Service

GROWTH PLANS/SPECIAL FEATURES:
Telefonica Deutschland O2 is the German subsidiary of Telefonica, which owns 69% of the company's stock. Following the E-Plus acquisition in 2014, O2 became one of the largest wireless operators in Germany. O2 is required to offer competitors, such as 1&1 Drillisch, access to its network as a condition of the regulator's approval of its merger with E-Plus. The firm does not have its own fixed-line network but resells capacity from Deutsche Telekom and Vodafone.

BRANDS/DIVISIONS/AFFILIATES:
Telefonica SA
Telefonica Germany GmbH & Co OHG
O2
AY YILDIZ
Blau
Ortel Mobile
AldiTalk
Tchibo mobil

CONTACTS: Note: Officers with more than one job title may be intentionally listed here more than once.
Markus Haas, CEO
Markus Rolle, CFO
Nicole Gerhardt, Chief Human Resources Officer
Mallik Rao, CTO

FINANCIAL DATA: Note: Data for latest year may not have been available at press time.

In U.S. $	2021	2020	2019	2018	2017	2016
Revenue	7,919,479,000	7,680,855,000	7,545,226,000	7,463,645,000	7,440,191,000	7,650,262,000
R&D Expense						
Operating Income	59,146,250	-45,889,340	-126,450,600	-186,616,600	-44,869,570	-415,043,600
Operating Margin %	.01%	-.01%	-.02%	-.03%	-.01%	-.05%
SGA Expense	300,830,100	244,743,100	243,723,400	258,000,000	296,751,000	346,719,400
Net Income	215,170,000	334,482,300	-216,189,800	-234,545,500	-388,529,700	-179,478,300
Operating Cash Flow	2,175,154,000	2,176,174,000	2,054,823,000	1,723,399,000	1,735,637,000	1,895,740,000
Capital Expenditure	1,190,063,000	1,019,763,000	985,091,100	998,348,000	1,057,494,000	1,054,435,000
EBITDA	2,704,411,000	2,737,044,000	2,340,356,000	1,834,554,000	1,826,396,000	2,117,028,000
Return on Assets %	.01%	.02%	-.01%	-.02%	-.03%	-.01%
Return on Equity %	.03%	.05%	-.03%	-.03%	-.04%	-.02%
Debt to Equity	.73%	0.616	0.64	0.265	0.153	0.183

CONTACT INFORMATION:
Phone: 49 8924421010   Fax: 49 8924421209
Toll-Free:
Address: Georg-Brauchle-Ring 50, Munich, BY 80992 Germany

STOCK TICKER/OTHER:
Stock Ticker: TELDY   Exchange: PINX
Employees: 8,271   Fiscal Year Ends: 12/31
Parent Company:

SALARIES/BONUSES:
Top Exec. Salary: $ Bonus: $
Second Exec. Salary: $ Bonus: $

OTHER THOUGHTS:
Estimated Female Officers or Directors:
Hot Spot for Advancement for Women/Minorities:

Telefonica SA
NAIC Code: 517110

TYPES OF BUSINESS:
Fixed Line & Mobile Telecommunications
Internet Access Service
Data Service
Digital Media

GROWTH PLANS/SPECIAL FEATURES:
Telefonica operates mobile and fixed networks in Spain (where
it is the incumbent telephone operator), U.K., Germany, Brazi
and other Latin American countries. The company derive
more than 30% of its revenue from Spain, close to 20% from
Germany and 20% from Brazil. Its U.K. operations are held
through a joint venture with Virgin Media. In Latin America
Telefonica operates in Mexico, Argentina, Chile, and Peru
among others.

BRANDS/DIVISIONS/AFFILIATES:
Telefonica Espana
Telefonica UK
Telefonica Deutschland
Telefonica Brasil
Telefonica HispanoAmerica
Telefonica
Movistar
O2

CONTACTS: *Note: Officers with more than one job title may be intentionally listed here more than once.*
Jose Maria Alvarez-Pallette, CEO
Angel Vila, COO
Laura Abasolo, CFO
Marta Machicot, Chief People Officer
Enrique Blanco, CTO
Ramiro Sanchez de Lerin Garcia-Ovies, General Counsel
Angel Vila Voix, Gen. Mgr.-Corp. Dev.
Angel Vila Voix, Gen. Mgr.-Finance
Guillermo Ansaldo Lutz, Gen. Mgr.-Global Resources
Matthew Key, Chmn.-Telefonica Digital
Eduardo Navarro, Gen. Mgr.-Strategy & Alliances
Eva Castillo, CEO-Telefonica Europe
Jose Maria Alvarez-Pallette, Chmn.
Santiago Fernandez Valbuena, Chmn.-Telefonica Latin America

FINANCIAL DATA: *Note: Data for latest year may not have been available at press time.*

In U.S. $	2021	2020	2019	2018	2017	2016
Revenue	40,053,230,000	43,927,310,000	49,378,960,000	49,655,320,000	53,035,840,000	53,064,390,000
R&D Expense						
Operating Income	2,610,593,000	4,824,499,000	3,712,957,000	6,776,325,000	6,787,543,000	5,503,661,000
Operating Margin %	.07%	.11%	.08%	.14%	.13%	.10%
SGA Expense	6,947,645,000	5,475,107,000	8,350,839,000	7,549,305,000	9,322,673,000	10,636,130,000
Net Income	8,297,811,000	1,613,265,000	1,164,569,000	3,396,831,000	3,193,898,000	2,415,818,000
Operating Cash Flow						
Capital Expenditure	6,285,819,000	7,158,736,000	7,810,364,000	8,754,665,000	9,169,709,000	9,368,563,000
EBITDA	22,965,060,000	14,642,780,000	16,413,080,000	17,652,100,000	17,699,010,000	17,713,280,000
Return on Assets %	.07%	.01%	.01%	.03%	.02%	.02%
Return on Equity %	.47%	.09%	.05%	.17%	.16%	.12%
Debt to Equity	1.79%	0.36	0.329			2.512

CONTACT INFORMATION:
Phone: 34 914823734   Fax: 34 914823768
Toll-Free:
Address: Distrito C, Ronda de la Comunicacion, Madrid, 28050 Spain

STOCK TICKER/OTHER:
Stock Ticker: TEF   Exchange: NYS
Employees: 104,150   Fiscal Year Ends: 12/31
Parent Company:

SALARIES/BONUSES:
Top Exec. Salary: $1,961,106   Bonus: $3,173,462
Second Exec. Salary: Bonus: $2,200,241
$1,631,621

OTHER THOUGHTS:
Estimated Female Officers or Directors: 1
Hot Spot for Advancement for Women/Minorities:

Telefonos de Mexico SAB de CV (Telmex)   www.telmex.com

NAIC Code: 517110

TYPES OF BUSINESS:
Local Telephone Service
Long-Distance Service
Internet Services
Telecom Equipment & Computer Retail
Internet of Things

GROWTH PLANS/SPECIAL FEATURES:
Telmex is the largest fixed-line provider in Mexico, and the firm holds a dominant position in the long-distance phone market. Telmex also is a strong player in Mexico's Internet access market, which represents a long-term growth opportunity. However, Internet access accounts for less than 20% of total sales; the firm still derives the majority of its revenue from mature, fixed-line operations.

BRANDS/DIVISIONS/AFFILIATES:
America Movil SAB de CV
Scitum
Infinitum

CONTACTS: *Note: Officers with more than one job title may be intentionally listed here more than once.*
Carlos Slim Domit, Pres.
Sergio Medina Noriega, Dir.-Legal
Arturo Elias Ayub, Dir.-Strategic Alliances & Institutional Rel.
Javier Mondragon Alarcon, Dir.-Regulation & Legal Affairs

FINANCIAL DATA: *Note: Data for latest year may not have been available at press time.*

In U.S. $	2021	2020	2019	2018	2017	2016
Revenue	4,984,049,623	4,603,903,851	5,091,563,848			
R&D Expense						
Operating Income						
Operating Margin %						
SGA Expense						
Net Income	223,563,640	-54,541,497	-90,396,738			
Operating Cash Flow						
Capital Expenditure						
EBITDA						
Return on Assets %						
Return on Equity %						
Debt to Equity						

CONTACT INFORMATION:
Phone: 52 5552221774   Fax: 52 5555455550
Toll-Free:
Address: Parque Via 190, Colonia Cuauhtemoc, Mexico City, DF 06599 Mexico

STOCK TICKER/OTHER:
Stock Ticker: Subsidiary   Exchange:
Employees: 54,317   Fiscal Year Ends: 12/31
Parent Company: America Movil SAB de CV

SALARIES/BONUSES:
Top Exec. Salary: $ Bonus: $
Second Exec. Salary: $ Bonus: $

OTHER THOUGHTS:
Estimated Female Officers or Directors:
Hot Spot for Advancement for Women/Minorities:

Telekom Austria AG

www.telekomaustria.com

NAIC Code: 517110

TYPES OF BUSINESS:

Wireless & Wireline Phone Services
Internet Services
Mobile Services
Data Services
Cloud
Internet of Things

BRANDS/DIVISIONS/AFFILIATES:

America Movil SAB de CV
A1 Telekom Austria AG
A1
A1 Bulgaria
velcom
Vipnet
Vip mobile
one.Vip

GROWTH PLANS/SPECIAL FEATURES:

Telekom Austria AG is a telecommunications provider.
generates revenue from four primary business units: fixed-lin
services, fixed-line equipment, mobile services, and the sale
mobile equipment. It operates across Austria, Croatia, Serbi
Slovenia, North Macedonia, Bulgaria, and Belarus. Th
company's revenue is generated from fixed-line and mobil
services across these regions. Austria generates the mo:
revenue across all four business units among all service
regions. The company is the owner of a telecommunication
infrastructure.

CONTACTS: *Note: Officers with more than one job title may be intentionally listed here more than once.*

Thomas Arnoldner, CEO
Alejandro Plater, COO
Siegfried Mayrhofer, CFO
Eve Zehetner, Dir.-Human Resources

FINANCIAL DATA: *Note: Data for latest year may not have been available at press time.*

In U.S. $	2021	2020	2019	2018	2017	2016
Revenue	4,758,197,000	4,552,104,000	4,557,658,000	4,429,295,000	4,469,094,000	4,206,352,000
R&D Expense						
Operating Income	767,636,800	651,569,400	626,969,700	442,943,200	452,686,000	498,642,700
Operating Margin %	.16%	.14%	.14%	.10%	.10%	.12%
SGA Expense	1,007,692,000	1,017,086,000	1,050,029,000	1,026,929,000	1,014,572,000	1,005,606,000
Net Income	463,439,400	396,097,300	333,424,800	248,106,300	351,270,600	420,989,800
Operating Cash Flow	1,616,893,000	1,510,329,000	1,486,841,000	1,256,692,000	1,198,028,000	1,219,158,000
Capital Expenditure	870,116,900	757,204,600	891,142,300	786,705,300	719,363,300	832,598,700
EBITDA	1,745,448,000	1,588,834,000	1,597,021,000	1,426,093,000	1,434,438,000	1,406,724,000
Return on Assets %	.05%	.05%	.04%	.03%	.04%	.05%
Return on Equity %	.15%	.14%	.13%	.09%	.11%	.15%
Debt to Equity	.53%	0.893	1.259	1.388	0.851	0.832

CONTACT INFORMATION:

Phone: 43 506640 Fax: 43 5917182100
Toll-Free:
Address: Lassallestrasse 9, Vienna, AT 1020 Austria

STOCK TICKER/OTHER:

Stock Ticker: TKAGY   Exchange: PINX
Employees: 17,856   Fiscal Year Ends: 12/31
Parent Company: America Movil SAB de CV

SALARIES/BONUSES:

Top Exec. Salary: $ Bonus: $
Second Exec. Salary: $ Bonus: $

OTHER THOUGHTS:

Estimated Female Officers or Directors: 2
Hot Spot for Advancement for Women/Minorities: Y

Telekom Malaysia Berhad

www.tm.com.my

NAIC Code: 517110

TYPES OF BUSINESS:

Wired Telecommunications Carriers
Telecommunications Services
Broadband
Mobile
TV Content
Smart Services

BRANDS/DIVISIONS/AFFILIATES:

Khazanah Nasional Berhad
unifi
TM One
TM Global
Multimedia University
Menara Kuala Lumpur
VADS Berhad
GITN

GROWTH PLANS/SPECIAL FEATURES:

Telekom Malaysia Bhd is a triple-play telecommunications company. It generates revenue from the provision of fixed-line voice services, data, and broadband, and multimedia services to businesses and individual households, and consumers. Broadband and multimedia services are the majority of company revenue. Data is composed of products such as Ethernet and Internet protocol services. Additionally, the company's voice product generates revenue from providing business and residential telephony services. The company owns telecommunications infrastructure. Its segments include unifi; TM ONE; TM WHOLESALE; and Shared Services/Others. It generates the vast majority of its revenue in Malaysia.

CONTACTS: *Note: Officers with more than one job title may be intentionally listed here more than once.*

Imri Mokhtar, CEO
Azizi A Haid, COO
Razidan Ghazalli, CFO
Shanti Jusnita Johari, CMO
Sarinah Abu Bakar, Chief Human Resources Officer
M. Umapathy Sivan, CIO

FINANCIAL DATA: *Note: Data for latest year may not have been available at press time.*

In U.S. $	2021	2020	2019	2018	2017	2016
Revenue	2,587,008,000	2,432,469,000	2,565,735,000	2,652,148,000	2,711,792,000	2,706,362,000
R&D Expense						1,907,326
Operating Income	417,928,900	363,626,200	390,755,100	294,760,500	263,166,200	257,197,400
Operating Margin %	.16%	.15%	.15%	.11%	.10%	.10%
SGA Expense	718,052	718,052				99,741,950
Net Income	200,875,100	227,981,600	141,972,400	34,376,750	208,616,600	174,127,700
Operating Cash Flow						
Capital Expenditure	427,173,800	322,472,800	336,317,700	510,737,200	744,687,600	826,545,500
EBITDA	943,902,200	913,317,600	838,281,200	611,601,000	856,322,200	857,511,500
Return on Assets %	.04%	.04%	.03%	.01%	.04%	.03%
Return on Equity %	.12%	.14%	.09%	.02%	.12%	.10%
Debt to Equity	.92%	1.177	1.267	1.108	0.896	0.996

CONTACT INFORMATION:

Phone: 60 322401211   Fax:
Toll-Free:
Address: Fl. 51, North Wing Menara TM, Jalan Pantai Baharu, Kuala Lumpur, 50672 Malaysia

STOCK TICKER/OTHER:

Stock Ticker: MYTEF   Exchange: PINX
Employees: 22,763   Fiscal Year Ends:
Parent Company: Khazanah Nasional Berhad

SALARIES/BONUSES:

Top Exec. Salary: $ Bonus: $
Second Exec. Salary: $ Bonus: $

OTHER THOUGHTS:

Estimated Female Officers or Directors:
Hot Spot for Advancement for Women/Minorities:

Sales, profits and employees may be estimates. Financial information, benefits and other data can change quickly and may vary from those stated here.

Telenet Group NV/SA

www.telenet.be/en

NAIC Code: 517110

TYPES OF BUSINESS:

Telecommunications Services
Telecommunications
Media Services
Internet
TV
Mobile
Digital Security
Smart Home

BRANDS/DIVISIONS/AFFILIATES:

Liberty Global

GROWTH PLANS/SPECIAL FEATURES:

Telenet Group Holding NV is a Belgian triple-pla
telecommunications company. Telenet generates revenu
from the provision of broadband services, fixed services
mobile services as well as pay-TV services. Of the subscribe
base, the majority receive all three solutions from the company
However, from a product revenue perspective, the compan
derives the majority of overall revenue from video an
broadband services. Both video and Internet services ar
delivered via the company's owned hybrid fiber-coaxia
infrastructure. Telenet operates in both Belgium an
Luxembourg.

CONTACTS: Note: Officers with more than one job title may be intentionally listed here more than once.

John Porter, CEO
Erik Van den Enden, CFO
Micha Berger, CTO

FINANCIAL DATA: Note: Data for latest year may not have been available at press time.

In U.S. $	2021	2020	2019	2018	2017	2016
Revenue	2,647,101,000	2,626,094,000	2,634,966,000	2,583,875,000	2,570,928,000	2,477,127,000
R&D Expense						
Operating Income	611,042,000	605,841,200	699,047,600	609,410,400	449,148,500	495,025,600
Operating Margin %	.23%	.23%	.27%	.24%	.17%	.20%
SGA Expense	728,008,800	590,952,700	555,872,800	545,573,200	500,647,500	503,508,000
Net Income	401,786,600	345,597,700	239,134,400	256,980,300	112,097,400	42,641,390
Operating Cash Flow	1,049,948,000	1,078,297,000	1,114,091,000	1,096,857,000	848,066,500	763,903,400
Capital Expenditure	488,568,400	481,124,200	420,040,400	411,678,300	489,455,600	493,456,200
EBITDA	1,451,633,000	1,374,029,000	1,320,695,000	1,338,031,000	1,140,036,000	948,398,000
Return on Assets %	.07%	.06%	.04%	.05%	.02%	.01%
Return on Equity %						
Debt to Equity						

CONTACT INFORMATION:

Phone: 32 15333000   Fax: 32 15333999
Toll-Free: 0800 66046
Address: Liersesteenweg 4, Mechelen, 2800 Belgium

STOCK TICKER/OTHER:

Stock Ticker: TLGHY   Exchange: PINX
Employees: 3,431   Fiscal Year Ends: 12/31
Parent Company: Liberty Global

SALARIES/BONUSES:

Top Exec. Salary: $ Bonus: $
Second Exec. Salary: $ Bonus: $

OTHER THOUGHTS:

Estimated Female Officers or Directors:
Hot Spot for Advancement for Women/Minorities:

Telenor ASA

www.telenor.com

IAIC Code: 517210

TYPES OF BUSINESS:

Mobile Telephone Services
Fixed-Line Telephone Services
Cable Services
Satellite Communications
Satellite Television Broadcasting
Internet of Things

BRANDS/DIVISIONS/AFFILIATES:

Telenor Norway
Telenor Sweden
Telenor Denmark
DNA
dtac
Digi
Grameenphone
Telenor Pakistan

GROWTH PLANS/SPECIAL FEATURES:

Telenor is an international provider of telecom, data, and media communication services. It is the incumbent dominant telecom operator in Norway and the Norwegian government holds an almost 54% stake in the firm. Telenor also operates in other Nordic countries and is an established player in faster-growing emerging markets. Telenor owns both fixed-line and mobile networks, though fixed-line services (telephony, Internet, TV, data services) provide around 10% of service revenue. The firm's fixed-line operations are in Sweden, Norway, Finland and Denmark. In the rest of its markets, Telenor offers only mobile services.

CONTACTS: *Note: Officers with more than one job title may be intentionally listed here more than once.*

Sigve Brekke, CEO
Tone Hegland Bachke, CFO
Cecilie Blydt Heuch, Chief People Officer
Ruza Sabanovic, CTO
Morten Karlsen Sorby, Head-Strategy & Regulatory Affairs
Rolv-Erik Spilling, Exec. VP
Hilde M. Tonne, Exec. VP
Berit Svendsen, Exec. VP
Bjorn Magnus Kopperud, Acting Head-Central & Eastern European Oper.
Gunn Waersted, Chmn.
Sigve Brekke, Head-Asia Oper.

FINANCIAL DATA: *Note: Data for latest year may not have been available at press time.*

In U.S. $	2021	2020	2019	2018	2017	2016
Revenue	11,343,850,000	11,920,000,000	11,696,290,000	10,899,630,000	11,532,160,000	12,903,220,000
R&D Expense						
Operating Income	2,246,543,000	2,478,483,000	2,602,582,000	2,177,497,000	2,483,320,000	2,627,073,000
Operating Margin %	.20%	.21%	.22%	.20%	.22%	.20%
SGA Expense	703,022,400	712,489,300	801,499,100	773,201,200	1,040,745,000	1,354,285,000
Net Income	157,233,300	1,784,413,000	799,852,700	1,515,841,000	1,233,068,000	291,416,800
Operating Cash Flow	4,349,848,000	4,509,139,000	3,521,492,000	3,744,994,000	4,190,454,000	4,093,212,000
Capital Expenditure	2,001,124,000	1,955,127,000	2,262,391,000	2,162,061,000	2,120,489,000	2,441,542,000
EBITDA	5,602,570,000	5,841,919,000	5,160,917,000	4,185,720,000	4,565,838,000	5,119,345,000
Return on Assets %	.01%	.07%	.04%	.07%	.06%	.01%
Return on Equity %	.05%	.45%	.18%	.28%	.22%	.05%
Debt to Equity	4.38%	3.495	3.04	1.118	0.888	1.173

CONTACT INFORMATION:

Phone: 47 81077000   Fax: 47 67890000
Toll-Free:
Address: Snaroyveien 30, Fornebu, N-1360 Norway

STOCK TICKER/OTHER:

Stock Ticker: TELNY   Exchange: PINX
Employees: 19,000   Fiscal Year Ends: 12/31
Parent Company:

SALARIES/BONUSES:

Top Exec. Salary: $ Bonus: $
Second Exec. Salary: $ Bonus: $

OTHER THOUGHTS:

Estimated Female Officers or Directors: 6
Hot Spot for Advancement for Women/Minorities: Y

Telephone and Data Systems Inc (TDS)

NAIC Code: 517110

www.tdsinc.com

TYPES OF BUSINESS:

Local Telephone Service
Cellular Telephone Services
Internet Access
Printing Services
Long-Distance Telephone Service
Data Networks
Broadband Service

BRANDS/DIVISIONS/AFFILIATES:

United States Cellular Corporation
TDS Telecommunications Corporation
OneNeck
Suttle-Straus Inc

GROWTH PLANS/SPECIAL FEATURES:

Telephone and Data Systems Inc is a diversifie
telecommunications operator that provides mobile, telephon
and broadband services to approximately 6 million customer
The firm's mobile operations are conducted by its 81%-owne
subsidiary, U.S. Cellular, which serves nearly 5 million wireles
customers. The firm's wireline operations are conducted by i
wholly owned subsidiary, TDS Telecom, which services near
1 million phone and Internet access lines in predominantly rur
and suburban areas.

TDS offers its employees comprehensive health benefits, li
and disability insurance, 401(k) and assistance programs.

CONTACTS: *Note: Officers with more than one job title may be intentionally listed here more than once.*

James Butman, CEO, Subsidiary
Laurent Therivel, CEO, Subsidiary
Leroy Carlson, CEO
Peter Sereda, CFO
Walter Carlson, Chairman of the Board
Anita Kroll, Chief Accounting Officer
Jane McCahon, Secretary
Daniel DeWitt, Senior VP, Divisional
Scott Williamson, Senior VP, Divisional
Joseph Hanley, Senior VP, Divisional
Kurt Thaus, Senior VP, Divisional

FINANCIAL DATA: *Note: Data for latest year may not have been available at press time.*

In U.S. $	2021	2020	2019	2018	2017	2016
Revenue	5,329,000,000	5,225,000,000	5,176,000,000	5,109,000,000	5,044,000,000	5,155,000,000
R&D Expense						
Operating Income	285,000,000	281,000,000	190,000,000	196,000,000	152,000,000	114,000,000
Operating Margin %	.05%	.05%	.04%	.04%	.03%	.02%
SGA Expense	1,677,000,000	1,681,000,000	1,717,000,000	1,694,000,000	1,689,000,000	1,762,000,000
Net Income	156,000,000	226,000,000	121,000,000	135,000,000	153,000,000	43,000,000
Operating Cash Flow	1,103,000,000	1,532,000,000	1,016,000,000	1,017,000,000	776,000,000	782,000,000
Capital Expenditure	1,151,000,000	1,368,000,000	962,000,000	778,000,000	685,000,000	636,000,000
EBITDA	1,348,000,000	1,365,000,000	1,308,000,000	1,276,000,000	892,000,000	1,112,000,000
Return on Assets %	.01%	.02%	.01%	.01%	.02%	.00%
Return on Equity %	.02%	.05%	.03%	.03%	.04%	.01%
Debt to Equity	.80%	0.908	0.698	0.53	0.571	0.587

CONTACT INFORMATION:

Phone: 312-630-1900   Fax: 312-630-1908
Toll-Free:
Address: 30 N. LaSalle St., Ste. 4000, Chicago, IL 60602 United States

STOCK TICKER/OTHER:

Stock Ticker: TDS
Employees: 9,400
Parent Company:

Exchange: NYS
Fiscal Year Ends: 12/31

SALARIES/BONUSES:

Top Exec. Salary: $1,352,700   Bonus: $1,270,000
Second Exec. Salary: $785,894   Bonus: $358,374

OTHER THOUGHTS:

Estimated Female Officers or Directors: 5
Hot Spot for Advancement for Women/Minorities: Y

Telesat Corporation

www.telesat.com

NAIC Code: 334220

TYPES OF BUSINESS:

Satellite Communications Equipment Manufacturing
Satellites
Telecommunication Services
Global Connectivity Services
Teleports
Low-Earth Orbit Network
Geostationary Satellite Fleet

BRANDS/DIVISIONS/AFFILIATES:

Loral Space & Communications

GROWTH PLANS/SPECIAL FEATURES:

Telesat Corp is a global satellite operator, providing its customers with mission-critical communications services. The company's segment includes Broadcast; Enterprise and Consulting and other. It generates maximum revenue from the Broadcast segment. The Broadcast segment includes Direct-to-home television, video distribution and contribution, and occasional use services. Geographically, it derives a majority of revenue from Canada.

Telesat offers its employees comprehensive health benefits, retirement and pension plans, performance-based compensation and bonuses and company perks.

CONTACTS: *Note: Officers with more than one job title may be intentionally listed here more than once.*

Daniel S. Goldberg, CEO
Andrew Browne, CFO
Glenn Katz, CCO
David Wendling, CTO

FINANCIAL DATA: *Note: Data for latest year may not have been available at press time.*

In U.S. $	2021	2020	2019	2018	2017	2016
Revenue	589,727,000	638,148,900	708,480,200			
R&D Expense						
Operating Income	320,998,700	315,403,300	372,677,200			
Operating Margin %	.54%	.49%	.53%			
SGA Expense	120,537,500	69,909,000	68,400,870			
Net Income	80,533,560	191,007,200	145,600,100			
Operating Cash Flow	230,529,700	289,679,600	292,149,800			
Capital Expenditure	27,830,750	13,292,370	27,955,200			
EBITDA	500,930,200	528,167,600	575,573,600			
Return on Assets %	.02%	.04%				
Return on Equity %	.11%	.17%				
Debt to Equity	9.19%	2.202				

CONTACT INFORMATION:

Phone: 613 748 8700   Fax:
Toll-Free:
Address: 160 Elgin St., Ste. 2100, Ottawa, ON K2P 2P7 Canada

STOCK TICKER/OTHER:

Stock Ticker: TSAT   Exchange: NAS
Employees: 471   Fiscal Year Ends: 12/31
Parent Company:

SALARIES/BONUSES:

Top Exec. Salary: $1,326,510   Bonus: $
Second Exec. Salary: Bonus: $
$698,775

OTHER THOUGHTS:

Estimated Female Officers or Directors:
Hot Spot for Advancement for Women/Minorities:

Sales, profits and employees may be estimates. Financial information, benefits and other data can change quickly and may vary from those stated here.

Telia Company AB
NAIC Code: 517110

www.teliacompany.com

TYPES OF BUSINESS:
Telephone Service
Telecommunications
Mobile
Fixed Voice
Broadband
Television

GROWTH PLANS/SPECIAL FEATURES:
Telia is the incumbent telecom operator in Sweden and one c
the dominant players in Finland and Norway. Its home marke
Sweden, represents more than 40% of its revenue. Beside
extensive operations in these countries, the firm also ha
assets in Denmark and in the Baltic region. The firm diveste
its outstanding operations in Eurasia in 2018 following allege
corruption linked to local partners and disappointing
performance that was exacerbated by tough economic an
market conditions. The firm's strategy has been to focus o
connectivity (high-quality, cost-efficient network) an
convergence (bundling of fixed and mobile offerings).

BRANDS/DIVISIONS/AFFILIATES:
Telia
Call me
Okarte
Ezys
MyCall
Phonero
C More
Halebop

CONTACTS: *Note: Officers with more than one job title may be intentionally listed here more than once.*
Allison Kirkby, CEO
Rainer Deutschmann, COO
Per Christian Morland, CFO
Markus Messerer, CCO
Cecilia Lundin, Chief People Officer
Jan H. Ahrnell, Head-Legal Affairs
Cecilia Edstrom, Sr. VP-Group Comm.
Christian Luga, Head-Finance
Tero Kivisaari, Pres., Mobility Svcs.
Malin Frenning, Pres., Broadband Svcs.
Lars-Johan Jarnheimer, Chmn.
Veysel Aral, Pres., Eurasia

FINANCIAL DATA: *Note: Data for latest year may not have been available at press time.*

In U.S. $	2021	2020	2019	2018	2017	2016
Revenue	8,724,296,000	8,808,039,000	8,489,456,000			
R&D Expense	35,452,970	29,428,930	15,010,730			
Operating Income	919,209,700	1,053,319,000	1,173,010,000			
Operating Margin %	.11%	.12%	.14%			
SGA Expense	2,023,288,000	2,102,786,000	1,983,194,000			
Net Income	1,153,456,000	-2,270,669,000	700,467,800			
Operating Cash Flow	2,703,512,000	2,824,782,000	2,725,040,000			
Capital Expenditure	1,545,216,000	1,332,202,000	1,503,443,000			
EBITDA	3,504,413,000	987,646,800	3,121,244,000			
Return on Assets %	.05%	- .09%	.03%			
Return on Equity %	.16%	- .30%	.08%			
Debt to Equity	.46%	0.715	1.102			

CONTACT INFORMATION:
Phone: 46 850455000   Fax: 46 850455001
Toll-Free:
Address: Stjarntorget 1, Stockholm, 169 94 Sweden

STOCK TICKER/OTHER:
Stock Ticker: TLSNF
Employees: 20,741
Parent Company:

Exchange: PINX
Fiscal Year Ends: 12/31

SALARIES/BONUSES:
Top Exec. Salary: $ Bonus: $
Second Exec. Salary: $ Bonus: $

OTHER THOUGHTS:
Estimated Female Officers or Directors: 7
Hot Spot for Advancement for Women/Minorities: Y

Telia Lietuva AB

www.telia.lt

NAIC Code: 517110

TYPES OF BUSINESS:

Local & Long-Distance Phone Service
Internet Services
Consulting Services
Managed IT Services
VoIP Services
Call Center Services
Network Construction
Next-Generation Services

BRANDS/DIVISIONS/AFFILIATES:

Telia Company AB

CONTACTS: *Note: Officers with more than one job title may be intentionally listed here more than once.*

Dan Stromberg, CEO
Arunas Linge, Dir.-Finance
Nerijus Ivanauskas, Chief Mktg. Officer

GROWTH PLANS/SPECIAL FEATURES:

Telia Lietuva AB provides telecommunications, information technology (IT) and television services from a single source throughout the country of Lithuania. The firm is part of the Telia Company AB, which operates in several countries, from Norway to Turkey. Telia Lietuva's range of products and services include smart TV; mobile telephone, music and entertainment; smart devices and solutions for home and/or office; office equipment maintenance; and IT security. In-late 2020, Telia Lietuva announced that it entered into a strategic partnership with Ericsson to modernize its mobile network and roll out 5G services throughout Lithuania. Therefore, Ericsson would be the sole partner to deliver radio access network technology (RAN) in that country. As a result, Telia Lietuva launched 11 base stations of the 5G technology, operating in Vilnius, Kaunas and Lkaipeda, and two other 5G stations were launched in the Klaipeda Free Economic Zone by the end of 2020. The firm announced plans to provide commercial 5G communication services in Lithuania during 2021.

FINANCIAL DATA: *Note: Data for latest year may not have been available at press time.*

In U.S. $	2021	2020	2019	2018	2017	2016
Revenue	433,701,855	488,943,000	434,842,000	430,631,000	430,388,000	221,862,054
R&D Expense						
Operating Income						
Operating Margin %						
SGA Expense						
Net Income		68,617,100	61,256,400	62,565,400	62,655,500	9,106,964
Operating Cash Flow						
Capital Expenditure						
EBITDA						
Return on Assets %						
Return on Equity %						
Debt to Equity						

CONTACT INFORMATION:

Phone: 370-5-262-15-11   Fax: 370-5-212-66-65
Toll-Free:
Address: Saltoniskiu g. 7, Vilnius, LT-03501 Lithuania

STOCK TICKER/OTHER:

Stock Ticker: TEO1L   Exchange: Vilnius
Employees: 2,000   Fiscal Year Ends: 12/31
Parent Company: Telia Company AB

SALARIES/BONUSES:

Top Exec. Salary: $ Bonus: $
Second Exec. Salary: $ Bonus: $

OTHER THOUGHTS:

Estimated Female Officers or Directors: 4
Hot Spot for Advancement for Women/Minorities: Y

Telkom SA SOC Limited

NAIC Code: 517110

www.telkom.co.za

TYPES OF BUSINESS:

Telephony Services
Information Technology
Telecommunications
Broadband Networks
Digital Solutions
Property Management
Data Connectivity
Advertising Solutions

BRANDS/DIVISIONS/AFFILIATES:

Openserve
Telkom Consumer
BCX
Gyro
Telkom Business
Yellow Pages

GROWTH PLANS/SPECIAL FEATURES:

Telkom SA SOC Ltd provides fixed-line voice and dat
communications services in South Africa. The fixed-lin
segment has more than 4.5 million access lines in South Afric
where fixed-line penetration is 9.3%. Telkom offers wireles
data services through subsidiary Swiftnet. The firm owns 75°
of Multi-Links, a mobile operation in Nigeria with 1.78 millic
subscribers. The firm also owns Africa Online, which has mor
than 17,000 Internet customers throughout the continent.

CONTACTS: *Note: Officers with more than one job title may be intentionally listed here more than once.*

Sipho Maseko, CEO
Dirk Reyneke, CFO
Xoliswa Makasi, Company Sec.
Miriam Altman, Head-Strategy
Praveen Naidoo, Group Exec.-Comm.
Robin Coode, Group Exec.-Acct.
Ouma Rasethaba, Chief Corp. & Regulatory Affairs Exec.
Bashier Sallie, Managing Dir.-Wholesale & Networks
Manelisa Mavuso, Managing Dir.-Consumer Svcs. & Retail
Attila Vitai, Managing Dir.-Telkom Mobile
Vuledzani Nemukula, Group Exec.-Procurement Svcs.

FINANCIAL DATA: *Note: Data for latest year may not have been available at press time.*

In U.S. $	2021	2020	2019	2018	2017	2016
Revenue	2,599,757,000	2,588,990,000	2,512,661,000	2,385,566,000	2,464,302,000	
R&D Expense						
Operating Income	345,796,200	226,400,600	299,722,100	295,451,500	305,917,500	
Operating Margin %	.13%	.09%	.12%	.12%	.12%	
SGA Expense	267,722,900	238,249,900	211,904,700	191,875,100	251,482,700	
Net Income	145,680,700	32,179,680	168,116,300	175,454,400	228,385,500	
Operating Cash Flow	658,089,400	515,055,300	343,209,800			
Capital Expenditure		463,868,500	456,169,500	466,515,100	510,002,800	
EBITDA	686,539,800	564,858,600	643,954,400	633,548,700	642,450,700	
Return on Assets %	.04%	.01%	.05%	.06%	.08%	
Return on Equity %	.08%	.02%	.10%	.11%	.14%	
Debt to Equity	.44%	0.47	0.164	0.266	0.171	

CONTACT INFORMATION:

Phone: 27 123115322   Fax: 27 813912016
Toll-Free:
Address: 61 Oak Ave., Highveld Techno Park, Centurion, SA 0157
South Africa

STOCK TICKER/OTHER:

Stock Ticker: TLKGY
Employees: 12,039
Parent Company:

Exchange: PINX
Fiscal Year Ends: 03/31

SALARIES/BONUSES:

Top Exec. Salary: $ Bonus: $
Second Exec. Salary: $ Bonus: $

OTHER THOUGHTS:

Estimated Female Officers or Directors: 5
Hot Spot for Advancement for Women/Minorities: Y

Tellabs Inc

www.tellabs.com

NAIC Code: 334210

TYPES OF BUSINESS:

Wireline & Wireless Products & Services
Wireline Products and Services
Wireless Voice Products
Data Products
Video Services
Broadband Solutions
Fiber-based Innovation

BRANDS/DIVISIONS/AFFILIATES:

Marlin Equity Partners LLC

CONTACTS: Note: Officers with more than one job title may be intentionally listed here more than once.

Rich Schroder, CEO
Norm Burke, CFO
Karen Leos, VP-Global Sales
Tom Dobozy, VP-Engineering
James M. Sheehan, Chief Admin. Officer
James M. Sheehan, General Counsel
John M. Brots, Exec. VP-Global Oper.
Kenneth G. Craft, Exec. VP-Product Dev.

GROWTH PLANS/SPECIAL FEATURES:

Tellabs, Inc. provides products and services that enable customers to deliver wireline and wireless voice, data and video services to business and residential customers. The firm operates in two segments: enterprise and broadband. The enterprise segment offers a passive optical local area network (LAN) infrastructure, which is secure, scalable and sustainable. This division serves the business enterprise, federal government, hospitality, higher education, K-12 education, healthcare and transportation industries. The broadband segment offers solutions to service providers that deliver stability and scalability while increasing flexibility. These broadband solutions help telecommunications companies grow HSI (high-speed internet) subscribers, extend service area coverage and offer faster internet service speeds. They also enable Ethernet business services while continuing to support time-division multiplexing (TDM) and automated teller machine (ATM) services. Tellabs offers services such as technical support, professional network services and training. The company is an innovation leader in fiber-based technologies, with many first-to-market accomplishments and investments in fiber optic research and development innovations. Tellabs is a subsidiary of Marlin Equity Partners, LLC.

FINANCIAL DATA: Note: Data for latest year may not have been available at press time.

In U.S. $	2021	2020	2019	2018	2017	2016
Revenue	1,640,000,000	1,610,256,375	1,533,577,500	1,460,550,000	1,391,000,000	1,372,000,000
R&D Expense						
Operating Income						
Operating Margin %						
SGA Expense						
Net Income						
Operating Cash Flow						
Capital Expenditure						
EBITDA						
Return on Assets %						
Return on Equity %						
Debt to Equity						

CONTACT INFORMATION:

Phone: 972-588-7000   Fax:
Toll-Free:
Address: 4240 International Pkwy, St. 105, Carrollton, TX 75007 United States

STOCK TICKER/OTHER:

Stock Ticker: Private   Exchange:
Employees: 9,400   Fiscal Year Ends: 12/31
Parent Company: Marlin Equity Partners LLC

SALARIES/BONUSES:

Top Exec. Salary: $ Bonus: $
Second Exec. Salary: $ Bonus: $

OTHER THOUGHTS:

Estimated Female Officers or Directors: 1
Hot Spot for Advancement for Women/Minorities: Y

Sales, profits and employees may be estimates. Financial information, benefits and other data can change quickly and may vary from those stated here.

Telstra Corporation Limited
NAIC Code: 517110

www.telstra.com.au

TYPES OF BUSINESS:
Wireless & Wireline Telephony Services
Telecommunications
Communications Technology
Mobile
Broadband
Contract Management
5G

BRANDS/DIVISIONS/AFFILIATES:

GROWTH PLANS/SPECIAL FEATURES:
Telstra is Australia's largest telecommunications entity, wit
material market shares in voice, mobile, data and interne
spanning retail, corporate and wholesale segments. Its fixed
line copper network will gradually be wound down as th
government-owned National Broadband Network rolls out to a
Australian households, but the group will be compensate
accordingly. Investments into network applications an
services, media, technology and overseas are being made t
replace the expected lost fixed-line earnings longer term, whil
continuing cost-cuts are also critical.

CONTACTS: *Note: Officers with more than one job title may be intentionally listed here more than once.*
Andrew Penn, CEO
Vicki Brady, CFO
Kate McKenzie, Managing Dir.-Prod. & Innovation
Carmel Mulhern, General Counsel
Will Irving, Managing Dir.-Telstra Bus.
Tony Warren, Managing Dir.-Corp. Affairs
Andrew Keys, Acting Dir.-Investor Rel.
Andrew Penn, Managing Dir.-Finance
Paul Geason, Group Managing Dir.-Enterprise & Govt
Damien Coleman, Sec.
Gordon Ballantyne, Chief Customer Officer
Stuart Lee, Managing Dir.-Telstra Wholesale
Timothy Y. Chen, Pres.

FINANCIAL DATA: *Note: Data for latest year may not have been available at press time.*

In U.S. $	2021	2020	2019	2018	2017	2016
Revenue	14,672,770,000	15,908,900,000	17,650,230,000	18,175,710,000	18,177,100,000	18,105,830,000
R&D Expense						
Operating Income	1,018,108,000	565,304,900	1,800,031,000	2,243,052,000	3,018,686,000	3,939,665,000
Operating Margin %	.07%	.04%	.10%	.12%	.17%	.22%
SGA Expense	3,635,700,000	3,962,725,000	5,091,238,000	5,330,217,000	5,193,957,000	4,873,222,000
Net Income	1,297,616,000	1,271,063,000	1,505,150,000	2,489,718,000	2,718,914,000	4,038,890,000
Operating Cash Flow						
Capital Expenditure	2,194,138,000	2,405,166,000	3,053,625,000	3,446,333,000	3,718,155,000	2,930,641,000
EBITDA	5,351,180,000	6,337,844,000	5,745,287,000	7,129,550,000	7,558,595,000	7,372,722,000
Return on Assets %	.04%	.04%	.05%	.08%	.09%	.14%
Return on Equity %	.13%	.13%	.15%	.24%	.26%	.39%
Debt to Equity	.91%	1.088	1.033	1.019	1.018	0.923

CONTACT INFORMATION:
Phone: 61 883081721   Fax: 61 396323215
Toll-Free:
Address: 242 Exhibition St., Level 41, Melbourne, VIC 3000 Australia

STOCK TICKER/OTHER:
Stock Ticker: TLSYY   Exchange: PINX
Employees: 27,015   Fiscal Year Ends: 06/30
Parent Company:

SALARIES/BONUSES:
Top Exec. Salary: $ Bonus: $
Second Exec. Salary: $ Bonus: $

OTHER THOUGHTS:
Estimated Female Officers or Directors: 6
Hot Spot for Advancement for Women/Minorities: Y

Sales, profits and employees may be estimates. Financial information, benefits and other data can change quickly and may vary from those stated here.

Telular Corporation

www.telular.com

NAIC Code: 334220

TYPES OF BUSINESS:

Fixed Wireless Terminals
Wireless Security Products
GPS Products

BRANDS/DIVISIONS/AFFILIATES:

AMETEK Inc
SkyBitz
Telguard

CONTACTS: *Note: Officers with more than one job title may be intentionally listed here more than once.*

Joseph A. Beatty, Pres.
Christopher Bear, VP-Prod. Dev.
George S. Brody, Sr. VP
Henry Popplewell, Sr. VP
Pat Barron, VP
David A. Zapico, Chmn.-AMETEK

GROWTH PLANS/SPECIAL FEATURES:

Telular Corporation designs, develops and distributes products and services that utilize wireless networks to provide data and voice connectivity among people and machines. The company's strategy consists of leveraging domain expertise to identify opportunities for Internet of Things (IoT) solutions to enterprise scale problems. Once identified, Telular utilizes hardware derived from internal designs that connect into the appropriate Telular analytics cloud via a wide range of connectivity options. The firm then rolls the solution out to targeted industries with the goal of delivering a one-year return on investment for customers. Examples for this strategy include trailer tracking, remote oilfield, intermodal asset tracking, local fleet telematics, petroleum logistics, industrial tank monitoring, residential security/home automation, commercial security, fire and personal emergency response systems. Connectivity partners include AT&T, Verizon, Telefonica, Iridium, Inmarsat, Globalstar and Rogers. Ecosystem partners include Sierra Wireless, Salesforce, CISCO Jasper, Windows Azure and Telit. Telular's own brands include: SkyBitz, offering asset tracking and information management services and solutions; and Telguard, offering home security and automation solutions. Telular is a subsidiary of AMETEK, Inc., a global manufacturer of electronic instruments and electromechanical devices.

FINANCIAL DATA: *Note: Data for latest year may not have been available at press time.*

In U.S. $	2021	2020	2019	2018	2017	2016
Revenue	167,310,000	160,875,000	165,000,000	152,500,000	150,000,000	145,000,000
R&D Expense						
Operating Income						
Operating Margin %						
SGA Expense						
Net Income						
Operating Cash Flow						
Capital Expenditure						
EBITDA						
Return on Assets %						
Return on Equity %						
Debt to Equity						

CONTACT INFORMATION:

Phone: 312 379-8397   Fax: 312 379-8310
Toll-Free: 800-835-8527
Address: 200 S. Wacker Dr., Ste. 1800, Chicago, IL 60606 United States

STOCK TICKER/OTHER:

Stock Ticker: Subsidiary
Employees: 155
Parent Company: AMETEK Inc

Exchange:
Fiscal Year Ends: 09/30

SALARIES/BONUSES:

Top Exec. Salary: $ Bonus: $
Second Exec. Salary: $ Bonus: $

OTHER THOUGHTS:

Estimated Female Officers or Directors:
Hot Spot for Advancement for Women/Minorities:

Sales, profits and employees may be estimates. Financial information, benefits and other data can change quickly and may vary from those stated here.

TELUS Corporation

NAIC Code: 517110

www.telus.com

TYPES OF BUSINESS:

Local & Long-Distance Telephone Service
Wireless Phone Service
Internet Service Provider
Paging Services
Mobile Media Services
EMR Technology
Start-up Investments

BRANDS/DIVISIONS/AFFILIATES:

TELUS SmartHome Security
Conservis

GROWTH PLANS/SPECIAL FEATURES:

Telus is one of the Big Three wireless service providers
Canada, with its 9 million mobile phone subscribers nationwic
constituting about 30% of the total market. It is the incumbe
local exchange carrier in the western Canadian provinces
British Columbia and Alberta, where it provides interne
television, and landline phone services. It also has a sma
wireline presence in eastern Quebec. In recent years Telus ha
moved to bring fiber to the home over most of its wirelir
footprint as it upgrades its legacy copper network, leaving
able to compete on more equal footing with cable provider
Telus' other businesses participate in the internation
business services, health, security, and agriculture industries

CONTACTS: *Note: Officers with more than one job title may be intentionally listed here more than once.*

Darren Entwistle, CEO
Doug French, CFO
Richard Auchinleck, Chairman of the Board
Francois Gratton, Chairman, Divisional
Andrea Wood, Chief Legal Officer
Eros Spadotto, Executive VP, Divisional
Sandy McIntosh, Executive VP, Divisional
Tony Geheran, Executive VP
Navin Arora, President, Divisional
Zainul Mawji, President, Divisional
Jim Senko, President, Divisional
Stephen Lewis, Senior VP

FINANCIAL DATA: *Note: Data for latest year may not have been available at press time.*

In U.S. $	2021	2020	2019	2018	2017	2016
Revenue	13,096,370,000	11,932,020,000	11,347,130,000	10,962,900,000	10,348,450,000	9,897,333,000
R&D Expense						
Operating Income	2,070,468,000	1,845,687,000	2,261,803,000	1,995,800,000	2,066,579,000	1,658,241,000
Operating Margin %	.16%	.15%	.20%	.18%	.20%	.17%
SGA Expense	3,320,370,000	2,878,588,000	2,359,804,000	2,252,470,000	2,017,578,000	2,285,914,000
Net Income	1,287,237,000	938,788,300	1,358,015,000	1,244,458,000	1,212,569,000	951,232,800
Operating Cash Flow	3,412,927,000	3,557,595,000	3,054,367,000	3,156,257,000	3,069,923,000	2,503,695,000
Capital Expenditure	4,134,713,000	2,194,914,000	3,028,700,000	2,236,136,000	2,396,360,000	2,253,247,000
EBITDA	4,899,277,000	4,253,714,000	4,284,047,000	3,938,711,000	3,818,154,000	3,270,592,000
Return on Assets %	.04%	.03%	.05%	.05%	.05%	.04%
Return on Equity %	.12%	.11%	.17%	.16%	.17%	.15%
Debt to Equity	1.19%	1.566	1.625	1.293	1.302	1.287

CONTACT INFORMATION:

Phone: 604 697-8044   Fax: 604 432-2984
Toll-Free: 800-667-4871
Address: 510 W. Georgia St., Fl. 7, Vancouver, BC V6B 0M3 Canada

STOCK TICKER/OTHER:

Stock Ticker: TU
Employees: 47,640
Parent Company:

Exchange: NYS
Fiscal Year Ends: 12/31

SALARIES/BONUSES:

Top Exec. Salary: $1,375,000   Bonus: $
Second Exec. Salary: Bonus: $
$650,000

OTHER THOUGHTS:

Estimated Female Officers or Directors: 1
Hot Spot for Advancement for Women/Minorities:

Sales, profits and employees may be estimates. Financial information, benefits and other data can change quickly and may vary from those stated here.

Telvue Corporation

www.telvue.com

NAIC Code: 334210

TYPES OF BUSINESS:
Video Broadcast Systems, Servers and Software

BRANDS/DIVISIONS/AFFILIATES:

CONTACTS: *Note: Officers with more than one job title may be intentionally listed here more than once.*
H. Lenfest, Chairman of the Board
John Fell, Controller
Jesse Lerman, Director
Paul Andrews, Senior VP, Divisional
Randy Gilson, Vice President, Divisional
Dan Pisarski, Vice President, Divisional

GROWTH PLANS/SPECIAL FEATURES:
TelVue Corporation operates two business segments. The first segment, TelVue Products and Services, includes equipment such as the TelVue Princeton Server Product Line (TVP) and services such as WEBUS and PEG.TV. The TVP includes video systems, servers and software that support capture, storage, manipulation and play-out of digital media in multiple popular formats. WEBUS is a broadcast digital signage system for displaying a fully automated TV station-like display on a cable system access channel using computer-based digital technology. PEG.TV is a live streaming and Video-on-Demand service for integrating video on the Internet. TelVue's second business segment is a Marketing and Service company which sells automatic number identification (ANI) telecommunications services to the cable and satellite television industry for the automated ordering of pay-per-view features and events. The Company co-markets its products and services to hyperlocal broadcasters using direct mail, e-mail, Internet banner ads, Search Engine Optimization, its website, online videos & webinars and telemarketing. In 2009 the Company introduced a new TVP broadcast server model (B100-IP) and new multi-channel high-definition TVP broadcast server model (B3000-PRO). The Company currently has competition from software and hardware suppliers to broadcasters including municipalities that operate PEG cable access channels.

FINANCIAL DATA: *Note: Data for latest year may not have been available at press time.*

In U.S. $	2021	2020	2019	2018	2017	2016
Revenue						
R&D Expense						
Operating Income						
Operating Margin %						
SGA Expense						
Net Income						
Operating Cash Flow						
Capital Expenditure						
EBITDA						
Return on Assets %						
Return on Equity %						
Debt to Equity						

CONTACT INFORMATION:
Phone: 856 273-8888   Fax: 856 866-7411
Toll-Free:
Address: 16000 Horizon Way, Mount Laurel, NJ 08054 United States

STOCK TICKER/OTHER:
Stock Ticker: Private
Employees: 26
Parent Company:

Exchange:
Fiscal Year Ends: 12/31

SALARIES/BONUSES:
Top Exec. Salary: $ Bonus: $
Second Exec. Salary: $ Bonus: $

OTHER THOUGHTS:
Estimated Female Officers or Directors:
Hot Spot for Advancement for Women/Minorities:

Sales, profits and employees may be estimates. Financial information, benefits and other data can change quickly and may vary from those stated here.

TESSCO Technologies Incorporated   www.tessco.com

NAIC Code: 423430

TYPES OF BUSINESS:

Wireless Communications Products Distributor
Technology Distribution
Base Station Infrastructure
Network Systems
Installation
Testing

BRANDS/DIVISIONS/AFFILIATES:

GROWTH PLANS/SPECIAL FEATURES:

Tessco Technologies Inc is a United States-based value
added technology distributor, manufacturer, and solution
provider. The company supplies wireless communication
products for network infrastructure, site support, and fixed an
mobile broadband networks. It offers products related to powe
systems, Wi-Fi Networks, Broadband, DAS((Distribute
Antenna Systems) for In-Building Cellular and Public Safet
Coverage, IoT (Internet of Things), Mobile Devices an
Accessories, and others. The operating segments of th
company are Carrier and Commercial, of which the majority c
the revenue is derived from the Commercial segment.

TESSCO offers its employees health and retirement benefits
life and disability insurance, product discounts and othe
assistance plans and programs.

CONTACTS: *Note: Officers with more than one job title may be intentionally listed here more than once.*

Sandip Mukerjee, CEO
Aric Spitulnik, Chief Accounting Officer
Joseph Cawley, Chief Information Officer
Douglas Rein, Senior VP, Divisional

FINANCIAL DATA: *Note: Data for latest year may not have been available at press time.*

In U.S. $	2021	2020	2019	2018	2017	2016
Revenue	373,340,700	409,014,400	419,044,800	580,274,700	533,295,100	
R&D Expense						
Operating Income	-17,791,500	-12,363,300	-10,942,300	7,901,700	3,351,500	
Operating Margin %	-.05%	-.03%	-.03%	.01%	.01%	
SGA Expense	85,507,100	92,005,200	95,347,000	112,326,700	108,416,300	
Net Income	-8,742,900	-21,568,900	5,545,800	5,195,400	1,445,100	
Operating Cash Flow	-684,200	908,200	8,246,700	-9,247,100	3,051,300	
Capital Expenditure	11,855,900	6,845,700	5,164,700	3,539,400	2,563,000	
EBITDA	-14,047,000	-17,933,800	-7,323,400	11,894,300	6,783,800	
Return on Assets %	-.04%	-.10%	.03%	.03%	.01%	
Return on Equity %	-.11%	-.22%	.05%	.05%	.01%	
Debt to Equity	.51%	0.137		0.00	0.00	

CONTACT INFORMATION:

Phone: 410 229-1000   Fax: 410 527-0005
Toll-Free:
Address: 11126 McCormick Rd., Hunt Valley, MD 21031 United States

STOCK TICKER/OTHER:

Stock Ticker: TESS   Exchange: NAS
Employees: 678   Fiscal Year Ends: 03/31
Parent Company:

SALARIES/BONUSES:

Top Exec. Salary: $550,000   Bonus: $
Second Exec. Salary: Bonus: $
$345,000

OTHER THOUGHTS:

Estimated Female Officers or Directors:
Hot Spot for Advancement for Women/Minorities:

Sales, profits and employees may be estimates. Financial information, benefits and other data can change quickly and may vary from those stated here.

Thuraya Telecommunications Company

www.thuraya.com

NAIC Code: 517410

TYPES OF BUSINESS:

Satellite Telecommunications
Mobile Satellite Services
Telecommunications
Mobile Satellite Handsets
Broadband Devices

BRANDS/DIVISIONS/AFFILIATES:

Mubadala Investment Company
Al Yah Satellite Communications Company (Yahsat)

CONTACTS: *Note: Officers with more than one job title may be intentionally listed here more than once.*

Sulaiman Al Ali, CEO
Khalid Al Kaf, COO
Nassem Nasser, CMO
Adnan Al Muhairi, CTO
Rashed Al Ghafri, Chmn.

GROWTH PLANS/SPECIAL FEATURES:

Thuraya Telecommunications Company is the mobile satellite services subsidiary of Al Yah Satellite Communications Company (Yahsat), a global satellite operator based in the United Arab Emirates. Yahsat itself is wholly-owned by Mubadala Investment Company. Thuraya offers innovative communications solutions to a variety of sectors, including energy, government, broadcast media, maritime, military, aerospace and humanitarian non-government organizations. The company's network enables communications and coverage across two-thirds of the globe by mobile satellite services (MSS), quasi-global very small aperture terminal (VSAT) coverage and through its GSM (global system for mobile) roaming capabilities. Thuraya has established roaming agreements in more than 160 countries, enabling the firm to provide roaming services for postpaid and prepaid customers onto worldwide GSM networks. Thuraya also offers mobile satellite handsets and broadband devices. Product lines are categorized into land voice, land data, marine, aero, machine-to-machine and certified.

FINANCIAL DATA: *Note: Data for latest year may not have been available at press time.*

In U.S. $	2021	2020	2019	2018	2017	2016
Revenue	187,218,435	180,017,726	184,444,391	175,661,325	167,296,500	159,330,000
R&D Expense						
Operating Income						
Operating Margin %						
SGA Expense						
Net Income						
Operating Cash Flow						
Capital Expenditure						
EBITDA						
Return on Assets %						
Return on Equity %						
Debt to Equity						

CONTACT INFORMATION:

Phone: 971 4-4488888   Fax: 971-4-4488999
Toll-Free:
Address: P.O. Box 283333, Dubai,   United Arab Emirates

STOCK TICKER/OTHER:

Stock Ticker: Private   Exchange:
Employees: Fiscal Year Ends:
Parent Company: Mubadala Investment Company

SALARIES/BONUSES:

Top Exec. Salary: $ Bonus: $
Second Exec. Salary: $ Bonus: $

OTHER THOUGHTS:

Estimated Female Officers or Directors:
Hot Spot for Advancement for Women/Minorities:

TIM SA

NAIC Code: 517210

TYPES OF BUSINESS:

Cell Phone Service
Telecommunications Infrastructure
Mobile Telephone
Fixed Telephone
Internet
Broadband

GROWTH PLANS/SPECIAL FEATURES:

TIM is one of the top three wireless carriers by subscribers
Brazil, with 52 million subscribers, equal to about 20% of th
market. TIM also owns some fixed-line assets, including a
extensive long-haul fiber network and local networks that read
about 6.7 million locations (a bit less than 10% of the country
The firm is investing to expand its fiber network to mo
customer locations, including through its minority ownership
an infrastructure partnership. TIM leases capacity on th
venture's network to serve retail customers. The company
67%-owned by Telecom Italia.

BRANDS/DIVISIONS/AFFILIATES:

Telecom Italia SpA

CONTACTS: *Note: Officers with more than one job title may be intentionally listed here more than once.*

Pietro Labriola, CEO
Leonardo de Carvalho Capdeville, CTO
Rogerio Tostes, Investor Rel. Officer
Lorenzo Federico Zanotti Lindner, Chief Commercial Officer
Mario Girasole, Chief Regulatory Affairs Officer
Antonino Ruggiero, Chief Wholesale Officer
Daniel Junqueira Pinto Hermeto, Chief Supplies Officer

FINANCIAL DATA: *Note: Data for latest year may not have been available at press time.*

In U.S. $	2021	2020	2019	2018	2017	2016
Revenue	3,533,791,000	3,379,154,000	3,400,559,000	3,323,091,000	3,176,838,000	3,056,186,000
R&D Expense						
Operating Income	575,344,200	542,787,600	874,581,600	464,927,900	370,913,700	262,512,900
Operating Margin %	.16%	.16%	.26%	.14%	.12%	.09%
SGA Expense	1,122,806,000	1,053,253,000	1,163,469,000	1,152,464,000	1,060,005,000	1,079,967,000
Net Income	578,692,000	357,772,700	708,817,300	498,053,100	241,581,800	146,851,700
Operating Cash Flow	1,972,190,000	1,697,398,000	1,382,503,000	1,199,465,000	1,057,535,000	976,937,500
Capital Expenditure	1,033,973,000	761,493,200	754,091,800	749,869,100	811,707,600	881,078,100
EBITDA	1,986,832,000	1,652,740,000	2,085,330,000	1,222,159,000	1,175,843,000	1,081,174,000
Return on Assets %	.06%	.04%	.10%	.08%	.04%	.02%
Return on Equity %	.12%	.08%	.17%	.13%	.07%	.04%
Debt to Equity	.44%	0.344	0.337	0.136	0.278	0.424

CONTACT INFORMATION:

Phone: 55 2141093742   Fax: 55 2141093314
Toll-Free:
Address: Joao Cabral de Melo Neto Ave, 850-South Tower, Fl. 12, Rio de Janeiro, RJ 22775-057 Brazil

STOCK TICKER/OTHER:

Stock Ticker: TIMB   Exchange: NYS
Employees: 9,337   Fiscal Year Ends: 12/31
Parent Company: Telecom Italia SpA

SALARIES/BONUSES:

Top Exec. Salary: $ Bonus: $
Second Exec. Salary: $ Bonus: $

OTHER THOUGHTS:

Estimated Female Officers or Directors:
Hot Spot for Advancement for Women/Minorities:

TKH Group NV

NAIC Code: 238210

www.tkhgroup.com

TYPES OF BUSINESS:
Network Wiring and Electrical Contracting
Telecommunications Infrastructure
Fiber Optic Cable
Security & Closed-Circuit Television Systems

BRANDS/DIVISIONS/AFFILIATES:

CONTACTS: Note: Officers with more than one job title may be intentionally listed here more than once.
Alexander van der Lof, CEO
Elling de Lange, CFO
Arne Dehn, Managing Dir.-Building Solutions
Alexandervan der Lof, Chmn.

GROWTH PLANS/SPECIAL FEATURES:
TKH Group NV is an internationally-operating group of companies involved in the development of systems and networks for providing information, electro technical engineering, telecommunication and industrial production solutions. The group offers its services under three primary divisions: telecom solutions, building solutions and industrial solutions. The telecom solutions division installs and upgrades fiber-optic cable networks, copper networks and indoor telecom systems, serving customers such as telecom operators, cable operators, service providers, telecom installers and telecom retailers. The building solutions division is focused on three main areas: building technologies, encompassing systems such as intercoms, programmable lighting and motion-activated lighting for public spaces; security systems, including closed-circuit television, control room systems and audio and video analysis technology; and connectivity systems, encompassing specialty cable for energy transmission in both home and industrial settings as well as pre-assembled cable systems engineered to reduce installation time and complexity. The industrial solutions division provides services in two primary areas: connectivity systems, including specialty cable installations for medical, automotive and industrial applications; and manufacturing systems, focused primarily on process-improvement technologies for tire manufacturers and automated handling systems for customers in both food and non-food packaging industries. TKH Group operates on a global scale, with primary activities in Europe, North America and Asia.

FINANCIAL DATA: Note: Data for latest year may not have been available at press time.

In U.S. $	2021	2020	2019	2018	2017	2016
Revenue			1,610,214,912	1,762,826,368	1,604,609,024	1,484,703,360
R&D Expense						
Operating Income						
Operating Margin %						
SGA Expense						
Net Income			123,279,144	117,337,208	94,392,080	94,892,608
Operating Cash Flow						
Capital Expenditure						
EBITDA						
Return on Assets %						
Return on Equity %						
Debt to Equity						

CONTACT INFORMATION:
Phone: 31-53-573-2900   Fax: 31-53-573-2180
Toll-Free:
Address: Spinnerstraat 15, Haaksbergen, 7481 KJ Netherlands

STOCK TICKER/OTHER:
Stock Ticker: TWKN
Employees: 5,980
Parent Company:

Exchange: Amsterdam
Fiscal Year Ends: 12/31

SALARIES/BONUSES:
Top Exec. Salary: $ Bonus: $
Second Exec. Salary: $ Bonus: $

OTHER THOUGHTS:
Estimated Female Officers or Directors: 1
Hot Spot for Advancement for Women/Minorities:

Sales, profits and employees may be estimates. Financial information, benefits and other data can change quickly and may vary from those stated here.

T-Mobile Polska SA

www.t-mobile.pl

NAIC Code: 517210

TYPES OF BUSINESS:

Mobile Phone Service
Mobile Telephone
Fixed Telephone
LTE Technologies
Wi-Fi
Fiber Optic Network
Devices
5G

BRANDS/DIVISIONS/AFFILIATES:

Deutsche Telekom AG

GROWTH PLANS/SPECIAL FEATURES:

T-Mobile Polska SA is a mobile and fixed telephone servic operator in Poland, serving more than 11 million customers The company provides mobile services using long-term evolution (LTE) technologies, covering nearly 100% of th country's population. T-Mobile Polska also provides voice ove LTE and voice over Wi-Fi; and provides 5G network service across select areas. Fixed-line services are provided throug T-Mobile Polska's owned and leased fiber-optic network. Th company's solutions and services span the entire spectrum o information and communications technology (ICT). Relate devices such as telephones, tablets, modems and routers, a well as smartwatches are sold through T-Mobile as well. T Mobile Polska SA is a subsidiary of Deutsche Telkom AG.

CONTACTS: *Note: Officers with more than one job title may be intentionally listed here more than once.*

Maciej Rogalski, Dir.-Legal Affairs, Data Protect & Compliance Mgmt.
Grzegorz Bors, Dir.-Private Mkt.
Igor Matejov, Dir.-Bus. Mkt.

FINANCIAL DATA: *Note: Data for latest year may not have been available at press time.*

In U.S. $	2021	2020	2019	2018	2017	2016
Revenue	1,615,934,800	1,784,640,000	1,664,110,000	1,744,280,000	1,807,580,000	1,663,757,612
R&D Expense						
Operating Income						
Operating Margin %						
SGA Expense						
Net Income	103,048,400	106,857,000	104,147,000	-631,373,000	-819,339,000	223,676,481
Operating Cash Flow						
Capital Expenditure						
EBITDA						
Return on Assets %						
Return on Equity %						
Debt to Equity						

CONTACT INFORMATION:

Phone: 48 22-413-6000   Fax: 4822-413-6889
Toll-Free:
Address: ul. Marynarska 12, Warsaw, 02-674 Poland

STOCK TICKER/OTHER:

Stock Ticker: Subsidiary   Exchange:
Employees: 3,895   Fiscal Year Ends: 12/31
Parent Company: Deutsche Telekom AG

SALARIES/BONUSES:

Top Exec. Salary: $ Bonus: $
Second Exec. Salary: $ Bonus: $

OTHER THOUGHTS:

Estimated Female Officers or Directors: 2
Hot Spot for Advancement for Women/Minorities:

T-Mobile US Inc   www.t-mobile.com

NAIC Code: 517210

TYPES OF BUSINESS:
Mobile Phone and Wireless Services
Wireless Services
Cellular
Mobile Devices
5G

BRANDS/DIVISIONS/AFFILIATES:
Deutsche Telekom AG
T-Mobile International AG
T-Mobile
Metro by T-Mobile
Sprint Corporation

GROWTH PLANS/SPECIAL FEATURES:
Deutsche Telekom merged its T-Mobile USA unit with prepaid specialist MetroPCS in 2013, creating T-Mobile us. Following the merger, the firm provided nationwide service in major markets but spottier coverage elsewhere. T-Mobile spent aggressively on low-frequency spectrum, well suited to broad coverage, and has substantially expanded its geographic footprint. This expansion, coupled with aggressive marketing and innovative offerings, produced rapid customer growth. With the Sprint acquisition, the firm's scale now roughly matches its larger rivals: T-Mobile now serves 71 million postpaid and 21 million prepaid phone customers, equal to around 30% of the U.S. retail wireless market. In addition, the firm provides wholesale service to resellers.

CONTACTS: Note: Officers with more than one job title may be intentionally listed here more than once.
G. Sievert, CEO
Peter Osvaldik, CFO
Timotheus Hottges, Chairman of the Board
Dara Bazzano, Chief Accounting Officer
Matthew Staneff, Chief Marketing Officer
Peter Ewens, Executive VP, Divisional
David Miller, Executive VP
Deeanne King, Executive VP
Neville Ray, President, Divisional

FINANCIAL DATA: Note: Data for latest year may not have been available at press time.

In U.S. $	2021	2020	2019	2018	2017	2016
Revenue	80,118,000,000	68,397,000,000	44,998,000,000	43,310,000,000	40,604,000,000	37,490,000,000
R&D Expense						
Operating Income	6,892,000,000	7,054,000,000	5,722,000,000	5,309,000,000	4,653,000,000	3,319,000,000
Operating Margin %	.09%	.10%	.13%	.12%	.11%	.09%
SGA Expense	20,238,000,000	18,926,000,000	14,139,000,000	13,161,000,000	12,259,000,000	11,378,000,000
Net Income	3,024,000,000	3,064,000,000	3,468,000,000	2,888,000,000	4,536,000,000	1,460,000,000
Operating Cash Flow	13,917,000,000	8,640,000,000	6,824,000,000	3,899,000,000	3,831,000,000	2,779,000,000
Capital Expenditure	21,692,000,000	12,367,000,000	7,358,000,000	5,668,000,000	11,065,000,000	8,670,000,000
EBITDA	23,096,000,000	20,411,000,000	12,354,000,000	11,760,000,000	10,816,000,000	10,300,000,000
Return on Assets %	.01%	.02%	.04%	.04%	.07%	.02%
Return on Equity %	.04%	.07%	.13%	.12%	.22%	.08%
Debt to Equity	1.39%	1.449	1.279	0.49	1.184	1.197

CONTACT INFORMATION:
Phone: 425-378-4000   Fax: 425-378-4040
Toll-Free: 800-318-9270
Address: 12920 SE 38th St., Bellevue, WA 98006-1350 United States

STOCK TICKER/OTHER:
Stock Ticker: TMUS   Exchange: NAS
Employees: 75,000   Fiscal Year Ends: 12/31
Parent Company: Deutsche Telekom AG

SALARIES/BONUSES:
Top Exec. Salary: $200,962   Bonus: $2,000,000
Second Exec. Salary: Bonus: $
$1,498,462

OTHER THOUGHTS:
Estimated Female Officers or Directors:
Hot Spot for Advancement for Women/Minorities:

TomTom International BV

NAIC Code: 334511

TYPES OF BUSINESS:

Communications Equipment-GPS-Based
Navigation Services
Mapping Technology
PDA & Smartphone Software

GROWTH PLANS/SPECIAL FEATURES:

TomTom is a software company that specializes in maintainin·
a robust digital mapping database. Its mapping software serve·
business-to-business needs, like auto infotainment system·
and third-party app integrations, as well as business-tc
consumer needs, such as mobile mapping apps and portabl·
navigation devices for functions such as navigation, traffi·
monitoring, and autonomous driving visualization.

BRANDS/DIVISIONS/AFFILIATES:

TomTom

CONTACTS: *Note: Officers with more than one job title may be intentionally listed here more than once.*

Harold Goddijn, CEO
Taco Titulaer, CFO
James Joy, General Counsel
Peter-Frans Pauwels, Mgr.-Online Strategy & e-commerce
Richard Piekaar, Mgr.-Investor Rel.
Taco Titulaer, Treas.
Corinne Vigreux, Managing Dir.-Consumer Bus. Unit
Giles Shrimpton, Managing Dir.-TomTom Automotive
Maarten van Gool, Managing Dir.-Licensing Bus. Unit
Lucien Groenhuijzen, Managing Dir.-Places Bus. Unit
Derk Haank, Chmn.
Kirvan Pierson, Head-Bus. Dev. China

FINANCIAL DATA: *Note: Data for latest year may not have been available at press time.*

In U.S. $	2021	2020	2019	2018	2017	2016
Revenue	516,944,400	538,623,500	714,608,100	700,371,200	753,894,400	1,006,842,000
R&D Expense	373,250,600	578,603,300	329,164,200	225,217,700	205,970,700	194,237,300
Operating Income	-95,032,740	-293,396,000	-235,521,400	2,962,412	-72,268,560	9,069,772
Operating Margin %	-.18%	-.54%	-.33%	.00%	-.10%	.01%
SGA Expense	136,932,800	146,551,200	165,385,200	146,202,400	207,803,200	280,776,400
Net Income	-96,523,620	-262,730,700	645,394,700	45,743,510	-197,807,500	12,223,900
Operating Cash Flow	37,509,940	-20,578,820	95,530,380	233,498,200	177,028,800	147,169,100
Capital Expenditure	13,536,330	6,422,468	24,535,500	85,815,100	107,512,600	119,908,800
EBITDA	-11,701,780	-7,599,274	61,066,470	169,308,200	111,580,400	143,590,800
Return on Assets %	-.10%	-.23%	.45%	.03%	-.13%	.01%
Return on Equity %	-.28%	-.49%	.88%	.06%	-.23%	.01%
Debt to Equity	.07%	0.074	0.034	0.033	0.046	0.056

CONTACT INFORMATION:

Phone: 31 20-757-5000   Fax:
Toll-Free:
Address: De Ruyterlade 154, Amsterdam, 1011 AC Netherlands

STOCK TICKER/OTHER:

Stock Ticker: TMOAF   Exchange: PINX
Employees: 4,373   Fiscal Year Ends: 12/31
Parent Company:

SALARIES/BONUSES:

Top Exec. Salary: $ Bonus: $
Second Exec. Salary: $ Bonus: $

OTHER THOUGHTS:

Estimated Female Officers or Directors: 3
Hot Spot for Advancement for Women/Minorities: Y

Toshiba Corporation

www.toshiba.co.jp

NAIC Code: 334413

TYPES OF BUSINESS:

Memory Chip Manufacturing
Infrastructure Systems
Digital Products
Electronic Devices
Power Systems
Retail
Identification
Semiconductor

BRANDS/DIVISIONS/AFFILIATES:

GROWTH PLANS/SPECIAL FEATURES:

Founded in 1875, Toshiba is Japan's largest semiconductor manufacturer and its second-largest diversified industrial conglomerate. After the accounting scandal in 2015, Toshiba reorganized into six major segments: energy systems and solutions; infrastructure systems and solutions; building solutions; retail and printing solutions; storage and electronic devices solutions; and digital solutions. Toshiba is the second-largest manufacturer of NAND flash memory with a market share of 16.5% in 2017, and it concentrates business resources in this area.

CONTACTS: Note: Officers with more than one job title may be intentionally listed here more than once.

Nobuaki Kurumatani, CEO
Norio Sasaki, Vice Chmn.
Hidejiro Shimomitsu, Sr. Exec. VP
Hideo Kitamura, Sr. Exec. VP
Makoto Kubo, Sr. Exec. VP
Satoshi Tsunakawa, Chmn.

FINANCIAL DATA: Note: Data for latest year may not have been available at press time.

In U.S. $	2021	2020	2019	2018	2017	2016
Revenue	22,651,850,000	25,139,950,000	27,392,010,000	29,276,150,000	29,989,140,000	
R&D Expense						
Operating Income	774,265,800	967,517,100	335,842,500	639,157,600	733,677,000	
Operating Margin %	.03%	.04%	.01%	.02%	.02%	
SGA Expense	5,333,410,000	5,839,581,000	6,412,711,000	6,514,187,000	6,894,178,000	
Net Income	845,305,500	-850,140,900	7,514,506,000	5,962,704,000	-7,161,547,000	
Operating Cash Flow	1,076,424,000	-1,054,198,000	925,949,300	277,121,100	994,979,200	
Capital Expenditure	1,031,986,000	1,002,433,000	1,025,193,000	1,482,557,000	1,340,366,000	
EBITDA	1,803,975,000	277,996,100	741,545,600	1,704,331,000	1,679,465,000	
Return on Assets %	.03%	- .03%	.23%	.18%	- .20%	
Return on Equity %	.11%	- .10%	.90%	6.99%		
Debt to Equity	.40%	0.306	0.053	0.499		

CONTACT INFORMATION:

Phone: 81 334572096   Fax: 81 354449202
Toll-Free:
Address: 1-1, Shibaura 1-chome, Minato-ku, Tokyo, 105-8001 Japan

STOCK TICKER/OTHER:

Stock Ticker: TOSBF   Exchange: PINX
Employees: 128,697   Fiscal Year Ends: 03/31
Parent Company:

SALARIES/BONUSES:

Top Exec. Salary: $ Bonus: $
Second Exec. Salary: $ Bonus: $

OTHER THOUGHTS:

Estimated Female Officers or Directors:
Hot Spot for Advancement for Women/Minorities:

Sales, profits and employees may be estimates. Financial information, benefits and other data can change quickly and may vary from those stated here.

TOT pcl

NAIC Code: 517110

www.tot.co.th

TYPES OF BUSINESS:

Local Telephone Services
International Calling Services
Payphone Installation & Operation
Internet Service Provider

BRANDS/DIVISIONS/AFFILIATES:

Thai Ministry of Finance

GROWTH PLANS/SPECIAL FEATURES:

TOT pcl, 100% owned by the Thai Ministry of Finance, wa
organized in 2002 as the national telecommunication
company of Thailand. TOT offers fixed line, mobile, public an
international voice services; internet (including fiber optic
data, international service and internet managed servic
multimedia and content services such as eLearning, net-ca
eConferencing, mail, smart SMS (short message service
cloud applications, eMarket, iptv (internet protocol televisio
and Netlog services; and other services such as payment, car
ad billing, contact center and procurement. TOT provides thes
services for small, medium and large enterprises in th
healthcare, finance, energy/utilities, retail, manufacturin
automotive, transportation/distribution, media/entertainmer
insurance, technology, education, travel/hospitality and re
estate sectors.

CONTACTS: *Note: Officers with more than one job title may be intentionally listed here more than once.*

Montchai Noosong, CEO
Morakot Thienmontree, Sr. Exec. VP-Prod. Dev.
Nutnucha Chaipresert, Sr. Exec. VP-Legal & Beneficial Assurance Mgmt.
Preeya Danchaivijit, Sr. Exec. VP-Finance
Phithakphong Tupinprom, Sr. Exec. VP-Enterprise Effectiveness
Chate Phanchan, Sr. Exec. VP-Mobile Bus.
Rungsun Channarukul, Sr. Exec. VP-Wireless & Universal Svcs. Obligation
Kajhonsak Eiamsopa, Sr. Exec. VP-Metropolitan Sale & Customer Svcs.

FINANCIAL DATA: *Note: Data for latest year may not have been available at press time.*

In U.S. $	2021	2020	2019	2018	2017	2016
Revenue						
R&D Expense						
Operating Income						
Operating Margin %						
SGA Expense						
Net Income						
Operating Cash Flow						
Capital Expenditure						
EBITDA						
Return on Assets %						
Return on Equity %						
Debt to Equity						

CONTACT INFORMATION:

Phone: 66-2240-0701   Fax:
Toll-Free:
Address: 89/2 Moo 3 Chaeng Wattana Rd, Thungsong-Hong, Laks, Bangkok, 10210 Thailand

STOCK TICKER/OTHER:

Stock Ticker: Government-Owned   Exchange:
Employees: Fiscal Year Ends: 12/31
Parent Company: Thai Ministry of Finance

SALARIES/BONUSES:

Top Exec. Salary: $ Bonus: $
Second Exec. Salary: $ Bonus: $

OTHER THOUGHTS:

Estimated Female Officers or Directors: 7
Hot Spot for Advancement for Women/Minorities: Y

Sales, profits and employees may be estimates. Financial information, benefits and other data can change quickly and may vary from those stated here.

Total Access Communication PCL

www.dtac.co.th

NAIC Code: 517210

TYPES OF BUSINESS:
Telecommunications Services
Telecommunications
Wireless
Internet
Wi-Fi
Voice
Smartphones
Tablets

BRANDS/DIVISIONS/AFFILIATES:
Telenor Group
dtac
dtac Reward

GROWTH PLANS/SPECIAL FEATURES:
Total Access Communication PCL is a telecommunications company. The company's operating segment includes Mobile telephone service & related services and Sales of handsets & starter kits. It serves both individual and business customers. The majority of revenue stems from providing mobile services. Within mobile services, a majority of revenue stems from both voice and data services. Revenue is on a subscription basis and in terms of subscribers, the majority are prepaid customers. It operates in a single geographical segment, which is Thailand.

CONTACTS: *Note: Officers with more than one job title may be intentionally listed here more than once.*
Sharad Mehrotra, CEO
Rajiv Bawa, Chief Business Officer
Nakul Sehgal, CFO
How Lih Ren, CMO
Nardrerdee Arj-Harnwongse, Chief People Officer
Prathet Tankuranun, CTO
Saranya Sanghiran, Chief Strategy Officer
Darmp Sukontasap, Chief Corp. Affairs Officer
Chaiyod Chirabowornkul, Chief Customer Officer
Boonchai Bencharongkul, Chmn.

FINANCIAL DATA: *Note: Data for latest year may not have been available at press time.*

In U.S. $	2021	2020	2019	2018	2017	2016
Revenue	2,288,770,000	2,218,351,000	2,284,468,000	2,119,045,000	2,203,062,000	2,321,351,000
R&D Expense						
Operating Income	194,599,200	238,520,000	270,930,200	-132,001,500	103,738,100	109,597,500
Operating Margin %	.09%	.11%	.12%	-.06%	.05%	.05%
SGA Expense	386,849,100	404,917,700	434,291,900	639,136,700	430,861,200	500,875,200
Net Income	94,453,500	143,741,000	152,600,400	-122,957,800	59,526,450	58,706,170
Operating Cash Flow	745,674,500	730,130,300	493,523,300	513,809,700	769,200,100	749,391,100
Capital Expenditure	622,797,900	585,738,800	501,240,100	714,732,900	562,258,100	534,485,800
EBITDA	827,961,000	843,310,400	839,632,500	578,149,400	851,415,800	765,507,200
Return on Assets %	.02%	.03%	.03%	-.03%	.02%	.02%
Return on Equity %	.15%	.21%	.23%	-.17%	.08%	.08%
Debt to Equity	3.20%	2.692	2.125	1.716	1.608	1.805

CONTACT INFORMATION:
Phone: 66 22028000   Fax: 66 22028102
Toll-Free:
Address: 319 Phayathai Rd., Chamchuri Sq. Bldg., Fl. 22-41, Bangkok, 10330 Thailand

STOCK TICKER/OTHER:
Stock Ticker: TACYY   Exchange: PINX
Employees: 5,573   Fiscal Year Ends: 12/31
Parent Company: Telenor Group

SALARIES/BONUSES:
Top Exec. Salary: $ Bonus: $
Second Exec. Salary: $ Bonus: $

OTHER THOUGHTS:
Estimated Female Officers or Directors: 4
Hot Spot for Advancement for Women/Minorities: Y

Sales, profits and employees may be estimates. Financial information, benefits and other data can change quickly and may vary from those stated here.

TPG Telecom Limited

NAIC Code: 517110

www.tpgtelecom.com.au

TYPES OF BUSINESS:

Wired Telecommunications Carriers
Telecommunications Services
Mobile Networks
Fixed Networks
Next-Generation Networks

GROWTH PLANS/SPECIAL FEATURES:

TPG Telecom is Australia's third-largest integrated teleco
services provider. It offers broadband, telephony, mobile a
networking solutions catering to all market segme
(consumer, small business, corporate and wholesa
government). The company has grown significantly since 20
both via organic growth and acquisitions, and in July 20
merged with Vodafone Australia. It owns an extensive stable
infrastructure assets.

BRANDS/DIVISIONS/AFFILIATES:

Vodafone
TPG
iiNet
AAPT
Internode
Lebara
felix

CONTACTS: *Note: Officers with more than one job title may be intentionally listed here more than once.*

Inaki Berroeta, CEO
Stephen Banfield, CFO
Rob James, CIO

FINANCIAL DATA: *Note: Data for latest year may not have been available at press time.*

In U.S. $	2021	2020	2019	2018	2017	2016
Revenue		3,373,816,576	1,747,855,232	1,760,413,440	1,615,135,360	1,742,019,584
R&D Expense						
Operating Income						
Operating Margin %						
SGA Expense						
Net Income		574,712,192	122,619,368	280,020,864	268,335,392	276,937,184
Operating Cash Flow						
Capital Expenditure						
EBITDA						
Return on Assets %						
Return on Equity %						
Debt to Equity						

CONTACT INFORMATION:

Phone: 61 299644646   Fax: 61 298889148
Toll-Free:
Address: Level 1, 177 Pacific Hwy., Sydney, NSW 2060 Australia

STOCK TICKER/OTHER:

Stock Ticker: TPPTY   Exchange: PINX
Employees: 5,000   Fiscal Year Ends: 07/31
Parent Company:

SALARIES/BONUSES:

Top Exec. Salary: $ Bonus: $
Second Exec. Salary: $ Bonus: $

OTHER THOUGHTS:

Estimated Female Officers or Directors:
Hot Spot for Advancement for Women/Minorities:

Trilogy International Partners Inc   www.trilogy-international.com

NAIC Code: 517210

TYPES OF BUSINESS:
Wireless Telecommunications Carriers (except Satellite)
Wireless Communications
LTE Coverage
Wireless Networks
Fixed Broadband
Wireless Mobile

BRANDS/DIVISIONS/AFFILIATES:
2degrees
NuevaTel
Trilogy International Partners LLC

GROWTH PLANS/SPECIAL FEATURES:
Trilogy International Partners Inc together with its subsidiaries in New Zealand and Bolivia, is a provider of wireless voice and data communications including local, international long distance and roaming services, for both subscribers and international visitors roaming on its networks. The company also provides fixed broadband communications to residential and enterprise customers in New Zealand. It has two reporting segments, New Zealand and Bolivia. Services are provided to subscribers on both a postpaid and prepaid basis. Its networks support several digital technologies: GSM (NuevaTel only), 3G and 4G LTE.

CONTACTS: Note: Officers with more than one job title may be intentionally listed here more than once.
Tomas Perez, CEO, Subsidiary
Mark Aue, CEO, Subsidiary
Bradley Horwitz, CEO
Erik Mickels, CFO
John Stanton, Chairman of the Board
Theresa Gillespie, Co-Founder
Scott Morris, Senior VP

FINANCIAL DATA: Note: Data for latest year may not have been available at press time.

In U.S. $	2021	2020	2019	2018	2017	2016
Revenue	653,564,000	610,299,000	693,927,000	798,175,000	778,900,000	952,689
R&D Expense						
Operating Income	-4,919,000	-7,943,000	17,489,000	22,931,000	34,866,000	278,793
Operating Margin %	-.01%	-.01%	.03%	.03%	.04%	.29%
SGA Expense	212,721,000	192,581,000	204,834,000	227,233,000	224,758,000	673,896
Net Income	-144,689,000	-47,787,000	2,878,000	-20,205,000	-15,337,000	-7,034,023
Operating Cash Flow	48,702,000	40,876,000	45,671,000	74,602,000	65,013,000	-529,215
Capital Expenditure	99,573,000	77,331,000	115,905,000	83,638,000	95,631,000	
EBITDA	-22,876,000	96,893,000	139,059,000	130,961,000	144,786,000	278,793
Return on Assets %	-.16%	-.05%	.00%	-.03%	-.02%	-.03%
Return on Equity %						-3.86%
Debt to Equity						

CONTACT INFORMATION:
Phone: 425 458-5900   Fax: 425 458-5999
Toll-Free:
Address: 155 - 108th Avenue NE, Bellevue, WA 98004 United States

STOCK TICKER/OTHER:
Stock Ticker: TRL   Exchange: TSE
Employees: 1,663   Fiscal Year Ends: 12/31
Parent Company:

SALARIES/BONUSES:
Top Exec. Salary: $ Bonus: $
Second Exec. Salary: $ Bonus: $

OTHER THOUGHTS:
Estimated Female Officers or Directors:
Hot Spot for Advancement for Women/Minorities:

Sales, profits and employees may be estimates. Financial information, benefits and other data can change quickly and may vary from those stated here.

Trimble Inc

NAIC Code: 334511

www.trimble.com

TYPES OF BUSINESS:

GPS Technologies
Surveying & Mapping Equipment
Navigation Tools
Autopilot Systems
Data Collection Products
Fleet Management Systems
Outdoor Recreation Information Service
Telecommunications & Automotive Components

BRANDS/DIVISIONS/AFFILIATES:

Applanix
AXIO-NET GmbH
Beena Vision Systems Inc
e-Builder
HHK Datentechnik GmbH
Innovative Software Engineering
MyTopo
Viewpoint

GROWTH PLANS/SPECIAL FEATURES:

Trimble Inc provides location-based solutions that are used
global positioning system (GPS), laser, optical and inerti
technologies. Its products portfolio includes 3D laser scannin
flow and application control systems, monitoring system
water management, and navigation infrastructure. It als
manufactures laser and optics-based products, and GP
products. The company serves various industries whic
include agriculture, architecture, civil engineering, survey ar
land administration, construction, geospatial among other
The company operates in four reportable segments namel
Buildings and Infrastructure, Geospatial, Resources ar
Utilities, and Transportation. It derives most of its revenue
from the US and Europe with the rest coming from the As
Pacific and other markets.

CONTACTS: Note: Officers with more than one job title may be intentionally listed here more than once.

Robert Painter, CEO
David Barnes, CFO
Steven Berglund, Chairman of the Board
Julie Shepard, Chief Accounting Officer
James Kirkland, General Counsel
Bryn Fosburgh, Senior VP, Divisional
Darryl Matthews, Senior VP, Divisional
Ronald Bisio, Senior VP, Divisional
James Langley, Senior VP, Divisional

FINANCIAL DATA: Note: Data for latest year may not have been available at press time.

In U.S. $	2021	2020	2019	2018	2017	2016
Revenue	3,659,100,000	3,147,700,000	3,264,300,000	3,108,400,000	2,646,500,000	2,362,100,000
R&D Expense	536,600,000	475,900,000	469,700,000	446,100,000	370,200,000	349,600,000
Operating Income	571,300,000	445,600,000	402,700,000	328,900,000	242,600,000	192,000,000
Operating Margin %	.16%	.14%	.12%	.11%	.09%	.08%
SGA Expense	875,900,000	767,900,000	834,800,000	829,600,000	701,800,000	630,700,000
Net Income	492,700,000	389,900,000	514,300,000	282,800,000	118,400,000	132,400,000
Operating Cash Flow	750,500,000	672,000,000	585,000,000	486,700,000	429,700,000	431,100,000
Capital Expenditure	46,100,000	56,800,000	69,000,000	67,600,000	43,700,000	26,000,000
EBITDA	819,900,000	670,100,000	634,400,000	566,400,000	456,800,000	389,800,000
Return on Assets %	.07%	.06%	.08%	.06%	.03%	.04%
Return on Equity %	.13%	.12%	.18%	.11%	.05%	.06%
Debt to Equity	.36%	0.389	0.557	0.64	0.325	0.212

CONTACT INFORMATION:

Phone: 408 481-8000   Fax: 408 481-2218
Toll-Free: 800-874-6253
Address: 935 Stewart Dr., Sunnyvale, CA 94085 United States

STOCK TICKER/OTHER:

Stock Ticker: TRMB
Employees: 11,931
Parent Company:

Exchange: NAS
Fiscal Year Ends: 12/31

SALARIES/BONUSES:

Top Exec. Salary: $882,692   Bonus: $
Second Exec. Salary: Bonus: $16,000
$800,000

OTHER THOUGHTS:

Estimated Female Officers or Directors: 5
Hot Spot for Advancement for Women/Minorities: Y

Sales, profits and employees may be estimates. Financial information, benefits and other data can change quickly and may vary from those stated here.

TruConnect Communications Inc   www.truconnect.com

NAIC Code: 517110

TYPES OF BUSINESS:
Local Exchange Carrier
Wireless Telecommunications Services
Internet
Broadband Services
Telephone Services
Telephone Devices

GROWTH PLANS/SPECIAL FEATURES:
TruConnect Communications, Inc. is a U.S. wireless telecommunications service provider. The company offers Lifeline and pay-as-you-go affordable plans. Lifeline is a state and federal government program that provides wireless internet and mobile broadband service to low-income consumers. This program is only available to consumers who can provide documentation for eligibility based on either income or participation in public assistance programs such as Supplemental Nutrition Assistance Program (SNAP) and Medicaid. Only one Lifeline service is allowed per household. TruConnect currently offers Lifeline service to residents in select states. Phones for sale by TruConnect range from $50 to $200, and consumers can also bring their own devices for service plans. Phones can be picked up at TruConnect corporate stores, branded partner locations and through authorized dealers.

BRANDS/DIVISIONS/AFFILIATES:

CONTACTS: Note: Officers with more than one job title may be intentionally listed here more than once.
Nathan Jonson, Co-CEO
Matthew Johnson, Co-CEO
Brian Kushner, Pres.
Lucy Sung, COO
Samuel Braff, Chief Product Officer
Aleksandr Gudkov, CTO
Danielle Perry, CIO
John Debus, Treas.
Shahin Sazej, Sr. VP-Systems
Nathan Johnson, Chmn.

FINANCIAL DATA: Note: Data for latest year may not have been available at press time.

In U.S. $	2021	2020	2019	2018	2017	2016
Revenue	153,972,000	148,050,000	141,000,000	138,000,000	132,300,000	126,000,000
R&D Expense						
Operating Income						
Operating Margin %						
SGA Expense						
Net Income						
Operating Cash Flow						
Capital Expenditure						
EBITDA						
Return on Assets %						
Return on Equity %						
Debt to Equity						

CONTACT INFORMATION:
Phone: 323-776-9880   Fax:
Toll-Free:
Address: 1149 S. Hill St., H-400, Los Angeles, CA 90015 United States

STOCK TICKER/OTHER:
Stock Ticker: Private   Exchange:
Employees: 325   Fiscal Year Ends: 12/31
Parent Company: TSC Acquisition Corporation

SALARIES/BONUSES:
Top Exec. Salary: $ Bonus: $
Second Exec. Salary: $ Bonus: $

OTHER THOUGHTS:
Estimated Female Officers or Directors:
Hot Spot for Advancement for Women/Minorities:

Sales, profits and employees may be estimates. Financial information, benefits and other data can change quickly and may vary from those stated here.

True Corporation Public Company Limited true.listedcompany.com
NAIC Code: 517110

TYPES OF BUSINESS:
Wired Telecommunications Carriers
Telecommunications
Mobile Services
Streaming Services
Broadband and Internet Services
Payment Solutions
Entertainment Services
Internet of Things

BRANDS/DIVISIONS/AFFILIATES:
TrueMove H
TrueVisions
TrueOnline
TrueMoney
TrueMusic
TrueID
Charoen Pokphand Group

GROWTH PLANS/SPECIAL FEATURES:
True Corp PCL is a triple-play telecommunications company. operates through three business segments: TrueOnlin TrueMove H, and TrueVisions. The majority of revenue derived from TrueMove H, the company's mobile servic product offering. The majority of the company's customer ba in its mobile services division is considered prepai TrueOnline comprises the company's fixed-line telephone a public phone services, Internet, and business data service TrueVisions is the group's pay-TV service. It is delivered v digital direct-to-home satellite and hybrid fiber-coaxial cab network platforms. It operates in a single geographical are which is Thailand.

CONTACTS: Note: Officers with more than one job title may be intentionally listed here more than once.
Yupa Leewongcharoen, CFO
Dhanin Chearavanont, Chmn.

FINANCIAL DATA: Note: Data for latest year may not have been available at press time.

In U.S. $	2021	2020	2019	2018	2017	2016
Revenue	4,043,210,000	3,890,018,000	3,966,887,000	4,554,470,000	3,926,801,000	3,510,250,000
R&D Expense						
Operating Income	349,374,000	299,639,600	149,194,400	574,328,600	115,889,100	-13,836,560
Operating Margin %	.09%	.08%	.04%	.13%	.03%	.00%
SGA Expense	725,258,600	757,899,300	794,592,800	1,002,952,000	937,395,600	833,412,900
Net Income	-40,202,900	29,507,400	158,647,100	197,990,200	15,516,210	-79,210,470
Operating Cash Flow	997,914,900	1,060,285,000	405,960,000	1,067,518,000	140,538,800	227,867,900
Capital Expenditure	1,656,475,000	2,106,531,000	1,306,276,000	1,964,689,000	1,339,178,000	1,322,416,000
EBITDA	1,771,238,000	1,755,480,000	1,254,683,000	1,547,015,000	1,239,625,000	856,569,100
Return on Assets %	.00%	.00%	.01%	.01%	.00%	-.01%
Return on Equity %	-.02%	.01%	.04%	.05%	.00%	-.03%
Debt to Equity	3.65%	3.20	1.42	0.511	0.392	0.477

CONTACT INFORMATION:
Phone: 66 2858-2515   Fax: 66 2858-2519
Toll-Free:
Address: 18 True Tower, Ratchadapisek Rd., Bangkok, 10310 Thailand

STOCK TICKER/OTHER:
Stock Ticker: TCPFF   Exchange: PINX
Employees: 3,862   Fiscal Year Ends: 12/31
Parent Company: Charoen Pokphand Group

SALARIES/BONUSES:
Top Exec. Salary: $ Bonus: $
Second Exec. Salary: $ Bonus: $

OTHER THOUGHTS:
Estimated Female Officers or Directors:
Hot Spot for Advancement for Women/Minorities:

TTEC Holdings Inc

www.ttec.com

NAIC Code: 541800

TYPES OF BUSINESS:

Customer Support Services
Customer Experience
Digital Consulting
Customer Acquisition

GROWTH PLANS/SPECIAL FEATURES:

TTEC Holdings provides customer engagement management tools and services. The company operates through four operating segments that are organized into two groups, TTEC Digital and TTEC Engage. TTEC Digital is engaged in building and implementing cloud-based and on-premises customer experience tools that enable clients to develop customer engagement strategies. TTEC Engage focuses on delivering sales and marketing solutions to help clients boost their revenue as well as on managing customer's front-to-back office processes to optimize the customer experience. TTEC Engage contributes the vast majority of the company's revenue, and most of the sales are derived from North America, followed by Asia-Pacific and India.

BRANDS/DIVISIONS/AFFILIATES:

TTEC Digital
TTEC Engage

CONTACTS: Note: Officers with more than one job title may be intentionally listed here more than once.

Kenneth Tuchman, CEO
Dustin Semach, CFO
Margaret McLean, Chief Risk Officer
Regina Paolillo, COO
Judi Hand, Executive VP
George Demou, President, Divisional

FINANCIAL DATA: Note: Data for latest year may not have been available at press time.

In U.S. $	2021	2020	2019	2018	2017	2016
Revenue	2,273,062,000	1,949,248,000	1,643,704,000	1,509,171,000	1,477,365,000	1,275,258,000
R&D Expense						
Operating Income	232,253,000	213,765,000	129,191,000	99,637,000	120,476,000	89,194,000
Operating Margin %	.10%	.11%	.08%	.07%	.08%	.07%
SGA Expense	239,994,000	203,902,000	202,540,000	182,428,000	182,314,000	175,797,000
Net Income	140,970,000	118,648,000	77,164,000	35,817,000	7,256,000	33,678,000
Operating Cash Flow	251,296,000	271,920,000	237,989,000	168,345,000	113,152,000	111,830,000
Capital Expenditure	60,358,000	59,772,000	60,776,000	43,450,000	51,958,000	50,832,000
EBITDA	316,974,000	266,619,000	198,610,000	154,091,000	167,128,000	126,916,000
Return on Assets %	.08%	.08%	.06%	.03%	.01%	.04%
Return on Equity %	.29%	.27%	.20%	.10%	.02%	.09%
Debt to Equity	1.64%	1.087	0.997	0.817	0.967	0.612

CONTACT INFORMATION:

Phone: 303 397-8100   Fax: 303 397-8668
Toll-Free: 800-835-3832
Address: 9197 S. Peoria St., Englewood, CO 80112 United States

STOCK TICKER/OTHER:

Stock Ticker: TTEC
Employees: 65,000
Parent Company:

Exchange: NAS
Fiscal Year Ends: 12/31

SALARIES/BONUSES:

Top Exec. Salary: $425,000   Bonus: $
Second Exec. Salary: $400,000   Bonus: $

OTHER THOUGHTS:

Estimated Female Officers or Directors: 4
Hot Spot for Advancement for Women/Minorities: Y

Sales, profits and employees may be estimates. Financial information, benefits and other data can change quickly and may vary from those stated here.

Turk Telekomunikasyon AS

NAIC Code: 517110

www.turktelekom.com.tr

TYPES OF BUSINESS:

Telecommunications Services

GROWTH PLANS/SPECIAL FEATURES:

Turk Telekomunikasyon AS is a multi-play telecommunication company that provides broadband, mobile, and TV services f companies and individuals. From a product perspectiv revenue comes from mobile and broadband products, whic are typically offered on a subscription basis. Within mobi however, there is roughly an even split between postpaid an prepaid customers. Additionally, Turk Telekomunikasyon own mobile and fiber infrastructure. The company generates th vast majority of its revenue in Turkey.

BRANDS/DIVISIONS/AFFILIATES:

Ojer Telekomunikasyon AS
TTNET
iNNOVA Bilisim Cozumleri AS
Sebit Egitim ve Bilgi Teknolojileri AS
Argela Software and Information Technologies
Avea
AssisTT
Turk Telekom International

CONTACTS: Note: *Officers with more than one job title may be intentionally listed here more than once.*

Rami Aslan, CEO
Necdet Mert Basar, CMO
Bahattin Aydin, Chief Human Resources Officer
Mehmet Komurcu, VP-Legal
Celalettin Dincer, VP-Oper.
Ramazan Demir, VP-Bus. Dev. & Strategy
Erem Demircan, VP-Corp. Comm.
Mustafa Uysal, Interim VP-Finance
Kamil Gokhan Bozkurt, CEO-Turk Telekom
Sukru Kutlu, VP-Regulation & Support Svcs.
Aydin Camlibel, VP-Sales
Nazif Burca, Head-Internal Audit
Mohammed Hariri, Chmn.
Mehmet Candan Toros, VP-Intl & Wholesale

FINANCIAL DATA: Note: *Data for latest year may not have been available at press time.*

In U.S. $	2021	2020	2019	2018	2017	2016
Revenue	1,910,654,000	1,577,054,000	1,318,841,000	1,138,986,000	1,011,247,000	898,025,100
R&D Expense	15,264,800	13,775,660	9,281,574	8,238,414	6,953,863	5,230,352
Operating Income	579,969,500	438,594,000	364,036,600	275,513,600	225,838,300	152,063,700
Operating Margin %	.30%	.28%	.28%	.24%	.22%	.17%
SGA Expense	299,906,300	265,379,400	268,613,700	233,826,100	229,204,400	245,271,200
Net Income	321,190,700	177,160,700	134,173,800	-77,560,300	63,303,860	-40,380,650
Operating Cash Flow	874,191,500	760,322,800	623,415,100	423,069,500	331,377,600	274,972,200
Capital Expenditure	483,649,200	388,122,700	275,705,100	226,633,500	235,454,200	260,191,700
EBITDA	783,412,700	641,696,600	547,606,900	156,686,400	283,078,900	169,848,400
Return on Assets %	.11%	.08%	.06%	- .04%	.04%	- .03%
Return on Equity %	.44%	.30%	.28%	- .23%	.29%	- .17%
Debt to Equity	1.59%	1.256	1.522	1.766	3.075	3.854

CONTACT INFORMATION:

Phone: 90-444-1-444   Fax: 90 312-306-07-32
Toll-Free:
Address: Turgut Ozal Bulvari, Ankara, 06103 Turkey

STOCK TICKER/OTHER:

Stock Ticker: TRKNF   Exchange: PINX
Employees: 33,417   Fiscal Year Ends: 12/31
Parent Company: Ojer Telekomunikasyon AS

SALARIES/BONUSES:

Top Exec. Salary: $ Bonus: $
Second Exec. Salary: $ Bonus: $

OTHER THOUGHTS:

Estimated Female Officers or Directors:
Hot Spot for Advancement for Women/Minorities:

Turkcell Iletisim Hizmetleri AS   www.turkcell.com.tr/en/aboutus
NAIC Code: 517210

TYPES OF BUSINESS:
Cell Phone Service
Value-Added Services
Mobile Media Content
Mobile Internet

GROWTH PLANS/SPECIAL FEATURES:
Turkcell Iletisim Hizmetleri AS provides mobile telephone services in Turkey. The firm provides mobile voice and data services over its global system for mobile communications networks. Currently, Turkcell provides service to subscribers in 208 countries through commercial roaming agreements with operators. Turkcell has 26 million prepaid subscribers and more than 9 million postpaid subscribers. The firm also has investments in Azerbaijan, Georgia, Kazakhstan, Moldova, and Ukraine.

BRANDS/DIVISIONS/AFFILIATES:
Turkcell Holding AS
Cukurova Telecom Holdings Limited
Telia Sonera Finland Oyj
Cukurova Finance International Limited
Alfa Telecom Turkey Limited

CONTACTS: *Note: Officers with more than one job title may be intentionally listed here more than once.*
Kaan TerzioÄŸlu, CEO
Osman Yilmaz, Exec. VP-Finance
Omer Barbaros Yis, Exec. VP-Marketing
Serkan Ozturk, Mngr.-IT & Communication
Tayfun Cataltepe, Exec. VP-Corp. Strategy & Regulations
Selen Kocabas, Assistant Gen. Mgr.-Corp. Mktg. & Sales Group
Burak Sevilengul, Assistant Gen. Mgr.-Retail Mktg.
Hulusi Acar, Assistant Gen. Mgr.-Retail Sales
Ekrem Yener, Assistant Gen. Mgr.-Intl Expansion
Iter Terzioglu, Assistant Gen. Mgr.-Strategic Projects
Bulent Aksu, Chmn.
Lale Saral Develioglu, Assistant Gen. Mgr.-Intl Affairs

FINANCIAL DATA: *Note: Data for latest year may not have been available at press time.*

In U.S. $	2021	2020	2019	2018	2017	2016
Revenue	2,002,506,000	1,622,481,000	1,401,350,000	1,187,017,000	982,955,800	796,394,300
R&D Expense						
Operating Income	436,869,900	370,453,100	318,807,800	270,178,000	204,459,400	146,480,700
Operating Margin %	.22%	.23%	.23%	.23%	.21%	.18%
SGA Expense	148,852,500	117,206,500	128,685,500	126,727,800	144,387,600	133,856,800
Net Income	280,474,700	236,209,900	180,985,800	112,670,700	110,332,900	83,181,220
Operating Cash Flow	1,112,023,000	729,899,800	503,214,200	325,003,900	172,890,000	33,846,800
Capital Expenditure	550,609,600	433,815,600	335,391,200	302,733,700	217,666,700	202,766,400
EBITDA	708,629,600	634,993,500	512,350,300	418,830,100	301,029,600	253,975,800
Return on Assets %	.08%	.09%	.07%	.05%	.06%	.05%
Return on Equity %	.23%	.22%	.19%	.13%	.13%	.10%
Debt to Equity	1.24%	0.787	0.702	0.824	0.551	0.433

CONTACT INFORMATION:
Phone: 90 2123131244   Fax: 90 2165044058
Toll-Free:
Address: Aydinevler Mahallesi Inonu Caddesi No. 20, B Block, Maltepe, Istanbul, 34854 Turkey

STOCK TICKER/OTHER:
Stock Ticker: TKC
Employees: 25,638
Parent Company:
Exchange: NYS
Fiscal Year Ends: 12/31

SALARIES/BONUSES:
Top Exec. Salary: $ Bonus: $
Second Exec. Salary: $ Bonus: $

OTHER THOUGHTS:
Estimated Female Officers or Directors: 4
Hot Spot for Advancement for Women/Minorities: Y

Sales, profits and employees may be estimates. Financial information, benefits and other data can change quickly and may vary from those stated here.

Ubiquiti Inc

www.ubnt.com

NAIC Code: 334220

TYPES OF BUSINESS:

Radio & Wireless Communication, Manufacturing

BRANDS/DIVISIONS/AFFILIATES:

GROWTH PLANS/SPECIAL FEATURES:

Ubiquiti Inc is a wireless and wireline network equipme
provider for small Internet service providers and small- a
midsize-business integrators. Its product is based on tw
primary categories namely Service Provider Technology a
Enterprise Technology. The company generates maximu
revenue from Enterprise Technology. Geographically,
derives a majority of revenue from North America and also h
a presence in Europe, the Middle East and Africa; Asia Paci
and South America.

CONTACTS: *Note: Officers with more than one job title may be intentionally listed here more than once.*

Robert Pera, CEO
Kevin Radigan, Chief Accounting Officer

FINANCIAL DATA: *Note: Data for latest year may not have been available at press time.*

In U.S. $	2021	2020	2019	2018	2017	2016
Revenue	1,898,094,000	1,284,500,000	1,161,733,000	1,016,861,000	865,268,000	666,395,000
R&D Expense	116,171,000	89,405,000	82,070,000	74,324,000	69,094,000	57,765,000
Operating Income	742,592,000	478,198,000	412,297,000	326,127,000	289,761,000	233,761,000
Operating Margin %	.39%	.37%	.35%	.32%	.33%	.35%
SGA Expense	53,513,000	40,569,000	43,237,000	43,121,000	36,853,000	33,269,000
Net Income	616,584,000	380,297,000	322,694,000	196,290,000	257,506,000	213,616,000
Operating Cash Flow	612,022,000	460,284,000	259,258,000	332,047,000	112,036,000	197,508,000
Capital Expenditure	18,325,000	30,619,000	51,684,000	9,115,000	7,232,000	6,248,000
EBITDA	754,692,000	485,893,000	401,853,000	333,437,000	296,864,000	239,798,000
Return on Assets %	.76%	.47%	.34%	.20%	.30%	.32%
Return on Equity %			1.56%	.43%	.49%	.50%
Debt to Equity	184.99%		4.681	1.458	0.402	0.435

CONTACT INFORMATION:

Phone: 408 942-3085   Fax: 408 351-4973
Toll-Free:
Address: 91 E. Tasman Drive, San Jose, CA 95134 United States

STOCK TICKER/OTHER:

Stock Ticker: UI   Exchange: NYS
Employees: 1,021   Fiscal Year Ends: 06/30
Parent Company:

SALARIES/BONUSES:

Top Exec. Salary: $420,000   Bonus: $100,000
Second Exec. Salary: $ Bonus: $

OTHER THOUGHTS:

Estimated Female Officers or Directors:
Hot Spot for Advancement for Women/Minorities:

Uniden Holdings Corporation

NAIC Code: 334220

TYPES OF BUSINESS:

Communications Equipment-Wireless Devices
Wireless Communications Devices
Electronics
Manufacture
Marine Electronics
Two-Way Radios
Digital Tuners
HD TV

BRANDS/DIVISIONS/AFFILIATES:

Uniden America Corporation
Uniden Australia Proprietary Limited
Uniden Vietnam Limited
Uniden Japan Corporation
Attowave Co Ltd

GROWTH PLANS/SPECIAL FEATURES:

Uniden Holdings Corporation, based in Japan, provides wireless communications devices through its global companies. The group manufactures and markets wireless communication equipment and consumer products such as CB radios, marine electronics, two-way radios, scanners, high-definition televisions, digital tuners for homes and cars and more. Group companies include Uniden America Corporation, Uniden Australia Proprietary Limited, Uniden Vietnam Limited, Uniden Japan Corporation, and Attowave Co. Ltd.

CONTACTS: Note: Officers with more than one job title may be intentionally listed here more than once.

Hidero Fujimoto, Chmn.

FINANCIAL DATA: Note: Data for latest year may not have been available at press time.

In U.S. $	2021	2020	2019	2018	2017	2016
Revenue	174,850,006	185,786,000	144,087,300	137,226,000	117,642,000	117,114,954
R&D Expense						
Operating Income						
Operating Margin %						
SGA Expense						
Net Income	33,191,558	-4,284,380	16,834,440	16,032,800	13,282,000	-42,439,908
Operating Cash Flow						
Capital Expenditure						
EBITDA						
Return on Assets %						
Return on Equity %						
Debt to Equity						

CONTACT INFORMATION:

Phone: 81 3-5543-2800   Fax: 81-3-5543-2921
Toll-Free:
Address: 2-12-7 Hatchobori, Chuo-ku, Tokyo, 104-8512 Japan

STOCK TICKER/OTHER:

Stock Ticker: 6815   Exchange: Tokyo
Employees: 833   Fiscal Year Ends: 03/31
Parent Company:

SALARIES/BONUSES:

Top Exec. Salary: $ Bonus: $
Second Exec. Salary: $ Bonus: $

OTHER THOUGHTS:

Estimated Female Officers or Directors:
Hot Spot for Advancement for Women/Minorities:

Sales, profits and employees may be estimates. Financial information, benefits and other data can change quickly and may vary from those stated here.

United Online Inc

NAIC Code: 517110

www.untd.com

TYPES OF BUSINESS:

Internet Service Provider

BRANDS/DIVISIONS/AFFILIATES:

B Riley Financial Inc
NetZero
Juno

CONTACTS: *Note: Officers with more than one job title may be intentionally listed here more than once.*

Edward Zinser, CFO
Howard Phanstiel, Director
Mark Harrington, Executive VP
Shahir Fakiri, General Manager, Divisional
Bryant Riley, Chmn.-B. Riley Financial

GROWTH PLANS/SPECIAL FEATURES:

United Online, Inc., a subsidiary of B. Riley Financial, Inc
provides consumer products and internet and media service
over the internet. These products feature value-priced intern
access through the NetZero and Juno brands. The brands off
a variety of plans, and customers can choose the option th
best meets their connection needs. NetZero offers mobile ar
home wireless broadband services. NetZero mobile broadbar
provides high-speed, affordable internet access to on-the-g
consumers. There are four value-priced NetZero mobi
broadband plans to choose from. Customers are not require
to sign a contract, cannot incur overage charges and ca
upgrade their data plan at any time. New customers can try th
service free for up to one year with the purchase of a NetZe
hotspot or NetZero stick. NetZero home wireless broadband
a secure, wireless internet access service designed for hon
or office use. NetZero and Juno each offer a variety of plan
including a full range of dial-up and digital subscriber line (DS
internet access services. Customers who purchase th
accelerated dial-up service (either NetZero HiSpeed or Jur
Turbo) can surf the web at up to 5 times the speed of standa
dial-up. The Toll-free service allows customers who live
hard-to-serve areas to connect to the internet using a toll-fre
800 access number. DSL service is available in select citie
and includes Norton Antivirus online and MegaMail with bui
in spam and email virus protection. Additionally, United Onlir
provides advertising solutions to marketers with brand ar
direct response objectives through a full suite of displa
search, email and text-link opportunities across its intern
properties.

FINANCIAL DATA: *Note: Data for latest year may not have been available at press time.*

In U.S. $	2021	2020	2019	2018	2017	2016
Revenue	196,716,000	189,150,000	194,000,000	193,000,000	191,000,000	184,400,000
R&D Expense						
Operating Income						
Operating Margin %						
SGA Expense						
Net Income						
Operating Cash Flow						
Capital Expenditure						
EBITDA						
Return on Assets %						
Return on Equity %						
Debt to Equity						

CONTACT INFORMATION:

Phone: 818 287-3000   Fax: 818 287-3001
Toll-Free:
Address: 21301 Burbank Blvd., Woodland Hills, CA 91367 United States

STOCK TICKER/OTHER:

Stock Ticker: Subsidiary
Employees: 625
Parent Company: B Riley Financial Inc

Exchange:
Fiscal Year Ends: 12/31

SALARIES/BONUSES:

Top Exec. Salary: $ Bonus: $
Second Exec. Salary: $ Bonus: $

OTHER THOUGHTS:

Estimated Female Officers or Directors:
Hot Spot for Advancement for Women/Minorities:

United States Cellular Corporation

www.uscellular.com

NAIC Code: 517210

TYPES OF BUSINESS:

Mobile Phone and Wireless Services
Wireless Services
Telecommunications
Voice
Data Services

BRANDS/DIVISIONS/AFFILIATES:

Telephone and Data Systems Inc

GROWTH PLANS/SPECIAL FEATURES:

U.S. Cellular is a regional wireless carrier that serves about 5 million customers spread across four major geographic clusters: the Midwest, mid-Atlantic, New England, and the Pacific Northwest. These service territories encompass a total population of about 32 million people. The vast majority of the markets the firm serves are rural or second/third-tier cities, with only the greater Milwaukee and Oklahoma City regions boasting populations greater than 1 million. U.S. Cellular also owns a 5.5% stake in Verizon Wireless' Los Angeles operations and, unlike its wireless carrier peers, owns most of its own towers.

U.S. Cellular offers its employees medical, dental and vision coverage; life insurance and AD&D; short- and long-term disability; a 401(k) and Roth IRA; a pension plan; and tuition reimbursement.

CONTACTS: *Note: Officers with more than one job title may be intentionally listed here more than once.*

Laurent Therivel, CEO
Douglas Chambers, CFO
Leroy Carlson, Chairman of the Board
Anita Kroll, Chief Accounting Officer
Michael Irizarry, Chief Technology Officer
Jeffrey Hoersch, Controller
Deirdre Drake, Director

FINANCIAL DATA: *Note: Data for latest year may not have been available at press time.*

In U.S. $	2021	2020	2019	2018	2017	2016
Revenue	4,122,000,000	4,037,000,000	4,022,000,000	3,967,000,000	3,890,000,000	3,990,000,000
R&D Expense						
Operating Income	191,000,000	193,000,000	130,000,000	150,000,000	60,000,000	51,000,000
Operating Margin %	.05%	.05%	.03%	.04%	.02%	.01%
SGA Expense	1,345,000,000	1,368,000,000	1,406,000,000	1,388,000,000	1,412,000,000	1,480,000,000
Net Income	155,000,000	229,000,000	127,000,000	150,000,000	12,000,000	48,000,000
Operating Cash Flow	802,000,000	1,237,000,000	724,000,000	709,000,000	469,000,000	501,000,000
Capital Expenditure	2,046,000,000	1,190,000,000	921,000,000	522,000,000	654,000,000	496,000,000
EBITDA	1,033,000,000	1,045,000,000	997,000,000	971,000,000	456,000,000	813,000,000
Return on Assets %	.02%	.03%	.02%	.02%	.00%	.01%
Return on Equity %	.03%	.05%	.03%	.04%	.00%	.01%
Debt to Equity	.80%	0.763	0.564	0.396	0.441	0.445

CONTACT INFORMATION:

Phone: 773 399-8900   Fax: 773 399-8936
Toll-Free: 888-944-9400
Address: 8410 W. Bryn Mawr Ave., Chicago, IL 60631 United States

STOCK TICKER/OTHER:

Stock Ticker: USM   Exchange: NYS
Employees: 4,800   Fiscal Year Ends: 12/31
Parent Company: Telephone and Data Systems Inc

SALARIES/BONUSES:

Top Exec. Salary: $785,894   Bonus: $358,374
Second Exec. Salary: $733,867   Bonus: $174,587

OTHER THOUGHTS:

Estimated Female Officers or Directors: 6
Hot Spot for Advancement for Women/Minorities: Y

Sales, profits and employees may be estimates. Financial information, benefits and other data can change quickly and may vary from those stated here.

Veon Ltd

NAIC Code: 517210

TYPES OF BUSINESS:

Cell Phone Service
Wireless Internet Service
IPTV
Fixed-Line Telephony
Digital

BRANDS/DIVISIONS/AFFILIATES:

Veon
Beeline
Kyivstar
banglalink
Jazz
Djezzy

GROWTH PLANS/SPECIAL FEATURES:

VEON Ltd is a global provider of connectivity and intern
services. The company provides more than 210 milli
customers with voice, fixed broadband, data and digi
services. Currently, the company offers services to custome
in 10 countries: Russia, Pakistan, Algeria, Uzbekistan, Ukrair
Bangladesh, Kazakhstan, Kyrgyzstan, Armenia and Georg
The reportable segments currently consist of the following
segments: Russia; Pakistan; Bangladesh; Ukrair
Uzbekistan; and HQ. The company provides services under t
Beeline, Kyivstar, banglalink, Jazz and Djezzy brands. T
maximum revenue derives from Russia.

CONTACTS: Note: Officers with more than one job title may be intentionally listed here more than once.

Sergi Herrero, Co-CEO
Kaan Terzioglu, Co-CEO
Serkan Okandan, CFO
Jeffrey D. McGhie, General Counsel
Dmitry G. Kromsky, Head-CIS Bus. Unit
Anton Kudryashov, Head-Russia Bus. Unit
Ahmed Abou Doma, Head-Africa & Asia Bus. Unit
Romano Righetti, Group Chief Regulatory Officer
Gennady Gazin, Chmn.
Maximo Ibarra, Head-Italy

FINANCIAL DATA: Note: Data for latest year may not have been available at press time.

In U.S. $	2021	2020	2019	2018	2017	2016
Revenue	7,788,000,000	7,291,000,000	8,089,000,000	9,086,000,000	9,474,000,000	8,885,000,000
R&D Expense						
Operating Income	1,508,000,000	1,465,000,000	2,070,000,000	1,501,000,000	1,618,000,000	1,354,000,000
Operating Margin %	.19%	.20%	.26%	.17%	.17%	.15%
SGA Expense	2,470,000,000	2,321,000,000	2,529,000,000	3,418,000,000	3,470,000,000	3,366,000,000
Net Income	674,000,000	-349,000,000	621,000,000	582,000,000	-505,000,000	2,328,000,000
Operating Cash Flow	2,639,000,000	2,443,000,000	2,948,000,000	2,515,000,000	2,475,000,000	1,875,000,000
Capital Expenditure	1,796,000,000	1,677,000,000	1,582,000,000	1,948,000,000	2,037,000,000	1,651,000,000
EBITDA	3,471,000,000	3,117,000,000	3,873,000,000	3,260,000,000	3,357,000,000	3,246,000,000
Return on Assets %	.04%	-.02%	.04%	.03%	-.02%	.08%
Return on Equity %	1.80%	-.50%	.25%	.15%	-.10%	.48%
Debt to Equity	16.05%	54.184	6.329	1.789	2.393	1.354

CONTACT INFORMATION:

Phone: 31 207977200   Fax: 31 207977201
Toll-Free:
Address: Claude Debussylaan 88, Amsterdam, 1082 MD Netherlands

STOCK TICKER/OTHER:

Stock Ticker: VEON   Exchange: NAS
Employees: 44,586   Fiscal Year Ends: 12/31
Parent Company:

SALARIES/BONUSES:

Top Exec. Salary: $6,155,568   Bonus: $6,124,678
Second Exec. Salary: Bonus: $4,215,842
$1,398,993

OTHER THOUGHTS:

Estimated Female Officers or Directors: 1
Hot Spot for Advancement for Women/Minorities:

Verio Inc

www.verio.com

NAIC Code: 517210

TYPES OF BUSINESS:

Internet Service Provider
Domain Name Registration
Web Site Hosting
Virtual Private Networks
Ecommerce Services

BRANDS/DIVISIONS/AFFILIATES:

Nippon Telegraph and Telephone Corporation (NTT)
NTT Communications

CONTACTS: *Note: Officers with more than one job title may be intentionally listed here more than once.*

Hideyuki Yamasawa, CEO
Fred Martin, Sr. VP-Oper.
Tomoyuki Sakae, VP-Corp. Svcs.
William Gunther, VP-Systems Dev. & Architecture
Fred Martin, Sr. VP-Customer Service
Fred White, VP-Product Mgmt.
Wataru Imajuku, VP-Global Service Dev. & Japanese Sales

GROWTH PLANS/SPECIAL FEATURES:

Verio, Inc. offers shared website hosting solutions, domain name registration, virtual private server (VPS) hosting and other online services to individuals and small-to-medium-sized businesses. The company's hosting plans include free site-building software, as well as access to more than 200 other tools and services. Domain name registrations have hundreds of new domains available, as well as pre-registration services for soon-to-be-released domain extensions. Domain name Trademark capabilities are also available. Verio offers secure customer data and payments with SSL Certificates to enhance customer confidence from the business' website, to help boost Google rankings and utilizing encryption at lower costs. Email services through Microsoft offers businesses the latest in business-class email management and collaboration, including email, calendar, contact and task management from anywhere, any time. Verio's marketing services provide solutions for generating more traffic to the business' website, including design services, marketing services, Google Workspace, search engine optimization, search engine marketing, email marketing and more. Marketing strategies include website, advertising and website traffic management and analytics. Verio operates as a wholly-owned subsidiary of NTT Communications, which itself is a subsidiary of Nippon Telegraph and Telephone Corporation.

FINANCIAL DATA: *Note: Data for latest year may not have been available at press time.*

In U.S. $	2021	2020	2019	2018	2017	2016
Revenue						
R&D Expense						
Operating Income						
Operating Margin %						
SGA Expense						
Net Income						
Operating Cash Flow						
Capital Expenditure						
EBITDA						
Return on Assets %						
Return on Equity %						
Debt to Equity						

CONTACT INFORMATION:

Phone: 855-765-0425   Fax:
Toll-Free:
Address: 10 Corporate Dr., Ste. 300, Burlington, MA 01803 United States

STOCK TICKER/OTHER:

Stock Ticker: Subsidiary   Exchange:
Employees: 1,960   Fiscal Year Ends: 03/31
Parent Company: Nippon Telegraph and Telephone Corporation (NTT)

SALARIES/BONUSES:

Top Exec. Salary: $ Bonus: $
Second Exec. Salary: $ Bonus: $

OTHER THOUGHTS:

Estimated Female Officers or Directors:
Hot Spot for Advancement for Women/Minorities:

Sales, profits and employees may be estimates. Financial information, benefits and other data can change quickly and may vary from those stated here.

VeriSign Inc

NAIC Code: 511210E

www.verisigninc.com

TYPES OF BUSINESS:
Computer Software: Network Security, Managed Access, Digital ID, Cybersecurity & Anti-Virus
Domain Name Registration

BRANDS/DIVISIONS/AFFILIATES:
Registry Services
Security Services

GROWTH PLANS/SPECIAL FEATURES:
Verisign is the sole authorized registry for several generic top level domains, including the widely utilized .com and .net top level domains. The company operates critical internet infrastructure to support the domain name system, including operating two of the world's 13 root servers that are used t route internet traffic. In 2018, the firm sold its Security Service business, signaling a renewed focus on the core registr business.

Employees of VeriSign receive a flexible benefits package tha includes health, dental, vision, disability and life insurance flexible spending accounts; a 401(k); an employee assistanc program; a group legal plan; domestic partner coverage; tuitio ass

CONTACTS: Note: Officers with more than one job title may be intentionally listed here more than once.
D. Bidzos, CEO
George Kilguss, CFO
Todd Strubbe, COO
Thomas Indelicarto, Executive VP

FINANCIAL DATA: Note: Data for latest year may not have been available at press time.

In U.S. $	2021	2020	2019	2018	2017	2016
Revenue	1,327,576,000	1,265,052,000	1,231,661,000	1,214,969,000	1,165,095,000	1,142,167,000
R&D Expense	80,529,000	74,671,000	60,805,000	57,884,000	52,342,000	59,100,000
Operating Income	866,803,000	824,201,000	806,127,000	767,392,000	707,722,000	686,572,000
Operating Margin %	.65%	.65%	.65%	.63%	.61%	.60%
SGA Expense	188,311,000	186,003,000	184,262,000	197,559,000	211,705,000	198,253,000
Net Income	784,830,000	814,888,000	612,299,000	582,489,000	457,248,000	440,645,000
Operating Cash Flow	807,152,000	730,183,000	753,892,000	697,767,000	702,761,000	693,007,000
Capital Expenditure	53,033,000	43,395,000	40,316,000	37,007,000	49,499,000	169,574,000
EBITDA	913,414,000	886,740,000	895,717,000	889,197,000	785,226,000	754,904,000
Return on Assets %	.42%	.45%	.32%	.24%	.17%	.19%
Return on Equity %						
Debt to Equity						

CONTACT INFORMATION:
Phone: 703 948-3200　　Fax:
Toll-Free: 800-922-4917
Address: 12061 Bluemont Way, Reston, VA 20190 United States

STOCK TICKER/OTHER:
Stock Ticker: VRSN　　　　　Exchange: NAS
Employees: 904　　　　　　　Fiscal Year Ends: 12/31
Parent Company:

SALARIES/BONUSES:
Top Exec. Salary: $925,000　　Bonus: $
Second Exec. Salary: $573,462　　Bonus: $

OTHER THOUGHTS:
Estimated Female Officers or Directors:
Hot Spot for Advancement for Women/Minorities:

Verizon Communications Inc

www.verizon.com

NAIC Code: 517110

TYPES OF BUSINESS:

Mobile Phone and Wireless Services
Communications Services
Mobile Services
Home Services
Business Services
Network Technologies
Wireless
Wireline

BRANDS/DIVISIONS/AFFILIATES:

Verizon Consumer Group
Verizon Business Group
Verizon Fios

GROWTH PLANS/SPECIAL FEATURES:

Verizon is primarily a wireless business (nearly 80% of revenue and nearly all operating income). It serves about 93 million postpaid and 24 million prepaid phone customers (following the acquisition of Tracfone) via its nationwide network, making it the largest U.S. wireless carrier. Fixed-line telecom operations include local networks in the Northeast, which reach about 25 million homes and businesses, and nationwide enterprise services.

Verizon offers comprehensive employee benefits.

CONTACTS: Note: Officers with more than one job title may be intentionally listed here more than once.

Ronan Dunne, CEO, Divisional
Tami Erwin, CEO, Divisional
Hans Vestberg, CEO
Matthew Ellis, CFO
Anthony Skiadas, Chief Accounting Officer
Craig Silliman, Chief Administrative Officer
Rima Qureshi, Chief Strategy Officer
Kyle Malady, Chief Technology Officer
Christine Pambianchi, Executive VP

FINANCIAL DATA: Note: Data for latest year may not have been available at press time.

In U.S. $	2021	2020	2019	2018	2017	2016
Revenue	133,613,000,000	128,292,000,000	131,868,000,000	130,863,000,000	126,034,000,000	125,980,000,000
R&D Expense						
Operating Income	32,448,000,000	28,798,000,000	30,564,000,000	26,869,000,000	29,199,000,000	30,256,000,000
Operating Margin %	.24%	.22%	.23%	.21%	.23%	.24%
SGA Expense	28,658,000,000	31,573,000,000	29,896,000,000	31,083,000,000	26,818,000,000	27,095,000,000
Net Income	22,065,000,000	17,801,000,000	19,265,000,000	15,528,000,000	30,101,000,000	13,127,000,000
Operating Cash Flow	39,539,000,000	41,768,000,000	35,746,000,000	34,339,000,000	24,318,000,000	21,689,000,000
Capital Expenditure	67,882,000,000	22,088,000,000	18,837,000,000	18,087,000,000	17,830,000,000	17,593,000,000
EBITDA	49,111,000,000	44,934,000,000	44,145,000,000	41,859,000,000	42,281,000,000	41,290,000,000
Return on Assets %	.06%	.06%	.07%	.06%	.12%	.05%
Return on Equity %	.29%	.28%	.34%	.32%	.92%	.67%
Debt to Equity	2.04%	2.081	1.94	1.992	2.637	4.681

CONTACT INFORMATION:

Phone: 212 395-1000   Fax:
Toll-Free: 800-837-4966
Address: 1095 Avenue of the Americas, New York, NY 10036 United States

STOCK TICKER/OTHER:

Stock Ticker: VZ
Employees: 118,400
Parent Company:

Exchange: NYS
Fiscal Year Ends: 12/31

SALARIES/BONUSES:

Top Exec. Salary: $588,462   Bonus: $3,000,000
Second Exec. Salary: Bonus: $
$1,500,000

OTHER THOUGHTS:

Estimated Female Officers or Directors: 5
Hot Spot for Advancement for Women/Minorities: Y

ViaSat Inc
NAIC Code: 334220

www.viasat.com

TYPES OF BUSINESS:
Telecommunications Equipment-Digital Satellite
Networking & Wireless Signal Processing
Satellite Broadband Internet Service Provider

GROWTH PLANS/SPECIAL FEATURES:
Viasat Inc provides bandwidth technologies and services i
three segments: satellite services, which provides satellite
based high-speed broadband services to consumer:
enterprises, and commercial airlines; commercial network:
which develops end-to-end communication and connectivit
systems; and government systems, which produces networl
centric Internet Protocol-based secure governmer
communication systems. A large majority of the firm's revenu
is generated in the United States, with the rest coming from th
Americas, Europe, Middle East, Africa, and Asia-Pacific.

ViaSat offers employee benefits such as health, 401(k) an
tuition reimbursement, depending on the location.

BRANDS/DIVISIONS/AFFILIATES:

CONTACTS: Note: Officers with more than one job title may be intentionally listed here more than once.
Richard Baldridge, CEO
Mark Dankberg, Chairman of the Board
Shawn Duffy, Chief Accounting Officer
Keven Lippert, Chief Administrative Officer
Krishna Nathan, Chief Information Officer
Mark Miller, Chief Technology Officer
Girish Chandran, Chief Technology Officer
Kevin Harkenrider, COO
Robert Blair, General Counsel
Melinda Kimbro, Other Executive Officer
Evan Dixon, President, Divisional
Craig Miller, President, Divisional
James Dodd, President, Divisional
David Ryan, President, Divisional

FINANCIAL DATA: Note: Data for latest year may not have been available at press time.

In U.S. $	2021	2020	2019	2018	2017	2016
Revenue	2,256,107,000	2,309,238,000	2,068,258,000	1,594,625,000	1,559,337,000	
R&D Expense	115,792,000	130,434,000	123,044,000	168,347,000	129,647,000	
Operating Income	58,233,000	38,421,000	-60,620,000	-92,187,000	36,459,000	
Operating Margin %	.03%	.02%	-.03%	-.06%	.02%	
SGA Expense	512,316,000	523,085,000	458,458,000	385,420,000	333,468,000	
Net Income	3,691,000	-212,000	-67,623,000	-67,305,000	23,767,000	
Operating Cash Flow	727,215,000	436,936,000	327,551,000	358,633,000	411,298,000	
Capital Expenditure	885,271,000	761,078,000	686,820,000	584,487,000	585,658,000	
EBITDA	455,775,000	382,247,000	258,142,000	154,208,000	283,389,000	
Return on Assets %	.00%	.00%	-.02%	-.02%	.01%	
Return on Equity %	.00%	.00%	-.04%	-.04%	.02%	
Debt to Equity	.90%	1.04	0.73	0.533	0.489	

CONTACT INFORMATION:
Phone: 760 476-2200   Fax: 760 929-3941
Toll-Free:
Address: 6155 El Camino Real, Carlsbad, CA 92009 United States

STOCK TICKER/OTHER:
Stock Ticker: VSAT
Employees: 7,000
Parent Company:

Exchange: NAS
Fiscal Year Ends: 03/31

SALARIES/BONUSES:
Top Exec. Salary: $1,300,000   Bonus: $
Second Exec. Salary: Bonus: $
$1,300,000

OTHER THOUGHTS:
Estimated Female Officers or Directors:
Hot Spot for Advancement for Women/Minorities:

Viavi Solutions Inc

www.viavisolutions.com

NAIC Code: 334220

TYPES OF BUSINESS:

Communications and Commercial Optical Products
Network Systems
Security Products
Systems Enablement and Consulting

BRANDS/DIVISIONS/AFFILIATES:

Optically Variable Pigment
Optically Variable Magnetic Pigment

GROWTH PLANS/SPECIAL FEATURES:

Viavi Solutions Inc. is a global provider of network test, monitoring and assurance solutions to communications service providers, enterprises, network equipment manufacturers, civil government, military and avionics customers. The company also offers high-performance thin-film optical coatings, providing light management solutions to anti-counterfeiting, 3D sensing, electronics, automotive, defense and instrumentation markets. Its operating segments include Network Enablement, Service Enablement, and Optical Security and Performance Products. Geographically, it derives a majority of revenue from the United States.

Viavi offers its employees medical, dental and vision benefits, wellness initiatives, retirement benefits, stock compensation and an employee stock purchase plan.

CONTACTS: Note: Officers with more than one job title may be intentionally listed here more than once.

Oleg Khaykin, CEO
Henk Derksen, CFO
Richard Belluzzo, Chairman of the Board
Paul McNab, Chief Marketing Officer
Pam Avent, Controller
Kevin Siebert, General Counsel
Luke Scrivanich, General Manager, Divisional
Gary Staley, Senior VP, Divisional
Ralph Rondinone, Senior VP, Divisional

FINANCIAL DATA: Note: Data for latest year may not have been available at press time.

In U.S. $	2021	2020	2019	2018	2017	2016
Revenue	1,198,900,000	1,136,300,000	1,130,300,000	875,700,000	805,000,000	906,300,000
R&D Expense	203,000,000	193,600,000	187,000,000	133,300,000	136,300,000	166,400,000
Operating Income	140,600,000	121,600,000	82,800,000	10,200,000	28,600,000	17,600,000
Operating Margin %	.12%	.11%	.07%	.01%	.04%	.02%
SGA Expense	337,500,000	315,000,000	343,500,000	323,900,000	300,100,000	351,100,000
Net Income	46,100,000	28,700,000	5,400,000	-48,600,000	160,200,000	-99,200,000
Operating Cash Flow	243,300,000	135,600,000	138,800,000	66,000,000	94,300,000	64,300,000
Capital Expenditure	52,100,000	31,900,000	45,000,000	42,500,000	38,600,000	35,500,000
EBITDA	247,800,000	235,500,000	185,800,000	95,000,000	280,900,000	60,400,000
Return on Assets %	.02%	.02%	.00%	-.02%	.08%	-.05%
Return on Equity %	.06%	.04%	.01%	-.06%	.22%	-.11%
Debt to Equity	.39%	0.907	0.833	0.796	1.22	0.888

CONTACT INFORMATION:

Phone: 408-404-3600   Fax: 408-404-4500
Toll-Free:
Address: 7047 E. Greenway Pkwy., Ste. 250, Scottsdale, AZ 85254 United States

STOCK TICKER/OTHER:

Stock Ticker: VIAV   Exchange: NAS
Employees: 3,600   Fiscal Year Ends: 06/27
Parent Company:

SALARIES/BONUSES:

Top Exec. Salary: $800,000   Bonus: $
Second Exec. Salary: $435,000   Bonus: $

OTHER THOUGHTS:

Estimated Female Officers or Directors:
Hot Spot for Advancement for Women/Minorities:

Sales, profits and employees may be estimates. Financial information, benefits and other data can change quickly and may vary from those stated here.

Virgin Media Business Ltd

www.virginmediabusiness.co.uk

NAIC Code: 517110

TYPES OF BUSINESS:

Fixed-line Telecommunications
High-Speed Internet Services
Telephony Services
Voice Services
Business Telecommunications
Data Services

BRANDS/DIVISIONS/AFFILIATES:

Virgin Media Limited

CONTACTS: *Note: Officers with more than one job title may be intentionally listed here more than once.*

Peter Kelly, Managing Dir.
Mike Smith, Managing Dir.-Business
Luke Milner, Dir.-Finance
Michelle King, Dir.-Dev. & Strategy
Alison Bawn, Dir.-People
Phil Stewart, Dir.-Customer Service
Brendan Lynch, Dir.-Bus. & Partner Markets
Lee Hull, Dir.-Public Sector

GROWTH PLANS/SPECIAL FEATURES:

Virgin Media Business Ltd., a division of Virgin Media Limited is a leading telecommunications provider for small- and medium-sized businesses as well as for enterprises and public sector entities within the U.K., Scotland and northern Ireland. The firm offers a national product portfolio of broadband internet, mobile, voice over internet protocol (VoIP) solutions, site connectivity, virtual private networks (VPNs) and phone lines. Virgin Media works with approximately 600 wholesale partners to service businesses and public sector organizations in the U.K. The company's wholly-owned network hosts 84% of Ethernet services entirely on their own, connects hundreds of data centers and points of presence, and carries over 35% of the U.K.'s broadband traffic. Virgin Media Business' office locations include: Reading, Berkshire, Hammersmith, Bristol, Peterborough, Nottingham, Staverton, Swansea, Bradford, Stockton on Tees, Liverpool, Manchester, Knowsley and Gateshead, England; Edinburgh, Scotland; and Belfast, Ireland.

FINANCIAL DATA: *Note: Data for latest year may not have been available at press time.*

In U.S. $	2021	2020	2019	2018	2017	2016
Revenue	1,100,000,000	1,084,820,000	1,110,900,000	1,058,000,000	1,030,000,000	1,020,432,000
R&D Expense						
Operating Income						
Operating Margin %						
SGA Expense						
Net Income						
Operating Cash Flow						
Capital Expenditure						
EBITDA						
Return on Assets %						
Return on Equity %						
Debt to Equity						

CONTACT INFORMATION:

Phone: 44-173-323-0666   Fax:
Toll-Free:
Address: Southgate   House,   Bakewell   Rd.,   Orton   Southgate, Peterborough, PE2 6YS United Kingdom

STOCK TICKER/OTHER:

Stock Ticker: Subsidiary
Employees: 1,978
Parent Company: Virgin Media Limited

Exchange:
Fiscal Year Ends: 12/31

SALARIES/BONUSES:

Top Exec. Salary: $ Bonus: $
Second Exec. Salary: $ Bonus: $

OTHER THOUGHTS:

Estimated Female Officers or Directors:
Hot Spot for Advancement for Women/Minorities: Y

VMED O2 UK Limited (Virgin Media O2)   www.virginmedia.com

NAIC Code: 517210

TYPES OF BUSINESS:

Wireless Service
Cable TV
Internet Service Provider
Fixed Line & Mobile Telephony
Digital
Broadband
Fiber Broadband

BRANDS/DIVISIONS/AFFILIATES:

Liberty Global plc
Telefonica SA

GROWTH PLANS/SPECIAL FEATURES:

VMED O2 UK Limited, doing business as Virgin Media O2, is a leading entertainment and communications company serving the U.K. The company provides broadband, TV, phone and mobile services to consumers and businesses, of which the services can be bundled. Broadband options include fiber broadband and gigabit fiber broadband. VMED O2 is a 50:50 joint venture between Liberty Global plc and Telefonica S.A. VMED O2 combines Virgin Media with Telefonica's O2 mobile operator to create a leading U.K. mobile platform, bringing together 5G technology with gigabit broadband.

CONTACTS: *Note: Officers with more than one job title may be intentionally listed here more than once.*

Lutz Schuler, CEO
Richard Williams, Dir.-Investor Relations

FINANCIAL DATA: *Note: Data for latest year may not have been available at press time.*

In U.S. $	2021	2020	2019	2018	2017	2016
Revenue	13,965,845,608	14,215,464,172	6,778,630,000	7,030,768,500	6,695,970,000	5,912,270,000
R&D Expense						
Operating Income						
Operating Margin %						
SGA Expense						
Net Income	24,685,602	-1,176,600,152	-442,404,000	-46,953,900	-117,239,000	-351,334,000
Operating Cash Flow						
Capital Expenditure						
EBITDA						
Return on Assets %						
Return on Equity %						
Debt to Equity						

CONTACT INFORMATION:

Phone: 44 800-052-0626   Fax:
Toll-Free:
Address: 160 Great Portland St., London, W1W 5QA United Kingdom

STOCK TICKER/OTHER:

Stock Ticker: Subsidiary   Exchange:
Employees: 1,700   Fiscal Year Ends: 12/31
Parent Company:

SALARIES/BONUSES:

Top Exec. Salary: $ Bonus: $
Second Exec. Salary: $ Bonus: $

OTHER THOUGHTS:

Estimated Female Officers or Directors:
Hot Spot for Advancement for Women/Minorities:

Vocera Communications Inc

NAIC Code: 511210C

www.vocera.com

TYPES OF BUSINESS:

Computer Software: Telecom, Communications & VOIP

BRANDS/DIVISIONS/AFFILIATES:

CONTACTS: *Note: Officers with more than one job title may be intentionally listed here more than once.*

Brent Lang, CEO
Steve Anheier, CFO
M. Duffy, Chief Medical Officer
Paul Johnson, Executive VP, Divisional
Douglas Carlen, General Counsel

GROWTH PLANS/SPECIAL FEATURES:

Vocera Communications, Inc. is a provider of mob
communication solutions focused on addressing criti
communication challenges facing hospitals today. It helps
customers improve patient safety and satisfaction, a
increase hospital efficiency and productivity through its Vo
Communication solutions and new Messaging and Ca
Transition solutions. Its Voice Communication solution, wh
includes a lightweight, wearable, voice-controll
communication badge and a software platform, enables use
to connect instantly with other hospital staff simply by say
the name, function or group name of the desired recipient.
Messaging solution securely delivers text messages and ale
directly to and from smartphones, replacing legacy pagers.
Care Transition solution is a hosted voice and text bas
software application that captures, manages and monit
patient information when responsibility for the patient
transferred or handed-off from one caregiver to another.
communication platform helps hospitals increase productiv
and reduce costs by streamlining operations, and improv
patient and staff satisfaction by creating a differentiated Voce
hospital experience. Its solutions provide the following k
benefits: improves patient safety; enhances pati
experience; improves caregiver job satisfaction & increas
revenue and reduce expenses. It has following key competiti
strengths: unified communication solutions focused on t
requirements of healthcare providers; comprehens
proprietary communication solutions; broad and loyal custom
base; recognized and trusted brand & experienc
management team.

FINANCIAL DATA: *Note: Data for latest year may not have been available at press time.*

In U.S. $	2021	2020	2019	2018	2017	2016
Revenue	200,000,000	198,420,000	180,500,992	179,630,000	162,548,000	127,696,000
R&D Expense						
Operating Income						
Operating Margin %						
SGA Expense						
Net Income		-9,656,000	-17,980,000	-9,674,000	-14,217,000	-17,267,000
Operating Cash Flow						
Capital Expenditure						
EBITDA						
Return on Assets %						
Return on Equity %						
Debt to Equity						

CONTACT INFORMATION:

Phone: 408 882-5100   Fax: 408 882-5101
Toll-Free:
Address: 525 Race Street, San Jose, CA 95126 United States

STOCK TICKER/OTHER:

Stock Ticker: Subsidiary   Exchange:
Employees: 688   Fiscal Year Ends: 12/31
Parent Company: Stryker Corp

SALARIES/BONUSES:

Top Exec. Salary: $ Bonus: $
Second Exec. Salary: $ Bonus: $

OTHER THOUGHTS:

Estimated Female Officers or Directors:
Hot Spot for Advancement for Women/Minorities:

Vocus Group Limited

www.vocusgroup.com.au

NAIC Code: 517110

TYPES OF BUSINESS:
Wired Telecommunications Carriers

BRANDS/DIVISIONS/AFFILIATES:
Dodo
iPrimus
Slingshot
Flip
Orcon
Australia Singapore Cable

CONTACTS: *Note: Officers with more than one job title may be intentionally listed here more than once.*
Kevin Russell, Managing Dir.
Ellie Sweeney, COO
Nitesh Naidoo, CFO
Amber Kristof, Dir.-People & Culture
Robert Mansfield, Chmn.

GROWTH PLANS/SPECIAL FEATURES:
Vocus Group Limited provides retail, wholesale and corporate telecommunications services across Australia and New Zealand through various brands. The company's range of products and services span connectivity, cloud, data center, voice and security. Vocus owns and operates the group's telecom network, which comprises approximately 18,640 miles (30,000 km) of fiber optic cable. Vocus' internet services include business internet, business nbn (national broadband network) satellite service, enterprise internet, IP transit carrier grade internet and add-on products such as distributed denial-of-service (DDoS) protection. Internet services also include connectivity solutions. The firm's Commander brand focuses solely on business communications, offering communications and technology solutions to Australian businesses. Dodo is a leading Australian brand that serves individuals and families with affordable and easy-to-use telecommunications and insurance products nation-wide. Dodo also provides power and gas in selected Australian states. iPrimus offers broadband, home phone and mobile phone products and services to the Australian market. Slingshot offers broadband services in New Zealand. Flip is a no-frills, budget-friendly internet provider in New Zealand. Orcon offers broadband services to New Zealand residential customers. Beyond Australia and New Zealand, Vocus built, owns and operates the Australia Singapore Cable, a 2,858 miles (4,600 km) fiber network that delivers connectivity via direct-current interconnects in Perth, Jakarta and Singapore, enabling low-latency connections to Asian hubs such as Hong Kong.

FINANCIAL DATA: *Note: Data for latest year may not have been available at press time.*

In U.S. $	2021	2020	2019	2018	2017	2016
Revenue	1,270,006,400	1,221,160,000	1,457,934,400	1,401,860,000	1,420,960,000	600,344,000
R&D Expense						
Operating Income						
Operating Margin %						
SGA Expense						
Net Income		-122,355,000	47,337,045	45,082,900	-1,143,340,000	46,319,300
Operating Cash Flow						
Capital Expenditure						
EBITDA						
Return on Assets %						
Return on Equity %						
Debt to Equity						

CONTACT INFORMATION:
Phone: 61-3-9923-3000   Fax: 61-3-9674-6599
Toll-Free:
Address: 452 Flinders St., Fl. 10, Melbourne, VIC 3000 Australia

STOCK TICKER/OTHER:
Stock Ticker: VOC   Exchange: ASX
Employees: 2,000   Fiscal Year Ends: 06/30
Parent Company:

SALARIES/BONUSES:
Top Exec. Salary: $ Bonus: $
Second Exec. Salary: $ Bonus: $

OTHER THOUGHTS:
Estimated Female Officers or Directors:
Hot Spot for Advancement for Women/Minorities:

Sales, profits and employees may be estimates. Financial information, benefits and other data can change quickly and may vary from those stated here.

Vodafone Group plc

www.vodafone.com

NAIC Code: 517210

TYPES OF BUSINESS:

Cell Phone Service
Mobile Communications
Fixed Communications
Unified Communications
Cloud Hosting
Internet of Things
Carrier Services

BRANDS/DIVISIONS/AFFILIATES:

TPG Telecom

GROWTH PLANS/SPECIAL FEATURES:

Vodafone operates mobile and fixed-line networks ar
businesses in more than 20 countries. Its largest market
Germany, where it is the second mobile operator aft
Deutsche Telekom and owns a cable network after acquirir
Kabel Deutschland in 2013 and Liberty Global Germany
2019. In the U.K. and Italy, it acts as a mobile operator, whi
in Spain, it offers converged services after the acquisition
cable operator Ono in 2014. Vodafone also has operations
several Central European and African countries, whic
combined represent around one third of revenue.

CONTACTS: Note: Officers with more than one job title may be intentionally listed here more than once.

Nick Read, CEO
Margherita Della Valle, CFO
Ahmed Essam, Chief Commercial Operations Officer
Leanne Wood, Chief Human Resources Officer
Johan Wibergh, Group Technology Officer
Rosemary Martin, General Counsel
Warren Finegold, Dir.-Strategy & Bus. Dev.
Matthew Kirk, Dir.-External Affairs
Morten Lundal, Chief Commercial Officer
Philipp Humm, CEO-Northern & Central Europe
Paulo Bertoluzzo, CEO-Southern Europe
Nick Jeffery, Dir-Group Enterprises
Gerard Kleisterlee, Chmn.
Nick Read, CEO-Asia-Pacific, Africa & Middle East Region

FINANCIAL DATA: Note: Data for latest year may not have been available at press time.

In U.S. $	2021	2020	2019	2018	2017	2016
Revenue	44,674,800,000	45,862,820,000	44,528,970,000	47,491,380,000	48,572,330,000	
R&D Expense						
Operating Income	5,526,095,000	9,125,859,000	4,137,178,000	4,982,562,000	4,351,329,000	
Operating Margin %	.12%	.20%	.09%	.10%	.09%	
SGA Expense	9,047,337,000	9,814,199,000	9,484,815,000	9,307,377,000	10,034,470,000	
Net Income	114,213,500	-938,182,000	-8,178,499,000	2,487,202,000	-6,421,448,000	
Operating Cash Flow	17,555,220,000	17,722,460,000	13,236,520,000	13,868,780,000	14,504,090,000	
Capital Expenditure	8,810,752,000	7,755,298,000	8,312,088,000	8,324,325,000	9,036,120,000	
EBITDA	21,476,210,000	17,850,950,000	8,968,816,000	15,783,890,000	15,895,050,000	
Return on Assets %	.00%	-.01%	-.06%	.02%	-.04%	
Return on Equity %	.00%	-.01%	-.12%	.03%	-.08%	
Debt to Equity	1.06%	1.025	0.782	0.487	0.478	

CONTACT INFORMATION:

Phone: 44 163533251   Fax: 44 1635238080
Toll-Free:
Address: Vodafone House, The Connection, Newbury, Berkshire RG14
2FN United Kingdom

STOCK TICKER/OTHER:

Stock Ticker: VOD   Exchange: NAS
Employees: 104,000   Fiscal Year Ends: 03/31
Parent Company:

SALARIES/BONUSES:

Top Exec. Salary: $1,268,898   Bonus: $1,754,704
Second Exec. Salary: Bonus: $1,169,803
$845,932

OTHER THOUGHTS:

Estimated Female Officers or Directors: 2
Hot Spot for Advancement for Women/Minorities: Y

Vonage Holdings Corp

www.vonage.com

NAIC Code: 517110

TYPES OF BUSINESS:
VOIP Telecommunications

BRANDS/DIVISIONS/AFFILIATES:

GROWTH PLANS/SPECIAL FEATURES:
Vonage Holdings Corp is a North American technology company that provides cloud communication services to businesses and consumers. For businesses, the company provides unified communications (as a service), which consists of integrated voice, text, video, data, and mobile applications over Voice over Internet Protocol network. Its reportable operating segments include Vonage Communications Platform and Consumer. For consumer service customers, there is a home telephone replacement service. This can include services such as voicemail, call waiting, and call forwarding. This service is delivered over the Internet. The company generates most of its revenue within the United States.

CONTACTS: Note: Officers with more than one job title may be intentionally listed here more than once.
Rory Read, CEO
Stephen Lasher, CFO
Jeffrey Citron, Chairman of the Board
David Levi, Chief Accounting Officer
Randy Rutherford, Chief Legal Officer
Joy Corso, Chief Marketing Officer
Vinod Lala, Chief Strategy Officer
Joseph Bellissimo, COO
Jay Bellissimo, COO
Savinay Berry, Executive VP, Divisional
Susan Quackenbush, Other Executive Officer
Sanjay Macwan, Other Executive Officer
Omar Javaid, Other Executive Officer
Rodolpho Cardenuto, President, Divisional

FINANCIAL DATA: Note: Data for latest year may not have been available at press time.

In U.S. $	2021	2020	2019	2018	2017	2016
Revenue	1,409,015,040	1,247,933,952	1,189,346,048	1,048,782,016	1,002,286,016	955,620,992
R&D Expense						
Operating Income						
Operating Margin %						
SGA Expense						
Net Income	-24,497,000	-36,212,000	-19,482,000	35,728,000	-33,933,000	17,907,000
Operating Cash Flow						
Capital Expenditure						
EBITDA						
Return on Assets %						
Return on Equity %						
Debt to Equity						

CONTACT INFORMATION:
Phone: 732 528-2600   Fax: 732 287-9119
Toll-Free: 800-980-1455
Address: 23 Main Street, Holmdel, NJ 07733 United States

STOCK TICKER/OTHER:
Stock Ticker: VG   Exchange: NAS
Employees: 2,082   Fiscal Year Ends: 12/31
Parent Company:

SALARIES/BONUSES:
Top Exec. Salary: $ Bonus: $
Second Exec. Salary: $ Bonus: $

OTHER THOUGHTS:
Estimated Female Officers or Directors: 1
Hot Spot for Advancement for Women/Minorities:

Sales, profits and employees may be estimates. Financial information, benefits and other data can change quickly and may vary from those stated here.

VTech Holdings Limited

www.vtech.com

NAIC Code: 334210

TYPES OF BUSINESS:
Cordless Telephone Sets
Electronic Learning Products
Contract Manufacturing Services
Data Networking Products
Cordless Phones
Telecommunications

GROWTH PLANS/SPECIAL FEATURES:

VTech Holdings Ltd is engaged in designing, manufacturing and distribution of consumer electronic products. It al provides contract manufacturing services. VTech's divers collection of products include Telephone Products, Busine Phones, Electronic Learning Toys, Baby Monitors a Hospitality Products. It offers turnkey services to customers a number of product categories. The company operate through four geographical segments, North America includi the United States and Canada, Europe, Asia Pacific a Others, which covers sales of electronic products to the rest the world. VTech generates a majority of its revenue from Nor America and Europe.

BRANDS/DIVISIONS/AFFILIATES:
Vtech
Leap Frog

CONTACTS: Note: Officers with more than one job title may be intentionally listed here more than once.
Allan Chi Yun Wong, CEO
King Fai Pang, Pres.
Andy Leung Hon Kwong, CEO-Contract Mfg. Svcs.
William To, Pres., Vtech Electronics North America
Gordon Chow, Pres., Vtech Technologies Canada
Nicholas Delany, Pres., Vtech Communications, Inc.
Allan Chi Yun Wong, Chmn.

FINANCIAL DATA: Note: Data for latest year may not have been available at press time.

In U.S. $	2021	2020	2019	2018	2017	2016
Revenue	2,372,300,000	2,165,500,000	2,161,900,000	2,130,100,000	2,079,300,000	
R&D Expense	86,400,000	81,700,000	77,200,000	77,600,000	77,200,000	
Operating Income	266,200,000	219,700,000	193,200,000	231,300,000	200,000,000	
Operating Margin %	.11%	.10%	.09%	.11%	.10%	
SGA Expense	378,200,000	368,100,000	371,900,000	397,300,000	412,200,000	
Net Income	230,900,000	190,700,000	171,300,000	206,300,000	179,000,000	
Operating Cash Flow	280,300,000	237,000,000	249,300,000	175,700,000	185,300,000	
Capital Expenditure	48,000,000	34,600,000	37,300,000	37,600,000	35,700,000	
EBITDA	325,900,000	276,900,000	231,200,000	267,400,000	234,800,000	
Return on Assets %	.17%	.17%	.16%	.19%	.18%	
Return on Equity %	.35%	.32%	.27%	.34%	.32%	
Debt to Equity	.26%	0.245			0.002	

CONTACT INFORMATION:
Phone: 852 26801000   Fax: 852 26801300
Toll-Free:
Address: 57 Ting Kok Rd., Tai Ping Ctr., Block 1, 23/Fl, Hong Kong, Hong Kong 999077 Hong Kong

STOCK TICKER/OTHER:
Stock Ticker: VTKLF   Exchange: PINX
Employees: 25,000   Fiscal Year Ends: 03/31
Parent Company:

SALARIES/BONUSES:
Top Exec. Salary: $ Bonus: $
Second Exec. Salary: $ Bonus: $

OTHER THOUGHTS:
Estimated Female Officers or Directors:
Hot Spot for Advancement for Women/Minorities:

Westell Technologies Inc

www.westell.com

NAIC Code: 334210

TYPES OF BUSINESS:

Telecommunications Equipment-High-Speed Data Transmission
Distributed Antenna Systems
Digital Repeaters
Passive System Components
Remote Units
Integrated Cabinets
Power Distribution
Fiber Network Connectivity

BRANDS/DIVISIONS/AFFILIATES:

GROWTH PLANS/SPECIAL FEATURES:

Westell Technologies Inc is a provider of wireless network infrastructure solutions. It is organized into the following operating segments, In-Building Wireless (IBW), Intelligent Site Management (ISM), and Communications Network Solutions (CNS). IBW segment solutions enable cellular and public safety coverage in stadiums, arenas, malls, buildings, and other indoor areas not served well or at all by the existing macro outdoor cellular network. ISM segment solutions include a suite of remote units, which provide machine-to-machine communications that enable operators to remotely monitor, manage, and control physical site infrastructure and support systems. The majority of revenue is derived from the CNS segment Geographically, it derives a majority of revenue from the US.

Westell offers its employees comprehensive medical benefits, a 401(k) plan and paid time off.

CONTACTS: *Note: Officers with more than one job title may be intentionally listed here more than once.*

Thomas Minichiello, CFO
Kirk Brannock, Chairman of the Board
Alfred John, President
Jesse Swartwood, Senior VP, Divisional

FINANCIAL DATA: *Note: Data for latest year may not have been available at press time.*

In U.S. $	2021	2020	2019	2018	2017	2016
Revenue	29,947,000	29,956,000	43,570,000	58,577,000	62,965,000	88,203,000
R&D Expense	4,032,000	5,346,000	6,790,000	7,375,000	12,367,000	19,317,000
Operating Income	-4,156,000	-9,281,000	-6,902,000	-1,289,000	-11,717,000	-16,008,000
Operating Margin %	-.14%	-.31%	-.16%	-.02%	-.19%	-.18%
SGA Expense	9,293,000	12,349,000	15,041,000	14,892,000	18,335,000	25,653,000
Net Income	-2,734,000	-10,102,000	-11,382,000	31,000	-15,941,000	-16,212,000
Operating Cash Flow	95,000	-2,272,000	-760,000	6,945,000	-7,011,000	-5,607,000
Capital Expenditure	72,000	2,135,000	290,000	408,000	596,000	1,932,000
EBITDA	-2,862,000	-7,381,000	-2,876,000	3,668,000	-5,573,000	-8,910,000
Return on Assets %	-.08%	-.23%	-.21%	.00%	-.21%	-.17%
Return on Equity %	-.10%	-.28%	-.24%	.00%	-.27%	-.22%
Debt to Equity	.16%	0.008				

CONTACT INFORMATION:

Phone: 630 898-2500   Fax: 630 375-4931
Toll-Free:
Address: 750 North Commons Drive, Aurora, IL 60504 United States

STOCK TICKER/OTHER:

Stock Ticker: WSTL   Exchange: PINX
Employees: 96   Fiscal Year Ends: 03/31
Parent Company:

SALARIES/BONUSES:

Top Exec. Salary: $253,962   Bonus: $
Second Exec. Salary: Bonus: $
$225,000

OTHER THOUGHTS:

Estimated Female Officers or Directors: 2
Hot Spot for Advancement for Women/Minorities:

Sales, profits and employees may be estimates. Financial information, benefits and other data can change quickly and may vary from those stated here.

WIND Hellas Telecommunications SA

www.wind.com.gr

NAIC Code: 517210

TYPES OF BUSINESS:

Mobile Telephone Service
Telecommunications
Mobile Services
Fixed Services
Internet Services
Broadband
Pay-TV
Telecommunication Devices

BRANDS/DIVISIONS/AFFILIATES:

BC Partners LLP
United Group
WIND
Nova

CONTACTS: Note: Officers with more than one job title may be intentionally listed here more than once.

Panagiotls Georglopoulos, CEO-WIND
Kiki Silvestriadou, CEO-Nova
George Lambrou, CFO
Michalis Anagnostakos, CMO
Gerasimos Karmirls, Dir.-Human Resources
Dimitris Papazafiropoulos, CTO
Antonis Tzortzakakis, Chief Commercial Officer
Yiannis Gavrielides, Exec. Dir.-Online Mktg. Comm. & Consumer Prepaid
George Tsaprounis, Exec. Dir.-Corp. Affairs
Nikos Babalis, Chief Network Officer
Christos Noulis, CCO
Nikos Panopoulos, Exec. Dir.-Supply Chain & Facilities Mgmt.

GROWTH PLANS/SPECIAL FEATURES:

WIND Hellas Telecommunications SA is one of the large telecommunications operators in Greece. WIND's princip business is the provision of mobile and intern telecommunications services, including voice, broadband, pa TV and related value-added services, to pre-paid and contra customers. The company also offers businesses with netwo services, such as a voice mailbox, online payment solution teleconference services, caller ID, call waiting, call barring, ca forwarding and mobile email. WIND offers 2G, 3G, 4G and 5 network coverage, including 4G LTE services in select Grecia cities. Internet and fixed programs are available. WIND als sells devices such as cell phones, tablets, mobile broadban wearables, smart home and related accessories, as well a fixed devices, routers, modems, power connectors an extenders. In January 2022, WIND Hellas was acquired t United Group, an alternative telecommunications provider i southeast Europe, operating both telecommunication platforms and mass media outlets. United Group itself is owne by BC Partners LLP. United announced plans to merge i Nova business with WIND.

FINANCIAL DATA: Note: Data for latest year may not have been available at press time.

In U.S. $	2021	2020	2019	2018	2017	2016
Revenue	602,436,800	642,369,420	611,780,400	582,648,000	591,745,000	501,495,000
R&D Expense						
Operating Income						
Operating Margin %						
SGA Expense						
Net Income						
Operating Cash Flow						
Capital Expenditure						
EBITDA						
Return on Assets %						
Return on Equity %						
Debt to Equity						

CONTACT INFORMATION:

Phone: 30 210-615-8000   Fax: 30-210-510-0001
Toll-Free:
Address: 66 Kifissias Ave., Athens, 15125 Greece

STOCK TICKER/OTHER:

Stock Ticker: Private   Exchange:
Employees: 877   Fiscal Year Ends: 12/31
Parent Company: BC Partners LLP

SALARIES/BONUSES:

Top Exec. Salary: $ Bonus: $
Second Exec. Salary: $ Bonus: $

OTHER THOUGHTS:

Estimated Female Officers or Directors: 1
Hot Spot for Advancement for Women/Minorities:

Windstream Holdings Inc
investor.windstream.com/home/default.aspx

NAIC Code: 517110

TYPES OF BUSINESS:
Telephone Service--Local Exchange Carrier & Diversified
Network Communications
Technology Solutions
Broadband
Internet
Cloud
Digital Transformation
Fiber Optic

BRANDS/DIVISIONS/AFFILIATES:
Kinetic by Windstream
Windstream Enterprise
Windstream Wholesale

CONTACTS: Note: Officers with more than one job title may be intentionally listed here more than once.
Anthony Thomas, CEO
Robert Gunderman, CFO
Alan Wells, Director
Kristi Moody, General Counsel
Jeffery Small, President, Divisional
Layne Levine, President, Divisional

GROWTH PLANS/SPECIAL FEATURES:
Windstream Holdings, Inc. is a provider of advanced network communications and technology solutions for residences and businesses throughout the U.S. The firm also offers broadband, entertainment and security solutions to consumers and small businesses primarily in rural areas within 18 states. For residential consumers, Windstream's products and services keep the home environment connected via high-speed internet, digital TV and phone. For small and mid-sized businesses, products and solutions include internet, voice, unified communications, cloud and security. For enterprises, the firm offers multi-location networking, an office suite platform, SD-WAN, cloud, security and unified communications solutions and services. Windstream also offers digital transformation services for wholesale customers, including content and media providers, cloud and data center operators, international carriers, resale partners, cable operators, wireless and traditional network service providers and more. Wholesale solutions include scalable fiber optic connectivity, 5G fixed wireless service, wavelengths and Ethernet. Industries served by Windstream span healthcare, retail, K-12 education, higher education, hospitality, banking, government, content and media. Windstream owns and operates a nationwide network with over 170,000 fiber route miles, as well as a proprietary cloud core architecture. Business units of Windstream Holdings include Kinetic by Windstream, Windstream Enterprise, and Windstream Wholesale.

FINANCIAL DATA: Note: Data for latest year may not have been available at press time.

In U.S. $	2021	2020	2019	2018	2017	2016
Revenue	5,000,000,000	4,987,515,000	5,115,400,000	5,713,099,776	5,852,899,840	5,386,999,808
R&D Expense						
Operating Income						
Operating Margin %						
SGA Expense						
Net Income			-3,517,000,000	-723,000,000	-2,116,600,064	-383,500,000
Operating Cash Flow						
Capital Expenditure						
EBITDA						
Return on Assets %						
Return on Equity %						
Debt to Equity						

CONTACT INFORMATION:
Phone: 501 748-7000　　Fax:
Toll-Free: 866-445-5880
Address: 4001 N. Rodney Parkham Rd., Little Rock, AR 72212 United States

STOCK TICKER/OTHER:
Stock Ticker: Private　　Exchange:
Employees: 11,080　　Fiscal Year Ends: 12/31
Parent Company:

SALARIES/BONUSES:
Top Exec. Salary: $　　Bonus: $
Second Exec. Salary: $　　Bonus: $

OTHER THOUGHTS:
Estimated Female Officers or Directors: 3
Hot Spot for Advancement for Women/Minorities: Y

XIUS

NAIC Code: 511210C

TYPES OF BUSINESS:

Computer Software, Telecom, Communications & VOIP
Payment Processing Solutions
Mobile Technology Innovation
Mobile Payments
Over-the-Top Solutions
Internet of Things Solutions

BRANDS/DIVISIONS/AFFILIATES:

Megasoft Limited
XIUS AMPLIO
XIUS Infinet
XIUS Inergy
XIUS PowerRoam
XIUS Payment Manager
XIUS Wireless Wallet
X-Connect

CONTACTS: *Note: Officers with more than one job title may be intentionally listed here more than once.*

G.V. Kumar, CEO
Shridhar Thathachary, CFO
Sridhar Lanka, VP-Eng.
Kevin Bresnahan, Exec. VP-Oper.
Derek Bowman, Dir.-Bus. Dev.
Umakanta Mansingh, Head-Quality
Sundaresan Nandyal, Head-Global Sales

GROWTH PLANS/SPECIAL FEATURES:

XIUS, a subsidiary of Indian company Megasoft Limited, is mobile technology specialist focused on real-time transacti processing in mobile infrastructure, mobile payments and ov the-top (OTT) solutions. The company's mobile infrastructu solutions division provides mobile virtual network opera (VNO) services, a mobile services platform, 4G LTE service Internet of Things (IoT) connectivity and control, solutions XIUS AMPLIO mobile infrastructure solutions platform, re time billing through its XIUS INfinet product, real-time mob services through its XIUS INergy application, and steering roaming services through its XIUS PowerRoam solution. T mobile commerce division provides domestic and internatior recharge services through XIUS Payment Manager, mob payment technology via its XIUS Wireless Wallet applicatic mobile banking as well as mobile kiosks that offer mob operator services via its Active Poster product; an end-to-e platform for payment banks to set up technology infrastructu and launch services, and a secure authentication online/mob platform called eCognito. The mobile-enabled solutio division consists of business intelligence and geo-referencir which allows the mining of data and creation of custom view reports and dashboards, along with the capability of viewi key information data on a map; and retail consum engagement, providing consumer mobile interaction a engagement data to brands and retailers. IoT solutions inclu X-Connect, X-Fleet, X-Care device management, X-Ass Tracking/Management, and X-Manufacturing. XIUS headquartered in Massachusetts, with a global delivery cent in Hyderabad, India, and operations in the U.S., Mexic Malaysia and India.

FINANCIAL DATA: *Note: Data for latest year may not have been available at press time.*

In U.S. $	2021	2020	2019	2018	2017	2016
Revenue	468,231,000	507,385,000	480,600,000	480,612,000	385,400,000	354,955,000
R&D Expense						
Operating Income						
Operating Margin %						
SGA Expense						
Net Income	3,979,000	-3,782,000	33,419,000	86,900,000	26,000,000	21,856,000
Operating Cash Flow						
Capital Expenditure						
EBITDA						
Return on Assets %						
Return on Equity %						
Debt to Equity						

CONTACT INFORMATION:

Phone: 781-904-5000   Fax: 781-904-5601
Toll-Free:
Address: 15 Tyngsboro Rd., Unit 8C, North Chelmsford, MA 01863
United States

STOCK TICKER/OTHER:

Stock Ticker: Subsidiary   Exchange:
Employees: 500   Fiscal Year Ends: 03/31
Parent Company: Megasoft Limited

SALARIES/BONUSES:

Top Exec. Salary: $ Bonus: $
Second Exec. Salary: $ Bonus: $

OTHER THOUGHTS:

Estimated Female Officers or Directors:
Hot Spot for Advancement for Women/Minorities:

Zoom Video Communications Inc

zoom.us

NAIC Code: 511210C

TYPES OF BUSINESS:
Computer Software, Telecom, Communications & VOIP

GROWTH PLANS/SPECIAL FEATURES:
Zoom Video Communications provides a communications platform that connects people through video, voice, chat, and content sharing. The company's cloud-native platform enables face-to-face video and connects users across various devices and locations in a single meeting. Zoom, which was founded in 2011 and is headquartered in San Jose, California, serves companies of all sizes from all industries around the world.

BRANDS/DIVISIONS/AFFILIATES:
Zoom
Zoom Room

CONTACTS: *Note: Officers with more than one job title may be intentionally listed here more than once.*
Eric Yuan, CEO
Kelly Steckelberg, CFO
Shane Crehan, Chief Accounting Officer
Aparna Bawa, Chief Legal Officer
Janine Pelosi, Chief Marketing Officer
Ryan Azus, Other Executive Officer
Velchamy Sankarlingam, President, Divisional

FINANCIAL DATA: *Note: Data for latest year may not have been available at press time.*

In U.S. $	2021	2020	2019	2018	2017	2016
Revenue	2,651,368,000	622,658,000	330,517,000	151,478,000	60,817,000	
R&D Expense	164,080,000	67,079,000	33,014,000	15,733,000	9,218,000	
Operating Income	659,848,000	12,696,000	6,167,000	-4,833,000		
Operating Margin %	.25%	.02%	.02%	-.03%		
SGA Expense	1,005,451,000	427,487,000	230,335,000	109,798,000	39,127,000	
Net Income	672,316,000	25,305,000	7,584,000	-3,822,000	-14,000	
Operating Cash Flow	1,471,177,000	151,892,000	51,332,000	19,426,000	9,361,000	
Capital Expenditure	85,815,000	38,225,000	30,450,000	9,738,000	4,824,000	
EBITDA	688,705,000	29,145,000	13,175,000	-2,047,000	1,219,000	
Return on Assets %	.20%	.03%	.03%	-.04%		
Return on Equity %	.29%	.04%	.11%			
Debt to Equity	.02%	0.078				

CONTACT INFORMATION:
Phone: 650-397-6096   Fax:
Toll-Free: 888-799-9666
Address: 55 Almaden Blvd., Fl. 6, San Jose, CA 95113 United States

STOCK TICKER/OTHER:
Stock Ticker: ZM   Exchange: NAS
Employees: 6,787   Fiscal Year Ends: 01/31
Parent Company:

SALARIES/BONUSES:
Top Exec. Salary: $427,452   Bonus: $
Second Exec. Salary: $402,308   Bonus: $

OTHER THOUGHTS:
Estimated Female Officers or Directors:
Hot Spot for Advancement for Women/Minorities:

Sales, profits and employees may be estimates. Financial information, benefits and other data can change quickly and may vary from those stated here.

ZTE Corporation
NAIC Code: 334210

TYPES OF BUSINESS:
Telecommunications Equipment Manufacturing
Optical Networking Equipment
Intelligent & Next-Generation Network Systems
Mobile Phones

GROWTH PLANS/SPECIAL FEATURES:
ZTE Corp offers a suite of telecommunications and informatic
technology equipment to carriers, businesses, and the publ
sector. The firm's product portfolio covers wireless network:
core networks, fixed access, terminals, and other teleco
verticals. ZTE generates a majority of its revenue fro
equipment supporting carriers' networks, but also provide
handset terminals and telecom software systems. The Asi
Pacific region accounts for a majority of the firm's revenue, b
ZTE also has a presence in Europe and the Americas.

BRANDS/DIVISIONS/AFFILIATES:

CONTACTS: *Note: Officers with more than one job title may be intentionally listed here more than once.*
Ziyang Xu, CEO
Junshi Xie, COO
Ying LI, CFO
Zixue Li, Chmn.

FINANCIAL DATA: *Note: Data for latest year may not have been available at press time.*

In U.S. $	2021	2020	2019	2018	2017	2016
Revenue	16,966,420,000	15,029,950,000	13,442,660,000	12,668,800,000	16,121,020,000	14,997,730,000
R&D Expense	2,785,821,000	2,192,184,000	1,858,975,000	1,615,666,000	1,920,361,000	1,890,703,000
Operating Income	1,264,308,000	827,959,800	1,378,410,000	881,158,300	1,154,202,000	378,700,600
Operating Margin %	.07%	.06%	.10%	.07%	.07%	.03%
SGA Expense	1,673,631,000	1,470,890,000	1,505,384,000	1,547,288,000	1,924,901,000	1,938,890,000
Net Income	1,009,340,000	633,489,100	814,305,000	-972,847,700	751,044,000	-274,984,500
Operating Cash Flow	2,329,446,000	1,515,971,000	1,103,209,000	-1,365,262,000	1,069,642,000	779,301,300
Capital Expenditure	842,437,400	958,777,400	970,476,600	723,251,000	886,532,400	592,965,800
EBITDA	2,158,257,000	1,560,189,000	1,846,945,000	-525,803,600	1,344,558,000	417,598,800
Return on Assets %	.04%	.03%	.04%	-.05%	.04%	-.01%
Return on Equity %	.14%	.11%	.17%	-.19%	.13%	-.05%
Debt to Equity	.59%	0.539	0.305	0.081	0.075	0.14

CONTACT INFORMATION:
Phone: 86 75526770000   Fax: 86 75526770286
Toll-Free:
Address: Hi-tech Rd. S., No. 55, Shenzhen, Guangdong 518057 China

STOCK TICKER/OTHER:
Stock Ticker: ZTCOF   Exchange: PINX
Employees: 73,709   Fiscal Year Ends: 12/31
Parent Company:

SALARIES/BONUSES:
Top Exec. Salary: $ Bonus: $
Second Exec. Salary: $ Bonus: $

OTHER THOUGHTS:
Estimated Female Officers or Directors:
Hot Spot for Advancement for Women/Minorities:

ADDITIONAL INDEXES

Contents:

Index of Firms Noted as "Hot Spots for Advancement" for Women/Minorities   **448**

Index by Subsidiaries, Brand Names and Selected Affiliations   **450**

INDEX OF FIRMS NOTED AS HOT SPOTS FOR ADVANCEMENT FOR WOMEN & MINORITIES

Advanced Info Service plc
Advanced Micro Devices Inc (AMD)
Akamai Technologies Inc
Alaska Communications
Altice USA Inc
Anixter International Inc
Asia Satellite Telecommunications Holdings Ltd
AT&T Inc
Axiata Group Berhad
Bangkok Cable Co Ltd
BCE Inc (Bell Canada Enterprises)
Belden Inc
Bell Aliant Inc
Bell MTS Inc
Bezeq-The Israel Telecommunication Corp Ltd
Bharti Airtel Limited
BlackBerry Limited
Bouygues SA
Broadcom Inc
BT Global Services plc
BT Group plc
Celestica Inc
Cellcom Israel Ltd
Ceske Radiokomunikace as
China Mobile Limited
China Telecom Corporation Limited
Chunghwa Telecom Co Ltd
Cincinnati Bell Inc
Cisco Systems Inc
Colt Technology Services Group Limited
Comcast Corporation
Corning Incorporated
COSMOTE Mobile Telephones SA
Cox Communications Inc
Cricket Wireless LLC
Deutsche Telekom AG
Digi.com Bhd
DirecTV LLC (DIRECTV)
EarthLink LLC
Elisa Corporation
Empresa Nacional de Telecommunicacions SA (Entel)
Equinix Inc
Extreme Networks Inc
F5 Inc
Flex Ltd
Frontier Communications Corporation
Garmin Ltd
Gilat Satellite Networks Ltd
GoDaddy Inc
Hellenic Telecommunications Organization SA
HP Inc
Huawei Technologies Co Ltd

Hutchison Telecommunications Hong Kong Holdings Limited
Iliad SA
Ingram Micro Mobility
Inmarsat Global Limited
Intel Corporation
Internap Corporation
International Business Machines Corporation (IBM)
Jabil Inc
Juniper Networks Inc
Koninklijke Philips NV (Royal Philips)
Kratos Defense & Security Solutions Inc
L3Harris Technologies Inc
Liberty Global plc
LM Ericsson Telephone Company (Ericsson)
Lumen Technologies Inc
M1 Limited
Maxis Berhad
McAfee Corp
Microsoft Corporation
Millicom International Cellular SA
MiTAC Holdings Corp
Momentum Telecom
NETGEAR Inc
Neustar Inc
Newfold Digital Inc
Nokia Corporation
Orange
Partner Communications Co Ltd
PCCW Limited
Plantronics Inc
PLDT Inc
Proximus Group
Purple Communications Inc
Qualcomm Incorporated
Rogers Communications Inc
Rostelecom PJSC
Shaw Communications Inc
Shenandoah Telecommunications Company
Siemens AG
Singapore Telecommunications Limited
Skype Technologies Sarl
SMTC Corporation
Spok Inc
Swisscom AG
Sykes Enterprises Incorporated
Syniverse Technologies LLC
TDC A/S
Tele2 AB
Telekom Austria AG
Telenor ASA
Telephone and Data Systems Inc (TDS)
Telia Company AB
Telia Lietuva AB
Telkom SA SOC Limited
Tellabs Inc
Telstra Corporation Limited
TomTom International BV

TOT pcl
Total Access Communication PCL
Trimble Inc
TTEC Holdings Inc
Turkcell Iletisim Hizmetleri AS
United States Cellular Corporation
Verizon Communications Inc
Virgin Media Business Ltd
Vodafone Group plc
Windstream Holdings Inc

INDEX OF SUBSIDIARIES, BRAND NAMES AND AFFILIATIONS

Brand or subsidiary, followed by the name of the related corporation

10-10-719; **Startec Global Communications Corporation**
128 Technology; **Juniper Networks Inc**
2degrees; **Trilogy International Partners Inc**
3; **Hutchison Telecommunications Hong Kong Holdings Limited**
800 Series; **Socket Mobile Inc**
A1; **Telekom Austria AG**
A1; **America Movil SAB de CV**
A1 Bulgaria; **Telekom Austria AG**
A1 Telekom Austria AG; **Telekom Austria AG**
a4; **Altice USA Inc**
AAPT; **TPG Telecom Limited**
Acacia Communications Inc; **Cisco Systems Inc**
ACX; **Juniper Networks Inc**
Adaptive Network; **Ciena Corporation**
Adesol SA; **Telecom Argentina SA**
Advent International; **Laird Connectivity**
AGC Networks Ltd; **Black Box Corporation**
AISWare; **AsiaInfo Technologies Limited**
Al Yah Satellite Communications Company (Yahsat); **Thuraya Telecommunications Company**
AldiTalk; **Telefonica Deutschland Holding AG**
Alfa Telecom Turkey Limited; **Turkcell Iletisim Hizmetleri AS**
All3Media; **Liberty Global plc**
Altice; **Altice Europe NV**
Altice; **Altice Portugal SA**
Altice Empresas; **Altice Portugal SA**
Altice Group; **Altice Portugal SA**
Altice Mobile; **Altice USA Inc**
Altice NV; **Altice Portugal SA**
Alvaria CX Suite; **Alvaria Inc**
Alvaria WEM Suite; **Alvaria Inc**
AMD; **Advanced Micro Devices Inc (AMD)**
America Movil SAB de CV; **Embratel Participacoes SA**
America Movil SAB de CV; **Telecomunicaciones de Puerto Rico Inc**
America Movil SAB de CV; **Telefonos de Mexico SAB de CV (Telmex)**
America Movil SAB de CV; **Telekom Austria AG**
AMETEK Inc; **Telular Corporation**
Amobee; **Singapore Telecommunications Limited**
Anixter Trakr; **Anixter International Inc**
Antenna Products Corporation; **Phazar Antenna Corp**
Apollo Global Management Inc; **Intrado Corporation**
Apollo Global Management LLC; **Rackspace Technology Inc**
AppDynamics Inc; **Cisco Systems Inc**
Applanix; **Trimble Inc**
Applied Optical Systems Inc; **Optical Cable Corporation**

AppNeta Inc; **Broadcom Inc**
Apstra; **Juniper Networks Inc**
AQCOLOR; **BenQ Corporation**
Arbor; **NetScout Systems Inc**
Argela Software and Information Technologies; **Turk Telekomunikasyon AS**
Arm Limited; **SoftBank Group Corp**
AS Watson Group (HK) Limited; **CK Hutchison Holdings Limited**
Asia Satellite Telecommunications Company Limited; **Asia Satellite Telecommunications Holdings Ltd**
AsiaSat 5; **Asia Satellite Telecommunications Holdings Ltd**
AsiaSat 6; **Asia Satellite Telecommunications Holdings Ltd**
AsiaSat 7; **Asia Satellite Telecommunications Holdings Ltd**
AsiaSat 8; **Asia Satellite Telecommunications Holdings Ltd**
AsiaSat 9; **Asia Satellite Telecommunications Holdings Ltd**
Aspera; **International Business Machines Corporation (IBM)**
Aspirasi; **Axiata Group Berhad**
Assistance Services Group; **Sykes Enterprises Incorporated**
AssisTT; **Turk Telekomunikasyon AS**
Astound Broadband; **enTouch Systems Inc**
AT&T Inc; **AT&T Mexico SAU**
AT&T Inc; **AT&T Mobility LLC**
AT&T Inc; **Cricket Wireless LLC**
AT&T Inc; **DirecTV LLC (DIRECTV)**
AT&T TV; **DirecTV LLC (DIRECTV)**
AT&T Wireless; **AT&T Mobility LLC**
Athlon; **Advanced Micro Devices Inc (AMD)**
ATI; **Advanced Micro Devices Inc (AMD)**
ATLAS P25; **EF Johnson Technologies Inc**
ATN International Inc; **Alaska Communications**
Attowave Co Ltd; **Uniden Holdings Corporation**
Australia Singapore Cable; **Vocus Group Limited**
Avaya Inc; **Avaya Holdings Corp**
AVC Continente Audiovisual SA; **Telecom Argentina SA**
Avea; **Turk Telekomunikasyon AS**
Axiata Digital Labs; **Axiata Group Berhad**
AXIO-NET GmbH; **Trimble Inc**
Axway AMPLIFY; **Axway Inc**
AY YILDIZ; **Telefonica Deutschland Holding AG**
Azteca America; **INNOVATE Corp**
B Communications; **Bezeq-The Israel Telecommunication Corp Ltd**
B Riley Financial Inc; **United Online Inc**
bangalink; **Global Telecom Holding SAE**
Bangkok Cable Ventures Co Ltd; **Bangkok Cable Co Ltd**
Bangkok Magnet Wire Co Ltd; **Bangkok Cable Co Ltd**
Bangkok Solar Power Co Ltd; **Bangkok Cable Co Ltd**
banglalink; **Veon Ltd**
Bare Metal; **Equinix Inc**

INDEX OF SUBSIDIARIES, BRAND NAMES AND AFFILIATIONS, CONT.

Barkadahan sa SmarTone; **SmarTone Telecommunications Holdings Limited**

Bbox; **Bouygues SA**

BC Card Co Ltd; **KT Corporation**

BC Partners LLP; **WIND Hellas Telecommunications SA**

BC Partners LLP; **Cyxtera Technologies Inc**

BCE Inc; **Bell MTS Inc**

BCE Inc; **Glentel Inc**

BCE Inc (Bell Canada Enterprises); **Bell Aliant Inc**

BCX; **Telkom SA SOC Limited**

Beeline; **Veon Ltd**

Beena Vision Systems Inc; **Trimble Inc**

Beijing Ericsson Potevio; **Potevio Corporation**

Beijing Hi Sunsray Information Technology Limited; **Hi Sun Technology (China) Limited**

Belden Inc; **OPTERNA**

Belkin; **Belkin International Inc**

Belkin International Inc; **Foxconn Technology Co Ltd**

Bell; **Bell Aliant Inc**

Bell Aliant; **Bell Aliant Inc**

BendBroadband; **TDS Telecommunications LLC**

BenQ America Corporation; **BenQ Corporation**

BenQ Asia Pacific Corporation; **BenQ Corporation**

BenQ China; **BenQ Corporation**

BenQ Europe BV; **BenQ Corporation**

BenQ Latin America Corporation; **BenQ Corporation**

Bezeq; **B Communications Ltd**

Bharthi Group; **OneWeb Ltd**

Bharti Enterprises Limited; **Bharti Airtel Limited**

BICS; **Proximus Group**

BioTelemetry Inc; **Koninklijke Philips NV (Royal Philips)**

Birdie; **SmarTone Telecommunications Holdings Limited**

BlackBerry Alert; **BlackBerry Limited**

BlackBerry AtHoc; **BlackBerry Limited**

BlackBerry Certicom; **Certicom Corp**

BlackBerry IVY; **BlackBerry Limited**

BlackBerry Limited; **Certicom Corp**

BlackBerry Persona; **BlackBerry Limited**

BlackBerry Protect; **BlackBerry Limited**

BlackBerry QNX; **BlackBerry Limited**

BlackBerry Spark; **BlackBerry Limited**

Blackbird; **TCI International Inc**

Blau; **Telefonica Deutschland Holding AG**

Blizz; **TeamViewer AG**

Blue Planet Automation; **Ciena Corporation**

Blue Yonder; **Panasonic Corporation**

bluehost inc.; **Newfold Digital Inc**

Boost; **Axiata Group Berhad**

BOSS Revolution; **IDT Corporation**

Bouygues Construction; **Bouygues SA**

Bouygues Immobilier; **Bouygues SA**

Brightstar Corporation; **Likewize Corp**

British Telecommunications plc; **BT Group plc**

BT; **BT Group plc**

BT Group plc; **BT Global Services plc**

Buoygues Telecom; **Bouygues SA**

C More; **Telia Company AB**

Calient Optical Components; **Calient Technologies Inc**

Call me; **Telia Company AB**

Cam IT Solutions; **Koninklijke KPN NV (Royal KPN NV)**

Campaigner; **Consensus Cloud Solutions Inc**

Carlyle Group (The); **Syniverse Technologies LLC**

Casanet; **Maroc Telecom SA**

Celcom; **Axiata Group Berhad**

Cellcom Fixed Line Communications LP; **Cellcom Israel Ltd**

CellConnect; **Startec Global Communications Corporation**

CellMate; **Lattice Incorporated**

Cenovus Energy Inc; **CK Hutchison Holdings Limited**

Centerbridge Partners LP; **Speedcast International Limited**

Centric Solutions LLC; **Optical Cable Corporation**

CenturyLink Inc; **Lumen Technologies Inc**

Cerato; **MedTel Services LLC**

Charoen Pokphand Group; **True Corporation Public Company Limited**

chatr; **Rogers Communications Inc**

Cheddar; **Altice USA Inc**

CHIC Holdings Inc; **Hikari Tsushin Inc**

China Mobile Communications Group Co Ltd; **China Mobile Limited**

China Telecom Corporation Limited; **China Communications Services Corp Ltd**

China United Network Communications Group Co Ltd; **China Unicom (Hong Kong) Limited**

Choice; **ATN International Inc**

Chunghwa System Integration Co Ltd; **Chunghwa Telecom Co Ltd**

Chunghwa Telecom Global Inc; **Chunghwa Telecom Co Ltd**

Ciel; **SES SA**

Cincinnati Bell Extended Territories LLC; **Cincinnati Bell Inc**

Cincinnati Bell Telephone Company LLC; **Cincinnati Bell Inc**

Cisco Systems Inc; **Cisco Webex**

CK Hutchison Group Telecom; **CK Hutchison Holdings Limited**

CK Hutchison Holdings Limited; **Hutchison Telecommunications Hong Kong Holdings Limited**

CK Hutchison Holdings Limited; **Indosat Ooredoo Hutchison**

Claro; **America Movil SAB de CV**

Claro; **Telecomunicaciones de Puerto Rico Inc**

Claro Puerto Rico; **Telecomunicaciones de Puerto Rico Inc**

ClearCurve; **Corning Incorporated**

INDEX OF SUBSIDIARIES, BRAND NAMES AND AFFILIATIONS, CONT.

Clearlake Capital Group LP; **Newfold Digital Inc**
clearlink; **Sykes Enterprises Incorporated**
Code School; **Orange**
Cognos; **International Business Machines Corporation (IBM)**
Colas; **Bouygues SA**
Colt IQ Network; **Colt Technology Services Group Limited**
Columbia; **Sony Corporation**
Commnet; **ATN International Inc**
CommScope NEXT; **CommScope Holding Company Inc**
Connect Bidco Limited; **Inmarsat Global Limited**
Connect IQ; **Garmin Ltd**
Conservis; **TELUS Corporation**
Content Wavve; **SK Telecom Co Ltd**
Cordiant Digital Infrastructure Limited; **Ceske Radiokomunikace as**
Corrections Operating Platform; **Lattice Incorporated**
CosmoOne; **Hellenic Telecommunications Organization SA**
COSMOTE; **Hellenic Telecommunications Organization SA**
COSMOTE eValue; **COSMOTE Mobile Telephones SA**
COSMOTE Payments; **COSMOTE Mobile Telephones SA**
COSMOTE TV PRODUCTIONS; **Hellenic Telecommunications Organization SA**
Cox Enterprises Inc; **Cox Communications Inc**
Cricket; **Cricket Wireless LLC**
Crypto Learning Center; **Technical Communications Corporation**
CsomoONE; **COSMOTE Mobile Telephones SA**
CT Cinetrade AG; **Swisscom AG**
Cukurova Finance International Limited; **Turkcell Iletisim Hizmetleri AS**
Cukurova Telecom Holdings Limited; **Turkcell Iletisim Hizmetleri AS**
CX Analytics; **Sitel Corporation**
CX Digital; **Sitel Corporation**
CX Learning; **Sitel Corporation**
CX Operations; **Sitel Corporation**
CX Technology; **Sitel Corporation**
Cylance Inc; **BlackBerry Limited**
D600; **Socket Mobile Inc**
DASAN Zhone Solutions Inc; **DZS Inc**
DBM Global Inc; **INNOVATE Corp**
DBM10; **DSP Group Inc**
Deutsche Telekom (UK) Ltd; **Deutsche Telekom AG**
Deutsche Telekom AG; **COSMOTE Mobile Telephones SA**
Deutsche Telekom AG; **Hellenic Telecommunications Organization SA**
Deutsche Telekom AG; **Magyar Telekom plc**
Deutsche Telekom AG; **T-Mobile Polska SA**
Deutsche Telekom AG; **T-Mobile US Inc**

Dialog; **Axiata Group Berhad**
Diamond; **Glentel Inc**
DiGi; **Digi.com Bhd**
Digi; **Telenor ASA**
Digi Telecommunications Sdn Bhd; **Digi.com Bhd**
DigitalBridge Group Inc; **Boingo Wireless Inc**
Direct Dial; **Startec Global Communications Corporation**
Djezzy; **Global Telecom Holding SAE**
Djezzy; **Veon Ltd**
DNA; **Telenor ASA**
docomo; **NTT DOCOMO Inc**
Dodo; **Vocus Group Limited**
dtac; **Total Access Communication PCL**
dtac; **Telenor ASA**
dtac Reward; **Total Access Communication PCL**
DuraCase; **Socket Mobile Inc**
DuraScan; **Socket Mobile Inc**
DVCOM Data Private Limited; **Pakistan Telecommunication Company Limited**
Eagle XG; **Corning Incorporated**
EarthLink Mobile; **EarthLink LLC**
e-Builder; **Trimble Inc**
Edge/640 Optical Circuit Switch; **Calient Technologies Inc**
EdgeQAM Collection; **Blonder Tongue Laboratories Inc**
Edotco; **Axiata Group Berhad**
EE; **BT Group plc**
eir Mobile; **eircom Limited**
Eleven Street; **SK Telecom Co Ltd**
Elisa Automate; **Elisa Corporation**
Elisa Estonia; **Elisa Corporation**
Elisa Santa Monica; **Elisa Corporation**
Elisa Saunalahti; **Elisa Corporation**
Elisa Smart Factory; **Elisa Corporation**
Elisa Videra; **Elisa Corporation**
Empresa Brasileira de Telecomunicacoes SA; **Embratel Participacoes SA**
Enghouse Systems Limited; **Dialogic Inc**
Enia Oy; **Elisa Corporation**
Entel Movil; **Empresa Nacional de Telecommunicacions SA (Entel)**
Entel PCS; **Empresa Nacional de Telecommunicacions SA (Entel)**
Entel Phone Local Telephone; **Empresa Nacional de Telecommunicacions SA (Entel)**
Entel Telefonia Personal; **Empresa Nacional de Telecommunicacions SA (Entel)**
Entrust; **Entrust Corporation**
Epic; **Sony Corporation**
EPYC; **Advanced Micro Devices Inc (AMD)**
Equinix Fabric; **Equinix Inc**
Equinix Internet Exchange; **Equinix Inc**
Equinix SmartKey; **Equinix Inc**
ESS; **EchoStar Corporation**

INDEX OF SUBSIDIARIES, BRAND NAMES AND AFFILIATIONS, CONT.

Essar Group; **Black Box Corporation**
Etihad Etisalat Company; **Emirates Telecommunications Corporation (Etisalat)**
Etisalat; **Emirates Telecommunications Corporation (Etisalat)**
Etisalat Software Solutions (P) Ltd; **Emirates Telecommunications Corporation (Etisalat)**
Etisalat Sri Lanka; **Emirates Telecommunications Corporation (Etisalat)**
Eurocom Group; **B Communications Ltd**
Eurocom Group; **Bezeq-The Israel Telecommunication Corp Ltd**
European Aviation Network; **Inmarsat Global Limited**
Eutelsat; **OneWeb Ltd**
Eutelsat SA; **Eutelsat Communications SA**
Everyday Health; **Consensus Cloud Solutions Inc**
E-Vision; **Emirates Telecommunications Corporation (Etisalat)**
eVoice; **Consensus Cloud Solutions Inc**
EX; **Juniper Networks Inc**
Exchange; **Microsoft Corporation**
Ezys; **Telia Company AB**
Fastweb; **Swisscom AG**
Federal Agency for State Property Management; **Rostelecom PJSC**
felix; **TPG Telecom Limited**
Fibe; **Bell MTS Inc**
FibreNation; **TalkTalk Telecom Group Limited**
Fidelity Investments Inc; **Colt Technology Services Group Limited**
Fido; **Rogers Communications Inc**
Fintech Telecom LLC; **Nortel Inversora SA**
FiOS; **Frontier Communications Corporation**
Fireminds; **ATN International Inc**
Flip; **Vocus Group Limited**
Fortive Corporation; **Tektronix Inc**
Fortress Investment Group LLC; **SoftBank Group Corp**
Foundation Technology Worldwide LLC; **McAfee Corp**
Foxconn Interconnect Technology Limited; **Belkin International Inc**
Free; **Iliad SA**
Free Mobile Plan; **Iliad SA**
Free Pro; **Iliad SA**
Freebox; **Iliad SA**
Freedom; **Shaw Communications Inc**
Freenet Digital GmbH; **Freenet AG**
Freenet Energy GmbH; **Freenet AG**
Freenet TV; **Freenet AG**
FT Group Co ltd; **Hikari Tsushin Inc**
Fujitsu Limited; **Fujitsu Network Communications Inc**
Fullrate; **TDC A/S**
Future Mobility Corporation; **Foxconn Technology Co Ltd**
Gabon Telecom; **Maroc Telecom SA**
Garmin Connect; **Garmin Ltd**

Geoverse; **ATN International Inc**
GITN; **Telekom Malaysia Berhad**
Global Business Power Corporation; **First Pacific Company Limited**
Global Telesat Communications Ltd; **NextPlat Corp**
Global Xpress; **Inmarsat Global Limited**
Globalstar System; **Globalstar Inc**
Gorilla; **Corning Incorporated**
GovSat; **SES SA**
Goyo Electronics Co Ltd; **Hitachi Kokusai Electric Inc**
Grameenphone; **Telenor ASA**
GRAVIS-Computervertriebsgesellschaft mbH; **Freenet AG**
GrayWolf; **INNOVATE Corp**
Gruppo TIM; **Telecom Italia SpA**
GTCTrack; **NextPlat Corp**
GTT+; **ATN International Inc**
Guardicore Ltd; **Akamai Technologies Inc**
Gyro; **Telkom SA SOC Limited**
Halebop; **Telia Company AB**
Hawaiian Telecom Holdco Inc; **Cincinnati Bell Inc**
HBC Telecom Co Ltd; **Bangkok Cable Co Ltd**
HBO Max; **AT&T Inc**
HC2 Broadcasting Holdings Inc; **INNOVATE Corp**
HD Plus GmbH; **SES SA**
HDClear; **DSP Group Inc**
Hellenic Telecommunications Organization SA; **COSMOTE Mobile Telephones SA**
HHK Datentechnik GmbH; **Trimble Inc**
HIG Capital LLC; **SMTC Corporation**
HiNet; **Chunghwa Telecom Co Ltd**
HIS; **Realtime Corporation**
Hitachi Kokusai Electric America Ltd; **Hitachi Kokusai Electric Inc**
Hitachi Kokusai Electric Asia (Singapore) Pte Ltd; **Hitachi Kokusai Electric Inc**
Hitachi Kokusai Electric Comark LLC; **Hitachi Kokusai Electric Inc**
Hitachi Kokusai Electric Europe GmbH; **Hitachi Kokusai Electric Inc**
Hitachi Kokusai Electric Turkey; **Hitachi Kokusai Electric Inc**
Hitachi Kokusai Linear Equip. Eletronicos S/A; **Hitachi Kokusai Electric Inc**
HKT; **PCCW Limited**
Home Box Office; **AT&T Inc**
Home Controller; **Mediacom Communications Corporation**
Hon Hai Precision Industry Co Ltd; **Belkin International Inc**
HP Labs; **HP Inc**
Huawei; **Huawei Technologies Co Ltd**
Huddle; **Net2Phone Inc**
Hughes; **EchoStar Corporation**

INDEX OF SUBSIDIARIES, BRAND NAMES AND AFFILIATIONS, CONT.

Hutchison 3G Hong Kong Holdings Limited; **Hutchison Telecommunications Hong Kong Holdings Limited**
Hutchison China MediTech; **CK Hutchison Holdings Limited**
Hutchison Port Holdings Limited; **CK Hutchison Holdings Limited**
Hutchison Telecom Macau; **Hutchison Telecommunications Hong Kong Holdings Limited**
Hutchison Telecommunications Hong Kong Holdings; **CK Hutchison Holdings Limited**
Hutchison Telephone Company Limited; **Hutchison Telecommunications Hong Kong Holdings Limited**
Hutchison Water; **CK Hutchison Holdings Limited**
Hutchison Whampoa; **CK Hutchison Holdings Limited**
HYS Engineering Service Inc; **Hitachi Kokusai Electric Inc**
i24NEWS; **Altice USA Inc**
Ibeo Automotive Systems GmbH; **AAC Technologies Holdings Inc**
IBM; **International Business Machines Corporation (IBM)**
IDT Carrier Services; **IDT Corporation**
IDT Corporation; **Net2Phone Inc**
IE Group Inc; **Hikari Tsushin Inc**
iG; **Realtime Corporation**
IGN; **Consensus Cloud Solutions Inc**
iiNet; **TPG Telecom Limited**
Iliad; **Iliad SA**
IM3; **Indosat Ooredoo Hutchison**
INAP; **Internap Corporation**
Indian Telephone Industries; **ITI Limited**
Indofood CBP; **First Pacific Company Limited**
Indra Industrial Park Co Ltd; **Bangkok Cable Co Ltd**
InfiniCor; **Corning Incorporated**
Infinitum; **Telefonos de Mexico SAB de CV (Telmex)**
INFO AG; **Q Beyond AG**
Ingram Micro Inc; **Ingram Micro Mobility**
iNNOVA Bilisim Cozumleri AS; **Turk Telekomunikasyon AS**
INNOVATE Corp; **INNOVATE Corp**
Innovative Software Engineering; **Trimble Inc**
Innovium Inc; **Marvell Technology Group Ltd**
Inphi Corporation; **Marvell Technology Group Ltd**
Inspark; **Koninklijke KPN NV (Royal KPN NV)**
Intellitube; **Energy Focus Inc**
Intelsat General Communications LLC; **Intelsat SA**
IntelsatOne; **Intelsat SA**
International Business Exchange (IBX); **Equinix Inc**
Internet Business Technologies; **Realtime Corporation**
Internet Gold-Golden Lines Ltd; **B Communications Ltd**
Internet Gold-Golden Lines Ltd; **Bezeq-The Israel Telecommunication Corp Ltd**
Internode; **TPG Telecom Limited**
InTouch; **Lattice Incorporated**
INWIT SpA; **Telecom Italia SpA**

Ipanema; **Extreme Networks Inc**
iPrimus; **Vocus Group Limited**
IPVanish; **Consensus Cloud Solutions Inc**
Iridium NEXT; **Iridium Communications Inc**
Iris; **Corning Incorporated**
Israeli Telecommunications Corp Ltd; **B Communications Ltd**
ITI Neovision; **Liberty Global plc**
Itissalat Al-Maghrib; **Emirates Telecommunications Corporation (Etisalat)**
ITV; **Liberty Global plc**
Jaguar Network; **Iliad SA**
Jazz; **Veon Ltd**
J-Communication Inc; **Hikari Tsushin Inc**
Jio; **Jio (Reliance Jio Infocomm Limited)**
JioFi; **Jio (Reliance Jio Infocomm Limited)**
JioFiber; **Jio (Reliance Jio Infocomm Limited)**
Juno; **United Online Inc**
JUST AI Limited; **Mobile TeleSystems PJSC**
JVCKENWOOD Corporation; **EF Johnson Technologies Inc**
KAIROS; **EF Johnson Technologies Inc**
KEB HanaCard Co Ltd; **SK Telecom Co Ltd**
Keithley; **Tektronix Inc**
KENWOOD Viking; **EF Johnson Technologies Inc**
Keppel Corporation; **M1 Limited**
Khazanah Nasional Berhad; **Telekom Malaysia Berhad**
Kinetic by Windstream; **Windstream Holdings Inc**
Klarmobil GmbH; **Freenet AG**
KPN; **Koninklijke KPN NV (Royal KPN NV)**
KPN; **America Movil SAB de CV**
KPN Security; **Koninklijke KPN NV (Royal KPN NV)**
Kuwait Investment Authority; **Mobile Telecommunications Company KSCP (Zain Group)**
Kyivstar; **Veon Ltd**
Kyndryl Holdings Inc; **International Business Machines Corporation (IBM)**
LEAF; **Corning Incorporated**
Leap Frog; **VTech Holdings Limited**
Lebara; **TPG Telecom Limited**
LG Chem; **LG Corporation**
LG Corporation; **LG Electronics Inc**
LG Corporation; **LG Electronics USA Inc**
LG Corporation; **LG Uplus Corp**
LG Display; **LG Corporation**
LG Electronics; **LG Corporation**
LG Electronics Inc; **LG Electronics USA Inc**
LG Energy Solution; **LG Corporation**
LG Household & Health Care; **LG Corporation**
LG Innotek; **LG Corporation**
LG Signature; **LG Electronics Inc**
LG ThinQ; **LG Electronics Inc**
LG U+; **LG Uplus Corp**
LG U+; **LG Corporation**

INDEX OF SUBSIDIARIES, BRAND NAMES AND AFFILIATIONS, CONT.

Liberty Broadband Corporation; **GCI Communication Corp**
Liberty Global; **Telenet Group NV/SA**
Liberty Global plc; **VMED O2 UK Limited (Virgin Media O2)**
Light Era Development Co Ltd; **Chunghwa Telecom Co Ltd**
LightConnect Fabric Manager; **Calient Technologies Inc**
Lingo Communications LLC; **Startec Global Communications Corporation**
LinkedIn; **Microsoft Corporation**
LiquidIO Server Adapter; **Marvell Technology Group Ltd**
LiveDrive; **Consensus Cloud Solutions Inc**
Loral Space & Communications; **Telesat Corporation**
Lumen; **Lumen Technologies Inc**
Lumentum Holdings Inc; **Lumentum Operations LLC**
LX Holdings Corp; **LG Corporation**
Macquarie European Infrastructure Fund 6; **KCOM Group Limited**
Macquarie Group Limited; **KCOM Group Limited**
Macquarie Group Limited; **TDC A/S**
Macquarie Infrastructure and Real Assets; **KCOM Group Limited**
MADA Bahrain; **Mobile Telecommunications Company KSCP (Zain Group)**
Magellan; **MiTAC Holdings Corp**
Mahanagar Telephone Mauritius Limited; **Mahanagar Telephone Nigam Limited**
Marlin Equity Partners LLC; **Tellabs Inc**
Mashable; **Consensus Cloud Solutions Inc**
Mauritel SA; **Maroc Telecom SA**
MaxisONE; **Maxis Berhad**
McAfee Global Threat Intelligence; **McAfee Corp**
McAfee LLC; **McAfee Corp**
Medina Capital; **Cyxtera Technologies Inc**
Megasoft Limited; **XIUS**
Members Mobile Inc; **Hikari Tsushin Inc**
Menara Kuala Lumpur; **Telekom Malaysia Berhad**
MEO; **Altice Portugal SA**
Metro by T-Mobile; **T-Mobile US Inc**
Metro Pacific Investments Corporation; **First Pacific Company Limited**
Meural; **NETGEAR Inc**
Mezzanine; **Oblong Inc**
MiCloud Connect CX; **Mitel Networks Corporation**
MiContact Center; **Mitel Networks Corporation**
Micro Sistemas SA; **Nortel Inversora SA**
Microsoft Corporation; **Skype Technologies Sarl**
Microsoft Teams; **Microsoft Corporation**
Ministry of Transportation and Communications; **Chunghwa Telecom Co Ltd**
Mio; **MiTAC Holdings Corp**
Mist; **Juniper Networks Inc**

MiTAC Computing Technology Corp; **MiTAC Holdings Corp**
MiTAC Digital Technology Corp; **MiTAC Holdings Corp**
MiTAC International Corp; **MiTAC Holdings Corp**
Mitel Teamwork; **Mitel Networks Corporation**
Mobbit; **Realtime Corporation**
Mobilcom Debitel GmbH; **Freenet AG**
Mobile Vikings; **Proximus Group**
Mobileye; **Intel Corporation**
Mobilink; **Global Telecom Holding SAE**
Mobily; **Emirates Telecommunications Corporation (Etisalat)**
MOCHE; **Altice Portugal SA**
Monitis; **TeamViewer AG**
Moov Benin; **Maroc Telecom SA**
Moov Ivory Coast; **Maroc Telecom SA**
Moov Togo; **Maroc Telecom SA**
Movistar; **Telefonica de Argentina SA**
Movistar; **Telefonica del Peru SAA**
Movistar; **Telefonica Chile SA**
Movistar; **Telefonica SA**
MTS; **Mobile TeleSystems PJSC**
MTS Bank; **Mobile TeleSystems PJSC**
MTS Systems Corporation; **Amphenol Corporation**
Mubadala Investment Company; **Thuraya Telecommunications Company**
Multimedia University; **Telekom Malaysia Berhad**
Multisys Technologies Corporation; **PLDT Inc**
MusicFreedom; **Digi.com Bhd**
MVISION Device; **McAfee Corp**
MX; **Juniper Networks Inc**
MyCall; **Telia Company AB**
MyCountry Number; **Startec Global Communications Corporation**
MyDigi; **Digi.com Bhd**
MyFax; **Consensus Cloud Solutions Inc**
MyTopo; **Trimble Inc**
N2P Remote; **Net2Phone Inc**
Name Jet; **Newfold Digital Inc**
Nanjing Ericsson Panda; **Potevio Corporation**
National Retail Solutions; **IDT Corporation**
Navman; **MiTAC Holdings Corp**
NCS Pte Ltd; **Singapore Telecommunications Limited**
net2phone; **IDT Corporation**
NETVIGATOR; **HKT Trust and HKT Limited**
NetVisit; **Lattice Incorporated**
Network Solutions; **Newfold Digital Inc**
NetZero; **United Online Inc**
New York Interconnect; **Altice USA Inc**
News 12 Networks; **Altice USA Inc**
Nexa; **Aware Inc**
NEXEDGE; **EF Johnson Technologies Inc**
Next Generation Network; **Q Beyond AG**
Next47; **Siemens AG**

INDEX OF SUBSIDIARIES, BRAND NAMES AND AFFILIATIONS, CONT.

Nexus; **Lattice Incorporated**
NFC Holdings Inc; **Hikari Tsushin Inc**
NFL Sunday Ticket; **DirecTV LLC (DIRECTV)**
Ngenius; **NetScout Systems Inc**
Nighthawk; **NETGEAR Inc**
Ningbo Electronics; **Potevio Corporation**
Nippon Telegraph and Telephone Corporation (NTT);
NTT DOCOMO Inc
Nippon Telegraph and Telephone Corporation (NTT);
Verio Inc
NITROX; **Marvell Technology Group Ltd**
NJJ Group; **eircom Limited**
NJJ Telecom Europe; **eircom Limited**
Nojoom; **Ooredoo QPSC**
Nokia Bell Labs; **Nokia Corporation**
Nova; **WIND Hellas Telecommunications SA**
Now TV; **PCCW Limited**
NTT Anode Energy Corporation; **Nippon Telegraph and Telephone Corporation (NTT)**
NTT Communications; **Verio Inc**
NTT Communications Corporation; **Nippon Telegraph and Telephone Corporation (NTT)**
NTT Data Corporation; **Nippon Telegraph and Telephone Corporation (NTT)**
NTT DoCoMo Inc; **Nippon Telegraph and Telephone Corporation (NTT)**
NTT East; **Nippon Telegraph and Telephone Corporation (NTT)**
NTT Ltd; **Nippon Telegraph and Telephone Corporation (NTT)**
NTT Urban Solutions Inc; **Nippon Telegraph and Telephone Corporation (NTT)**
NTT West; **Nippon Telegraph and Telephone Corporation (NTT)**
Nucleo SA; **Nortel Inversora SA**
Nucleo SAE; **Telecom Argentina SA**
NuevaTel; **Trilogy International Partners Inc**
Nuuday; **TDC A/S**
O2; **Telefonica Deutschland Holding AG**
O2; **Telefonica SA**
Oasis Smart SIM Europe SAS; **Tata Communications Limited**
OCTEON; **Marvell Technology Group Ltd**
OCTEON Fusion-M; **Marvell Technology Group Ltd**
Office 365; **Microsoft Corporation**
Oi TV; **Oi SA**
Ojer Telekomunikasyon AS; **Turk Telekomunikasyon AS**
Okarte; **Telia Company AB**
Olivetti; **Telecom Italia SpA**
Omantel; **Mobile Telecommunications Company KSCP (Zain Group)**
Omnis; **NetScout Systems Inc**
Onatel; **Maroc Telecom SA**
One; **ATN International Inc**

one.Vip; **Telekom Austria AG**
OneDrive; **Microsoft Corporation**
OneNeck; **Telephone and Data Systems Inc (TDS)**
ONERetail; **Maxis Berhad**
Onica; **Rackspace Technology Inc**
Ooredoo Money; **Ooredoo QPSC**
Ooredoo QPSC; **Indosat Ooredoo Hutchison**
Open eir; **eircom Limited**
Openreach; **BT Group plc**
Openserve; **Telkom SA SOC Limited**
Optically Variable Magnetic Pigment; **Viavi Solutions Inc**
Optically Variable Pigment; **Viavi Solutions Inc**
Optimum; **Altice USA Inc**
Optus; **Singapore Telecommunications Limited**
Orang Fab; **Orange**
Orange; **Orange Polska SA**
Orange Business Services; **Orange**
Orange Communications Luxembourg; **Orange Belgium**
Orange Customer Service Sp z o o; **Orange Polska SA**
Orange Foundation; **Orange Polska SA**
Orange SA; **Orange Belgium**
Orange Ventures; **Orange**
Orascom Telecom Algeria SpA; **Global Telecom Holding SAE**
Orbi Voice; **NETGEAR Inc**
Orbital Satcom Corp; **NextPlat Corp**
Orbsat Corp; **NextPlat Corp**
ORCHESTRA; **Inmarsat Global Limited**
Orcon; **Vocus Group Limited**
Ortel Mobile; **Koninklijke KPN NV (Royal KPN NV)**
Ortel Mobile; **Telefonica Deutschland Holding AG**
OTE Estate; **Hellenic Telecommunications Organization SA**
OTE Rural; **Hellenic Telecommunications Organization SA**
OTE SAT-MARITEL; **Hellenic Telecommunications Organization SA**
OTEGlobe; **Hellenic Telecommunications Organization SA**
OTN Systems NV; **Belden Inc**
Otter Media Holdings; **AT&T Inc**
Outlook.com; **Microsoft Corporation**
Pacific Century Premium Developments; **PCCW Limited**
Pak Telecom Mobile Ltd; **Pakistan Telecommunication Company Limited**
Pakistan Mobile Communications Ltd; **Global Telecom Holding SAE**
Panseed Life Sciences LLC; **INNOVATE Corp**
Partner; **Partner Communications Co Ltd**
PayPay Corporation; **SoftBank Group Corp**
PCCW Global; **HKT Trust and HKT Limited**
PCCW Limited; **HKT Trust and HKT Limited**
PCCW Media; **PCCW Limited**
PCCW Solutions; **PCCW Limited**
PCCW Teleservices; **HKT Trust and HKT Limited**

INDEX OF SUBSIDIARIES, BRAND NAMES AND AFFILIATIONS, CONT.

Peacock; **Comcast Corporation**
PEM SAU; **Telecom Argentina SA**
Personal Envios SA; **Nortel Inversora SA**
Philadelphia Flyers; **Comcast Corporation**
Philex Mining Corporation; **First Pacific Company Limited**
Phonero; **Telia Company AB**
Pineapple Energy LLC; **Pineapple Energy Inc**
Platinum Equity LLC; **Ingram Micro Mobility**
PLDT Inc; **First Pacific Company Limited**
Plusnet; **BT Group plc**
Poly; **Plantronics Inc**
Pongtipa Co Ltd; **Bangkok Cable Co Ltd**
POP; **Purple Communications Inc**
Potevio Capitel; **Potevio Corporation**
Potevio Designing & Planning Institute; **Potevio Corporation**
Potevio Eastcom; **Potevio Corporation**
Potevio Taili; **Potevio Corporation**
PowerMarketing; **Realtime Corporation**
Poynt; **GoDaddy Inc**
PPF Group NV; **O2 Czech Republic AS**
Premium Water Holdings Co Ltd; **Hikari Tsushin Inc**
Proxim ClearConnect; **Proxim Wireless Corporation**
Proxim FastConnect; **Proxim Wireless Corporation**
Proxim SmartConnect; **Proxim Wireless Corporation**
Proximus; **Proximus Group**
PT Indofood Sukses Makmur Tbk; **First Pacific Company Limited**
PTC Systems Pte Ltd; **Internet Initiative Japan Inc**
PTX; **Juniper Networks Inc**
PXP Energy; **First Pacific Company Limited**
Qisda Corporation; **BenQ Corporation**
QSC AG; **Q Beyond AG**
QuetzSat; **SES SA**
Radeon; **Advanced Micro Devices Inc (AMD)**
Rapid Fire; **Anixter International Inc**
RCA; **Sony Corporation**
RCN Telecom Services LLC; **enTouch Systems Inc**
RealPresence; **Plantronics Inc**
Realtime DMC; **Realtime Corporation**
Red Hat OpenShift; **International Business Machines Corporation (IBM)**
Redu Space Services; **SES SA**
Register.com; **Newfold Digital Inc**
Registry Services; **VeriSign Inc**
Reliance Industries Limited; **Jio (Reliance Jio Infocomm Limited)**
RIFT; **DZS Inc**
RL Drake Holdings LLC; **Blonder Tongue Laboratories Inc**
Robi; **Axiata Group Berhad**
Rogers; **Rogers Communications Inc**
Rogers Communication Inc; **Glentel Inc**

Rohde & Schwarz USA Inc; **Rohde & Schwarz GmbH & Co KG**
Rozgar Microfinance Bank Limited; **Pakistan Telecommunication Company Limited**
Ryzen; **Advanced Micro Devices Inc (AMD)**
S B Israel Telecom Ltd; **Partner Communications Co Ltd**
S Series Optical Circuit Switch; **Calient Technologies Inc**
S550; **Socket Mobile Inc**
S700; **Socket Mobile Inc**
Sahabat Setia SmarTone; **SmarTone Telecommunications Holdings Limited**
salesforce.com Inc; **Slack Technologies Inc**
Samsung Canada; **Glentel Inc**
Samsung Group; **Samsung Electronics Co Ltd**
SAPO; **Altice Portugal SA**
SB Northstar LP; **SoftBank Group Corp**
Scaleway; **Iliad SA**
Scarlet; **Proximus Group**
Scitum; **Telefonos de Mexico SAB de CV (Telmex)**
Searchlight Capital Partners LP; **Mitel Networks Corporation**
Sebit Egitim ve Bilgi Teknolojileri AS; **Turk Telekomunikasyon AS**
Security Services; **VeriSign Inc**
SES Government Solutions; **SES SA**
SES Techcom; **SES SA**
SharePoint; **Microsoft Corporation**
Shaw; **Shaw Communications Inc**
Shaw Business; **Shaw Communications Inc**
Shaw Direct; **Shaw Communications Inc**
Sheba Telecom (Pvt) Limited; **Global Telecom Holding SAE**
Siemens Advanta; **Siemens AG**
Siemens Financial Services; **Siemens AG**
Siemens Healthineers AG; **Siemens AG**
Siemens Real Estate; **Siemens AG**
Simply Mac; **Simply Inc**
Simyo; **Koninklijke KPN NV (Royal KPN NV)**
Singapore Telecommunications Limited; **Singtel Optus Pty Limited**
Singapore Telecommunications Limited; **Bharti Airtel Limited**
Singtel Digital Media; **Singapore Telecommunications Limited**
Singtel Innov8; **Singapore Telecommunications Limited**
Siris Capital Group LLC; **Newfold Digital Inc**
Sitel Group; **Sykes Enterprises Incorporated**
SK Hynix Inc; **SK Telecom Co Ltd**
SK Telecom Co Ltd; **SK Broadband Co Ltd**
Sky Limited; **Comcast Corporation**
Sky News; **Comcast Corporation**
Sky Sports; **Comcast Corporation**
SkyBitz; **Telular Corporation**
Skype for Business; **Microsoft Corporation**

INDEX OF SUBSIDIARIES, BRAND NAMES AND AFFILIATIONS, CONT.

Slack; **Slack Technologies Inc**
Slack Enterprise Grid; **Slack Technologies Inc**
Slingshot; **Vocus Group Limited**
Smart Communications Inc; **First Pacific Company
Limited**
Smart Sky Private Limited; **Pakistan Telecommunication
Company Limited**
SmartOne; **Globalstar Inc**
SmarTone Solutions; **SmarTone Telecommunications
Holdings Limited**
SocketScan; **Socket Mobile Inc**
SoftBank Group Capital Limited; **SoftBank Group Corp**
SoftBank Investment Advisers; **SoftBank Group Corp**
SoftBank Vision Fund LP; **SoftBank Group Corp**
Solcon; **Koninklijke KPN NV (Royal KPN NV)**
Solidarity FabLab; **Orange**
Sony Classical; **Sony Corporation**
Sony Corporation of America; **Sony Corporation**
Sony Group Corporation; **Sony Corporation**
Sony Music Nashville; **Sony Corporation**
Sotelma; **Maroc Telecom SA**
SparkLab.co.nz; **Spark New Zealand Limited**
Spectra; **NetScout Systems Inc**
Spectrum; **Charter Communications Inc**
Spectrum Community Solutions; **Charter
Communications Inc**
Spectrum Enterprise Solutions; **Charter Communications
Inc**
Spectrum Internet Gig; **Charter Communications Inc**
Spectrum Mobile; **Charter Communications Inc**
Spectrum Reach; **Charter Communications Inc**
Spectrum TV; **Charter Communications Inc**
Spectrum Voice; **Charter Communications Inc**
Speedpull; **Anixter International Inc**
Spok Go; **Spok Inc**
Spok Holdings Inc; **Spok Inc**
SPOT; **Globalstar Inc**
Sprint Corporation; **T-Mobile US Inc**
SPX Corporation; **TCI International Inc**
SRA Holdings Inc; **Proxim Wireless Corporation**
SRX; **Juniper Networks Inc**
Star One SA (Embratel Star One); **Embratel
Participacoes SA**
STINGR; **Globalstar Inc**
Straight Talk; **America Movil SAB de CV**
STX-3; **Globalstar Inc**
Suddenlink; **Altice USA Inc**
Sun Hung Kai Properties Limited; **SmarTone
Telecommunications Holdings Limited**
Sunrise Communications AG; **Liberty Global plc**
Suttle-Straus Inc; **Telephone and Data Systems Inc
(TDS)**
Suzhou Chunxing Precision Mechanical Co Ltd; **Calient
Technologies Inc**
SWIPGLOBAL Limited; **Mobile TeleSystems PJSC**

Swisscom AG; **FASTWEB SpA**
Swisscom Italia Srl; **FASTWEB SpA**
Swisscom Switzerland; **Swisscom AG**
Sykes Digital Services; **Sykes Enterprises Incorporated**
Synercy Communications Management; **Hello Direct Inc**
TACSAT-214-SATCOM; **Phazar Antenna Corp**
TalkTalk; **TalkTalk Telecom Group Limited**
TalkTalk Business; **TalkTalk Telecom Group Limited**
Tango; **Proximus Group**
Tata Capital; **Tata Group**
Tata Communications; **Tata Group**
Tata Consultancy Services; **Tata Group**
Tata Group; **Tata Communications Limited**
Tata Group; **Tata Teleservices Limited**
Tata Investment Corporation; **Tata Group**
Tata Motors; **Tata Group**
Tata Oil Mills Company (TOMCO); **Tata Group**
Tata Sons Private Limited; **Tata Group**
Tata Sons Private Limited; **Tata Teleservices Limited**
Tata Steel; **Tata Group**
Tata Teleservices (Maharashtra) Limited; **Tata
Teleservices Limited**
Tawana Container Co Ltd; **Bangkok Cable Co Ltd**
tBooth Wireless; **Glentel Inc**
Tchibo mobil; **Telefonica Deutschland Holding AG**
TDC Erhverv; **TDC A/S**
TDC NET; **TDC A/S**
TDS Broadband Service LLC; **TDS Telecommunications
LLC**
TDS Telecommunications Corporation; **Telephone and
Data Systems Inc (TDS)**
TeamViewer; **TeamViewer AG**
TeamViewer Frontline; **TeamViewer AG**
TeamViewer Pilot; **TeamViewer AG**
TeamViewer Remote Management; **TeamViewer AG**
TeamViewer Tensor; **TeamViewer AG**
Telcel; **America Movil SAB de CV**
Tele Danmark Commmunications; **TDC A/S**
Telecom Argentina SA; **Nortel Inversora SA**
Telecom Argentina USA Inc; **Telecom Argentina SA**
Telecom Argentina USA Inc; **Nortel Inversora SA**
Telecom Italia SpA; **TIM SA**
Telecom Italia Sparkle SpA; **Telecom Italia SpA**
Telecom Personal SA; **Nortel Inversora SA**
Telecom Service Co Ltd; **Hikari Tsushin Inc**
Telefonica; **Telefonica SA**
Telefonica Brasil; **Telefonica SA**
Telefonica Deutschland; **Telefonica SA**
Telefonica Empresas Chile SA; **Telefonica Chile SA**
Telefonica Espana; **Telefonica SA**
Telefonica Germany GmbH & Co OHG; **Telefonica
Deutschland Holding AG**
Telefonica Gestion de Servicios Compartidos; **Telefonica
Chile SA**
Telefonica HispanoAmerica; **Telefonica SA**

INDEX OF SUBSIDIARIES, BRAND NAMES AND AFFILIATIONS, CONT.

Telefonica Ingenieria de Seguridad; **Telefonica del Peru SAA**

Telefonica Moviles Chile Larga Distancia SA; **Telefonica Chile SA**

Telefonica Moviles Peru SAC; **Telefonica del Peru SAA**

Telefonica Multimedia; **Telefonica Chile SA**

Telefonica SA; **Telefonica Brasil SA**

Telefonica SA; **Telefonica Chile SA**

Telefonica SA; **Telefonica de Argentina SA**

Telefonica SA; **Telefonica del Peru SAA**

Telefonica SA; **Telefonica Deutschland Holding AG**

Telefonica SA; **VMED O2 UK Limited (Virgin Media O2)**

Telefonica UK; **Telefonica SA**

Telekom Hotel Balatonkenese; **Magyar Telekom plc**

Telenet; **Liberty Global plc**

Telenor Denmark; **Telenor ASA**

Telenor Group; **Digi.com Bhd**

Telenor Group; **Total Access Communication PCL**

Telenor Norway; **Telenor ASA**

Telenor Pakistan; **Telenor ASA**

Telenor Sweden; **Telenor ASA**

Telephone and Data Systems Inc; **United States Cellular Corporation**

Telephone and Data Systems Inc (TDS); **TDS Telecommunications LLC**

Telfonica Moviles Chile SA; **Telefonica Chile SA**

Telguard; **Telular Corporation**

Telia; **Telia Company AB**

Telia Company AB; **Telia Lietuva AB**

Telia Sonera Finland Oyj; **Turkcell Iletisim Hizmetleri AS**

Telindus; **Proximus Group**

Telkom Business; **Telkom SA SOC Limited**

Telkom Consumer; **Telkom SA SOC Limited**

Telmex; **America Movil SAB de CV**

Telmore; **TDC A/S**

TELUS SmartHome Security; **TELUS Corporation**

Telxius Torres Peru; **Telefonica del Peru SAA**

Temasek Holdings Pvt Limited; **Singapore Technologies Telemedia Pte Ltd**

Terra Networks Perus SA; **Telefonica del Peru SAA**

TF1; **Bouygues SA**

Tgestiona; **Telefonica del Peru SAA**

Thai Copper Rod Co Ltd; **Bangkok Cable Co Ltd**

Thai Ministry of Finance; **TOT pcl**

Thermo Capital Partners LLC; **Globalstar Inc**

Threadripper; **Advanced Micro Devices Inc (AMD)**

Three; **Indosat Ooredoo Hutchison**

ThunderX; **Marvell Technology Group Ltd**

Thuraya; **Emirates Telecommunications Corporation (Etisalat)**

Tigo; **Millicom International Cellular SA**

Tigo Business; **Millicom International Cellular SA**

TIM SpA; **Telecom Italia SpA**

TM Global; **Telekom Malaysia Berhad**

TM One; **Telekom Malaysia Berhad**

TMC; **Bouygues SA**

T-Mobile; **Deutsche Telekom AG**

T-Mobile; **Magyar Telekom plc**

T-Mobile; **T-Mobile US Inc**

T-Mobile International AG; **T-Mobile US Inc**

Tofane Global; **iBasis Inc**

TomTom; **TomTom International BV**

Toscafund Asset Management LLP; **TalkTalk Telecom Group Limited**

TPG; **TPG Telecom Limited**

TPG Capital; **enTouch Systems Inc**

TPG Capital; **DirecTV LLC (DIRECTV)**

TPG Telecom; **Vodafone Group plc**

TPG Telecom Limited; **AAPT Limited**

TracFone; **America Movil SAB de CV**

TransUnion; **Neustar Inc**

Trilogy International Partners LLC; **Trilogy International Partners Inc**

Trive Capital; **EarthLink LLC**

TrueID; **True Corporation Public Company Limited**

TrueMoney; **True Corporation Public Company Limited**

TrueMove H; **True Corporation Public Company Limited**

TrueMusic; **True Corporation Public Company Limited**

TrueOnline; **True Corporation Public Company Limited**

TrueVisions; **True Corporation Public Company Limited**

Trust Networks Inc; **Internet Initiative Japan Inc**

Trustwave; **Singapore Telecommunications Limited**

TTEC Digital; **TTEC Holdings Inc**

TTEC Engage; **TTEC Holdings Inc**

TTM Technologies Inc; **Anaren Inc**

TTNET; **Turk Telekomunikasyon AS**

Turk Telekom International; **Turk Telekomunikasyon AS**

Turkcell Holding AS; **Turkcell Iletisim Hizmetleri AS**

TYAN; **MiTAC Holdings Corp**

U Microfinance Bank Limited; **Pakistan Telecommunication Company Limited**

U+ homeBoy; **LG Uplus Corp**

Ubimax GmbH; **TeamViewer AG**

Ufone; **Pakistan Telecommunication Company Limited**

UHP Networks Inc; **Comtech Telecommunications Corp**

UK Department for Business Energy and Industrial; **OneWeb Ltd**

Uniden America Corporation; **Uniden Holdings Corporation**

Uniden Australia Proprietary Limited; **Uniden Holdings Corporation**

Uniden Japan Corporation; **Uniden Holdings Corporation**

Uniden Vietnam Limited; **Uniden Holdings Corporation**

INDEX OF SUBSIDIARIES, BRAND NAMES AND AFFILIATIONS, CONT.

unifi; **Telekom Malaysia Berhad**
Union of Huawei Investment & Holding Co; **Huawei Technologies Co Ltd**
United Group; **WIND Hellas Telecommunications SA**
United States Cellular Corporation; **Telephone and Data Systems Inc (TDS)**
United Telecom Limited; **Mahanagar Telephone Nigam Limited**
Universal Pictures; **Comcast Corporation**
Universal Studios; **Comcast Corporation**
UPC; **Liberty Global plc**
U-verse; **DirecTV LLC (DIRECTV)**
VADS Berhad; **Telekom Malaysia Berhad**
Vantage; **Frontier Communications Corporation**
Vascade; **Corning Incorporated**
Vector Capital; **Alvaria Inc**
velcom; **Telekom Austria AG**
Veon; **Veon Ltd**
VEON Ltd; **Global Telecom Holding SAE**
Verizon Business Group; **Verizon Communications Inc**
Verizon Consumer Group; **Verizon Communications Inc**
Verizon Fios; **Verizon Communications Inc**
Vesper Marine; **Garmin Ltd**
Vibrant Energy; **ATN International Inc**
VideoFreedom; **Digi.com Bhd**
Viewpoint; **Trimble Inc**
Viking P25; **EF Johnson Technologies Inc**
Vip mobile; **Telekom Austria AG**
Vipnet; **Telekom Austria AG**
Virgin Media; **Liberty Global plc**
Virgin Media Limited; **Virgin Media Business Ltd**
Virtual1; **TalkTalk Telecom Group Limited**
Vivendi SA; **Telecom Italia SpA**
Viya; **ATN International Inc**
Vodafone; **TPG Telecom Limited**
Vodafone Egypt; **Telecom Egypt SAE**
Vodafone Ziggo; **Liberty Global plc**
Volterra; **F5 Inc**
Voyager Innovations Holdings Pte Ltd; **PLDT Inc**
Vrio; **AT&T Inc**
Vtech; **VTech Holdings Limited**
waipu.tv; **Freenet AG**
WanerMedia; **AT&T Inc**
Warner Bros; **AT&T Inc**
Watson; **International Business Machines Corporation (IBM)**
Wayra Peru Acelerador de Proyectos SAC; **Telefonica del Peru SAA**
Web.com Contractor Services; **Newfold Digital Inc**
Web.com Online Marketing; **Newfold Digital Inc**
WebSpectator; **Realtime Corporation**
Wemo; **Belkin International Inc**
WESCO International Inc; **Anixter International Inc**
Westcoast Group Holdings Limited; **Data Select Limited**
WIND; **WIND Hellas Telecommunications SA**

Windstream Enterprise; **Windstream Holdings Inc**
Windstream Wholesale; **Windstream Holdings Inc**
WIRELESS etc; **Glentel Inc**
Wireless Zone; **Glentel Inc**
WIRELESSWAVE; **Glentel Inc**
WORP; **Proxim Wireless Corporation**
www.bluewin.ch; **Swisscom AG**
www.simplymac.com; **Simply Inc**
Xandr; **AT&T Inc**
X-Connect; **XIUS**
XFINITY; **Comcast Corporation**
XIUS AMPLIO; **XIUS**
XIUS Inergy; **XIUS**
XIUS Infinet; **XIUS**
XIUS Payment Manager; **XIUS**
XIUS PowerRoam; **XIUS**
XIUS Wireless Wallet; **XIUS**
XL Axiata; **Axiata Group Berhad**
Xperia; **Sony Corporation**
XS4ALL; **Koninklijke KPN NV (Royal KPN NV)**
xScale; **Equinix Inc**
Xtream; **Mediacom Communications Corporation**
YahLive; **SES SA**
Yellow Pages; **Telkom SA SOC Limited**
YouSee; **TDC A/S**
Z Holdings; **SoftBank Group Corp**
Zain Ventures; **Mobile Telecommunications Company KSCP (Zain Group)**
Zoom; **Zoom Video Communications Inc**
Zoom Room; **Zoom Video Communications Inc**
ZOWIE; **BenQ Corporation**
ZP Better Together LLC; **Purple Communications Inc**

A Short Telecommunications Industry Glossary

3G Cellular: Short for third generation, this term refers to high speed enhancements to mobile telephone service. 3G enables wireless e-mail, Internet browsing and data transfer. 3G will be largely replaced by advanced 4G and 5G technologies.

3GPP: Third Generation Partnership Project. It is an organization set up to create and monitor advanced 3G wireless standards.

4G Cellular: An advancement in speed and capabilities over 3G wireless networks. 4G not only features high data transfer speeds, it also has an enhanced ability to support interactive multimedia, internet access, mobile video and other vital tasks. It will eventually be surpassed by 5G and higher networks, with 5G beginning to rollout on a major basis in the 2020s.

5G Cellular: A wireless technology that is expected to produce blinding download speeds of one gigabyte per second (Gbps), and perhaps as high as 10 Gbps. The first specifications for 5G were agreed to by the global wireless industry from 2017 to 2019. Significant rollout was expected to begin in the early 2020s. While certain 5G features can be used to boost speeds of earlier 4G networks, a true rollout requires major investment in new cellular infrastructure and systems.

802.11: See "Wi-Fi."

802.11n (MIMO): Multiple Input Multiple Output. MIMO is a standard in the series of 802.11 Wi-Fi specifications for wireless networks. It can provide very high speed network access. 802.11n also boasts better operating distances than many networks. MIMO uses spectrum more efficiently without any loss of reliability. The technology is based on several different antennas all tuned to the same channel, each transmitting a different signal. Advancements include MU-MIMO (Multi-User MIMO) and OFDMA (Orthogonal Frequency-Division Multiple Access), each of which improves network throughput.

802.15: See "Ultrawideband (UWB)." For 802.15.1, see "Bluetooth."
802.15.1: See "Bluetooth."

802.16: See "WiMAX."

Access Network: The network that connects a user's telephone equipment to the telephone exchange.

Active Server Page (ASP): A web page that includes one or more embedded programs, usually written in Java or Visual Basic code. See "Java."

Active X: A set of technologies developed by Microsoft Corporation for sharing information across different applications.

ADN: See "Advanced Digital Network (ADN)."

ADSL: See "Asymmetrical Digital Subscriber Line (ADSL)."

Advanced Digital Network (ADN): See "Integrated Digital Network (IDN)."

Analog: A form of transmitting information characterized by continuously variable quantities. Digital transmission, in contrast, is characterized by discrete bits of information in numerical steps. An analog signal responds to changes in light, sound, heat and pressure.

Analog IC (Integrated Circuit): A semiconductor that processes a continuous wave of electrical signals based on real-world analog quantities such as speed, pressure, temperature, light, sound and voltage.

Analytics: Generally refers to the deep examination of massive amounts of data, often on a continual or real-time basis. The goal is to discover deeper insights, make recommendations or generate predictions. Advanced analytics includes such techniques as big data, predictive analytics, text analytics, data mining, forecasting, optimization and simulation.

ANSI: American National Standards Institute. Founded in 1918, ANSI is a private, non-profit organization that administers and coordinates the U.S. voluntary standardization and conformity assessment system. Its mission is to enhance both the global competitiveness of U.S. business and the quality of U.S. life by promoting and facilitating voluntary consensus standards and conformity assessment systems, and safeguarding their integrity. See www.ansi.org.

APAC: Asia Pacific Advisory Committee. A multi-country committee representing the Asia and Pacific region.

Applets: Small, object-based applications written in Java that net browsers can download from the Internet on an as-needed basis. These may be software, accessories (such as spell checkers or calculators), information-packed databases or other items. See "Object Technology."

Application Service Provider (ASP): A web site that enables utilization of software and databases that reside permanently on a service company's remote web server, rather than having to be downloaded to the user's computer. Advantages include the ability for multiple remote users to access the same tools over the Internet and the fact that the ASP provider is responsible for developing and maintaining the software. (ASP is also an acronym for "active server page," which is not related.) For the latest developments in ASP, see "Software as a Service (SaaS)."

Applied Research: The application of compounds, processes, materials or other items discovered during basic research to practical uses. The goal is to move discoveries along to the final development phase.

ARPANet: Advanced Research Projects Agency Network. The forefather of the Internet, ARPANet was developed during the latter part of the 1960s by the United States Department of Defense.

ARPU: See "Average Revenue Per User (ARPU)."

Artificial Intelligence (AI): The use of computer technology to perform functions somewhat like those normally associated with human intelligence, such as reasoning, learning and self-improvement.

ASCII: American Standard Code for Information Exchange. There are 128 standard ASCII codes that represent all Latin letters, numbers and punctuation. Each ASCII code is represented by a seven-digit binary number, such as 0000000 or 0000111. This code is accepted as a standard throughout the world.

ASEAN: Association of Southeast Asian Nations. A regional economic development association established in 1967 by five original member countries: Indonesia, Malaysia, Philippines, Singapore, and Thailand. Brunei joined on 8 January

1984, Vietnam on 28 July 1995, Laos and Myanmar on 23 July 1997, and Cambodia on 30 April 1999.

ASP: See "Application Service Provider (ASP)."

Asymmetrical Digital Subscriber Line (ADSL): High-speed technology that enables the transfer of data over existing copper phone lines, allowing more bandwidth downstream than upstream.

Asynchronous Communications: A stream of data routed through a network as generated instead of in organized message blocks. Most personal computers use this format to send data.

Asynchronous Transfer Mode (ATM): A digital switching and transmission technology based on high speed. ATM allows voice, video and data signals to be sent over a single telephone line at speeds from 25 million to 1 billion bits per second (bps). This digital ATM speed is much faster than traditional analog phone lines, which allow no more than 2 million bps. See "Broadband."

ATCA: Advanced Telecommunications Computing Architecture. It is a set of standards widely used in telecommunications equipment due to the rapid growth of VOIP. ATCA technology increases performance and reliability by optimizing the architecture specifically for communications servers by eliminating proprietary communications port specifications.

Average Revenue Per User (ARPU): A measure of the average monthly billing revenue of a wireless company on a per user basis.

B2B: See "Business-to-Business."

B2C: See "Business-to-Consumer."

Backbone: Traditionally the part of a communications network that carries the heaviest traffic: the high-speed line or series of connections that forms a large pathway within a network or within a region. The combined networks of AT&T, MCI and other large telecommunications companies make up the backbone of the Internet.

Bandwidth: The data transmission capacity of a network, measured in the amount of data (in bits and bauds) it can transport in one second. A full page of text is about 15,000 to 20,000 bits. Full-motion, full-screen video requires about 10 million bits per second, depending on compression.

Basic Cable: Primary level or levels of cable service offered for subscription. Basic cable offerings may include retransmitted broadcast signals as well as local and access programming. In addition, regional and national cable network programming may be provided.

Basic Research: Attempts to discover compounds, materials, processes or other items that may be largely or entirely new and/or unique. Basic research may start with a theoretical concept that has yet to be proven. The goal is to create discoveries that can be moved along to applied research. Basic research is sometimes referred to as "blue sky" research.

Baud: Refers to how many times the carrier signal in a modem switches value per second or how many bits a modem can send and receive in a second.

Beam: The coverage and geographic service area offered by a satellite transponder. A global beam effectively covers one-third of the earth's surface. A spot beam provides a very specific high-powered downlink pattern that is limited to a particular geographical area to which it may be steered or pointed.

Binhex: A means of changing non-ASCII (or non-text) files into text/ASCII files so that they can be used, for example, as e-mail.

Bit: A single digit number, either a one or a zero, which is the smallest unit of computerized data.

Bits Per Second (Bps): An indicator of the speed of data movement.

Blog (Web Log): A web site consisting of a personal journal, news coverage, special-interest content or other data that is posted on the Internet, frequently updated and intended for public viewing by anyone who might be interested in the author's thoughts. Short for "web log," blog content is frequently distributed via RSS (Real Simple Syndication). Blog content has evolved to include video files (VLOGs) and audio files (Podcasting) as well as text. Also, see "Real Simple Syndication (RSS)," "Video Blog (VLOG)," "Moblog": "Podcasting," and "User Generated Content (UGC)."

Bluetooth: An industry standard for a technology that enables wireless, short-distance infrared connections between devices such as cell phone headsets, Palm Pilots or PDAs, laptops, printers and Internet appliances.

BPL: See "Broadband Over Power Lines (BPL)."

BPO: See "Business Process Outsourcing (BPO)."

Bps: See "Bits Per Second (Bps)."

Brand: A marketing strategy that places a focus on the brand name of a product, service or firm in order to increase the brand's market share, increase sales, establish credibility, improve satisfaction, raise the profile of the firm and increase profits. Also, see "Brand."

Branding: A marketing strategy that places a focus on the brand name of a product, service or firm in order to increase the brand's market share, increase sales, establish credibility, improve satisfaction, raise the profile of the firm and increase profits. Also, see "Brand."

Bring Your Own Device (BYOD): The trend of employees bringing their own laptops, tablets and cellphones into an office setting, as opposed to using equipment issued by the company. Similar phrases include BYOP (Bring Your Own Phone) and BYOT (Bring Your Own Technology). This practice presents serious network security issues.

Broadband: The high-speed transmission range for telecommunications and computer data. Broadband generally refers to any transmission at 2 million bps (bits per second) or higher (much higher than analog speed). A broadband network can carry voice, video and data all at the same time. Internet users enjoying broadband access typically connect to the Internet via DSL line, cable modem or T1 line. Several wireless methods offer broadband as well.

Broadband Over Power Lines (BPL): Refers to the use of standard electric power lines to provide fast Internet service. Internet data is converted into radio frequency signals, which are not affected by electricity. Subscribers utilize special modems.

Browser: A program that allows a user to read Internet text or graphics and to navigate from one page to another. The most popular browsers are

Microsoft Internet Explorer and Netscape Navigator. Firefox is an open source browser introduced in 2005 that is rapidly gaining popularity.

Business Process Outsourcing (BPO): The process of hiring another company to handle business activities. BPO is one of the fastest-growing segments in the offshoring sector. Services include human resources management, billing and purchasing and call centers, as well as many types of customer service or marketing activities, depending on the industry involved. Also, see "Knowledge Process Outsourcing (KPO)" and Business Transformation Outsourcing (BTO)."

Business Transformation Outsourcing (BTO): A segment within outsourcing in which the client company revamps its business processes with the goal of transforming its business by following a collaborative approach with its outsourced services provider.

Business-to-Business: An organization focused on selling products, services or data to commercial customers rather than individual consumers. Also known as B2B.

Business-to-Consumer: An organization focused on selling products, services or data to individual consumers rather than commercial customers. Also known as B2C.

Byte: A set of eight bits that represent a single character.

Cable Modem: An interface between a cable television system and a computer or router. Most cable modems are external devices that connect to the PC through a standard 10Base-T Ethernet card and twisted-pair wiring. External Universal Serial Bus (USB) modems and internal PCI modem cards are also available.

CAFTA-DR: See "Central American-Dominican Republic Free Trade Agreement (CAFTA-DR)."

Call Automation: Part of the telephone equipment revolution, including voice mail, automated sending and receiving of faxes and the ability for customers to place orders and gather information using a touch-tone telephone to access sophisticated databases. See "Voice Mail."

Call Center: A department within a company or a third-party organization that manages inbound and outbound telephone calls. This organization usually processes orders, provides technical support and/or provides marketing support. Call centers are frequently provided on an outsourced basis by service firms in lower-cost nations.

Capex: Capital expenditures.

Captive Offshoring: Used to describe a company-owned offshore operation. For example, Microsoft owns and operates significant captive offshore research and development centers in China and elsewhere that are offshore from Microsoft's U.S. home base. Also see "Offshoring."

CAR: See "Committed Access Rate (CAR)."

Carrier: In communications, the basic radio, television or telephony center of transmit signal. The carrier in an analog signal is modulated by varying volume or shifting frequency up or down in relation to the incoming signal. Satellite carriers operating in the analog mode are usually frequency-modulated.

CDMA: See "Code Division Multiple Access (CDMA)."

CDMA 1xRTT: See "Code Division Multiple Access (CDMA)."

CDMA2000: The commercial name for a high-speed version of 3G CDMA, based on and compatible with current 2G CDMA networks. See "EV-DO (CDMA 2000 1xEV-DO)."

CDMAOne: See "Code Division Multiple Access (CDMA)."

Cellular Mobile Telephone Service: Refers to the method in which advanced mobile telephone systems hand off calls to the nearest "cells" as the users travel. Cells represent the range of fixed antenna. In the best systems, cells overlap so that there is less possibility of service interruptions.

CEM: Contract electronic manufacturing. See "Contract Manufacturing."

Central American-Dominican Republic Free Trade Agreement (CAFTA-DR): A trade agreement signed into law in 2005 that aimed to open up the Central American and Dominican Republic markets to American goods. Member nations include Guatemala, Nicaragua, Costa Rica, El Salvador, Honduras and the Dominican Republic. Before the law was signed, products from those countries could enter the U.S. almost tariff-free, while American goods heading into those countries faced stiff tariffs. The goal of this agreement was to create U.S. jobs while at the same time offering the non-U.S. member citizens a chance for a better quality of life through access to U.S.-made goods.

CGI: See "Common Gateway Interface (CGI)."

CGI-BIN: The frequently used name of a directory on a web server where CGI programs exist.

Channel Definition Format (CDF): Used in Internet-based broadcasting. With this format, a channel serves as a web site that also sends an information file about that specific site. Users subscribe to a channel by downloading the file.

Churn Rate: The percentage of customers of subscribers who terminate contracts in a given time period.

CLEC: See "Competitive Local Exchange Carrier (CLEC)."

Client/Server: In networking, a way of running a large computer setup. The server is the host computer that acts as the central holding ground for files, databases and application software. The clients are all of the PCs connected to the network that share data with the server. This represents a vast change from past networks, which were connected to expensive, complicated "mainframe" computers.

Cloud: Refers to the use of outsourced servers to store and access data, as opposed to computers owned or managed by one organization. Firms that offer cloud services for a fee run clusters of servers networked together, often based on open standards. Such cloud networks can consist of hundreds or even thousands of computers. Cloud services enable a client company to immediately increase computing capability without any investment in physical infrastructure. (The word "cloud" is also broadly used to describe any data or application that runs via the Internet.) The concept of cloud is also increasingly linked with software as a service.

CMOS: Complementary Metal Oxide Semiconductor. The technology used in making modern silicon-based microchips.

Coaxial Cable: A type of cable widely used to transmit telephone and broadcast traffic. The distinguishing feature is an inner strand of wires surrounded by an insulator that is in turn surrounded by another conductor, which serves as the ground. Cable TV wiring is typically coaxial.

Code Division Multiple Access (CDMA): A cellular telephone multiple-access scheme whereby stations use spread-spectrum modulations and orthogonal codes to avoid interfering with one another. IS-95 (also known as CDMAOne) is the 2G CDMA standard. CDMA2000 is the 3G standard. CDMA in the 1xEV-DO standard offers data transfer speeds up to 2.4 Mbps. CDMA 1xRTT is a slower standard offering speeds of 144 kbps.

Codec: Hardware or software that converts analog to digital and digital to analog (in both audio and video formats). Codecs can be found in digital telephones, set-top boxes, computers and videoconferencing equipment. The term is also used to refer to the compression of digital information into a smaller format.

Co-Location: Refers to the hosting of computer servers at locations operated by service organizations. Co-location is offered by firms that operate specially designed co-location centers with high levels of security, extremely high-speed telecommunication lines for Internet connectivity and reliable backup electrical power systems in case of power failure, as well as a temperature-controlled environment for optimum operation of computer systems.

Committed Access Rate (CAR): A premium level of access from the Internet carrier to the customer.

Common Gateway Interface (CGI): A set of guidelines that determines the manner in which a web server receives and sends information to and from software on the same machine.

Communications Satellite Corporation (COMSAT): Serves as the U.S. Signatory to INTELSAT and INMARSAT.

Competitive Local Exchange Carrier (CLEC): A newer company providing local telephone service that competes against larger, traditional firms known as ILECs (incumbent local exchange carriers).

Compression: A technology in which a communications signal is squeezed so that it uses less bandwidth (or capacity) than it normally would. This saves storage space and shortens transfer time. The original data is decompressed when read back into memory.

COMSAT: See "Communications Satellite Corporation (COMSAT)."

Contract Manufacturing: A business arrangement whereby a company manufactures products that will be sold under the brand names of its client companies. For example, a large number of consumer electronics, such as laptop computers, are manufactured by contract manufacturers for leading brand-name computer companies such as Dell and Apple. Many other types of products, such as shoes and apparel, are made under contract manufacturing. Also see "Original Equipment Manufacturer (OEM)" and "Original Design Manufacturer (ODM)."

Cookie: A piece of information sent to a web browser from a web server that the browser software saves and then sends back to the server upon request. Cookies are used by web site operators to track the actions of users returning to the site.

CRM: See "Customer Relationship Management (CRM)."

Customer Relationship Management (CRM): Refers to the automation, via sophisticated software, of business processes involving existing and prospective customers. CRM may cover aspects such as sales (contact management and contact history), marketing (campaign management and telemarketing) and customer service (call center history and field service history). Well known providers of CRM software include Salesforce, which delivers via a Software as a Service model (see "Software as a Service (Saas)"), Microsoft and Oracle.

Cyberspace: Refers to the entire realm of information available through computer networks and the Internet.

D-AMPS: See "Time Division Multiple Access (TDMA)."

Dark Fiber: A reference to fiber optic bandwidth that is not being utilized.

Data Over Cable Service Interface Specification (DOCSIS): A set of standards for transferring data over cable television. DOCSIS 3.0 will enable very high-speed Internet access that may eventually reach 160 Mbps.

Datanets: Private networks of land-based telephone lines, satellites or wireless networks that allow corporate users to send data at high speeds to remote locations while bypassing the speed and cost constraints of traditional telephone lines.

Decompression: See "Compression."

Dedicated Internet Access (DIA): A high speed Internet service with dedicated access from the carrier to the customer.

Demographics: The breakdown of the population into statistical categories such as age, income, education and sex.

Development: The phase of research and development (R&D) in which researchers attempt to create new products from the results of discoveries and applications created during basic and applied research.

DIA: "Dedicated Internet Access (DIA)."

Digital: The transmission of a signal by reducing all of its information to ones and zeros and then regrouping them at the reception end. Digital transmission vastly improves the carrying capacity of the spectrum while reducing noise and distortion of the transmission.

Digital Local Telephone Switch: A computer that interprets signals (dialed numbers) from a telephone caller and routes calls to their proper destinations. A digital switch also provides a variety of calling features not available in older analog switches, such as call waiting.

Digital Rights Management (DRM): Restrictions placed on the use of digital content by copyright holders and hardware manufacturers. DRM for Apple, Inc.'s iTunes, for example, allows downloaded music to be played only on Apple's iPod player and iPhones, per agreement with music production companies Universal Music Group, SonyBMG, Warner Music and EMI.

Digital Signal Processor: A chip that converts analog signals such as sound and light into digital signals.

Digital Subscriber Line (DSL): A broadband (high-speed) Internet connection provided via telecommunications systems. These lines are a cost-effective means of providing homes and small businesses with relatively fast Internet access. Common variations include ADSL and SDSL. DSL competes with cable modem access and wireless access.

Digital Transformation (DX): The implementation of digital technologies into as many areas of a business as reasonably possible Goals may include: to fundamentally change how the enterprise operates: how data is gathered and tracked: how innovation is launched: and how value is delivered to customers. The hoped-for result is to create new operating efficiencies and develop new revenue or profit opportunities, while better positioning the enterprise for the future. Also abbreviated as DX or DT.

Direct Broadcast Satellite (DBS): A high-powered satellite authorized to broadcast television programming directly to homes. Home subscribers use a dish and a converter to receive and translate the TV signal. An example is the DirecTV service. DBS operates in the 11.70- to 12.40-GHz range.

Disaster Recovery: A set of rules and procedures that allow a computer site to be put back in operation after a disaster has occurred. Moving backups off-site constitutes the minimum basic precaution for disaster recovery. The remote copy is used to recover data if the local storage is inaccessible after a disaster.

Discrete Semiconductor: A chip with one diode or transistor.

Disk Mirroring: A data redundancy technique in which data is recorded identically on multiple separate disk drives at the same time. When the primary disk is off-line, the alternate takes over, providing continuous access to data. Disk mirroring is sometimes referred to as RAID.

Distributor: An individual or business involved in marketing, warehousing and/or shipping of products

manufactured by others to a specific group of end users. Distributors do not sell to the general public. In order to develop a competitive advantage, distributors often focus on serving one industry or one set of niche clients. For example, within the medical industry, there are major distributors that focus on providing pharmaceuticals, surgical supplies or dental supplies to clinics and hospitals.

Domain: A name that has server records associated with it. See "Domain Name."

Domain (Top-Level): Either an ISO country code or a common domain name such as .com, .org or .net.

Domain Name: A unique web site name registered to a company, organization or individual (e.g., plunkettresearch.com).

DS-1: A digital transmission format that transmits and receives information at a rate of 1,544,000 bits per second.

DSL: See "Digital Subscriber Line (DSL)."

Duplicate Host: A single host name that maps to duplicate IP addresses.

DX: See "Digital Transformation (DX)."

Dynamic HTML: Web content that changes with each individual viewing. For example, the same site could appear differently depending on geographic location of the reader, time of day, previous pages viewed or the user's profile.

Electronic Data Interchange (EDI): An accepted standard format for the exchange of data between various companies' networks. EDI allows for the transfer of e-mail as well as orders, invoices and other files from one company to another.

E-Mail (eMail): The use of software that allows the posting of messages (text, audio or video) over a network. E-mail can be used on a LAN, a WAN or the Internet, as well as via online services or wireless devices that are Internet enabled. It can be used to send a message to a single recipient or may be broadcast to a large group of people at once.

EMEA: The region comprised of Europe, the Middle East and Africa.

EMS: Electronics Manufacturing Services. See "Contract Manufacturing."

Enhanced 911 (E911): A Federal Communication Commission rule that all wireless carriers must be able to identify a 911 caller by telephone number and location to within 100 meters.

Enhanced Data Rate for Global Evolution (EDGE): Technology that uses enhanced TDMA to achieve 3G transmission speeds and is compatible with GSM and TDMA networks.

Enterprise Resource Planning (ERP): An integrated information system that helps manage all aspects of a business, including accounting, ordering and human resources, typically across all locations of a major corporation or organization. ERP is considered to be a critical tool for management of large organizations. Suppliers of ERP tools include SAP and Oracle.

ERP: See "Enterprise Resource Planning (ERP)."

Ethernet: The standard format on which local area network equipment works. Abiding by Ethernet standards allows equipment from various manufacturers to work together.

eTOM: A business process flow standard created by the TeleManagement Forum.

EU: See "European Union (EU)."

EU Competence: The jurisdiction in which the European Union (EU) can take legal action.

European Community (EC): See "European Union (EU)."

European Union (EU): A consolidation of European countries (member states) functioning as one body to facilitate trade. Previously known as the European Community (EC). The EU has a unified currency, the Euro. See europa.eu.int.

EV-DO (CDMA 2000 1xEV-DO): A 3G (third generation) cellular telephone service standard that is an improved version of 1xRTT. The EV-DO (Evolution-Data Optimized) standard introduced in 2004 allows data download speeds of as much as 2.4 Mbps. A version introduced in 2006 allows up to 14.7 Mbps data download speeds. EV-DO is also

known as CDMA 2000 1xEV-DO. EV-DO's capabilities are used by the entertainment industry to enable video via cell phone.

Extensible Markup Language (XML): A programming language that enables designers to add extra functionality to documents that could not otherwise be utilized with standard HTML coding. XML was developed by the World Wide Web Consortium. It can communicate to various software programs the actual meanings contained in HTML documents. For example, it can enable the gathering and use of information from a large number of databases at once and place that information into one web site window. XML is an important protocol to web services. See "Web Services."

Extranet: A computer network that is accessible in part to authorized outside persons, as opposed to an intranet, which uses a firewall to limit accessibility.

FASB: See "Financial Accounting Standards Board (FASB)."

FCC: See "Federal Communications Commission (FCC)."

FDDI: See "Fiber Distributed Data Interface (FDDI)."

Federal Communications Commission (FCC): The U.S. Government agency that regulates broadcast television and radio, as well as satellite transmission, telephony and all uses of radio spectrum.

Femtocell: A device used to boost performance of cell phones on a local basis, such as in a consumer's home or office. It utilizes nearby licensed wireless spectrum. The femtocell, in the form of a small box, routes wireless phone calls from a cell phone handset to the central office of a cellular service provider via a consumer's high speed Internet line.

Fiber Distributed Data Interface (FDDI): A token ring passing scheme that operates at 100 Mbps over fiber-optic lines with a built-in geographic limitation of 100 kilometers. This type of connection is faster than both Ethernet and T-3 connections. See "Token Ring."

Fiber Optics (Fibre Optics): A type of telephone and data transmission cable that can handle vast amounts of voice, data and video at once by carrying them along on beams of light via glass or plastic threads embedded in a cable. Fiber optics are rapidly replacing older copper wire technologies. Fiber optics offer much higher speeds and the ability to handle extremely large quantities of voice or data transmissions at once.

Fiber to the Home (FTTH): Refers to the extension of a fiber-optic system through the last mile so that it touches the home or office where it will be used. This can provide high speed Internet access at speeds of 15 to 100 Mbps, much faster than typical T1 or DSL line. FTTH is now commonly installed in new communities where telecom infrastructure is being built for the first time. Another phrase used to describe such installations is FTTP, or Fiber to the Premises.

Fiber to the Node (FTTN): Refers to the extension of a fiber-optic system through the last mile so that it touches a central neighborhood junction close to the home or office where it will be used. The remaining distance is covered by existing copper phone line that uses DSL (digital subscriber line) technology to speed data transfer.

File Transfer Protocol (FTP): A widely used method of transferring data and files between two Internet sites.

Financial Accounting Standards Board (FASB): An independent organization that establishes the Generally Accepted Accounting Principles (GAAP).

Firewall: Hardware or software that keeps unauthorized users from accessing a server or network. Firewalls are designed to prevent data theft and unauthorized web site manipulation by hackers.

Firmware: A software program or string of code programmed on a hardware device. Firmware is usually stored in the read only memory (ROM) of the device, and essentially provides the control program for that device.

Fixed Wireless: Refers to the use of Wi-Fi, WiMAX or other wireless receivers that remain fixed in a stationary place, to provide Internet service.

FOMA: Freedom of Mobile Multimedia Access. See "Wideband CDMA."

Fourth-Party Logistics (4PL): A service that integrates a company's third-party logistics providers into a single entity for ease of use. Often formed by a telecommunications company, a 4PL is also called a lead logistics provider. A 4PL service provider provides a top layer of business processes, generally technology-driven, to the client's supply chain. Also see "Third-Party Logistics (3PL)."

Frame Relay: An accepted standard for sending large amounts of data over phone lines and private datanets. The term refers to the way data is broken down into standard-size "frames" prior to transmission.

Free Space Optics (FSO): A cost-effective alternative to fiber-optic broadband access, FSO uses lasers, or light pulses, to send packetized data in the terahertz spectrum range. Air, rather than fiber, is the transport medium.

Frequency: The number of times that an alternating current goes through its complete cycle in one second. One cycle per second is referred to as one hertz: 1,000 cycles per second, one kilohertz: 1 million cycles per second, one megahertz: and 1 billion cycles per second, one gigahertz.

Frequency Band: A term for designating a range of frequencies in the electromagnetic spectrum.

FTP: See "File Transfer Protocol (FTP)."

FTTC: Fiber to the curb. See "Fiber to the Home (FTTH)."

FTTP: Fiber to the premises. See "Fiber to the Home (FTTH)."

Fuzzy Logic: Recognizes that some statements are not just "true" or "false," but also "more or less certain" or "very unlikely." Fuzzy logic is used in artificial intelligence. See "Artificial Intelligence (AI)."

GAAP: See "Generally Accepted Accounting Principles (GAAP)."

Gateway: A device connecting two or more networks that may use different protocols and media. Gateways translate between the different networks and can connect locally or over wide area networks.

GDP: See "Gross Domestic Product (GDP)."

General Packet Radio Service (GPRS): A system for mobile communications, GPRS providers faster data transmission than GSM networks. See "Global System for Mobile Communications (GSM)."

Generally Accepted Accounting Principles (GAAP): A set of accounting standards administered by the Financial Accounting Standards Board (FASB) and enforced by the U.S. Security and Exchange Commission (SEC). GAAP is primarily used in the U.S.

Geostationary: A geosynchronous satellite angle with zero inclination, making a satellite appear to hover over one spot on the earth's equator.

Gigabyte: 1,024 megabytes.

Gigahertz (GHz): One billion cycles per second. See "Frequency."

Global Positioning System (GPS): A satellite system, originally designed by the U.S. Department of Defense for navigation purposes. Today, GPS is in wide use for consumer and business purposes, such as navigation for drivers, boaters and hikers. It utilizes satellites orbiting the earth at 10,900 miles to enable users to pinpoint precise locations using small, electronic wireless receivers.

Global System for Mobile Communications (GSM): The standard cellular format used throughout Europe, making one type of cellular phone usable in every nation on the continent and in the U.K. In the U.S., Cingular and T-Mobile also run GSM networks. The original GSM, introduced in 1991, has transfer speeds of only 9.6 kbps. GSM EDGE offers 2.75G data transfer speeds of up to 473.6 kbps. GSM GPRS offers slower 2.5G theoretical speeds of 144 kbps.

Globalization: The increased mobility of goods, services, labor, technology and capital throughout the world. Although globalization is not a new development, its pace has increased with the advent of new technologies.

GPRS: A system for mobile communications, GPRS providers faster data transmission than GSM networks. See "Global System for Mobile Communications (GSM)."

GPS: See "Global Positioning System (GPS)."

Graphic Interchange Format (GIF): A widely used format for image files.

Gross Domestic Product (GDP): The total value of a nation's output, income and expenditures produced with a nation's physical borders.

Gross National Product (GNP): A country's total output of goods and services from all forms of economic activity measured at market prices for one calendar year. It differs from Gross Domestic Product (GDP) in that GNP includes income from investments made in foreign nations.

GSM: See "Global System for Mobile Communications (GSM)."

GSM EDGE: See "Global System for Mobile Communications (GSM)."

GSM GPRS: See "Global System for Mobile Communications (GSM)."

Handheld Devices Markup Language (HDML): A text-based markup language designed for display on a smaller screen (e.g., a cellular phone, PDA or pager). Enables the mobile user to send, receive and redirect e-mail as well as access the Internet (HDML-enabled web sites only).

HDML: See "Handheld Devices Markup Language (HDML)."

HDSL: See "High-Data-Rate Digital Subscriber Line (HDSL)."

HDSPA: See "High Speed Packet Access (HSPA)."

Headend: A facility that originates and distributes cable service in a given geographic area. Depending on the size of the area it serves, a cable system may be comprised of more than one headend.

Hertz: A measure of frequency equal to one cycle per second. Most radio signals operate in ranges of megahertz or gigahertz.

HFC: Hybrid Fiber Coaxial. A type of cable system.

High Speed Packet Access (HSPA): A 3G wireless standard introduced in 2007 that encompasses two protocols: High Speed Downlink Packet Access (HSDPA) and High Speed Uplink Packet Access (HSUPA). Downlink speeds can reach up to 14.4 Mbps while uplink speeds of up to 5.76 Mbps. (These are theoretical speeds.) HSPA+ is an advanced technology with the following theoretical download speeds: Release 8, 42 Mbps: Release 9, 84 Mbps: Release 10, 168 Mbps.

High-Data-Rate Digital Subscriber Line (HDSL): High-data-rate DSL, delivering up to T1 or E1 speeds.

Homes Passed: Households that have the ability to receive cable service and may opt to subscribe.

HSDPA: See "High Speed Packet Access (HSPA)."

HSPA: See "High Speed Packet Access (HSPA)."

HTML: See "Hypertext Markup Language (HTML)."

HTML5: A specification for Internet development that represents the fifth major revision of the Hypertext Markup Language, or HTML. HTML5 is designed to better handle the types of Internet content that are rapidly growing in popularity, such as online video, audio and interactive documents and pages. For example, HTML5 enables the designer to embed images, audio and video directly into a web-based document.

HTTP: See "Hypertext Transfer Protocol (HTTP)."

Hypertext Markup Language (HTML): A language for coding text for viewing on the World Wide Web. HTML is unique because it enables the use of hyperlinks from one site to another, creating a web.

Hypertext Transfer Protocol (HTTP): The protocol used most frequently on the World Wide Web to move hypertext files between clients and servers on the Internet.

ICANN: The Internet Corporation for Assigned Names and Numbers. ICANN acts as the central coordinator for the Internet's technical operations.

ICT: See "Information and Communication Technologies (ICT)."

iDEN: A packet switched cellular mobile telephone standard. It is a proprietary standard of Nextel. The maximum data speed is 19.2 kbps. This standard is considered 2.5G. It was introduced in 2001.

IDN: See "Integrated Digital Network (IDN)."

IEEE: See "Institute of Electrical and Electronic Engineers (IEEE)."

IFRS: See "International Financials Reporting Standards (IFRS)."

ILEC: See "Incumbent Local Exchange Carrier (ILEC)."

IM: See "Instant Messaging (IM)."

IMT-Advanced: See "Long-Term Evolution (LTE)."

Incumbent Local Exchange Carrier (ILEC): A traditional telephone company that was providing local service prior to the establishment of the Telecommunications Act of 1996, when upstart companies (CLECs, or competitive local exchange carriers) were enabled to compete against the ILECS and were granted access to their system wiring.

Industry Code: A descriptive code assigned to any company in order to group it with firms that operate in similar businesses. Common industry codes include the NAICS (North American Industrial Classification System) and the SIC (Standard Industrial Classification), both of which are standards widely used in America, as well as the International Standard Industrial Classification of all Economic Activities (ISIC), the Standard International Trade Classification established by the United Nations (SITC) and the General Industrial Classification of Economic Activities within the European Communities (NACE).

Information and Communication Technologies (ICT): A term used to describe the relationship between the myriad types of goods, services and networks that make up the global information and communications system. Sectors involved in ICT include landlines, data networks, the Internet, wireless communications, (including cellular and remote wireless sensors) and satellites.

Information Technology (IT): The systems, including hardware and software, that move and store voice, video and data via computers and telecommunications.

Infrastructure: 1) The equipment that comprises a system. 2) Public-use assets such as roads, bridges, water systems, sewers and other assets necessary for public accommodation and utilities. 3) The underlying base of a system or network. 4) Transportation and shipping support systems such as ports, airports and railways.

Infrastructure (Telecommunications): The entity made up of all the cable and equipment installed in the worldwide telecommunications market. Most of today's telecommunications infrastructure is connected by copper and fiber-optic cable, which represents a huge capital investment that telephone companies would like to continue to utilize in as many ways as possible.

Initial Public Offering (IPO): A company's first effort to sell its stock to investors (the public). Investors in an up-trending market eagerly seek stocks offered in many IPOs because the stocks of newly public companies that seem to have great promise may appreciate very rapidly in price, reaping great profits for those who were able to get the stock at the first offering. In the United States, IPOs are regulated by the SEC (U.S. Securities Exchange Commission) and by the state-level regulatory agencies of the states in which the IPO shares are offered.

INMARSAT: The International Maritime Satellite Organization. INMARSAT operates a network of satellites used in transmissions for all types of international mobile services, including maritime, aeronautical and land mobile.

Instant Messaging (IM): A type of e-mail that is viewed and then deleted. IM is used between opt-in networks of people for leisure or business purposes.

Institute of Electrical and Electronic Engineers (IEEE): An organization that sets global technical standards and acts as an authority in technical areas including computer engineering, biomedical technology, telecommunications, electric power, aerospace and consumer electronics, among others. www.ieee.org.

Integrated Circuit (IC): Another name for a semiconductor, an IC is a piece of silicon on which

thousands (or millions) of transistors have been combined.

Integrated Digital Network (IDN): A network that uses both digital transmission and digital switching.

Integrated Services Digital Networks (ISDN): Internet connection services offered at higher speeds than standard "dial-up" service. While ISDN was considered to be an advanced service at one time, it has been eclipsed by much faster DSL, cable modem and T1 line service.

Intellectual Property (IP): The exclusive ownership of original concepts, ideas, designs, engineering plans or other assets that are protected by law. Examples include items covered by trademarks, copyrights and patents. Items such as software, engineering plans, fashion designs and architectural designs, as well as games, books, songs and other entertainment items are among the many things that may be considered to be intellectual property. (Also, see "Patent.")

INTELSAT: The International Telecommunications Satellite Organization. INTELSAT operates a network of 20 satellites, primarily for international transmissions, and provides domestic services to some 40 countries.

Interactive TV (ITV): Allows two-way data flow between a viewer and the cable TV system. A user can exchange information with the cable system—for example, by ordering a product related to a show he/she is watching or by voting in an interactive survey.

Interexchange Carrier (IXC or IEC): Any company providing long-distance phone service between LECs and LATAs. See "Local Exchange Carrier (LEC)" and "Local Access and Transport Area (LATA)."

Interface: Refers to (1) a common boundary between two or more items of equipment or between a terminal and a communication channel, (2) the electronic device that interconnects two or more devices or items of equipment having similar or dissimilar characteristics or (3) the electronic device placed between a terminal and a communication channel to protect the network from the hazard of excess voltage levels.

International Financials Reporting Standards (IFRS): A set of accounting standards established by the International Accounting Standards Board (IASB) for the preparation of public financial statements. IFRS has been adopted by much of the world, including the European Union, Russia and Singapore.

International Telecommunications Union (ITU): The international body responsible for telephone and computer communications standards describing interface techniques and practices. These standards include those that define how a nation's telephone and data systems connect to the worldwide communications network.

Internet: A global computer network that provides an easily accessible way for hundreds of millions of users to send and receive data electronically when appropriately connected via computers or wireless devices. Access is generally through HTML-enabled sites on the World Wide Web. Also known as the Net.

Internet Appliance: A non-PC device that connects users to the Internet for specific or general purposes. A good example is an electronic game machine with a screen and Internet capabilities.

Internet of Things (IoT): A concept whereby individual objects, such as kitchen appliances, automobiles, manufacturing equipment, environmental sensors or air conditioners, are connected to the Internet. The objects must be able to identify themselves to other devices or to databases. The ultimate goals may include the collection and processing of data, the control of instruments and machinery, and eventually, a new level of synergies, artificial intelligence and operating efficiencies among the objects. The Internet of Things is often referred to as IoT. Related technologies and topics include RFID, remote wireless sensors, telecommunications and nanotechnology.

Internet Protocol (IP): A set of tools and/or systems used to communicate across the World Wide Web.

Internet Protocol Television (IPTV): Television delivered by Internet-based means such as fiber to the home (FTTH) or a very high speed DSL. Microsoft is a leading provider of advanced IPTV software. SBC and BT are two leading telecom firms that are using Microsoft's new software to offer television services over high speed Internet lines.

Internet Protocol Version 6 (IPv6): The next-generation of IP standard. IPv6 is intended to first work with, and eventually replace, IPv4. Version 6 will enable a vastly larger number of devices to each utilize one internet address (an IP address) at one time. Specifically, it will allow for 340 trillion, trillion, trillion addresses.

Internet Service Provider (ISP): A company that sells access to the Internet to individual subscribers. Leading examples are MSN and AOL.

Internet Telephony: See "Voice Over Internet Protocol (VOIP)."

Intranet: A network protected by a firewall for sharing data and e-mail within an organization or company. Usually, intranets are used by organizations for internal communication.

IoT: See "Internet of Things (IoT)."

IP: See "Intellectual Property (IP)."

IP Number/IP Address: A number or address with four parts that are separated by dots. Each machine on the Internet has its own IP (Internet protocol) number, which serves as an identifier.

IP VOD: See "VOD-Over-IP."

IPL: International Private Line.

IPv6: See "Internet Protocol Version 6 (IPv6)."

IS-95: See "Code Division Multiple Access (CDMA)."

ISDN: See "Integrated Services Digital Networks (ISDN)."

ISO 9000, 9001, 9002, 9003: Standards set by the International Organization for Standardization. ISO 9000, 9001, 9002 and 9003 are the highest quality certifications awarded to organizations that meet exacting standards in their operating practices and procedures.

IT: See "Information Technology (IT)."

IT-Enabled Services (ITES): The portion of the Information Technology industry focused on providing business services, such as call centers, insurance claims processing and medical records transcription, by utilizing the power of IT, especially the Internet. Most ITES functions are considered to be back-office procedures. Also, see "Business Process Outsourcing (BPO)."

ITES: See "IT-Enabled Services (ITES)."

ITU: See "International Telecommunications Union (ITU)."

ITV: See "Interactive TV (ITV)."

Java: A programming language developed by Sun Microsystems that allows web pages to display interactive graphics. Any type of computer or operating systems can read Java.

Joint Photographic Experts Group (JPEG): A widely used format for digital image files.

Ka-Band: The frequency range from 18 to 31 GHz. The spectrum allocated for satellite communications is 30 GHz for the up-link and 20 GHz for the downlink.

Kbps: One thousand bits per second.

Kilobyte: One thousand (or 1,024) bytes.

Kilohertz (kHz): A measure of frequency equal to 1,000 Hertz.

Knowledge Process Outsourcing (KPO): The use of outsourced and/or offshore workers to perform business tasks that require judgment and analysis. Examples include such professional tasks as patent research, legal research, architecture, design, engineering, market research, scientific research, accounting and tax return preparation. Also, see "Business Process Outsourcing (BPO)."

LAC: An acronym for Latin America and the Caribbean.

Landline: Refers to standard telephone and data communications systems that use in-ground and telephone pole cables, as opposed to wireless cellular and satellite services.

LATA: See "Local Access and Transport Area (LATA)."

LDCs: See "Least Developed Countries (LDCs)."

Leased Line: A phone line that is rented for use in continuous, long-term data connections.

Least Developed Countries (LDCs): Nations determined by the U.N. Economic and Social Council to be the poorest and weakest members of the international community. There are currently 50 LDCs, of which 34 are in Africa, 15 are in Asia Pacific and the remaining one (Haiti) is in Latin America. The top 10 on the LDC list, in descending order from top to 10th, are Afghanistan, Angola, Bangladesh, Benin, Bhutan, Burkina Faso, Burundi, Cambodia, Cape Verde and the Central African Republic. Sixteen of the LDCs are also Landlocked Least Developed Countries (LLDCs) which present them with additional difficulties often due to the high cost of transporting trade goods. Eleven of the LDCs are Small Island Developing States (SIDS), which are often at risk of extreme weather phenomenon (hurricanes, typhoons, Tsunami): have fragile ecosystems: are often dependent on foreign energy sources: can have high disease rates for HIV/AIDS and malaria: and can have poor market access and trade terms.

LEC: See "Local Exchange Carrier (LEC)."

Li-Fi: Optical wireless systems that operate somewhat like Wi-Fi, but they utilize light to transfer data.

LINUX: An open, free operating system that is shared readily with millions of users worldwide. These users continuously improve and add to the software's code. It can be used to operate computer networks and Internet appliances as well as servers and PCs.

LMDS: Local Multipoint Distribution Service. A fixed, wireless, point-to-multipoint technology designed to distribute television signals.

Local Access and Transport Area (LATA): An operational service area established after the breakup of AT&T to distinguish local telephone service from long-distance service. The U.S. is divided into over 160 LATAs.

Local Area Network (LAN): A computer network that is generally within one office or one building. A LAN can be very inexpensive and efficient to set up when small numbers of computers are involved. It may require a network administrator and a serious investment if hundreds of computers are hooked up to the LAN. A LAN enables all computers within the office to share files and printers, to access common databases and to send e-mail to others on the network.

Local Exchange Carrier (LEC): Any local telephone company, i.e., a carrier, that provides ordinary phone service under regulation within a service area. Also see "Incumbent Local Exchange Carrier (ILEC)" and "Competitive Local Exchange Carrier (CLEC)."

Long Term Evolution (LTE): An advanced wireless technology expected to deliver wireless Internet access speeds equal to ultrafast DSL speeds. The technology may be useful not only for cell phones, but also for connecting items like digital cameras to the Internet. LTE implementations are planned by such operators as Verizon Wireless, AT&T and Vodafone. It will compete with WiMAX. Practical LTE data transfer speeds may reach 100 to 1,000 Mbps.

LTE: See "Long-Term Evolution (LTE)."

M2M: See "Machine-to-Machine (M2M)."

Machine Learning (ML): The ability of a computer or computerized device to learn based on the results of previous actions or the analysis of a stream of related data. It is a vital branch of Artificial Intelligence that uses advanced software in order to identify patterns and make decisions or predictions.

Machine-to-Machine (M2M): Refers to communications from one device to another (or to a collection of devices). It is typically through wireless means such as Wi-Fi or cellular. Wireless sensor networks (WSNs) will be a major growth factor in M2M communications, in everything from factory automation to agriculture and transportation. In logistics and retailing, M2M can refer to the advanced use of RFID tags. See "Radio Frequency Identification (RFID)." The Internet of Things is based on the principle of M2M communications. Also, see "Internet of Things (IoT)."

MAN: See "Metropolitan Area Network (MAN)."

Managed Service Provider (MSP): An outsourcer that deploys, manages and maintains the back-end software and hardware infrastructure for Internet businesses.

Market Segmentation: The division of a consumer market into specific groups of buyers based on demographic factors.

Mbps (Megabits per second): One million bits transmitted per second.

M-Commerce: Mobile e-commerce over wireless devices.

Media Oriented Systems Transport (MOST): A standard adopted in 2004 by the Consumer Electronics Association for the integration of or interface with consumer electronics (such as iPods) into entertainment systems in automobiles.

Megabytes: One million bytes, or 1,024 kilobytes.

Megahertz (MHz): A measure of frequency equal to 1 million Hertz.

Mesh Network: A network that uses multiple Wi-Fi repeaters or "nodes" to deploy a wireless Internet access network. Typically, a mesh network is operated by the users themselves. Each user installs a node at his or her locale, and plugs the node into his/her local Internet access, whether DSL, cable or satellite. Other users within the mesh can access all other nodes as needed, or as they travel about. A mesh network can provide access to an apartment complex, an office building, a campus or an entire city. Meraki is a leading node brand in this sector.

Metropolitan Area Network (MAN): A data and communications network that operates over metropolitan areas and recently has been expanded to nationwide and even worldwide connectivity of high-speed data networks. A MAN can carry video and data.

Microprocessor: A computer on a digital semiconductor chip. It performs math and logic operations and executes instructions from memory. (Also known as a central processing unit or CPU.)

Microwave: Line-of sight, point-to-point transmission of signals at high frequency. Microwaves are used in data, voice and all other types of information transmission. The growth of fiber-optic networks has tended to curtail the growth and use of microwave relays.

Miles of Plant: The number of cable plant miles laid or strung by a cable system: the cable miles in place.

MIME: See "Multipurpose Internet Mail Extensions (MIME)."

MIMO: See "802.11n (MIMO)."

MMS: See "Multimedia Messaging System (MMS)."

Mobile Virtual Network Operator (MVNO): A seller of cellular service that doesn't own its own cellular network. Instead, it buys all access and network services from a major provider such as Sprint Nextel, and then resells that service under its own brand.

Moblog: Mobile blog. This is a blog created by cell phone or other mobile device. It often consists largely of photos taken by a cell phone's built-in camera. Also, see "Blog (Web Log)."

Modem: A device that allows a computer to be connected to a phone line, which in turn enables the computer to receive and exchange data with other machines via the Internet.

Modulator: A device that modulates a carrier. Modulators are found in broadcasting transmitters and satellite transponders. The devices are also used by cable TV companies to place a baseband video television signal onto a desired VHF or UHF channel. Home video tape recorders also have built-in modulators that enable the recorded video information to be played back using a television receiver tuned to VHF channel 3 or 4.

MOST: See "Media Oriented Systems Transport (MOST)."

MP3: A subsystem of MPEG used to compress sound into digital files. It is the most commonly used format for downloading music and audio books. MP3 compresses music significantly while retaining CD-like quality. MP3 players are personal, portable devices used for listening to music and audio book files. See "MPEG."

MPEG, MPEG-1, MPEG-2, MPEG-3, MPEG-4: Moving Picture Experts Group. It is a digital standard for the compression of motion or still video for transmission or storage. MPEGs are used in digital cameras and for Internet-based viewing.

MPLS: See "Multi-Protocol Label Switching (MPLS)."

MSO: See "Multi-System Operator (MSO)."

MSO: See "Multi-System Operator (MSO)."

Multicasting: Sending data, audio or video simultaneously to a number of clients. Also known as broadcasting.

Multimedia Messaging System (MMS): See "Text Messaging."

Multiple System Operator (MSO): An individual or company owning two or more cable systems.

Multipoint Distribution System (MDS): A common carrier licensed by the FCC to operate a broadcast-like omni-directional microwave transmission facility within a given city. MDS carriers often pick up satellite pay-TV programming and distribute it, via their local MDS transmitter, to specially installed antennas and receivers.

Multi-Protocol Label Switching (MPLS): A technology that enables network operators to route Internet traffic around network failures and bottlenecks.

Multipurpose Internet Mail Extensions (MIME): A widely used method for attaching non-text files to e-mails.

Multi-System Operator (MSO): Refers to companies that operate cable TV (and/or direct-broadcast satellite TV) systems in multiple cities or regions.

MU-MIMO: Mulit-User, Mutiple-Inut, Multiple-Output. See "802.11n (MIMO)."

MVNO: See "Mobile Virtual Network Operator (MVNO)."

NAICS: North American Industrial Classification System. See "Industry Code."

Nanosecond (NS): A billionth of a second. A common unit of measure of computer operating speed.

National Telecommunications and Information Administration (NTIA): A unit of the Department of Commerce that addresses U.S. government telecommunications policy, standards setting and radio spectrum allocation. www.ntia.doc.gov.

Network: In computing, a network is created when two or more computers are connected. Computers may be connected by wireless methods, using such technologies as 802.11b, or by a system of cables, switches and routers.

Network Numbers: The first portion of an IP address, which identifies the network to which hosts in the rest of the address are connected.

Node: Any single computer connected to a network or a junction of communications paths in a network.

NS: See "Nanosecond (NS)."

NTIA: See "National Telecommunications and Information Administration (NTIA)."

Object Technology: By merging data and software into "objects," a programming system becomes object-oriented. For example, an object called "weekly inventory sold" would have the data and programming needed to construct a flow chart. Some new programming systems–including Java–contain this feature. Object technology is also featured in many Microsoft products. See "Java."

OC3, up to OC768: Very high-speed data lines that run at speeds from 155 to 39,813.12 Mbps.

ODM: See "Original Design Manufacturer (ODM)."

OEC: See "Organisation for Economic Co-operation and Development (OECD)."

OEM: See "Original Equipment Manufacturer (OEM)."

OFDM: See "Orthogonal Frequency Division Multiplexing (OFDM)."

OFDMA: Orthogonal Frequency-Division Multiple Access. See "802.11n (MIMO)."

Offshoring: The rapidly growing tendency among U.S., Japanese and Western European firms to send knowledge-based and manufacturing work overseas. The intent is to take advantage of lower wages and operating costs in such nations as China, India, Hungary and Russia. The choice of a nation for offshore work may be influenced by such factors as language and education of the local workforce, transportation systems or natural resources. For example, China and India are graduating high numbers of skilled engineers and scientists from their universities. Also, some nations are noted for large numbers of workers skilled in the English language, such as the Philippines and India. Also see "Captive Offshoring" and "Outsourcing."

Onshoring: The opposite of "offshoring." Providing or maintaining manufacturing or services within or nearby a company's domestic location. Sometimes referred to as reshoring.

Open Source (Open Standards): A software program for which the source code is openly available for modification and enhancement as various users and developers see fit. Open software is typically developed as a public collaboration and grows in usefulness over time. See "LINUX."

Optical Fiber (Fibre): See "Fiber Optics (Fibre Optics)."

Organisation for Economic Co-operation and Development (OECD): A group of more than 30 nations that are strongly committed to the market economy and democracy. Some of the OECD members include Japan, the U.S., Spain, Germany, Australia, Korea, the U.K., Canada and Mexico. Although not members, Estonia, Israel and Russia are invited to member talks: and Brazil, China, India, Indonesia and South Africa have enhanced engagement policies with the OECD. The Organisation provides statistics, as well as social and economic data: and researches social changes, including patterns in evolving fiscal policy, agriculture, technology, trade, the environment and other areas. It publishes over 250 titles annually: publishes a corporate magazine, the OECD Observer: has radio and TV studios: and has centers in Tokyo, Washington, D.C., Berlin and Mexico City that distributed the Organisation's work and organizes events.

Original Design Manufacturer (ODM): A contract manufacturer that offers complete, end-to-end design, engineering and manufacturing services. ODMs design and build products, such as consumer electronics, that client companies can then brand and sell as their own. For example, a large percentage of laptop computers, cell phones and PDAs are made by ODMs. Also see "Original Equipment Manufacturer (OEM)" and "Contract Manufacturing."

Original Equipment Manufacturer (OEM): 1) A company that manufactures a component (or a completed product) for sale to a customer that will integrate the component into a final product. The OEM's customer will put its own brand name on the end product and distribute or resell it to end users. 2) A firm that buys a component and then incorporates it into a final product, or buys a completed product and then resells it under the firm's own brand name. This usage is most often found in the computer industry, where OEM is sometimes used as a verb. Also see "Original Design Manufacturer (ODM)" and "Contract Manufacturing."

Orthogonal Frequency Division Multiplexing (OFDM): An alternative to CDMA cell phone technology backed by Flarion Technologies, Inc. OFDM is a frequency-division multiplexing system that uses closely spaced, eight-sided subcarriers to transmit data.

Outsourcing: The hiring of an outside company to perform a task otherwise performed internally by the company, generally with the goal of lowering costs and/or streamlining work flow. Outsourcing contracts are generally several years in length. Companies that hire outsourced services providers often prefer to focus on their core strengths while sending more routine tasks outside for others to perform. Typical outsourced services include the running of human resources departments, telephone call centers and computer departments. When outsourcing is performed overseas, it may be referred to as offshoring. Also see "Offshoring."

P2P Network: See "Peer-to-Peer (P2P) Network."

Packet Switching: A higher-speed way to move data through a network, in which files are broken down into smaller "packets" that are reassembled electronically after transmission.

PAS: See "Personal Access System (PAS)."

Passive Optical Network (PON): A telecommunications network that brings high speed fiber optic cable all the way (or most of the way) to the end user. Also, see "Fiber to the Home (FTTH)."

Patent: An intellectual property right granted by a national government to an inventor to exclude others from making, using, offering for sale, or selling the invention throughout that nation or importing the invention into the nation for a limited time in exchange for public disclosure of the invention when the patent is granted. In addition to national patenting agencies, such as the United States Patent and Trademark Office, and regional organizations such as the European Patent Office, there is a cooperative international patent organization, the World Intellectual Property Organization, or WIPO, established by the United Nations.

Pay Cable: A network or service available for an added monthly fee. Also called premium cable. Some services, called mini-pay, are marketed at an average monthly rate below that of full-priced premium.

Pay Cable Unit: Each premium service to which a household subscribes.

Pay-Per-View (PPV): A service that enables television subscribers, including cable and satellite viewers, to order and view events or movies on an individual basis. PPV programming may include sporting events.

PBX: A central telephone system within a large business office used to route incoming and outgoing calls to various employees and onto long-distance networks. PBX functions are typically enhanced by the application of computer functions, such as voice mail and call forwarding.

PC: See "Personal Computer (PC)."

PCS: See "Personal Communication Service (PCS)."

Peer-to-Peer (P2P) Network: Refers to a connection between computers that creates equal status between the computers. P2P can be used in an office or home to create a simple computer network. However, P2P more commonly refers to networks of computers that share information online. For example, peer-to-peer music sharing networks enable one member to search the hard drives of other members to locate music files and then download those files.

Personal Access System (PAS): A type of mobile phone service that is very popular in China. Unlike traditional cellular mobile phone service, a PAS customer typically cannot roam beyond their home service area. However, monthly charges tend to be much lower.

Personal Communication Service (PCS): A type of cellular mobile telephone service.

Personal Computer (PC): An affordable, efficient computer meant to be used by one person. The device may be a desktop computer or a laptop. Frequently, the PC is connected to a local area network (LAN), or uses wireless methods such as Wi-Fi to access the Internet. PCs are used both in the home and in the office. There is no firm agreement on whether tablets should be regarded as PCs.

Personal Handyphone System (PHS): A type of mobile phone service that is very popular in Japan. Unlike traditional cellular mobile phone service, a PHS customer typically cannot roam beyond their home service area. However, monthly charges tend to be much lower.

PHS: See "Personal Handyphone System (PHS)."

Podcasting: The creation of audio files as webcasts. Podcasts can be anything from unique radio-like programming to sales pitches to audio press releases. Audio RSS (Real Simple Syndication) enables the broadcast of these audio files to appropriate parties. Also see "Real Simple Syndication (RSS)," "Video Blog (VLOG)" and "Blog (Web Log)."

Point-to-Point Protocol (PPP): A protocol that enables a computer to use the combination of a standard telephone line and a modem to make TCP/IP connections.

PON: See "Passive Optical Networking (PON)."

POP: An acronym for both "Point of Presence" and "Post Office Protocol." Point of presence refers to a location that a network can be connected to (generally used to count the potential subscriber base of a cellular phone system). Post office protocol refers to the way in which e-mail software obtains mail from a mail server.

Port: An interface (or connector) between the computer and the outside world. The number of ports

on a communications controller or front-end processor determines the number of communications channels that can be connected to it. The number of ports on a computer determines the number of peripheral devices that can be attached to it.

Portal: A comprehensive web site for general or specific purposes.

Positioning: The design and implementation of a merchandising mix, price structure and style of selling to create an image of the retailer, relative to its competitors, in the customer's mind.

Powerline: A method of networking computers, peripherals and appliances together via the electrical wiring that is built in to a home or office. Powerline competes with 802.11b and other wireless networking methods.

Predictive Analytics: See "Analytics."

Product Lifecycle (Product Life Cycle): The prediction of the life of a product or brand. Stages are described as Introduction, Growth, Maturity and finally Sales Decline. These stages track a product from its initial introduction to the market through to the end of its usefulness as a commercially viable product. The goal of Product Lifecycle Management is to maximize production efficiency, consumer acceptance and profits. Consequently, critical processes around the product need to be adjusted during its lifecycle, including pricing, advertising, promotion, distribution and packaging.

Product Lifecycle Management (PLM): See "Product Lifecycle (Product Life Cycle)."

Proportionate Subscribers: The number of subscribers, usually to a mobile telecommunications system, relative to the interest owned in the system by an investor or investors. For example, a firm with a 100% ownership interest in a wireless business with 100,000 subscribers would have 100,000 proportionate subscribers, but a firm with a 25% interest in a system with 100,000 subscribers would be attributed 25,000 proportionate subscribers for that system.

Protocol: A set of rules for communicating between computers. The use of standard protocols allows products from different vendors to communicate on a common network.

PSTN: See "Public Switched Telephone Network (PSTN)."

Public Switched Telephone Network (PSTN): A term that refers to the traditional telephone system.

QoS: See "Quality of Service (QoS)."

Quality of Service (QoS): The improvement of the flow of broadband information on the Internet and other networks by raising the data flow level of certain routes and restricting it on others. QoS levels are supported on robust, high-bandwidth technologies such as 4G.

R&D: Research and development. Also see "Applied Research" and "Basic Research."

Radio Frequency Identification (RFID): A technology that applies a special microchip-enabled tag to an individual item or piece of merchandise or inventory. RFID technology enables wireless, computerized tracking of that inventory item as it moves through the supply chain from factory to transport to warehouse to retail store or end user. Also known as radio tags.

RBOC: See "Regional Bell Operating Company (RBOC)."

Real Simple Syndication (RSS): Uses XML programming language to let web logs and other data be broadcast to appropriate web sites and users. Formerly referred to as RDF Site Summary or Rich Site Summary, RSS also enables the publisher to create a description of the content and its location in the form of an RSS document. Also useful for distributing audio files. See "Podcasting."

Regional Bell Operating Company (RBOC): Former Bell system telephone companies (or their successors), created as a result of the breakup of AT&T by a Federal Court decree on December 31, 1983 (e.g., Bell Atlantic, now part of Verizon).

Request for Bids (RFB): A request for pricing and supporting details, sent by a firm that requires products or services, outlining all the firm's requirements. Proposing companies are asked to place a bid based on the requested goods or services.

Request for Quotation (RFQ): A proposal that asks companies to submit pricing for goods or a described level of services. See "Request for Bids (RFB)."

Reshoring: See "Onshoring."

RF: Radio Frequency.

RFID: See "Radio Frequency Identification (RFID)."

RoHS Compliant: A directive that restricts the total amount of certain dangerous substances that may be incorporated in electronic equipment, including consumer electronics. Any RoHS compliant component is tested for the presence of Lead, Cadmium, Mercury, Hexavalent chromium, Polybrominated biphenyls and Polybrominated diphenyl ethers. For Cadmium and Hexavalent chromium, there must be less than 0.01% of the substance by weight at raw homogeneous materials level. For Lead, PBB, and PBDE, there must be no more than 0.1% of the material, when calculated by weight at raw homogeneous materials. Any RoHS compliant component must have 100 ppm or less of mercury and the mercury must not have been intentionally added to the component. Certain items of military and medical equipment are exempt from RoHS compliance.

Router: An electronic device that enables networks to communicate with each other. For example, the local area network (LAN) in an office connects to a router to give the LAN access to an Internet connection such as a T1 or DSL. Routers can be bundled with several added features, such as firewalls.

RSS: See "Real Simple Syndication (RSS)."

SaaS: See "Software as a Service (SaaS)."

Satellite Broadcasting: The use of Earth-orbiting satellites to transmit, over a wide area, TV, radio, telephony, video and other data in digitized format.

Scalable: Refers to a network that can grow and adapt as the total customer count increases and as customer needs increase and change. Scalable websites (and the hardware and software behind them) are extremely vital to rapidly-growing online services. Scalable networks can easily manage increasing numbers of workstations, servers, user workloads and added functionality.

SDSL: See "Digital Subscriber Line (DSL)."

Semiconductor: A generic term for a device that controls electrical signals. It specifically refers to a material (such as silicon, germanium or gallium arsenide) that can be altered either to conduct electrical current or to block its passage. Carbon nanotubes may eventually be used as semiconductors. Semiconductors are partly responsible for the miniaturization of modern electronic devices, as they are vital components in computer memory and processor chips. The manufacture of semiconductors is carried out by small firms, and by industry giants such as Intel and Advanced Micro Devices.

Serial Line Internet Protocol (SLIP): The connection of a traditional telephone line, or serial line, and modem to connect a computer to an Internet site.

Server: A computer that performs and manages specific duties for a central network such as a LAN. It may include storage devices and other peripherals. Competition within the server manufacturing industry is intense among leaders Dell, IBM, HP and others.

Server-Based SVOD Programming: Programming that is delivered directly to the customer's TV from where it is stored on the content provider's servers. In contrast, non-server-based SVOD (satellite TV) needs a storage device at the customer's location (such as a PVR or DVR) to store and play VOD content for the viewer's TV. Server-based SVOD surpasses non-server-based SVOD in its ability to simultaneously send or receive more than one video stream to or from the customer.

Service Level Agreement (SLA): A detail in a contract between a service provider and the client. The agreement specifies the level of service that is expected during the service contract term. For example, computer or Internet service contracts generally stipulate a maximum amount of time that a system may be unusable.

Set-Top Box: Sits on top of a TV set and provides enhancement to cable TV or other television reception. Typically a cable modem, this box may enable interactive enhancements to television viewing. For example, a cable modem is a set-top box that enables Internet access via TV cable. See "Cable Modem."

Short Messaging System (SMS): See "Text Messaging."

SIC: Standard Industrial Classification. See "Industry Code."

Simple Mail Transfer Protocol (SMTP): The primary form of protocol used in the transference of e-mail.

Simple Network Management Protocol (SNMP): A set of communication standards for use between computers connected to TCP/IP networks.

SIP (SIPphone): SIP stands for Session Initiated Protocol. An SIP telephone, or SIPphone, is a telephone that can make calls at no cost to any other SIPphone, anywhere in the world, via the Internet. Both telephones must be equipped with a SIPphone adapter or an Internet-connected softphone.

Six Sigma: A quality enhancement strategy designed to reduce the number of products coming from a manufacturing plant that do not conform to specifications. Six Sigma states that no more than 3.4 defects per million parts is the goal of high-quality output. Motorola invented the system in the 1980s in order to enhance its competitive position against Japanese electronics manufacturers.

SLA: See "Service Level Agreement (SLA)."

SLIP: See "Serial Line Internet Protocol (SLIP)."

Smartphones: Mobile devices that have the capability to perform complex tasks and run user-generated programs. Newer devices include high-speed Internet access by connecting to wireless data services. Examples include Apple's iPhone, Research in Motion's BlackBerry and various devices with Google's Android operating system.

SMDS: See "Switched Multimegabit Data Service (SMDS)."

SMS: See "Short Messaging System (SMS)."

SMTP: See "Simple Mail Transfer Protocol (SMTP)."

Social Media: Sites on the Internet that feature user generated content (UGC). Such media include wikis, blogs and specialty web sites such as MySpace.com,

Facebook, YouTube, Yelp and Friendster.com. Social media are seen as powerful online tools because all or most of the content is user-generated.

Software as a Service (SaaS): Refers to the practice of providing users with software applications that are hosted on remote servers and accessed via the Internet. Excellent examples include the CRM (Customer Relationship Management) software provided in SaaS format by Salesforce. An earlier technology that operated in a similar, but less sophisticated, manner was called ASP or Application Service Provider.

SONET: See "Synchronous Optical Network Technology (SONET)."

Spam: A term used to refer to generally unwanted, solicitous, bulk-sent e-mail. In recent years, significant amounts of government legislation have been passed in an attempt to limit the use of spam. Also, many types of software filters have been introduced in an effort to block spam on the receiving end. In addition to use for general advertising purposes, spam may be used in an effort to spread computer viruses or to commit financial or commercial fraud.

Streaming Media: One-way audio and/or video that is compressed and transmitted over a data network. The media is viewed or heard almost as soon as data is fed to the receiver: there is usually a buffer period of a few seconds.

Subscriber: A term used interchangeably with household in describing cable, Internet access or telephone customers.

Subscription Video On Demand (SVOD): Allows subscribers unlimited access to selected VOD television programming for a fixed monthly fee.

Subsidiary, Wholly-Owned: A company that is wholly controlled by another company through stock ownership.

Supply Chain: The complete set of suppliers of goods and services required for a company to operate its business. For example, a manufacturer's supply chain may include providers of raw materials, components, custom-made parts and packaging materials.

SVOD: See "Subscription Video On Demand (SVOD)."

Switch: A network device that directs packets of data between multiple ports, often filtering the data so that it travels more quickly.

Switched Multimegabit Data Service (SMDS): A method of extremely high-speed transference of data.

Synchronous Optical Network Technology (SONET): A mode of high-speed transmission meant to take full advantage of the wide bandwidth in fiber-optic cables.

System (Cable): A facility that provides cable television service in a given geographic area, consisting of one or more headends.

T1: A standard for broadband digital transmission over phone lines. Generally, it can transmit at least 24 voice channels at once over copper wires, at a high speed of 1.5 Mbps. Higher speed versions include T3 and OC3 lines.

T3: Transmission over phone lines that supports data rates of 45 Mbps. T3 lines consist of 672 channels, and such lines are generally used by Internet service providers. They are also referred to as DS3 lines.

Tablet: A mobile computing device that offers similar functionality as a smartphone, except with a larger viewing area and a more complex processor. However, some tablets do not have the ability to make phone calls. Examples include the Apple iPad and Samsung Galaxy tablet. Tablets are designed to interact with the user primarily through a touchscreen rather than a keyboard. While tablets offer many PC-like functions, there is no firm agreement as to whether they should be counted as part of the PC market.

TCP/IP: Transmission Control Protocol/Internet Protocol. The combination of a network and transport protocol developed by ARPANet for internetworking IP-based networks.

TDMA: See "Time Division Multiple Access (TDMA)."

Telecommunications: Systems and networks of hardware and software used to carry voice, video and/or data within buildings and between locations around the world. This includes telephone wires, satellite signals, wireless networks, fiber networks, Internet networks and related devices.

Telepresence: The use of highly sophisticated digital video cameras, microphones and high speed Internet connections to create a video conference for remote participants that is nearly life-like. Conference participants may consult with each other from specially-equipped rooms that can be almost anywhere in the world. With the most advanced equipment, such as that produced by Cisco, the images on screens can be near life-size and the results can be of almost face-to-face quality.

Telnet: A terminal emulation program for TCP/IP networks like the Internet, which runs on a computer and connects to a particular network. Directions entered on a computer that is connected using Telnet will be read and followed just as if they had been entered on the server itself. Through Telnet, users are able to control a server and communicate with other servers on the same network at the same time. Telnet is commonly used to control web servers remotely.

Text Messaging: The transmission of very short, text messages in a format similar to e-mail. Generally, text messaging is used as an additional service on cell phones. The format has typically been SMS (Short Messaging System), but a newer standard is evolving: MMS (Multimedia Messaging System). MMS can transmit pictures, sound and video as well as text.

Third-Party Logistics (3PL): A specialist firm in logistics, which may provide a variety of transportation, warehousing and logistics-related services to buyers or sellers. These tasks were previously performed in-house by the customer. When 3PL services are provided within the client's own facilities, it can also be referred to as insourcing. Also see "Fourth-Party Logistics (4PL)."

TIME: Telecommunications, Information Technology, Media and Electronics.

Time Division Multiple Access (TDMA): A 2G digital service for relatively large users of international public-switched telephony, data, facsimile and telex. TDMA also refers to a method of multiplexing digital signals that combines a number of signals passing through a common point by transmitting them sequentially, with each signal sent

in bursts at different times. TDMA is sometimes referred to as IS-136 or D-AMPS.

Token Ring: A local area network architecture in which a token, or continuously repeating frame, is passed sequentially from station to station. Only the station possessing the token can communicate on the network.

Transactional VOD: Allows VOD customers to pay a single price for a single VOD program or a set of programs rather than paying a set fee for a set amount of VOD programming (as in SVOD services).

Transistor: A device used for amplification or switching of electrical current.

TV over IP: See "Internet Protocol Television (IPTV)."

U-Commerce (U Commerce): Ubiquitous Commerce, Universal Commerce or Ultimate Commerce (ubiquitous meaning ever-present), depending on whom you ask. It describes the concept that buyers and sellers have the potential to interact anywhere, anytime thanks to the use of wireless devices, such as cell phones, by buyers to connect with sellers via the Internet where orders can be placed online and payments can be made via credit card or PayPal. The Association for Information Systems states that the qualities of U-Commerce include ubiquity, uniqueness, universality and unison.

Ultra 3G Cellular: See "4G Cellular."

Ultrawideband (UWB): A means of low-power, limited-range wireless data transmission that takes advantage of bandwidth set aside by the FCC in 2002. UWB encodes signals in a dramatically different way, sending digital pulses in a relatively secure manner that will not interfere with other wireless systems that may be operating nearby. It has the potential to deliver very large amounts of data to a distance of about 230 feet, even through doors and other obstacles, and requires very little power. Speeds are scalable from approximately 100 Mbps to 2Gbps. UWB works on the 802.15.3 IEEE specification.

UMA: See "Unlicensed Mobile Access (UMA)."

UMTS: Universal Mobile Telecommunications System. See "Wideband CDMA."

Unified Communications: The use of advanced technology to replace traditional telecommunications infrastructure such as PBX, fax and even the desktop telephone. Special software operating on a local or remote server enables each office worker to have access, via the desktop PC, to communications tools that include VOIP phone service, email, voice mail, fax, instant messaging (IM), collaborative calendars and schedules, contact information such as address books, audio conferencing and video conferencing.

Uniform Resource Locator (URL): The address that allows an Internet browser to locate a homepage or web site.

Universal Mobile Telecommunications System (UMTS): An overarching standard based on W-CDMA technology. See "Wideband CDMA."

UNIX: A multi-user, multitasking operating system that runs on a wide variety of computer systems, from PCs to mainframes.

Unlicensed Mobile Access (UMA): A standard technology developed for GSM cellular phone networks. UMA is designed to enable access to the Internet and to VOIP telephony for dual mode cellular phone handsets that are capable of switching from cellular to Wi-Fi and back.

URL: See "Uniform Resource Locator (URL)."

User Generated Content (UGC): Data contributed by users of interactive web sites. Such sites can include wikis, blogs, entertainment sites, shopping sites or social networks such as Facebook. UGC data can also include such things as product reviews, photos, videos, comments on forums, and how-to advice. Also see "Social Media."

UWB: See "Ultrawideband (UWB)."

Value Added Tax (VAT): A tax that imposes a levy on businesses at every stage of manufacturing based on the value it adds to a product. Each business in the supply chain pays its own VAT and is subsequently repaid by the next link down the chain: hence, a VAT is ultimately paid by the consumer, being the last link in the supply chain, making it comparable to a sales tax. Generally, VAT only applies to goods bought for consumption within a given country: export goods are exempt from VAT, and purchasers

from other countries taking goods back home may apply for a VAT refund.

V-Chip: A system built into TV sets that helps parents screen out programs with questionable parental guideline ratings. Consumers can purchase a special set-top box that performs the same function.

VDSL: Very high-data-rate digital subscriber line, operating at data rates from 55 to 100 Mbps.

Vendor: Any firm, such as a manufacturer or distributor, from which a retailer obtains merchandise.

Very Small Aperture Terminal (VSAT): A small Earth station terminal, generally 0.6 to 2.4 meters in size, that is often portable and primarily designed to handle data transmission and private-line voice and video communications.

Video Blog (VLOG): The creation of video files as webcasts. VLOGs can be viewed on personal computers and wireless devices that are Internet-enabled. They can include anything from unique TV-like programming to sales pitches to music videos, news coverage or audio press releases. Online video is one of the fastest-growing segments in Internet usage. Leading e-commerce companies such as Microsoft, through its MSN service, Google and Yahoo!, as well as mainstream media firms such as Reuters, are making significant investments in online video services. Real Simple Syndication (RSS) enables the broadcasting of these files to appropriate parties. Also see "Real Simple Syndication (RSS)," "Podcasting" and "Blog (Web Log)."

Video On Demand (VOD): A system that allows customers to request programs or movies over cable or the Internet. Generally, the customer can select from an extensive list of titles. In some cases, a set-top device can be used to digitally record a broadcast for replay at a future date.

Virtual Private Network (VPN): Cordons off part of a public network to create a private LAN. It is a common way of increasing login security.

VLOG: See "Video Blog (VLOG)."

VOD: See "Video On Demand (VOD)."

VOD-Over-IP: VOD (video on demand) television viewing that is distributed via the Internet.

Voice Mail: A sophisticated electronic telephone answering service that utilizes a computer. Voice mail enables users to receive faxes and phone messages and to access those messages from remote sites.

Voice Over Internet Protocol (VOIP): The ability to make telephone calls and send faxes over IP-based data networks, i.e., real-time voice between computers via the Internet. Leading providers of VOIP service include independent firms Skype and Vonage. However, all major telecom companies, such as SBC are planning or offering VOIP service. VOIP can offer greatly reduced telephone bills to users, since toll charges, certain taxes and other fees can be bypassed. Long-distance calls can pass to anywhere in the world using VOIP. Over the mid-term, many telephone handsets, including cellular phones, will have the ability to detect wireless networks offering VOIP connections and will switch seamlessly between landline and VOIP or cellular and VOIP as needed.

VOIP: See "Voice Over Internet Protocol (VOIP)."

VPN: See "Virtual Private Network (VPN)."

WAN: See "Wide Area Network (WAN)."

WAP: See "Wireless Access Protocol (WAP)."

W-CDMA (WCDMA): See "Wideband CDMA."

Web 2.0: The second stage of development of the World Wide Web. Services include collaborative sites that emphasize dynamic user-generated content such as those for social media.

Web of Things: See "Internet of Things (IoT)."

Web Services: Self-contained modular applications that can be described, published, located and invoked over the World Wide Web or another network. Web services architecture evolved from object-oriented design and is geared toward e-business solutions. Microsoft Corporation is focusing on web services with its .NET initiative. Also see "Extensible Markup Language (XML)."

Webmaster: Any individual who runs a web site. Webmasters generally perform maintenance and upkeep.

Website Meta-Language (WML): A free HTML generation toolkit for the Unix operating system.

WFH: Work from home.

Wide Area Network (WAN): A regional or global network that provides links between all local area networks within a company. For example, Ford Motor Company might use a WAN to enable its factory in Detroit to talk to its sales offices in New York and Chicago, its plants in England and its buying offices in Taiwan. Also see "Local Area Network (LAN)."

Wideband CDMA: A high-speed version of CDMA based on Qualcomm technology. Significantly modified by Japanese and European manufacturers, it is used as a 3G technology outside the U.S. with Japan's Freedom of Multimedia Access (FOMA) and Europe's Universal Mobile Telecommunications System (UMTS). It is also known as "IMT-2000 Direct Spread."

WiFi: See "Wi-Fi."

Wi-Fi: Wireless Fidelity. Refers to 802.11 wireless network specifications. The 802.XX standards are set by the IEEE (Institute of Electrical and Electronics Engineers). Wi-Fi enables very high speed local networks in homes, businesses, factories, industrial and transportation infrastructure, public spaces and vehicles. Wi-Fi networks enable computing devices of all types to connect to each other and to the internet, including smartphones, laptops, desktops and tablet computers. In addition, Wi-Fi enables machine-to-machine (M2M) communication between devices, providing a backbone for the Internet of Things. These networks can be made reasonably secure when strong passwords are required and additional cybersecurity measures are in place. (Also, see 'Internet of Things".)

WiMAX: An advanced wireless standard with significant speed and distance capabilities, WiMAX is officially known as the 802.16 standard. Using microwave technologies, it has the theoretical potential to broadcast at distances up to 30 miles and speeds of up to 70 Mbps. The 802.XX standards are set by the IEEE (Institute of Electrical and Electronics Engineers).

Wireless: Transmission of voice, video or data by a cellular telephone or other wireless device, as opposed to landline, fiber or cable. It includes Bluetooth, Cellular, Wi-Fi, WiMAX and other local or long-distance wireless methods.

Wireless Access Protocol (WAP): A technology that enables the delivery of internet pages in a smaller format readable by screens on smartphones.

Wireless LAN (WLAN): A wireless local area network. WLANs frequently operate on 802.11-enabled equipment (Wi-Fi).

Wireline (in telecommunications): See "Landline."

WLAN: See "Wireless LAN (WLAN)."

WML: See "Website Meta-Language (WML)."

Workstation: A high-powered desktop computer, usually used by engineers.

World Trade Organization (WTO): One of the only globally active international organizations dealing with the trade rules between nations. Its goal is to assist the free flow of trade goods, ensuring a smooth, predictable supply of goods to help raise the quality of life of member citizens. Members form consensus decisions that are then ratified by their respective parliaments. The WTO's conflict resolution process generally emphasizes interpreting existing commitments and agreements, and discovers how to ensure trade policies to conform to those agreements, with the ultimate aim of avoiding military or political conflict.

World Wide Web: A system (the internet) that provides enhanced access to various sites on the Internet through the use of hyperlinks. Clicking on a link displayed in one document takes you to a related document. The World Wide Web is governed by the World Wide Web Consortium, located at www.w3.org. Also known as the web.

WoT: Web of Things. See "Internet of Things."

WPA: Wireless Protected Access. A basic security standard for wireless networking, including Wi-Fi.

WTO: See "World Trade Organization (WTO)."

XML: See "Extensible Markup Language (XML)."